Using AutoCAD® 2011

# Using AutoCAD® 2011

**RALPH GRABOWSKI**

Autodesk

Australia • Canada • Mexico • Singapore • Spain • United Kingdom • United States

Using AutoCAD® 2011
Ralph Grabowski

Vice President, Career and Professional Editorial: Dave Garza

Director of Learning Solutions: Sandy Clark

Acquisitions Editor: Stacy Masucci

Managing Editor: Larry Main

Senior Product Manager: John Fisher

Senior Editorial Assistant: Dawn Daugherty

Editorial Assistant: Andrea Timpano

Vice President, Career and Professional Marketing: Jennifer McAvey

Marketing Director: Deborah Yarnell

Marketing Manager: Katie Hall

Associate Marketing Manager: Mark Pierro

Production Director: Wendy Troeger

Art Director: David Arsenault

Senior Content Project Manager: Angela Sheehan

Technology Project Manager: Joe Pliss

Designer and Typesetter: Ralph Grabowski

© 2011 Delmar Cengage Learning.

ALL RIGHTS RESERVED. No part of this work covered by the copyright herein may be reproduced, transmitted, stored, or used in any form or by any means graphic, electronic, or mechanical, including but not limited to photocopying, recording, scanning, digitizing, taping, Web distribution, information networks, or information storage and retrieval systems, except as permitted under Section 107 or 108 of the 1976 United States Copyright Act, without the prior written permission of the publisher.

> For product information and technology assistance, contact us
> at **Career & Professional Group Customer Support,**
> **1-800-648-7450**
>
> For permission to use material from this text or product, submit all requests online at **www.cengage.com/permissions**
> Further permissions questions can be emailed to
> **permissionrequest@cengage.com**

Autodesk, AutoCAD, and the Autodesk logo are registered trademarks of Autodesk. Delmar Cengage Learning uses "Autodesk Press" with permission from Autodesk for certain purposes.

Library of Congress Control Number: 2010925289
ISBN-13: 978-1-1111-2514-1
ISBN-10: 1-1111-2514-7

**Delmar Cengage Learning**
5 Maxwell Drive
Clifton Park, NY 12065
USA

Cengage Learning is a leading provider of customized learning solutions with office locations around the globe, including Singapore, the United Kingdom, Australia, Mexico, Brazil and Japan. Locate your local office at: **international.cengage.com/region**

Cengage Learning products are represented in Canada by Nelson Education, Ltd.

For your course and learning solutions, visit
**delmar.cengage.com**

Purchase any of our products at your local college store or at our preferred online store **www.ichapters.com**

**Notice to the Reader**
Publisher does not warrant or guarantee any of the products described herein or perform any independent analysis in connection with any of the product information contained herein. Publisher does not assume, and expressly disclaims, any obligation to obtain and include information other than that provided to it by the manufacturer. The reader is expressly warned to consider and adopt all safety precautions that might be indicated by the activities described herein and to avoid all potential hazards. By following the instructions contained herein, the reader willingly assumes all risks in connection with such instructions. The publisher makes no representations or warranties of any kind, including but not limited to, the warranties of fitness for particular purpose or merchantability, nor are any such representations implied with respect to the material set forth herein, and the publisher takes no responsibility with respect to such material. The publisher shall not be liable for any special, consequential, or exemplary damages resulting, in whole or part, from the readers' use of, or reliance upon, this material.

Printed in the United States of America
1 2 3 4 5 XXX 14 13 12 11 10

# BRIEF CONTENTS

INTRODUCTION ...................................................................................................................LI

## UNIT I — SETTING UP DRAWINGS ............................................................................. 1
  1  QUICK TOUR OF AUTOCAD 2011 ............................................................................3
  2  UNDERSTANDING CAD CONCEPTS ......................................................................41
  3  SETTING UP DRAWINGS .........................................................................................87

## UNIT II — DRAFTING ESSENTIALS ......................................................................... 127
  4  DRAWING BASIC OBJECTS ...................................................................................129
  5  DRAWING WITH PRECISION ................................................................................185
  6  DRAWING WITH EFFICIENCY ...............................................................................247

## SECTION III — EDITING DRAWINGS ..................................................................... 319
  7  PROPERTIES AND LAYERS .....................................................................................321
  8  CONSTRUCTING OBJECTS ...................................................................................373
  9  ADDITIONAL EDITING METHODS .......................................................................417

## UNIT IV — TEXT AND DIMENSIONS ...................................................................... 479
  10  PLACING AND EDITING TEXT ............................................................................481
  11  PLACING DIMENSIONS .......................................................................................563
  12  EDITING DIMENSIONS ........................................................................................617
  13  APPLYING DIMENSIONAL CONSTRAINTS .........................................................663
  14  USING GEOMETRIC CONSTRAINTS ...................................................................685
  15  REPORTING ON DRAWINGS ..............................................................................709

## UNIT V — 3D MODELING ......................................................................................... 751
  16  INTRODUCTION TO 3D SOLID MODELING .......................................................753
  17  INTRODUCTION TO 3D MESH MODELING ........................................................787
  18  ADDITIONAL 3D TECHNIQUES ...........................................................................811

## UNIT VII — PLOTTING DRAWINGS ............ 851
- 19 WORKING WITH LAYOUTS ............ 853
- 20 PLOTTING DRAWINGS ............ 901

## UNIT VII — APPENDICES ............ 965
- A AUTOCAD, COMPUTERS, AND WINDOWS ............ 967
- B AUTOCAD COMMANDS, ALIASES, AND KEYBOARD SHORTCUTS ............ 981
- C AUTOCAD SYSTEM VARIABLES ............ 1009
- D SUMMARY OF DIMENSION VARIABLES ............ 1067

INDEX ............ 1079

# CONTENTS

## Introduction .................................................................................. LI

### BENEFITS OF COMPUTER-AIDED DRAFTING ................................................. LI

#### CAD APPLICATIONS ........................................................................ LII
- Architecture .................................................................................. liii
- Engineering ................................................................................... liv
- Interior Design ................................................................................ lv
- Manufacturing ................................................................................. lv
- Facilities Management ......................................................................... lvi
- Mapping ...................................................................................... lvi
- Entertainment ................................................................................. lvii
- Online Applications ........................................................................... lviii
- OTHER BENEFITS ............................................................................... LIX

### AUTOCAD OVERVIEW .......................................................................... LX

#### TERMINOLOGY .............................................................................. LX
- Coordinates ................................................................................... lx
- Drawing Files ................................................................................. lx

### WHAT'S NEW IN AUTOCAD 2011 ................................................................ LXI
- USER INTERFACE ............................................................................... LXI
- 2D DRAFTING .................................................................................. LXII
- PARAMETRICS .................................................................................. LXII
- 3D MODELING ................................................................................. LXIII

The icon indicates the command is new to AutoCAD 2011.

| | |
|---|---|
| **USING THIS BOOK** | **LXIV** |
|    **CONVENTIONS** | **LXIV** |
|       Keys | lxiv |
|          Control and Alternate Keys | lxiv |
|          Flip Screen | lxiv |
|       Command Nomenclature | lxiv |
| **COMPANION FILES** | **LXV** |
|    **ONLINE COMPANION** | **LXVI** |
|    **WE WANT TO HEAR FROM YOU!** | **LXVII** |
| **ACKNOWLEDGMENTS** | **LXVII** |
|    **ABOUT THE AUTHOR** | **LXVII** |

# UNIT 1 — SETTING UP DRAWINGS .......... 1

## 1  Quick Tour of AutoCAD 2011 .......... 3

### STARTING AUTOCAD .......... 4
- TUTORIAL: STARTING AUTOCAD .......... 4
- COMMANDS IN AUTOCAD .......... 8
  - Command Bar .......... 8
    - Text Window .......... 9
    - Dynamic Input .......... 9
  - Mouse .......... 9
  - RIBBON .......... 11
  - TUTORIAL: EXECUTING COMMANDS WITH THE RIBBON .......... 11
  - TUTORIAL: ACCESSING FLYOUTS .......... 15
  - TUTORIAL: ACCESSING DROPLISTS .......... 16
  - QUICK ACCESS TOOLBAR .......... 17
  - ACCESSING THE APPLICATION MENU .......... 17
    - Recent Drawings .......... 18
    - Open Documents .......... 19
    - Displaying the Menu Bar .......... 19
  - CHANGING WORKSPACES .......... 20
    - AutoCAD Classic Workspace .......... 21
      - Toolbars .......... 21

### EXITING AUTOCAD .......... 22

### 2D TUTORIAL .......... 23
- SETTING UP NEW DRAWINGS .......... 24
- DRAWING THE BASE PLATE .......... 25
- DRAWING THE SHAFT .......... 26
- DRAWING HOLES .......... 27
- COPYING CIRCLES .......... 27
- SAVING THE DRAWING .......... 28
- PLOTTING THE DRAWING .......... 29

### 3D TUTORIAL .......... 32
- MODELING THE SHAFT .......... 33
- MODELING HOLES .......... 35
- COPYING CYLINDERS .......... 36
- SUBTRACTING CYLINDERS TO CREATE HOLES .......... 37

### EXERCISES .......... 38

### CHAPTER REVIEW .......... 39

# 2  Understanding CAD Concepts .......................... 41

## AUTOCAD USER INTERFACE .......................... 42
### USER INTERFACE CHANGES IN AUTOCAD 2011 .......................... 42

### TOURING THE USER INTERFACE .......................... 43
#### TITLE BAR .......................... 43
- Info Center .......................... 43
- Window Controls .......................... 43

#### RIBBON, MENU BAR, AND TOOLBARS .......................... 44
#### DRAWING AREA .......................... 44
- Cursors .......................... 44
  - Cursor Icons .......................... 45
- View Cube .......................... 45
- Navigation Bar .......................... 45
- UCS Icon .......................... 46
- Other Screen Markings .......................... 47
  - Highlights and Grips .......................... 47
  - Rollover Tooltips .......................... 47
    - Selection Cycling .......................... 48
  - Dynamic Input .......................... 48
  - Grid .......................... 49
  - Tracking Lines .......................... 49

#### LAYOUT TABS AND SCROLL BARS .......................... 49
- Scroll Bars .......................... 49

#### STATUS BAR .......................... 50
  - X,Y Coordinates and Elevation .......................... 50
- Mode Indicators .......................... 50
- Layouts .......................... 51
- Annotation Scaling .......................... 52
- Workspaces .......................... 52
- Tray .......................... 52
  - Tray Options .......................... 53
- Cleanscreen and Resize Window .......................... 53

## ENTERING COMMANDS .......................... 54

### USING THE KEYBOARD .......................... 54
- Command Input Considerations .......................... 55
- Repeating Commands .......................... 55
  - Recent Input .......................... 55
- Dynamic Input .......................... 56
  - Entering Commands .......................... 56
  - Entering Coordinates .......................... 57
  - Selecting Options .......................... 57
  - Dynamic Dimensions .......................... 58
  - Selecting a Default Value .......................... 58
  - Dynamic Angles .......................... 58

| | |
|---|---|
| Error Messages and Warnings | 58 |
| Modifying Dynamic Input | 59 |
| Keyboard Shortcuts | 60 |
| AutoComplete | 60 |
| Function Keys | 61 |
| Aliases | 61 |
| Other Keys | 62 |

**USING MOUSE BUTTONS** .................................................................. 62

**USING DIALOG BOXES** ..................................................................... 63

| | |
|---|---|
| Hyphen Prefix | 64 |
| System Variables | 64 |
| FileDia | 64 |
| Expert | 65 |
| AttDia | 65 |
| Alternate Commands | 65 |

## THE DRAFT-EDIT-PLOT CYCLE ............................................................ 66

**DRAFTING** ........................................................................................ 66

Constructing Objects ........................................................................... 66

**BLOCKS AND TEMPLATES** ............................................................... 67

**EDITING** .......................................................................................... 67

**PLOTTING** ....................................................................................... 68

ePlots ................................................................................................. 68

**PROTECTING DRAWINGS** ................................................................ 68

## SCREEN POINTING ............................................................................ 68

Showing Points by Window Corners .................................................... 68

## COORDINATE SYSTEMS ..................................................................... 69

**AUTOCAD COORDINATES** ............................................................... 70

| | |
|---|---|
| Cartesian Coordinates | 71 |
| Polar Coordinates | 72 |
| Cylindrical Coordinates | 73 |
| Spherical Coordinates | 74 |
| Relative Coordinates | 75 |
| User-defined Coordinate Systems | 76 |

**ENTERING COORDINATES** ................................................................ 76

| | |
|---|---|
| Screen Picks | 77 |
| Keyboard Coordinates | 77 |
| Snaps, Etc. | 77 |
| Point Filters | 78 |
| Snap Spacing | 78 |
| Direct Distance Entry | 78 |
| Tracking | 78 |
| Object Snaps | 79 |

## EXERCISES ......................................................................................... 80

## CHAPTER REVIEW ............................................................................. 82

# 3 Setting Up Drawings .................................................................. 87

## NEW DRAWINGS ........................................................................................................ 88
### TUTORIAL: STARTING NEW DRAWINGS ................................................................ 88
#### About Templates ............................................................................................................. 89

## QNEW ..................................................................................................................... 90
### TUTORIAL: STARTING NEW DRAWINGS WITH TEMPLATES ...................................... 90
### STARTUP DIALOG BOX .............................................................................................. 90
#### Starting From Scratch ...................................................................................................... 90
#### Using Templates ............................................................................................................... 91
#### Using Wizards .................................................................................................................. 92
### TUTORIAL: USING THE ADVANCED WIZARD ........................................................... 92

## OPEN ...................................................................................................................... 96
### TUTORIAL: OPENING DRAWINGS ............................................................................. 96
#### Files List ........................................................................................................................... 98
#### Preview ............................................................................................................................ 99
#### Files of Type .................................................................................................................... 99
##### Read-Only .................................................................................................................. 100
##### Partial Open ............................................................................................................... 100
##### Select Initial View ...................................................................................................... 101
##### Making Dialog Boxes Bigger ....................................................................................... 101
#### Standard Folders Sidebar ............................................................................................... 101
#### Tools .............................................................................................................................. 102
##### Finding Files .............................................................................................................. 102

## CHANGING SCREEN COLORS .................................................................................. 103
#### Default Background Colors ............................................................................................ 103
### TUTORIAL: CHANGING COLORS ............................................................................ 104

## UNITS .................................................................................................................... 105
### TUTORIAL: SETTING UNITS ..................................................................................... 105
### SETTING SCALE FACTORS ...................................................................................... 107
#### Scale-dependent Objects ................................................................................................ 108
#### Annotation Scaling ......................................................................................................... 109

## SAVE AND SAVEAS ................................................................................................ 110
### TUTORIAL: SAVING DRAWINGS BY OTHER NAMES .............................................. 110
### SAVING DRAWINGS: ADDITIONAL METHODS ....................................................... 111
#### My Network Places ........................................................................................................ 111
### TUTORIAL: SAVING DRAWINGS ON OTHER COMPUTERS .................................... 111
#### FTP ................................................................................................................................ 112
### TUTORIAL: SENDING DRAWINGS BY FTP .............................................................. 113
#### Files of Type .................................................................................................................. 114
### TUTORIAL: SAVING DRAWINGS IN OTHER FORMATS ........................................... 114
#### Options .......................................................................................................................... 115

## QSAVE ............................................................................................................................. 117
### TUTORIAL: SAVING DRAWINGS ...................................................................... 117
### AUTOMATIC BACKUPS .................................................................................... 117
### TUTORIAL: MAKING BACKUP COPIES ........................................................... 117
### MAKING BACKUPS: ADDITIONAL METHODS ............................................ 119
#### SaveFilePath .................................................................................................................. 119
#### Accessing Backup Files ................................................................................................ 119

## QUIT ................................................................................................................................. 120
### TUTORIAL: QUITTING AUTOCAD ................................................................... 120

## EXERCISES .................................................................................................................... 121

## CHAPTER REVIEW .................................................................................................... 123

# UNIT II — DRAFTING ESSENTIALS ................................................................ 127

## 4 Drawing Basic Objects ............................................................. 129

### LINE .......................................................................................................... 130
- TUTORIAL: DRAWING LINE SEGMENTS ........................................................ 131
- DRAWING LINE SEGMENTS: ADDITIONAL METHODS .................................. 132
  - Close ............................................................................................................. 133
  - Undo ............................................................................................................. 133
  - Relative Coordinates and Angles .................................................................. 133
  - Continue ....................................................................................................... 134

### RECTANG ................................................................................................ 134
- TUTORIAL: DRAWING RECTANGLES ............................................................. 134
- ADDITIONAL METHODS FOR DRAWING RECTANGLES ................................ 135
  - Dimensions ................................................................................................... 135
  - Area .............................................................................................................. 136
  - Rotation ........................................................................................................ 137
  - Width ............................................................................................................ 137
  - Elevation ....................................................................................................... 138
  - Thickness ...................................................................................................... 138
  - Fillet .............................................................................................................. 139
  - Chamfer ........................................................................................................ 140

### POLYGON ................................................................................................ 140
- TUTORIAL: DRAWING POLYGONS ................................................................ 141
- ADDITIONAL METHODS FOR DRAWING POLYGONS ................................... 141
  - Edge .............................................................................................................. 142
  - Inscribed in Circle ......................................................................................... 143
  - Circumscribed About Circle ......................................................................... 143

### CIRCLE ..................................................................................................... 144
- TUTORIAL: DRAWING CIRCLES .................................................................... 144
- ADDITIONAL METHODS FOR DRAWING CIRCLES ....................................... 144
  - Diameter ....................................................................................................... 145
  - 2P .................................................................................................................. 145
  - 3P .................................................................................................................. 145
  - Ttr (tan tan radius) ........................................................................................ 146
  - Tan, Tan, Tan ................................................................................................ 146

### ARC .......................................................................................................... 147
- TUTORIAL: DRAWING ARCS ......................................................................... 148
- ADDITIONAL METHODS FOR DRAWING ARCS ............................................ 148
- START, CENTER, END & CENTER, START, END ............................................. 149
- START, CENTER, ANGLE & START, END, ANGLE & CENTER, START, ANGLE ........... 149
- START, CENTER, LENGTH & CENTER, START, LENGTH ................................ 150
- START, END, RADIUS .................................................................................... 150

| | |
|---|---|
| START, END, DIRECTION | 151 |
| CONTINUE | 151 |

## DONUT .................................................................................................................. 152

### TUTORIAL: DRAWING DONUTS ........................................................................ 152

## PLINE .................................................................................................................... 153

### TUTORIAL: DRAWING POLYLINES ................................................................... 153
### ADDITIONAL METHODS FOR DRAWING POLYLINES ................................. 154
- Undo .............................................................................................................. 154
- Close .............................................................................................................. 154
- Arc ................................................................................................................. 154
- Width ............................................................................................................. 154
- Halfwidth ....................................................................................................... 155
- Length ............................................................................................................ 156

## ELLIPSE ................................................................................................................. 156

### TUTORIAL: DRAWING ELLIPSES ....................................................................... 157
### ADDITIONAL METHODS FOR DRAWING ELLIPSES ..................................... 157
- Center ............................................................................................................ 158
- Rotation ......................................................................................................... 158
- Elliptical Arc ................................................................................................... 158
- Isocircle .......................................................................................................... 159

## POINT .................................................................................................................. 160

### ADDITIONAL METHODS FOR DRAWING POINTS ....................................... 160
- Multiple .......................................................................................................... 160
- DdPType ........................................................................................................ 161

## U AND UNDO ..................................................................................................... 162

### TUTORIAL: UNDOING COMMANDS .............................................................. 162
### UNDOING COMMANDS: ADDITIONAL METHODS .................................... 163
- Undo .............................................................................................................. 163
  - Auto ........................................................................................................... 163
  - Control ....................................................................................................... 163
  - BEgin/End .................................................................................................. 163
  - Mark/Back .................................................................................................. 164
- Combined Zooms and Pans ........................................................................ 164
- Undo, Previous Options and Cancel Buttons .......................................... 164
  - Undo Option ............................................................................................. 164
  - Previous Option ........................................................................................ 164
  - Cancel Button ............................................................................................ 164

## REDO AND MREDO ........................................................................................... 165

### TUTORIAL: REDOING UNDOS ......................................................................... 165
- MRedo ........................................................................................................... 165

CONSTRUCTING OTHER OBJECTS ............................................................................. 166
   EQUILATERAL TRIANGLES ............................................................................... 166
   ISOSCELES TRIANGLES ..................................................................................... 167
   SCALENE TRIANGLES ....................................................................................... 170
   RIGHT ANGLE TRIANGLES ............................................................................... 171

EXERCISES ........................................................................................................... 172

CHAPTER REVIEW ............................................................................................. 182

# 5 Drawing with Precision ........................................................... 185

GRID ....................................................................................................................... 186
   TUTORIAL: TOGGLING THE GRID ................................................................. 186
   CONTROLLING THE GRID DISPLAY: ADDITIONAL METHODS ............. 187
     DSettings Command ........................................................................................... 187
     Grid Spacing ......................................................................................................... 189
     LIMITS .................................................................................................................. 189

SNAP ....................................................................................................................... 189
   TUTORIAL: TOGGLING SNAP MODE ............................................................. 189
   CONTROLLING THE SNAP: ADDITIONAL METHODS ............................... 190
     DSettings ............................................................................................................... 190
     Snap ....................................................................................................................... 191
   TUTORIAL: CHANGING THE GRID ............................................................... 192
   TUTORIAL: DRAWING WITH SNAP MODE .................................................. 192

ORTHO .................................................................................................................. 193
   TUTORIAL: TOGGLING ORTHO MODE ....................................................... 193
   TUTORIAL: DRAWING IN ORTHO MODE .................................................... 194

POLAR .................................................................................................................... 194
   TUTORIAL: TOGGLING POLAR MODE ......................................................... 194
   CONTROLLING POLAR MODE: ADDITIONAL METHODS ......................... 195
     DSettings ............................................................................................................... 195
     PolarAng .............................................................................................................. 197
     PolarDist ............................................................................................................... 197
     PolarMode ............................................................................................................ 197
   TUTORIAL: DRAWING WITH POLAR MODE ............................................... 198

OSNAP .................................................................................................................... 200
   APERTURE ......................................................................................................... 200
   RUNNING AND TEMPORARY OSNAPS ........................................................ 201
   SUMMARY OF ALL OBJECT SNAP MODES ................................................. 202
     APParent intersection ......................................................................................... 202
     CENter ................................................................................................................. 202
     ENDpoint ............................................................................................................. 203

| EXTension | 203 |
| INSERTION | 203 |
| INTersection | 204 |
| MIDpoint | 204 |
| NEArest | 204 |
| NODe | 205 |
| PARallel | 205 |
| PERpendicular | 205 |
| QUAdrant | 205 |
| TANgent | 206 |
| ADDITIONAL OBJECT SNAP MODES | 206 |
| From | 206 |
| NONe | 207 |
| TT | 207 |
| QUIck | 207 |
| OFF | 207 |
| TUTORIAL: TOGGLING AND SELECTING OSNAP MODE | 207 |
| CONTROLLING OSNAP MODES: ADDITIONAL METHODS | 208 |
| OSnap (DSettings) | 208 |
| -OSnap | 211 |
| TUTORIAL: USING OBJECT SNAP MODES | 211 |

## OTRACK ............................................................................................................... 214

| OTRACK OPTIONS | 214 |
| AutoTrack Settings | 215 |
| Alignment Point Acquisition | 215 |
| TUTORIAL: TOGGLING OTRACK MODE | 215 |

## RAY AND XLINE ................................................................................................. 216

| TUTORIAL: DRAWING CONSTRUCTION LINES | 216 |
| DRAWING CONSTRUCTION LINES: ADDITIONAL METHODS | 217 |
| TUTORIAL: DRAWING WITH CONSTRUCTION LINES | 218 |

## REDRAW AND REGEN ....................................................................................... 221

| TUTORIAL: USING THE REDRAW COMMAND | 221 |
| TRANSPARENT REDRAWS | 222 |

## ZOOM .................................................................................................................. 222

| TUTORIAL: USING THE ZOOM COMMAND | 222 |
| ZOOM COMMAND: ADDITIONAL METHODS | 223 |
| Zoom All | 223 |
| Zoom Extents | 224 |
| Zoom Object | 224 |
| Zoom Previous | 224 |
| Zoom Scale | 224 |
| Zoom Window | 225 |
| Zoom Realtime | 226 |
| Rarely Used Zoom Modes | 226 |
| Transparent Zoom | 226 |

| PAN | 227 |
|---|---|
| TUTORIAL: PANNING THE DRAWING | 227 |
| TRANSPARENT PANNING | 227 |
| SCROLL BARS | 227 |
| -PAN | 228 |
| NAVIGATION BAR | 229 |
| NAVBAR AND NAVBARDISPLAY | 229 |
| CLEANSCREENON | 231 |
| TUTORIAL: TOGGLING THE DRAWING AREA | 231 |
| VIEW | 232 |
| PREDEFINED VIEWS | 232 |
| TUTORIAL: STORING AND RECALLING NAMED VIEWS | 232 |
| TUTORIAL: RENAMING AND DELETING NAMED VIEWS | 234 |
| TUTORIAL: OPENING DRAWINGS WITH NAMED VIEWS | 235 |
| TUTORIAL: STARTING AUTOCAD WITH NAMED VIEWS | 235 |
| TUTORIAL: CHANGING WINDOWED VIEWS | 236 |
| EXERCISES | 238 |
| CHAPTER REVIEW | 243 |

# 6 Drawing with Efficiency ............ 247

| BLOCK AND INSERT | 248 |
|---|---|
| Editing the Content of Blocks | 249 |
| TUTORIAL: CREATING BLOCKS | 249 |
| Suggestions for Designing Blocks | 252 |
| CREATING BLOCKS: ADDITIONAL METHODS | 252 |
| Objects | 252 |
| Behavior | 253 |
| Settings | 254 |
| Description | 254 |
| INSERTING BLOCKS | 255 |
| TUTORIAL: PLACING BLOCKS | 255 |
| INSERTING BLOCKS: ADDITIONAL METHODS | 256 |
| Basepoint | 257 |
| Scale | 257 |
| Rotation Angle | 258 |
| Explode | 259 |
| Preview | 260 |
| DESIGNCENTER AND TOOL PALETTES | 260 |
| Inserting Blocks with DesignCenter | 260 |
| Inserting Blocks with Tool Palettes | 261 |

## CREATING AND INSERTING BLOCKS: ADDITIONAL METHODS ........... 262
### Base ........... 262
### WBlock ........... 262
#### Source ........... 262
#### Destination ........... 262
### REDEFINING BLOCKS ........... 263

## BEDIT ........... 264
### TUTORIAL: EDITING BLOCKS WITH THE BLOCK EDITOR ........... 264

## HATCH ........... 266
### ABOUT HATCHING ........... 266
#### Full Sections ........... 266
#### Revolved, Offset, and Removed Sections ........... 266
#### How Hatching Works ........... 267
### BASIC TUTORIAL: PLACING HATCH PATTERNS ........... 268
### PLACING HATCH PATTERNS: ADDITIONAL METHODS ........... 270
#### Boundaries Panel ........... 270
##### Defining Boundaries by Picking Areas ........... 272
##### Defining Boundaries by Selecting Objects ........... 273
##### Removing Boundaries ........... 273
##### Recreating Boundaries ........... 273
#### Additional Boundary Options ........... 274
##### Boundary Retention ........... 274
##### Selecting Boundary Sets ........... 274
#### Select a Hatch ........... 274
##### ISO Patterns ........... 275
##### User Defined Patterns ........... 275
#### Hatch Properties ........... 275
##### Pattern ........... 275
##### Hatch and Background Color ........... 276
##### Hatch Transparency ........... 276
##### Angle and Scale ........... 276
##### Layer ........... 277
##### Relative to Paper Space ........... 277
##### Double Hatching ........... 277
##### ISO Pen Width ........... 277
#### Origin Panel ........... 277
##### Additional Origin Options ........... 278
#### Options Panel ........... 279
##### Associative and Non-Associative Hatch Patterns ........... 279
##### Annotative Scaling ........... 279
##### Matching Properties ........... 279
##### Additional Options ........... 280
##### Gap Tolerance ........... 280

| | |
|---|---|
| Create Separate Hatches | 280 |
| Island Detection | 280 |
| Draw Order | 281 |
| Close | 281 |

## GRADIENT .................................................................................................................. 282

## HATCHEDIT ................................................................................................................ 284
### Direct Editing of Hatch Boundaries ................................................................ 284

## BOUNDARY ................................................................................................................ 285
### TUTORIAL: CREATING BOUNDARIES ........................................................... 285

## SELECT ...................................................................................................................... 286
### TUTORIAL: SELECTING OBJECTS BY LOCATION ......................................... 286
- ALL (Select All Objects) .................................................................... 288
- Pick (Select Single Object) ................................................................ 288
- Visual Selection Cycling .................................................................... 288
- W (Select Within a Window) ............................................................. 288
- C (Select with a Crossing Window) ................................................... 289
- BOX (Select with a Box) .................................................................... 290
- F (Select with a Fence) ..................................................................... 290
- AU (Select with the Automatic Option) ............................................. 291
- WP (Select with a Windowed Polygon) ............................................. 291
- CP (Select with a Crossing Polygon) ................................................. 291
- SI (Select a Single Object) ................................................................ 292
- SU and O (Select Subobjects and Objects) ...................................... 292
- M (Select Through the Multiple Option) ........................................... 292
- P (Select the Previous Selection Set) ............................................... 292
- L (Select the Last Object) ................................................................. 292
- G (Select the Group) ......................................................................... 292

### Changing Selected Items ................................................................................ 292
- U (Undo the Selected Option) ........................................................... 293
- R (Remove Objects from the Selection Set) ..................................... 293
- A (Add Objects to the Selection Set) ................................................ 293

### Canceling the Selection Process .................................................................... 293
### Selecting Objects During Commands ............................................................ 293

## QSELECT ................................................................................................................... 294
### TUTORIAL: SELECTING OBJECTS BY PROPERTIES ..................................... 294
### CONTROLLING THE SELECTION: ADDITIONAL METHODS ....................... 295
#### Pickbox ........................................................................................................... 295
#### DdSelect ......................................................................................................... 296
- Pickbox Size ...................................................................................... 296
- Selection Preview .............................................................................. 296
- Noun/Verb Selection .......................................................................... 297
- Use Shift to Add to Selection ............................................................ 298
- Press and Drag .................................................................................. 298

- Implied Windowing ........................................................................................... 298
- Object Grouping ............................................................................................. 298
- Associative Hatch ........................................................................................... 298
- Filter .................................................................................................................... 299
- SELECTSIMILAR ............................................................................................. 301
- TUTORIAL: SELECTING SIMILAR OBJECTS ............................................. 301
- ADDSELECTED ............................................................................................... 302
- TUTORIAL: DRAWING SIMILAR OBJECTS .............................................. 302
- HIDEOBJECTS AND ISOLATEOBJECTS .................................................... 303
- UnisolateObjects ............................................................................................. 304
- TUTORIAL: HIDING AND ISOLATING OBJECTS ................................... 304
  - Isolating Objects ........................................................................................... 305
  - Unisolating Objects ...................................................................................... 305

# DRAWORDER .................................................................................................... 306
- TUTORIAL: SPECIFYING DRAWORDER .................................................... 306
- CHANGING THE DRAW ORDER: ADDITIONAL METHODS ............... 306
- HpDrawOrder ................................................................................................... 306
  - HatchToBack .................................................................................................. 307
- TextToFront ...................................................................................................... 307

# EXERCISES ........................................................................................................... 308

# CHAPTER REVIEW ............................................................................................ 316

## SECTION III — EDITING DRAWINGS .................................................................................... 319

# 7  Properties and Layers .................................................................. 321

### MATCHPROP ........................................................................................................... 322
- TUTORIAL: MATCHING PROPERTIES ........................................................................ 322
  - Settings ............................................................................................................... 323

### COLOR ..................................................................................................................... 324
- TUTORIAL: CHANGING COLORS ................................................................................ 324
- UNDERSTANDING AUTOCAD'S COLOR SYSTEM ........................................................ 325
  - Color Numbers and Names ................................................................................ 326
  - Background Color 0 ............................................................................................ 326
    - Color 7 .......................................................................................................... 326
  - Colors ByLayer, ByBlock, and ByEntity ............................................................. 326
    - ByLayer Color #256 ..................................................................................... 326
    - ByBlock Color #0 ......................................................................................... 326
    - ByEntity Color #257 .................................................................................... 327
  - True Color ........................................................................................................... 327
    - RGB .............................................................................................................. 327
    - HSL .............................................................................................................. 328
    - Color Books ................................................................................................. 329
- CHANGING COLORS: ADDITIONAL METHODS ........................................................... 330
- TUTORIAL: SETTING AND CHANGING COLORS ......................................................... 330

### ⊕ TRANSPARENCYDISPLAY ..................................................................................... 331

### LWEIGHT .................................................................................................................. 332
- TUTORIAL: CHANGING LINEWEIGHTS ...................................................................... 333
- AUTOCAD'S DEFAULT LINEWEIGHTS ........................................................................ 333
  - Lineweight Control ............................................................................................ 334

### LINETYPE .................................................................................................................. 335
- TUTORIAL: CHANGING LINETYPES ........................................................................... 335
  - Global, Local, and Annotative Linetype Scales ................................................. 338
    - Local (Object) Linetype Scaling ................................................................... 338
    - Annotative Scaling ...................................................................................... 339
- LINETYPE SCALING (LTSCALE) ................................................................................... 339
  - Linetype Control ................................................................................................. 340

### PROPERTIES ............................................................................................................. 340
- TUTORIAL: CHANGING ALL PROPERTIES ................................................................... 340
- CHANGING PROPERTIES: ADDITIONAL METHODS .................................................... 343
- QuickProperties .................................................................................................... 343
- TUTORIAL: USING QUICK PROPERTIES ..................................................................... 343
  - QuickProperties Settings .................................................................................... 345
- Rollover Tips ........................................................................................................ 345

## LAYER ............ 346
### TUTORIAL: CREATING LAYERS ............ 347
#### Shortcut Keystrokes ............ 348
#### Working with Layers ............ 348
- Status ............ 348
- Turning Layers On and Off ............ 348
- Freezing and Thawing Layers ............ 348
- Locking and Unlocking Layers ............ 348
- Setting Layer Colors ............ 349
- Setting Layer Linetypes ............ 349
- Setting Layer Lineweights ............ 349
- Setting Transparency ............ 350
- Setting the Layer Plot Style ............ 350
- Setting Layer Print Toggles ............ 350
- Description ............ 352

#### Layer Controls in Layouts ............ 352
- Layout Freeze Modes ............ 352
- Layout Property Modes ............ 352
- Resetting Overridden Layout Properties ............ 353

#### Selecting Layers ............ 353
- Sorting by Columns ............ 353
- Searching for Layer Names ............ 355
- Creating Group Filters ............ 356

### TUTORIAL: CREATING GROUP FILTERS ............ 357
### TUTORIAL: CREATING PROPERTY FILTERS ............ 359
- Deleting Layers ............ 361

#### Layer States Manager ............ 361
### TUTORIAL: CREATING AND APPLYING LAYER STATES ............ 361
### CONTROLLING LAYERS: ADDITIONAL METHODS ............ 363
#### Layers on the Ribbon ............ 363
- Changing the Current Layer ............ 364
- Changing the Layer's Status ............ 364

#### LayMCur ............ 365
#### CLayer ............ 365
#### LayerP and LayerPMode ............ 365
- Layer Tools ............ 365

## LAYISO ............ 366

## EXERCISES ............ 368

## CHAPTER REVIEW ............ 370

# 8 Constructing Objects ... 373

## COPY ... 374
### TUTORIAL: MAKING COPIES ... 374
### MAKING COPIES: ADDITIONAL METHODS ... 375
#### Displacement ... 375
##### More on Displacements ... 375

## MIRROR ... 376
### TUTORIAL: MAKING MIRRORED COPIES ... 377
### MAKING MIRRORED COPIES: ADDITIONAL METHOD ... 377

## OFFSET ... 378
### TUTORIAL: MAKING OFFSET COPIES ... 379
### MAKING OFFSET COPIES: ADDITIONAL METHODS ... 380
### TUTORIAL: THROUGH ... 380
#### Erase ... 380
#### Layer ... 380
#### OffsetGapType ... 381

## DIVIDE AND MEASURE ... 381
### TUTORIAL: DIVIDING OBJECTS ... 382
### TUTORIAL: "MEASURING" OBJECTS ... 382
### DIVIDING AND MEASURING: ADDITIONAL METHODS ... 383
#### Block ... 383
#### DdPType ... 384

## ARRAY ... 384
### TUTORIAL: LINEAR AND RECTANGULAR ARRAYS ... 385
### TUTORIAL: POLAR AND SEMICIRCULAR ARRAYS ... 387

## FILLET AND CHAMFER ... 389
### TUTORIAL: FILLETING OBJECTS ... 389
### DIFFERENT FILLET RESULTS FOR DIFFERENT OBJECTS ... 390
#### Intersecting Lines ... 390
#### Parallel Lines ... 390
#### Polylines ... 390
#### Arcs and Circles ... 391
### CONSTRUCTING FILLETS: ADDITIONAL METHODS ... 392
#### Undo ... 392
#### Polyline ... 392
#### Radius ... 392
#### Trim ... 392
#### Multiple ... 392
### CHAMFER ... 393
### TUTORIAL: CHAMFERING OBJECTS ... 393
### CONSTRUCTING CHAMFERS: ADDITIONAL METHODS ... 394

| | |
|---|---|
| Undo | 394 |
| Polyline | 394 |
| Angle | 394 |
| Trim | 394 |
| mEthod | 395 |
| Multiple | 395 |

### JOIN .................................................................................................. 395
- TUTORIAL: JOINING OBJECTS ........................................................ 396
- JOINING OBJECTS: ADDITIONAL METHODS .................................. 396

### REVERSE ........................................................................................... 397
- TUTORIAL: REVERSING OBJECTS .................................................. 397

### REVCLOUD ....................................................................................... 398
- TUTORIAL: REDLINING OBJECTS ................................................... 398
- REDLINING OBJECTS: ADDITIONAL METHODS .............................. 399
  - Arc length .................................................................................... 399
  - Object ......................................................................................... 399
  - Style ........................................................................................... 400

### MARKUP ........................................................................................... 400
- Design Review ............................................................................... 400
- Markup Options ............................................................................. 403

### EXERCISES ....................................................................................... 404

### CHAPTER REVIEW ........................................................................... 414

## 9 Additional Editing Methods .................................. 417

### ERASE ............................................................................................... 418
- TUTORIAL: ERASING OBJECTS ....................................................... 418
- RECOVERING ERASED OBJECTS .................................................... 418
  - Oops ........................................................................................... 418
  - U ................................................................................................. 419

### BREAK ............................................................................................... 419
- TUTORIAL: BREAKING OBJECTS ..................................................... 419
- BREAKING OBJECTS: ADDITIONAL METHOD ................................. 420
  - First ............................................................................................. 420

### TRIM, EXTEND, AND LENGTHEN .................................................. 421
- TUTORIAL: SHORTENING OBJECTS .............................................. 422
  - Trimming Closed Objects ............................................................. 422
  - Trimming Polylines ...................................................................... 423
- TRIMMING OBJECTS: ADDITIONAL METHODS .............................. 423
  - Select All ..................................................................................... 423
  - Fence and Crossing ..................................................................... 423
  - Project ........................................................................................ 424

- Edge .................................................................................................................................. 424
- eRase ................................................................................................................................ 424
- Undo ................................................................................................................................ 424
- Extend (SHIFT-select) ...................................................................................................... 424
- EXTEND .............................................................................................................................. 425
- TUTORIAL: EXTENDING OBJECTS ................................................................................. 425
- Extending Polylines ........................................................................................................ 426
- EXTEND OPTIONS ............................................................................................................ 426
- LENGTHEN ......................................................................................................................... 426
- TUTORIAL: LENGTHENING OBJECTS ............................................................................ 426
- LENGTHENING OBJECTS: ADDITIONAL METHODS ................................................... 427
- Delta ............................................................................................................................... 428
- Percent ........................................................................................................................... 428
- Total ............................................................................................................................... 428
- Dynamic .......................................................................................................................... 429

## STRETCH ............................................................................................................................ 429
- STRETCHING — THE RULES ........................................................................................... 429
- TUTORIAL: STRETCHING OBJECTS ............................................................................... 430

## MOVE .................................................................................................................................. 431
- TUTORIAL: MOVING OBJECTS ....................................................................................... 431
- Command-less Moving ................................................................................................... 431

## ROTATE ............................................................................................................................... 432
- TUTORIAL: ROTATING OBJECTS .................................................................................... 432
- ROTATING OBJECTS: ADDITIONAL METHODS ............................................................ 433
- Reference ........................................................................................................................ 433
- Copy ............................................................................................................................... 434
- Dragging ......................................................................................................................... 434

## SCALE ................................................................................................................................. 434
- TUTORIAL: CHANGING SIZE .......................................................................................... 434
- CHANGING SIZE: ADDITIONAL METHODS .................................................................. 435
- Copy ............................................................................................................................... 435
- Reference ........................................................................................................................ 435

## CHANGE ............................................................................................................................. 437
- TUTORIAL: CHANGING OBJECTS .................................................................................. 437
- CHANGING OBJECTS: ADDITIONAL METHODS .......................................................... 437
- Lines ............................................................................................................................... 437
- Circles ............................................................................................................................. 438
- Blocks ............................................................................................................................. 438
- Text ................................................................................................................................ 438
- Attribute Definitions ...................................................................................................... 438

## EXPLODE AND XPLODE ........... 439

### TUTORIAL: EXPLODING COMPOUND OBJECTS ........... 439
### XPLODE ........... 441
### TUTORIAL: CONTROLLED EXPLOSIONS ........... 441
All ........... 441
Color ........... 441
Layer ........... 441
LType ........... 441
Inherit from Parent Block ........... 441
Explode ........... 441

## PEDIT ........... 442

### TUTORIAL: STARTING THE PEDIT COMMAND ........... 442
Close/Open ........... 443
Join ........... 443
Width ........... 444
Edit vertex ........... 445
Fit ........... 445
Spline ........... 445
Decurve ........... 446
Ltype gen ........... 446
Reverse ........... 446
Undo ........... 446

### EDITING POLYLINES: ADDITIONAL METHODS ........... 446
Select ........... 447
Mutiple ........... 447
Jointype ........... 447

### CONTROLLING POLYLINE EDITING ........... 448
PEditAccept ........... 448
PLINECONVERTMODE ........... 448
SPLINESEGS ........... 448
SPLINETYPE ........... 448

## GRIPS ........... 449

### DIRECT EDITING WITH GRIPS ........... 449
Grip Color and Size ........... 450

### TUTORIAL: EDITING WITH GRIPS ........... 450
### GRIP EDITING MODES ........... 451
### GRIP EDITING OPTIONS ........... 452
Stretch Mode ........... 452
Move Mode ........... 452
Rotate Mode ........... 453
Scale Mode ........... 453
Mirror Mode ........... 453
Grip Editing Options ........... 454
    Base point ........... 454
    Copy ........... 454

| | |
|---|---|
| Undo | 454 |
| Reference | 454 |
| eXit | 455 |
| Other Grip Editing Commands | 455 |
| Grips on Arcs | 455 |
| Grips on Polyline | 456 |
| Grips on Hatches | 456 |

## RIGHT-CLICK MOVE/COPY/INSERT .................................................................. 457

## DIRECT DISTANCE ENTRY ............................................................................... 458
TUTORIAL: DIRECT DISTANCE ENTRY ............................................................. 458

## TRACKING ............................................................................................................ 459
TUTORIAL: TRACKING ......................................................................................... 459
OSNAP WITH OBJECT TRACKING .................................................................... 460

## M2P ........................................................................................................................ 462
TUTORIAL: M2P ..................................................................................................... 462

## EXERCISES ............................................................................................................ 463

## CHAPTER REVIEW .............................................................................................. 476

# UNIT IV — TEXT AND DIMENSIONS ............... 479

## 10  Placing and Editing Text .................. 481

### TEXT IN DRAWINGS ............... 482
#### FONTS ............... 483

### TEXT ............... 483
#### TUTORIAL: PLACING LINES OF TEXT ............... 483
#### PLACING TEXT ALL OVER DRAWINGS ............... 485
#### PLACING TEXT: ADDITIONAL METHODS ............... 485
##### Justify ............... 486
###### Align ............... 488
###### Fit ............... 488
##### Style ............... 489
##### Height ............... 489
###### Determining the Height of Text ............... 489
###### Annotative Scaling ............... 490
##### Rotation Angle ............... 490
##### Control Codes ............... 491
###### ASCII Characters ............... 492
##### DTextEd ............... 492
#### TEXT SHORTCUT MENU ............... 493
##### Editor Settings ............... 493
###### Always Display as WYSIWYG ............... 493
##### Opaque Background ............... 493
##### Check Spelling ............... 494
##### Check Spelling Options and Dictionaries ............... 494
##### Text Highlight Color ............... 494
##### Insert Field ............... 494
###### Update Field ............... 495
###### Convert Field to Text ............... 495
##### Find and Replace ............... 495
##### Select All ............... 495
##### Change Case ............... 495

### MTEXT ............... 496
#### TUTORIAL: PLACING PARAGRAPH TEXT ............... 496
#### RIBBON MTEXT TAB ............... 497
##### Style Panel ............... 497
###### Annotative Property ............... 498
###### Text Height ............... 498
##### Formatting Panel ............... 498
###### Bold, Italic, Underline, and Overline ............... 498
##### UPPER and lower Case ............... 498
###### Font ............... 499
###### Color ............... 499

- Background Mask .................................................................................................................. 499
- Oblique Angle, Tracking, and Width Factor ........................................................................ 500
- Stack ............................................................................................................................................. 500
- Paragraph Panel .................................................................................................................................. 502
  - Justification ................................................................................................................................. 502
  - Popular Justifications ................................................................................................................ 502
  - Bullets .......................................................................................................................................... 503
  - Line Spacing ................................................................................................................................ 504
  - Combine Paragraphs ................................................................................................................. 504
  - Paragraph Options ..................................................................................................................... 504
- Insert Panel ......................................................................................................................................... 505
  - Columns ....................................................................................................................................... 505
  - Insert Symbol .............................................................................................................................. 505
  - Insert Fields ................................................................................................................................. 506
- Spell Check Panel ............................................................................................................................... 507
  - Spell Check .................................................................................................................................. 507
  - Edit Dictionaries ......................................................................................................................... 507
  - Spell Check Settings .................................................................................................................. 507
- Tools Panel .......................................................................................................................................... 508
  - Find and Replace ........................................................................................................................ 508
  - Import Text ................................................................................................................................. 508
  - AutoCAPS .................................................................................................................................... 508
- Options Panel ..................................................................................................................................... 508
  - More Options ............................................................................................................................. 509
  - Character Set .............................................................................................................................. 509
  - Remove Formatting ................................................................................................................... 509
  - Editor Settings ............................................................................................................................ 509
  - Toggle Ruler ................................................................................................................................ 509
  - Undo and Redo .......................................................................................................................... 509
- Close Panel ......................................................................................................................................... 510

## MTEXT SHORTCUT MENU .................................................................................................................. 510
- Paste Special ............................................................................................................................... 510

## MTEXT TEXT ENTRY AREA .................................................................................................................. 511
- First Line Indent ................................................................................................................................. 511
- Paragraph Indent ................................................................................................................................ 512
- Tabs ....................................................................................................................................................... 512
  - Set Mtext Width and Set Mtext Height ................................................................................ 512

## PLACING PARAGRAPH TEXT: ADDITIONAL METHODS ................................................................. 513
- Height .................................................................................................................................................. 513
- Justify ................................................................................................................................................... 513
- Linespacing ......................................................................................................................................... 513
- Rotation ............................................................................................................................................... 513
- Style ...................................................................................................................................................... 514
- Width ................................................................................................................................................... 514
- Columns ............................................................................................................................................... 514

| | |
|---|---|
| TEXTTOFRONT | 514 |

## STYLE ........................................................................................................................ 515

### TUTORIAL: DEFINING TEXT STYLES ................................................................ 515
### USING TEXT STYLES ............................................................................................ 519
Styles Panel .................................................................................................................... 519
### DEFINING STYLES: ADDITIONAL METHODS ................................................ 519
FontAlt ............................................................................................................................ 520
TextFill ............................................................................................................................ 520
TextQlty .......................................................................................................................... 520

## QLEADER .................................................................................................................. 521

### TUTORIAL: PLACING LEADERS .......................................................................... 521
### PLACING LEADERS: ADDITIONAL METHODS ............................................... 522
Annotation Type ........................................................................................................... 523
MText Options .............................................................................................................. 523
Annotation Reuse ......................................................................................................... 524
Leader Line ................................................................................................................... 524
Number of Points ......................................................................................................... 524
Arrowhead ..................................................................................................................... 525
Angle Constraints ......................................................................................................... 525
Multiline Text Attachment .......................................................................................... 525

## MLEADER .................................................................................................................. 526

### TUTORIAL: DRAWING MULTILINE LEADERS WITH TEXT (MLEADER) ..... 526
MLeader Command Options ...................................................................................... 527
### TUTORIAL: ADDING LEADERS TO MLEADERS (MLEADEREDIT) .............. 528
### TUTORIAL: DRAWING MLEADERS WITH BUBBLES (MLEADERSTYLE) .... 529
### TUTORIAL: COLLECTING BUBBLES INTO ONE MLEADER (MLEADERCOLLECT) ............. 532
MLeaderCollect Options ............................................................................................. 532
### TUTORIAL: ALIGNING MLEADERS (MLEADERALIGN) ................................. 534
MLeaderAlign Options ................................................................................................ 535

## TEXTEDIT ................................................................................................................... 536

### TUTORIAL: EDITING TEXT ................................................................................... 536
Single-line Text Editor ................................................................................................. 537
Multiline and Leader Text Editor ............................................................................... 537
Attribute Text Editor .................................................................................................... 537
Enhanced Attribute Editor .......................................................................................... 538
### EDITING TEXT: ADDITIONAL METHODS ........................................................ 538

## SPELL .......................................................................................................................... 539

### TUTORIAL: CHECKING SPELLING ...................................................................... 539

## FIND ............................................................................................................................ 540

### TUTORIAL: FINDING TEXT .................................................................................. 540

## SCALETEXT AND JUSTIFYTEXT .................................................................................. 541
### TUTORIAL: RESIZING TEXT ............................................................................. 541
### TUTORIAL: REJUSTIFYING TEXT ..................................................................... 542

## ANNOTATION SCALING ........................................................................................ 542
### TUTORIAL: AUTOMATICALLY SCALING TEXT ................................................ 543
### ANNOTATION SCALING: ADDITIONAL METHODS ........................................ 547
#### ObjectScale ........................................................................................................ 548
#### SelectionAnnoDisplay and XFadeCtl ............................................................. 548
#### AnnoUpdate ...................................................................................................... 550
#### AnnoReset .......................................................................................................... 550
#### AnnotativeDwg ................................................................................................. 550
#### DimAnno ........................................................................................................... 550
#### MsLtScale ........................................................................................................... 550
#### SaveFidelity ....................................................................................................... 551

## EXERCISES ............................................................................................................... 552
### Creating an Annotative Text Style ..................................................................... 556
### Controlling Annotation Visibility ...................................................................... 556
### Resetting and Updating Annotative Text ......................................................... 556

## CHAPTER REVIEW .................................................................................................. 558

# 11 Placing Dimensions ...................................................... 563

## INTRODUCTION TO DIMENSIONS ..................................................................... 564
### THE PARTS OF DIMENSIONS ........................................................................... 564
#### Dimension Lines ............................................................................................... 565
#### Extension Lines ................................................................................................ 565
#### Arrowheads ...................................................................................................... 566
#### Dimension Text ................................................................................................ 566
##### Annotative Scaling ..................................................................................... 567
##### Tolerance Text ............................................................................................ 568
##### Limits Text .................................................................................................. 568
##### Alternate Units ........................................................................................... 568
### DIMENSIONING OBJECTS .............................................................................. 568
#### Lines and Polyline Segments .......................................................................... 568
#### Arcs, Circles, and Polyarcs .............................................................................. 569
#### Wedges and Cylinders ..................................................................................... 570
#### Cones and Pyramids ........................................................................................ 571
#### Holes .................................................................................................................. 572
#### Vertices .............................................................................................................. 572
#### Single Points ..................................................................................................... 572
#### Chamfers and Tapers ....................................................................................... 573
### DIMENSION MODES ......................................................................................... 573
#### Dimension Variables and Styles ..................................................................... 574
#### Dimension Standards ...................................................................................... 574
#### Object Snaps ..................................................................................................... 574

## DIMLINEAR ............................................................................................................................. 575
### TUTORIAL: PLACING DIMENSIONS HORIZONTALLY ........................................................... 575
### TUTORIAL: PLACING DIMENSIONS VERTICALLY AND ROTATED ........................................ 577
#### Dimensioning Objects ............................................................................................................ 578
### TUTORIAL: DIMENSIONING OBJECTS ..................................................................................... 578
### LINEAR DIMENSIONS: ADDITIONAL METHODS .................................................................. 579
#### MText ........................................................................................................................................ 579
##### Special Symbols .................................................................................................................. 580
#### Text ............................................................................................................................................ 580
#### Angle .......................................................................................................................................... 580
#### Horizontal, Vertical and Rotated ............................................................................................. 580

## DIMALIGNED ........................................................................................................................ 581
### TUTORIAL: PLACING ALIGNED DIMENSIONS ....................................................................... 582

## DIMBASELINE AND DIMCONTINUE ................................................................................ 583
### TUTORIAL: PLACING ADDITIONAL DIMENSIONS FROM A BASELINE .............................. 583
### TUTORIAL: CONTINUING DIMENSIONS ............................................................................... 584

## QDIM .................................................................................................................................... 585
### TUTORIAL: QUICK CONTINUOUS DIMENSIONS ................................................................. 585
### QUICK DIMENSIONS: ADDITIONAL METHODS .................................................................. 586
#### Continuous ................................................................................................................................ 586
#### Staggered ................................................................................................................................... 586
#### Baseline ...................................................................................................................................... 587
#### Ordinate .................................................................................................................................... 587
#### Radius ........................................................................................................................................ 587
#### Diameter .................................................................................................................................... 588
#### datumPoint ................................................................................................................................ 588
#### Edit ............................................................................................................................................. 588
#### seTtings ..................................................................................................................................... 588

## DIMRADIUS AND DIMDIAMETER ..................................................................................... 589
### TUTORIAL: PLACING RADIAL DIMENSIONS .......................................................................... 590
### DIMJOGGED .................................................................................................................................. 591
### TUTORIAL: PLACING JOGGED RADIAL DIMENSIONS .......................................................... 591
### DIMARC ......................................................................................................................................... 592
### TUTORIAL: PLACING ARC LENGTH DIMENSIONS ............................................................... 592
### ARC LENGTH DIMENSIONS: ADDITIONAL METHODS ...................................................... 592
#### Partial .......................................................................................................................................... 593
#### Leader ........................................................................................................................................ 593

## DIMANGULAR ..................................................................................................................... 593
### TUTORIAL: PLACING ANGULAR DIMENSIONS ..................................................................... 594
#### Quadrant ................................................................................................................................... 595
### DIMCENTER .................................................................................................................................. 595
### TUTORIAL: PLACING CENTER MARKS ................................................................................... 595

DimCen .................................................................................................................. 596
   TUTORIAL: DIMENSIONING ............................................................................ 597
   TUTORIAL: ANNOTATIVE DIMENSIONING ...................................................... 599
     Changing Linetype Scaling ............................................................................ 601
     Changing the Annotation Scale .................................................................... 601

**EXERCISES** ............................................................................................................ 603

**CHAPTER REVIEW** ............................................................................................... 613

# 12 Editing Dimensions ............................................................... 617

## AIDIMFLIPARROW ............................................................................................. 618
   TUTORIAL: FLIPPING ARROWHEADS .............................................................. 618

## DIMJOGLINE ....................................................................................................... 619
   TUTORIAL: JOGGING DIMENSION LINES ........................................................ 619

## DIMBREAK ......................................................................................................... 620
   TUTORIAL: BREAKING DIMENSION AND EXTENSION LINES ........................... 620

## DIMSPACE .......................................................................................................... 621
   TUTORIAL: EVENLY SPACING DIMENSIONS .................................................... 621

## DIMEDIT, DIMTEDIT AND DDEDIT ..................................................................... 622
   TUTORIAL: OBLIQUING EXTENSION LINES ..................................................... 623
   TUTORIAL: REPOSITIONING DIMENSION TEXT ............................................... 623
   TUTORIAL: EDITING DIMENSION TEXT ............................................................ 624

## DIMINSPECT ....................................................................................................... 625
   TUTORIAL: ADDING INSPECTION TEXT TO DIMENSIONS ............................... 626

## AIDIM AND AI_DIM ............................................................................................ 627
   AIDIMPREC ........................................................................................................ 628
   TUTORIAL: CHANGING DISPLAY PRECISION .................................................. 628
   AIDIMTEXTMOVE .............................................................................................. 629
   TUTORIAL: MOVING DIMENSION TEXT ........................................................... 629
   Quick Dimension Text Moves ............................................................................ 630
     Ai_Dim_TextCenter .................................................................................... 630
     Ai_Dim_TextHome ..................................................................................... 630
     Ai_Dim_TextAbove ..................................................................................... 630

## DIMENSION VARIABLES AND STYLES ................................................................. 631
   CONTROLLING DIMVARS ................................................................................ 631
   SOURCES OF DIMSTYLES ................................................................................. 631
   Dimstyles from Other Sources ........................................................................... 632
     DesignCenter .............................................................................................. 632
     Express Tools ............................................................................................... 632
     Binding External References ...................................................................... 633

**DIMSTYLE** ................................................................................................................. 633
    WORKING WITH DIMENSION STYLES ........................................................... 633
    TUTORIAL: NEW — CREATING NEW DIMSTYLES ...................................... 635
    MODIFY — MODIFYING DIMSTYLES ............................................................. 635
    LINES TAB ............................................................................................................. 636
        Dimension Lines .................................................................................................. 636
            Color ................................................................................................................ 636
            Linetype .......................................................................................................... 636
            Lineweight ..................................................................................................... 636
            Extend Beyond Ticks .................................................................................. 637
            Baseline Spacing ......................................................................................... 637
            Suppress ......................................................................................................... 637
        Extension Lines ................................................................................................... 637
            Color, Linetype, and Lineweight ............................................................. 637
            Suppress ......................................................................................................... 637
            Additional Options ..................................................................................... 637
    SYMBOLS AND ARROWS .................................................................................. 638
        Arrowheads ........................................................................................................ 638
            First, Second, and Leader ........................................................................ 638
            Arrow Size ..................................................................................................... 638
    TUTORIAL: CREATING CUSTOM ARROWHEADS ....................................... 639
        Center Marks for Circles ................................................................................. 640
        Dimension Break ................................................................................................ 641
        Arc Length Symbol ............................................................................................ 641
        Radius Dimension Jog ...................................................................................... 641
        Linear Jog Dimension ....................................................................................... 641
    TEXT ....................................................................................................................... 642
        Text Appearance ................................................................................................ 642
            Text Style ....................................................................................................... 642
            Text Color ...................................................................................................... 642
            Fill Color ......................................................................................................... 642
            Text Height ................................................................................................... 643
            Fraction Height Scale ................................................................................ 643
            Draw Frame Around Text ......................................................................... 643
        Text Placement .................................................................................................. 643
            Vertical ........................................................................................................... 643
            Horizontal ..................................................................................................... 643
            View Direction ............................................................................................. 643
            Offset From Dim Line ................................................................................ 643
        Text Alignment ................................................................................................... 643
    FIT ........................................................................................................................... 644
        Fit Options ........................................................................................................... 644
        Text Placement .................................................................................................. 645
        Scale for Dimension Features ........................................................................ 645
        Fine Tuning .......................................................................................................... 645

## PRIMARY UNITS ............ 646
### Linear Dimensions ............ 646
#### Units Format ............ 646
#### Precision ............ 646
#### Fraction Format ............ 647
#### Decimal Separator ............ 647
#### Round Off ............ 647
#### Prefix and Suffix ............ 647
### Measurement Scale ............ 647
#### Scale Factor ............ 647
#### Apply to Layout Dimensions Only ............ 647
### Zero Suppression ............ 647
#### Leading and Trailing ............ 648
#### Sub-units ............ 648
#### 0 Feet and 0 Inches ............ 648
### Angular Dimensions ............ 648
#### Units Format ............ 648
#### Precision ............ 648
#### Zero Suppression ............ 648

## ALTERNATE UNITS ............ 649
### Alternate Units ............ 649
### Zero Suppression ............ 649
### Placement ............ 649

## TOLERANCES ............ 650
### Tolerance Format ............ 650
#### Method ............ 650
#### Precision ............ 651
#### Upper Value and Lower Value ............ 651
#### Scaling for Height ............ 651
#### Vertical Position ............ 651
### Tolerance Alignment ............ 651
### Zero Suppression ............ 651
### Alternate Unit Tolerance ............ 651

## SETTING THE CURRENT DIMSTYLE ............ 652
### Renaming and Deleting Dimstyles ............ 652

## MODIFYING AND OVERRIDING DIMSTYLES ............ 652

## TUTORIAL: GLOBAL RETROACTIVE DIMSTYLE CHANGES ............ 653

## TUTORIAL: GLOBAL TEMPORARY DIMSTYLE CHANGES ............ 653

# STYLE OVERRIDES, DIMOVERRIDE, AND PROPERTIES ............ 655

## TUTORIAL: LOCAL DIMSTYLE CHANGES ............ 655

## TUTORIAL: LOCAL DIMVAR OVERRIDES ............ 655
### Clearing Local Overrides ............ 656

## PROPERTIES ............ 656

# DRAWORDER FOR DIMENSIONS ............ 656

GRIPS EDITING ............................................................................................................ 656
    TUTORIAL: EDITING DIMENSIONS THROUGH OBJECTS .................................. 657
    GRIPS EDITING OPTIONS ................................................................................... 657
EXERCISES ................................................................................................................ 658
CHAPTER REVIEW ................................................................................................... 660

# 13 Applying Dimensional Constraints .................. 663

ABOUT DIMENSIONAL CONSTRAINTS ................................................................ 664
    DC... COMMANDS .............................................................................................. 665
    Types of Dimensional Constraints ......................................................................... 665
    TUTORIAL: PLACING DIMENSIONAL CONSTRAINTS ...................................... 666
    Converting Associative Dimensions to Constraints ............................................. 667
    Placing Dimensional Constraints ........................................................................... 667
    Changing Object Size Through Constraints ........................................................ 668
    TUTORIAL: LINKING DIMENSIONAL CONSTRAINTS ....................................... 670
    TUTORIAL: LINKING OBJECTS WITH CONSTRAINTS ...................................... 671
DELCONSTRAINT .................................................................................................... 673
    TUTORIAL: ERASING CONSTRAINTS ................................................................. 673
PARAMETERS ........................................................................................................... 674
    TUTORIAL: EDITING PARAMETER VALUES ....................................................... 674
    Parameters Manager Palette .................................................................................. 676
    Headers .................................................................................................................... 677
    Shortcut Menu ........................................................................................................ 677
CONSTRAINTSETTINGS .......................................................................................... 679
    CONSTRAINTS SETTINGS: ADDITIONAL METHODS ..................................... 680
EXERCISES ................................................................................................................ 681
CHAPTER REVIEW ................................................................................................... 682

# 14 Using Geometric Constraints ......................... 685

ABOUT GEOMETRIC CONSTRAINTS .................................................................... 686
    AUTOCONSTRAIN ............................................................................................... 687
    TUTORIAL: CONSTRAINING GEOMETRY AUTOMATICALLY .......................... 687
    TUTORIAL: MODIFYING GEOMETRIC CONSTRAINTS .................................... 690
GEOMCONSTRAINT ............................................................................................... 691
    TUTORIAL: APPLYING CONSTRAINTS MANUALLY .......................................... 691
    GEOMETRIC CONSTRAINT CATALOG ............................................................. 692
    GcCoincident ........................................................................................................... 692
    GcCollinear .............................................................................................................. 693

- GcConcentric ............................................................................................................................ 693
- GcEqual .................................................................................................................................... 694
- GcFix ........................................................................................................................................ 694
- GcHorizontal ........................................................................................................................... 694
- GcParallel ................................................................................................................................. 695
- GcPerpendicular ..................................................................................................................... 695
- GcSmooth ............................................................................................................................... 696
- GcSymmetric .......................................................................................................................... 696
- GcTangent .............................................................................................................................. 697
- GcVertical ............................................................................................................................... 697

## CONSTRAINTINFER ................................................................................................................. 697
### TUTORIAL: INFERRING CONSTRAINTS ................................................................................. 698
Inferred Constraints on Lines .................................................................................................. 698
Inferred Constraints on Rectangles ......................................................................................... 699
Inferred Constraints on Fillets .................................................................................................. 700

## CONSTRAINTSETTINGS .......................................................................................................... 701
Geometric Tab ......................................................................................................................... 701
AutoConstrain Tab .................................................................................................................. 702
### TUTORIAL: COMBINING DIMENSIONAL AND GEOMETRIC CONSTRAINTS ..................... 703

## EXERCISES ................................................................................................................................. 707

## CHAPTER REVIEW .................................................................................................................... 708

# 15  Reporting on Drawings ........................................... 709

## ID ................................................................................................................................................ 710
### TUTORIAL: REPORTING POINT COORDINATES .................................................................. 710
LastPoint and @ ...................................................................................................................... 710

## MEASUREGEOM ....................................................................................................................... 711
### TUTORIAL: REPORTING DISTANCE AND ANGLE MEASUREMENTS ................................. 711
Angle Measurement ................................................................................................................ 712
### TUTORIAL: DETERMINING AREAS AND PERIMETERS ........................................................ 713
Adding and Subtracting Areas ................................................................................................ 714
### TUTORIAL: REPORTING VOLUMES ....................................................................................... 716
Volumes of 3D Objects ............................................................................................................ 717

## MASSPROP ................................................................................................................................ 717
### TUTORIAL: REPORTING INFORMATION ABOUT REGIONS ............................................... 717
Finding the Center ................................................................................................................... 718

## LIST ............................................................................................................................................. 718
### TUTORIAL: REPORTING INFORMATION ABOUT OBJECTS ................................................ 719
PROPERTIES ............................................................................................................................. 720

## TABLE .................................................................................................................721
### BASIC TUTORIAL: PLACING TABLES ..................................................................722
### ADVANCED TUTORIALS: INSERTING TABLES .......................................................724
#### Insertion Behavior ..................................................................................724
#### Column & Row Settings ...........................................................................724
#### Table and Cell Styles ...............................................................................724
#### Set Cell Styles .......................................................................................725
#### Data Source ..........................................................................................725
#### Preview ................................................................................................725
### EDITING TABLES ............................................................................................726
#### TablEdit Command .................................................................................726
#### TInsert Command ..................................................................................726
#### Properties Command ..............................................................................728
#### Grips Editing ........................................................................................729
#### Managing Cell Content ............................................................................729
#### TableExport Command ............................................................................729

## TABLESTYLE ...................................................................................................731
### BASIC TUTORIAL: CREATING TABLE STYLES .......................................................731

## DWGPROPS ....................................................................................................736
### TUTORIAL: REPORTING INFORMATION ABOUT DRAWINGS ....................................736

## SETVAR .........................................................................................................739
### TUTORIAL: LISTING SYSTEM VARIABLES ............................................................739

## TIME .............................................................................................................740
### TUTORIAL: REPORTING ON TIME .....................................................................740
### REPORTING ON TIME: ADDITIONAL METHODS ..................................................740
#### Display .................................................................................................740
#### ON, OFF, and Reset ................................................................................740

## EXERCISES .....................................................................................................741
## CHAPTER REVIEW ...........................................................................................748

UNIT V — 3D MODELING ..................................................................................... 751

# 16  Introduction to 3D Solid Modeling .................... 753

## HISTORY OF 3D IN AUTOCAD ........................................................................ 754

## PARTS OF 3D SOLID MODELS ........................................................................ 755

## TUTORIAL: PREPARING TO MODEL IN THREE DIMENSIONS ........................ 757
### TUTORIAL: DRAWING SOLID PRIMITIVES ............................................. 761
### TUTORIAL: MODIFYING 3D MODELS ..................................................... 762
### TUTORIAL: CREATING HOLES .............................................................. 764

## REVOLVED MODELS ...................................................................................... 768
### TUTORIAL: REVOLVING OBJECTS ......................................................... 769

## UNDERSTANDING THE UCS ........................................................................ 771
### WORLD COORDINATE SYSTEM (WCS) .................................................. 771
### User-defined Coordinate Systems (UCS) ................................................. 771
#### UCS Icons .......................................................................................... 771
### TUTORIAL: CHANGING FROM WCS TO UCS ....................................... 772
### TUTORIAL: USING DYNAMIC UCS ....................................................... 774

## JOINING SOLID MODELS .............................................................................. 776
### TUTORIAL: UNIONING SOLIDS ............................................................. 776

## EXERCISES .................................................................................................... 779
### 3D SOLID MODEL PRIMITIVES ............................................................ 779
### OTHER SOLIDS OPERATIONS ................................................................ 782

## CHAPTER REVIEW ........................................................................................ 784

# 17  Introduction to 3D Mesh Modeling .................... 787

## PARTS OF 3D MESH MODELS ...................................................................... 788
### MULTIPLYING MESHES .......................................................................... 789
### Increasing Smoothness .............................................................................. 789
### Manipulating Mesh Models ...................................................................... 790
### Creating and Editing 3D Mesh Models .................................................... 791
### TUTORIAL: CREATING 3D MESH PRIMITIVES (MESH) ......................... 791
### Setting Tessellation Lines ......................................................................... 793
#### MeshOptions .................................................................................... 793
### MESHSMOOTHMORE AND MESHSMOOTHLESS .................................. 794
### TUTORIAL: SMOOTHING MESH MODELS ............................................. 795
### MOVING, ROTATING, AND SCALING MESH FACES, EDGES, AND VERTICES ....... 796
### TUTORIAL: MOVING FACES .................................................................. 797

MESHCREASE AND MESHUNCREASE ................................................................. 799
TUTORIAL: ADDING CREASES ........................................................................... 799
MESHCAP ................................................................................................................ 802
TUTORIAL: CAPPING LEAKING MESH OBJECTS ........................................... 802
EXERCISES .................................................................................................................... 804
3D SURFACE MODEL PRIMITIVES ..................................................................... 804
OTHER MESH OPERATIONS ................................................................................ 806
CHAPTER REVIEW .................................................................................................... 809

# 18 Additional 3D Techniques ............................................ 811

ADDITIONAL 3D DESIGN AIDS .............................................................................. 812
3D OBJECT SNAPS ................................................................................................. 812
CULLED SELECTIONS ............................................................................................ 813
TUTORIAL: CULLING 3D OBJECT SELECTION ............................................... 813
3D MOUSE SUPPORT ............................................................................................. 814
3D Mouse Navigation Bar ......................................................................................... 815
2D Commands That Don't Work ............................................................................. 816
AutoCAD LT .............................................................................................................. 816
EDGESURF .................................................................................................................... 817
TUTORIAL: MESHING BETWEEN FOUR EDGES ............................................. 817
DRAWING SURFACES: ADDITIONAL METHODS ........................................... 818
RULESURF .................................................................................................................... 820
Defining Starting Points ............................................................................................. 820
TABSURF ....................................................................................................................... 821
REVSURF ....................................................................................................................... 822
SWEEP AND EXTRUDE ............................................................................................. 823
TUTORIAL: SWEEPING OBJECTS INTO SOLIDS ............................................ 825
UNION ............................................................................................................................ 826
TUTORIAL: JOINING 3D MESHES ....................................................................... 827
Smoothed Conversion ............................................................................................... 829
SUBTRACT .................................................................................................................... 829
TUTORIAL: SUBTRACTING SOLIDS .................................................................. 829
INTERSECT ................................................................................................................... 831
TUTORIAL: INTERSECTING SOLIDS ................................................................. 831
INTERFERE ................................................................................................................... 832
TUTORIAL: CHECKING INTERFERENCE BETWEEN SOLIDS ..................... 832

## CHAMFER AND CHAMFEREDGE ........................................................................................... 833
### TUTORIAL: CHAMFERING OBJECTS ............................................................................ 834
Chamfering Meshes ...................................................................................................... 835
### INTERACTIVE CHAMFERING ....................................................................................... 836
## FILLET AND FILLETEDGE ................................................................................................ 837
### TUTORIAL: CHAMFERING AND FILLETING 3D OBJECTS .......................................... 837
### INTERACTIVE FILLETING ............................................................................................. 839
## MASSPROP ........................................................................................................................ 840
### TUTORIAL: REPORTING PROPERTIES ........................................................................ 840
## EXERCISES ........................................................................................................................ 842
### ADDING BEVELS TO SOLIDS' EDGES ......................................................................... 846
### ADDING COUNTERSINKS ............................................................................................ 848
## CHAPTER REVIEW ........................................................................................................... 850

## UNIT VII — PLOTTING DRAWINGS ... 851

# 19 Working with Layouts ... 853

### ABOUT LAYOUTS ... 854
- Exploring Layouts ... 855
- "Hiding" Layout Tabs ... 857
- TUTORIAL: WORKING WITH LAYOUTS ... 858
- Hiding Viewport Borders ... 861
- Switching from Paper to Model Mode ... 861
- SCALING MODELS IN VIEWPORTS ... 863
- VP Scale ... 863
- Zoom XP ... 863
- SELECTIVELY DISPLAYING DETAILS IN VIEWPORTS ... 865
- Freezing Layers ... 865
- Viewport Border Properties ... 866
- Viewport Content Properties ... 866
- Other Viewport Commands ... 866
- Overriding Layer Properties ... 867
  - VpOverrideMode — Toggling Properties Overrides ... 869
  - Reverting Properties to ByLayer ... 869
- INSERTING TITLE BLOCKS ... 870
- USING TILEMODE ... 872

### VPMAX AND VPMIN ... 872
- TUTORIAL: MAXIMIZING AND MINIMIZING VIEWPORTS ... 872

### QVLAYOUT ... 874
- TUTORIAL: SWITCHING LAYOUTS THROUGH QUICK VIEWS ... 874

### SPACETRANS ... 875
- TUTORIAL: DETERMINING TEXT HEIGHT IN LAYOUTS ... 875

### PSLTSCALE ... 877
- TUTORIAL: MAKING LINETYPE SCALES UNIFORM ... 877

### LAYOUTWIZARD ... 878
- TUTORIAL: MANAGING LAYOUTS ... 878

### LAYOUT ... 884
- TUTORIAL: MANAGING LAYOUTS ... 884

### VIEWPORTS ... 886
- TUTORIAL: CREATING RECTANGULAR VIEWPORTS ... 886

- VPORTS AND VPCLIP ........................................................................................... 888
  - TUTORIAL: CREATING POLYGONAL VIEWPORTS ............................................ 888
  - TUTORIAL: CREATING VIEWPORTS FROM OBJECTS ........................................ 889
  - VPCLIP ............................................................................................................ 890
- EXPORTLAYOUT ................................................................................................... 891
- EXERCISES ............................................................................................................. 893
- CHAPTER REVIEW ................................................................................................ 897

# 20 Plotting Drawings .................................................................. 901

- ALL ABOUT PLOTTERS AND PRINTERS ............................................................. 902
  - SYSTEM PRINTERS ........................................................................................... 902
    - Nontraditional Printers ................................................................................... 902
  - LOCAL AND NETWORK PRINTER CONNECTIONS ......................................... 903
    - Differences Between Local and Network Printers ........................................... 903
  - ABOUT DEVICE DRIVERS ................................................................................ 903
    - Drivers Today .................................................................................................. 904
- PREVIEW ............................................................................................................... 905
  - TUTORIAL: PREVIEWING PLOTS ..................................................................... 905
  - PREVIEWING PLOTS: ADDITIONAL METHODS .............................................. 906
    - Controlling the Preview Image ....................................................................... 906
- PLOT ...................................................................................................................... 907
  - TUTORIAL: PLOTTING DRAWINGS ................................................................. 907
    - Select a Plotter/Printer .................................................................................... 909
    - Plot to File ...................................................................................................... 909
    - Select the Paper Size ....................................................................................... 910
    - Select the Plot Area ........................................................................................ 910
    - Plot Offset ....................................................................................................... 911
    - Number of Copies ........................................................................................... 911
    - Plot Scale ........................................................................................................ 911
    - Preview the Plot .............................................................................................. 912
      - Partial Preview ............................................................................................ 912
      - Full Preview ................................................................................................ 912
    - Save the Settings ............................................................................................. 912
    - Plot the Drawing ............................................................................................. 913
  - VIEWPLOTDETAILS .......................................................................................... 914
  - PLOTTING DRAWINGS: ADDITIONAL METHODS .......................................... 915
    - About Plot Style Tables .................................................................................. 915
      - Selecting a Plot Style .................................................................................. 915
    - Select Viewport Shade Options ...................................................................... 916
      - Shade Plot Options ..................................................................................... 916
      - Quality and DPI Options ............................................................................ 918
    - Plot Options .................................................................................................... 918

| | |
|---|---|
| Plot in Background | 918 |
| Plot Object Lineweights and Plot with Plot Styles | 918 |
| Plot Transparency | 919 |
| Plot Paperspace Last | 919 |
| Hide Paperspace Objects | 919 |
| Plot Stamp On | 919 |
| Save Changes to Layout | 919 |
| Drawing Orientation | 920 |
| AUTOSPOOL | 921 |

## OPTIONS ............................................................................................................................. 922

### PLOT AND PUBLISH TAB ............................................................................................. 922

| | |
|---|---|
| Default Plot Settings for New Drawings | 922 |
| Plot To File | 922 |
| Background Processing Options | 923 |
| Plot and Publish Log File | 923 |
| AutoPublish | 923 |
| General Plot Options | 923 |
| Specify Plot Offset Relative To | 924 |
| Other Options | 924 |

### FILE TAB ............................................................................................................................ 924

## SCALELISTEDIT ................................................................................................................ 925

### HIDEXREFSCALES ........................................................................................................... 926

## PLOTSTAMP ...................................................................................................................... 927

### TUTORIAL: STAMPING PLOTS .................................................................................... 927

### STAMPING PLOTS: ADVANCED OPTIONS ............................................................. 929

| | |
|---|---|
| Location and Offset | 929 |
| Text Properties | 929 |
| Plot Stamp Units | 929 |
| Log File Location | 929 |

## PAGESETUP ....................................................................................................................... 930

### TUTORIAL: PREPARING LAYOUTS FOR PLOTTING ............................................. 930

## PUBLISH ............................................................................................................................. 932

### TUTORIAL: PUBLISHING DRAWING SETS ............................................................. 932

### SHOW DETAILS ............................................................................................................... 934

| | |
|---|---|
| Publish Output | 936 |

### PUBLISH OPTIONS ......................................................................................................... 936

| | |
|---|---|
| Current User | 937 |
| Default Output Location | 937 |
| General DWF/PDF Options | 937 |
|     Type | 937 |
|     Naming | 937 |
|     Name | 937 |
|     Layer Information | 937 |
|     Merge Control | 937 |

DWF Data Options ............................................................................................................ 937
    Password Protection ................................................................................................ 937
    Password ................................................................................................................. 938
    Block Information .................................................................................................... 938
    Block Template File ................................................................................................. 938
3D DWF Options .............................................................................................................. 938

## PLOTTERMANAGER .......................................................................................... 939

### TUTORIAL: CREATING PLOTTER CONFIGURATIONS ............................................ 939
### TUTORIAL: EDITING PLOTTER CONFIGURATIONS ................................................ 944
### THE PLOTTER CONFIGURATION EDITOR OPTIONS .............................................. 946

Media Source and Size ...................................................................................................... 946
    Media Type ............................................................................................................. 946
    Duplex Printing ....................................................................................................... 946
    Physical Pen Configuration .................................................................................... 947
    Physical Pen Characteristics ................................................................................... 947
Graphics .......................................................................................................................... 947
    Resolution and Color Depths / Vector Graphics .................................................. 947
Custom Properties ........................................................................................................... 948
Initialization Strings ......................................................................................................... 948
User-Defined Paper Sizes and Calibration ...................................................................... 949
Save As ............................................................................................................................ 949

## STYLESMANAGER ............................................................................................... 950

### COLOR-DEPENDENT PLOT STYLE TABLES (.*CTB* FILES) ...................................... 950
### NAMED PLOT STYLE TABLES (.*STB* FILES) .......................................................... 951
### CONVERTCTB AND CONVERTPSTYLES .............................................................. 951

From Color-dependent to Named .................................................................................. 951
From Named to Color-dependent .................................................................................. 952
Setting Default Plot Styles ............................................................................................... 952
Assigning Plot Styles ........................................................................................................ 953
    Assigned Plot Styles to Layouts .............................................................................. 953
    Assigned Plot Styles to Objects .............................................................................. 954
    Assigned Plot Styles to Layers ................................................................................ 955

### TUTORIAL: CREATING NEW PLOT STYLES ........................................................... 956
### TUTORIAL: EDITING PLOT STYLES ....................................................................... 959

## EXERCISES ........................................................................................................... 961

## CHAPTER REVIEW .............................................................................................. 963

## UNIT VII — APPENDICES .................................................................................. 965

## A  AutoCAD, Computers, and Windows ............... 967

### HARDWARE OF AN AUTOCAD SYSTEM ................................................................ 967
#### COMPONENTS OF COMPUTERS .................................................................... 968
System Board .................................................................................................... 968
Central Processing Unit .................................................................................... 969
Memory ............................................................................................................. 969
Disk Drives ........................................................................................................ 969
#### DISK CARE ..................................................................................................... 970
Write-Protecting Data ....................................................................................... 970

### PERIPHERAL HARDWARE ................................................................................ 971
#### PLOTTERS ...................................................................................................... 971
Inkjet Plotters .................................................................................................... 971
Laser Printers ................................................................................................... 972
#### GRAPHICS BOARDS ....................................................................................... 972
Monitors ............................................................................................................ 972
#### INPUT DEVICES .............................................................................................. 973
Mouse ............................................................................................................... 973
3D Mouse ......................................................................................................... 973
Digitizer ............................................................................................................. 973
Scanners ........................................................................................................... 974

### DRIVES AND FILES ........................................................................................... 974
#### FILE NAMES, EXTENSIONS, AND PATHS ...................................................... 974
Wild-Card Characters ....................................................................................... 975
#### DISPLAYING FILES ......................................................................................... 975
Windows Explorer ............................................................................................. 975
File-Related Dialog Boxes ................................................................................ 975
DOS Session .................................................................................................... 976

### AUTOCAD FILE EXTENSIONS .......................................................................... 977
Drawing Files .................................................................................................... 977
Support Files .................................................................................................... 977
AutoCAD Program Files ................................................................................... 978
Plotting Support Files ....................................................................................... 978
Import-Export Files ........................................................................................... 978
Miscellaneous Files .......................................................................................... 978
LISP and ObjectARX Programming Files ........................................................ 979
REMOVING TEMPORARY FILES ..................................................................... 979

## B AutoCAD Commands, Aliases, and Keyboard Shortcuts ... 981
### AUTOCAD COMMANDS ... 981
### COMMAND ALIASES ... 1000
### KEYBOARD SHORTCUTS ... 1005
### MOUSE AND DIGITIZER BUTTONS ... 1008
### MTEXT EDITOR SHORTCUT KEYS ... 1008

## C AutoCAD System Variables ... 1009
### CONVENTIONS ... 1009

## D Summary of Dimension Variables ... 1067
#### General ... 1067
#### ALPHABETICAL LISTING OF DIMVARS ... 1073

**INDEX** ... 1079

# INTRODUCTION

With more than four million users around the world, AutoCAD offers engineers, architects, drafters, interior designers, and many others, a fast, accurate, and versatile drafting and modeling tool.

Now in its 18th edition, *Using AutoCAD 2011* makes using AutoCAD a snap, by presenting easy-to-master, step-by-step tutorials covering AutoCAD's commands.

## BENEFITS OF COMPUTER-AIDED DRAFTING

AutoCAD is efficient and versatile for creating drawings. Some of its advantages include the following:

**Accuracy** — AutoCAD drawings are created and plotted to an accuracy of up to fourteen digits.

**Speed** — AutoCAD operators can easily copy and array objects, and otherwise edit the drawing. When operators customize the system for specific tasks, the speed of work increases even more markedly.

**Consistency** — AutoCAD is consistent in its methodology, and so the problem of individual style is eliminated. Companies can have a number of drafters working on the same project and produce a consistent set of drawings, when they follow established CAD standards.

**Popularity** — AutoCAD is the most popular software for creating and editing drawings. After learning to use AutoCAD, students are more likely to be hired; firms that use AutoCAD are likewise more likely to find suitable new employees.

## CAD APPLICATIONS

AutoCAD is being applied to many industries. The flexibility of this software has a major impact on how tasks are performed in architecture, engineering, interior design, manufacturing, mapping, piping design, and entertainment. A brief discussion of each of these applications follows.

### Architecture

CAD allows architects to formulate designs in a shorter period of time than by traditional techniques. The work is neater and more uniform. The designer can use 3D modeling to help the client better visualize the finished design. Changes can be made and resubmitted in a very short time. Architects can assemble construction drawings using stored details.

AutoCAD's database allows architects to extract information from the drawings, perform cost estimates, and prepare bills of materials.

Many third-party applications help architects customize AutoCAD for their discipline. Software is available for doing quick 3D conceptual designs, providing building details, generating automatic stairs, and much more. www.autodesk.com/autocad

### Revit

Autodesk promotes its Revit product as the future of architectural design. Architects create the building in 3D, and then let Revit generate 2D section views. Worksets allow team members to work together on the same model, fully coordinating their work. Detailed graphic control and view-specific graphics permit drawings to show more or fewer details.

*China Construction Design International used Revit to model the initial architectural, structural, water, HVAC, and electrical design of a new international cruise terminal in Tianjin, China in nine days. The first phase of this seven million-square-foot facility includes two cruise berths handling 500,000 passengers a year, as well as includes shipping agencies, hotels, exhibition centers, and commercial areas.*

Image courtesy of CCDI Group.

*All figures in this chapter are courtesy of Autodesk, Inc.*

The software can be configured to office standards for graphic style, data and layer export, as well as to additional drafting and CAD standards. The software package includes thousands of building components, and can output its data to ODBC-compliant databases. www.autodesk.com/revit

## AutoCAD Architecture

Autodesk has an AutoCAD add-on called Architecture (formerly Architectural Desktop), which customizes AutoCAD for architectural design. Its Content Browser accesses catalogs, tool palettes, and design content in the form of blocks and multiview blocks. With direct manipulation, you can modify designs directly, using either grip manipulation or in-place object editing.

Materials provide graphic and nongraphic attributes for design, documentation, and visualization. You can attach materials to building model objects and to their respective components for greater visual detail.

Scheduling allows tracking of any object in drawings. Because schedules are linked to data, they automatically update as the design changes. You can create schedule tables, objects tagged through xrefs, and schedules of external drawings.

Third-party developers have add-on software for roadway design, digital terrain modeling, and more. www.autodesk.com/architecture

*The Green Bay School District's Preble High School offers a curriculum dedicated to preparing students for careers in architecture, engineering, and manufacturing. Students learn AutoCAD Architecture and other Autodesk software, from the fundamentals of design through to the visualization and rendering of residential and commercial buildings. Students put their skills to practical use by providing a local engineering firm with professional renderings of a medical facility.*

### Engineering

Engineers use AutoCAD for many different kinds of design; in addition, they use programs that interact with AutoCAD to perform calculations that would take more time using traditional techniques. Among the many engineering disciplines benefitting from AutoCAD are civil, structural, and mechanical.

### Building Systems

Autodesk's Building Systems software is meant for designing mechanical, electrical, and plumbing systems for buildings, such as offices. In this case, "mechanical" means heating, cooling, and fire protection. The Building Systems add-on helps you design splined flex ducts with editable grips; single-line, 2D, or 3D pipes for creating chilled water, hot water, steam, and other piping systems; rise/drop symbology for connecting ductwork; and fire protection content, such as sprinklers, deluge valves, and retard chambers.

The electrical design capabilities of Building Systems software include automatic calculation of service and feeder sizes, as well as panel schedules and single-, two-, and three-pole intelligent circuit objects that work across all referenced files.

For plumbing, the software includes schematic tools for piping layout, built-in plumbing code systems for automatically calculating pipe sizing and flow requirements, as well as 3D plumbing content and 3D piping tools. www.autodesk.com/buildingsystems

### Civil 3D

Autodesk's Civil 3D (formerly Land Desktop) performs COGO (coordinate geometry) tasks, and creates maps, models terrain, designs alignments, and parcels for land planning. Other features include topographic analysis, real-world coordinate systems, volume totals, and roadway geometry.

Civil 3D integrates parcels in a single topology. Changes to one parcel update neighboring parcels. It extracts profiles of multiple surfaces based on alignment geometry, and then automatically creates the profile based on styles. The software generates dynamic models of any road, rail, or corridor project based on design elements, including alignments, profiles, superelevation, and criteria included in design subassemblies. Changes to any element update the corridor volumes, surfaces, and sections.

www.autodesk.com/civil3d

*Civil 3D contains tools to design more efficient roads and highways, such as a best-fit alignment and profile. It can design new alignments from parts of existing alignments, applying rule-based curve widening automatically.*

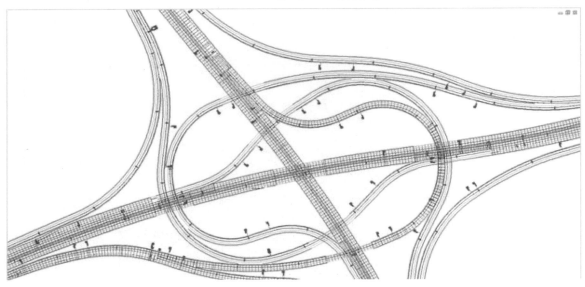

### Interior Design

AutoCAD is a valuable tool for interior designers. Its 3D capabilities can model interiors for clients, allowing 3D layouts, such as kitchens and baths, to be drawn and modified quickly with third-party add-ons.

### Manufacturing

Manufacturing uses for CAD are legion. One main advantage is integrating the program with a database for record keeping and tracking. Maintaining information in a central database simplifies much of the work required in manufacturing. Technical drawings used in manufacturing are constructed quickly and legibly.

Autodesk has several applications for mechanical engineers. For 2D drafting, there is AutoCAD Mechanical, and for 3D design, there is the Inventor Series.

### AutoCAD Mechanical

AutoCAD Mechanical is an add-on program that runs on AutoCAD. It is meant for 2D drafting of mechanical parts. Its parts-tracking feature and large pre-drawn content make it easy for mechanical engineers to create, manage, and reuse their 2D drawing data.

### Inventor Series

The Inventor software uses a paradigm different from "regular" AutoCAD, because mechanical parts can be very complex. You begin by drawing parts that make up assemblies. The assembly is the finished device, whether an MP3 music player or a plastic injection molding machine. The parts, as the name suggests, make up the assembly. Programs like Inventor are designed to handle tens of thousands of parts.

Once the assembly is created, Inventor can generate 2D plans, parts lists, and assembly instructions. Commands help place standard bolts, route wires and tubing, and determine whether the parts can withstand stresses from pressure and temperature.

The Inventor Series includes AutoCAD Mechanical and Inventor. (Inventor is independent of AutoCAD,

*Autodesk Mechanical analyzes and simulates 2D mechanical design under varying loads to identify possible areas of failure. The software allows you to add movable and fixed supports, as well as stress points, lines, and areas.*

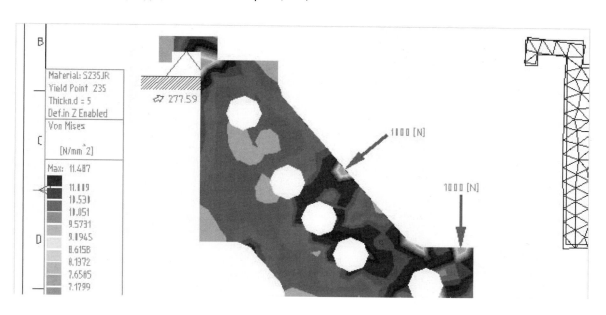

but can read and write its drawings.) Inventor Professional adds discipline-specific functions, such as cable routing and PCB (printed circuit board) design. Inventor LT is a low-cost version that leaves out assemblies and add-on applications.

### Facilities Management

After tenants move into the newly-constructed building, facilities managers can track how the building is used. When staff move to different offices, for example, they need to know whether the correct desks, chairs, cupboards, and telecommunications connections are available in their new office, or whether their existing furniture has to move with them.

Shopping malls and airports use AutoCAD linked to a database to maximize the profit from leased stores. Owners of these complexes need to track the mix of stores, and the revenue they produce. Some stores don't want to be located near competitors; others want to be in prime locations, such as near the entrance or at a corner. Other data stored in the database include agreements with cleaning and maintenance contractors, even details such as floor coverings and air conditioner capacities.

### Mapping

Map makers can use CAD to construct maps that store huge amounts of data. For example, a municipal map can contain information about all buildings, transportation systems, and underground services (such as storm sewer and water pipes). By using search criteria, the CAD system returns a map showing, for example, all water pipes older than 20 years. This map then shows where municipal engineers should replace old piping.

*Inventor uses dynamic sketch blocks so that you can evaluate different concepts, and then convert them to fully constrained 3D assembly models. Design Accelerators add the remaining components, such as drive mechanisms, fasteners, steel frames, and hydraulic systems. Inventor Routed Systems Suite designs complex electrical wiring, tube, and pipe runs. It automates aspects of routed system design to minimize errors, and validates the complete digital prototype. Inventor Simulation Suite performs motion simulation, and static and modal finite element analysis (FEA) of parts and assemblies. Inventor Tooling Suite designs injection molds for plastic parts. It creates and validates complete mold designs.*

GPS (global positioning system) generates the data used to create maps in CAD. This system consists of 24 satellites that always circle the earth, broadcasting data about their location. A handheld GPS receiver fixes your location by analyzing data broadcast from at least three satellites.

## Map 3D

Autodesk has a program specifically for AutoCAD users called Map 3D. It assigns properties to objects according to classification, and then identifies, manipulates, and selects objects by their description (e.g., road, river) rather than by their simple primitives (lines and arcs). Topology functions define how nodes, links, and polygons connect to each other. This permits network tracing, shortest-path routing, polygon overlaying, and polygon buffer generating.

Map has tools that help correct errors due to incorrect surveying, digitizing, or scanning. It connects to an external database, and can import data from other GIS products. www.autodesk.com/map3d

The GIS drawing above uses layers and colors to differentiate amongst geographic elements, such as homes, properties, streets, and underground services.

## Entertainment

The entertainment industry is largely based on digital media, so CAD graphics are a perfect match. What you view on television and in the newspaper is produced by a form of CAD graphics. Movie and theater set designs are often done in CAD. Movie makers are turning to forms of electronic graphics for manipulations and additions to their filmed and animated work. Autodesk also sells media and entertainment software, such as Maya and MotionBuilder.

*AutoCAD Map 3D reads in point cloud data generated by 3D laser scanners. The data consists of hundreds of millions of points, and are used as the basis of new 3D features designs.*

## Online Applications

Autodesk is rolling out a series of applications that operate inside Web browsers. Some programs are like AutoCAD, while others assist with tasks that occur before and after drafting, such as job planning and rendering. These applications are available from Autodesk Labs at labs.autodesk.com.

Much of this software works in Web browsers, and generally requires Flash, JavaScript, or Google Gears. The software is free to try while in the lab, although some require registration. This collection of software gives you an idea of the future of CAD.

### Project Butterfly

Butterfly allows one or more users to draw and edit AutoCAD *.dwg* files in their Web browsers. (You don't need AutoCAD on your computer.) The software also provides online meetings for discussing the project and its drawings. The Timeline slider bar lets you view all earlier changes and comments.

### Project Draw

Draw is a Visio-like vector drawing environment for creating structured diagrams, such as organizational charts, simple floor plans, electronic circuits, network diagrams, and user interface mock-ups. It runs in JavaScript-enabled Internet Explorer or Firefox Web browsers.

*Project Butterfly lets anyone operate a simplified version of AutoCAD in any Web browser that supports Flash. More than one person can edit the drawing at the same time.*

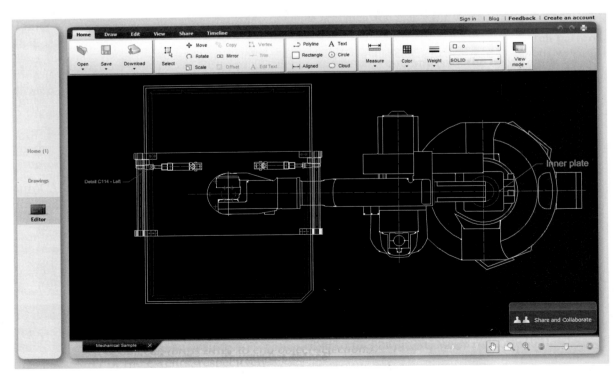

## OTHER BENEFITS

Using a computer to draft and design also opens a new world of information technology. Many CAD systems now access email and the Web. There are Web sites specific to AutoCAD, as well as discussion groups and email newsletters.

Autodesk maintains many kinds of information at its Web site, www.autodesk.com. You can learn about updates to AutoCAD, download useful utilities, on read up tips for using AutoCAD, and access large libraries of symbols.

For help with bugs in AutoCAD, visit Autodesk's support site at support.autodesk.com.

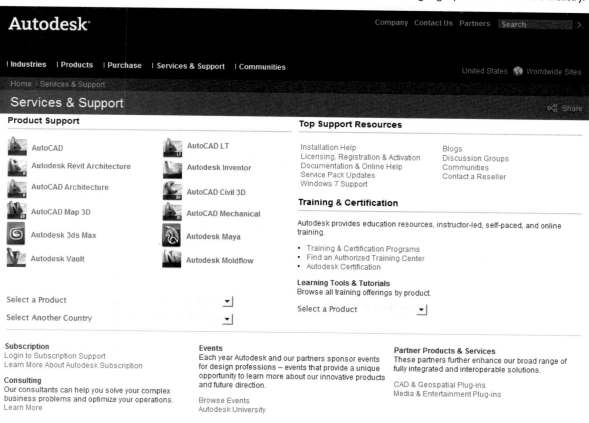

*Autodesk provides a lot of free support for its many software packages. Online support is available through FAQs (frequently-asked questions), discussion groups, Weblogs, bug fixes, and service packs. In addition, there many independent sites that provide additional tips, how-to tutorials, and ongoing information about the industry.*

## AUTOCAD OVERVIEW

AutoCAD displays your drawings on the screen of your computer's monitor. All the additions and changes to your drawing are shown on the screen as you perform them. You can work with one or more monitors, placing the drawing on the main monitor and tool palettes on the secondary monitors.

Drawings are made up of 2D objects, such as lines, arcs, circles, text, as well as 3D objects, such as boxes, cylinders, and tori.

You place objects in drawings by entering commands and selecting options. You issue commands by several methods: typing the names of commands on the keyboard, selecting icons from the ribbon, or choosing options from floating palettes. Some commands start automatically when you double-click objects in drawings.

Once a command starts, you are often asked to choose among its options. After you identify all the information that AutoCAD requests, the new objects, or changes to objects, are shown on the screen.

If you need additional help, refer to Appendix B, "AutoCAD Command Summary."

## TERMINOLOGY

To use AutoCAD properly, this book contains terms and concepts that you need to understand. Some of the terms are explained briefly in this chapter. In addition, you may consult the index, which contains many terms.

### Coordinates

AutoCAD uses the *Cartesian coordinate system*. The horizontal is represented by the x axis; the vertical, by the y axis. Any point on the plane can be represented by an x and y value shown in the form of x, y. For example **2,10** represents a point 2 units in the x direction and 10 units in the y direction.

The *origin* of the x and y axes is the **0,0** point. This point is normally at the lower left corner of the screen. You can, however, specify a different point for the lower left corner.

When AutoCAD works with 3D (three-dimensional) drawings, the third axis is called the "z axis," which normally points out of the screen at you.

### Drawing Files

Drawing files contain the information used to store the objects you draft. Drawing files automatically have the file extension of *.dwg* added to them.

**WHAT'S NEW IN AUTOCAD 2011**

Throughout this book, the New in 2011 logo alerts you to features new in AutoCAD 2011. This book includes information on these new features.

**USER INTERFACE**

AutoCAD 2011 changes the look of the 2D drafting environment by changing the background color to dark blue and using lines for the grid, instead of dots. The new navigation bar provides quick access to panning, zooming, and other navigation tools.

New to workspaces is 3D Basics, which reduces the number of commands displayed by the ribbon. Shown below is AutoCAD 2011's user interface with the simplified 3D modeling workspace.

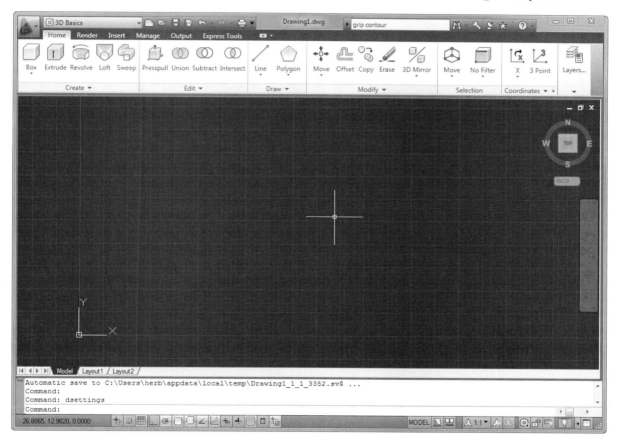

The Workspace droplist was added to the Quick Access toolbar. The new WSAUTOSAVE system variable automatically saves changes to workspaces.

The default state for the Quick Properties palette is now turned off.

The status bar gains new buttons for toggle inferred constraints, 3D object snaps, transparency, selection cycling, hardware acceleration, and object isolation.

Visual styles now include five more default settings: shaded, shaded with edges, gray, X-ray, and sketchy.

Two new system variables imply that Autodesk is getting AutoCAD ready for multi-touch interfaces. These are DIGITIZER and MAXTOUCHES.

Help can link to Autodesk Web pages, but this does not work well when the Internet connection is slow or does not exist.

## 2D DRAFTING

The HATCH and GRADIENT commands now show patterns in just two steps: (1) enter **h** at the command prompt, and then (2) pass the cursor over a closed area. The commands are interactive: selecting different patterns and parameters (scale, rotation, and so on) from the ribbon instantly shows the revised pattern. Hatches can have their own transparency level, and can sport different foreground and background colors.

Definitions for complex linetypes have a new 'U' parameter that forces text and symbols to remain upright.

The SKETCH command now draws splines when system variable SKPOLY is set to 2. The command's old control options (pen up, erase, connect, and so on) are replaced by mouse clicks: press to draw; let go to stop.

The SPLINE command now draws splines with control vertices, knots, and kinks. Starting and ending tangency is now ignored by default. The display of control vertices is controlled by the new CVSHOW and CVHIDE system variables.

The JOIN command now joins 3D polylines to other objects, such as helices, splines, and 2D polylines. A similar **Join** option is added to the PEDIT command.

The PEDIT command can now convert polyline segments between lines and arcs, as well as add and remove vertices. Cursor icons report the current editing mode. Splines can be converted to polylines, and vice versa.

You no longer need to hold down the CTRL key to select subobjects. Selection cycling now pops up a mini dialog box listing the names of overlapping objects. The new culling mode displays only the topmost 3D object during object selection.

The new SELECTSIMILAR command selects all objects that match specified properties. The new ADDSELECTED command creates new objects based on the type and properties of a selected object. The new HIDEOBJECTS and ISOLATEOBJECTS commands hide or isolate objects independent of layer settings; the related UNISOLATEOBJECTS command reveals them again.

The undocumented OPTCHPROP command changes styles, for example text, dimensions, and tables.

Transparency can be applied to any layer and any object. Transparent objects can be plotted, but the process may take longer, because AutoCAD must first rasterize the drawing. Commands such as PROPERTIES, MATCHPROP, and QSELECT now support the new transparency property.

The STYLE command's dialog box now reports missing font files.

## PARAMETRICS

The DIMCONSTRAINT command has been unbundled into individual commands, such as DCLINEAR for linear dimensional constraints. Similarly, the GEOMCONSTRAINT command is unbundled into commands, such as GCCOLLINEAR for placing collinear constraints. New constraint tags distinguish between point and object constraints. Coincident constraints are shown as a small blue square.

The PARAMETERS command's dialog box now supports filters, which allow you to create groups of constraints. The new search field lets you find specific constraints in large drawings.

The new CONSTRAINTINFER system variable automatically places geometric constraints as you draw and edit. The AUTOCONSTRAIN command now supports the Equal constraint.

## 3D MODELING

The new **3DOSNAP** command places 3D object snaps: ZVERtex (snaps to vertex or control vertex), ZMIDpoint (midpoint of a face edge), ZCENter (face center), ZKNOt (spline knot), ZPERpendicular (perpendicular to planar face), ZNEAr (nearest to face), and ZNONe (off).

The new **CHAMFEREDGE** and **FILLETEDGE** commands provide interactive chamfering and filleting of 3D objects.

A series of visual analysis commands superimposes zebra, curvature, and draft analyses colors onto 3D objects.

Commands like **EXTRUDE** and **REVOLVE** can now select subobjects as profiles. Many of these 3D construction commands have two new options: **Mode** (creates 3D solid or surface) and **Expression** (allows use of parametric formulas).

New editing commands for 3D mesh objects place caps on open meshes, merge vertices or faces, extrude and spin faces. The new **CONVTOMESH** command converts surfaces to 3D mesh objects; the new **CONVTONURBS** command converts 3D solids and surfaces to NURBS surfaces.

Many new commands were added for creating and editing surfaces, such as adding a continuous blend between two surfaces, extending and offsetting surfaces, filleting two surfaces, and creating a surface from several curves.

AutoCAD 2011 can now import point cloud data, which are generated by laser scanners.

For rendering, the materials palette is redesigned, and is now compatible with the materials used in all other Autodesk software.

## USING THIS BOOK

Each chapter in *Using AutoCAD 2011* builds on the previous chapters. The basic use for each command is given, along with one or more tutorials that help you understand how the command works. This is followed by a comprehensive look at the command and its variations.

The exercises at the end of a chapter are specifically designed to use the commands covered in the book to that point. Review questions reinforce the concepts explored in the chapter. This method allows you to pace learning. Remember, not everyone grasps each command in the same amount of time.

### CONVENTIONS

This book uses the following conventions.

#### Keys

Several references are made to keyboard keystrokes in this book, such as ENTER, CTRL, TAB, and function keys (F1, F2, and so on).

#### Control and Alternate Keys

Some commands are executed by holding down one key, while pressing a second key. The control key is labeled CTRL, and is always used in conjunction with another key or a mouse button.

To access menu commands from the keyboard, hold down the ALT key while pressing the underlined letter. The ALT key is also used in conjunction with other keys.

#### Flip Screen

The text and graphics windows can be alternately displayed by using the Flip Screen key. The F2 key is used for this function.

#### Command Nomenclature

When a command sequence is shown, the following notations are used:

> Command: **line** *(Press ENTER.)*
> Specify first point: *(Pick a point.)*
> Specify next point or [Undo]: *(Pick point 2, or enter an option.)*

**Boldface** text designates user input. This is what you type at the keyboard. If the command and its options can be entered by other methods, these are described in the book. For example, commands can be selected from menus and toolbars, or entered as keyboard shortcuts.

*(Press ENTER)* means you press the ENTER key; you do not type "enter."

*(Pick a point)* means that you enter a point in the drawing to show AutoCAD where to place the object. To pick the point, you can either click on the point, or type the x, y, z coordinates.

*(Pick point 2)* The book often shows you a point on a drawing. The points are designated, such as "Point 1."

<Default> is how AutoCAD indicates default values. The text or numbers are enclosed in angle brackets. The default value is executed when you press ENTER.

**[Undo]** is how AutoCAD lists command options. The words are surrounded by square brackets. At least one letter is always capitalized. This means you enter the capital letter as the response to select the option. For example, if the option is **Undo**, simply entering **U** chooses the option.

Additional options are separated by the slash mark. When two options start with the same letter, then two letters are capitalized, and you must enter both.

## COMPANION FILES

This book includes access to companion files needed for completing end-of-chapter exercises. These files include AutoCAD drawing files, which are identified by the *.dwg* suffix in their file names, as well as other types of files. When an exercise requires a file, it plainly tells you the file name, such as the following:

1. Open the *5-47.dwg* drawing file.

You can obtain the complete set of companion files from the Student Companion site at CengageBrain by following these steps:

1. With your computer's Web browser, go to www.cengagebrain.com.
2. Enter the name of the author, title of the book, or the book's ISBN (international standard book number) in the Search window:

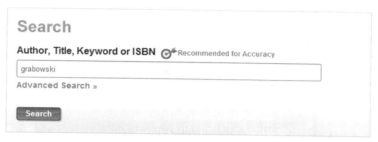

| Author: | Ralph Grabowski |
| Title: | Using AutoCAD 2011 |
| ISBN | 978-1-1111-2514-7 |

3. Locate the desired book, and then click on its title.
4. When you arrive at the Product Page, click on the **Free Stuff** tab.

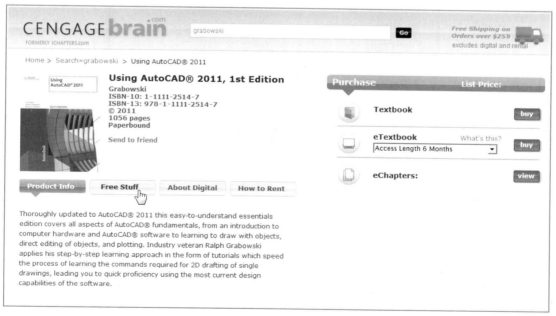

5. Choose the **Click Here** link to arrive at the Companion site. (If this is your first time at this site, you will have to complete a registration form.)
6. To access the resources, click the **Student Resources** link in the left navigation pane.

## Instructor Resources

Instructor Resource™ is an educational resource that creates a truly electronic classroom. It is a CD containing tools and instructional resources to enrich the classroom and reduce preparation time. The elements of Instructor Resource link directly to the text, and combine to provide a unified instructional system. With Instructor Resource, you can spend your time teaching, not preparing to teach. IB

Features contained in Instructor Resource (ISBN 1111125155) include the following:

**Instructor Syllabus:** This lists goals, topics covered, reading materials, required texts, lab materials, grading plan and terms. Additional resources and web sites covering descriptive geometry, manual drafting, AutoCAD, and Inventor are also listed.

**Student Syllabus:** This lists goals, topics covered, required texts, grading plans and terms.

**Lesson Plans:** These contain goals, discussion topics, suggested readings, and suggested homework assignments. You have the option of using these lesson plans with your own course information.

**Answers to Review Questions:** These solutions help you grade and evaluate end-of-chapter tests.

**PowerPoint® Presentation:** These slides help you present concepts and material. Key points and concepts can be graphically highlighted for student retention.

**Exam View Computerized Test Bank:** This includes over 800 questions of varying levels of difficulty in true/false and multiple-choice formats that assess student comprehension.

**AVI Files:** These movies, listed by topic, illustrate and explain key concepts.

**DWG Files:** This list of .dwg files matches many of the figures in the textbook. These files can be used to stylize the PowerPoint presentations.

### WE WANT TO HEAR FROM YOU!

Many of the changes to the look and feel of this new edition came by way of requests from and reviews by users of our previous editions. We'd like to hear from you as well! If you have any questions or comments, please contact:

The CADD Team
c/o Autodesk Press
5 Maxwell Drive
Clifton Park NY 12065-8007

or visit our Web site at www.autodeskpress.com

## ACKNOWLEDGMENTS

We would like to thank and acknowledge the professionals who reviewed the manuscript to help us publish this AutoCAD text:

**Technical Editor:** Bill Fane, retired from British Columbia Institute of Technology, Burnaby BC, Canada

**Copy Editor:** Stephen Dunning, Trinity Western University, Langley BC, Canada

## ABOUT THE AUTHOR

Ralph Grabowski has been writing about AutoCAD since 1985, and is the author of over 100 books on computer-aided design. He received his B.A.Sc. degree in Civil Engineering from the University of British Columbia in 1980.

Mr. Grabowski publishes upFront.eZine, the weekly email newsletter about the business of CAD. He is the author of a series of CAD e-books under the eBooks.onLine imprint. Mr. Grabowski is the former Senior Editor of CADalyst magazine, the original magazine for AutoCAD users. You can visit his Web site at www.upfrontezine.com and his Weblog at worldcadaccess.typepad.com.

# I

# Setting Up Drawings

# CHAPTER 1

## Quick Tour of AutoCAD 2011

Welcome to *Using AutoCAD 2011*!

As a new user, you need to experience the feel of AutoCAD firsthand. The concepts behind computer-aided design software can be disorienting to a first-time user, and so this quick-start chapter gets your feet wet. It introduces you to some of the features discussed in detail in later chapters.

In this chapter, you learn these commands:

> **NEW** starts fresh drawings.
> **LINE** draws straight line segments.
> **U** undoes mistakes.
> **QUIT** exits AutoCAD.
> **CLOSE** and **CLOSEALL** close drawings.
> **SAVEAS** renames drawings, and then saves them.
> **PLOT** prints the drawing on paper.

Let's get started!

## STARTING AUTOCAD

Before you can install and run AutoCAD, your computer must have Microsoft® Windows™ XP, Server 2003, Vista, or 7 installed and running.

### TUTORIAL: STARTING AUTOCAD

1. To start AutoCAD 2011, use one of these methods:
   - On the Windows desktop, double-click the icon labeled **AutoCAD 2011**. *(Double-click means to press the left mouse button twice, quickly.)*

   - Alternatively, you can also start AutoCAD from the Windows taskbar by following these steps:
     a. In Windows Vista and 7, click the round Windows logo. (In Windows XP, click the **Start** button.)
     b. Choose **All Programs**, the **Autodesk** folder, the **AutoCAD 2011** folder, and finally **AutoCAD 2011**. See figure below.

   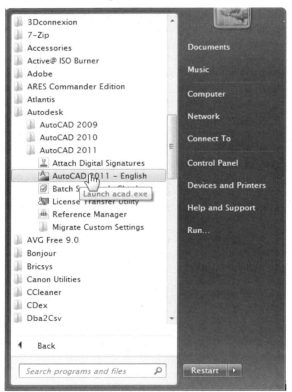

   - Or, in the Windows Explorer, double-click a *.dwg* drawing file. (If AutoCAD add-ons, such as Mechanical Desktop or Architecture, are installed, they may start instead of AutoCAD.)
   - Or, in the Run dialog box, enter *acad.exe*.

        Run: **acad.exe** *(Press* ENTER.*)*

     (This procedure bypasses command-line startup switches that may be associated with the icon.)

## ADDITIONAL INITIAL DIALOG BOXES

AutoCAD sometimes shows several dialog boxes before it finishes starting up. Here are some you can expect to see — and instruction on what to do with them.

- If the Getting Started Videos dialog box appears, click **X** to close it.

- If the Create New Drawing dialog box appears, choose the **Start from Scratch** button, and then select **Imperial (feet and inches)**. Click **OK**.

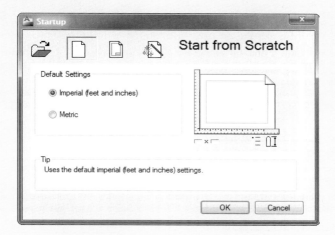

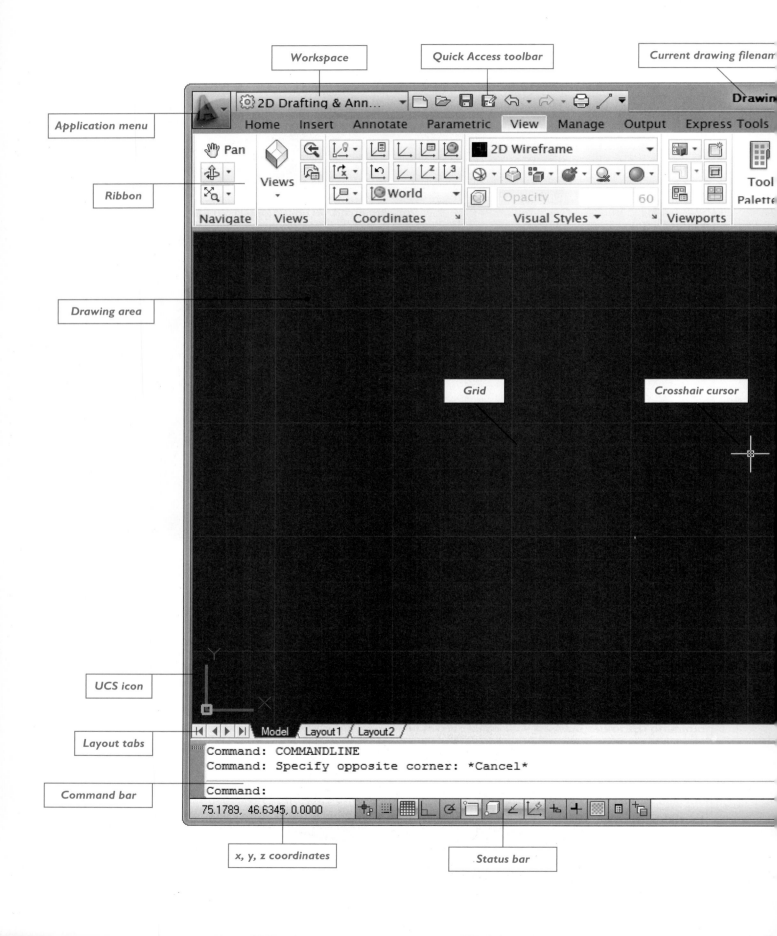

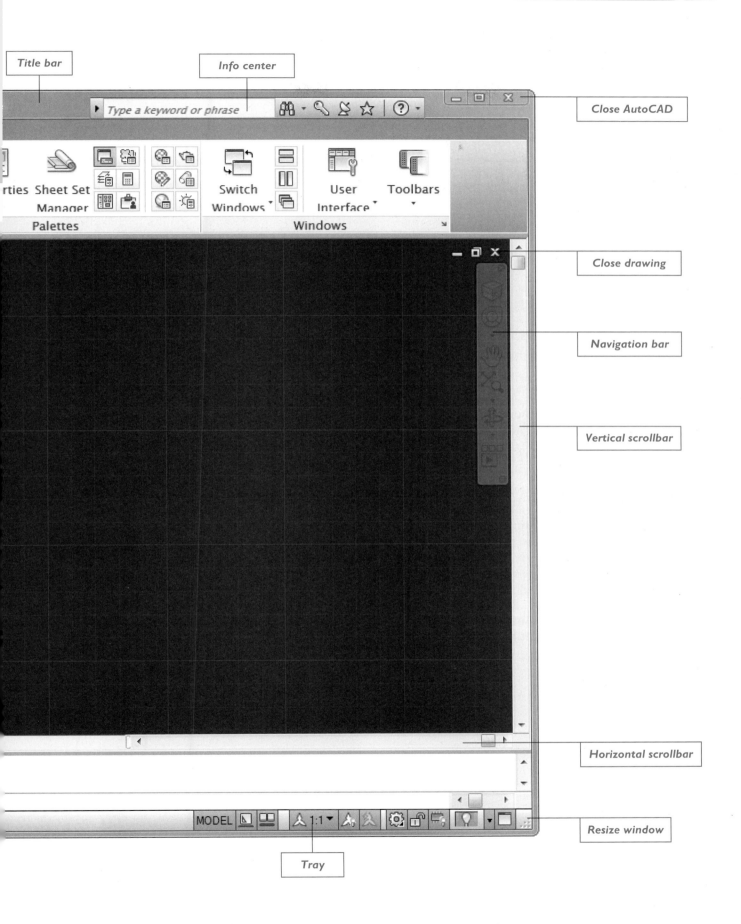

Quick Tour of AutoCAD 2011   7

2. In all cases, Windows opens the AutoCAD software. An opening screen is displayed, called the "splash screen."

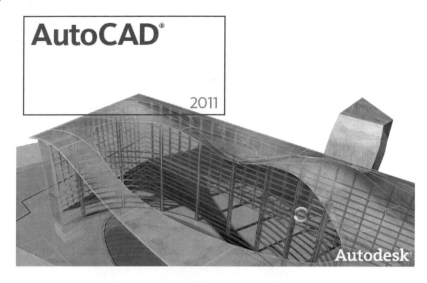

After the splash screen disappears, you see AutoCAD appear (as illustrated by the figure on the previous pages).

## COMMANDS IN AUTOCAD

You make drawings in AutoCAD by creating and editing *objects*; you instruct AutoCAD to draw and edit objects through *commands*. Often, these are simple words such as LINE, CIRCLE, MOVE, and ERASE.

You can command AutoCAD by a variety of input methods. Commands can be executed by several different methods, such as typing them on the keyboard, selecting them from ribbon panels, clicking a mouse button, or by entering shortcut keystrokes. As you work with AutoCAD, you will come to prefer one method over the others.

Despite the variety of input methods, AutoCAD, however, is always command-driven. Whether you click an item in a menu, on the ribbon or toolbars, or click a mouse button, AutoCAD translates the action into a command name. Because Autodesk changes AutoCAD's user interface every year or two, longtime users prefer typing command names, because that's the one interface that never changes.

### Command Bar

At the bottom of the AutoCAD window is the *command bar*. (If you do not see the command bar, press CTRL+9 to make it appear.)

The command bar is where you type commands, and one of the places where AutoCAD responds back to you. You should see the word 'Command:' on the bottom line now, like this:

Command:

You enter command names at this prompt, and then read instructions from AutoCAD for the steps to follow.

### Text Window

AutoCAD has a larger version of the command bar called the "text window." It displays only text. You can see it by pressing function key F2.

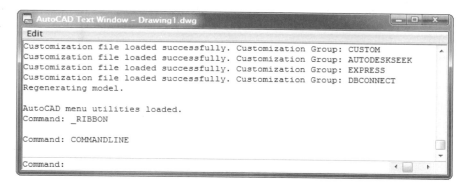

Press F2 a second time to hide the text window.

### Dynamic Input

In addition to the command bar, you can enter commands in the drawing area itself, and receive feedback from AutoCAD. This is known as *dynamic input*. When dynamic input is turned on, the command prompts and options appear in small rectangles near the cursor.

In the figure below, the "line" command is being entered.

Normally, dynamic input is turned on by default; if not, click the ⊞ (or DYN) **Dynamic Input** button on the status bar, and then ensure the cursor is in the drawing area. You learn more about dynamic input in Chapter 2, "Understanding CAD Concepts."

### Mouse

The alternative to typing commands is using the mouse to select commands. To see how the mouse interacts with AutoCAD, follow the next tutorial.

As alternatives to the mouse, you can also use digitizing tablets or other input device with AutoCAD. (At time of writing, Autodesk has implemented, but not yet turned on, touch screen controls.) From this point forward, however, we refer to the mouse only, being the most common of pointing devices.

In the following sections, we look at how the ribbon, application menu, and other user interface elements operate. All of these are accessed with the mouse.

## ABOUT RIBBON TABS AND PANELS

The ribbon is segregated into tabs and panels. *Tabs* represent major groupings of tasks in AutoCAD. They are identified by names such as Home, Visualize, and View. To access a tab, click its name. The Home tab is shown below. It contains the most commonly used AutoCAD commands, such as those that draw lines, move and copy objects, control layers, and place text.

Each tab consists of one or more *panels*. Each panel contains a group of related commands. The Draw panel contains commands for drawing, the Modify panel for modifying objects, and so on.

Each panel contains *buttons*, each typically labeled with an *icon*, the Greek word for "small picture." The pictures represent the functions of the buttons. When you click a button, AutoCAD executes the related command.

Sometimes, AutoCAD automatically displays additional tabs in reaction to a command, such as when editing text and creating hatch patterns; when the command ends, the tab disappears. These are known as *contextual tabs*.

AutoCAD provides ribbon sets that differ for 2D, 3D, and other workspaces. (*Workspaces* customize AutoCAD's user interface, and are described later in this chapter.) The ribbon for the 3D workspace is illustrated below.

### Ribbon Shortcuts

Some panels aren't big enough to hold all commands, and so you can expand them to see additional ones. To expand a panel, click the ▼ triangle icon found at the bottom edge of each panel. To keep the panel expanded, click the 📌 thumbtack icon.

You can drag panels away from the ribbon to keep them open all the time. Indeed, you can *undock* the entire ribbon, and then drag it to a second monitor. To do so, right-click a tab name, and then select **Undock** from the shortcut menu.

Some panels have a related dialog box that's used for selecting options. You access these dialog boxes by clicking the ↘ arrow icon.

If the ribbon takes up too much room on your computer's screen, you can minimize it by clicking the ⬒ arrow button found along the top edge of the ribbon.

Right-click the button to see a list of other collapsible states:

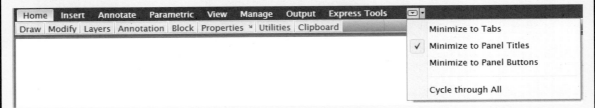

Panels appear again when you hover the cursor over their names.

## RIBBON

The ribbon is the primary interface. It spans the upper portion of AutoCAD. Through the use of tabs and panels, it segregates commands into logical groupings. (The ribbon replaces the Dashboard of earlier only two releases.)

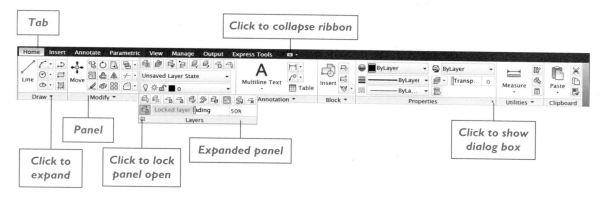

## TUTORIAL: EXECUTING COMMANDS WITH THE RIBBON

Commands are executed in AutoCAD by clicking buttons, selecting options from droplists, or dragging sliders. For example, to start drawing lines you can click the button for the Line command. (Make sure the Home tab is active, the "2D Drafting and Design" workspace is current, and Ortho and Polar modes are turned off.) Follow these steps:

1. Move the mouse, and notice how a crosshair cursor moves around the screen. (When the cursor looks like the crosshair illustrated below, you can use it to specify points in drawings.)

2. Move the crosshair cursor out of the drawing area. Notice that the cursor changes to an arrow. In this state, the cursor lets you select commands from the ribbon, application menu, and other areas.

3. Move the cursor up to the ribbon, and then pause it over the large ⬚ button. The icon on the button tells you that it starts the **LINE** command. But don't press the mouse button yet!

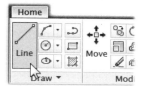

4. If you wait a couple of seconds with the cursor on the button, a *tooltip* appears that tells you the purpose of the button:

5. Wait a few seconds longer, and the tooltip expands to give you further details on the **LINE** command:

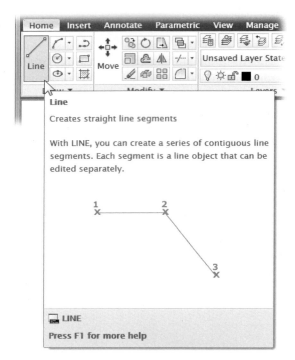

6. With the cursor on the **LINE** button, click the left button on your mouse. Notice that the following text appears in the command bar at the bottom of the window. This is called a *prompt:*

   Command: _line Specify first point:

   In the command bar, AutoCAD tells you what it expects from you. Here, AutoCAD asks you to specify the point at which the line starts, which it calls the "first point."

7. Move the crosshair cursor into the drawing area. Notice that the prompt is duplicated near the crosshair cursor. This is called the "dynamic prompt," because it moves with the cursor.

Next to the dynamic prompt, AutoCAD reports the current x and y coordinates.

8. Click the left mouse button to specify the starting point of the line.
9. Continue to move the crosshair cursor across the screen. Notice how a line "sticks" to the intersection of the cursor crosshairs. This is called the "rubber band." It stretches and follows your movement to preview the location of the line, when you click again.

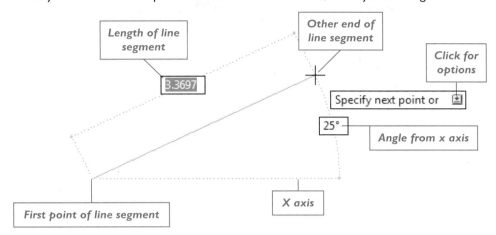

AutoCAD uses dynamic prompts to report the current length and angle from the x axis. In the illustration above, the line is 3.3697 units in length and is at an angle of 25 degrees from the axis.

10. To enter a specific length:
    a. Enter a value at the keyboard, such as **5**. Notice that it appears in the length box.
    b. Press the **TAB** key.

Notice that a 🔒 padlock icon appears. The padlock tells you that the length is *locked*; it no longer changes when you move the cursor — although you can certainly override it by entering another length.

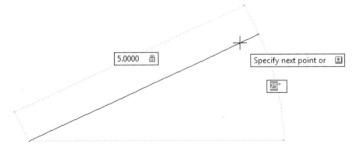

11. Move the cursor. Notice that the length remains the same, even as the angle changes. To indicate a specific angle, follow these steps:
    a. Enter the value, such as **20**.
    b. Press **TAB**.

Now the angle is also locked, as indicated by the second padlock.

12. Draw another line segment by moving the cursor and picking another point.

Notice that in the command prompt area, the prompt has expanded with the addition of the words "or [Undo]":

    Specify next point or [Undo]:

The word **Undo** is an *option*. AutoCAD indicates options by surrounding them by square brackets, such as [Undo]. Options are alternative actions that can be performed by the command. In this case, **Undo** "undraws" the previous line segment, which can be useful when you make mistakes or change your mind.

13. The **LINE** command remains active until you cancel it. While it is active, it allows you to draw as many connected segments as you require — without reselecting the command from the ribbon.

    Continue to enter line segments. As you do, the prompt changes:

    Specify next point or [Close/Undo]:

**Close** is another option of the **LINE** command. When you enter **Close**, it closes the polyline by drawing a line segment from the end of the last segment to the start of the first segment automatically.

14. Another way to end the command is by clicking the right mouse button. Notice the shortcut menu, as illustrated below. Notice that the shortcut menu gives you access to several other options of the **LINE** command (Close and Undo), as well as other AutoCAD commands, such as **PAN**, **ZOOM**, and **QUICKCALC**.

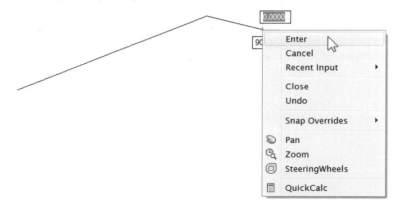

15. Choose **Enter**, which ends the command; you can also press **ENTER** (or the spacebar) on the keyboard. (Alternately, press **ESC** on the keyboard; for some commands, you will need to press **ESC** twice before they end.)

If you wish to erase lines you have drawn, click the ↶ **Undo** button in the Quick Access toolbar. (Alternatively, enter the u command at the 'Command:' prompt, short for "undo.")

    Command: **u** (*Press* ENTER.)

**Notes** If you wish to draw lines that are perfectly horizontal and vertical, hold down the **SHIFT** key while drawing.

Another way to access the options of the **LINE** command is through dynamic input. Press the ⬇ down arrow on the cursor keypad to see a list of options, as illustrated below.

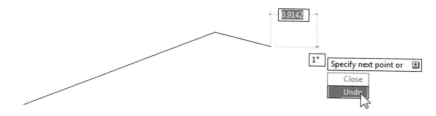

Frankly, I find it quicker to enter the options at the keyboard. You may prefer to draw on an uncluttered screen. Turn off dynamic input by clicking its ⊞ button on the status bar. When off, the on-screen prompts do not appear, looking like this:

## TUTORIAL: ACCESSING FLYOUTS

Some buttons on the ribbon hide additional buttons known as *flyouts*. You can recognize them by the tiny ▼ black triangle adjacent the button: ⊙ ▼ . To access flyouts, follow these steps:

1. Position the cursor over the tiny triangle next to a button, such as the Circle button illustrated below. (If you were to click the **Circle** button itself, AutoCAD would execute the **CIRCLE** command.)

2. Click the left mouse button. Notice that AutoCAD displays a flyout that lists additional ways to draw circles.

3. Drag the cursor over the buttons on the flyout. As the cursor passes over a button, it gives the illusion of being depressed.

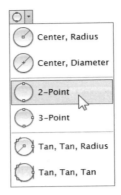

4. When you reach the button you want, click the mouse button. AutoCAD starts the command associated with the button.

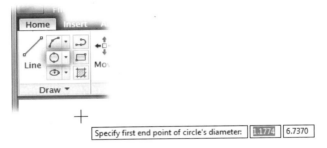

Notice that the icon associated with the option you selected now appears in the ribbon, becoming the default option the next time the CIRCLE command is executed.

### TUTORIAL: ACCESSING DROPLISTS

The ribbon compresses multiple choices through flyouts and *droplists*. For example, colors are "hidden" by a droplist. To access the droplist, follow these steps:

1. Position the cursor over the ● Color droplist, as found in the Home tab's Properties panel.

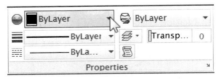

2. Click the droplist. Notice a list of colors extends downwards.

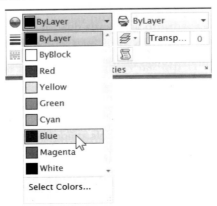

3. Move the cursor down the list, and then choose an option, such as the color **Blue**.
4. Click to choose the color. Notice that the list disappears, and that the option you chose now appears in the droplist.

While the Color droplist is extended, there are two controls worth noting:

Clicking **Select Colors** displays the Select Colors dialog box, which gives you access to all of AutoCAD's millions of colors and color books. Colors you select from the dialog box are added to the end of this droplist.

The size of the droplist can be lengthened by dragging the triangular grip located at its lower right corner. This is handy when you add colors from the Select Colors dialog box.

### QUICK ACCESS TOOLBAR

A toolbar that gives you "quick access" to system commands is found to the right of the title bar, at the top of the AutoCAD window. The Quick Access toolbar contains a selection of common functions. From left to right, these are Workspaces, New, Open, Save, SaveAs, Undo, Redo, and Print.

You can add any other command to this toolbar, as follows:

1. Right-click a button in the ribbon, such as **Line.** Notice the shortcut menu.

2. From the shortcut menu, choose **Add to Quick Access Toolbar**. Notice that the LINE command's icon is added to the end of the toolbar.

Clicking the **Line** button starts the LINE command.

### ACCESSING THE APPLICATION MENU

A large red A resides in the upper left corner of the AutoCAD window.

It is more than just the AutoCAD logo. It "hides" the *application menu* with its utility commands. (This menu replaces the menu browser of AutoCAD 2009.) To see the menu, click the **A**.

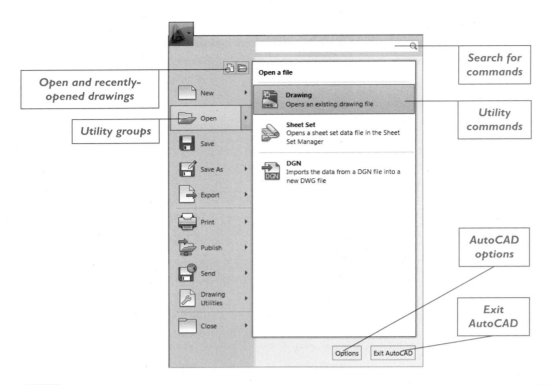

The Options **Options** button displays the Options dialog box (OPTIONS command), and the Exit AutoCAD **Exit AutoCAD** button exits AutoCAD (EXIT command). Alternatively, you can double-click the **A** to close all drawings and exit AutoCAD.

### Recent Drawings

The application menu lists the names of previously opened drawings — useful when you want a drawing file but can't remember its name or location. (Autodesk calls this feature "Recent Documents," even though AutoCAD does not open Office-style documents.)

Click the **A**, and then choose the **Recent Documents** button (if it is not already selected by default). Notice the list of previously opened drawing file names in the right-hand pane.

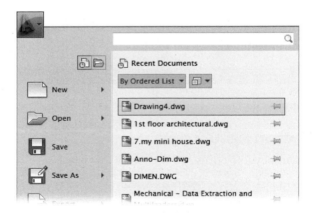

To open a drawing, click on its file name.

The pane contains several useful options. The **By Ordered List** droplist lets you sort drawings by order (the order in which they were opened), their file size, file type, or the date last accessed. The other droplist displays drawing names as icons or images.

When the list gets long enough, the oldest file names are removed automatically. Click the pushpin to keep them from falling off the end of the list.

## Open Documents

The application menu also lists the names of drawings currently open in AutoCAD. Click the **A**, and then click the **Open Documents** button. Notice the list of open drawing file names in the right-hand pane.

Click the droplist to display drawings by their preview images. (Unsaved drawings show an X as their preview image.) Alternatively, you can press ALT+TAB to switch between open drawings.

## Displaying the Menu Bar

Some users prefer the dropdown menu system to the ribbon. Normally, the *menu bar* is turned off, but when turned on it appears above the ribbon. In the figure below, the menu bar contains the words File, Edit, and so on.

To display the menu bar, enter the following at the 'Command:' prompt:

    Command: **menubar**

    Enter new value for MENUBAR <0>: **1**

Enter the number **1**, and then press ENTER. (In this case, "1" means on.)

(Alternatively, you can toggle the menu bar by selecting it from the droplist accessed by the drop arrow at the right end of the Quick Access tooblar.)

Like the ribbon, the menu bar groups commands by function. For instance, the **Line** command is found in the Draw category.

## CHANGING WORKSPACES

The way that AutoCAD looks — its menus, toolbars, ribbon, and palettes — is controlled by workspaces. Selecting another workspace changes the number and position of these user interface elements. The workspaces provided by Autodesk are selected from the Workspaces button on the status bar or the Quick Access menu.

1. Move the cursor to the Quick Access toolbar, and then click the **Workspaces** droplist. Notice the list of workspace names and options:

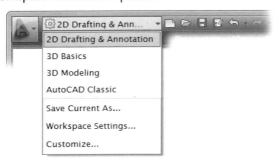

2. From the menu, choose **3D Modeling**. Notice that AutoCAD's user interface changes: the ribbon has different panels, and a tool palette appears on the right.

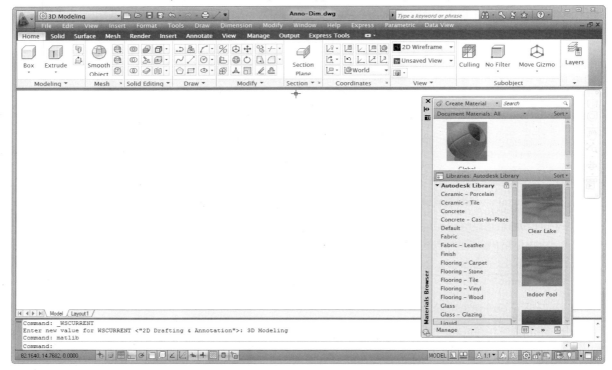

3. This book works with AutoCAD's default workspace, "2D Drafting & Annotation." To return AutoCAD to its default look, select this name from the Workspace droplist.

Workspaces are described in greater detail in Chapter 3, "Setting Up Drawings." For now, it is sufficient to know that if you do not see a user interface element, your workspace may have changed.

## AutoCAD Classic Workspace

If you are familiar with the menu bar and toolbars (as in AutoCAD 2008 and earlier), then choose **AutoCAD Classic** as the workspace. Notice that the menu bar and toolbars appear. If you also want the ribbon, you can redisplay it with the RIBBON command.)

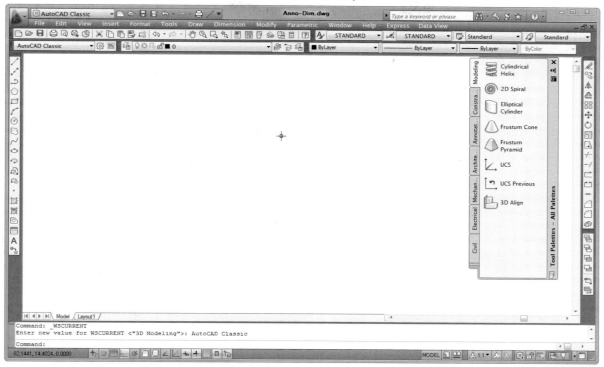

### Toolbars

The Classic workspace predominately features toolbars. They consist of collections of buttons labeled by icons, and group similar commands together. For example, many drawing commands appear on the Draw toolbar.

Toolbars operate like the ribbon: click a button to execute a command. Some toolbars feature flyouts and droplists, like the ribbon.

Not all toolbars are displayed at once. To access additional ones, follow these steps:

1. Right-click any toolbar button.
2. From the shortcut menu, select its name from the list.

Leave at least one toolbar visible so that you can easily access the other toolbars. If all of them are closed (such as in the default workspace), you can use the View menu's **Toolbars** item reopen them.

## EXITING AUTOCAD

AutoCAD provides several different ways to close drawings and exit AutoCAD. The following sections outline the possibilities.

To save your work and remain in AutoCAD, do the following:

1. On the Quick Access toolbar, click the 💾 **Save** button.
2. If the drawing has not yet been named, the Save Drawing As dialog box appears.

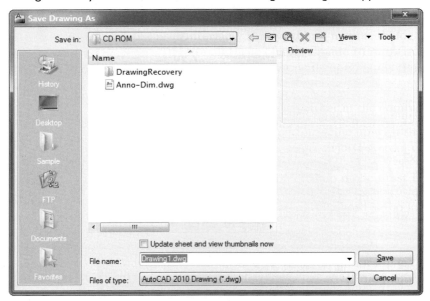

3. Name the drawing in the **File name** text box, and then choose **Save.**

You have these options to close just the drawing and remain in AutoCAD:

- Enter the **CLOSE** command to close the current drawing. If necessary, AutoCAD asks if you wish to save the drawing. AutoCAD then closes the drawing.
- To close all drawings at once, enter the **CLOSEALL** command.

To save your work and exit AutoCAD:

1. Click the **X** button found in the upper right corner of the AutoCAD window, adjacent to the Info Center.

2. When the "Save changes?" dialog box appears, choose **Yes** to save your work. AutoCAD saves your work, and then exits to the Windows desktop.
3. If this is a new drawing that has not be saved previously, then AutoCAD also displays the Save Drawing As dialog box. Enter a file name and then click **Save.**

To discard the drawing and exit AutoCAD:

1. Click the **X** button. Notice that AutoCAD displays a dialog box to confirm whether you want to save the drawing — it's your last chance!

2. Move the cursor to choose the **No** button to exit without saving your work. AutoCAD does not record your work to disk, and exits to the Windows desktop.

## 2D TUTORIAL

With the basics of navigating AutoCAD under your belt, you can now practice drawing in AutoCAD. In the following tutorial, you draw a two-dimensional object.

The figure below illustrates the drawing you construct. It is the plan view of a base with four holes. In the center is a shaft with rounded edges. This tutorial draws it in 2D; the following tutorial draws it 3D.

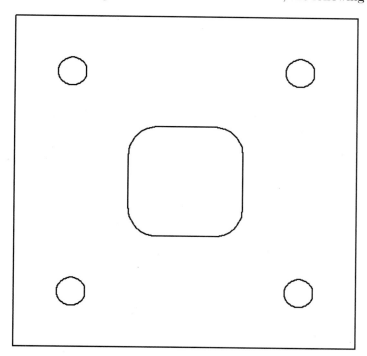

The tutorial describes the commands that you enter to create the drawing, which are shown by **boldface** text. (Responses can be in either UPPERCASE or lowercase.) Items enclosed in parentheses (such as these) are instructions and are not typed. If ENTER is shown, press the ENTER key on the keyboard. For example:

    Command: **line** *(Press ENTER.)*

Type the LINE command, and then press ENTER.

If you make a mistake, click the **Undo** button on the Quick Access toolbar. You may use it several times to undo each step in reverse order.

Before you enter a command, the prompt on the command line must be "empty," with nothing following the colon:

    Command:

If not, press ESC. This cancels the current command, and makes AutoCAD ready for your command.

## SETTING UP NEW DRAWINGS

1. Start AutoCAD, and then open a new drawing by clicking the **New** button on the Quick Access toolbar. Notice that AutoCAD opens the Select Template dialog box.

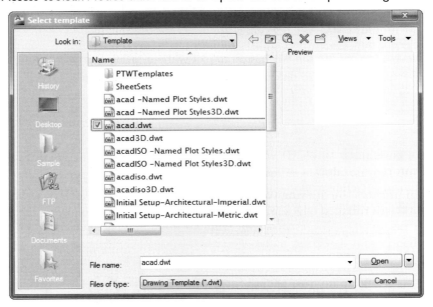

2. Choose *acad.dwt*, and then click **Open**.

To draw accurately, you use drawing aids like grid and snap. The grid is a visual guide that shows distances, like graph paper. When the grid spacing is set to 1 unit, for example, the drawing area is covered with lines that are spaced one unit apart. The grid is only displayed on the screen; it is not printed.

The snap is similar to drawing resolution. When on, snap mode causes the cursor to move in precise increments. For example, when the snap is set to 5 units, the cursor moves in a five-unit increment across the drawing. (The grid and snap can have either the same or different spacing from each other.)

You turn on these aids through the status bar, as follows:

3. On the status bar, click the **Snap** button to turn it on. When status bar buttons are turned on, they look blue; gray when off.
4. If necessary, click the **Grid** button to turn it on. Notice that a grid of lines fills the drawing area.

The default spacing for snap and grid is 0.5 units. (You can change these values for other drawings through the Drafting Settings dialog box by entering the DSETTINGS command.)

AutoCAD should look similar to the figure below.

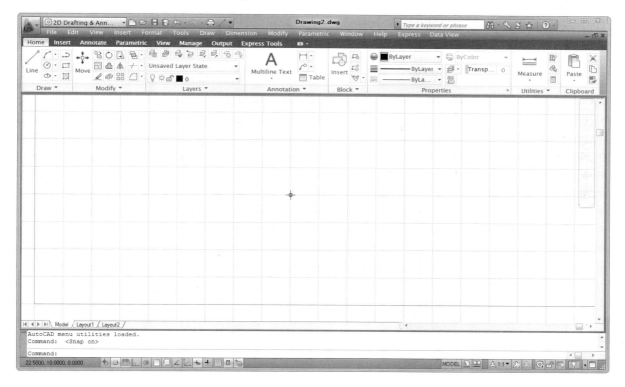

5. Move the mouse to see the crosshairs cursor move in increments of half units.

### DRAWING THE BASE PLATE

The base plate is a square. You can draw it with the RECTANGLE command, which is AutoCAD's command for drawing rectangles and squares:

1. From the Home tab's Draw panel, click the ▭ **Rectangle** button.

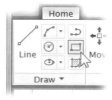

   As an alternative, enter the RECTANGLE command at the keyboard:
   Command: **rectangle** (Press ENTER.)

2. The command wants to know where to place the rectangle (square). The first corner is at the x, y coordinates of 3,2:
   Specify first corner point or [Chamfer/Elevation/Fillet/Thickness/Width]: **3,2** (Press ENTER.)

3. The opposite corner is located at x, y coordinates of 9,8:
   Specify other corner point or [Dimensions]: **9,8** (Press ENTER.)

Unlike the LINE command, the RECTANGLE command stops by itself; there is no need to press ESC. Your drawing should look similar to the figure below.

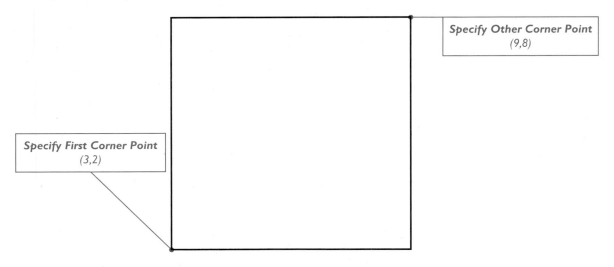

### DRAWING THE SHAFT

The shaft is also a square, but with rounded corners. The RECTANGLE command can draw squares with squared-off corners as above, rounded off, or cut off. Repeat the RECTANGLE command to draw the square shaft. In addition to specifying the two corners, you need to tell AutoCAD the radius of the rounded corners.

1. Press the spacebar. This repeats the previous command, the RECTANGLE command. This is a handy shortcut that saves you typing the name a second time.
   Command: *(Press SPACEBAR.)*
   RECTANGLE

2. Rounded corners are known in drafting as "fillets," and the size is specified by a radius. To specify the radius, enter the **Fillet** option:
   Specify first corner point or [Chamfer/Elevation/Fillet/Thickness/Width]: **fillet** *(Press ENTER.)*

3. These fillets are 0.5 inches in radius:
   Specify fillet radius for rectangles <0.0000>: **.5** *(Press ENTER.)*

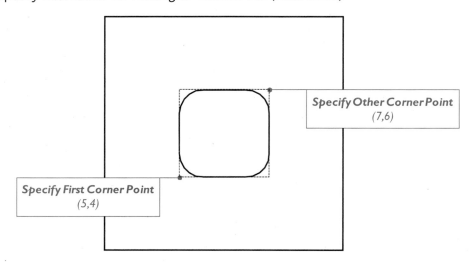

4. With the fillet radius set, you can now specify the corners of the square:
   Specify first corner point or [Chamfer/Elevation/Fillet/Thickness/Width]: **5,4** *(Press ENTER.)*
   Specify other corner point or [Dimensions]: **7,6** *(Press ENTER.)*

## DRAWING HOLES

Now draw the "holes" in the base plate. Holes are drawn with circles.

To draw a circle, AutoCAD needs to know two things from you: (1) its position in the drawing, based on a center point; and (2) the radius of the circle.

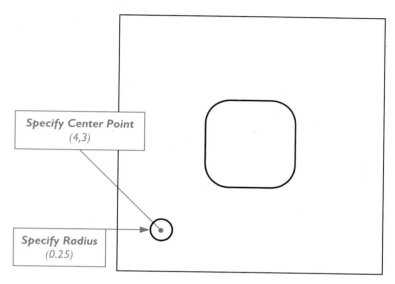

1. From the Home tab's Draw panel, click the ⊙ **Circle** button.
2. The first circle is located at 4,3 in the drawing, and so you specify its x,y coordinates like this:
   Specify center point for circle or [3P/2P/Ttr (tan tan radius)]: **4,3** *(Press ENTER.)*
3. The circle has a radius of 0.25 units:
   Specify radius of circle or [Diameter]: **0.25** *(Press ENTER.)*

## COPYING CIRCLES

To draw the other three circles, you don't repeat the CIRCLE command; instead, it is faster to make copies of the first one with the COPY command.

AutoCAD needs to know two things: (1) which object(s) to copy, and (2) the location(s) to place copies. Often, the location is specified as the distance between the original and the copied object.

1. Select the object to be copied using AutoCAD's COPY command together with an object selection method called "Last" (or "L" for short). It selects the last-drawn object visible on the screen, as follows:
   Command: **copy** *(Press ENTER.)*
   Select objects: **L** *(Press ENTER.)*

   Notice that the circle changes, and is now made of dashed lines.

   ⌒
   ( ⌣ )

   This effect is called "highlighting," how AutoCAD indicates selected objects.

2. Press ENTER to tell AutoCAD you have finished selecting objects.
   Select objects: *(Press ENTER to end object selection.)*

3. To specify the distance that the circle should be copied, you tell AutoCAD the *base point* and a *displacement*. The base point is the point from which the copying takes place:
   Specify base point or [Displacement] <Displacement>: **4,3** *(Press ENTER.)*

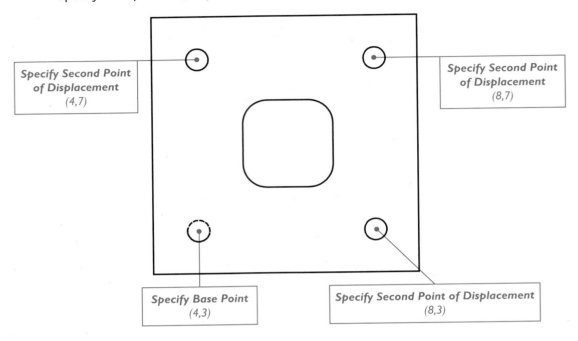

4. The displacement is the point at which the copy is placed:
   Specify second point of displacement or <use first point as displacement>: **8,3** *(Press ENTER.)*

5. The COPY command repeats its displacement prompt so that you can make multiple copies:
   Specify second point of displacement...: **8,7** *(Press ENTER.)*
   Specify second point of displacement...: **4,7** *(Press ENTER.)*

6. Press ESC to tell AutoCAD you have finished making copies:
   Specify second point of displacement...: *(Press ESC to end the command.)*

   Your drawing should look similar to the figure illustrated above.

### SAVING THE DRAWING

With the drawing complete, save your valuable work.

1. On the Quick Access toolbar, click the 💾 **Save** button.
2. In the Save Drawing As dialog box, enter "tutorial" for the **File name**.
3. Click **Save** to save the drawing.
   On the title bar, notice that the name changes from the generic *drawing1.dwg* to *tutorial.dwg*, perhaps prefixed by the path name.

## PLOTTING THE DRAWING

With the drawing saved, print a copy to show your fine work to your instructor, family, and friends! This is done with the PLOT command.

Before printing, AutoCAD needs to know: (1) the printer on which the drawing will be plotted, (2) the view to plot, and (3) the size. There are many other options, but these are the most important.

1. Click the big red **A**, and then choose **Print**.

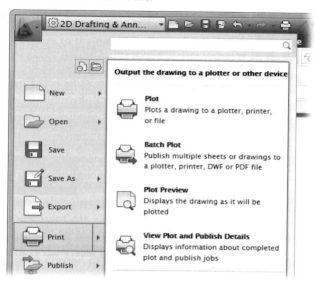

Notice that AutoCAD displays the Plot dialog box with its many options. It can be a little overwhelming, so I've highlighted the five areas you'll need for this tutorial.

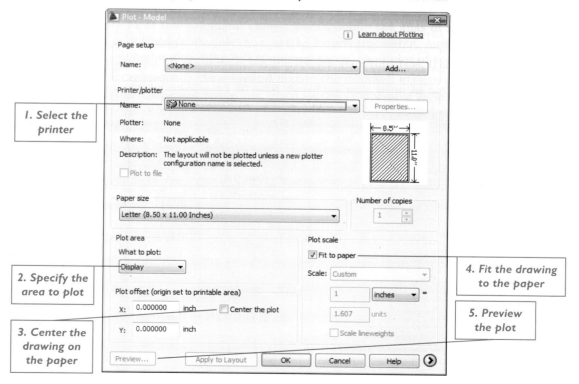

2. In the Printer/plotter area, select a printer from the **Name** droplist. "Default Windows System printer.pc3" is usually the safe choice. This plots the drawing on the same printer your computer uses for all other printing tasks.

3. In the Plot Area section, select **Extents** from the What to Plot droplist. This option ensures the entire drawing is plotted.

4. In the Plot Offset section, select **Center the plot** to center the drawing on the paper.

5. In the Plot Scale area, select **Fit to Paper**. This ensures the drawing fits the paper, no matter the size of paper or drawing.

At this point, the dialog box's options should look like those illustrated below.

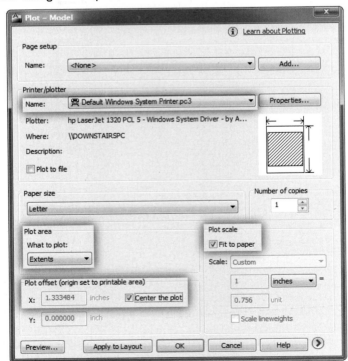

6. To ensure the plot will work out correctly, click the **Preview** button. This lets you check for errors before committing (or wasting) paper. AutoCAD shows you what the plot will look like, as illustrated by the figure below.

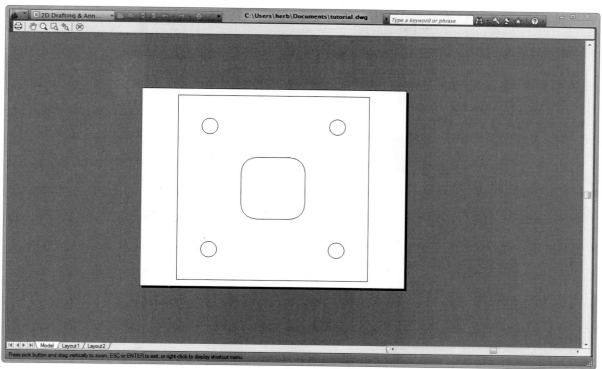

7. Click the 🖶 **Print** button. AutoCAD displays a dialog box showing the progress of the plot. When done, a yellow alert balloon appears at the right end of the status bar, reporting on the success (or failure) of the plot.

After a moment, the drawing should emerge from your printer.

8  Save the drawing with the **QSAVE** command, and then exit AutoCAD, as you learned earlier in this chapter.

## 3D TUTORIAL

Let's repeat the tutorial to draw the same part but in three dimensions. The figure below illustrates the result: the base plate with its four holes and the shaft with rounded edges.

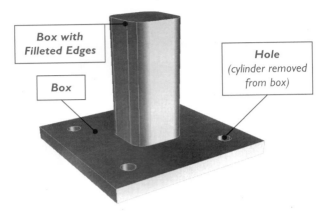

Modeling objects in 3D is very different from drafting in 2D, because 3D objects are made of solid primitives, which are elements like boxes and cylinders. For instance, the base is made of a 3D box, and the shaft is a second 3D box with four of its edges filleted. Holes are made of 3D cylinders that are then removed from the base.

Begin your work in the 3D modeling workspace.

1. From the Workspaces button on the status bar, choose **3D Modeling**.
2. Click the **New** button on the Quick Start toolbar, and then select the *acad3d.dwt* template file. Notice that AutoCAD's environment changes.

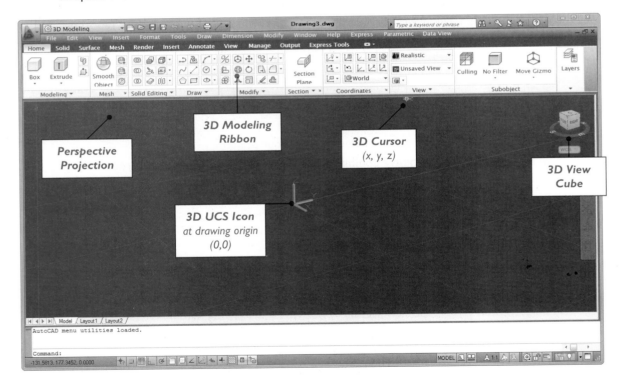

Ensure dynamic input is turned off; otherwise your results may differ from this tutorial: click the **DYN** button on the status bar.

3. Enter the **BOX** command at the keyboard:
   Command: **box** *(Press ENTER.)*

4. AutoCAD next needs to know where to place the box. One corner is at the x,y coordinates of 3,2:
   Specify first corner or [Center]: **3,2** *(Press ENTER.)*

5. The opposite corner is located at x, y coordinates of 9,8:
   Specify other corner or [Cube/Length]: **9,8** *(Press ENTER.)*

6. The height is 0.5:
   Specify height or [2Point]: **0.5** *(Press ENTER.)*

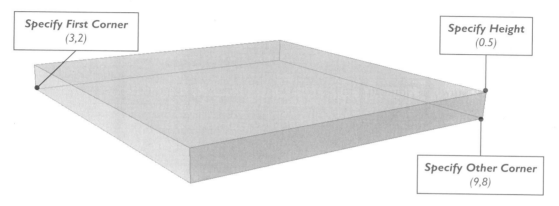

7. Your drawing should look similar to the figure above. If necessary, enter the **ZOOM** command, and select the **All** option:
   Command: **zoom** *(Press ENTER.)*
   Enter option [All/Extents/Window/Previous] <real time>: **all** *(Press ENTER.)*

## MODELING THE SHAFT

The shaft is also a box, but with rounded edges. Unlike RECTANGLE, the BOX command doesn't have a built-in filleting option, so you'll use the FILLET command later.

1. To make the boxes easier to see and work with, turn on transparency. In the ribbon's **View** tab, click the  **Opacity** button (found in the **Visual Styles** panel).

2. Enter the **BOX** command.
   Command: **box**

3. Specify the corners and height of the box by typing x, y coordinates, as follows:
   Specify first corner or [Center]: **5,4** *(Press ENTER.)*
   Specify other corner or [Cube/Length]: **7,6** *(Press ENTER.)*
   Specify height or [2Point]: **6** *(Press ENTER.)*

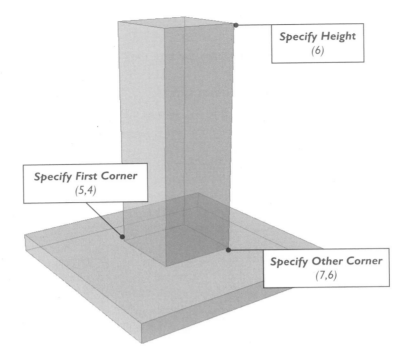

4. To round the edges of the box, enter the **FILLET** command:
   Command: **fillet**
   Current settings: Mode = TRIM, Radius = 0.0000

5. Enter the "m" option (short for **Multiple**), because you will be filleting multiple (four) edges:
   Select first object or [Undo/Polyline/Radius/Trim/Multiple]: **m**

6. The fillets are 0.5 inches in radius:
   Select first object or [Undo/Polyline/Radius/Trim/Multiple]: **r**
   Enter fillet radius <0.0000>: **0.5** *(Press* ENTER.*)*

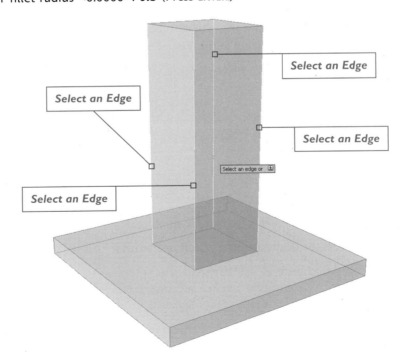

Notice that the cursor changes to a small square; this is called the "pick" cursor.

7. Now tell AutoCAD which edges to fillet by picking each one: you must position the square pick cursor over each edge, and then click the left mouse button:
   Select an edge or [Chain/Radius]: *(Pick an edge.)*

8. As you do, the edge turns white. Repeat for the other three edges:
   Select an edge or [Chain/Radius]: *(Pick another edge.)*
   Select an edge or [Chain/Radius]: *(Pick a third edge.)*
   Select an edge or [Chain/Radius]: *(Pick the last edge.)*

8. With the four edges selected, press **ENTER** to exit selection mode:
   Select an edge or [Chain/Radius]: *(Press ENTER to exit selection mode.)*
   4 edge(s) selected for fillet.

9. Press **ENTER** a second time to exit the command.
   Select first object or [Undo/Polyline/Radius/Trim/Multiple]: *(Press ENTER to exit the command.)*

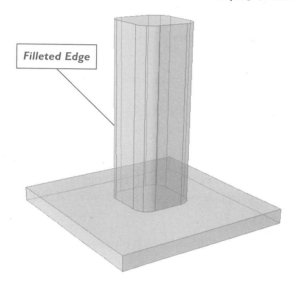

Notice that the column now has rounded edges.

## MODELING HOLES

The next step is to draw the holes in the base plate. In AutoCAD, holes are constructed by drawing, then removing, cylinder shapes.

1. Enter the **CYLINDER** command:
   Command: **cylinder**

2. The cylinder is located at 4,3 in the drawing:
   Specify center point of base or [3P/2P/Ttr/Elliptical]: **4,3** *(Press ENTER.)*

3. The circle has a radius of 0.25 units:
   Specify base radius or [Diameter]: **.25** *(Press ENTER.)*

4. The height does not matter, just as long as it is taller than the base:
   Specify height or [2Point/Axis endpoint] <6.0>: **1** *(Press ENTER.)*

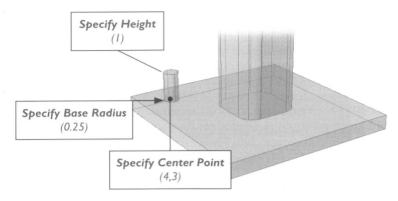

**COPYING CYLINDERS**

To draw the other three cylinders, use the COPY command to copy the first cylinder to the other locations on the base plate.

1. Start the **COPY** command, and then select the last-drawn cylinder with the **L** object selection option, as follows:
   Command: **copy** *(Press ENTER.)*
   Select objects: **L** *(Press ENTER.)*
   1 found Select objects: *(Press ENTER to end object selection.)*

2. Recall that the *base point* is the point from which the copying takes place:
   Specify base point or [Displacement] <Displacement>: **4,3** *(Press ENTER.)*

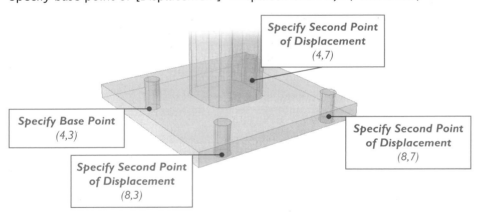

3. The *displacement* is the point at which each copied cylinder is placed:
   Specify second point of displacement or <use first point as displacement>: **8,3** *(Press ENTER.)*
   Specify second point of displacement...: **8,7** *(Press ENTER.)*
   Specify second point of displacement...: **4,7** *(Press ENTER.)*
   Specify second point of displacement...: *(Press ESC to end the command.)*

## SUBTRACTING CYLINDERS TO CREATE HOLES

With the four cylinders in place, you now use the **SUBTRACT** command to turn them into holes.

1. Enter the **SUBTRACT** command:
   Command: **subtract**

   AutoCAD first asks you for the object(s) to subtract *from*; that would be the base plate:
   Select solids and regions to subtract from ..
   Select objects: *(Select the base plate.)*
   Select objects: *(Press ENTER to end object selection.)*

2. Now AutoCAD needs to know which objects to subtract; that would be the four cylinders:
   Select solids and regions to subtract ..
   Select objects: *(Select one cylinder.)*
   Select objects: *(Select another cylinder.)*
   Select objects: *(Select a third cylinder.)*
   Select objects: *(Select the fourth cylinder.)*
   Select objects: *(Press ENTER to end object selection.)*

   Notice that AutoCAD removes the cylinders from the base plate, creating four holes.

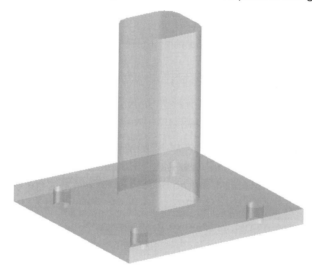

3. The final step is to join the shaft with the base plate. This is done with the **UNION** command.
   Command: **union**
   Select objects: **all**
   Select objects: *(Press ENTER to end object selection.)*

   The model does not look any different, but it is now a single object.

4. Save the drawing as "3d Tutorial," and then print it, if you wish.

So, that was a quick start to using AutoCAD.

Welcome to the rest of *Using AutoCAD 2011*. The remaining chapters take you step by step through the program. Before long, you'll be using AutoCAD to create drawings like a pro.

*Enjoy!*

## EXERCISES

1. For this exercise, start AutoCAD.
   (If the Start New Drawing dialog box appears, click **Cancel**.)
   With the **LINE** command, draw several shapes:
   a. Rectangle.
   b. Triangle.
   c. Irregular polygon (any shape you like).
2. Use the **U** command. What happens to the last object you drew?
3. Press function key **F9**, and then look at the status line.
   Does the button look blue (turned on)?
4. Repeat the **LINE** command, and again try drawing these shapes:
   a. Rectangle.
   b. Triangle.
   c. Irregular polygon.
   Do you find it easier?
5. Draw a rectangle with the **RECTANGLE** command using these parameters:
   Corner           2,3
   Other corner     6,7
6. Draw a circle with the **CIRCLE** command using these parameters:
   Center point     4,5
   Radius           0.75

   Is the circle drawn "inside" the box?
7. Print your drawing with the **PLOT** command.
   Does the plot look like the drawing on your computer screen?

## CHAPTER REVIEW

1. What are AutoCAD drawings constructed with?
2. Name three ways in which commands can be entered:
   a.
   b.
   c.
3. Describe the purpose of the 'Command:' prompt.
4. How do you exit from print preview mode?
5. What is the purpose of the LINE command?
6. How is the U command helpful?
7. Describe how to cancel commands.
8. What does the mouse control?
9. What is an *icon*?
10. Name two methods by which to determine the function of ribbon buttons:
    a.
    b.
11. What are *tooltips*?
12. What are *flyouts*?
13. Can toolbars be moved around the AutoCAD window?
14. Which commands close drawings without exiting AutoCAD?
15. Which commands exit AutoCAD?
16. Which command saves drawings?
17. Which command starts new drawings?
18. Describe how snap and grid are useful:
    Snap
    Grid
19. Is the grid plotted?
20. List three things AutoCAD needs to know, as a minimum, before plotting drawings:
    a.
    b.
    c.
21. Explain the advantage of the PLOT command's **Fit to Scale** option.
22. Why is the **Preview** option environmentally friendly?
23. How do you exit from print preview mode?

24. Identify the user interface elements illustrated below:
    a.
    b.
    c.
    d.
    e.
    f.
    g.
    h.
    i.

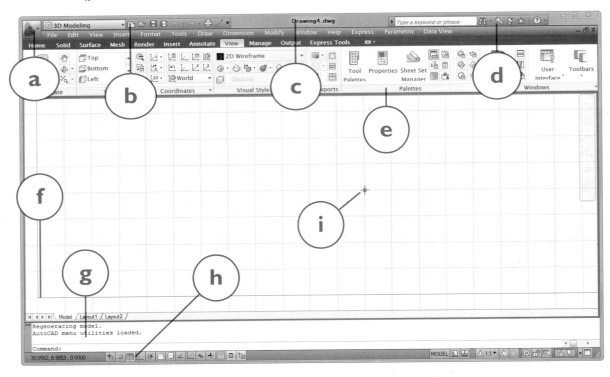

# CHAPTER 2
## Understanding CAD Concepts

Software used for computer-aided designs employe concepts that distinguish it from other software programs. In part, the uniqueness stems from CAD's basis on *vectors*. These are geometric objects that have length and direction; using AutoCAD means knowing how to manipulate vectors. So, before you can manipulate vectors successfully, you need to understand how AutoCAD operates.

This chapter covers the following activities:

> Touring the AutoCAD user interface.
> Saving and restoring user interface configurations.
> Understanding the information AutoCAD presents to you.
> Entering commands.
> Learning shortcut keystrokes and command aliases.
> Picking points on the screen.
> Using mouse buttons and menus.
> Identifying the elements of dialog boxes.
> Understanding the draft-edit-plot cycle.
> Recognizing AutoCAD's coordinate systems.
> Entering coordinates from the keyboard and on the screen.

---

**NEW TO AUTOCAD 2011** IN THIS CHAPTER
- Some aspects of AutoCAD's user interface have changed.
- The navigation bar was added to the drawing area.
- There are new commands, system variables, and aliases.

## AUTOCAD USER INTERFACE

Your drawing sessions begin by starting the AutoCAD program. As described in Chapter 1, "Quick Start in AutoCAD," double click the AutoCAD 2011 icon to start the program.

(If there is no icon on the Windows desktop, click the taskbar's **Start** button, and then choose **Programs**. Choose the **Autodesk** folder, the **AutoCAD 2011** folder, and then the **AutoCAD 2011** program.)

When AutoCAD loads, you will probably see a blank drawing. (If any dialog boxes appear, close them.) The look of AutoCAD may differ, depending on which workspace is loaded. For most of this book, you will work with the "2D Drafting & Annotation" workspace, which looks like this:

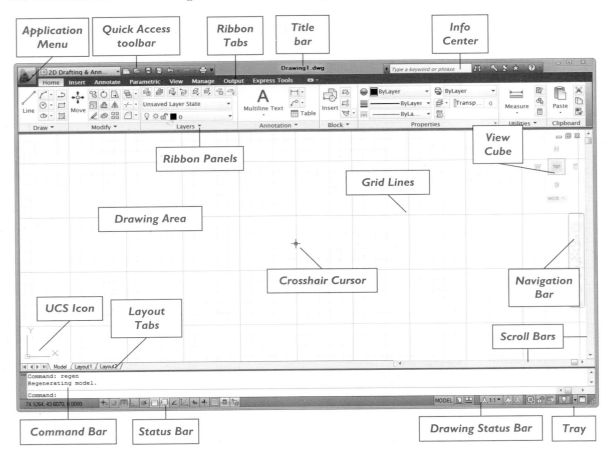

## USER INTERFACE CHANGES IN AUTOCAD 2011

If you are used to earlier releases of AutoCAD, then the changes to AutoCAD 2011's user interface might surprise you:

The default background color for the model tab is dark gray.

The grid is turned on by default in the 2D workspace, and uses the same adaptive lines as do 3D workspaces.

The ViewCube is also now turned on in 2D workspace, by default.

Quick properties are now turned off, by default.

Brand-new is the Navigation Bar, which collects the pan, zoom, and other navigation commands.

Workspaces can now be selected from the Quick Access toolbar, as well as from the status bar.

## TOURING THE USER INTERFACE

AutoCAD displays information in several areas of its window. You learned about some of them in the previous chapter; now let's look at the remaining areas of interest, starting at the top and then moving down the window.

### TITLE BAR

At the very top of the AutoCAD window sits the *title bar*. The center of the title bar tells you the name of the software — AutoCAD 2011 — and the file name of the current drawing, such as "Drawing1. dwg."

 At the left is a large red A; it hides the *application menu* described in detail in Chapter 1. Next to it is the *Quick Access toolbar*  , also described in Chapter 1.

### Info Center

To the right of the title bar is the *Info Center*. When you enter terms into its search field, it accesses the help system. (This replaces the Info palette from earlier releases of AutoCAD.)

From left to right, the Info Center's buttons perform the following functions:

| Button | Meaning |
|---|---|
| | Collapses the InfoCenter. |
| | Searches the help files, as well as resources on Autodesk's Web site. |
| | Displays a list of search options. |
| | Accesses Autodesk's subscription center by Web browser. |
| | Lists information from Autodesk's blogs and online learning. |
| | Collects "favorites," often used as help resources. |
| | Displays help. |

AutoCAD 2011 changes the help system from CHM (appearing in its own window) to HTML, which means the help topics now appear in your computer's default Web browser. By default, it accesses the latest help information from Autodesk's Web site.

### Window Controls

At the far right are three window controls common to every Windows application. (A similar set of buttons is available for drawing windows.)

From left to right, the buttons perform the following functions:

| Button | Meaning | Related Commands |
|---|---|---|
| | Minimizes AutoCAD, or the drawing window. | ... |
| | Restores the AutoCAD or drawing window. | ... |
| | Closes the AutoCAD window, and | EXIT command. |
| | closes the drawing window. | CLOSE command. |

You can quickly exit AutoCAD by double-clicking the red A button.

## RIBBON, MENU BAR, AND TOOLBARS

Below the title bar is the *ribbon*, which replaces the menu bar and toolbars from earlier releases of AutoCAD. You can toggle the display of the ribbon with the RIBBON and RIBBONCLOSE commands.

You can bring back the menu bar by choosing **Show Menu Bar** from the Quick Access toolbar's options list.

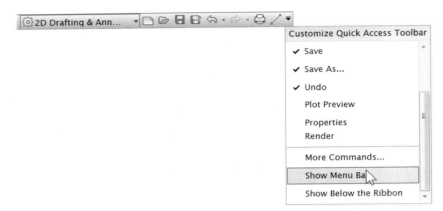

The menu appears in its traditional location, below the title bar:

A toolbar can be brought back initially through the -TOOLBAR command (be sure to enter the dash):

    Command: **-toolbar**

    Enter toolbar name or [ALL]: **standard**

    Enter an option [Show/Hide/Left/Right/Top/Bottom/Float] <Show>: *(Press* ENTER.*)*

Once one toolbar is visible, you can right-click it to easily access the name of all other toolbars.

## DRAWING AREA

The drawing area is where you draw, and it displays much information.

### Cursors

The *crosshair cursor* is the most important element of the drawing area, because it shows you where you are in the drawing. Large crosshair lines help you locate the cursor easily.

There is a small square at the cursor's center, which indicates that you can *select* (or "pick") objects in the drawing for editing. This square is called the "pickbox." Pressing the left mouse button selects the object under the pickbox. Although its size can be varied (through a system variable named PICK-BOX), I find the default of 10 pixels to be exactly the right size.

Moving the crosshair cursor out of the drawing area changes its shape to an arrow, with which you are probably familiar from other Windows software.

The arrow cursor lets you select items outside of the drawing area, such as buttons in the ribbon and options in dialog boxes.

## Cursor Icons

From time to time, AutoCAD displays small icons near the crosshair cursor — or in place of it. Here are just a few of the cursor icons you can expect to see:

| Icon | Meaning | Related Command |
|---|---|---|
|  | Specifies orthographic drawing mode is in effect. | ORTHO |
|  | Specifies that an object or its layer is locked. | LAYER |
|  | Specifies that the table is linked to an external data file. | DATALINK |
|  | Specifies that the drawing is being panned. | PAN |
|  | Specifies that the drawing has panned to its extents. | PAN |
|  | Specifies that the drawing is in real time zoom mode. | ZOOM |
|  | Specifies that the object has an annotative scale factor. | CANNOSCALE |
|  | Specifies that the object has two or more annotative scales. | ... |

### View Cube

In the upper right corner of the drawing area is the *view cube* (or "navigation cube"). Its primary purpose is to rotate the view in three dimensions. You do so by clicking the corners of the cube.

Although the view cube is not new to AutoCAD 2001, it is new to the 2D model space environment. Because the view cube is not particularly useful for 2D drafting, you can turn it off with the NAVVCUBE command, as follows:

Command: **navvcube**

Enter an option [ON/OFF/Settings] <ON>: **off**

### Navigation Bar

Below the view cube is the *navigation bar*, a toolbar that collects many of AutoCAD's viewing commands into one handy location.

The buttons perform the tasks listed below:

| Icon | Meaning | Related Command |
|---|---|---|
|  | Turns on the Navigation Wheel. | NAVSWHEEL |
|  | Pans the drawing. | PAN |
|  | Zoom the drawing. | ZOOM |
|  | Rotates the drawing view in 3D. | 3DORBIT |
|  | Opens the Show Motion interface. | NAVSMOTION |

The navigation bar has a number of options available through ▾ dropdown buttons:

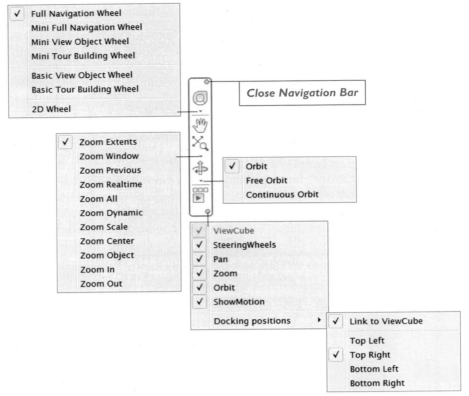

## UCS Icon

In the lower-left corner of the drawing area is an L shape with the letters Y and X. It is called the "UCS icon," *UCS* being short for "user coordinate system."

The arms of the Y and X point in the direction of the positive y and x axes, respectively. The UCS icon is, however, useful primarily for drawing in three dimensions, because it helps to orient you in three-dimensional space. It is used by a few 2D operations, such as determining the default location for hatch patterns, ordinate dimensions, and mass property calculations. Most of the time, however, the icon is not needed for 2D drafting and so you can turn it off with the UCSICON command.

Type the command name, and then enter the **Off** option, as follows:

Command: **ucsicon** *(Press* ENTER.*)*

Enter an option [ON/OFF/All/Noorigin/ORigin/Properties] <ON>: **off** *(Press* ENTER.*)*

## Other Screen Markings

The drawing area often displays additional markings that appear as you work with objects.

### Highlights and Grips

When you select objects during editing commands, AutoCAD *highlights* them with a dotted pattern. The highlighting is feedback that confirms which object(s) you selected. The figure below shows a circle that isn't selected (at left) and one that is (at right).

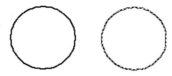

**Left:** Unselected circle.
**Right:** Circle selected during an editing command.

When you select objects without first entering a command, the objects also show *grips*. These are small squares that you use to edit objects directly without entering commands. You can move or copy them using the grips, as well as stretch, mirror, and rotate them, and sometimes even other operations.

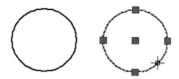

**Left:** Unselected circle.
**Right:** Selected circle with grips and highlighting.

Grips editing is detailed in a later chapter.

### Rollover Tooltips

In addition to displaying highlighting and grips, AutoCAD may also display a couple of floating panels around objects. One is called "Rollover Tooltips," and the other, "Quick Properties."

When you pause the cursor over an object, AutoCAD displays a tooltip that reports its type (such as "Circle") and some of its properties, such as color, layer, and linetype.

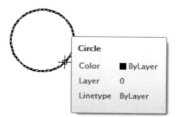

This is called a "rollover" tooltip, because it appears when you *roll* the cursor *over* objects. But it doesn't appear right away; there is a delay (of about two seconds) so that these tooltips aren't popping up all over the screen while you move the cursor. A better name might be "pause over" tooltip.

The content of this tooltip can be customized with the CUI command. These tooltips can be turned off with the **Show Rollover Tooltips** option in the OPTIONS command's Display tab.

## Quick Properties

The Quick Properties palette appears when you select an object. The palette displays the object's properties and allows you to change them.

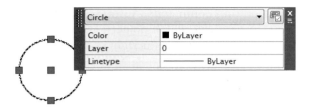

If the palette does not appear, then it has been turned off, the default setting as of AutoCAD 2011. To turn on the functionality, click the  (or QP) Quick Properties button on the status bar. You learn more about this palette in a later chapter.

## Selection Cycling

When you try to select one object of several lying on top of one another, it can be hard for AutoCAD to determine which one you want. In this situation, AutoCAD does a couple of things to help you out.

When you pass the cursor over objects that overlap, AutoCAD displays an icon near the cursor to alert you to the fact.

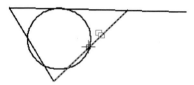

When you select the overlapping objects, AutoCAD next displays a palette listing the names of the overlapping objects.

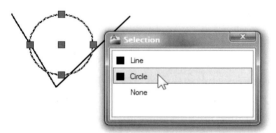

When you choose a name from the list, AutoCAD highlights the related object. To turn the functionality on or off, click the (or SC) Selection Cycling button on the status bar.

## Dynamic Input

As detailed in Chapter 1, AutoCAD can display prompts at the cursor. This is called "dynamic input."

Dynamic prompts are turned on and off through the (or DYN) Dynamic Input button on the status bar, and are on by default. (Press **F12** to turn off temporarily.)

Grid

The *grid* is a visual aid that helps you gauge distances in drawings. (It was described in Chapter 1.) It is turned on and off through the ▦ (or GRID) Grid button on the status bar.

Tracking Lines

A *tracking line* is a dotted line that radiates from the cursor. It shows you geometric relationships that exist between objects. There are two types of tracking: *object snap* and *polar*.

**Object snap tracking** shows you the relationships between the cursor and the geometric features of objects, such as their end points, centers, and intersections.

In the figure below, a line is being drawn and its end point is lined up with the circle's center — as shown by the dotted line. The accompanying tooltip reports the distance to the "Center" of the circle; the angle of 90 degrees is measured counterclockwise from the positive x-axis.

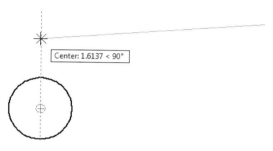

Object snap tracking is turned on with the ∠ (or OTRACK) Object Tracking button.

**Polar tracking** lets you know when the cursor is at discrete angles from the horizontal, such as in 15-degree increments (0, 15, 30, 45, and so on). In the figure below, the tooltip reports the line's length and its precise angle is 15 degrees. When the line is not at an increment of 15 degrees, the tooltip does not appear.

Polar tracking is turned on with the ⦟ (or POLAR) Polar Tracking button on the status bar. You can have both osnap and polar tracking turned on at the same time.

## LAYOUT TABS AND SCROLL BARS

Below the drawing area are three or more tabs labeled Model, Layout1, and Layout2. These tabs let you switch quickly between model space and paper space. You learn more about layouts later in this book.

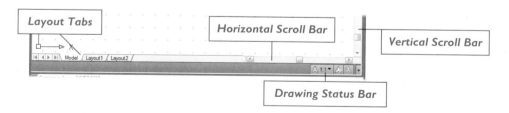

### Scroll Bars

Scroll bars let you move the drawing around the screen — called "panning." You can click or drag the bars to pan the drawing horizontally and vertically.

Scroll bars are turned off by default, so to turn them on, enter the OPTIONS command, choose the **Display** tab, and then select the **Display Scroll Bars in Drawing Window** option.

## STATUS BAR

At the very bottom of the AutoCAD window is the status bar. This bar reports the coordinates of the cursor and the status of the drawing through buttons.

Many settings are turned on and off simply by clicking the related button. Additional controls are available by right-clicking buttons or selecting options from shortcut menus. For example, to change the way coordinates are displayed, right-click the x, y, elevation display, and choose another format.

AutoCAD 2011 moved the visual controls from the status bar to the navigation bar.

### X, Y Coordinates and Elevation

At the left of the status bar is a set of three numbers. They represent the coordinates of the crosshair cursor: the x, y coordinates plus the current elevation (not the z coordinate).

-1.1936, -0.0736, 0.0000

The format of the numbers and angles is controlled by the UNITS command, as you will learn in Chapter 3, "Setting Up Drawings."

The x, y coordinates constantly update as you move the cursor. You can change the display to relative (*distance<angle*) readout through the shortcut menu described above.

4.0561<155, 0.0000

When the coordinates look gray and do not update as you move the cursor, then they are turned off; click the coordinate display to turn them on.

### Mode Indicators

Next on the status bar are mode indicators which look like buttons. These are also known as "toggles," because they turn modes on and off — like light switches. Clicking a button turns the mode on or off. When on, the mode's button looks bluish; when off, grayish.

From left to right, the buttons have the following meaning:

| Icon | Text | Shortcut Keystroke | Meaning |
|---|---|---|---|
| | INFER | ... | Constraint inference is on or off. |
| | SNAP | F9 | Cursor snap is on or off. |
| | GRID | F7 | Grid display is on or off. |
| | ORTHO | F8 | Orthogonal mode is on or off. |
| | POLAR | F10 | Polar snap mode is on or off. |
| | OSNAP | F3 | Object snap modes are on or off. |
| | 3DOSNAP | F4 | 3D object snap modes are on or off. |
| | OTRACK | F11 | Object tracking mode is on or off. |
| | DUCS | F6 | Dynamic user-coordinate system is on or off. |
| | DYN | F12 | Dynamic input mode is on or off. |
| | LWT | ... | Lineweights are displayed or not. |
| | TPY | ... | Object transparency is on or off. |
| | QP | ... | Quick Properties palette is on or off. |
| | SC | ... | Selection cycling is on or off. |

As an alternative to clicking buttons on the status bar, you can press the shortcut keystrokes listed by the table above.

The buttons display icons (as shown above) or text labels (below). To change between them, right-click any button, and then choose **Use Icons** from the shortcut menu.

## Layouts

Next to the toggle buttons are two controls that switch visually among drawings and layouts. (Layouts are described later in this book.) The two buttons have the following meaning:

| Icon | Meaning | Related Command |
|---|---|---|
| | Switch visually between drawings. | QVDRAWING |
| | Switch visually between layouts. | QVLAYOUT |

Click one of the buttons, and AutoCAD displays a filmstrip-like series of preview images of the open drawings and layouts. Click a thumbnail to switch to that drawing or layout.

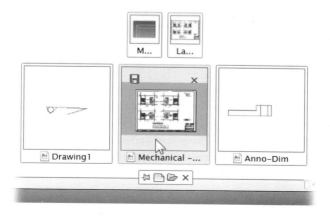

### Annotation Scaling

Most objects in AutoCAD can be drawn and plotted at any size, but some cannot. Text, hatching, and linetypes must look the correct size when plotted. Annotation scaling ensures that these scale-dependent objects display and plot at the correct scale factor.

From left to right, the buttons have the following meaning:

| Icon | Meaning | Related Command |
|---|---|---|
| | Specifies the annotation scale; lists scale factors when clicked. | CANNOSCALE OBJECTSCALE |
| | Toggles visibility of annotation objects. | ANNOALLVISIBLE |
| | Adds annotative scale factors automatically. | ANNOAUTOSCALE |

Annotative scaling is discussed in detail in Chapter 10, "Placing and Editing Text."

### Workspaces

Workspaces change aspects of the user interface, such as the ribbon, toolbars, and palettes. To change, click the **Workspace** button on the status bar or the Quick Access toolbar, and then choose another one.

You can create workspaces customized to your drafting needs.

### Tray

At the right end of the status bar are icons, many of which appear only when AutoCAD needs to tell you something. For some reason, this part of the status bar gets its own name: the tray. The one icon that is always present is illustrated below — unless turned off via Tray Options. As needed, other icons appear in the tray, as follows:

| Icon | Meaning | Related Commands |
|---|---|---|
| | Locks palettes and toolbars into position. | LOCKUI |
| | Indicates that drawing has CAD standards attached. | STANDARDS, CHECKSTANDARDS |
| | Indicates that tables contain linked attribute data. | DATAEXTRACTION |
| | Indicates that tables are linked to external data files. | DATALINK |
| | Indicates that externally-referenced drawings are attached. | XATTACH |
| | Indicates that drawing is signed digitally. | SIGVALIDATE |
| | Indicates that drawing is created by Autodesk software. | DWGCHECK |
| | Indicates that plots are underway or complete. | PLOT, VIEWPLOTDETAILS |
| | Warns that layers were added from xref drawings. | LAYERNOTIFY |
| | Indicates that Communications Center is active. | ... |

Sometimes, icons display yellow alert balloons to warn you of completed actions or problems.

Click the blue underlined text for more information, or click the x in the upper-right corner to dismiss the balloon.

To keep the toolbars and palettes from shifting on the screen, you can lock them in place. At the far right end of the status bar, click the padlock icon, and then choose which user interface elements to lock: floating and docked toolbars, panels, "windows," or all. (*Windows* is the old term for palettes, such as the Properties palette — and not the AutoCAD window itself.) Hold down the CTRL key to move locked items temporarily.

Tray Options

Click the arrow button for a menu of options for the status bar and tray.

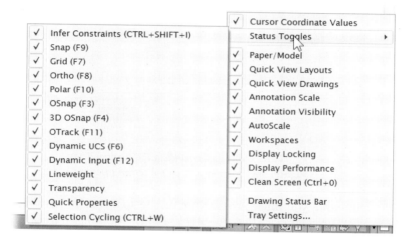

### Cleanscreen and Resize Window

The **Clean Screen** button maximizes the AutoCAD window, and turns off the title bar and ribbon. Some users find this mode handy for making screen grabs of the drawing. Click the button a second time to return the window to its former size.

The **Resize AutoCAD** button changes the size of the AutoCAD window; drag the corner with the mouse. (The button is not present when AutoCAD is in full-screen mode.)

## ENTERING COMMANDS

Commands can be entered in AutoCAD by several means, such with the keyboard, through dynamic input, and by keyboard shortcuts. Let's look at each.

### USING THE KEYBOARD

The original method for entering commands is with the keyboard. The earliest versions of AutoCAD did not rely on a mouse, which was expensive and rare at the time. (The technical editor notes that he used to use a joystick — until it fell out of favor due to its association with games.) All commands had to be entered with the keyboard.

Over the years, Autodesk expanded the user interface, slowly adding menus, dialog boxes, toolbars, right-click shortcut menus, and ribbons. Despite all these changes made to AutoCAD, you still need to deal with the keyboard, such as for overriding dimensions or using commands that work only with the keyboard.

When you type commands at the keyboard, your input is displayed either in the command prompt area (illustrated below) or on the screen — depending on where the cursor is located.

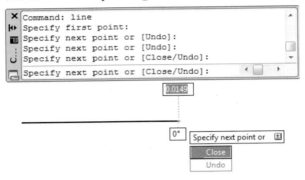

*Left:* Entering commands in the command line.
*Right:* Entering commands on the screen (dynamic input).

After typing the command, you must press ENTER or the spacebar to activate the command. In addition to ENTER, AutoCAD recognizes these keystrokes in the command line:

| Keystroke | Meaning |
| --- | --- |
| ESC | Cancels the current command. |
| ← | Moves the cursor one character to the left. |
| → | Moves the cursor one character to the right. |
| HOME | Moves the cursor to the beginning of the command text. |
| END | Moves the cursor to the end of the line. |
| DEL | Deletes the character to the right of the cursor. |
| BACKSPACE | Deletes the character to the left of the cursor. |
| INS | Switches between insert and typeover modes. |
| ↑ | Displays the previous line in the command history. |
| ↓ | Displays the next line in the command history. |
| PGUP | Displays the previous screen of command text. |
| PGDN | Displays the next screen of command text. |
| CTRL+V | Pastes text from the Clipboard into the command line. |
| TAB | Cycles through command names. |

## Command Input Considerations

Before typing a command, you must clear the command line. That is, another command must not be in progress. If the 'Command' prompt is not clear, you clear it by pressing ESC.

    Specify next point or [Undo]: *(Press* ESC.*)*
    Command:

*Transparent commands* are the exception. These commands are regular commands that operate during other commands when prefixed with the ' (apostrophe) character.

When you enter a command, AutoCAD doesn't care if you use UPPERCASE or lowercase characters, or a combination. If you make a mistake typing on the command line, press the BACKSPACE key to correct it, or edit the text on the command line using the keys listed in the table.

You can press ESC at any time. If you press ESC while an operation is in progress, it terminates the command.

## Repeating Commands

Some commands repeat automatically until you press ESC, but most commands do not. To repeat the command you previously entered, just press ENTER (or the spacebar) at the 'Command:' prompt. As an alternative, right-click to display the cursor menu, and then select the **Repeat** option.

To force a command to repeat automatically, type MULTIPLE before entering the command name, as in:

    Command: **multiple**
    Enter command name to repeat: **circle**

This forces the CIRCLE command to repeat itself until you press ESC.

## Recent Input

AutoCAD remembers the last twenty commands and options you enter at the 'Command:' prompt, or select from a menu or toolbar. Autodesk calls this "Recent Input."

To access command history, press the ↑ key at any time, except when you are in a dialog box. AutoCAD shows the previous commands you entered. When you see the one you need, press the ENTER key, and the command executes.

During commands you can also press the ↑ key. Instead of listing previous commands, however, AutoCAD shows you the coordinates you previously specified, whether entered at the keyboard or picked with the cursor.

If you prefer to see a list of the command history, right-click, and then select **Recent Input** from the shortcut menu. AutoCAD lists the previous commands and/or coordinate inputs in a submenu.

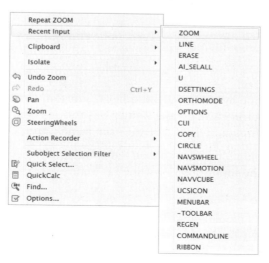

**Note** If you are unsure of how to spell a command, enter as much as you know, and then press the TAB key. AutoCAD displays the rest of the command name.

When you keep pressing tab, AutoCAD shows the names of additional commands and related aliases, in alphabetical order.

Command: **scale** *(Press TAB.)*
SCALE *(Press TAB.)*
SCALELISTEDIT *(Press TAB.)*
SCALETEXT *(Press TAB.)*
SCALE *(Press TAB.)*

### Dynamic Input

A second way to view commands typed at the keyboard is in the drawing area, as introduced in Chapter 1. Autodesk calls this "dynamic input," because it displays tooltips, distances, angles, and tracking lines dynamically on the screen during drawing and editing commands.

(*History*: The "dynamic" name goes back to Release 9, when Autodesk introduced the first dialog boxes to AutoCAD with the name of "dynamic dialogs.")

There are three aspects to dynamic input, all of which you can turn on or off:

**Dynamic Prompts** — display commands and prompts in tooltips near the cursor.
**Pointer Input** — displays coordinates and angles in tooltips near the cursor. They can be edited.
**Dimension Input** — displays distances and angles in tooltips near the cursor. These appear only at the 'Specify next point:' prompt.

You turn on overall dynamic input like this: on the status bar, turn on dynamic mode by clicking the ▭ (or DYN) **Dynamic Input** button. Next, move the cursor into the drawing area, and then type commands, as described below.

If you find you prefer using dynamic input, then you can hide the command bar with the COMMAND-LINEHIDE command. (Alternatively, press CTRL+9 to toggle the display of the command prompt bar.)

### Entering Commands

When you enter commands during dynamic input, the command name appears in the drawing area near the cursor, such as "circle" in the figure below.

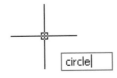

You can edit the command name by using the same keys as at the 'Command:' prompt. For example, you can press the ↑ key to scroll through the names of previous commands, or use the TAB key to step through the names of commands that start with the same letter(s).

Press BACKSPACE to erase the name, and ESC to exit the command. When you enter an incorrect command name or invalid value, the input box is outlined in red.

## Entering Coordinates

When you enter the command name and press ENTER, the command's prompts appear at the cursor. For the CIRCLE command, the first prompt is 'Specify center point for circle:'. Pick a point in the drawing, or interact with the dynamic prompt, as described next.

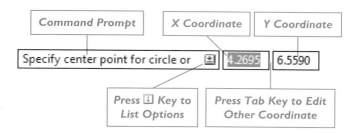

A pair of numbers appears next to the prompt. In many cases, the number represents the x and y coordinates of the cursor's current position. As the cursor moves, the values update.

In the figure above, the x coordinate (4.2695) is highlighted, because it can be edited. To edit the y coordinate, press the TAB key. (Pressing the TAB key a second time returns you to the x coordinate.)

You can press the ↓ key repeatedly to cycle through earlier inputs. As you do, older x, y coordinates appear next to the prompt. In addition, an orange marker appears in the drawing showing the location of the coordinate.

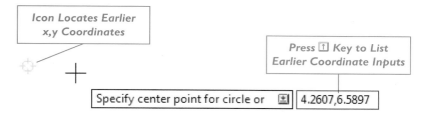

Press ENTER to fix the x, y-coordinate.

## Selecting Options

Next to the prompt text is a small icon ▣ that looks like an arrow on a keycap. This is a reminder to press the ↓ key to see the available options. (When the arrow-key icon does not appear, the command has no options at this time.)

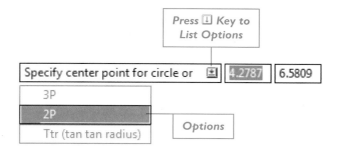

To select an option, move the cursor to it, and then click, such as **2P** illustrated above. (The 2P option draws two-point circles.)

## Dynamic Dimensions

After placing the first point, AutoCAD displays the CIRCLE command's next prompt, 'Specify radius of circle:'. It also shows a *dynamic dimension*, which in this case displays the size of the circle's radius (0.0462, in the figure below).

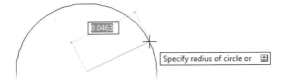

Technically, dynamic dimensions show the *relative* distance from the previous point to the current cursor location. The same distance is shown on the status bar when relative coordinates are selected.

## Selecting a Default Value

At this point, you can press the ↧ key to select an option, as before. This time, the previous radius (or diameter) is shown with a • black dot next to it. The black dot, as in • 0.0367, indicates this is the *default value*, and is equivalent to seeing '<0.0367>:' in the command prompt area.

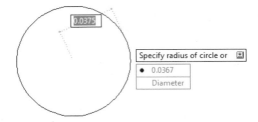

To select the default, press ENTER, or pick the value with the cursor.

## Dynamic Angles

Dynamic dimensions also display angles. To see the m, start the LINE command, and then pick two points. After picking the first point, notice that the dynamic angle appears (28 degrees):

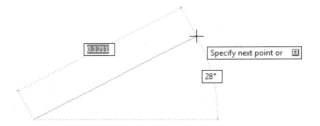

The angle is measured counterclockwise from the positive x axis.

(An angle wasn't shown while constructing the circle, because angles does not matter for circles specified by a radius.)

## Error Messages and Warnings

Dynamic inputs also display error messages and warnings at the cursor, such as when you enter a command name not recognized by AutoCAD. Here, I accidently entered "autocad":

⚠ Unknown command "AUTOCAD". Press F1 for help.

AutoCAD displays a red box when you enter incorrect values for coordinates, such as the "w" below:

## Modifying Dynamic Input

AutoCAD allows you to modify many aspects of dynamic input. To make changes, right-click the **DYN** button, and then select **Settings** from the shortcut menu.

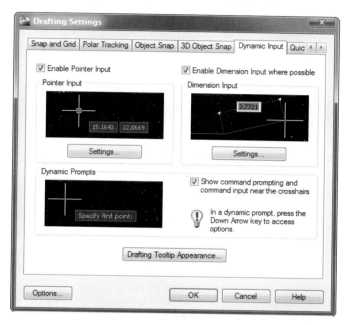

Notice that you can toggle pointer input, dimension input, and command prompting. Click the **Settings** buttons to determine how and when dynamic input appears.

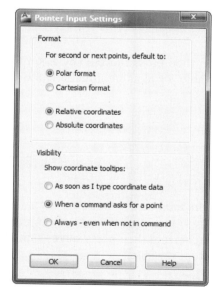
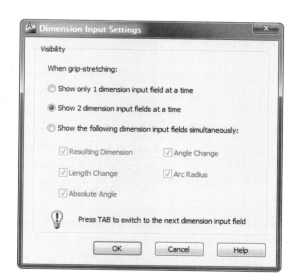

*Left: Settings for dynamic input.*
*Right: Settings for dimension input.*

To change the color and size of the tooltips, click the **Drafting Tooltip Appearance** button.

### Keyboard Shortcuts

In addition to accepting the full name of commands, AutoCAD allows you to enter *shortcuts*. AutoCAD recognizes several types of shortcuts, including aliases, function keys, and control keys. The advantage to shortcuts is that you execute commands more quickly.

### AutoComplete

AutoCAD has approximately 1,200 commands and system variables, and it can be difficult to remember the correct spelling of each. For example, do you enter POLYLINE or PLINE? POLYGON or PGON? One way to deal with the confusion is to select command names from the menu or ribbon; however, you have to know where the command resides. And the names and locations of commands in menus and ribbons change from release to release.

An alternative is to type the start of a command, such as "**p**," and then press the TAB key. Each time you press TAB, AutoCAD displays the next command, system variable, or alias in alphabetical order. Autodesk calls this "AutoComplete."

For example, enter **p**, and then press TAB:

    Command:  **p** *(Press TAB.)*

                        P *(Alias for PLOT command; press TAB.)*

                        PA *(Alias for PASTESPEC command; press TAB.)*

                        PAGESETUP *(Press ENTER to accept this command.)*

Press SHIFT+TAB to go backwards through the list. For example, enter **p** and then press SHIFT+TAB. The next to appear is the PURGE command, and then the PUCSBASE system variable.

When you reach the command you need, press ENTER to execute it.

## Function Keys

*Function keys* appear on the top row of your keyboard; they are all prefixed with the letter F (short for "function). AutoCAD assigns meanings to many of the function keys:

| Function Key | Meaning |
|---|---|
| F1 | Calls up the help window. |
| F2 | Toggles between the graphics and text windows. |
| F3 | Toggles object snap on and off. |
| F4 | Toggles constraint inference on and off. |
| F5 | Switches to the next isometric plane when in iso mode; planes are displayed left, top, right, and then repeated. |
| F6 | Toggles dynamic UCS on and off. |
| F7 | Toggles grid display on and off. |
| F8 | Toggles ortho mode on and off. |
| F9 | Toggles snap mode on and off. |
| F10 | Toggles polar tracking on and off. |
| F11 | Toggles object snap tracking on and off. |
| F12 | Toggles dynamic input mode. |
| CTRL+F4 | Closes the current drawing. |
| ALT+F4 | Exits AutoCAD. |
| ALT+F8 | Displays the Macros dialog box. |
| ALT+F11 | Starts the Visual Basic for Applications editor *. |

*) *The Visual Basic for Applications programming language is no longer included with AutoCAD, and support for VBA macros will be removed from AutoCAD 2012.*

## Aliases

*Aliases* are abbreviations of full command names. For example, the alias of the LINE command is L. To execute the LINE command more quickly, press L, followed by ENTER. AutoCAD executes the LINE command just as if you had entered the full command name. Examples of common aliases include:

| Alias | Command |
|---|---|
| a | ARC draws arcs. |
| aa | AREA finds the area. |
| adc | ADCENTER opens the Design Center. |
| b | BLOCK creates blocks (symbols). |
| c | CIRCLE draws circles. |
| co | COPY copies objects. |
| h | HATCH opens the Hatch dialog box. |
| i | INSERT inserts blocks. |
| l | LINE draws lines. |
| la | LAYER opens the Layers palette. |
| le | LEADER draws leaders. |
| m | MOVE moves objects. |
| t | TEXT places text in drawings. |
| z | ZOOM magnifies the drawing. |

There are many other command aliases, and you can create your own. The full list provided with AutoCAD is described in Appendix B.

## Other Keys

AutoCAD uses the SHIFT key to override drafting settings temporarily. For some, there are two shortcuts, one for left-hand and another for right-handed mouse users.

You use control-key shortcuts by holding down the CTRL key, and then pressing another key. You may already be familiar with some of them, such as CTRL+C to copy to Clipboard, and CTRL+Z for undo.

AutoCAD displays its control-key shortcuts in the drop-down menus, next to the related command. These keystrokes can also be customized (changed) with the CUI command. See Appendix B for the complete list of keyboard shortcuts.

### USING MOUSE BUTTONS

AutoCAD recognizes as many as sixteen buttons on mice and digitizer pucks. Most mouse devices, of course, are limited to two, three, or four buttons, but digitizer pucks commonly have four, twelve, or sixteen buttons.

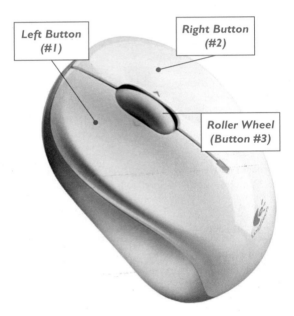

By default, AutoCAD defines the first ten buttons found on input devices, as follows:

| Button # | Mouse Button | Meaning |
| --- | --- | --- |
| 1 | Left | Selects commands or objects. |
| 2 | Right | Displays shortcut menus. |
| 3 | Center | Pans the drawing (when MBUTTONPAN = 1). |
| SHIFT+1 | SHIFT+Left | Toggles cycle mode. |
| SHIFT+2 | SHIFT+Left | Displays object snap shortcut menu. |
| CTRL+2 | CTRL+Right | Displays object snap shortcut menu. |
| ... | Wheel | Zooms or pans. |
| 4 | ... | Cancels command. |
| 5 | ... | Toggles snap mode. |
| 6 | ... | Toggles orthographic mode. |
| 7 | ... | Toggles grid display. |
| 8 | ... | Toggles coordinate display. |
| 9 | ... | Switches to next isometric plane. |
| 10 | ... | Toggles tablet mode. |

Buttons 11 through 16 are not predefined by AutoCAD. It is possible to change the meaning of all the buttons except the first (left) button, which is always the Select button. (The Windows mouse driver can reverse the meaning of the left and right buttons for left-handed users.)

**Left Mouse Button.** Earlier, I noted that you can pick points in the drawing with the cursor. By moving the mouse, you control the movement of the cursor. To pick a point, press the mouse's left button.

**Right Mouse Button.** In many cases, pressing the right mouse button (when inside the AutoCAD window) displays a context-sensitive shortcut menu. "Context-sensitive" means that the content of the menu changes, depending on where you right-click and which command is in effect.

The shortcut menu (cursor menu) appears at the cursor when you press the right-mouse button.

**Roller Wheel.** The wheel on a mouse can zoom and pan the drawing without invoking the ZOOM and PAN commands.

To zoom in or out, roll the wheel forward (zoom in) and backward (zoom out). To zoom the drawing to its extents, double-click the wheel button.

To pan about the drawing, hold down the wheel and drag the mouse.

In some cases, you might prefer that the wheel act as a button. You can do this by setting the value of the MBUTTONPAN system variable to 0. Now when you click the wheel, it acts like the middle button of a three-button mouse. Some mouse brands come with more than three buttons. This allows you to access additional AutoCAD functions.

## USING DIALOG BOXES

Many commands require you to specify options in dialog boxes. The benefit of dialog boxes is that they display all your options at the same time, unlike the command-line, which typically displays one option at a time. The dialog box and the command-line for creating arrays are compared below:

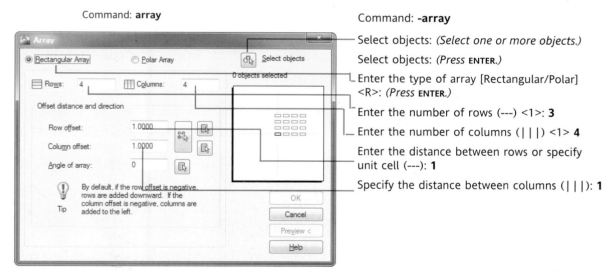

The drawback to dialog boxes is that they obscure drawings. The benefit, though, is that dialog boxes present options grouped logically; command-line prompts guide you through the steps needed to complete the action, but lack the previews and hints found in dialog boxes. The command-line steps you through options serially; dialog boxes allow you to jump to options directly.

A number of commands can display either a dialog box or prompts on the command line. There are several reasons for this. One is historical: until Release 9, AutoCAD didn't use dialog boxes at all, so all commands had to be entered at the command prompt. Also, many commands don't need a dialog box, such as drawing and editing commands like LINE and COPY. A third reason is that you may prefer using the command line, because it can be faster than using dialog boxes. For example, I find it faster to type the -VPORTS command and its options than to use the Viewports dialog box. A final reason is technical: script files and AutoLISP routines cannot control dialog boxes, and so must use the command-line version. There are two ways to suppress dialog boxes.

### Hyphen Prefix

Prefixing some commands with a *hyphen* ( – ) suppresses the related dialog box, displaying prompts at the command line instead. For example, the ARRAY command normally displays the Array dialog box:

>   Command: **array**
>
>   *(AutoCAD displays dialog box, shown above.)*

Prefixing the command with a hyphen displays the command line options:

>   Command: **-array**
>
>   Select objects: *(Select one or more objects.)*
>
>   Select objects: *(Press* ENTER.*)*
>
>   *(Command continues with the prompts listed earlier.)*

The following table lists some of the commands that accept the hyphen prefix (not all commands do):

| Commands | Hyphenated |
|---|---|
| ARCHIVE | -ARCHIVE |
| ARRAY | -ARRAY |
| ATTDEF | -ATTDEF |
| ATTEDIT | -ATTEDIT |
| ATTEXT | -ATTEXT |
| BEDIT | -BEDIT |
| BLOCK | -BLOCK |
| BOUNDARY | -BOUNDARY |
| COLOR | -COLOR |
| DIMSTYLE | -DIMSTYLE |
| HATCH | -HATCH |
| HATCHEDIT | -HATCHEDIT |

There are some curious exceptions: -PARTIALOPEN is the true name, while PARTIALOPEN is its alias. Similarly, -SHADEMODE is the true name, while SHADEMODE is an alias for the VSCURRENT command.

### System Variables

Whether or not certain commands display dialog boxes is controlled by *system variables*. System variables store the current state of AutoCAD. Here are a few common examples.

### FileDia

When the FILEDIA system variable is set to 0, AutoCAD suppresses the display of file-related dialog boxes, such as those associated with the NEW, OPEN, SAVEAS, and VSLIDE commands. Prompts are displayed at the command line:

>   Command: **filedia**
>
>   Enter new value for FILEDIA <1>: **0**
>
>   Command: **open**
>
>   Enter name of drawing to open: *(Enter path and name of .dwg drawing file.)*

To force the display of the dialog box, type the *tilde* ( ~ ) character when AutoCAD prompts for the file name:

> Command: **open**
>
> Enter name of drawing to open: ~
>
> *(AutoCAD displays the Select File dialog box.)*

## Expert

When the EXPERT system variable is set to a value other than zero, it suppresses warning dialog boxes displayed by AutoCAD. Examples include "About to regen, proceed?" and "Really want to turn the current layer off?" and "File [block] already exists. Overwrite it?"

## AttDia

When the ATTDIA system variable is set to 0, the INSERT command displays prompts for attribute data at the command line; when set to 1, a dialog box is displayed instead. (Curiously, though, in this age of dialog boxes and palettes, 0 continues to be the default.)

### Alternate Commands

In a very few cases, AutoCAD uses one command name for the command line, and a different one for dialog boxes. Examples include the following:

| Command Line | Dialog Box |
|---|---|
| FILEOPEN | OPEN |
| CHANGE | PROPERTIES |
| UCS | UCSMAN |

## THE DRAFT-EDIT-PLOT CYCLE

Once you set up a drawing, working with AutoCAD involves a cycle that consists of drafting, editing, and plotting. You draft drawings, edit portions of them, and then plot them. The contents of this book are arranged in a similar manner.

### DRAFTING

Drafting consists of placing and constructing objects in the drawing. AutoCAD comes with a set of objects, from which everything else in the drawing must be made:

| *Objects* | *Drawn with Command(s)* |
| --- | --- |
| Points | POINT |
| Lines | LINE, TRACE, PLINE, and 3DPLINE |
| Parallel lines | MLINE |
| Construction lines | RAY and XLINE |
| Circles | CIRCLE and DONUT |
| Ellipses | ELLIPSE |
| Arcs | ARC and PLINE |
| Elliptical arcs | ELLIPSE |
| Regular Polygons | POLYGON |
| Rectangles | RECTANG |
| Splines | SPLINE and HELIX |
| Irregular boundaries | BOUNDARY |
| Revision clouds | REVCLOUD |
| 3D meshes | MESH, EDGESURF, RULESURF, and TABSURF |
| 3D surfaces | PLANESURF, LOFT, SWEEP, THICKEN |
| 3D solids | BOX, CONE, CYLINDER, SPHERE, TORUS, PYRAMID, POLYSOLID, and WEDGE |
| Text | TEXT, MTEXT, and FIELD |
| Tables | TABLE |
| Dimensions | All commands starting with DIM, plus LEADER, QDIM, QLEADER, TOLERANCE, and MLEADER |
| Parametrics | PARAMETERS, DIMCONSTRAINT, and AUTOCONSTRAIN |
| Constraints | GEOMCONSTRAINT and GC variants. |

## Constructing Objects

When you cannot draw an object with one of AutoCAD's commands, you must use other commands to construct it. For example, AutoCAD has no command for drawing right-angle triangles or irregular polygons, so you construct them with the LINE or PLINE commands.

Here are some of the commands that are used to construct objects:

| *Construction Type* | *Constructed with Command(s)* |
| --- | --- |
| Hatch patterns | HATCH, GRADIENT, and HATCHEDIT |
| Filled areas | SOLID and HATCH |
| Blank areas | WIPEOUT |
| 3D revolutions | REVOLVE |
| 3D extrusions | EXTRUDE |
| 3D solids | INTERFERE, SLICE, SECTION, IMPRINT, PRESSPULL, THICKEN, SWEEP, and LOFT |
| 3D meshes | MESHSMOOTH |

In addition, AutoCAD has a collection of commands to construct copies of objects. Examples include the following:

| Copy Method | Copied with Command(s) |
|---|---|
| Copy objects | COPY |
| Copy blocks | INSERT, XBIND and ADCENTER |
| Offset copies | OFFSET |
| Mirror copies | MIRROR and MIRROR3D |
| Array copies | ARRAY and 3DARRAY |
| Array as block | MINSERT |
| Copy faces, edges | SOLIDEDIT, FLATSHOT and IMPRINT |
| Copy along paths | DIVIDE and MEASURE |

## BLOCKS AND TEMPLATES

One of Autodesk's cofounders, John Walker, said that we should never have to draw the same object twice in CAD. The best way to reuse drawing details is to turn them into *blocks*. Blocks can be used repeatedly in drawings, shared between drawings and offices, and can also hold database information, called "attributes." Sometimes you start drawings with *templates*, other drawing files that already contain previously-created settings and objects. The most common templates consist of a drawing border, title block, and perhaps layers. Many firms also use templates for standard drawings of details.

## EDITING

Editing consists of modifying objects in the drawing, correcting mistakes, moving objects, changing your mind, having your boss his mind, and so on.

| Editing Method | Edited with Command(s) |
|---|---|
| Move objects | MOVE, STRETCH, ALIGN and 3DMOVE |
| Rotate objects | ROTATE, ROTATE3D and 3DROTATE |
| Resize objects | SCALE |
| Stretch objects | STRETCH, LENGTHEN, EXTEND |
| Mirror objects | MIRROR |
| Erase objects | ERASE |
| Trim objects | TRIM and BREAK |
| Fillet objects | FILLET |
| Chamfer objects | CHAMFER |
| Undo editing | U, UNDO, REDO and MREDO |
| Edit splines | SPLINEDIT |
| Edit multilines | MLEDIT |
| Edit text | TEXTEDIT |
| Change text size | SCALETEXT and STYLE |
| Align text | JUSTIFYTEXT and STYLE |
| Edit attributes | EATTEDIT and ATTIPEDIT |
| Edit polylines | PEDIT |
| Edit 3D solids | UNION, INTERSECT, SUBTRACT and SOLIDEDIT |
| Reduce objects | EXPLODE and XPLODE |
| Change properties | CHANGE and PROPERTIES |
| Reverse directions | REVERSE |

In addition to using commands for editing objects, you can edit them *directly*. Single-clicking objects produces grips (handles), allowing you to stretch, move, copy, rotate, mirror, and resize them directly with the cursor. Double-clicking most objects brings up the Properties palette, which lets you change their properties; double-clicking text brings up the appropriate text editor.

## PLOTTING

When your drawings are complete, you *print* them on paper (a process also called "plotting"), or send them through the Internet by means of an "e-plot," short for *electronic plot*. Sometimes plots are created before drawings are complete, in which case they are known as "check plots."

AutoCAD's **preview** command lets you see plots before you commit them to paper. Another command, **plotstamp**, labels plots with the drawing file name, date and time of plot, and other information.

### ePlots

In addition to plotting drawings on paper, you can also plot drawings to file. This lets you send copies of drawings without sending the original *.dwg* files. Electronic plots are most often made with *.dwf* and *.pdf* files, which look just like the original drawings and cannot be edited.

To save drawings in Autodesk's DWF format (short for "design web format"), use the **exportdwf** command; in Adobe's PDF format (short for "portable document format"), use the **exportpdf** command. The **publishtoweb** command generates Web pages and images of the drawings, while the **3ddwf** command creates 3D versions of DWF format files.

To view *.dwf* files, use the Design Review software that was installed with AutoCAD. (Non-AutoCAD users can download the software free from www.autodesk.com/designreview.)

To view *.pdf* files, use the Acrobat Reader from Adobe, which you can download free from www.adobe.com/acrobat.

## PROTECTING DRAWINGS

To protect drawings, AutoCAD includes password protection and digital signatures. When you save drawings with passwords (through the **securityoptions** command), they cannot be opened unless the correct passwords are entered.

Be careful: If you forget the password, the drawing is rendered useless. In this regard, it's best not to use passwords at all.

A different level of security is the digital signature. When drawings are "signed" digitally, a warning appears after they have been edited or altered. This alerts you that the drawings may not be originals; the system does not, however, tell you what has been changed. You must purchase a digital ID from a certificate authority in order to apply digital certificates.

## SCREEN POINTING

You enter points, distances, and angles in two ways: by "showing" AutoCAD the information on the screen, or by entering coordinates and angles at the keyboard.

Picking two points in the drawing could indicate, for example, a distance, an angle, or both — depending on the command and what it expects as input. With AutoCAD, you press the left mouse button to pick points in the drawing.

### Showing Points by Window Corners

Some commands require input of both a horizontal and a vertical displacement. This is shown by a rectangle on the screen, called the "window." You pick a point in the drawing, and then move the cursor to form the rectangle. AutoCAD determines the x and y distances by measuring from the lower left corner to the upper right corner.

For example, the window in the figure starts at (2,3) and rises to (7,7). To AutoCAD, this indicates a displacement of (5,4) by using subtraction:

```
Horizontal displacement (x)   = 7 - 2 = 5
Vertical displacement (y)     = 7 - 3 = 4
```

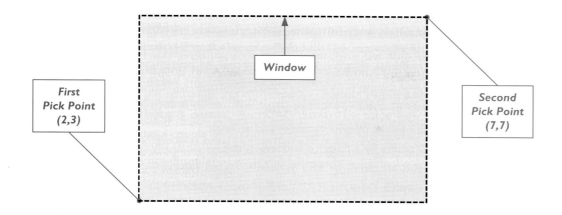

## COORDINATE SYSTEMS

AutoCAD works with coordinate systems. Coordinates locate objects in the drawing. (Technically, AutoCAD does not record lines and other objects, but rather their points and properties.) All 2D drafting is performed on the x, y-plane; for 3D drafting, you can redefine this plane to any orientation in space through UCSs.

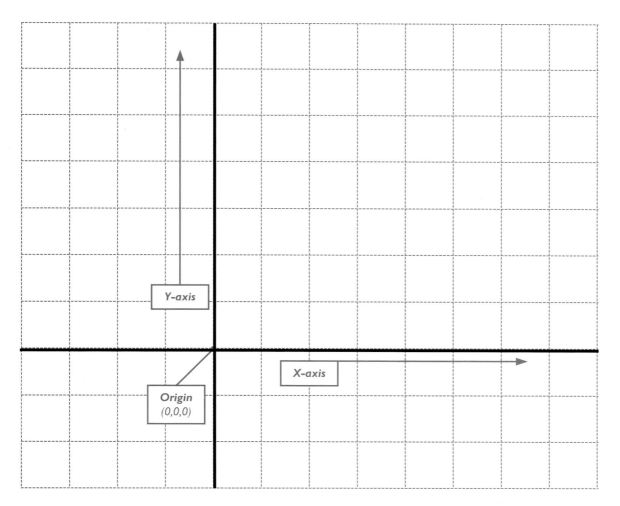

## AUTOCAD COORDINATES

AutoCAD works with several coordinate systems, most of which measure distances relative to the origin, which is located at 0,0,0:

**Cartesian** define 2D and 3D points by two or three distances along the x, y, and (sometimes) z axes.

**Polar** define 2D points by a distance and an angle.

**Cylindrical** define 3D points by two distances and an angle.

**Spherical** define 3D points by a distance and two angles.

**Relative** define 2D and 3D distances (and optionally angles) relative to the last point.

**User-defined coordinate systems** (UCS) define an x, y-plane anywhere in 3D space.

You enter coordinates in the format of your choice: decimal (metric), engineering (decimal inches), architectural (Imperial units), fractional (inches only), and scientific notation (exponents). Accuracy ranges from zero to eight decimal places (whole inches to 1/256th of an inch).

Angles can be entered in degrees-minutes-seconds (360 degrees in a circle), radians (2*pi), grads (400), and in surveyor's units (NdE).

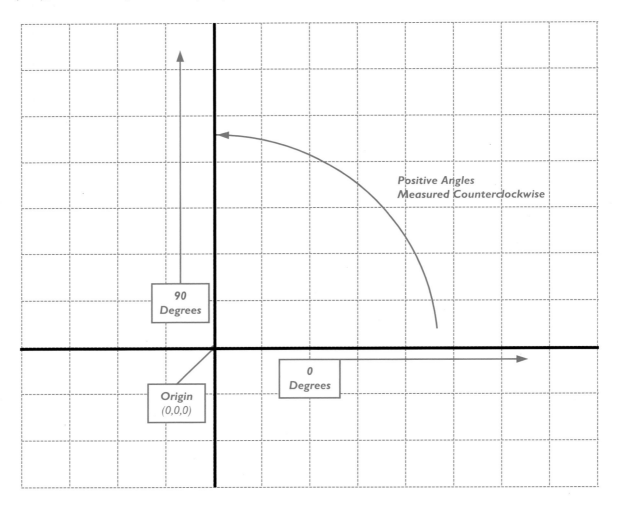

By default, AutoCAD measures angles counterclockwise, with 0 degrees pointing from the origin to the East, or 3 o'clock. Using this convention, 90 degrees points North (12 o'clock), 180 degrees points West (9 o'clock), and 270 degrees points South (6 o'clock).

To measure an angle clockwise, use negative angles. Prefix the angle with a dash, as in -45 degrees. Because there are 360 degrees in a circle, -45 degrees is the same as 315 degrees.

In the same way, you specify negative distances by prefixing numbers with the dash. For example, -2,-3 means two units in the negative x-direction, and three units in the negative y-direction.

Recall that absolute coordinates are measured relative to the origin (0,0,0), while relative coordinates are measured relative to the last point. Similarly, absolute angles are measured from the x-axis counter-clockwise, while relative angles are measured from the value stored in the LASTANGLE system variable.

If nothing else, AutoCAD is flexible in its measurement systems. For example, surveyors can change the UNITS command so that angles are measured clockwise and 0 degrees points in any direction.

### Cartesian Coordinates

Cartesian coordinates define 2D and 3D points by measuring two or three distances: each distance is along one of the three axes: x, y, and z, either positive or negative. In AutoCAD, the x-direction is horizontal, at 0 degrees; positive x is measured to the right. The y-direction is vertical along 90 degrees; positive y is measured upward. The z-direction comes out of the screen; positive z points at your face.

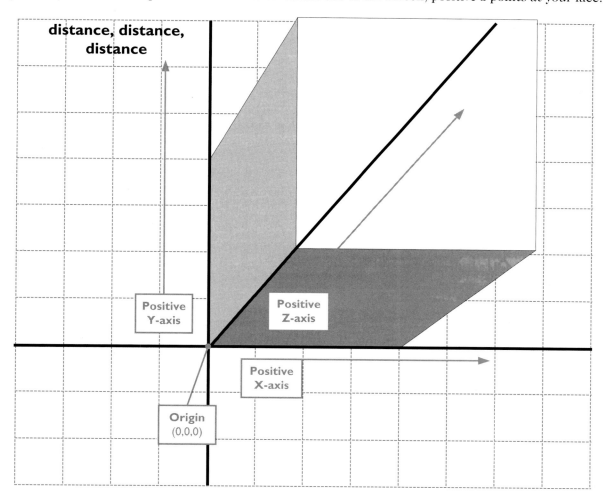

To draw a line using Cartesian coordinates with x, y (2D) and x, y, z (3D) coordinates:

    Command: **line**
    Specify first point: **1,2**
    Specify next point or [Undo]: **3,4,5**

## Polar Coordinates

Polar coordinates define 2D points by measuring one distance and one angle in the x, y-plane. The distance is the straight-line distance measured from the origin to the point, while the angle is measured from 0 degrees. To indicate the angle to AutoCAD, you prefix the angle with an angle bracket ( < ), such as <45 for 45 degrees.

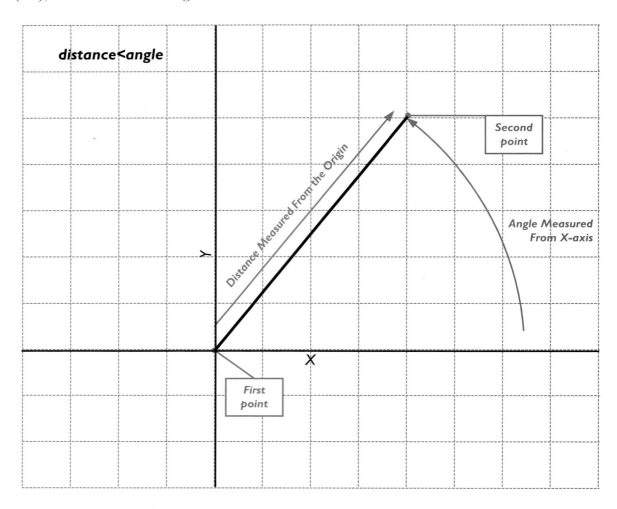

When you draw a line using polar coordinates, all points are relative to the origin (0,0):

    Command: **line**

    Specify first point: **1<23**

    Specify next point or [Undo]: **4<55**

## Cylindrical Coordinates

Cylindrical coordinates define 3D points by measuring two distances and one angle in 3D space. They are like polar coordinates, but with the z-distance added. The distances are measured from the origin to the point, while the angle is measured from 0 degrees.

The first distance lies in the x, y-plane; the second distance measures the distance along the z-axis. To indicate the measurement to AutoCAD, you use this notation: 1<23,4 — where 1 is the polar distance, <23 is angle from 0 degrees, and 4 is the height in the z direction.

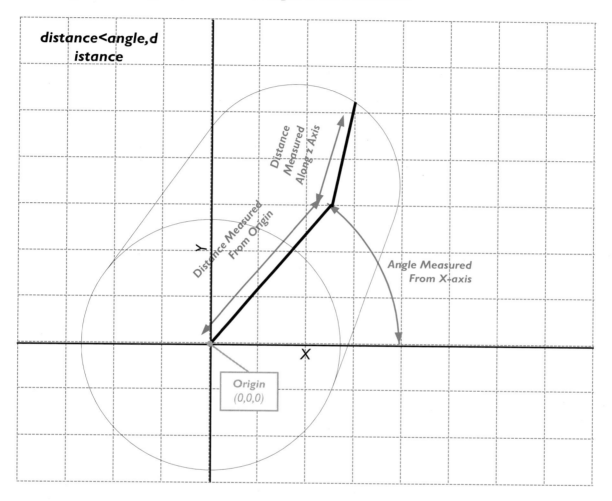

To draw a line using cylindrical coordinates:

    Command: **line**
    Specify first point: **1<23,4**
    Specify next point or [Undo]: **5<67,8**

## Spherical Coordinates

Spherical coordinates define 3D points by measuring a distance and two angles from the origin. The first angle lies in the x, y-plane, while the second angle is up (or down) from the x, y-plane. The distance lies, of necessity, in the x, y, z-plane.

To indicate the angles to AutoCAD, you prefix each with angle brackets, such as <15<45 for 15 and 45 degrees.

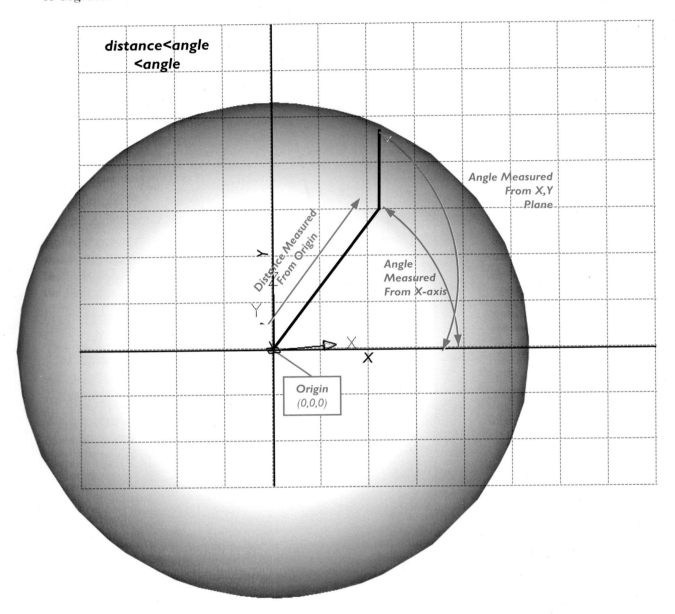

Here is how you draw a line using spherical coordinates:

    Command: **line**
    Specify first point: **1<23<45**
    Specify next point or [Undo]: **6<78<90**

## Relative Coordinates

All coordinate systems can use *relative coordinates*, which measure the distance (and angle) from the last point. (The opposite of relative is *absolute*, which measures distances from the origin.) To indicate relative coordinates to AutoCAD, you prefix them with the *at* symbol ( @ ), such as @2,3.

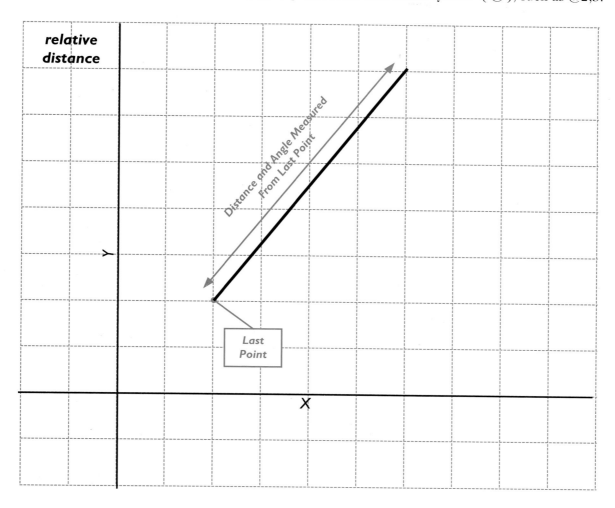

To draw a line using relative Cartesian, polar, cylindric, and spherical coordinates

    Command: **line**
    Specify first point: **@1,2**
    Specify next point or [Undo]: **@3,4,5**
    Specify next point or [Undo]: **@6<78**
    Specify next point or [Undo]: **@9<10<11**
    Specify next point or [Undo]: **@9<10<11**

Note that the first point can also be relative. It is measured from whatever last point was specified in the drawing with a previous drawing command. AutoCAD stores the coordinates of the last point in a system variable called LASTPOINT.

Similarly, relative angles are measured from the value stored in the LASTANGLE system variable.

## User-defined Coordinate Systems

As I noted earlier, all drawing in AutoCAD takes place in the x, y-plane, also called the "working plane." How, then, do you draw in 3D space, which involves the z axis, along with the x, z and y, z planes? You could specify the z coordinate, but this is not possible for all commands. In any case, specifying z coordinates becomes cumbersome when drafting details, say, on the side of a slanted roof, where the value of z is constantly changing.

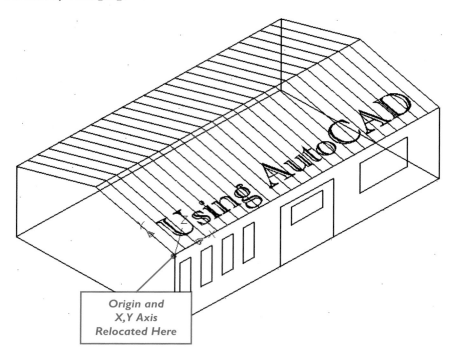

*Origin and X,Y Axis Relocated Here*

AutoCAD takes care of that through the user-defined coordinate system, or "UCS" for short. Through UCSs, you can define the working plane anywhere in 3D space, such as on the sloped roof of a barn. (The UCS command lets you set the working plane by making it relative to the current view, or to three points picked in the drawing, or by other methods.) UCSs can be saved by name, so that you can return to previously-created working planes.

### ENTERING COORDINATES

Usually, of working with AutoCAD involves specifying points on the screen. Typical are the prompts of the LINE command:

>Command: **line**
>
>Specify first point: *(Pick a point, or enter x, y, z coordinates.)*
>
>Specify next point or [Undo]: *(Pick another point, or enter more coordinates.)*

You can do this in three ways:

- Picking points on the screen.
- Entering coordinates at the keyboard.
- Using object snaps, tracking, direct distance entry, and point filters.

### Screen Picks

When AutoCAD prompts you to specify a point, you can use the crosshair cursor to pick a point in the drawing. Move the mouse to move the cursor; press the left mouse button to pick the point. AutoCAD determines the x, y coordinates of the pick point.

Screen picks are strictly 2D (two dimensional), even when working in 3D (three dimensional) drawings. "2D" means AutoCAD records the x and y coordinates only, and assumes the z coordinate is 0. If you wish to specify a z coordinate, you may do this with the ELEVATION command (which sets the distance along the z axis), with the .XY point filter (which forces AutoCAD to prompt you for the z distance), or with an x, y, z triplet.

### Keyboard Coordinates

You can enter coordinates at the keyboard in a variety of methods, as I described earlier. The formats permitted by AutoCAD are:

| Absolute | Relative | Meaning |
| --- | --- | --- |
| x,y | @x,y | 2D Cartesian coordinates. |
| x,y,z | @x,y,z | 3D Cartesian coordinates. |
| d<a | @d<a | 2D polar coordinates. |
| d<a,z | @d<a,z | 3D cylindrical coordinates. |
| d<a<r | @d<a<r | 3D polar coordinates. |

**x, y,** and **z** = distance along the x, y, and z axis, respectively.
**d** = distance in x, y plane.
**a** = angle from x axis (in x, y plane).
**r** = angle from x, y plane (in z direction).

Recall that absolute coordinates are measured relative to the origin (0,0,0), while relative coordinates are measured relative to the last point. Similarly, absolute angles are measured from the x-axis counterclockwise, while relative angles are measured from the value stored in the LASTANGLE system variable.

### Snaps, Etc.

Entering coordinates at the keyboard can get tedious. Thus, there are alternatives and aids:

**Point filters** combine screen picks with keyboard entry.

**Snap spacing** restricts cursor movement to increments.

**Polar snap** and **ortho mode** restrict cursor movement to specific angles.

**Direct distance entry** allows you to show angles and enter distances.

**Tracking** shows distances.

**Object snaps** snap the cursor to the nearest geometric feature.

**Object tracking** shows geometric relationships to nearby objects.

You learn more about snap mode, polar snaps, objects snaps, and object tracking in Chapter 5, "Drawing with Precision."

### Point Filters

Point filters combine screen picks with keyboard entry to specify, for example, the x coordinate with the keyboard and the y coordinate on the screen. AutoCAD recognizes these point filters:

| Point Filter | AutoCAD Asks For... |
|---|---|
| .x | y and z coordinates. |
| .y | z and z coordinates. |
| .z | x and y coordinates. |
| .xy | z coordinate. |
| .xz | y coordinate. |
| .yz | x coordinate. |

Here is an example of using point filters with the LINE command. Notice that you can enter coordinates at the keyboard or on the screen at any time:

>Command: **line**
>Specify first point: **.x**
>of *(Pick a point; AutoCAD reads just the x-coordinate.)*
>(need YZ): **2,3** *(AutoCAD reads these as the y and z coordinates.)*
>
>Specify next point or [Undo]: **.y**
>of **2** *(AutoCAD reads this as the y coordinate.)*
>(need XZ): *(Pick a point; AutoCAD reads the x and z coordinates.)*

### Snap Spacing

*Snap mode* restricts cursor movement. To restrict the cursor to moving in increments of 0.5, for example, set the snap spacing to 0.5 with the SNAP command. This allows you to pick points easily on the screen, accurate to the nearest 0.5 units. If you measured the rooms in your house to the nearest one inch, when drawing the floor plan in AutoCAD, you would set the snap spacing to 1".

Snap is often used in conjunction with the grid; when the grid is set to the same spacing as the snap, then you can "see" the snap points.

### Direct Distance Entry

Direct distance entry allows you to see angles and enter distances. It allows you to draw and edit by showing a relative distance: you move the cursor to indicate the angle, and then type a number to specify the distance. If you need a precise angle, then ensure that polar tracking or ortho mode is first turned on.

>Specify first point: *(Move the cursor, and then enter a distance, such as **2.5**.)*

### Tracking

Tracking shows distances with the "pen up" during drawing and editing commands. You can think of it as the opposite of direct distance entry, which draws; tracking moves the cursor without drawing.

During a prompt, such as "Specify first point:", enter **tk** to invoke tracking mode. Move the cursor, in the direction to track, and then pick a point or enter a distance, such as **2.5**. You can continue tracking as far as you need.

>Specify first point: **tk**
>First tracking point: *(Move cursor, and then enter a distance or pick a point.)*
>Next point (Press ENTER to end tracking): *(Press* ENTER.*)*

## Object Snaps

*Object snaps* snap the cursor to the nearest geometric feature, such as the endpoints of lines, the center points of circles, and the insertion points of text. To help you find these geometric features, AutoCAD displays icons and tooltips.

The figure below illustrates the line being drawn to the center of the circle. Because object snapping is turned on, AutoCAD displays the center icon (a circle), as well as the tooltip, "Center."

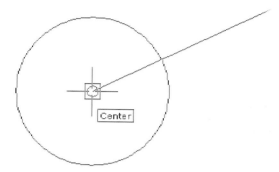

To use objects snaps, you can turn them on with the OSNAP command, or enter them at prompts, such as:

Specify first point: **center**
of *(Pick a circle or arc.)*

## EXERCISES

1. In the following exercises, watch how the coordinates and cursor react to different modes. Move the cursor around with your mouse. Do the coordinates at the lower left corner of the screen move?
   Press **F6**. Move the cursor again, and watch the coordinates. What is the difference?

2. Press **F8**. Check the status bar: is ortho mode turned on?
   Move the cursor. Do you notice a difference?

3. Press **F9** to turn on snap mode.
   Move the cursor. Now is there a difference in cursor movement?

4. In this exercise, you practice using ribbon buttons and keystrokes that affect commands. Move the crosshair cursor to the top of the drawing area and into the ribbon. Does the cursor change its shape?
   In the Home tab, position the cursor over the arrow next to **Circle**, and then click. Does a flyout drop down?

5. In the ribbon, choose **Line**.
   Draw some lines.
   To exit the command, right-click and select **Cancel** from the shortcut menu.

6. Press the **ENTER** key. Did the **Line** command repeat?
   Clear the command line with **ESC**.
   Press the spacebar. What happens?

7. Press **F2**. Do you see the Text window?
   Press **F2** again to return to the drawing window.

8. Press **F1**. Do you see the Help window? Familiarize yourself with the Help window.

9. Use the **LINE** command to draw a line graph of insurance premiums for nonsmoking men. Enter the coordinates listed by the table below:

   | X (Age) | Y (Rate $) |
   |---------|------------|
   | 25      | 14         |
   | 30      | 15         |
   | 35      | 19         |
   | 40      | 23         |
   | 45      | 37         |
   | 50      | 59         |
   | 55      | 88         |
   | 60      | 130        |
   | 65      | 194        |

   Are these coordinates relative or absolute? (If you cannot see the drawing, use the **ZOOM Extents** command.)

10. Using the figure below, label each axis and the origin.

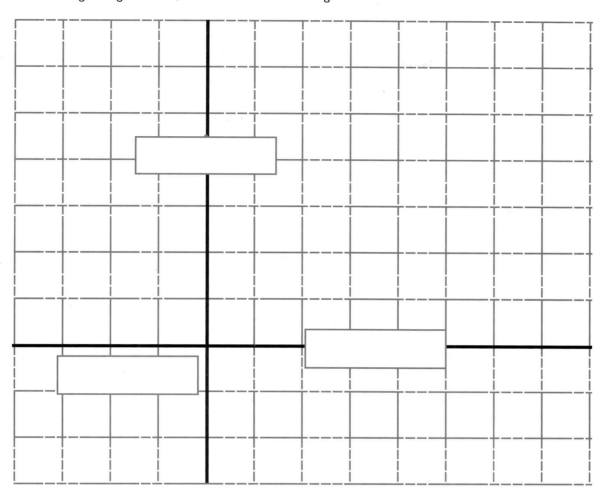

## CHAPTER REVIEW

1. List two pieces of information found on AutoCAD's title bar.
   a.
   b.
2. What are the benefits of using *template* drawings?
3. Where are the mode indicators located?
4. How do you access command options during dynamic input?
5. Describe what happens when you enter the first few letters of a command name, and then press the TAB key.
6. What happens when you press the F2 key?
7. List the control and/or function key associated with the following actions:
   Turns ortho mode on and off.
   Selects all objects in the drawing.
   Toggles between graphic and text screens.
   Cancels all the characters on the command line.
   Toggles to the next isometric plane in iso mode.
   Toggles the screen coordinate display modes.
   Toggles the grid on and off.
   Toggles snap on and off.
8. After a command ends, what happens when you press ENTER?
9. Name two ways to invoke the on-line help facility:
   a.
   b.
10. What is the purpose of the *prompt* area?
11. Which command turns the ribbon on? Off?
12. What does the color BYLAYER mean?
13. What are *linetypes*?
14. Explain the purpose of the small square at the center of the crosshair cursor.
15. "UCS" is short for:
16. Can the UCS icon be turned off? If so, why?
17. Explain the meaning of these numbers on the status line:
    32'-11 3/8", 1'-2 1/2", 0'-0"
18. What does it mean when the following buttons on the status line look bluish?
    SNAP
    OSNAP
    GRID
19. Where is the *tray* located?
20. Can you "show" AutoCAD an angle by picking points in the drawing?
21. How do you show AutoCAD x and y displacement simultaneously?
22. Describe how *transparent* commands work.
23. Can any command be transparent?

24. What is an *alias*?
25. Write out the alias for each of the following commands:
    **ARC**
    **LINE**
    **BLOCK**
    **CIRCLE**
    **COPY**
    **ZOOM**
26. Where are *function keys* located?
    How can you identify function keys?
27. Write out the function key you would use for the following actions:
    Toggle snap mode
    Toggle ortho mode
    Toggle grid display
    Exit AutoCAD
28. What does *toggle* mean?
29. Write out the **CTRL** key you would press for the following actions:
    Display Properties palette
    Toggle clean screen mode
    Copy objects to the Clipboard
    Save the drawing
    Switch to the next drawing window
30. Describe two methods to repeat commands.
31. Explain in two or three words the function of each of these mouse buttons:
    Left button
    Right button
    Scroll wheel
    **SHIFT**+Left button
32. How do you access the application menu?
33. Describe one way to display the menu bar.
34. List some pros and cons to using dialog boxes.
    Pro:
    Con:
35. Do all commands use dialog boxes?
36. Explain the difference between the **ARRAY** and the **-ARRAY** commands.
37. How do you force file-related commands to display their dialog boxes, when the function has been turned off?
38. What is *dynamic input*?
39. How is dynamic input activated?
40. Can the x, y coordinates be changed during dynamic input?
    If so, how?
41. What is the meaning of the dot ( • ) next to an option in dynamic input?
42. From where are dynamic angles measured?
43. From where are dynamic dimensions measured?

44. Describe the function of these commands:
    - **LINE**
    - **CIRCLE**
    - **ARC**
    - **POLYGON**
    - **TEXT**

    Are these commands for drawing or editing?

45. Describe the function of these commands:
    - **COPY**
    - **MIRROR**
    - **INSERT**
    - **MATCHPROP**

46. Give two reasons why you might need to edit drawings?

47. Describe the function of these commands:
    - **MOVE**
    - **ERASE**
    - **ROTATE**
    - **DDEDIT**
    - **PEDIT**
    - **PROPERTIES**

48. What happens when you double-click a line of text?

    Objects other than text?

49. Describe what happens when you press the [↑] key at the 'Command:' prompt.

50. List five ways to enter commands:
    a.
    b.
    c.
    d.
    e.

51. List one way that the **PREVIEW** command is useful.

52. What is an *eplot*?

53. How would you email a drawing without sending the *.dwg* file itself?

54. What is "DWF" short for?

    Name a benefit to *.dwf* files.

55. What happens if you forget the password to drawings?

56. Briefly explain these coordinate systems:
    - Cartesian
    - Polar
    - Spherical
    - Cylindrical

57. Describe the meaning of the following symbols as used by AutoCAD:

    <

    @

    , (comma)

58. Name the coordinate system used by the following:

    25<45

    25,45,60

    25<45<60

    25<45,60

59. Write out relative coordinates for a distance of 5 units and an angle of 75 degrees.
60. Can coordinates be negative?

    If yes, when?
61. What are the coordinates of the *origin*?
62. Are relative coordinates measured relative to the origin?

    Are absolute coordinates measured relative to the origin?
63. In which direction are positive angles measured in AutoCAD, usually?

    Can the direction be changed?
64. In which direction is 0 degrees in AutoCAD, usually?

    Can the direction be changed?
65. Fill in the missing Cartesian coordinates to draw a line from the origin to 10,15.

    Command: **line**

    Specify first point:

    Specify next point:
66. Fill in the missing relative coordinates to draw a vertical line 10 units long from 20,15.

    Command: **line**

    Specify first point:

    Specify next point:

    What are the coordinates of the line's endpoint?
67. What happens when you press ENTER at the LINE command's "Specify first point" prompt?
68. What is the likely z coordinate when you pick a point in the drawing?

    When might the z coordinate change?
69. When might you use *cylindrical* and *spherical* coordinates?
70. When might you use UCSs?
71. Describe the distances and angles employed by spherical coordinates:

    Distance

    Angle 1

    Angle 2

    Write down an example of a spherical coordinate.
72. How are *point filters* helpful?
73. When you enter **.xy**, what does AutoCAD ask for?

74. Explain the purpose of:
    Snap spacing
    Object snaps
    Polar snaps.
75. What is the difference between *direct distance entry* and *tracking*?
76. How do you enter tracking mode?
77. Can angles be negative?
    If so, when?

# CHAPTER 3

## Setting Up Drawings

After starting a new drawing, you must prepare it for use. This chapter covers the following items to help you set up drawings:

**NEW** starts new drawings from scratch, from a template, or through the Startup dialog box.
**QNEW** starts new drawings based on templates.
**Advanced Wizard** sets up measurement systems for new drawings.
**OPEN** opens existing drawings and manages files.
**OPTIONS** changes user interface colors, specifies backup options, and much more.
**UNITS** specifies unit and angle formats.
**Scale factors** affect the size of text, linetypes, hatch patterns, and dimensions.
**SAVE** and **SAVEAS** save drawings by other names and in other formats.
**QSAVE** saves drawings quickly.
**QUIT** exits AutoCAD.

> **NEW TO AUTOCAD 2011** IN THIS CHAPTER
> - **OPTIONS** command has new options.

## NEW DRAWINGS

When you start AutoCAD, you either begin with a new drawing or call up previously-saved drawings. Let's begin with a new drawing.

When AutoCAD launches, it displays a new blank drawing. (At the same time, it may also display a variety of startup dialog boxes. If the New Features Workshop dialog box appears, cancel it.)

You can use the NEW or QNEW commands at any time to start additional new drawings, as described below. You can have many drawings open in AutoCAD at the same time.

### TUTORIAL: STARTING NEW DRAWINGS

1. To open a new drawing when AutoCAD is already running, start the **NEW** command using one of the following methods:
   - In the application menu, choose **New**.
   - Alternatively, press the **CTRL+N** keyboard shortcut.
   - Or, at the 'Command:' prompt, enter the **new** command.

      Command: **new** *(Press* ENTER.*)*

   Notice that AutoCAD displays the Select Template dialog box. (In some cases, AutoCAD may display the Startup dialog box; if so, click **Cancel** to close it.)
2. Select a template file from the list in the dialog box. This file determines the look of new drawings. Some produce drawings that look blank, while others will contain drawing borders and other elements.

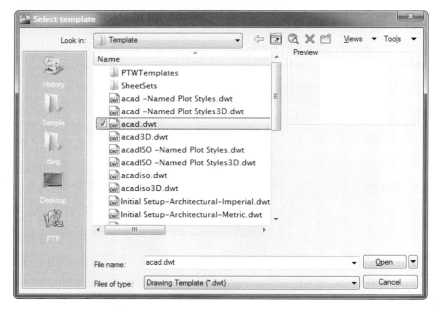

   If you are unsure of which template to choose, then select *acad.dwt*, the most generic of templates. As you select a file name, AutoCAD displays its preview image.
3. Click **Open**. Notice that AutoCAD displays a new drawing; if it is based on *acad.dwt*, then new drawing will look blank.

## About Templates

Each time you begin a new drawing, AutoCAD makes can exact copy of a template drawing. These drawing files are identified with the extension of *.dwt*. AutoCAD copies the settings from template files, which have names such as *acad.dwt*, *acadiso.dwt*, and *acad3d.dwt*.

Template drawings contain settings that new drawings use; sometimes, templates also include drawing elements. New drawings are identical to the template, except for the file name. You can change any setting in the drawings, regardless of how it is defined in the template.

AutoCAD includes a number of template drawings for a variety of sizes and standards. Template files are stored in a folder with the rather long path of */documents and settings/<login>/local settings/application data/autodesk/autocad 2011/r18.1/enu/template*. (Replace <login> with the name by which you started Windows, such as "administrator." By default, the Application Data folders are hidden, and you may need to use Windows Explorer's **Tools | Folder Options | View | Show Hidden Files** option.)

Clicking **Open** opens the template, and then this file becomes the starting point for your new drawing.

 **Notes** Whether AutoCAD displays the Select Template dialog box is determined by a setting in the Options dialog box. (From the Application Menu, click **Options**, choose the **Files** tab, and then look for the Template Settings section.) The **Default Template File Name for QNEW** item has two options:

- **None** means that the **QNEW** command will display the Select Template dialog box, just like **NEW**.
- A file name with a *.dwt* extension means that **QNEW** starts new drawings with the specified *.dwt* template file. This is the "quick" version of the command.

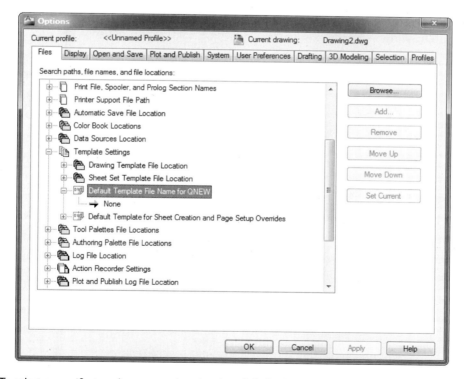

To select a specific template or another drawing, click the **Browse** button, and then choose the file.

 QNEW

The QNEW command (short for "quick new") is the more flexible version of NEW: depending on how AutoCAD is set up, the command either starts with a preselected *.dwt* file, or else prompts you to select the template.

### TUTORIAL: STARTING NEW DRAWINGS WITH TEMPLATES

To start a new drawing with a specific template file, follow these steps:

1. To open a new drawing based on a template, start the **QNEW** command:
   - On the Quick Access toolbar, click the **New** button.
   - Or, at the 'Command:' prompt, enter the **qnew** command.

       Command: **qnew** *(Press* ENTER.*)*

2. If a template file is defined as described above, then AutoCAD displays a new, blank drawing based on that preselected *.dwt* file.

### STARTUP DIALOG BOX

When AutoCAD is set up to display the Startup dialog box, it presents you with the following options:

**Start from Scratch** — starts new drawings in metric or Imperial units.
**Use a Template** — starts new drawings based on *.dwt* template files.
**Use a Wizard** — starts new drawings based on two wizards, Quick and Advanced.
**Open a Drawing** — opens existing drawings.

(The dialog box discussed here is displayed when AutoCAD first starts; the nearly-identical Create New Drawing dialog box is displayed when you enter the NEW command. The only difference between the two, other than the title, is that the Create New Drawing dialog box grays out the **Open a Drawing** button, because this function is handled by the OPEN command.)

 **Note** Whether AutoCAD displays the Create New Drawing dialog box is determined by the value of the STARTUP system variable:
  **1** — displays the Startup dialog box.
  **0** — does not display the dialog box (default setting).

### Starting From Scratch

The **Start from Scratch** option is one way to start new drawings. It presents just two options: measuring the drawing in either Imperial (English) or metric units

Select **Imperial** to use feet and inches. The drawing limits are set to 12" x 9", and dimension styles to inches. You can change these settings later. (Whether AutoCAD copies the settings stored in the *acad.dwt* template file seems to change from release to release. The technical editor corrects me: "Not 'seems to'; does!")

Select **Metric** for metric units. Limits are set to 429 mm x 297 mm, and the dimension styles to metric. AutoCAD copies the settings stored in the *acadiso.dwt* template file.

Behind the scenes, AutoCAD stores your selection in the MEASUREINIT system variable, which has a value of 0 or 1:

**0** (Imperial) — AutoCAD reads the hatch patterns and linetypes defined by the ANSIHatch and ANSILinetype settings in the Windows registry.

**1** (Metric) — AutoCAD reads the hatch patterns and linetypes defined by the ISOHatch and ISOLinetype registry settings.

## Using Templates

In the Setup dialog box, choose the **Use a Template** button. AutoCAD displays the same list of template drawings as it did with the NEW command.

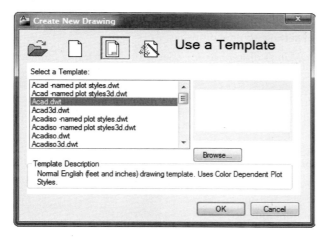

1. Select a file name, and AutoCAD displays its preview image.
2. Click **Open** to open the template, which is the starting point for your new drawing.

Although template drawings have the *.dwt* extension, you are in fact free to use *any* existing AutoCAD drawing as the template. Click **Browse** to find other files, such as those ending with *.dwg*.

## Using Wizards

AutoCAD includes two "wizards" that step you through the stages of setting up some of the many parameters that define a drawing. They are named Quick and Advanced.

The Quick Wizard takes you through just two steps — setting the units and the area — in creating new drawings. (You previously worked with the Quick Wizard in Chapter 1, "Quick Start in AutoCAD.")

### TUTORIAL: USING THE ADVANCED WIZARD

In this tutorial, you work through the Advanced Wizard's steps to set up the units, angles, and area for new drawings. Curiously, no help is available for this early encounter with AutoCAD.

1. To start a new drawing with the advanced wizard, click the **Use a Wizard** button.
2. Select **Advanced Wizard**, which will present a series of dialog boxes that leads you through the steps of setting up new drawings.
3. Click **OK**. Notice the Units dialog box for specifying the drawing's units.

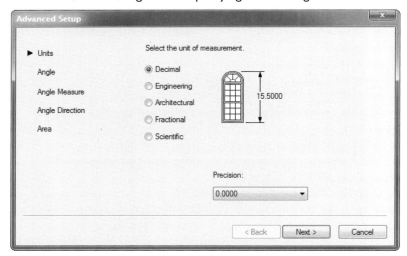

Select a measurement unit. (If you are not sure, select **Decimal**. You change the units later with the **UNITS** command.)

Technically, AutoCAD does not work with Imperial or metric units; internally, the software uses scientific units. To help human operators, however, AutoCAD accepts from us and displays to us units in the various formats listed below, converting to and from scientific format on the fly.

| AutoCAD Units | Example | Comments |
|---|---|---|
| Decimal | 14.5 | Decimal and metric units. |
| Engineering | 1'-2.5" | Feet and decimal inches. |
| Architectural | 1'-2 1/2" | Feet and fractional inches. |
| Fractional | 14 1/2 | Fractional inches, no feet. |
| Scientific | 1.45E+01 | Exponent notation. |

4. *Precision* refers to the number of decimal places or fractions of an inch. From the Precision droplist, select the number of decimal places (or fractional accuracy). If you are not sure, select 2 decimal places (**0.00**) or 1/8 fractional. The maximums are eight places or 1/256th.

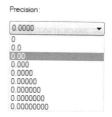 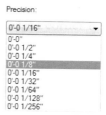

**Left:** *Selecting decimal precision.*
**Right:** *Selecting fractional precision.*

Precision affects the *display* of units only. No matter which precision you select, AutoCAD continues to calculate to 16 digits internally, and then displays the result rounded to the level of precision you select here.

Decimal, engineering, and scientific units: 0 to 8 decimal places.
Architectural and fractional units: 1/1, 1/2, 1/4, 1/8, 1/16, 1/32, 1/64, 1/128, or 1/256.

5. Click **Next**. Notice the Angle dialog box, which specifies the style of angle measurement.

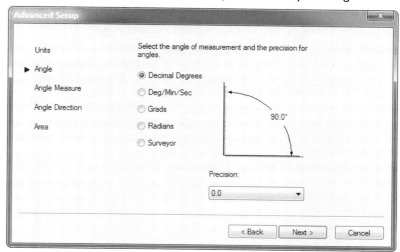

6. Select an angle measurement unit. If you are not sure, select **Decimal Degrees**. Internally, AutoCAD works with radians, but accepts and displays angles in the following formats:

| Angle Formats | Example | Comments |
|---|---|---|
| Decimal Degrees | 45.00 | 360 degrees in a circle. |
| Degrees-minutes-seconds | 45d00'00.00" | Fractions of seconds in decimals. |
| Grads | 50.00g | 400 grads in a circle. |
| Radians | 0.79r | 2pi radian in a circle. |
| Surveyor | N45d0'0.00"E | Fractions of seconds in decimals. |

7. From the Precision droplist, select the number of decimal places. (There is no fractional accuracy for angles.) If you are not sure, select 1 (**0.0**) decimal place. As with linear units, Precision affects only the *display* of angles:

    Decimal degrees, grads, and radians: 0 to 8 decimal places.
    Deg/min/sec and surveyor angles: 45d, 45d00' through to 45d00'00.00000".

8. Click **Next**. Notice the Angle Measure dialog box.

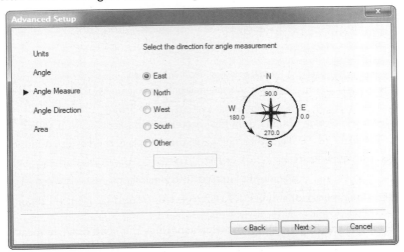

9. Although the dialog box asks you to "Select the direction for angle measurement," it would be more accurate if it read, "Select the direction for 0 degrees."

    The default is **East**, which is the correct setting for most drawings, because measuring angles from "East" is the same as measuring them from the positive x axis. If your drawings use a direction other than 0 degrees, select from **North**, **West**, **South**, or **Other**.

    In the **Other** text box, you enter an angle in degrees between 0 and 360 — even when you selected grads or radians in the previous dialog box.

10. Click **Next**. Notice the Angle Direction dialog box.

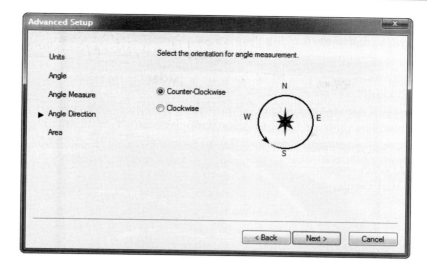

11. You are asked for the direction in which you wish to measure positive angles: (a) counter-clockwise or (b) clockwise. For most drawings, you keep the direction counterclockwise.
12. Click **Next.** Notice the Area dialog box.
13. You are asked to specify the limits of the drawing. The default is 12 units wide (x direction) by 9 units tall (long, y direction).

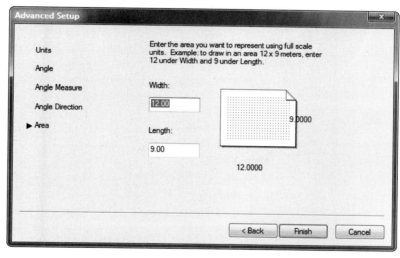

If you know how large the drawing will be, enter values in the **Width** and **Length** text boxes. Clear the current values using the **DEL** or **BACKSPACE** key.

AutoCAD uses these values to set the limits of the drawing, which affect the extent of the 2D grid display (as shown in the preview image above) and the **ZOOM All** command. You can always change the limits later with the **LIMITS** command.

(*History*: The earliest releases of AutoCAD prevented you from drawing outside the area defined by the limits; this didn't make much sense, and Autodesk dropped the restriction with 1986's version 2.17g, and added the LIMCHECK system variable.)

14. Click **Finish**. Notice that AutoCAD opens the new drawing with the settings you specified.

**OPEN**

The OPEN command opens existing drawings.

This command opens drawings for editing. It only opens files previously saved in AutoCAD formats. (Other commands are available to open drawings that are not in AutoCAD formats.)

**TUTORIAL: OPENING DRAWINGS**

1. To open drawings, start the **OPEN** command:
   - On the Quick Access toolbar, click the **Open** button.
   - From application menu, choose **Open**.
   - Alternatively, press the **CTRL+O** keyboard shortcut.
   - Or, at the 'Command:' prompt, enter the **open** command.

      Command: **open** *(Press ENTER.)*

   In most cases, AutoCAD displays the Select File dialog box.

2. Select a *.dwg* file from the list, and then choose **Open**. Notice that AutoCAD opens the drawing.

More than just opening drawings, the Select File dialog box accesses all kinds of file-related functions:

- Opens *.dwg* (drawing), *.dxf* (interchange), *.dws* (standards) and *.dwt* (template) files.
- Selects one or more files to open at the same time.
- Previews drawings before opening, via thumbnail representation.
- Sorts files by name, date, type, and size.
- Views names of drawings recently opened (History), and most often used (Favorites).
- Opens drawings protected against change (their status set to "read-only").
- Loads all or portions of drawings.

## TRUSTED DWG & DWGCHECK

When AutoCAD opens a drawing, it reports on its source. When the drawing was last saved by Autodesk software, the following prompt appears in the command bar:

> Autodesk DWG. This file is a TrustedDWG last saved by an Autodesk application or Autodesk licensed application.

When the drawing comes from a non-Autodesk source, however, a warning appears in this dialog box:

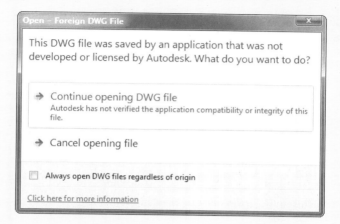

**Continue** opens the drawing, but then repeats the warning at the command line:

> Non Autodesk DWG. This DWG file was saved by a software application that was not developed or licensed by Autodesk. Autodesk cannot guarantee the application compatibility or integrity of this file.

**Cancel Opening File** does not open the drawing.
☐ **Always Open DWG Files Regardless of Origin** prevents this dialog box from appearing again.

AutoCAD 2007 introduced *Trusted DWG*, because Autodesk felt that drawings edited and saved by non-Autodesk software may corrupt data and destabilize applications used with AutoCAD. Trusted DWG does not repair defects in files, but allows you to choose whether to load these "foreign" *.dwg* files.

The stern wording of the dialog box changed in AutoCAD 2008 after Autodesk launched a law suit against the Open Design Alliance, after ODA began using the registered term "Trusted DWG" without permission from Autodesk. As part of the settlement, ODA agreed to stop using the term, and Autodesk agreed to tone down warning.

## DWGCHECK

The **DWGCHECK** system variable determines how and when these warnings are displayed.
- **0** — When the drawing has a possible problem, the warning appears in a dialog box.
- **1** — When the drawing was saved by software other than AutoCAD or AutoCAD LT or when it has a possible problem, the warning appears in a dialog box.
- **2** — When the drawing has a possible problem, the warning appears on the command line.
- **3** — When the drawing was saved by software other than AutoCAD or AutoCAD LT or when it has a possible problem, the warning appears on the command line.

Turning off **DWGCHECK** prevents the warnings from appearing in the future:

> Command: **dwgcheck**
> Enter new value for DWGCHECK <1>: **0**

- Accesses drawings stored on your computer, on any other computer located on your network, or from any location accessible through the Internet (FTP).
- Searches for drawings anywhere on your computer (Find).
- Searches the Internet for information.
- Creates a list of frequently-accessed folders.
- Creates new folders and manipulates files and folders, including renaming and deleting files.
- Sends drawings as email messages, or to removable drives.
- Resizes the Select Drawing dialog box.
- Accesses Autodesk's Buzzsaw Web site for managing construction projects.

Moving counterclockwise, the Select File dialog box consists of five primary sections:

**Files List** lists the names of files and folders.
**Preview** displays a thumbnail-size image of the drawing.
**Files of Type** specifies the name and type of file to be opened.
**Standard Folders Sidebar** provides shortcuts to folders and Web locations.
**Toolbar** holds tools useful for manipulating files and folders.

**Note** While the **OPEN** command is the most common method to open files, it is not the only way:

**ATTACH** attaches images and files in DWF, DWFX, DGN, and PDF formats as underlays.
**INSERT** inserts other drawings as blocks into the current drawing.
**OPENDWFMARKUP** attaches marked-up .dwf files.
**OPENSHEETSET** opens .dst sheet set data files and related drawings.
**DXBIN** opens .dxb (drawing exchange binary) files created by CAD\camera (obsolete).
**DXFIN** opens .dxf (drawing interchange format) files created by AutoCAD and other CAD programs.
**ACISIN** imports .sat format solid models created by ACIS-compatible CAD programs.
**WMFIN** inserts .wmf (Windows meta format) files.
**FBXIN** imports .fbx (filmbox) files.
**XOPEN** opens externally-referenced drawings in their own windows.
**3DSIN** opens .3ds models created by 3D Studio.

And the Clipboard pastes vector and raster graphics and text into drawings from almost any program.

### Files List

To open drawings:

1. Select the name from the Files list. Notice that the name appears in the **File name** field.
2. Click **Open**.

As a shortcut, you can double-click the file name. This is equivalent to selecting the file name, and then clicking the **Open** button.

(*History*: The first several releases of AutoCAD presented a numbered menu of choices. To open a drawing, users pressed **2**, and then ENTER. Option 2 was labeled "Edit an EXISTING drawing." AutoCAD then prompted for the name of the drawing; users had to memorize the drawing's file name, because they could not select file names from a list. Release 9 introduced the basic dialog box that allowed users to select drawing names, as well as to choose the drive and folder names. AutoCAD 2000i greatly expanded the options through the dialog box you see today.)

## Preview

When you select a drawing in the Files list, AutoCAD displays its preview image.

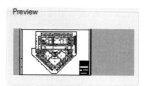

The image is small, but provides sufficient detail to distinguish it from most other drawings. The preview image, also called a "thumbnail," is generated automatically by AutoCAD when you save the drawing. Because the thumbnail shows the view when the drawing was last saved, the view may not include the entire drawing.

**Notes** Drawings in 3D are shown rendered when the drawing was saved with a visual style turned on.

If you see no preview image, the **RASTERPREVIEW** system variable was turned off (set to 0) when the drawing was saved. Turn it on by changing the value to 1:

    Command: **rasterpreview**

    Enter new value for RASTERPREVIEW <0>: **1** *(Press* **ENTER**.*)*

The other possibility is that the drawing files were created by old versions of AutoCAD or other types of CAD software that don't include preview images.

## Files of Type

By default, the dialog box displays the names of *.dwg* files, which are AutoCAD drawings. In addition, you can select other kinds of files to open:

To select a file type other than from *.dwg*, click the **Files of type** list:

    **DWG** (drawing) — for creating AutoCAD drawing files.

    **DWS** (drawing standards) — for ensuring drawings use standardized elements, such as linetypes and layers, against which drawings are checked.

    **DXF** (drawing interchange format) — for creating drawings that can be exchanged between otherwise incompatible programs.

    **DWT** (drawing template) — for creating preset frameworks of new drawings, such as dimension styles and drawing borders. When you select "Drawing Template (*.DWT)", the dialog box automatically switches to the *\template* folder.

If you are not sure, select *.dwg*.

There is more than one way to open a drawing file. Click the down-arrow next to the **Open** button.

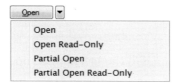

### Read-Only

Notice the menu that appears:

> **Open** — loads all of the drawing; you can edit the drawing and save the changes.
>
> **Open Read-Only** — loads all of the drawing, and does not allow you to save changes unless saved by a different file name.
>
> **Partial Open** — loads parts of the drawing, based on layer and view names.
>
> **Partial Open Read-Only** — loads parts of the drawing, and does not save editing changes.

When you select one of the read-only options, AutoCAD displays the "Read Only" message on the title bar, as illustrated below.

If you want to save a drawing that's read-only, then the SAVE command displays the Save As dialog box. When you attempt to save the drawing with the same name, however, a warning appears, as illustrated below.

AutoCAD prevents you from saving the drawing to the original file. (Give it a different file name, as described later in this chapter.)

### Partial Open

When you select one of the partial-open options, AutoCAD displays the Partial Open dialog box. (Partial-open means that only part of the drawing is loaded, useful for very large drawings.)

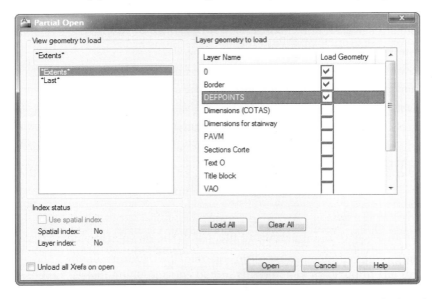

Select a view from the left hand column and the layers to be opened in the right hand column. When done, click **Open**.

### Select Initial View

Getting back to the Select File dialog box, the **Select Initial View** check box determines whether the drawing is opened with a saved view:

☐ S<u>e</u>lect Initial View

**Off** — when the drawing is opened, AutoCAD displays the same view as when the drawing was last saved.

**On** — before the drawing is opened, AutoCAD displays the Select Initial View dialog box, which lets you select a named view.

If no views have been saved, then the drawing opens normally. Views are described in Chapter 5, "Drawing with Precision."

### Making Dialog Boxes Bigger

When you first see the dialog box, it is at its smallest. You can make it larger to see more file names at a time. With the cursor, grab an edge of the dialog box, and stretch the dialog box, revealing more file names.

 **Note** Press function key **F5** to update the files list. You may need to do this when files appear to be missing or incorrectly named.

### Standard Folders Sidebar

A vertical list of folder names, called the "sidebar," is located on the left of the dialog box. It directly accesses folders on your computer, networks, and the Internet. From top to bottom, these are:

**History** — displays a list of the most-recently opened drawings.

**Desktop** — displays the files and folders found on the Windows "desktop."

**FTP** — browses FTP (file transfer protocol) sites on the Internet; you can add more sites with Tools | Add/Modify FTP Locations.

**Documents** — displays files stored in your computer's \documents folder.

**Favorites** — displays files stored in the \windows\favorites folder; you can add files to this folder with Tools | Add to Favorites.

**Buzzsaw** — connects to Autodesk's www.buzzsaw.com Web site, after you install their ProjectPoint client software.

To add folders to the list, simply drag them from the Files list. For example, I like to have AutoCAD's \sample folder on the list for fast access to the sample drawings, so I drag it onto the sidebar.

Commands for editing the sidebar are available by right-clicking the folder sidebar, as illustrated by the figure.

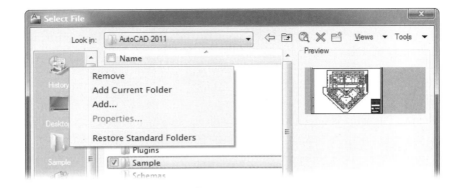

## Tools

On the Select File dialog box's toolbar, there is an item named **Tools** that contains a number of handy functions.

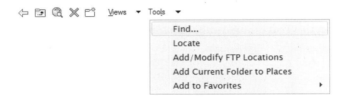

## Finding Files

Sometimes you cannot remember the names of drawing files or where they are located. When you choose **Find** from the **Tools** item, AutoCAD displays a Windows dialog box for finding files.

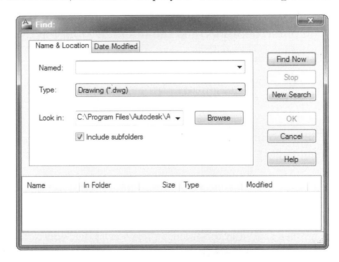

It is able to look at every folder on every drive in your computer. Drives include the hard drives, external drives, CD/DVD drives, and other storage devices. In addition, if your computer is connected to a network of other computers, the Find feature can search every folder of every networked drive your computer has permission to read.

The **Named** text box allows you to enter as much of the name as you can remember. Use wildcard characters to specify part of a name. Recall that ? is a placeholder for any character, while * searches for all files that match the rest of the criteria you supply. For example, door* searches for all files that start with **Door**:

    door   door36   doorstop

The **Type** droplist provides a filter for selecting *.dwg* drawing, interchange *.dxf*, template *.dwt*, and standards *.dws* files.

The **Look In** droplist restricts the search to specific drives, folders, and paths.

The **Date Modified** item narrows the search to specific dates. This is useful when you know roughly the date the drawing was created. For example, if the drawing was created last month, specify "Between these dates," and then enter the month.

After you enter the search parameters, choose **Find Now**. Windows spends a bit of time rummaging through your computer's drives, and produces a list of matching files. At any time, choose **Stop** to bring the process to a premature halt.

## CHANGING SCREEN COLORS

When first installed on your computer, AutoCAD is designed to start up in a workspace called "2D Drafting & Annotation." Workspaces change the look of AutoCAD's user interface. The workspaces provided by Autodesk are as follows:

**2D Drafting & Annotation** — sets the default workspace for AutoCAD 2011.
**3D Basics** — presents a simplified interface for 3D modeling.
**3D Modeling** — sets up interface elements for 3D modeling.
**AutoCAD Classic** — makes AutoCAD look like it did before 2009.

You can change workspaces through the Workspaces button on the status bar or Quick Access toolbar.

### Default Background Colors

In the 2D workspaces, AutoCAD displays the drawing area of model space in dark gray. Some drafters prefer black or white as the background color. Black makes the colors vibrant, while white simulates paper.

In the "3D Modeling" workspace, AutoCAD displays the drawing area of model space in a variety of shades of gray to provide a sense of depth. Notice that the cursor and UCS icon are 3D and colored. (If your copy of AutoCAD does not look like the figure below, start a new drawing with the *acad3d.dwt* template file.)

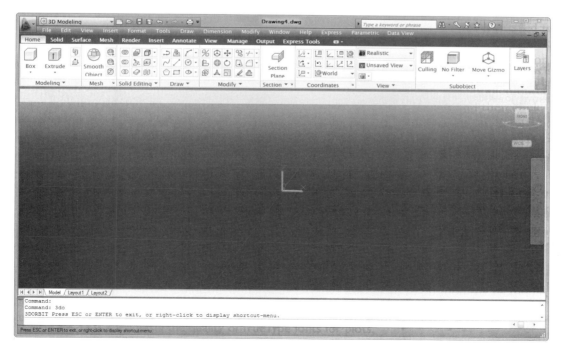

(*History*: In the past, computer screens had black backgrounds, displaying the text and graphics in monochrome colors, such as green, amber, or white — hence the name, "monochrome display." With the advent of Macintosh and its paper metaphor, the background color of choice became white, because it is the color of paper upon which drawings are printed. Still, some drafters prefer the black background — the default background color until AutoCAD 2008 — because it makes lighter colors, such as yellow and green, stand out better; others like it because black can create less eyestrain. And, added the technical editor, black grid dots are nearly invisible on a white background. AutoCAD 2009 switched the color to pale yellow, and then AutoCAD 2011 switched it to dark gray.)

### TUTORIAL: CHANGING COLORS

In this tutorial, you learn how to change the background color of the drawing area and many other parts of AutoCAD's user interface.

1. To change the colors of the AutoCAD user interface, start the **OPTIONS** command:
   - In the application menu, click the **Options** button.
   - At the 'Command:' prompt, enter the **options** command.

   Command: **options** (*Press* ENTER.)

2. In the Options dialog box, choose the **Display** tab, and then click **Colors**. Notice that AutoCAD displays the Drawing Window Colors dialog box.

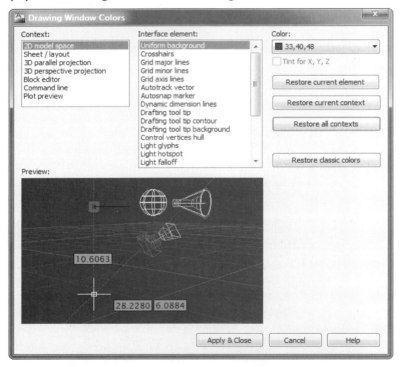

 **Note**  This dialog box has several buttons that reset colors:

   **Restore Current Element** — resets the color of the selected item (Interface Element).
   **Restore Current Context** — resets the colors for the selected environment (Context).
   **Restore All Contexts** — resets all colors to their defaults.
   **Restore Classic Colors** — matches the color scheme of AutoCAD 2008 and earlier, turning the background color of model space to black and resetting all other colors to their defaults.

3. From the Context column, select a user interface "context," such as **2D model space**.
4. From the Interface Element column, select an element, such as **Uniform background**.

Setting Up Drawings | 105

5. From the Color droplist, assign a color to the element.

If the basic eight colors are insufficient, click **Select Color**. AutoCAD displays the Select Color dialog box, from which you can access the full palette of 16.7 million colors.

6. If you mess up, click **Restore current element** to change the element back to its default color. Or, choose **Restore current context** to change all elements back to their default colors.
7. Click **Apply & Close** to see the change in colors.
8. Click **OK** to exit the Options dialog box.

## UNITS

The UNITS command specifies the units of measurement, angles, and precision.

(If you did not to use the Advanced Wizard, then you can change the drawing units with this command.)

### TUTORIAL: SETTING UNITS

1. To set the units for drawings, start the **UNITS** command:
   - At the 'Command:' prompt, enter the **units** command.

     Command: **units** *(Press ENTER.)*

   - Alternatively, enter the aliases **un** or **ddunits** (the old name for this command) at the 'Command:' prompt.

   Notice that AutoCAD displays the Drawing Units dialog box.

2. In the **Length** area, select the type of unit and precision.

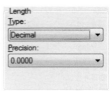

Watch the Sample Output area for examples of the units you select.

(To learn more about the units used by AutoCAD, see the discussion earlier in this chapter for the Advanced Wizard.)

3. In the **Angle** area, select the type and precision of angles.

4. In the **Insertion scale** area (named "Drag-and-drop scale" in earlier releases of Auto-CAD), select the units for blocks dragged into drawings from the DesignCenter, Tool palettes, or from a Web site using i-drop:

| Unit | Equivalent |
|---|---|
| Microinches | 0.000001 inches |
| Mils | 0.001 inches |
| Yards | 3 feet |
| Inches | 25.4mm |
| Feet | 12 inches |
| Miles | 5,280 feet |
| Angstroms | 0.1 nanometers |
| Nanometers | 10E-9 meters |
| Microns | 10E-6 meters |
| Millimeters | 0.0393 inches |
| Centimeters | 10 mm |
| Decimeters | 0.1 meter |
| Meters | 100 cm |
| Kilometers | 1000 m |
| Decameters | 10 meters |
| Hectometers | 100 meters |
| Gigameters | 10E+9 meters |
| Astronomical Units | 149.597E+8 kilometers |
| Light Years | 9.4605E+9 kilometers |
| Parsecs | 3.26 light years |
| Unitless | Inserted blocks are not scaled; units match the drawing's units. |

5. The Lighting area specifies a different kind of unit, one that assigns the units for measuring the intensity of photometric lights — and one that you can safely ignore, at least until you get involved in advanced photorealistic renderings of 3D drawings. (To use photometric lights, select any unit other than Generic.)

6. To change the direction of angle measurement, click **Direction**. AutoCAD displays the Direction Control dialog ox.

   a. Select a direction for the angle. If you wish to specify an angle that isn't listed, select **Other**, and then enter the angle in the **Angle** text box. (Alternatively, choose the **Pick an Angle** button to pick the angle in the drawing:

   Pick angle: *(Pick a point in the drawing.)*

   Specify second point: *(Pick a second point.)*

   b. Click **OK** to dismiss the Direction Control dialog box.
7. Click **OK** to dismiss the Drawing Units dialog box.
8. Move the cursor, and watch the coordinate display on the status bar. It should match the units you selected.

## SETTING SCALE FACTORS

One of the hardest concepts for the new CAD user to grasp is that of scale. Think of when you sketch a picture of your house on a piece of paper. You draw the house small enough to fit the paper. (There isn't a piece of paper big enough for you to draw the house full size! "I don't know about that," rejoins the technical editor. "I've seen paper rolls 50' wide and several thousand feet long. A full-size house drawing would be difficult to fold up, however, and carry under your arm")

A typical house is 50 feet long. To fit a drawing of it on a 10"-wide sheet of paper, you sketch the house about 60 times smaller. Where does "60" come from? The math works out like this:

Convert house width from feet to inches: 50 feet x 12 inches/foot = 600 inches.

Divide the house size (600 inches) by the paper size (10 inches) = 60.

The scale factor is 1:60. One inch on the drawing is 60 inches on the house.

Here are additional examples of scale factors:

A truck is 20 feet long. The paper is 10 inches wide. The scale factor works out to 1:24.

The sailboat is 30 m long. The paper is 1m wide. The scale factor is 1:30. (It's much easier in metric!)

The gear is 1 inch in diameter. The paper is 8 inches wide. The scale factor is 8:1. (Scale factor numbers are "reversed" when objects are smaller than the paper.)

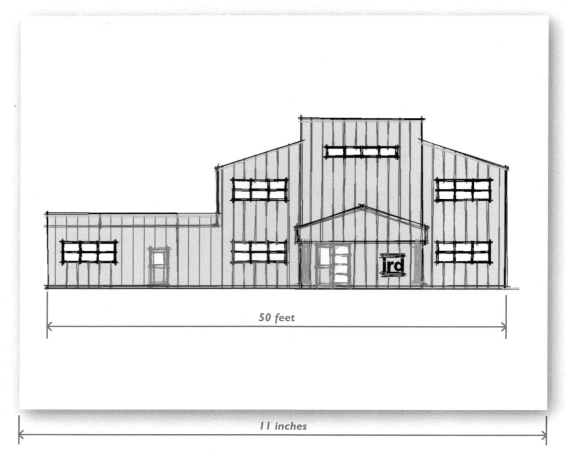

In CAD, scaling is done late in the process. Instead of drawing the house to scale, you create the CAD drawing full-size. That means the 50-foot long house is drawn 50 feet long; the one-inch gear is drawn one inch in diameter. That's much easier, isn't it? (AutoCAD can handle drawings of extreme size. The technical editor notes that the largest object would be a circle with a radius of 1 000 000 000 000 000 000 E+99 units, although strange things may occur above 1E+99.)

You could draw the entire known universe, full size if you want. Full size is shown as 1:1. There is no need to work out scale factors — or is there?

Scale comes into the scene when you plot your drawing. That big drawing, whether of your house or the solar system, must be made small enough to fit the paper it is plotted on. Thus, scale factor comes into play during the PLOT command, where you specify that AutoCAD should print the house drawing 60 times smaller to fit the paper.

### Scale-dependent Objects

There is, however, a catch. Some items in the drawing cannot be 60 times smaller. These items include text, hatch patterns, linetypes, and dimensions. These are called scale-dependent. If AutoCAD plots the text 60 times smaller, you would not be able to read it. The solution is to do the inverse: draw the text 60 times larger. When it is plotted, the text becomes the correct size.

You may be puzzled. How big should text be when plotted on a drawing? The standard in drafting is to draw text 1/8" tall for "normal" sizes. When you place text in your house drawing, specify a height of 7.5" tall (= 1/8" x 60). This may seem much too high to you, but trust me: when plotted, it looks correct.

You must apply the scale factor to other scale-dependent objects as well, including dimensions, linetypes, and hatch patterns. Scale factors for hatch patterns, linetypes, and dimensions are stored in these system variables:

| System Variable | Scale Factor |
| --- | --- |
| HPSCALE | Hatch patterns. |
| LTSCALE | Linetype patterns. |
| DIMSCALE | Dimension text height, arrowhead size, and leaders; does not affect dimension values. |

To change the scale factor, enter the name of the related system variable. For example, to set the hatch pattern scale to 60:

Command: **hpscale**

Enter new value for HPSCALE <1.0000>: **60**

Fortunately, you can use the same scale factor for all. For text, you specify a scaled height during the TEXT command:

Command: **text**

Current text style: "TECHNICLIGHT" Text height: 0'-2"

Specify start point of text or [Justify/Style]: *(Pick a point.)*

Specify height <0'-2">: **7.5**

Alternatively, use the STYLE command to create a text style with a fixed height of 7.5". If the text is the wrong size, you can use the SCALETEXT command to change it.

 **Note** Be aware of this inconsistency: HPSCALE and DIMSCALE do not apply to objects already in the drawing, but LTSCALE does.

## Annotation Scaling

Until recently, there was no master scale factor that applied to all scale-dependent objects, and there was no global scale factor for text at all. AutoCAD 2008 introduced *annotation scaling*, a way to ensure that text, dimensions, linetypes, and hatch patterns are displayed and plotted at the correct size.

Annotation scaling works like this: when the annotative scale of objects matches the view and plot scales, then the objects appear; when not, they do not appear. If the drawing is plotted at several sizes, then you need to assign several annotation scale factors to each object. (Fortunately, AutoCAD makes this automatic.) When multiple annotation scales are assigned to objects, AutoCAD takes care of automatically selecting the representation that appears at the correct scale.

Annotation is a property of the following objects: text, mtext, fields, dimensions, tolerances, attributes, leaders, quick leaders, multi leaders, and hatch patterns. In addition, blocks and dynamic blocks can have annotation scaling.

Linetypes are handled differently. They don't have individual annotative scales; instead, the new MSLTYPE system variable forces all linetypes in the drawing to take on the current annotative scale factor.

You learn more about annotation scaling in Chapter 10, "Placing and Editing Text," as well as in other chapters that deal with scale-dependent objects (hatches, dimensions, and so on).

## SAVE AND SAVEAS

The SAVE and SAVEAS commands operate identically — both display the Save Drawing As dialog box.

The purpose of the commands is to save drawings by other names and in a few other formats. You may want to give drawings different names when you open them for editing, but don't want to save the changes to the original file. These commands also allow you to save the drawings in other folders, on other drives, and over networks on other computers.

Drawing files created by AutoCAD 2011 cannot be read by versions of AutoCAD earlier than 2010, unless the drawings are *translated*. In contrast, AutoCAD 2011 reads just about any drawing created by earlier releases of AutoCAD. For this reason, the SAVEAS command includes options to translate drawings to certain earlier formats.

### TUTORIAL: SAVING DRAWINGS BY OTHER NAMES

1. To save drawings under different names, start the **SAVEAS** command:
   - From application menu, choose **Save As**.
   - In the Quick Access toolbar, click the **Save As** button.
   - Alternatively, use the **CTRL+SHIFT+S** keyboard shortcut.
   - Or, at the 'Command:' prompt, enter the **saveas** or **save** commands.

        Command: **saveas** *(Press* ENTER.*)*

   Notice that AutoCAD displays the Save Drawing As dialog box.

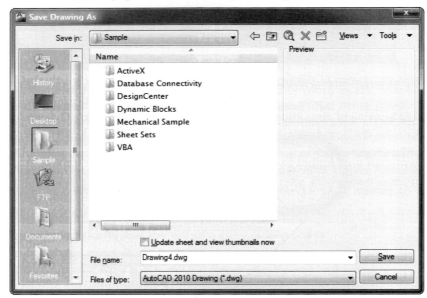

2. From the **Save in** droplist, select the drive and folder in which to save the drawing.
3. In the **File name** text box, enter a different name for the drawing. There is no need to include the *.dwg* extension as AutoCAD adds it automatically.
4. Click **Save**.

When you enter the name of a drawing that already exists in the folder, AutoCAD displays the message box shown belw.

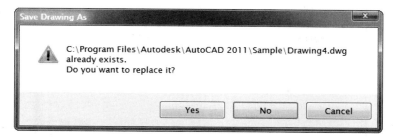

To replace the existing drawing with the new one, choose **Yes**. If not, choose **No**, and then rename the drawing.

### SAVING DRAWINGS: ADDITIONAL METHODS

The SAVEAS command allows you to save drawings on computers connected to networks, including the Internet. It also saves drawings in other formats. The dialog box provides these options:

> **My Network Places** — saves drawings over a network.
>
> **FTP** — saves drawings on the Internet.
>
> **Files of type** — saves drawings in other file formats.
>
> **Options** — specifies options for saving .dwg and .dxf files.

Let's look at each option.

### My Network Places

The **My Network Places** option saves drawings to other computers, provided two conditions are met: (1) the computers are connected via a network, and (2) you have access rights.

### TUTORIAL: SAVING DRAWINGS ON OTHER COMPUTERS

In this tutorial, you learn how to save AutoCAD drawings to networked computers.

1. Enter the SAVEAS command.
2. In the Save Drawing As dialog box, click the **Save in** droplist.

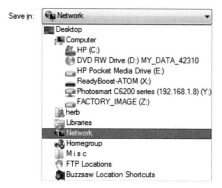

2. From the droplist, select **Network**. (Older dialects of Windows call this "My Network Places.")

Notice that the dialog box now displays the names of computers connected to yours.

3. Double-click a computer name to view the drives; double-click a drive to view its folders. The figure below shows folder *Public Documents* inside the *Public* folder on the *DownstairsPC* computer.

4. Click **Save**. Depending on the speed of the network, you may notice it takes longer to save drawings through the network.

 **Note** If you regularly access a folder on another computer, drag the folder to the Places list.

## FTP

*FTP* (short for "file transfer protocol") is a method of sending and receiving files over the Internet. You may be familiar with sending files through email as attachments. The drawback to using email is that it was not designed to handle very large files. Some email providers limit the attachment size to 5MB or smaller.

FTP allows you to send files of any size, even entire DVDs worth of files — gigabytes worth. FTP is also the method used to upload Web pages to Web sites. The drawback to using FTP is that you first need to set up the FTP site with your user name, password, and other data.

Clicking on **FTP** in the Places list results in an empty list of FTP Locations. You first need to add a site. There are two types of FTP sites: *anonymous* and *password*.

**Anonymous** — allows anyone to access them (also known as "public FTP sites"). You need only your email address as the password; a user name is unnecessary.

**Password** — requires a user name and a password, usually provided by the person running the FTP site.

## TUTORIAL: SENDING DRAWINGS BY FTP

In this tutorial, you learn how to set up AutoCAD to access a public FTP site so that it can open and save files.

1. In the Save Drawing As dialog box, choose **Tools**, and then **Add/Modify FTP Locations**.

Notice that AutoCAD displays the Add/Modify FTP Locations dialog box.

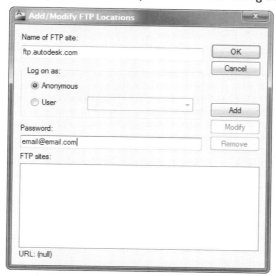

2. Enter the details for the ftp site. For example, enter the following parameters to access Autodesk's public FTP site:

   | | |
   |---|---|
   | Name of FTP site: | **ftp.autodesk.com** |
   | Logon as: | **Anonymous** |
   | Password: | **email@email.com** |

3. Click **Add**.
4. Click **OK**. Notice that AutoCAD adds the FTP site to the list in the dialog box.

   The FTP item has been create.

5. To save the drawing to the FTP site, follow these steps:
   a. Double-click the name of the FTP site (*ftp.autodesk.com*). Notice that the dialog box displays a list of folders found on the remote site.

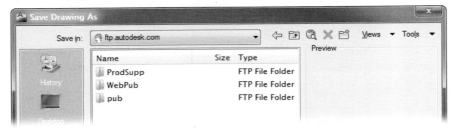

   b. Select a folder.
   c. Click **Save**. Notice the dialog box shows the progress as the file is transferred to the file to the site. Expect it to be significantly slower than saving to disk.

Some sites require a user name and password before you can save files. If so, a dialog box appears requesting the information. Fill in the missing information, and click **OK**.

### Files of Type

To save the drawing in an earlier release of AutoCAD, or in other formats, choose the **File of type** list box.

#### TUTORIAL: SAVING DRAWINGS IN OTHER FORMATS

1. To save drawings in other formats, start the **SAVEAS** command:
   Command: **saveas** (*Press* ENTER.)

   Notice that AutoCAD displays the Save Drawing As dialog box.
2. In the **Files of type** droplist, select a file format:

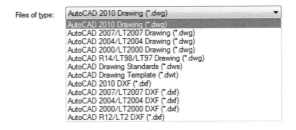

   **AutoCAD 2010 (*.dwg)** — saves drawings in 2010 format, which can also be read by AutoCAD 2011, as well as by AutoCAD LT 2010 and 2011.

   **AutoCAD 2007/LT 2007 (*.dwg)** — saves drawings in 2007 format, which can also be read by AutoCAD 2008/9 and AutoCAD LT 2007/8/9.

   **AutoCAD 2004/LT 2004 Drawing (*.dwg)** — saves drawings in 2004 format, which can also be read by AutoCAD 2005 and 2006.

   **AutoCAD 2000/LT 2000 Drawing (*.dwg)** — saves drawings in 2000 format, which can also be read by AutoCAD 2000i and 2002, as well as by AutoCAD LT 2000i and 2002.

   **AutoCAD R14/LT 98/LT 97 Drawing (*.dwg)** — saves drawings in Release 14 format.

   **AutoCAD Drawing Standards (*.dws)** — saves drawings as standards files.

   **AutoCAD Drawing Template (*.dwt)** — saves drawings as templates.

**AutoCAD 2010 DXF (*.dxf)** — saves drawings in DXF format compatible with AutoCAD 2010. This saves the drawing in a format called DXF (short for "drawing interchange format").

**AutoCAD 2007 DXF (*.dxf)** — saves drawings in DXF format compatible with AutoCAD 2007 through 2009.

**AutoCAD 2004 DXF (*.dxf)** — saves drawings in DXF format compatible with AutoCAD 2004 through 2006.

**AutoCAD 2000/LT 2000 DXF (*.dxf)** — saves drawings in DXF format compatible with AutoCAD 2000 and 2000i.

**AutoCAD R12/LT R2 DXF (*.dxf)** — saves the drawing for use with AutoCAD Release 11 and 12.

3. Click **Save**.

**Notes** To open drawings in Release 12, Release 11, and AutoCAD LT Release 1 and 2, use the **DXFIN** command. *Caution!* Some objects may be lost or changed in translation to older formats.

To save drawings in...

TIF, JPG, or PNG formats, use the **TIFOUT**, **JPGOUT**, and **PNGOUT** commands, respectively.
PDF format, use the **EXPORTPDF** command.
ESP (encapsulated PostScript) format, use the **PSOUT** command.
FBX (filmbox) format, use the **FBXOUT** command.
SLD slide format, use the **MSLIDE** command.
DWF or DWFX formats, use the **EXPORTDWF** or **EXPORTDWFX** commands, respectively.
3D DWF format, use the **3DDWF** command.
MicroStation V8/V7 format, use the **DGNEXPORT** command.
WMF, SAT, STL, BMP, or 3DS, use the **EXPORT** command.

## Options

The **Options** option allows you to specify how *.dwg* and *.dxf* files are saved.

1. In the Save Drawing As dialog box, choose **Tools**, and then **Options**.

Notice that AutoCAD displays the Saveas Options dialog box.

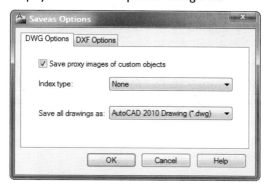

2. The **DWG Options** tab has the following settings:
   - ☑ **Save proxy images of custom objects** stores images of *custom objects* in the drawing file. Custom objects are created by ObjectARX programs, and are not understood by AutoCAD when the programs are not present. *Proxy images* are graphical representations of custom objects, which allow you to see, but not edit them.
   - ☐ When the option is off, AutoCAD displays rectangles in place of custom objects. It usually makes sense to leave this option turned on.

   **Index type** specifies whether AutoCAD creates indices when saving drawings. *Indices* improve performance during demand loading, but may increase save time. The options are:
   - **None** creates no indices.
   - **Layer** loads layers that are on and thawed.
   - **Spatial** loads only drawing parts within clipped boundaries.
   - **Layer & Spatial** combines both options for optimal partial loading performance.

   **Save all drawings as** determines the default format when drawings are saved with the **SAVEAS** and **QSAVE** commands. Keep this set to "AutoCAD 2010 Drawing (*.dwg)" to preserve all features unique to AutoCAD 2010 - 2011. If, however, you regularly exchange drawings with clients who use earlier releases, you may want to change the setting here.

3. The **DXF Options** tab has these settings:

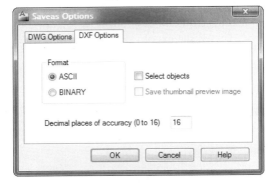

   **Format** selects either ASCII or binary format:
   - ⦿ **ASCII** .dxf files are readable by a larger number of non-Autodesk software applications than binary, and also by humans when files are opened in a text editor.
   - ○ **Binary** files are created and read faster than ASCII, and have smaller file sizes.
   - ☑ **Select Objects** prompts you to select objects from the drawing.
   - ☐ When option is off, AutoCAD outputs the entire drawing to the .dxf file.
   - ☑ **Save Thumbnail Preview Image** includes a preview image in the .dxf file.

   **Decimal Places of Accuracy (0 to 16)** specifies the accuracy of .dxf files. Higher values, such as 16, create larger files. CNC machines typically work to four decimal places, and get confused when files contain five or more decimal places. This option is available only for ASCII .dxf files; binary files are always saved with 16 decimal places of accuracy.

## QSAVE

The QSAVE command (short for "quick save") saves drawings as *.dwg* files on disc.

When you start a new drawing, AutoCAD gives it the generic name of *drawing1.dwg*. The first time you use QSAVE, AutoCAD displays the Save Drawing As dialog box, so that you can name the drawing, and file it in the correct folder. Subsequent uses of QSAVE unobtrusively save the drawing without the dialog box.

### TUTORIAL: SAVING DRAWINGS

In this tutorial, you learn how to save drawings.

1. To save drawings to disc, start the **QSAVE** command:
   - In the Quick Access toolbar, click the **Save** button.
   - From application menu, choose **Save**.
   - As an alternative, press the **CTRL+S** keyboard shortcut.
   - Or, at the 'Command:' prompt, enter the **qsave** command.

     Command: **qsave** *(Press ENTER.)*

2. Notice that AutoCAD saves the drawing.

   (If the drawing's name is *drawing1.dwg*, then AutoCAD displays the Save Drawing As dialog box. See **SAVEAS** command earlier in this chapter.)

### AUTOMATIC BACKUPS

Windows has been notorious for *crashing*, where the software stops working for no apparent reason. Even with improvements to Windows 7, AutoCAD can crash unexpectedly due to bugs (accidental errors in the programming code) and conflicts with the operating system. (Technically, computers do not crash; the application software conflicting with the operating system causes the crash.) When this happens, you can lose work.

To guard against loss of work, AutoCAD includes two functions to back up drawings: one makes backup copies of drawing files; the other creates a second set of backup copies at set time intervals. Autodesk warns that it is possible to lose drawing data when the power fails during a save. I recommend plugging your computers into uninterruptable power supplies that provide a minimum of ten minutes of emergency power, giving you time to save drawings during power outages.

It is therefore wise to save your work periodically, in addition to having AutoCAD automatically save open drawings to temporary files.

(If you choose to discard your work with the QUIT command, only the part of the drawing changed since the last file save is discarded.)

### TUTORIAL: MAKING BACKUP COPIES

In this tutorial, you learn how to set up AutoCAD to make back up copies of open drawings.

1. To ensure your drawings are backed up, open the Options dialog box:
   - Click the red "A," and then choose **Options** from the application menu.
   - At the 'Command:' prompt, enter the **options** command.

     Command: **options** *(Press ENTER.)*

2. Notice that AutoCAD displays the Options dialog box. In the Options dialog box, click the **Open and Save** tab.

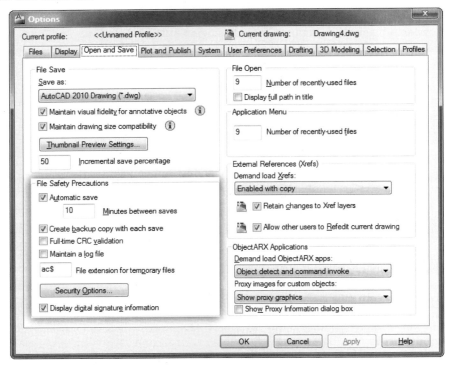

3. Notice the items in the **File Safety Precautions** section:

   ☑ **Automatic save** turns on the automatic backup facility. *This setting should always be turned on!* The only time to turn it off is when your computer's hard drive is low on free space — which is rare in this day of huge, cheap hard drives.

   **10 Minutes between saves** means that AutoCAD automatically backs up the drawing every ten minutes. You can change this value to any number you like. If your drawings are small, and your computer fast, then you may want to set this value to five minutes or less. That way, at most, only five minutes' work is lost.

   A setting of 0 (zero) means that no backups are made.

   ☑ **Create backup copy with each save** creates a *.bak* backup copy each time you use **QSAVE** and **CTRL+S**. *This option also should always be turned on,* unless the hard drive is low on free space.

4. Click **OK** to save the settings, and exit the dialog box.

AutoCAD starts the SAVETIME timer when you first change the drawing, not when you first open the drawing. The timer resets each time you use the QSAVE, SAVE, and SAVEAS commands. This prevents unnecessary hard drive activity.

### MAKING BACKUPS: ADDITIONAL METHODS

In addition to instructing AutoCAD to back up copies of your drawings, you can specify the folder in which backup files are stored.

- **SAVEFILEPATH** specifies the folder for storing backup copies.

Let's look at how it works.

## SaveFilePath

AutoCAD normally saves backup files in the following folder:

*C:\Documents and Settings\\<login>\Local Settings\Temp\\*

This is the same folder used by all other software on your computer. On my computer, for example, this folder holds 1,873 backup files (!) from a variety of sources — word processors, Web browsers, paint programs, and AutoCAD.

With all that clutter, you may prefer to store AutoCAD's backup files in a more suitable location. I recommend that you use another drive and a specially named folder, such as *\autocad 2011\dwgbackups*.

Further, I recommend that the drive be a portable, removable one, so that it can be quickly removed from the office in case of disaster. The only important part of your computer is the data; all else can be replaced easily, both hardware and software. Ideally, backups are kept outside the building, such as taken home each evening.

1. Use Windows Explorer to create the *dwgbackups* folder in \autocad 2011. (AutoCAD 2011 already has a folder called *\backup* for its own purposes.)
2. In AutoCAD, change the setting of the **SAVEFILEPATH** system variable, as follows:
   Command: **savefilepath**
   
   Enter new value for SAVEFILEPATH, or . for none <>: *(Enter a new path, such as* **"c:\autocad 2011\dwgbackups"**.*)*

Backup files are now stored in the new folder.

When the path name includes spaces, as does the one shown above, you must surround it with quotation marks.

If the path name is not valid, AutoCAD complains, "Cannot set SAVEFILEPATH to that value. *Invalid*." Ensure the folders exist, and that the full path is correct — full path means all folder names, from the drive name down to the file folder.

## Accessing Backup Files

The first time you save a drawing, AutoCAD adds the *.dwg* extension to the file name. With the second save, AutoCAD renames the file with the *.bak* file extension. Each time you use the QSAVE command, AutoCAD updates the backup file.

When AutoCAD crashes, it attempts to rename the *.bak* file as *.bk1*. (If *.bk1* exists, then AutoCAD renames it *.bk2*, and, if necessary, continues with *.bka* and then all the way through to *.bkz*.) This prevents AutoCAD from replacing previous backup files.

Backup files are full drawing files, just with a different file extension. If necessary, you can open the backup files in AutoCAD, following these steps:

1. Use Windows Explorer to copy the *.bak* files to a different folder.
2. Rename the *.bak* files to *.dwg* files.
3. Open the renamed *.dwg* files in AutoCAD.

(These steps are also carried out by the DRAWINGRECOVERY command.)

When AutoCAD crashes, it sometimes leaves behind temporary files; if your computer's hard drive is getting low on free space, you can erase these leftover files — provided AutoCAD is not running at the time. Drawings saved by the automatic backup process are given the extension *.ac$*.

When drawings are in use, AutoCAD locks the drawings to prevent other AutoCAD users from editing them. The lock status is indicated by the presence of *.dwl* and *.dwl2* (drawing lock) files. These files are used by the WHOHAS command to determine which user is editing specific drawings.

The *acminidump.dmp* file is created by AutoCAD when it crashes, to help programmers determine the reason. When AutoCAD is not running, all these files can be erased with Windows Explorer or a utility program, such as CCleaner.

## QUIT

Occasionally you create drawings that you do not wish to keep. To exit the current drawing and discard the changes, use the QUIT command.

### TUTORIAL: QUITTING AUTOCAD

In this tutorial, you learn how to exit AutoCAD.

1. To exit AutoCAD, without saving changes to drawings, start the QUIT command:
   - From the application menu, choose **Exit AutoCAD**.
   - Double-click the red A that identifies the application menu.
   - At the 'Command:' prompt, enter the **quit** or **exit** commands.
   - Alternatively, use the **CTRTL+Q** or **ALT+F4** keyboard shortcuts.

   Command: **quit** *(Press ENTER.)*

2. If nothing has changed in the drawing since it was opened, AutoCAD exits. In all other cases, AutoCAD displays the dialog box:

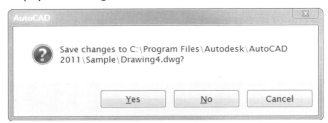

3. To save the drawing, choose **Yes**. AutoCAD displays the Save Drawing As dialog box. Enter a name, and then click **Save**.

   To not save the drawing, choose **No**. AutoCAD exits.

   To not save the drawing and return to AutoCAD, choose **Cancel**.

**Note** The CLOSEALL command closes all drawings. QUIT closes all drawings *and* exits AutoCAD.

## EXERCISES

1. In the following exercises, you start new drawings by a variety of methods.
   Start AutoCAD.
   Which do you see: a blank drawing, or the Startup dialog box?
   If you do not see the Startup dialog box, follow the instruction given in this chapter to turn it on.

2. Once the Startup dialog box is on the screen, click the **Use a Wizard** button.
   Select **Advanced Setup**, and then enter the settings for an engineering drawing with four decimal places of precision.
   Write down a sample length expressed in the units you set: _____.
   Select Degrees-minutes-seconds with 0 decimal places of precision.
   Write down a sample angle expressed in the units you set: _____.
   Select the 0 angle direction as North and clockwise.
   Draw a sketch below showing the direction of North and the direction of positive angle measurement:

   Specify a drawing size of 11" wide and 17" long.

3. On Quick Access toolbar, click the **New** icon. What happens?
   If AutoCAD displays the Create New Drawing dialog box, select the **Use a Template** button.
   In either the Create New Drawing dialog box, or the Select Template dialog box, watch the preview window while clicking on the names of template drawings.
   When you see a template that looks interesting, open it.

4. In the following exercises, you change the units of measurement.
   a. From the **Format** menu, select **Units.**
   b. In the Length droplist of the dialog box, select **Fractional Units**.
   c. Set the precision to **1/16.**
   d. Click **OK** to exit the dialog box.
   e. Move the cursor, and watch the coordinates on the status line.
   f. What do they look like?

5. Press the spacebar to return to the Units dialog box.
    a. In the **Angle** droplist, select **Surveyor's Units**.
    b. Change the precision to two decimal places.
    c. After exiting the dialog box, start the **LINE** command, and then click a point.
    d. Move the cursor, and watch the status line. What do the coordinates look like?
    e. Move the cursor to the coordinates, and click. Does the display change?
    f. Press **CTRL+D**, and then move the cursor. Is the coordinate display different?

6. Start a new drawing, and then draw a few lines.
    a. Use the **QSAVE** command to save the drawing.
    b. Does AutoCAD ask for a name? If so, enter *save.dwg*.
    c. Repeat the **QSAVE** command. Does AutoCAD ask for a name?
    d. This time, use the **SAVE** command. Does AutoCAD ask for a name? If so, press **Cancel**.
    e. Finally, use the **SAVEAS** command. Does this work identically to the **SAVE** command?

7. If your computer is connected to others through a network, ask your instructor for the following information:

    Name of another computer.

    Name of the drive and folder in which to store drawings.

    Use the **SAVEAS** command to save the drawing from exercise #6 using the network.

8. If your computer has a connection to the Internet, ask your instructor for the following information:

    Address of the school's FTP site.

    User name and password (or email address, if an anonymous FTP site).

    Name of a folder in which to store drawings.

    Use the **SAVEAS** command to save the drawing from exercise #6 to the Web site using FTP transfer.

## CHAPTER REVIEW

1. Is it necessary to add a *.dwg* extension to the drawing name when you save drawing files?
2. What is the danger of saving an AutoCAD 2011 drawing to an earlier version?
3. Which file types does the **OPEN** command open?
4. Write out the meaning of the following file extensions:

    DWG

    DXF

    DWF

    DWT
5. Does the Select Drawing File dialog box allow you to access drawing files located on the hard drive of your coworkers' computers and at sites located on the Internet?
6. Can the **OPEN** command access Web pages?
7. Is AutoCAD limited to working with just one drawing at a time?
8. Describe how to open more than one drawing in the Select Files dialog box.
9. How can you open a 3D model provided to you in SAT format?
10. Is AutoCAD able to open files other than DWG and DXF?

    If so, how?
11. Describe how to rename, move, and delete files with the **OPEN** command.
12. When can you erase a *.ac$ file?
13. AutoCAD starts up, but does not display the Startup dialog box. Describe how to enable this dialog box.
14. Explain the difference between the **NEW** and **QNEW** commands.
15. What is the importance of the *acad.dwt* file?
16. Describe how *template* drawings are useful.
17. What is the purpose of *wizards* in AutoCAD?
18. Write out the unit of measurement for each example shown below:

    1'-2 1/2"

    12.3456

    2'-2.2"

    6.54E+03

    43 3/4
19. Which unit of measurement is used for metric drawings?
20. Which unit of measurement does AutoCAD use internally?
21. How many decimal places can AutoCAD display?

    When would you not want to work with that many decimal places?
22. When you specify two decimal places, does this restrict the accuracy of AutoCAD's calculations?
23. Write out the angle of measurement for each example shown below:

    45.00

    45d00'00.00;

    50.00g

    0.79r

    N45d0'0.00"E

24. How many *degrees* are there in a circle?

    How many *grads*?

    How many *radians*?

25. In which direction does AutoCAD think of 0 degrees, by default?
26. In which direction are *positive* angles measured, by default?
27. Can you change the direction of 0 degrees?
28. What does AutoCAD use *limits* for?
29. Does the Select File dialog box allow you to preview drawings?
30. How would you open a drawing, but prevent changes from being saved?
31. What happens when you double-click a drawing file name in the Select Files dialog box?

    What happens when you click a drawing file name twice?

32. Explain the meaning of:

    Write-protected

    Read-only

    Read-write

33. Describe how to save drawings that have been opened *read-only*.
34. What are *.bak* files?

    Why are they useful?

35. Describe how to open a *.bak* file.
36. How can you guard against losing drawing data from a power outage?
37. What happens when drawings are "saved"?
38. Which command saves drawings quickly?

    Can AutoCAD save drawings on its own?

39. Describe the difference between the **QSAVE** and **SAVE** commands.

    And the difference between the **SAVE** and **SAVEAS** commands.

40. In the Select File dialog box, displayed by the **OPEN** command, does the **Folders** sidebar provides quick access to:

    My Documents

    History

    FTP

    Desktop

    Buzzsaw

41. Can you create new folders with the Select Files dialog box?
42. Explain when you would use the **Find** tool.
43. What is the difference between *anonymous* and *private* FTP sites?
44. Describe the difference between the two types of *.dxf* file:

    ASCII

    Binary

45. Name the command that belongs to each keyboard shortcut:

    **CTRL+S**

    **CTRL+SHIFT+S**

    **CTRL+O**

    **CTRL+N**

    **CTRL+Q**

46. Can you change the background color of the drawing area?
    If so, why might you want to?
47. Describe the purpose of the **UNITS** command.
48. How does the **UNIT** command's **Unitless** setting help the operator insert blocks into drawings?
49. Work out the scale factors for drawing these objects on a sheet of paper that is 10 inches wide. Show all your calculations.
    a. An automobile and trailer 12 feet long.
    b. A tool shed 10 feet wide.
    c. A telephone 9 inches long.
50. Work out the scale factors for the following objects drawn on a sheet of paper that is 100cm wide. Show all your calculations.
    a. A pen holder 7 cm in diameter.
    b. A stereo system 42 cm wide.
    c. A house 16m long.
51. Which system variable sets the scale factor for the following objects:
    Hatch patterns
    Linetypes
    Dimensions
    Text
52. Explain why text cannot be drawn full-size in drawings scaled at 1:50.
53. How tall should text be drawn in drawings with the following scale factors. (Show all your calculations.)
    1:50
    1:1
    10:1
    1:2500
54. If the scale factor for a drawing is 1:50, what should the scale factor be for hatch patterns?
55. If the scale factor for linetypes is 10:10, what should the scale factor be for hatch patterns?
56. Under what conditions can drawings be saved to others' computers?
57. What happens when you use the **SAVE** command to save a drawing, when another drawing of the same name already exists?
58. Describe what happens when you quit AutoCAD without saving the drawing.
59. Explain a benefit and drawback to FTP.
    What is FTP short for? When might you use FTP?

# II

## Drafting Essentials

# CHAPTER 4
## Drawing Basic Objects

Now that you know how to set up new drawings, you can begin creating them. Creating drawings with AutoCAD involves using a primary set of commands. These drafting commands construct the objects basic to most drawings — lines, arcs, circles, and so on.

In this chapter, you learn how to draw basic objects with these commands:

- **LINE** draws line segments.
- **RECTANG** constructs squares and rectangles.
- **POLYGON** constructs regular polygons, from 3 to 1,024 sides.
- **CIRCLE** constructs circles by several methods.
- **ARC** constructs arcs by many methods.
- **DONUT** draws thick and solid-filled circles.
- **PLINE** draws connected lines, arcs, and curves.
- **ELLIPSE** constructs ellipses and elliptical arcs.
- **POINT** draws point objects.
- **U** and **UNDO** reverse the effects of most commands.
- **REDO** and **MREDO** reverse the effect of the undoing.

**LINE**

The LINE command draws line segments.

Constructing lines in drawings is the most basic CAD operation. In AutoCAD, you can draw many types of lines (as listed below) and apply a variety of options to them, but most lines are drawn with the LINE command.

A single line is sometimes called a "segment." At each end, the line has endpoints.

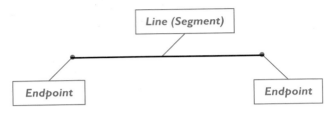

Let's talk about "first points" and "next points," because you see these terms a lot in AutoCAD. To draw lines and other objects, you first must show AutoCAD the point from which to start. AutoCAD calls this the "first point." Think of this as the place where you initially put pencil to paper.

Next, determine the point at which the line segment should end. AutoCAD calls this is the "next point."

The line segment is drawn. AutoCAD then repeats the 'Specify next point:' prompt so that you can draw another line segment without the inconvenience of restarting the LINE command.

AutoCAD repeats the 'Specify next point:' prompt until you press ESC to end the command. Each segment is an independent line.

 **Note** AutoCAD has commands that draw different types of lines. You encounter some of these commands in this chapter:

**MLINE** (multiline) draws as many as 16 parallel lines as a single object.

**PLINE** (polyline) draws lines in the same manner as the LINE command, but a polyline can include arcs, splines, and variable widths. The 3DPOLY command draws three-dimensional polylines.

**SKETCH** draws freehand lines.

**SPLINE** draws splined curves.

**XLINE** (construction line) and **RAY** draw infinite construction lines (xlines) and semi-infinite construction lines (rays).

Drawing Basic Objects    131

## TUTORIAL: DRAWING LINE SEGMENTS

1. To draw line segments, start the **LINE** command with one of these methods:
   - In the ribbon's 2D Home tab, choose the **Line** button in the Draw panel.
   - Or, at the 'Command:' prompt, enter the **line** command.

     Command: **line** *(Press ENTER.)*
   - Alternatively, enter the **l** alias at the 'Command:' prompt.

2. Specify the starting point of the line segment:

   Specify first point: *(Pick point 1, or specify coordinates.)*
   - With the cursor, pick a point on the screen.
   - Or, at the keyboard, enter x, y coordinates, such as **2,3**.

3. Specify the next point(s):

   Specify next point or [Undo]: *(Pick point 2, or specify coordinates.)*
   - Move the cursor, and then pick another point.
   - Or, enter another set of x, y coordinates, such as **3,4**.

4. Press **ENTER** to end the command.

   Specify next point or [Undo]: *(Press ENTER to exit the command.)*

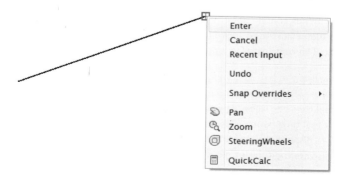

Alternatives to pressing **ENTER** include:
- Pressing **ESC**.
- Right-clicking, and selecting **Enter** from the shortcut menu.

5. To repeat the **LINE** command, press **ENTER**.

   AutoCAD restarts the **LINE** command and displays "LINE" at the 'Command:' prompt.

   Command: LINE Specify first point:

6. This time, draw several lines. Notice that each new segment connects precisely at the endpoint of the previous segment. The connection is called a "vertex."

7. Press **ENTER** to exit the command.

   In AutoCAD, pressing **ENTER** ends (most) commands; pressing it again restarts the same command.

## DRAWING LINE SEGMENTS: ADDITIONAL METHODS

The LINE command provides additional options for drawing lines and ending the command:

- **Close** draws another line to the start of the first one automatically.
- **Undo** undraws the last segment.
- **@** and **<** draw with relative coordinates and angles.
- **ENTER** continues from the last segment.

Shortcut menus (below, at left) and drafting tooltips (at right) list some (but not all) of the options found in commands.

**Note** Shortcut menus contain options specific to the command, as well as options that can apply to any command. In the menu illustrated above (at left), four options belong to the **LINE** command: **Enter**, **Cancel**, **Close**, and **Undo**. The remaining five options apply to any command:

**Recent Input** — lists the last 20 commands and options you entered.
**Snap Overrides** — displays a submenu of object snap modes. (See a later chapter.)
**Pan** — enters real time pan, allowing you to shift the drawing around the viewport. (See a later chapter.)
**Zoom** — enters real time zoom, allowing you to change the apparent size of the drawing.
SteeringWheels — displays the steering wheel interface for navigating drawings.
**QuickCalc** — displays the Quick Calc window, which allows you to carry out calculations, the results of which are used as input for the active command.

Let's look at each of the options specific to the LINE command.

### Close

You can use the LINE command to construct polygons (multi-sided objects, like the outline of a stop sign). In polygons, the last line segment connects to the first one, but aligning them can be tedious. The **Close** option does this for you automatically. (This option does not appear until you have picked three points, because a minimum of two line segments are needed before a polygon can be created by the third closing segment.)

To see how this works, consider drawing a right angle triangle. The illustration shows the line segment at the final intersection being connected with the **Close** option.

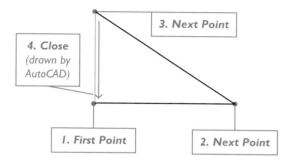

Command: **line**

Specify first point: *(Pick point 1.)*

Specify next point or [Undo]: *(Pick point 2.)*

Specify next point or [Undo]: *(Pick point 3.)*

Specify next point or [Close/Undo]: **c**

### Undo

As you draw line segments with the LINE command, you may make a mistake. Instead of canceling the command and starting over, use the **Undo** option to "undraw" the last segment. You can undo all the way back to before the first point was placed.

### Relative Coordinates and Angles

The @ and < symbols allow you to specify relative coordinates and angles. When you prefix x, y-coordinates with @, AutoCAD reads the numbers as distances, not coordinates. When you include the < symbol, AutoCAD reads the number following it as an angle.

Here are two examples of using relative coordinates:

Command: **line**

Specify first point: @2,3    *Draws the line 2 units right and 3 units up from the previous point.*

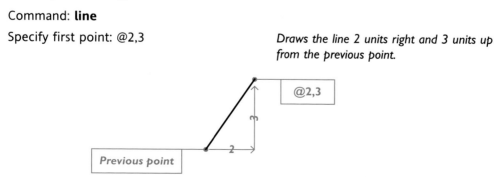

Specify next point or [Undo]: 10<45    *Draws the line 10 units long at a 45-degree angle from the last point.*

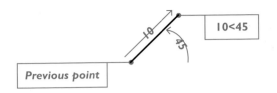

## Continue

When you terminate the LINE command, you usually begin another line somewhere else, but sometimes you want to continue from the last segment. To connect precisely, use the LINE command's hidden **Continue** option: when you next start the LINE command, press ENTER at the "Specify first point:" prompt. The start point of the new segment is located precisely at the endpoint of the last-drawn segment.

# RECTANG

The RECTANG command draws rectangles and squares by a variety of methods and in a variety of styles.

This command can draw rectangles with thin or fat lines, tilt them at an angle, and add rounded or cutoff corners. The primary parts of a rectangle are its length and width:

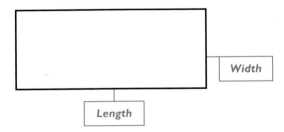

This command does not draw geometric variations of rectangles, such as parallelograms or rhomboids.

### TUTORIAL: DRAWING RECTANGLES

To draw a basic rectangle, pick two points that define the opposite corners of the rectangle.

1. Start the RECTANG command by one of these methods:
   - In the ribbon's 2D Home tab, choose the **Rectangle** button in the Draw panel.
   - Or, at the 'Command:' prompt, enter the **rectang** command.

     Command: **rectang** *(Press* ENTER.*)*
   - Alternatively, enter the aliases **rec** or **rectangle** at the 'Command:' prompt.

2. Pick a point for one corner of the rectangle, such as the lower-left corner:
   Specify first corner point or [Chamfer/Elevation/Fillet/Thickness/Width]: *(Pick point 1.)*

3. And pick a point for the opposite corner:
   Specify other corner point or [Area/Dimensions/Rotation]: *(Pick point 2.)*

**Notes** The sides of the rectangle are parallel to the x and y axes of the current UCS (user-defined coordinate system). To draw rectangles at an angle, use the command's **Rotation** option.

To draw squares, pick two points that create the square. As an alternative, you can use the POLYGON command to draw squares.

## ADDITIONAL METHODS FOR DRAWING RECTANGLES

The RECTANG command provides additional options for drawing rectangles:

- **Dimensions** lets you specify the width and height of the rectangle.
- **Area** lets you specify the area and one side of the rectangle.
- **Rotation** lets you specify the angle of the rectangle.
- **Width** lets you specify the width of the four lines making up the rectangle.
- **Elevation** draws the rectangle a specific height above the x,y-plane.
- **Thickness** draws the rectangle with a thickness in the z-direction.
- **Fillet** rounds off the corners of the rectangle.
- **Chamfer** cuts off the corners of the rectangle.

Let's look at each.

### Dimensions

The basic way to draw a rectangle is to specify the points of opposite corners. If you prefer, however, you can specify the rectangle's size by its length and width. In the tutorial below, you draw a 3" x 2" box.

1. Start the RECTANG command, pick the starting point, and then enter the **Dimension** option:
   Command: **rectang**
   Specify first corner point or [Chamfer/Elevation/Fillet/Thickness/Width]: *(Pick point 1.)*
   Specify other corner point or [Dimensions]: **d**

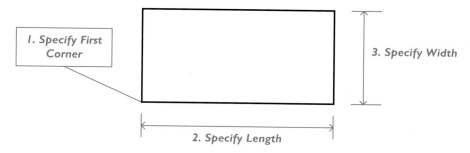

2. Enter the length, which is in the x-direction (to the right):
   Specify length for rectangles <0.0000>: **3**

3. And enter the width, which is in the y-direction (up):
   Specify width for rectangles <0.0000>: **2**

4. At this point, AutoCAD asks you to specify the orientation of the rectangle. Depending on where you pick the "other corner point," you place the rectangle in one of four positions around the first corner point:
   Specify other corner point or [Dimensions]: *(Move the cursor to position the rectangle, and then pick point 4.)*

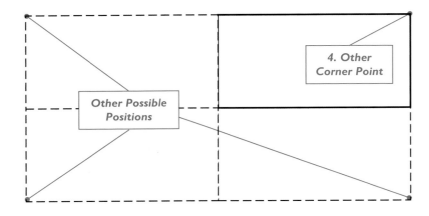

### Area

The **Area** option specifies the area of the rectangle and the length of one of its sides. After you enter the **Area** option, the command prompts you:

> Enter area of rectangle in current units <100>: *(Enter a value.)*
>
> Calculate rectangle dimensions based on [Length/Width] <Length>: *(Type **L** or **W**.)*

When you enter **L**, AutoCAD prompts:

> Enter rectangle length <10>: *(Enter a value smaller than the area.)*

If you enter **W**, AutoCAD prompts:

> Enter rectangle width <10>: *(Enter a value smaller than the area.)*

AutoCAD then repeats the earlier prompt:

> Specify other corner point or [Area/Dimensions/Rotation]: *(Pick a point or enter an option.)*

You can define the other corner of the rectangle, or specify an option, such as the rotation angle.

When the **Chamfer** or **Fillet** option is active, AutoCAD accounts for the subtracting effect of the chamfers or fillets on the area; that is, the area is made "larger" to compensate. Both rectangles illustrated below have a length of 10 units and an area of 50 square units. The black rectangle seems larger, because the fillets at each corner would otherwise reduce the area.

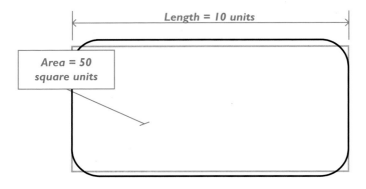

## Rotation

The **Rotation** option specifies the angle of the rectangle. AutoCAD prompts you:

> Specify rotation angle or [Points] <0>: *(Enter the angle or type **P**.)*

When you enter the angle, AutoCAD ghosts the rotated rectangle, and then prompts you for the other corner point:

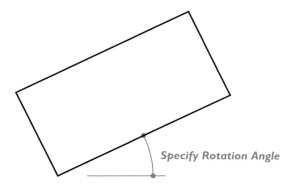

> Specify other corner point or [Area/Dimensions/Rotation]: *(Pick a point, or enter an option.)*

When you enter **P** (for the **Points** option), AutoCAD prompts you to pick two points that define the angle:

> Specify first point: *(Pick a point.)*
>
> Specify second point: *(Pick another point.)*

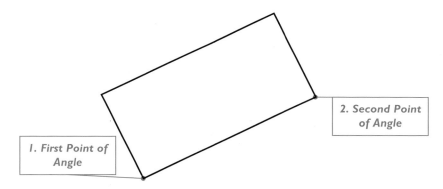

And then asks for the location of the other corner.

## Width

You can specify the width of the four lines making up the rectangle. (Technically, the "lines" are a polyline, which are discussed later in this chapter.) Once you set the width, the same value is used by subsequent RECTANG commands, until you change it.

To change the width, enter the **Width** option, and then specify a width in units:

> Command: **rectang**
>
> Specify first corner point or [Chamfer/Elevation/Fillet/Thickness/Width]: **w**
>
> Specify line width for rectangles <0.0000>: **.1**

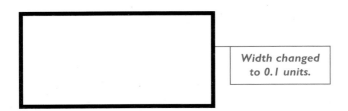

*Width changed to 0.1 units.*

### Elevation

You can elevate rectangles by specifying their height above (or below) the x, y-plane — the distance in the z-direction. Once you set the elevation, the same value is used by subsequent RECTANG commands, until you change it. To set the elevation, enter the **Elevation** option, and then specify it in units:

    Command: **rectang**

    Specify first corner point or [Chamfer/Elevation/Fillet/Thickness/Width]: **e**

    Specify the elevation for rectangles <0.0000>: **2**

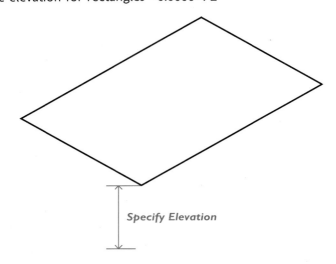

*Specify Elevation*

A positive value for elevation draws the rectangle above the x, y plane. A negative value pushes the rectangle below the x, y-plane.

### Thickness

You can turn rectangles into 3D box-like objects by specifying a thickness. This is not the recommended way to create 3D boxes, because these lack tops and bottoms. (Use the BOX command, to draw solid boxes.) Once the thickness is set, the same value is used by subsequent RECTANG commands, until you change it. To set the thickness, enter the **Thickness** option, and then specify it in units:

    Command: **rectang**

    Specify first corner point or [Chamfer/Elevation/Fillet/Thickness/Width]: **t**

    Specify thickness for rectangles <0.0000>: **2**

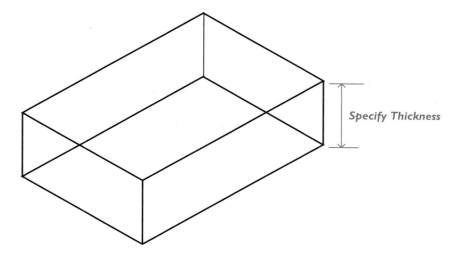

*Specify Thickness*

A positive value for thickness draws the rectangle upward from the base, while a negative value draws the rectangle downward.

### Fillet

To round the corners of rectangles, enter the **Fillet** option:

> Command: **rectang**
> Current rectangle modes: Elevation=2.0000 Thickness=2.0000 Width=0.1000
> Specify first corner point or [Chamfer/Elevation/Fillet/Thickness/Width]: **f**
> Specify fillet radius for rectangles <0.0000>: **.25**

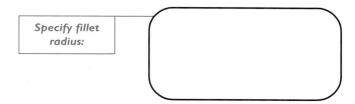

*Specify fillet radius:*

The single fillet radius applies to all four corners; you cannot specify a different radius for each corner. If the fillet radius is too large for the rectangle, AutoCAD does not draw the fillets; instead, the rectangle has square corners. Once you set the fillet, the same value is used until you change it.

You can enter a negative radius, which produces this interesting effect:

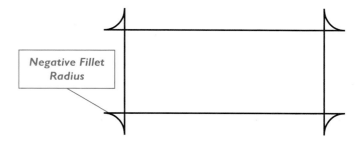

*Negative Fillet Radius*

After the rectangle is drawn, you can use grips to move the fillets around, which can result in quite a mess of a rectangle. "But," warns the technical editor, "just because something can be done doesn't always mean it should be done."

### Chamfer

To cut off the corners of the rectangle, enter the **Chamfer** option:

> Command: **rectang**
>
> Specify first corner point or [Chamfer/Elevation/Fillet/Thickness/Width]: **c**
>
> Specify first chamfer distance for rectangles <0.2500>: *(Press* ENTER.*)*
>
> Specify second chamfer distance for rectangles <0.2500>: *(Press* ENTER.*)*

The chamfer distances are measured from the corners of the rectangle. You can enter a different value for the first and second chamfer distances, as illustrated below.

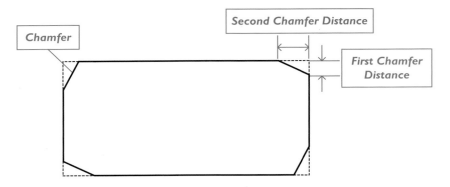

The pair of chamfer distances applies to all four corners; you cannot specify a different chamfer for each corner. If the chamfer distances are too large for the rectangle, AutoCAD does not draw the chamfers; instead, the rectangle has square corners. Once you set the chamfers, the same value is used by subsequent RECTANG commands, until you change it.

You can enter negative and positive distances, which produces this pinwheel effect:

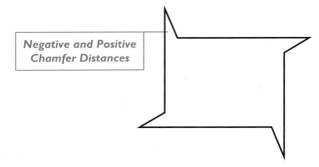

## POLYGON

The POLYGON command draws regular polygons, from 3 to 1,024 sides.

A regular polygon has all the sides the same length; in contrast, an irregular polygon has sides of different lengths. With this command, you can draw equilateral triangles, squares, pentagons, hexagons, octagons, and the like. AutoCAD draws polygons from polylines, so you can change their widths with the PEDIT command.

Polygons are constructed by any of three methods. By specifying the length of one edge, the polygon is defined, because all lengths are the same. By fitting it inside ("inscribed" within) or outside ("circumscribed" by) an imaginary circle, the polygon's size is defined by the circle's radius.

The primary parts of polygons are their vertices, edges, radii of inscribed circles, and center points, as illustrated below:

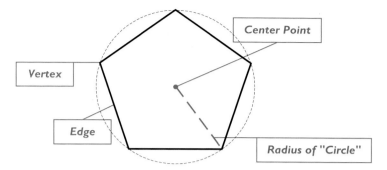

In drawing regular polygons, the word "circle" is used, because each vertex is the same distance from the center point — as is every point on a circle's circumference.

This command cannot draw irregular polygons, such as acute triangles or parallelograms; you need to construct these using AutoCAD's other commands, as described at the end of this chapter.

### TUTORIAL: DRAWING POLYGONS

1. To draw regular polygons, start the **POLYGON** command by one of these methods:
   - In the ribbon's 2D Home tab, choose the **Polygon** button in the Draw panel.
   - Or, at the 'Command:' prompt, enter the **polygon** command.

       Command: **polygon** *(Press ENTER.)*
   - Alternatively, enter the **pol** alias at the 'Command:' prompt.

2. Specify the number of sides. For example, enter **3** for a triangle, **4** for a square, **5** for a pentagon, and so on.
   Enter number of sides <4>: *(Enter a value between 3 and 1024.)*

3. Pick a point for the center of the polygon:
   Specify center of polygon or [Edge]: *(Pick a point.)*

4. Decide if the polygon fits inside (is inscribed within) or outside (circumscribes) an imaginary circle:
   Enter an option [Inscribed in circle/Circumscribed about circle] <I>: *(Type I or C.)*

5. Specify the radius of the circle, which determines the size of the polygon:
   Specify radius of circle: *(Enter a radius, or pick a point.)*

### ADDITIONAL METHODS FOR DRAWING POLYGONS

The **POLYGON** command provides options for drawing polygons:

- **Edge** defines the polygon by the length of one of its edges.
- **Inscribed in circle** fits the polygon inside an imaginary circle.
- **Circumscribed about circle** fits the imaginary circle inside the polygon.

Let's look at each option.

### Edge

Perhaps the easiest method is to construct polygons by specifying the length of one edge. Pick two points that define the length of the edge. Here is how to draw a pentagon with the **Edge** option.

1. Start the **POLYGON** command:
   Command: **polygon**

2. To draw a pentagon, enter **5** for the number of sides:
   Enter number of sides <4>: **5**

3. Select the **Edge** option by typing **e**:
   Specify center of polygon or [Edge]: **e**

4. Pick a point (or enter x, y coordinates) for the start of the edge:
   Specify first endpoint of edge: *(Pick a point.)*

5. And pick another point for the end of the edge:
   Specify second endpoint of edge: *(Pick another point.)*

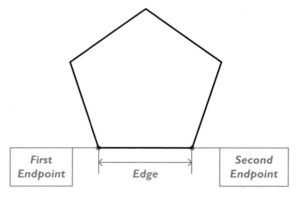

The angle of the two edge points determines the angle of the polygon. For example, to draw a diamond shape, pick the two points at a 45-degree angle (below).

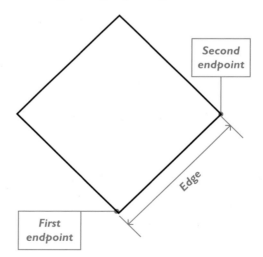

### Inscribed in Circle

Inscribed polygons are constructed inside a circle of a specified radius. (The circle itself is not drawn, but is imaginary.) The vertices of the polygon fall on the circle. Here is an example of how to draw a triangle inside a circle:

>  Command: **polygon**
>
>  Enter number of sides <4>: **3**
>
>  Specify center of polygon or [Edge]: *(Pick a point.)*
>
>  Enter an option [Inscribed in circle/Circumscribed about circle] <I>: **i**
>
>  Specify radius of circle: *(Pick a point, or enter a radius.)*

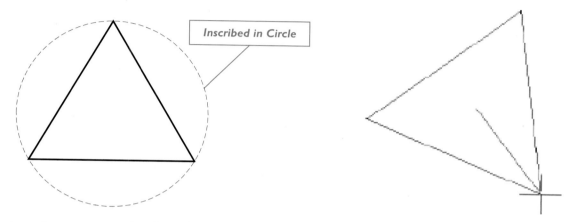

Just as the two points for the **Edge** option determine the rotation of the polygon, so too does the radius specification. As you move the cursor during the "Specify radius of circle" prompt, notice that one vertex moves with the cursor (above, right).

### Circumscribed About Circle

Circumscribed polygons are constructed outside a circle of a specified radius. The midpoints of the polygon's edges are placed on the circle's circumference. As with the inscribed option, the rotation of the polygon is determined at the "Specify radius of circle" prompt.

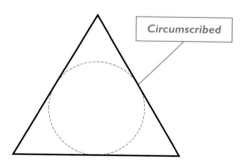

# CIRCLE

The CIRCLE command draws circles by several methods.

The important parts of the circle are its center point, the radius or diameter, and its circumference.

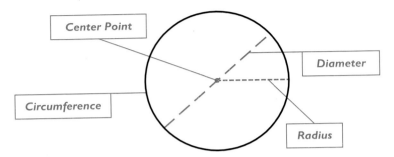

(This command does not draw isometric circles; instead, use the **Isocircle** option of the ELLIPSE command, after turning on **Iso** mode with the SNAP command.)

## TUTORIAL: DRAWING CIRCLES

1. To draw circles, start the **CIRCLE** command by one of these methods:
   - In the ribbon's 2D Home tab, choose the **Circle** button in the Draw panel.
   - Or, at the 'Command:' prompt, enter the **circle** command:

     Command: **circle** *(Press ENTER.)*
   - Alternatively, enter the **c** alias at the 'Command:' prompt.

2. Pick a point for the center of the circle:

   Specify center point for circle or [3P/2P/Ttr (tan tan radius)]: *(Pick point 1, or enter x,y coordinates.)*

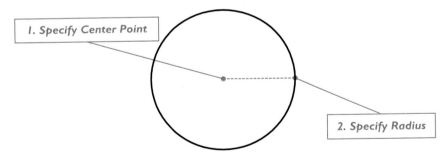

3. Enter the radius of the circle:

   Specify radius of circle or [Diameter]: *(Specify the radius, or pick point 2.)*

## ADDITIONAL METHODS FOR DRAWING CIRCLES

The CIRCLE command provides additional options for drawing circles:

- **Diameter** draws circles based on a center point and diameter.
- **2P** draws circles based on two diameter points.
- **3P** draws circles based on three points along the circumference.
- **Ttr (tan tan radius)** draws circles touching two tangent points and a radius.
- **Tan, Tan, Tan** draws circles touching three tangent points.

Let's look at each option.

### Diameter

Instead of specifying the circle's center point and radius, you can specify the diameter. (The diameter is twice the radius.)

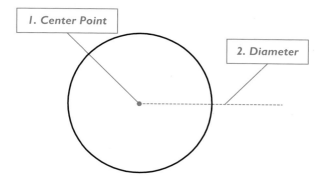

Command: **circle**
Specify center point for circle or [3P/2P/Ttr (tan tan radius)]: *(Pick point 1.)*
Specify radius of circle or [Diameter] <1.0>: **d**
Specify diameter of circle <2.0>: *(Pick point 2, or enter a value for the diameter.)*

### 2P

Instead of specifying a center point and diameter, you can specify only the diameter by picking two points on the circumference.

Command: **circle**
Specify center point for circle or [3P/2P/Ttr (tan tan radius)]: **2p**
Specify first end point of circle's diameter: *(Pick point 1.)*
Specify second end point of circle's diameter: *(Pick point 2.)*

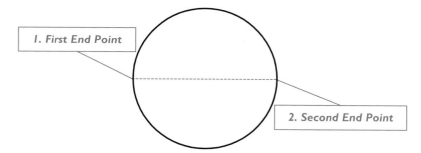

### 3P

As an alternative, you can pick any three points on the circumference, and AutoCAD constructs the circle.

Command: **circle**
Specify center point for circle or [3P/2P/Ttr (tan tan radius)]: **3p**
Specify first point on circle: *(Pick point 1.)*
Specify second point on circle: *(Pick point 2.)*

Specify third point on circle: *(Pick point 3.)*

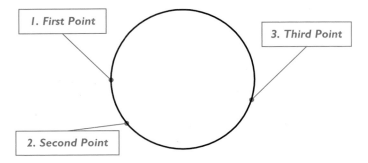

### Ttr (tan tan radius)

Often, you need to draw a circle that's precisely tangent to other objects. The **Ttr** option automatically invokes **Tangent** object snap. (A tangent is the point at which the circle touches another object.) You can draw circles tangent to other circles, arcs, lines, polylines, and so on.

Command: **circle**

Specify center point for circle or [3P/2P/Ttr (tan tan radius)]: **t**

Specify point on object for first tangent of circle: *(Pick point 1.)*

Specify point on object for second tangent of circle: *(Pick point 2.)*

Specify radius of circle <0.9980>: *(Pick point 3, or enter a value for the radius.)*

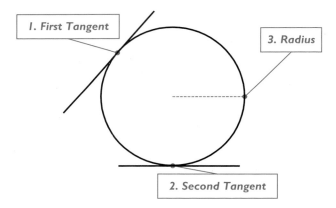

You may find it tricky placing the circle with tangents, because the radius determines where the tangents are placed.

### Tan, Tan, Tan

AutoCAD has a sixth way of drawing circles that's not available through the CIRCLE command. Instead, you have to access it from the ribbon by selecting **Circle | Tan Tan Tan.** This hidden option constructs circles tangent to three objects. After you select the command from the menu, notice that AutoCAD fills in several options for you, and automatically invokes **Tangent** object snap (displayed as "_tan").

All you do is pick three objects, as follows:

Command: _circle Specify center point for circle or [3P/2P/Ttr (tan tan radius)]: _3p Specify first point on circle: _tan to *(Pick point 1.)*

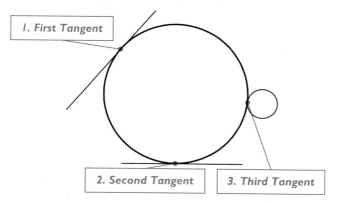

Specify second point on circle: _tan to *(Pick point 2.)*

Specify third point on circle: _tan to *(Pick point 3.)*

 ARC

The ARC command draws arcs by a variety of methods.

An arc is a portion of a circle. Like a circle, an arc has a center point and a radius (or diameter). Like lines, arcs have starting and ending points. The parts of an arc are shown below:

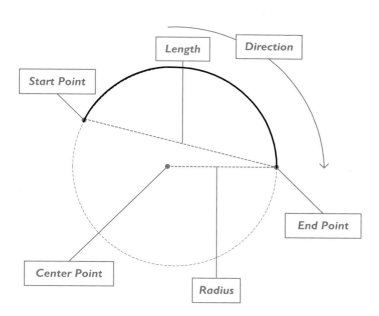

Some disciplines, such as railroad track design, define arcs by chords, the straight distance between the start and endpoints. (AutoCAD calls this distance the "length.")

If AutoCAD does not draw arcs as you expect, it may be because you are trying to draw it clockwise. It can be easy for beginners to forget that AutoCAD usually draws arcs counterclockwise from the starting point. (The exceptions are the three-point arc and the start-end-direction arc.) Sometimes, you may find it easier to draw a circle, and then use the BREAK command to create the arc.

AutoCAD constructs arcs by many methods — too many, you might think by the end of this section. This flexibility allows you to place arcs in many different situations.

**TUTORIAL: DRAWING ARCS**

1. To draw arcs, start the **ARC** command by one of these methods:
   - In the ribbon's 2D Home tab, choose the **Arc** button in the Draw panel.
   - Or, at the 'Command:' prompt, enter the **arc** command:

     Command: **arc** *(Press ENTER.)*
   - Alternatively, enter the **a** alias at the 'Command:' prompt.

2. Pick the starting point of the arc:
   Specify start point of arc or [Center]: *(Pick point 1.)*

3. Pick a point that lies on the arc:
   Specify second point of arc or [Center/End]: *(Pick point 2.)*

4. Pick a point at the end of the arc:
   Specify end point of arc: *(Pick point 3.)*

You can pick points in the drawing, or enter x ,y coordinates. The arc drawn in this tutorial is called the "three-point arc." Unlike other arcs, this one is drawn in the direction determined by the first and second points.

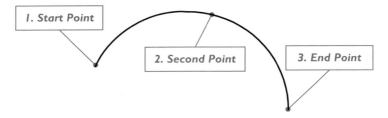

**ADDITIONAL METHODS FOR DRAWING ARCS**

The ARC command provides additional options for drawing arcs. Several of the variations are similar; thus, I have grouped them together:

- **Start, Center, End** constructs arcs from the start, center, and endpoints.
- **Center, Start, End** constructs arcs from the center, start, and endpoints.

- **Start, Center, Angle** constructs arcs from the start and center points, and the included angle.
- **Center, Start, Angle** constructs arcs from the center and start points, and the included angle.
- **Start, End, Angle** constructs arcs from the start and endpoints, and the included angle.

- **Start, Center, Length** constructs arcs from the start and center points, and the chord length.
- **Center, Start, Length** constructs arcs from the start, center, and endpoints.

- **Start, End, Radius** constructs arcs from the start and endpoints, and the radius.
- **Start, End, Direction** constructs arcs from the start and endpoints, and the direction.
- **Continue** continues the arc tangent to a line or another arc.

Let's look at each option.

## START, CENTER, END & CENTER, START, END

The most common method of drawing arcs is to specify their start, center, and endpoints. AutoCAD constructs the arc in the counterclockwise direction, from the start and to the endpoint. AutoCAD calculates the arc's radius as the distance from the start point to the center point.

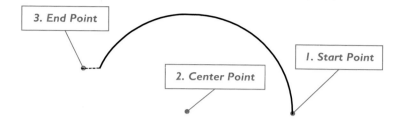

If the arc is too big or looks wrong, it could be you started the arc at the point where it should end. From the **Draw** menu, select **Arcs | Start, Center, End** or **Center, Start, End**.

    Command: **arc**

    Specify start point of arc or [Center]: *(Pick point 1, or enter x, y coordinates.)*

    Specify second point of arc or [Center/End]: **c**

    Specify center point of arc: *(Pick point 2, or enter x, y coordinates.)*

    Specify end point of arc or [Angle/chord Length]: *(Pick point 3, or enter x, y coordinates.)*

After you pick the center point, notice that AutoCAD ghosts the endpoint as you move the cursor.

The endpoint that you pick need not lie on the arc.

## START, CENTER, ANGLE & START, END, ANGLE & CENTER, START, ANGLE

Arcs are sometimes specified by their included angle. This is the angle formed between the start and endpoints, with the angle's vertex at the arc's center point.

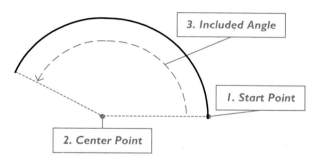

Keep in mind that AutoCAD measures the arc counterclockwise, starting from the start to the endpoint. When you enter a negative angle, AutoCAD draws the arc clockwise from the start point.

    Command: **arc**

    Specify start point of arc or [Center]: *(Pick point 1.)*

    Specify second point of arc or [Center/End]: **c**

    Specify center point of arc: *(Pick point 2.)*

    Specify end point of arc or [Angle/chord Length]: **a**

    Specify included angle: *(Enter an angle, 3.)*

## START, CENTER, LENGTH & CENTER, START, LENGTH

When you need to draw an arc with a specific chord, use the **Length** option. The length is the distance from the start point to the endpoint.

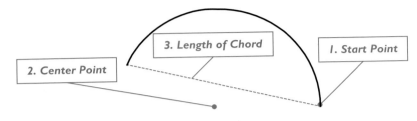

Command: **arc**
Specify start point of arc or [Center]: *(Pick point 1.)*
Specify second point of arc or [Center/End]: **c**
Specify center point of arc: *(Pick point 2.)*
Specify end point of arc or [Angle/chord Length]: **l**
Specify length of chord: *(Enter a length, 3.)*

AutoCAD does not draw the arc if the chord length does not work with the start and center points you pick. When you see "*Invalid*" on the command prompt area, the chord is too long or too short. Note that the center point is not usually on the chord.

AutoCAD constructs a major arc or a minor arc, depending on the chord length. (Given that the chord divides a circle into two parts, the major arc is the larger arc, while the minor arc is the smaller one.) When the chord is positive, AutoCAD draws the minor arc; when negative, the major arc.

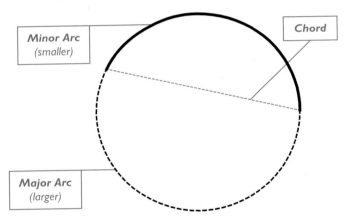

## START, END, RADIUS

This option is most like drawing an arc like a circle: you specify two points and the radius.

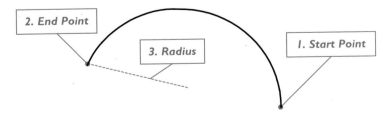

Command: **arc**

Specify start point of arc or [Center]: *(Pick point 1.)*
Specify second point of arc or [Center/End]: **e**
Specify end point of arc: *(Pick point 2.)*
Specify center point of arc or [Angle/Direction/Radius]: **r**
Specify radius of arc: *(Enter a positive or negative radius, or pick point 3.)*

By default, AutoCAD draws the minor arc. If, however, you enter a negative value for the radius, AutoCAD draws the major arc.

### START, END, DIRECTION

This option is perhaps the trickiest to understand. AutoCAD starts the arc tangent in the direction you specify.

Command: **arc**
Specify start point of arc or [Center]: *(Pick point 1.)*
Specify second point of arc or [Center/End]: **e**
Specify end point of arc: *(Pick point 2.)*
Specify center point of arc or [Angle/Direction/Radius]: **d**
Specify tangent direction for the start point of arc: *(Pick point 3.)*

At the "Specify tangent direction" prompt, move the cursor. Notice how AutoCAD ghosts the arc, depending in the cursor's location.

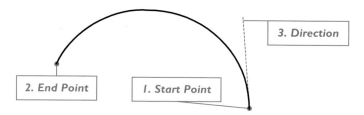

### CONTINUE

Among the ARC command's many options, one manages to stay hidden. The **Continue** option is designed to draw an arc tangent to the last-drawn object. Here's how it works:

Command: **arc**
Specify start point of arc or [Center]: *(Press ENTER.)*

Notice that AutoCAD automatically selects the end of the last line, polyline, or arc as the starting point for the new arc, which is drawn tangent to the last-drawn object. As you move the cursor, AutoCAD ghosts it in the drawing. There is just one option:

Specify end point of arc: *(Pick point 1.)*

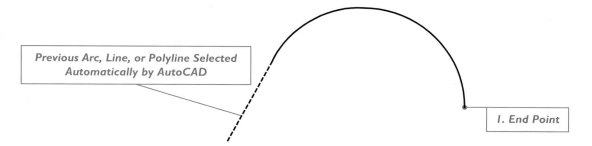

**Note** AutoCAD can convert arcs into circles, and back again. Here's how:
- To convert arcs to circles, use the **JOIN** command's **Close** option.
- To convert circles to arcs, use the **BREAK** command to remove the unwanted part of the circle.

# DONUT

The DONUT command draws circles with thick walls and solid-filled circles. These kinds of circles are useful for PCB (printed circuit board) designs. Donuts are drawn as polylines.

This is one of AutoCAD's few drawing commands that keeps on repeating itself until you press ESC or ENTER.

### TUTORIAL: DRAWING DONUTS

1. To draw donuts, start the **DONUT** command by one of these methods:
   - In the ribbon's 2D Home tab, choose the **Donut** button in the Draw panel.
   - Or, at the 'Command:' prompt, enter the **donut** command:
   
     Command: **donut** (Press ENTER.)
   - Alternatively, enter the **doughnut** alias at the 'Command:' prompt.

2. Enter a value for the donut's "hole," its inside diameter:
   Specify inside diameter of donut <0.5000>: *(Specify value, or pick two points, 1.)*

3. Enter a value for the outside of the donut:
   Specify outside diameter of donut <1.0000>: *(Specify value, or pick two points, 2.)*

4. Pick a point to place the donut.
   Specify center of donut or <exit>: (Pick point 3.)

5. This command repeats until you exit the command. Press ENTER to end it:
   Specify center of donut or <exit>: *(Press ENTER to exit command.)*

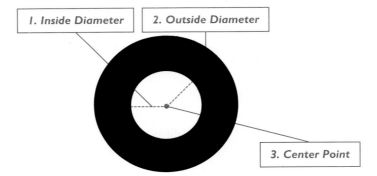

**Note** When the inside diameter equals the outside diameter, AutoCAD draws donuts as circles, albeit ones made from polylines. This is a "circle" that can be edited with the **PEdit** command. When the inside diameter is zero, AutoCAD draws solid-filled donuts, as illustrated below.

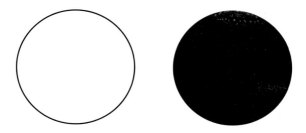

## PLINE

The PLINE command draws polylines.

We have referred to the "polyline" in some commands earlier, specifically POLYGON and DONUT. The PLINE command draws polylines, perhaps the most unique and flexible object created by any CAD program.

A polyline is a single object that consists of connected lines and arcs. It can be curved, splined, and open or closed (like an irregular polygon). You can specify the width of each segment, or give each segment a tapered width.

(*History*: until lineweights were introduced to AutoCAD, polylines were the primary way to draw objects with width.)

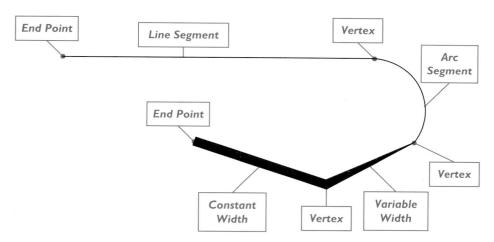

### TUTORIAL: DRAWING POLYLINES

1. To draw polylines, start the **PLINE** command.
   - In the ribbon's 2D Home tab, choose the **Polyline** button in the Draw panel.
   - Or, at the 'Command:' prompt, enter the **pline** command.

      Command: **pline** *(Press ENTER.)*
   - Alternatively, enter the **pl** alias at the 'Command:' prompt.

2. Pick a point from which to start drawing the polyline:
   Specify start point: *(Pick a point.)*

   AutoCAD reminds you of the current line width, which is 0, unless you change it with the **Width** option:
   Current line-width is 0.0000

3. Pick the next point, or select an option:
   Specify next point or [Arc/Halfwidth/Length/Undo/Width]: *(Pick a point, or enter an option.)*

4. Continue picking points, and then press **ENTER** to exit the command:

   Specify next point or [Arc/Close/Halfwidth/Length/Undo/Width]: *(Press* **ENTER** *to exit the command.)*

### ADDITIONAL METHODS FOR DRAWING POLYLINES

The **PLINE** command provides additional options for drawing polylines:

- **Undo** removes the previous segment.
- **Close** draws a segment to the start point.
- **Arc** switches to arc-drawing mode.
- **Width** specifies the width of the polyline.
- **Halfwidth** specifies the halfwidth.
- **Length** draws a tangent segment of specific length.

Let's look at each option.

### Undo

As you draw polyline segments, you may sometimes make a mistake. Instead of canceling the command and starting over, use the **Undo** option to "undraw" the last segment. You can undo all the way back to the first segment.

### Close

When you construct irregular polygons using the **PLINE** command, the last segment connects to the first segment. Aligning the ends of segments can be tedious, so the **Close** option automatically performs this for you. The **Close** option appears after you pick three points, because AutoCAD needs a minimum of two segments before it can close the polygon with a third segment.

### Arc

The **Arc** option switches the **PLINE** command to arc-drawing mode. This mode has options similar to those of the **ARC** command, with some additions:

Specify next point or [Arc/Halfwidth/Length/Undo/Width]: **a**

[Angle/CEnter/Direction/Halfwidth/Line/Radius/Second pt/Undo/Width]:

The **Width** and **Halfwidth** options specify the width of the arc segments, as described below. See the **ARC** command for the meaning of the other options.

The **Line** option switches out of arc-drawing mode and back to line drawing mode.

### Width

The **Width** option allows you to specify the width of each segment. In addition, you can specify a different starting and ending width, which creates tapers. By default, the width is 0 in new drawings. When changed, AutoCAD remembers the previous width setting.

1. Start the **PLINE** command, and pick a starting point:

   Command: **pline**

   Specify start point: *(Pick a point.)*

2. Notice the current line width. Enter **w** to access the **Width** option:

   Current line-width is 0.1000

   Specify next point or [Arc/Halfwidth/Length/Undo/Width]: **w**

3. Enter a value for the "starting width." This is the width at the starting end of the polyline.
   Specify starting width <0.1000>: *(Press ENTER to accept the default, or enter another value.)*
4. Press ENTER if you want the ending width to be the same as the starting width. For a taper, enter a different value for the ending width:
   Specify ending width <0.1000>: **.25**
5. Pick the next point, or select another option:
   Specify next point or [Arc/Halfwidth/Length/Undo/Width]:

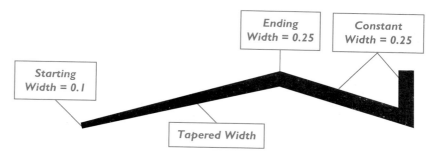

**Notes** When you provide different values for the starting and ending widths, AutoCAD draws a taper but for that segment only! The ending width becomes the next starting width.

When two polyline segments have a width other than 0, AutoCAD automatically bevels the vertex between them. To see the beveled vertices, set the **FILLMODE** system variable to 0, followed by the **REGEN** command.

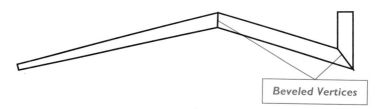

To select a polyline of width other than 0, you must pick its edge. Picking the centerline of the polyline fails. For some reason, AutoCAD is unable to detect the inside of polylines.

## Halfwidth

The **Halfwidth** option is identical to **Width**, except that you specify the width from the edge of the polyline to its centerline.

Specify next point or [Arc/Halfwidth/Length/Undo/Width]: **h**
Specify starting half-width <0.1250>: *(Enter a value, or press ENTER to keep the default.)*
Specify ending half-width <0.1250>: *(Press ENTER for constant width; enter a different value for tapered width.)*

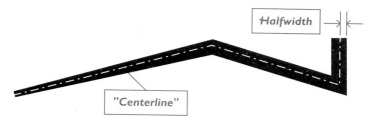

### Length

The **Length** option draws the next segment at the same angle as the previous one. You specify the length. This option is most useful when the previous segment is an arc, because the segment is drawn tangentially to the arc's endpoint.

Specify endpoint of arc or
[Angle/CEnter/CLose/Direction/Halfwidth/Line/Radius/Second pt/Undo/Width]: **line**
Specify next point or [Arc/Close/Halfwidth/Length/Undo/Width]: **length**
Specify length of line: **3**

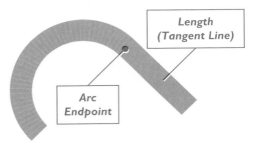

## ELLIPSE

The **ELLIPSE** command draws ellipses and elliptical arcs. When isometric mode is turned on, this command adds the **Isocircle** option for drawing isometric circles.

Ellipses are elongated circles drawn with two diameters called "axes." The major axis is the longer axis; the minor axis, the shorter.

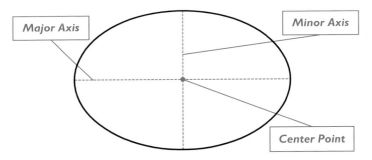

An ellipse is sometimes referred to by its rotation angle. A circle is an ellipse that has not been rotated: its rotation is zero degrees; the major and minor axes have the same length.

An ellipse is a circle that is rotated, or viewed at an angle. For instance, tilt this page away from you and the circle looks like an ellipse. A 40-degree ellipse is a circle that has been rotated by 40 degrees about the major axis. AutoCAD allows you to specify ellipse rotations between 0.0 (a circle) and 89.4 degrees (a very thin ellipse).

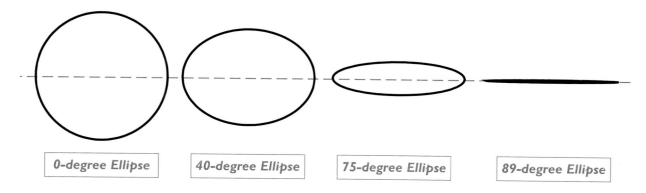

## TUTORIAL: DRAWING ELLIPSES

1. To draw ellipses, start the **ELLIPSE** command.
   - In the ribbon's 2D Home tab, choose the **Ellipse** button in the Draw panel.
   - Or, at the 'Command:' prompt, enter the **ellipse** command.

     Command: **ellipse** *(Press* ENTER.*)*
   - Alternatively, enter the **el** alias at the 'Command:' prompt.

2. Pick a point to indicate an endpoint of one axis.
   Specify axis endpoint of ellipse or [Arc/Center]: *(Pick point 1.)*

3. Pick another point for the other end of the axis.
   Specify other endpoint of axis: *(Pick point 2.)*

4. Show the half-distance to the other axis.
   Specify distance to other axis or [Rotation]: *(Pick point 3.)*

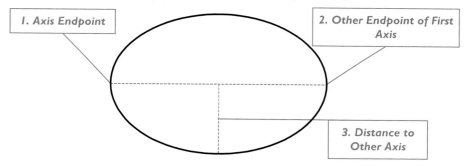

The "distance to other axis" does not need to be perpendicular to the first axis.

## ADDITIONAL METHODS FOR DRAWING ELLIPSES

The ELLIPSE command contains options that provide additional methods for drawing ellipses:

- **Center** starts with the center point of the ellipse, followed by the axes.
- **Rotation** starts with the rotation angle about the major axis.
- **Arc** constructs an elliptical arc.

### Center

The **Center** option allows you to pick the center of the ellipse, and then to specify distance from the center point to the endpoints of the axes.

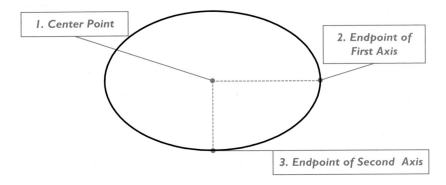

```
Command: ellipse
Specify axis endpoint of ellipse or [Arc/Center]: c
Specify center of ellipse: (Pick point 1.)
Specify endpoint of axis: (Pick point 2.)
Specify distance to other axis or [Rotation]: (Pick point 3.)
```

As you move the cursor during the "Specify distance to other axis" prompt, AutoCAD ghosts in the ellipse. Note that AutoCAD doesn't care which axis you create first.

### Rotation

The **Rotation** option rotates the ellipse about the major axis; it replaces the prompts for drawing the minor axis. (This option is also available when constructing ellipses with the **Center** option.) As you move the cursor during the "Specify rotation" prompt, AutoCAD ghosts the image of the ellipse.

```
Command: ellipse
Specify axis endpoint of ellipse or [Arc/Center]: (Pick point 1.)
Specify other endpoint of axis: (Pick point 2.)
Specify distance to other axis or [Rotation]: r
Specify rotation around major axis: (Pick point 3, or enter an angle.)
```

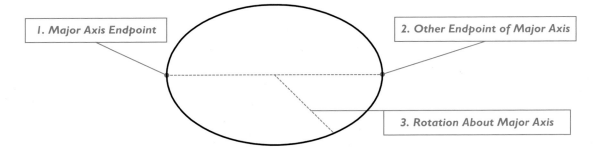

### Elliptical Arc

The **Arc** option draws elliptical arcs. The first few times you try drawing an elliptical arc, it might not turn out as desired. It takes some practice! Or, draw an ellipse and then trim or break it to an arc.

Command: **ellipse**
Specify axis endpoint of ellipse or [Arc/Center]: **a**
Specify axis endpoint of elliptical arc or [Center]: *(Pick point 1.)*
Specify other endpoint of axis: *(Pick point 2.)*
Specify distance to other axis or [Rotation]: *(Pick point 3.)*
Specify start angle or [Parameter]: *(Pick point 4.)*
Specify end angle or [Parameter/Included angle]: *(Pick point 5.)*

As you move the cursor during the "Start angle" and "End angle" prompts, AutoCAD ghosts a preview of the elliptical arc. The figure below shows an arc that consists of the upper half of an ellipse.

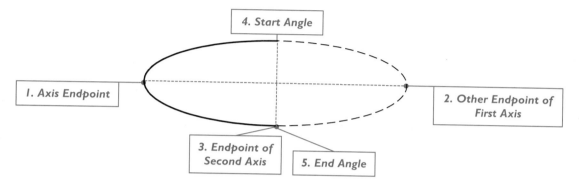

The **Arc** option has three methods of determining the arc: **Angle**, **Included Angle**, and **Parameter**.

The **Angle** option specifies the start angle and the end angle of the arc. The two angles are measured relative to the first axis endpoint. The start angle determines where the arc starts; the end angle where it ends. When you specify 0 degrees for the start angle, the arc starts at the first axis endpoint. AutoCAD draws the arc counterclockwise. If you run into trouble, it's probably because you are trying to draw the arc clockwise.

The **Included Angle** option defines the arc by an angle that starts from the start angle (described above). The angle is relative to the start angle. AutoCAD prompts:

Specify end angle or [Parameter/Included angle]: **i**
Specify included angle for arc <180>: *(Enter an angle, or pick a point.)*

The **Parameter** option uses this formula to construct the elliptical arc :

$p(u) = c + a \times \cos(u) + b \times \sin(u)$

where:
a is the major axis.
b is the minor axis.
c is the center point of the ellipse.

Does anyone use this formula in the real world? Probably not. In any case, AutoCAD prompts you:

Specify start angle or [Parameter]: **p**
Specify start parameter or [Angle]: *(Pick a point.)*
Specify end parameter or [Angle/Included angle]: *(Pick a point.)*

Notice how you can switch back and forth between the three arc definitions – Angle, Included Angle, and Parameter.

## Isocircle

The **Isocircle** option appears in the ELLIPSE command's prompts only when isometric drafting mode is turned on with the DSETTINGS command.

## POINT

The POINT command constructs point objects.

Points have no height or width: they are dots. Points are the smallest object that output devices can produce: single pixels on the screen, single dots on printed paper.

Because points are so small, they can be hard to see. Use the pdmode and pdsize system variables to change the look and size of points. As illustrated below, points can be invisible, or combinations of lines, circles, and squares. The effect of the two system variables is retroactive, meaning changing their values changes the look and size of all previously-drawn points.

### TUTORIAL: DRAWING POINTS

1. To draw points, start the **POINT** command.
   - In the ribbon's 2D Home tab, choose the **Point** button in the Draw panel's flyout.
   - Or, at the 'Command:' prompt, enter the **point** command:

     Command: **point** *(Press ENTER.)*
   - Alternatively, enter the **po** alias at the 'Command:' prompt.

2. Notice that AutoCAD displays the values of the **PDMODE** and **PDSIZE** system variables.
   Current point modes: PDMODE=0 PDSIZE=0.0000

3. Pick a point to place the point.
   Specify a point: *(Pick a point.)*

### ADDITIONAL METHODS FOR DRAWING POINTS

The POINT command does not contain any options, but there are related commands that provide additional methods for drawing points:

- **MULTIPLE** command repeats the **POINT** command, as well as many other commands.
- **DDPTYPE** command changes the look of points.

### Multiple

The MULTIPLE command repeats the POINT command, which is handy when you want to place more than one point at a time quickly. At the command prompt, enter:

Command: **multiple**
Enter command name to repeat: **point**
Current point modes: PDMODE=0 PDSIZE=0.0000
Specify a point: *(Pick a point.)*
POINT Current point modes: PDMODE=0 PDSIZE=0.0000
Specify a point: *(Press ESC or ENTER to exit repeating command.)*

The command repeats until you press ESC or ENTER.

## DdPType

The DDPTYPE command displays a dialog box that allows you to select the style and size of points. (DDPTYPE is short for "dynamic dialog point type.")

There are limitations: you can only choose from the styles listed in the dialog box, which excludes custom point styles; all points in the drawing take on the same style and size, which means different points cannot have different styles.

Enter DDPTYPE at the 'Command:' prompt to display the dialog box:

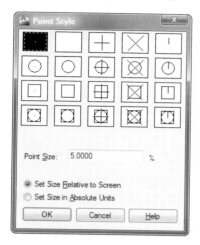

The upper half of the dialog box displays the point styles you can choose from. Note that the first style (the single dot) cannot be resized; the second style (the blank square) creates invisible points, which also cannot be sized.

The lower half changes the size of the point. The **Point Size** option sets the size as a percentage relative to the screen size or in units. The defaults are 5% and 5 units. I recommend you stick with the **Set Size Relative to Screen** (percentage) setting, because the **Absolute Units** setting can create points that are too large to see (when zoomed in) and too small to see (when zoomed out).

Click **OK** to accept the changes; AutoCAD automatically regenerates the drawing so that you can see the changes to the point display.

## U AND UNDO

The U and UNDO commands reverse the effect of many (not all) commands.

> **U** — undoes the last action; this is the quick command to use when just one or two changes need to be undone.
>
> **UNDO** — provides many options for undoing the effects of commands; this is the advanced command that provides utter control.

You can draw, edit, and then undo your work — then redo it. This is useful if you have just performed an operation that you wish to reverse, either a mistake or a what-if scenario. After undoing one or more operations, you can use the REDO command once to reverse the undo, while MREDO redoes multiple undoes.

The U command reverses the effect of the most recent command. You can execute a series of U commands to back up through a string of changes. (The U command is not an alias for the more advanced UNDO command, although it does function identically to the UNDO command's **1** option.)

Undoing a command restores the drawing to the state before the command was executed. For example, erase an object, and then execute the U command: the object is restored. When you scale an object, and then undo it, the object is scaled back to its original size. Undoing a just-completed BLOCK command restores the block, and deletes the block definition that was created, leaving the drawing exactly as it was before the block was inserted.

At the command prompt, the U command lists the command that is undone to alert you to the type of command that was affected.

### TUTORIAL: UNDOING COMMANDS

1. Open a drawing, and erase a part of it with the **ERASE** command.
2. To reverse the erasure, start the **U** command with one of these methods:
   - From the Edit menu, choose **Undo**.
   - On the Quick Action toolbar, choose **Undo**.
   - At the keyboard, press **CTRL+Z**.
   - At the 'Command:' prompt, enter the **u** command.

     Command: **u** *(Press* ENTER.*)*

3. AutoCAD undoes the effect of the **ERASE** command, and then displays its name:
   ERASE

   When you undo all the way back to the first command, AutoCAD reports:
   Everything has been undone

**Note** *Several commands cannot be undone.* **SAVE**, **PLOT**, *and* **WBLOCK**, *for example, are unaffected, because AutoCAD cannot "unsave" or "unplot." If you attempt to use the* **U** *or* **UNDO** *command after these commands, the name of the command is displayed, but the command's action is not undone.*

(The way to undo a save is to retrieve the backup copy, as described later in this chapter. The way to undo a plot is to fold up the paper and then throw it in the recycling bin, suggests the tech editor.)

**UNDO** and **U** have no effect on commands that change the arrangement of windows (as opposed to viewports), and redraw or regenerate (such as **REGEN**) the drawing. The two undo commands also have no effect on commands that took place before the drawing was opened. For example, you work on a drawing, save it, and then close the drawing. When you open the drawing again, AutoCAD cannot undo the commands of the previous editing session.

### UNDOING COMMANDS: ADDITIONAL METHODS

AutoCAD has several additional methods to undo the effects of commands:

- **UNDO** command controls the undo process at the command line.
- **Undo** and **Previous** options undo actions in commands.

Let's look at each.

### Undo

The UNDO command provides fine control over the undo process.

> Command: **undo**
>
> Enter the number of operations to undo or [Auto/Control/BEgin/End/Mark/Back] <1>: *(Enter a number or an option.)*

By default, the UNDO command operates like the U command: press ENTER at the "Enter the number of operations to undo" prompt, and AutoCAD undoes the last command (if possible). Enter a number, such as **4**, to undo the last four commands. This is the same as entering U four times.

### Auto

The **Auto** option groups all the actions of a single command into a single undo. The **Auto** option is not available when the **Control** option is turned off.

> Enter UNDO Auto mode [ON/OFF] <On>: *(Enter **ON** or **OFF**.)*

### Control

The **Control** option limits the effect of the UNDO command:

> Enter an UNDO control option [All/None/One] <All>: *(Enter an option.)*

The **All** option turns on the UNDO and U commands, and allows them to undo all the way to the beginning of the drawing session.

The **None** option turns off the UNDO and U commands, grays out the **Undo** button on the Standard toolbar, and discards the undo history. Use this option only if your computer is low on disk space.

The **One** option restricts the UNDO command to a single undo, and makes the **Auto**, **Begin**, and **Mark** options unavailable.

### BEgin/End

The **BEgin** option groups several commands into a set.

> Command: **undo**
>
> Enter the number of operations to undo or [Auto/Control/BEgin/End/Mark/Back] <1>: **be**

After you enter the **BEgin** option, the UNDO command ends, but AutoCAD starts recording all the commands you enter, until you enter the **End** option:

> Command: **undo**
>
> Enter the number of operations to undo or [Auto/Control/BEgin/End/Mark/Back] <1>: **e**

The UNDO, REDO, and U commands now treat the set of commands as a single undo:

> Command: **u**
>
> GROUP

### Mark/Back

The **Mark** option places a marker in the undo collection:

> Command: **undo**
>
> Enter the number of operations to undo or [Auto/Control/BEgin/End/Mark/Back] <1>: **m**

The **Back** option undoes all commands back to the marker:

> Enter the number of operations to undo or [Auto/Control/BEgin/End/Mark/Back] <1>: **b**
>
> Mark encountered

You may place as many marks as you require; the **Back** option moves back through the undo collection one mark at a time, removing each mark.

 **Note** It is dangerous to use the UNDO command's **Back** option without marks, because it undoes every action in the current editing session:

> Command: **undo**
>
> Enter the number of operations to undo or [Auto/Control/BEgin/End/Mark/Back] <1>: **b**
>
> This will undo everything. OK? <Y> (Enter **Y**.)
>
> Everything has been undone

### Combined Zooms and Pans

When you perform several zooms and pans in a row, undoing them could be annoying if AutoCAD were to redraw each view change. Fortunately, AutoCAD groups all sequential view changes into a single undo through the **Combine zoom and pan commands** option. You find it in the Undo/Redo section of the User Preferences tab of the Options dialog box.

### Undo, Previous Options and Cancel Buttons

Several commands include methods of undoing operations within them.

### Undo Option

Commands that execute more than one action often include the **Undo** option. This allows you to reverse the effect of the last action without leaving the command. For example, the LINE command's **Undo** option erases the last-drawn segment. Other commands that have the undo option include PLINE, PEDIT, and 3DPOLY.

### Previous Option

Instead of an undo option, other commands have an equivalent option called "previous." The ZOOM command's **Previous** option, for example, displays the previous view, whether created by the ZOOM, PAN, or -SHADEMODE command. This option remembers up to ten previous views.

Other commands with a **Previous** option include SELECT and UCS. One command is itself a "previous" command: LAYERP restores previous layer states.

### Cancel Button

Many dialog boxes have a button labeled **Cancel**. Click this button after making changes to the dialog box, if you don't want to keep the changes.

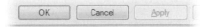

# REDO AND MREDO

The REDO and MREDO (short for "multiple redo") commands are the antidotes to the U and UNDO commands: they reverse the undoing.

- **REDO** — redoes the last undo; this is the quick command to use when just one or two changes need to be reversed.
- **MREDO** — provides a couple of options for redoing the effects of undo; this advanced command is also available as a droplist on the toolbar.

The REDO command must be used immediately after one of the undo commands.

### TUTORIAL: REDOING UNDOS

1. To undo an undo command, start the **REDO** command with one of these methods:
   - From the Edit menu, choose **Redo**.
   - At the keyboard, press **CTRL+Y**.
   - On the Quick Action toolbar, choose **Redo**.
   - At the 'Command:' prompt, enter the **redo** command.

   Command: **redo** *(Press ENTER.)*

2. AutoCAD redoes the previous undo, and then displays the name of the redone command:
   ARC

   When you redo back to the first undo, AutoCAD reports:
   Everything has been redone

## MRedo

The MREDO command provides additional control over the redo process.

Command: **mredo**

Enter number of actions or [All/Last]: *(Enter a number or an option.)*

Enter the number of undone commands you want reversed.

The **All** option reverses all previous undo actions.

The **Last** option reverses the last undo action only; this is like the REDO command.

## CONSTRUCTING OTHER OBJECTS

This chapter has introduced you to AutoCAD's basic drawing commands. These allow you easily to draw lines, circles, rectangles, regular polygons, arcs, ellipses, elliptical arcs, polylines, and points. They don't, however, make it easy to draw other common 2D geometric shapes, such as isosceles and right triangles, spirals, parallelograms; as well as 3D objects, such as cubes and cylinders. You learn about 3D objects in a later chapter of this book.

To draw other kinds of 2D shapes, for example the four kinds of triangle, you need to construct them, using one or more commands available in AutoCAD:

>**Equilateral** — all three sides the same length.
>
>**Isosceles** — two sides the same length.
>
>**Scalene** — all three sides at different lengths.
>
>**Right Angle** — one angle at 90 degrees.

As you may recall from your high school geometry class, triangles are drawn with a combination of lengths and angles, and sometimes require the use of sine, cosine, and tangent calculations. Although it's not obvious, AutoCAD provides all the tools you need to draw any kind of triangle — without getting out the scientific calculator.

### EQUILATERAL TRIANGLES

All three sides of equilateral triangles have the same length. (Again, if you recall your high school geometry class, the three interior angles are all 60 degrees.) This kind of triangle is most easily drawn with the POLYGON command set to draw 3 sides. When you know the length of the sides, use the **Edge** option. For example, if the sides are 2.5 units long, use direct distance entry to specify the length, as follows:

1. Start the **POLYGON** command:
   Command: **polygon**

2. To draw triangles, specify 3 sides:
   Enter number of sides <4>: **3**

3. By using the **Edge** option, you specify the length of one side; AutoCAD draws the other two sides to match.
   Specify center of polygon or [Edge]: **e**
   Specify first endpoint of edge: *(Pick point 1.)*
   Specify second endpoint of edge: *(Move cursor in a direction, and then enter length.)* **2.5**

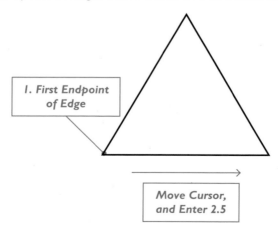

AutoCAD draws the equilateral triangle for you.

## ISOSCELES TRIANGLES

Two sides of isosceles triangles have the same length (the two opposite angles also being the same). One method to draw this triangle is to use the LINE, MIRROR, and TRIM commands. For example, consider a triangle with a base of 3 units and isosceles angles of 63 degrees.

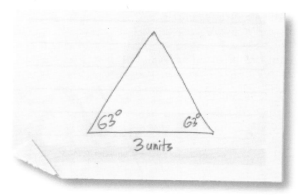

We don't know the length of the other two sides, so we draw them an arbitrary length, and then trim or extend to fit. For the best result, turn on polar mode.

1. Start the **LINE** command to draw the base and one angled leg:

    Command: **line**

    Specify first point: *(Pick point 1.)*

2. Use direct distance entry to draw the base 3 units long:

    Specify next point or [Undo]: *(Move cursor to the right, and then enter length.)* **3**

3. Use relative coordinates to draw one leg 10 units long at 63 degrees. The 10 units is arbitrary, because we do not know its length; we edit the length later.

    Specify next point or [Undo]: **10<63**

    Specify next point or [Close/Undo]: *(Press ENTER to exit command.)*

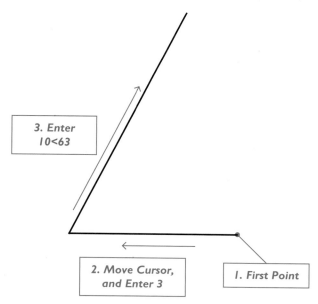

4. Start the **MIRROR** command to make a mirror copy of the angled leg:

    Command: **mirror**

5. Enter L to select the last-drawn object:

   Select objects: **L**

   1 found Select objects: *(Press* **ENTER** *to end object selection.)*

6. Use MIDpoint object snap to create an imaginary vertical mirroring line:

   Specify first point of mirror line: **mid**

   of *(Pick point 4.)*

   Specify second point of mirror line: *(Pick point 5.)*

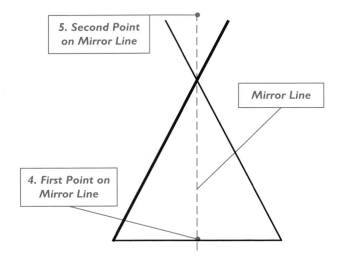

7. Keep the source object, and exit the command:

   Delete source objects? [Yes/No] <N>: *(Press* **ENTER** *to accept default of* **No**, *and then exits the command.)*

8. Use the **TRIM** command to trim back the two angled lines to their intersection point:

   Command: **trim**

   Current settings: Projection=UCS, Edge=None

   Select cutting edges ...

   (Filleting with radius = 0 also works.)

9. Select all the lines. They become each other's cutting edge:

   Select objects or <select all>: *(Press* **ENTER** *to select all objects.)*

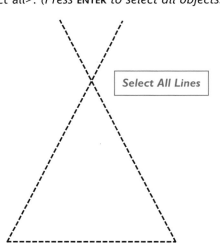

10. Select the two line extensions to trim them. Be sure to pick the portion of the lines to be trimmed away:

   Select object to trim or shift-select to extend or
   [Fence/Crossing/Project/Edge/eRase/Undo]: *(Select end of one angled line, 6.)*
   Select object to trim or shift-select to extend or
   [Fence/Crossing/Project/Edge/eRase/Undo]: *(Select end of the other line, 7.)*

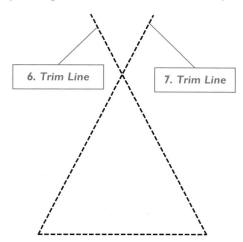

   Select object to trim or shift-select to extend or
   [Fence/Crossing/Project/Edge/eRase/Undo]: *(Press ENTER to exit command.)*

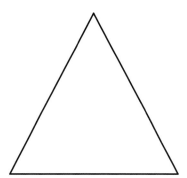

The isosceles triangle is complete. The technical editor found another method that requires just one application of the LINE command:

1. Set running object snap to MIDpoint. Turn on object tracking.
2. Start the **LINE** command to draw the base and one angled leg:

   Command: **line**

   Specify first point: *(Pick point 1.)*
3. Use direct distance entry to draw the base 3 units long:

   Specify next point or [Undo]: *(Move cursor to the right, and then enter length.)* **3**
4. Draw one leg any length at 63 degrees.

   Specify next point or [Undo]: **<63**
5. Pick up the tracking point at the midpoint of the first line.
6. Track to and select the upper vertex.
7. Enter **c** to close the polygon.

## SCALENE TRIANGLES

All three sides and corners of scalene triangles have different lengths and angles. This kind of triangle is best drawn with the LINE command, using a combination of direct distance entry and polar coordinates, as required. Take, for example, a triangle with the following specifications:

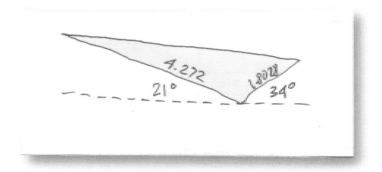

1. Start the LINE command:
   Command: **line**
   Specify first point: *(Pick point 1).*

2. Use polar coordinates (which are also relative) to draw the short leg at right:
   Specify next point or [Undo]: **1.8028<34**

3. End the LINE command, and start over to draw the next leg (at left):
   Specify next point or [Undo]: *(Press ENTER.)*
   Command: *(Press ENTER.)*
   LINE Specify first point: **endp**
   of *(Again, pick point 1).*

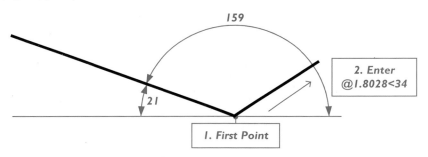

4. Use polar coordinates to draw the other leg. The angle of 159 degrees = 180 - 21 = 159.
   Specify next point or [Undo]: **4.272<159**

5. Use ENDpoint object snap to draw the third leg:

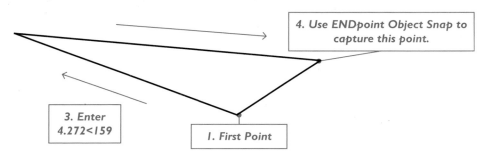

Specify next point or [Undo]: **end**

of *(Pick point 4.)*

Specify next point or [Close/Undo]: *(Press ENTER.)*

### RIGHT ANGLE TRIANGLES

One corner of right angle triangles is at 90 degrees; the two remaining corners have other angle that add up to 90 degrees, such as the same angle (45 degrees each, making them isosceles triangles), or two different angles, making them scalene.

This kind of triangle is easily drawn with the LINE command. When you know the length of two sides, such as 3 and 2 units, follow these steps:

1. Turning on ortho mode helps you draw the 90-degree angle easily:

    *(Press F8 to turn on ortho mode.)*

2. Start the LINE command:

    Command: **line**

    Specify first point: *(Pick point 1.)*

3. Notice that it is easier to draw this kind of triangle when you start at a corner away from the right angle, and then draw toward the right angle.

    Specify next point or [Undo]: *(Move cursor to the left, and then enter length.)* **3**

    Specify next point or [Undo]: *(Move cursor up, and then enter length.)* **2**

4. By using the **Close** option, you don't have to work out the length of angle of the third side.

    Specify next point or [Close/Undo]: **c**

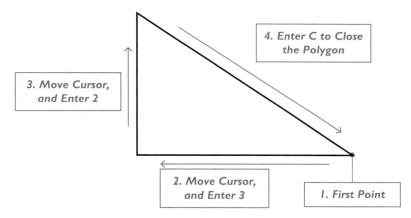

The right angle triangle is complete.

## EXERCISES

1. With the **LINE** command, draw the house shape shown below:

    Command: **line**

    Specify first point: *(Enter point 1.)*

    Specify next point: *(Enter point 2.)*

    Specify next point: *(Enter point 3.)*

    Specify next point: *(Enter point 4.)*

    Specify next point: *(Enter point 5.)*

    Specify next point: *(Enter point 1 again.)*

    Notice that the line stretched behind the crosshairs. This is called "rubber banding."

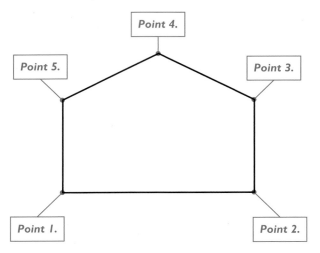

2. Using the figure below, fill in the missing absolute coordinates. Each side of the floor plan is dimensioned. Place the answers in the boxes provided. Assume the origin is at the lower left corner.

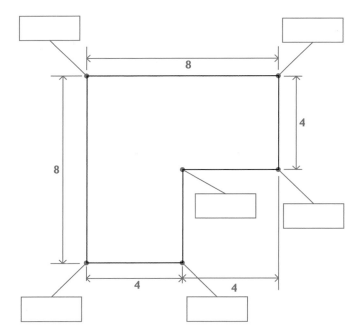

3. Using the LINE command, connect the following points designated by absolute coordinates.

   Point 1:   1,1
   Point 2:   5,1
   Point 3:   5,5
   Point 4:   1,5
   Point 5:   1,1

   What shape did you draw?

4. Determine the length of each side of the floor plan in the figure below. (Calculate the lengths from the absolute coordinates given.)

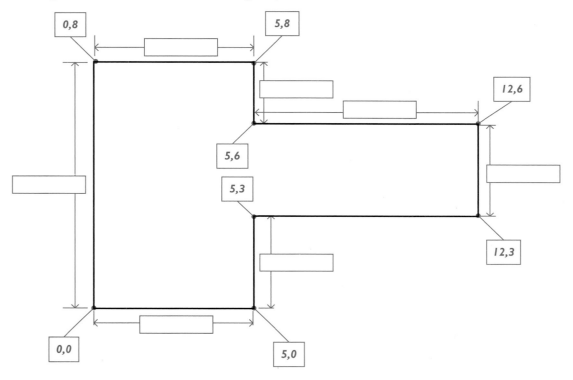

5. Use the following absolute and relative coordinates to draw an object.

   Point 1:   0,0
   Point 2:   @3,0
   Point 3:   @0,1
   Point 4:   @−2,0
   Point 5:   @0,2
   Point 6:   @−1,0
   Point 7:   0,0

6. List the relative coordinates used to draw the following base plate.

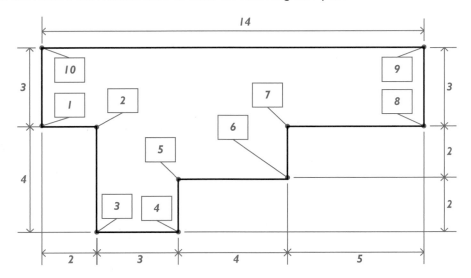

1. _____ , _____
2. _____ , _____
3. _____ , _____
4. _____ , _____
5. _____ , _____
6. _____ , _____
7. _____ , _____
8. _____ , _____
9. _____ , _____
10. _____ , _____

7. List the polar coordinates used to draw the base plate above.

8. Write a list of the absolute coordinates used to construct the shim shown in the figure below. The origin is at the lower left corner.

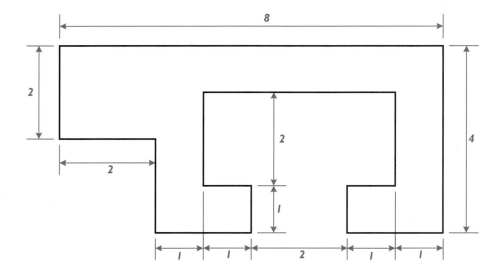

9. Use the following relative distances and angles to draw an object.

   Point 1: **0,0**
   Point 2: **@4<0**
   Point 3: **@4<120**
   Point 4: **@4<240**

10. Write a list of the relative coordinates used to construct the support plate shown in the figure below. The origin is at the lower left corner.

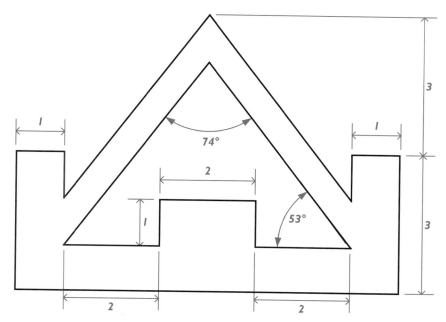

11. Write a list of polar coordinates used to construct the corner shelf shown in the figure below.

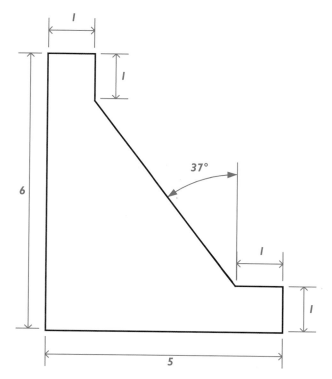

12. Set up a point type of 66, and then place it in the drawing:
    Command: **pdmode**
    New value for PDMODE <default>: **66**

    Recall that **PDMODE** affects all points in the drawing. Points placed previously are updated with the next command that causes a regeneration, such as **REGEN**. To place the point in the drawing, enter:
    Command: **point**
    Current point modes:  PDMODE=66  PDSIZE=0.0000
    Specify a point: *(Pick a point.)*

13. Construct a circle with a *radius* of 5. From the Home tab's **Draw** panel, choose **Circle | Center, Radius**.
    Command: _circle
    Specify center point for circle or [3P/2P/Ttr (tan tan radius)]: *(Pick a point.)*
    Specify radius of circle or [Diameter]: **5**

14. Construct a circle using a center point and a *diameter* of 3.
    Command: **circle**
    Specify center point for circle or [3P/2P/Ttr (tan tan radius)]: (Enter point 1.)
    Specify radius of circle or [Diameter]: **d**
    Specify diameter of circle <5.0000>: **3**

15. With the **RECTANG** command, draw two rectangles. One rectangle represents a B-size drawing sheet (17" x 11"); the second represents the title block in the lower right corner (4" x 2").

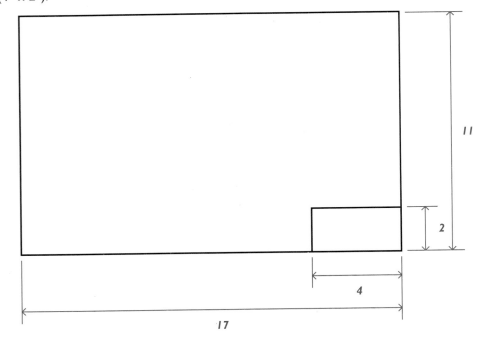

16. With the **RECTANGLE** command, draw the outline of a standard 24" x 36" speed limit sign; do not draw the text.

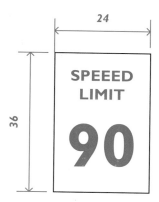

17. Use the **POLYGON** command to draw the outline of a standard 30" warning sign; do not draw the text:

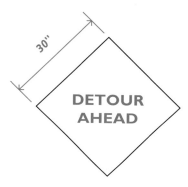

18. Use the **POLYGON** command to draw the outline of a standard 30" Stop sign; do not draw the text:

19. With the **POLYGON** command, draw the outline of a standard 36" Yield sign:

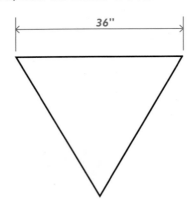

20. Draw donuts with the following diameters:
    a. ID = 1.0; OD = 1.5
    b. ID = OD = 2.5
    c. ID = 0.0; OD = 1.0

21. Use drawing commands discussed in this chapter to draw the following baseplates:
    a. Rectangular base plate with two holes:

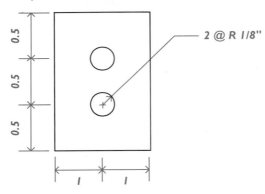

    b. Rectangular base plate with four holes:

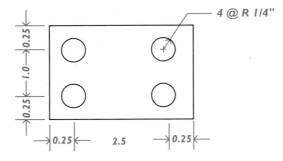

22. Use drawing commands discussed in this chapter to draw the following baseplates:
    a. Circular base plate with two square holes:

    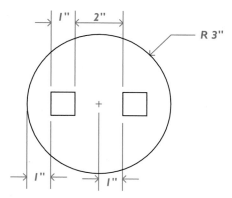

    b. Elliptical base plate with two square holes. The minor radius is 3", and the major radius is 4.5":

    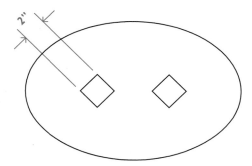

23. Use the **ELLIPSE** command to draw the face of an alien. Use the figure below to help you. Don't worry about the size.

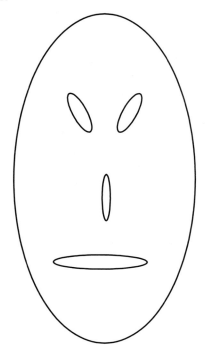

24. Draw a compact disc, which has an outer diameter of 4.75" and an inner diameter of 0.6". What are the equivalent measurements as radii?

25. Draw arcs with the following specifications:

    a.   Start point    **7,5**
         Second point  **9.5,3.5**
         Endpoint     **6,2**

    b.   Center point  **4,4**
         Start point    **4,2**
         Angle         **270**

    c.   Start point    **10.5,5.5**
         Center point  **9,6**
         Length       **3**

26. Draw circles with the following specifications:

    a.   Center point  **5,5**
         Radius       **3.5**

    b.   Center point  **5,5**
         Diameter    **3.5**

    c.   2P
         First point    **10,5**
         Second point  **3,2**

    d.   3P
         First point    **4,7**
         Second point  **6,5**
         Third point   **5,2**

27. Draw the 4.5" x 2.75" electrical cover plate shown below. The screw holes are 1/8" in diameter, and 2-5/16" apart. The light switch cover plate has an opening of 0.95" x 0.4".

28. With the **Line** and **Arc** options of the **PLINE** command, draw the outline of a standard credit card — 3.35" wide by 2.15" high, with a 0.1"-radius arc at each corner.

29. Draw the inserts for CD cases:
    a. The cover insert is 12cm x 12cm.
    b. The back cover insert is 15cm wide x 11.7cm tall; the two spine strips are 6.4mm wide.

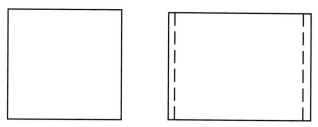

*Left:* Cover Insert = 12x12cm
*Right:* Back Cover Insert = 15x11.7cm

30. Draw the bicycle shown in the figure below. Use polylines for the frame, lines for the spokes, and donuts for the tires.

## CHAPTER REVIEW

1. What is another name for a single line?
2. Is using the **Close** option of the **LINE** command more accurate at closing the polygon than attempting to line up endpoints manually?
3. What does the **@** symbol mean?

    The **<** symbol?
4. How do you continue drawing a line *tangent* to a previously-drawn arc?
5. Name the best command for drawing the following triangles:

    Equilateral triangle

    Scalene triangle
6. What is AutoCAD's default method of constructing arcs?
7. What command can be used to change the look and size of points?
8. When rectangles are drawn with the **RECTANG** command, can each corner have a different fillet radius?
9. The **POLYGON** command uses three methods to determine the size of polygons. Explain the meaning of the methods:

    Edge

    Inscribed

    Circumscribed
10. The **PDMODE** and **PDSIZE** system variables are *retroactive*. What does this mean?
11. There are six methods of constructing circles. What four circle properties are used in various combinations to comprise these methods?

    a.

    b.

    c.

    d.
12. Describe the meaning of these options for drawing circles:

    2P

    3P

    Ttr
13. Describe a method of drawing arcs that does not involve the **ARC** command.
14. List three methods for ending the **LINE** command?

    a.

    b.

    c.
15. In which direction does AutoCAD draw arcs (by default).
16. What is the difference between a *minor* arc and a *major* arc?
17. What is the *chord* of an arc?
18. How do you continue drawing an arc tangent to a previously-drawn arc or line?
19. What does the **ARC** command's **Direction** option specify?
20. What must you do to end the **DONUT** command?

21. What is the alias for the following commands:

    **ARC**

    **LINE**

    **CIRCLE**

    **PLINE**

22. Where on the ribbon do you find commands for drawing objects?
23. Name the three kinds of circles drawn by the **DONUT** command:

    a.

    b.

    c.

24. What do the **POLYLINE, DONUT,** and **POLYGON** commands have in common?
25. Can polylines have varying width?
26. What is the purpose of the **PLINE** command's **Arc** option?
27. When selecting a wide polyline, where must you pick it?
28. Describe the purpose of the **PLINE** command's **Length** option.
29. What is the difference between the *minor* axis and the *major* axis of an ellipse?
30. What is another name for an ellipse with a rotation of 0 degrees?
31. Which must you specify first, the minor axis or the major axis of an ellipse?
32. Describe the purpose of the **MULTIPLE** command.
33. List two methods of drawing circles without using the **CIRCLE** command:

    a.

    b.

34. What is the effect of chamfers and fillets on the **RECTANGLE** command's **Area** option?
35. Does the **POLYGON** command draw scalene triangles?
36. Does the **RECTANGLE** command draw rhomboids?

# CHAPTER 5
## Drawing with Precision

One major advantage of CAD over hand drafting is precision. Instead of drawing with an accuracy to the nearest pencil (or pen) width, AutoCAD draws with an accuracy to 14 digits*. In this chapter, you learn to use some of AutoCAD's commands for drawing with precision, and to view the drawing in different ways:

**GRID** displays a grid of dots or lines (changed in AutoCAD 2011).
**DSETTINGS** controls drafting settings (changed in AutoCAD 2011).
**SNAP** specifies the cursor increment and additional drafting options.
**ORTHO** constrains cursor movement to the horizontal and vertical.
**POLAR** helps draw at specific angles.
**OSNAP** helps accurately select geometric features.
**OTRACK** implements object snap tracking.
**RAY** and **XLINE** place construction lines in the drawing.
**REDRAW** and **REGEN** clean up and update the drawing.
**ZOOM** enlarges and reduces the view of the drawing.
**PAN** moves the view around.
**NAVBAR** toggles the navigation bar (new in AutoCAD 2011).
**CLEANSCREENON** minimizes the user interface to maximize the drawing.
**VIEW** stores and recalls views of drawings by name, and controls viewport backgrounds.

---

**NEW TO AUTOCAD 2011 IN THIS CHAPTER**
- New **NAVBAR** command accesses the zoom, pan, navigation wheel, orbit, and show motion functions.
- Grid now displays as lines in model space, block editor, and layouts.
- **DSETTINGS** command's dialog box has new options for grids, 3D object snap, and selection cycling.

GRID

The GRID command toggles (turns on and off) the grid, a display of dots or lines. The grid is like a sheet of graph paper that helps you see the horizontal and vertical frames of reference. Normally, it displays a grid of lines; for compatibility with older releases of AutoCAD, it can displays dots (the intersections of the lines).

The grid gives you a feeling for the size of the drawing and for distances within the drawing. Because it is a visual aid, the grid is not plotted.

You can change the spacing of major and minor grid lines, their angle, style (standard or isometric), extent, and color.

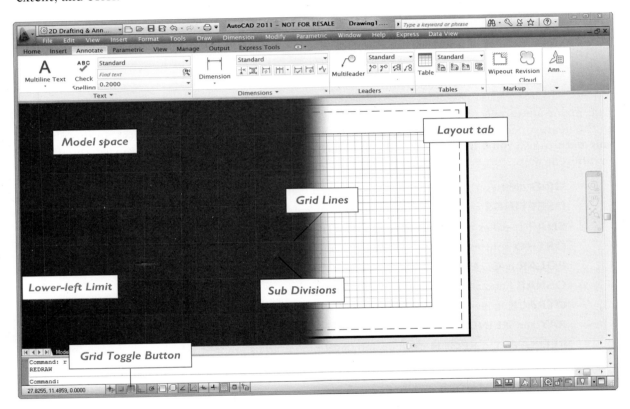

*Left:* Grid in model space.
*Right:* Grid in layout tabs.

## TUTORIAL: TOGGLING THE GRID

1. To turn the grid on and off, use the **GRID** command by one of these methods:
   - From the status bar, click the **GRID** or ▦ button (turned on when button looks blue).
   - On the keyboard, press function key **F7** or **CTRL+G**.
   - Or, at the 'Command:' prompt, enter the **grid** command:
   
   Command: **grid** *(Press ENTER.)*

2. If using the **GRID** command, enter the **ON** option to turn on the grid:
   Specify grid spacing(X) or [ON/OFF/Snap/Major/aDaptive/Limits/Follow/Aspect] <0.5>: **on**
   
   Notice that the grid is displayed.

3. To turn *off* the grid, repeat the command with the **OFF** option:
   Command: **grid** *(Press ENTER.)*
   Specify grid spacing(X) or [ON/OFF/Snap/Major/aDaptive/Limits/Follow/Aspect] <0.5>: **off**

## CONTROLLING THE GRID DISPLAY: ADDITIONAL METHODS

When you start new drawings in AutoCAD, the grid spacing is 0.5 units. You can increase and decrease the spacing, make the spacing the same as the snap distance (as detailed later in this chapter), and change the spacing in the x and y directions separately.

To effect these changes, you can use a dialog box or the command line:

- **DSETTINGS** displays a dialog box for changing grid settings.
- **GRID** specifies the grid spacing and aspect ratio via the command line.
- **LIMITS** controls the extent of the grid.

Let's look at how each command affects the grid display.

### DSettings Command

The DSETTINGS command (short for "drafting settings") displays a dialog box with settings that control many aspects of the grid display. The dialog box lists some of the options found in the GRID and SNAP commands.

1. To access the Drafting Settings dialog box, so the following:
   - The easiest way is to right-click **GRID** (or  ) on the status bar. Notice the shortcut menu; select **Settings**.

   - At the 'Command:' prompt, enter the **dsettings** command.
     Command: **dsettings**
   - Or, enter one of the aliases — **ds**, **se** (short for "settings"), or **ddrmodes** (the old name for this command) at the 'Command:' prompt.

   Notice that AutoCAD displays the Drafting Settings dialog box. (If necessary, choose the **Snap and Grid** tab.)

2. In the right half of this dialog box, you can change the grid settings:
   **Grid On (F7)** — toggles the grid display. The "(F7)" notation reminds you that you can also press function key **F7** to turn the grid on and off.
   **Grid Style** — section determines when to display the legacy dotted grid (new to AutoCAD 2011): in 2D model space, the block editor, and/or in layouts.
   **Grid X spacing** — specifies the horizontal spacing between grid lines or dots. When you enter a value for the x spacing, AutoCAD automatically changes the y-spacing to match, if the **Equal X and Y spacing** option is turned on (a check mark shows).
   **Grid Y spacing** — specifies the vertical spacing between lines. This is useful for those disciplines, such as highway design, that use a different scale in the x-direction than in the y-direction. To make the vertical spacing different from the horizontal, you must turn off the **Equal X and Y spacing** option (under Snap spacing).

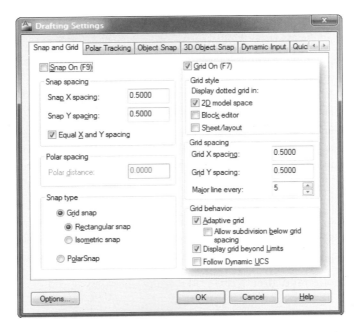

**Major line every** — specifies the number of subdivision lines per major line; subdivisions are not used by the dotted grid.

**Adaptive grid** — makes the grid spacing wider when zoomed out. (AutoCAD no longer complains that the grid is too dense to display as it did in earlier releases.)

**Allow subdivision below grid spacing** — displays more and more minor grid lines as you zoom into the drawing.

**Display grid beyond limits** — allows the grid display to go beyond the limits set by the LIMITS command.

**Follow Dynamic UCS** — forces the grid to match the x, y-orientation of the dynamic UCS, when activated.

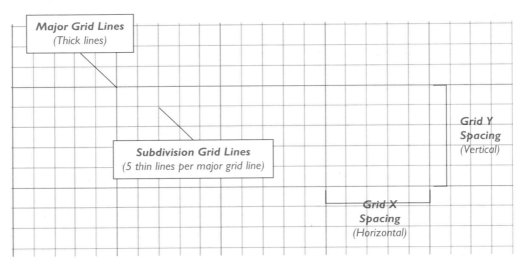

3. To accept the changes, click **OK**.

### Grid Spacing

The GRID command lets you relate the grid spacing to that of the snap distance:

1. To make the grid spacing the same as the snap distance, use the command's **Snap** option:

    Command: **grid** *(Press ENTER.)*

    Specify grid spacing(X) or [ON/OFF/Snap/Major/aDaptive/Limits/Follow/Aspect] <0.5>: **s**

2. To set the grid spacing as a multiple of the snap distance, add an "x" to the GRID command's default option. For example, to make the grid spacing twice that of the snap spacing, enter **2x**, as follows:

    Command: **grid** *(Press ENTER.)*

    Specify grid spacing(X) or [ON/OFF/Snap/Major/aDaptive/Limits/Follow/Aspect] <0.5>: **2x**

### LIMITS

The grid could extend nearly infinitely in all directions, and so the extent of the grid is limited by the LIMITS command. Typically, you want the grid covering the area in which you are drafting. Often, the lower-left corner is kept at 0,0 and only the location of the upper-right corner is changed, as follows:

> Command: **limits**
>
> Reset Model space limits:
>
> Specify lower left corner or [ON/OFF] <0.0000,0.0000>: *(Press ENTER, or enter x, y coordinates.)*
>
> Specify upper right corner <12.0000,9.0000>: *(Enter x, y coordinates.)*

**Notes** The grid can be set independently in each viewport, layout, and drawing.

Grid lines are colored gray, but you can change the color in the Display tab of the OPTIONS dialog box: click the **Colors** button. The grid lines that lie along the x and y axes are colored differently: red for x and green for y, which matches the color coding of the UCS icon.

## SNAP

The SNAP command specifies the cursor increment.

For example, when you are drafting a drawing to the nearest one inch, it's useful to set the snap to 1". When turned on, snap restricts the cursor to the interval you specify — 1" in this example. As you move the cursor, it seems to jump from point to point.

Snap is often used in conjunction with the grid. As with the grid, you can change the snap's spacing, aspect ratio, and so on. Similarly, the snap can be set individually for each viewport, layout, and drawing. (The LIMITS command does not apply to snap.)

### TUTORIAL: TOGGLING SNAP MODE

1. To turn snap on and off, invoke the **SNAP** command by one of these methods:
   - On the status bar, click the **SNAP** or [icon] button (turned on when buttons looks blue).
   - On the keyboard, press function key **F9** or **CTRL+B**.
   - Or, at the 'Command:' prompt, enter the **snap** command.

        Command: **snap** *(Press ENTER.)*

   - Alternatively, enter the **sn** alias at the 'Command:' prompt.

2. When using the **SNAP** command, use the **ON** option to turn on snap:

   Specify snap spacing or [ON/OFF/Aspect/Rotate/Style/Type] <0.5000>: **on**

   When you move the mouse, notice that the cursor jumps.

3. To turn off snap, repeat the command with the **OFF** option:

   Command: **grid** *(Press* ENTER.*)*

   Specify snap spacing or [ON/OFF/Aspect/Rotate/Style/Type] <0.5000>: **off**

   Notice that the cursor no longer jumps.

### CONTROLLING THE SNAP: ADDITIONAL METHODS

When you start new drawings in AutoCAD, the default snap spacing is set to 0.5 units — the same as the grid. You can increase and decrease the spacing, make the spacing the same as the grid distance, or specify different spacing in the x and y directions. To make these changes, you can use a dialog box, or enter the changes at the command line, using these commands:

- **DSETTINGS** displays a dialog box for changing snap settings.
- **SNAP** specifies the snap spacing, aspect ratio, and modes via the command line.

Let's look at how each command affects the snap.

## DSettings

The **DSETTINGS** command displays a dialog box that controls all settings affecting the snap. (The **Snap and Grid** tab combines the **SNAP** and **GRID** commands.)

1. To access the Drafting Settings dialog box, right-click **SNAP** on the status line.

   From the shortcut menu, select **Settings**.

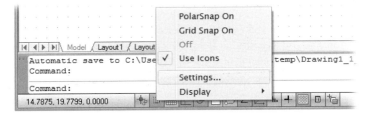

**Note** The shortcut menu (shown above) has settings that you can access without going through the dialog box:

**Polar Snap On** turns on polar snap mode, where the cursor aligns with distances and angles.
**Grid Snap On** turns on both the snap and the grid.
**Off** turns off snap mode.

2. AutoCAD displays the Drafting Settings dialog box, showing the Snap and Grid tab. In this dialog box, you can make the following changes to the snap settings:

   **Snap On (F9)** — toggles the snap on and off. The "(F9)" reminds you that you can press **F9** to turn snap mode on and off.

   **Snap X spacing** — specifies the horizontal snap increment. When you enter a value for the x-spacing, AutoCAD automatically changes the y-spacing to match, if **Equal X and Y spacing** is turned on. (The old press-the-TAB-key trick no longer works.)

   **Snap Y spacing** — specifies the vertical snap increment.

   **Equal X and Y spacing** — forces the y distance to equal that of x.

   **Grid Snap** — sets the snap spacing equal to the grid spacing, and ignores the values set earlier in the **Snap X and Y spacing** fields.

   **Rectangular Snap** — keeps standard snap, as opposed to isometric snap.

**Isometric Snap** — turns on isometric snap, which is useful for drawing isometric objects.

**Polar Snap** — sets snap to polar angles.

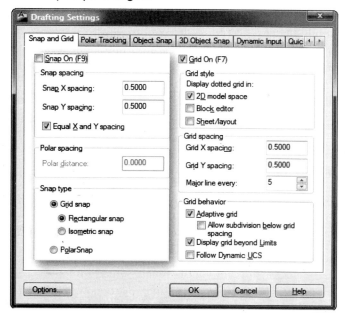

3. To accept the changes, click **OK**.

## Snap

The SNAP command has these options for changing the snap spacing:

1. To set the snap increment, use the **SNAP** command's default option. For example, to change the increment to 12 units (or inches), enter 12, as follows:

    Command: **snap** *(Press ENTER.)*

    Specify snap spacing or [ON/OFF/Aspect/Style/Type] <A>: **12**

2. To make the vertical (y) snap increment 10 times that of the horizontal (x) direction, use the **Aspect** option:

    Command: **snap** *(Press ENTER.)*

    Specify snap spacing or [ON/OFF/Aspect/Style/Type] <A>: **a**

    Specify the horizontal spacing <0.5000>: **1**

    Specify the vertical spacing <0.5000>: **10**

3. To select the type of snap mode, use the **Style** and **Type** options:
    - **Style** selects between regular and isometric modes.
    - **Type** selects between rectangular and polar modes. Here is how to enter polar mode:

        Command: **snap** *(Press ENTER.)*

        Specify snap spacing or [ON/OFF/Aspect/Style/Type] <A>: **t**

        Enter snap type [Polar/Grid] <Grid>: **p**

## TUTORIAL: CHANGING THE GRID

Although the grid is often used together with snap, the two can be used independently of each other. It is common to set the snap spacing to the drawing resolution (such as 1/8" or 1 cm), and then to set the grid spacing to a larger value (such as 1' or 1 m), so that the grid does not clutter the screen.

In this tutorial, you turn the grid display on and off, and then change its spacing.

1. Press **F7**. This key toggles the grid off and on. If the grid is on, turn it off by pressing **F7**.
2. On AutoCAD's status bar, right-click **GRID** (or ), and then choose **Settings** from the shortcut menu. Notice the Drafting Settings dialog box; choose the Snap and Grid tab, if necessary.
3. Select **Grid On** to turn on the grid display. (A check mark appears in the box.) The grid does not appear until after the dialog box closes.
4. Set the **Grid X** and **Y Spacing** to 1.0.
5. Click **OK**. Notice the grid appears as vertical and horizontal lines.
6. Press the spacebar to return to the Snap and Grid tab of the dialog box.
7. Turn off **Equal X and Y spacing**.
8. Set the **X Grid Spacing** to 10, and then the **Y Grid Spacing** to 1.

   Notice that it is not necessary for the vertical and horizontal values to be the same.
9. Click **OK**, and notice the grid spacing has changed, as illustrated below.

   (Depending on the mode AutoCAD is in, you may see grid dots instead of lines.)

## TUTORIAL: DRAWING WITH SNAP MODE

In the following tutorial, you draw a few lines with snap mode turned on.

1. Right-click the word **SNAP** (or ) on the status bar, and then choose **Settings**.
2. In the Snap and Grid tab, choose **Snap On** to turn on snap mode.

   Ensure the **Equal X and Y spacing** option is turned on, and then change the **Snap X** and **Y Spacing** to 0.25.

   Click **OK**. You won't notice any difference in the display until you start drawing.
3. Start the **LINE** command, and then pick endpoints on the screen. Notice the crosshair cursor "snapping" to specific points on the screen.

Next, you make the grid spacing equal to the snap spacing.

4. Right-click **GRID** on the status line, and then choose **Settings**.

   Set the grid spacing to 0.25. The grid display has the same spacing as the snap.

Click **OK** to exit the dialog box.

5. Use the **LINE** command to draw some lines. Notice that the crosshair cursor now lines up with the grid points.

Now you practice drawing rectangles, one with snap turned off, and another with it on.

6. Use **F9** to turn the snap mode off and on.
7. Draw rectangles with the **LINE** command — one with snap mode on and one with snap mode off. Notice how the endpoints are easier to line up with snap mode on.

Finally, rotate the snap angle with the **SNAPANG** system variable to see how much easier it is for drawing at angles. (Even though you enter system variables at the 'Command:' prompt just like commands, system variables are different from commands: they contain settings and do not execute commands.)

8. At the 'Command:' prompt, change the snap angle to 45 degrees, as follows:
   Command: **snapang**
   Enter new value for SNAPANG <0>: **45**

   Notice that the **SNAPANG** system variable affects the snap and crosshair cursor. If the grid is composed of dots, then it is also rotated. (Grid lines do not rotate, but disappear.)
9. Draw a rectangle with the rotated snap grid and cross hairs. This is an excellent method of drawing objects that have many lines at the same angle.

## ORTHO

When turned on, ortho mode (short for "orthographic") constrains cursor movement to the horizontal and vertical directions. This mode is very useful, because much drafting takes place at right angles — 0, 90, 180, and 270 degrees — as evidenced by the T-squares and right-triangles used in manual drafting.

Movement does not need to be only in the horizontal and vertical directions. Use the **SNAPANG** system variable again to change the angle, as detailed above. Alternatively, you can use the polar option, as described later.

### TUTORIAL: TOGGLING ORTHO MODE

1. To turn ortho mode on and off, use the **ORTHO** command by one of these methods:
   - From the status bar, click the **ORTHO** or ⌐ button (turned on when button looks blue).
   - On the keyboard, press function key **F8** or **CTRL+L**.
   - At the 'Command:' prompt, enter the **ortho** command.

   Command: **ortho** *(Press* ENTER.*)*
2. If you use the **ORTHO** command, enter the **ON** option to turn on ortho:
   Enter mode [ON/OFF] <ON>: **on**

   You don't notice any difference in the cursor movement until you use drawing and editing commands. Then the cursor moves only vertically or horizontally.

3. To turn off ortho mode, repeat the command with the **OFF** option:
   Command: **ortho** *(Press* ENTER.*)*
   Enter mode [ON/OFF] <ON>: **off**

Unlike the GRID and SNAP commands, the ORTHO command has no additional options. Ortho mode is, however, affected by these snap options: angle, base point, and isometric style.

**Notes** While you can rotate the ortho angle, drawing still takes place at 90 degree increments in most cases. The exception is when isometric mode is turned on: then the ortho angle is 120 degrees.

Ortho has generally been supplanted by polar mode, which is more flexible. To draw at other angles with polar tracking, also turn on Polar Snap. Ortho mode and polar tracking, however, cannot both be on at the same time: turning on one turns off the other. AutoCAD ignores ortho mode in 3D perspective views.

### TUTORIAL: DRAWING IN ORTHO MODE

1. If ortho mode is on, click **ORTHO** on the status line to turn it off.
2. Draw a rectangle with the **LINE** command. Notice that it is hard to draw lines that are perfectly horizontal and vertical.
3. Press **F8** to turn on ortho mode again.
4. Draw another rectangle with the **LINE** command. After picking the first endpoint, notice how the cursor moves precisely horizontally or vertically until you pick the next point.

**POLAR**

If ortho mode is like manual drafting with a T-square and right-triangle, then polar mode is like drafting with a set of 30-60-90 and 45-45-90 triangles. Polar mode is more flexible than ortho mode in that it guides you in drawing at any angle, such as 45 degrees or 22.5 degrees. It's like using a Vemco drafting machine to set specific angles.

When you turn on polar mode, AutoCAD automatically turns off ortho mode. That's because ortho limits cursor movement to 90 degrees, while polar is any angle,

### TUTORIAL: TOGGLING POLAR MODE

Curiously, there is no "POLAR" command, even though AutoCAD has shortcut keystrokes that turn polar mode on and off. Polar mode can be controlled by system variables and dialog boxes.

1. To turn on polar mode, use one of these methods:
   - From the status bar, click **POLAR** or [icon] (turned on when button looks blue).
   - On the keyboard, press function key **F10** or CTRL+**U**.
2. To initiate polar mode, start the **LINE** command, and then pick a point.
3. At the "Specify next point" prompt, move the cursor around. Notice that every so often a *tooltip* appears, together with an *x-marker* and an *alignment path*, as illustrated below:

*Toolips* are small rectangles with explanatory text. In this case, the tooltip displays the polar distance, such as:

**Polar: 8.8504 < 180°**

The content of the tooltip reports that the line is 8.8504 units long, at an angle of 180 degrees.

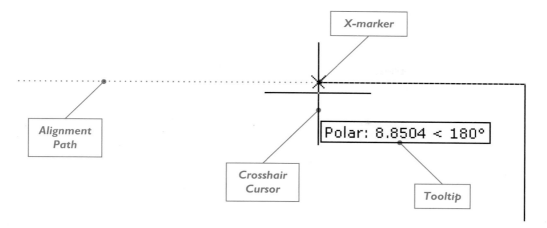

The *X-marker* emphasizes the end of the line and the start of the alignment path. It marks the point where the line would end if you were to click the mouse button.

The *alignment path* is the thin dotted line that shows you where the line would be drawn, if you were to continue in that direction — almost like a preview.

### CONTROLLING POLAR MODE: ADDITIONAL METHODS

In new drawings, the default polar angle is 90 degrees, the same as that of ortho mode. You can add angles, as well as specify whether angle measurements are absolute or relative to the last angle. The changes are made through a dialog box, or at the command line using system variables:

- **DSETTINGS** command displays a dialog box for changing snap settings.
- **POLARANG** system variable specifies the increment for polar angles.
- **POLARDIST** system variable specifies the polar snap increment, but only when the **SNAPSTYL** system variable is set to 1.
- **POLARMODE** system variable specifies a number of variables for polar snap.

## DSettings

The DSETTINGS command controls polar mode, replacing the function of the nonexistent "POLAR" command.

1. To access the Drafting Settings dialog box:
   - At the 'Command:' prompt, enter the **dsettings** command.
   - Alternatively, right-click **POLAR** on the status bar From the shortcut menu, select **Settings**.

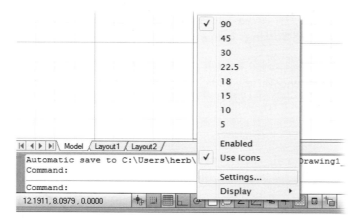

**Note** The shortcut menu (shown above) has angle settings that you can access without going to the dialog box. The numbers **90, 45, 30,** and so on, set the polar increment angle to 90, 45, 30 and other degrees.

2. In all cases, AutoCAD displays the Drafting Settings dialog box with the **Polar Tracking** tab, in which you can make the following changes to the polar settings:

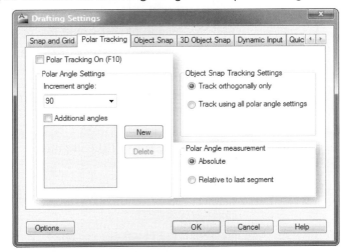

**Polar Tracking On (F10)** — toggles the snap on and off. The "(F10)" reminds you that you can press function key **F10** to turn on and off polar mode.

**Increment Angle** — specifies the polar tracking angle. AutoCAD presets a number of common angles, ranging from 5 to 90 degrees.

**Additional Angles** — allows you to add any angle not provided by the Increment Angle list, such as 7.5. Adding angles is a bit awkward:
   a. Click the **New** button.
   b. Enter a value, such as **35**, and then press ENTER.
   c. Repeat for additional angles.

**Polar Angle Measurement** — switches between absolute and relative angle measurements:

> **Absolute** means that the angle is measured from AutoCAD's 0-degree point, which usually is the positive x axis.
>
> **Relative to last segment** means the angle is measured relative to the last-drawn segment. The tooltip changes to read "Relative Polar."

(**Object Snap Tracking Settings** is an option described later in this chapter.)

3. Not all polar settings are listed in this tab of the dialog box, and so you need to click the **Snap and Grid** tab to uncover the additional settings.

4. Under Snap Type, select **PolarSnap**. This action causes all the Snap settings to gray out, meaning they are unavailable.

5. Instead, you now specify a distance next to **Polar distance**, such as 2. This sets a snap spacing in the polar direction, i.e., in the direction of the angle specified in the previous tab. (Snap must be turned on.)

6. Click **OK** to exit the dialog box.

7. With the LINE command, draw some lines, noticing the action of the cursor: you draw lines at specific angles, and the lines have lengths that are multiples of 2.

In general, you use the Drafting Settings dialog box to control polar mode, but you may at times prefer to change its settings by entering system variables at the command line:

## PolarAng

The POLARANG system variable specifies the increments for polar angles:

Command: **polarang**

Enter new value for POLARANG <90>: *(Enter a new value, such as 15.)*

## PolarDist

The POLARDIST system variable specifies the polar snap increment (distance), but only when the SNAPTYPE system variable is set to 1. Recall that polar measurements consist of a distance and an angle; this system variable specifies the distance:

Command: **polardist**

Enter new value for POLARDIST <0.0>: *(Enter a new value, such as 1.0.)*

## PolarMode

The POLARMODE system variable specifies a number of variables for polar snap. It handles eight situations using a single number by means of *bit codes*. Bit codes are added up to create a single number.

When all options are turned on, for example, the value of POLARMODE is 15, which comes from 1 + 2 + 4 + 8). Zero means no, off, or not.

| PolarMode | Meaning |
|---|---|
| **Measurement Mode:** | |
| 0 | Absolute mode: Bases polar angle measurements on the current UCS. |
| 1 | Relative mode: Measures polar angles from the last-drawn object. |
| **Additional Polar Tracking Angles:** | |
| 0 | Does not use additional angles. |
| 4 | Uses additional angles. |

Some values used by this system variable are meant for object tracking, covered later in this chapter.

| | |
|---|---|
| **Object Snap Tracking:** | |
| 0 | Tracks using orthogonal angles only. |
| 2 | Tracks using polar angle settings. |
| **Acquire Object Snap Tracking Points:** | |
| 0 | Acquires points automatically. |
| 8 | Acquires points when user presses SHIFT. |

## TUTORIAL: DRAWING WITH POLAR MODE

In the previous chapter, you saw that the POLYGON command draws regular polygons but not irregular ones. An example of an irregular polygon is a 45-degree isosceles triangle, which has two angles at 45 degrees and the third at 90 degrees.

In this tutorial, you draw such a triangle with two sides 2.5 units long.

1. Before starting to draw, set up polar mode using the DSETTINGS command. In the Drafting Settings dialog box, make these changes:

    *In the **Snap and Grid** tab*

    | | |
    |---|---|
    | Snap On | **On** |
    | Snap Type | **PolarSnap** |
    | Polar Distance | **2.5** |

    *In the **Polar Tracking** tab*

    | | |
    |---|---|
    | Polar Tracking On | **On** |
    | Increment Angle | **45 degrees** |
    | Polar Angle Measurement | **Absolute** |

    The dialog box's tabs should look like the ones illustrated below.

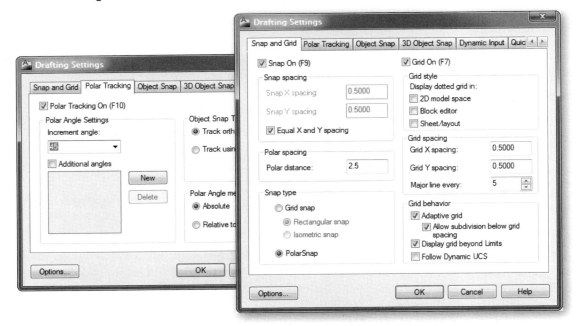

2. Click **OK** to exit the dialog box.
3. Start the LINE command, and then pick a point anywhere in the drawing:

    Command: **line**

    Specify first point: *(Pick point 1.)*

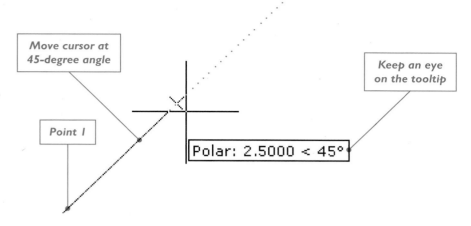

4. Move the cursor at a 45-degree angle to the upper-right. You may need to move the cursor around until the tooltip appears, reporting **Polar: 2.5000 < 45°**.

   Specify next point or [Undo]: *(Pick point 2.)*

5. Now move the cursor at a 45-degree angle to the lower-right. Click when the tooltip reports **Polar: 2.5000 < 315°**. (The angle of 315 comes from 45 + 270 — or 360 - 45.)

   Specify next point or [Undo]: *(Pick point 3.)*

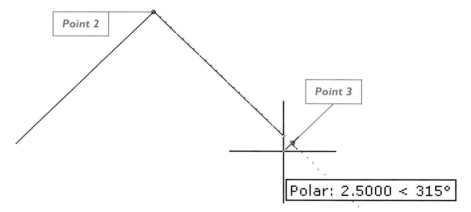

6. Close the triangle using the **Close** option:

   Specify next point or [Close/Undo]: **c**

   The isosceles triangle is complete. AutoCAD determines the length of the third side.

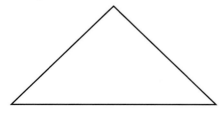

 **OSNAP**

Object snap is the *most* important tool for drawing accurately in AutoCAD. You might be able to do without the precise drawing aids described up to this point, but not without object snap. As implied by its name, object snap causes AutoCAD to snap to objects. Object snaps are often called "osnaps" for short.

More precisely, AutoCAD's cursor selects *geometric* features of objects. These include the ends of arcs, centers of circles, and the intersection of two lines.

In all, there are thirteen types of object snaps available in AutoCAD. When osnap mode is active, AutoCAD displays temporary icons at the geometric snap point in the drawing. (See figure at right.)

You can preset osnap modes or activate them temporarily through keyboard shortcuts or shortcut menus. You can have just one osnap mode turned on, or some or all of them. In most cases, you work with a few common modes, such as endpoint, midpoint, intersection, and center. They are used during drawing and editing commands.

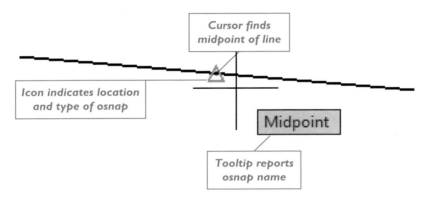

### APERTURE

A powerful aspect of object snap is that you don't have to be right on the geometric feature — you just have to be close enough (as in horseshoes, hand grenades, and dancing, notes the technical editor) — and AutoCAD finds the point to which to snap.

This close-enough distance is called the *aperture*, which is a square ten pixels in size centered on the crosshair cursor. AutoCAD examines all objects that lie within (or cross the aperture) for geometry that might match the current object snap mode(s).

In the figure above, the cursor has located the midpoint of the line.

You can change the size of the aperture, from as large as 50 pixels to as small as 1 pixel with the **APERTURE** command; I find that the default of 10 pixels works best. When it is too large, AutoCAD has to examine too much geometry; when too small, it is hard to see.

 **Note** AutoCAD has a separate set of objects snaps designed for 3D modeling and editing (new to AutoCAD 2011). These are handled by the **3DOSNAP** command, and are described later in this book.

## RUNNING AND TEMPORARY OSNAPS

Object snaps are used in two ways: running and temporary. *Running* object snaps are set with the OSNAP or -OSNAP commands; in effect, running object snaps run on and on, until you turn them off.

*Temporary* object snaps are in effect for the next object selection only. At prompts, such as "Select first point:" or "Specify next point", you enter the abbreviations of one or more object snap modes. When entering more than one mode, separate them with commas.

In the following command sequence, a line is drawn from the ENDpoint of one object to the MIDpoint or CENter point of another. Note how "end" and "mid,cen" are entered as abbreviations for the endpoint, midpoint, and center object snaps:

> Command: **line**
> Specify first point: **end**
> of *(Pick a point.)*
> Specify next point or [Undo]: **mid,cen**
> of *(Pick a point.)*
> Specify next point or [Undo]: *(Press ENTER to end command.)*

If you prefer to select osnap modes from a list, you can do so by accessing a shortcut menu, as follows:

1. Start a drawing or editing command, and then hold down the **CTRL** key.
2. Press the right mouse button. (Alternatively, you can press **SHIFT**+right-click.)
3. Notice that the shortcut menu appears, listing all osnap modes:

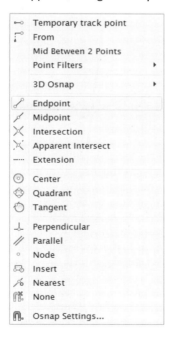

4. Choose an osnap mode from the shortcut menu, and then carry on with the command.

## SUMMARY OF ALL OBJECT SNAP MODES

In this review of object snap modes, notice that the first three letters of each mode are capitalized. These are the letters that abbreviate each mode, so that you do not need to type in the entire name.

In alphabetical order, AutoCAD's object snap modes are:

 **APParent intersection**

The apparent intersection object snap mode (called "APPint" for short) works two ways, depending on whether you are working with z coordinates.

In 2D drawings, APPint turns on an extended intersection mode, so that AutoCAD snaps to the intersection of two objects that don't physically intersect. AutoCAD creates an imaginary extension to the two objects, and then determines if an intersection could occur.

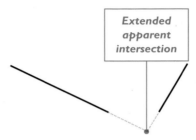

In 3D drawings, APPint snaps to the point where two objects appear to intersect from your viewpoint; the objects need not physically intersect.

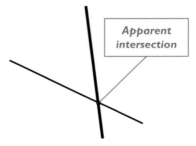

This object snap works with arcs, circles, ellipses, elliptical arcs, lines, multilines, polylines, rays, splines, and xlines.

 **CENter**

The center osnap snaps to the center point of arcs, circles, ellipses, and elliptical arcs.

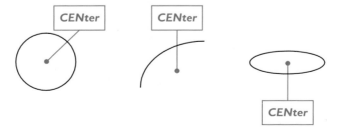

**ENDpoint**

The endpoint osnap snaps to the closest endpoint of arcs, elliptical arcs, lines, multilines, polyline segments, splines, regions, rays, and the closest corner of traces and 2D solids.

Use ENDpoint for the corners of 3D faces and solids, because INTersection and EXTended intersection don't work with them.

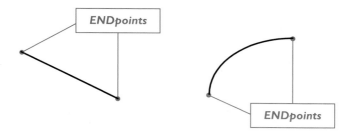

(History: Early versions of AutoCAD used the END command to exit the program. However, users found that they mistakenly exited AutoCAD when they intended to toggle endpoint object snap by entering 'end.' Autodesk solved the problem by removing the END command, and replacing it with QUIT. Today, the same exit problem can occur when you double-click the big red A.)

 **EXTension**

The extension osnap displays an extension line when the cursor passes over the endpoints of objects. This allows you to draw objects that start and end along the extension line.

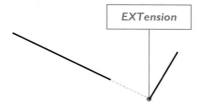

In the figure above, AutoCAD displays the extension line (shown dotted) from the endpoint of the line. (Extensions are similar to, but a more limited form of, *object snap tracking*, discussed later in this chapter.)

**INSERTION**

The insertion osnap snaps to the insertion point of attributes, blocks, shapes, and text.

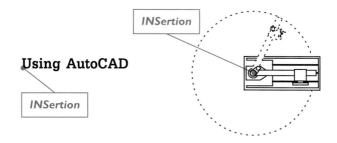

### INTersection

The intersection osnap snaps to the intersection of arcs, circles, ellipses, elliptical arcs, lines, multilines, polylines, rays, regions, splines, and xlines. Sometimes, more than one intersection is possible, such as when two circles intersect.

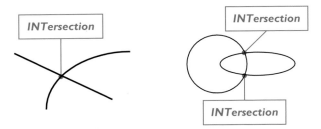

If extended intersection mode is turned on, AutoCAD can snap to the intersection of two objects that don't physically intersect. An imaginary extension to the two objects is created to determine if an intersection could occur.

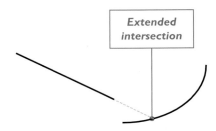

### MIDpoint

The midpoint osnap snaps to the midpoint of arcs, elliptical arcs, lines, multilines, polyline segments, edges of regions, solids, splines, and the first point that defines xlines.

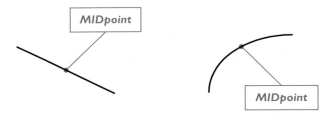

### NEArest

The nearest osnap snaps to the nearest point on the nearest arc, circle, ellipse, elliptical arc, line, multiline, point, polyline, ray, spline, or xline.

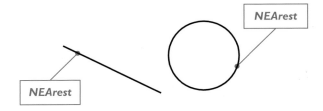

### NODe

The node osnap snaps to points, dimension definition points, and dimension text origins.

### PARallel

The parallel osnap displays an alignment path parallel to a straight line segment, such as a line or polyline.

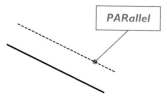

### PERpendicular

The perpendicular osnap snaps to a point perpendicular to an arc, circle, ellipse, elliptical arc, line, multiline, polyline, ray, region, solid, spline, or xline. Objects can be at any angle.

If necessary, AutoCAD turns on deferred perpendicular mode automatically when you need to pick more than one point to determine perpendicularity, such as making a line perpendicular to an arc.

### QUAdrant

The quadrant osnap snaps to the quadrant points of arcs, circles, ellipses, and elliptical arcs.

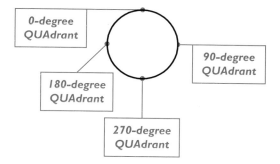

The quadrant points are located at the 0-, 90-, 180-, and 270-degree points of these curves. Thus, circles and ellipses always have four possible quadrant points, while arcs and elliptical arcs can have anywhere from one to four quadrant points.

## TANgent

The tangent osnap snaps to the tangent points of arcs, circles, ellipses, elliptical arcs, and splines.

If necessary, AutoCAD turns on deferred tangent mode automatically when you draw more than one tangent, such as a line tangent to two arcs.

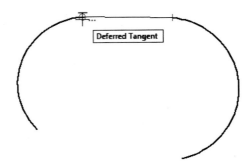

### ADDITIONAL OBJECT SNAP MODES

There are several additional object snap modes that don't work with geometry. These find distances, or else turn off snap modes.

## From

The "from" mode allows you to specify a temporary reference or base point, from which to locate the next point.

Command: **line**
Specify first point: **from**
Base point: *(Pick point 1.)*

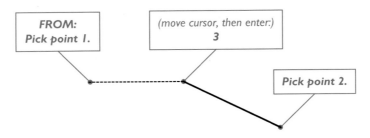

<Offset>: (Move the cursor to show the distance, or enter a distance:) **3**
Specify next point or [Undo]: *(Pick point 2.)*
Specify next point or [Undo]: *(Press* ENTER *to end command.)*

**NONe**

The none mode turns off all object snap modes temporarily for the next pick point.

**TT**

The temporary tracking mode allows you to specify a temporary tracking point. AutoCAD places a + marker at the point, and then, as you move the crosshair cursor, it displays horizontal and vertical alignment paths relative to the + marker.

Command: **line**

Specify first point: **tt**

Specify temporary OTRACK point: *(Pick point 1.)*

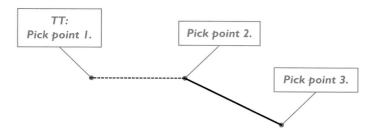

Specify first point: *(Pick point 2.)*

Specify next point or [Undo]: *(Pick point 3.)*

Specify next point or [Undo]: *(Press* ENTER *to end command.)*

### QUIck

The **Quick** object snap snaps to the first snap point AutoCAD finds. Quick mode works only when you have two or more other snap modes turned on. For example, when you have intersection and endpoint turned on, AutoCAD returns the first object snap point it finds that is either an intersection or endpoint; this is significant in large, complex drawings. When Quick is off, however, AutoCAD seeks all osnaps, and returns the one closest to the center of the aperture.

### OFF

The **Off** option turns off all object snap modes.

### TUTORIAL: TOGGLING AND SELECTING OSNAP MODE

Object snap probably has the most complete set of command variations:

1. To toggle ortho mode, use one of these methods:
   - On the status bar, click **OSNAP** or ▢ (turned on when button looks blue).
   - At the keyboard, press function key **F3** or CTRL+F.

2. To toggle ortho mode *and* specify one or more object snaps, use one of these methods:
   - During another command, hold down the CTRL (or SHIFT) key, and then press the right mouse button to display a shortcut menu of object snap modes.
   - Similarly, you can use the Object Snap toolbar during a command: from the toolbar, select a button corresponding to the object snap mode you wish to employ.
   - On the keyboard, enter the **-osnap** command, followed by one or more object snap mode names. (The hyphen prefix forces AutoCAD to display the command at the 'Command:' prompt; without the hyphen, the **OSNAP** command displays the dialog box.)

- Alternatively, type the aliases **os** or **ddosnap** (the old command name) for the dialog box, and the **-os** alias for the command line.
- And finally, the **OSMODE** system variable can be used to set object snap modes.

### CONTROLLING OSNAP MODES: ADDITIONAL METHODS

In new drawings, AutoCAD remembers the object snap modes set previously. You change the modes through a dialog box, or at the command-line:

- **OSNAP** command displays a dialog box for changing osnap settings.
- **-OSNAP** command changes osnap settings at the command line.
- Several system variables affect object snaps.

## OSnap (DSettings)

The OSNAP command displays a dialog box that allows you to select running object snap modes. (Strictly speaking, the OSNAP command is a shortcut for using the DSETTINGS command, followed by selecting the Object Snap tab in the dialog box.)

1. To access the Drafting Settings dialog box, do one of the following:
   - At the 'Command:' prompt, enter the **osnap** command.
   - Alternatively, right-click **OSNAP** (or ) on the status bar. From the shortcut menu, select **Settings**.
   
   Notice that AutoCAD displays the Drafting Settings dialog box. Click the **Object Snap** tab.

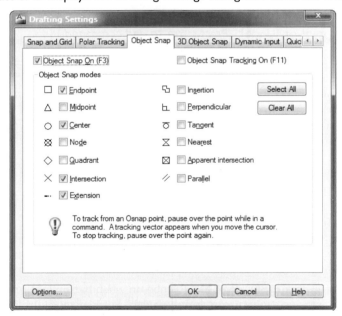

2. In this dialog box, you can make the following changes to the snap settings:

   **Object Snap On (F3)** — toggles the snap on and off. The "(F3)" reminds you that you can press **F3** to turn on and off osnaps.

   **Object Snap Modes** — lists most of the modes available. Notice the icon next to each mode; this is the same icon AutoCAD displays in the drawing when it finds a geometric feature matching the selected mode(s).

## OSNAP SYSTEM VARIABLES

The following system variables affect how AutoCAD performs its object snaps.

OSNAPNODELEGACY determines whether NODe osnap snaps to mtext grips:

| OSnapNodeLegacy | Comments |
|---|---|
| 0 | NODe snaps to mtext grips (default). |
| 1 | NODe ignores mtext grips. |

OSNAPHATCH determines whether osnap works with hatch objects:

| OSnapHatch | Comments |
|---|---|
| 0 | Ignores hatch objects (default). |
| 1 | Snaps to hatch objects. |

DWFOSNAP determines if osnaps works with geometry in DWF files:

| DwfOsnap | Comments |
|---|---|
| 0 | Ignores DWF entities. |
| 1 | Snaps to entities in attached DWF files (default). |

DGNOSNAP determines if osnap works with geometry in attached DGN design files:

| DgnOsnap | Comments |
|---|---|
| 0 | Ignores MicroStation elements. |
| 1 | Snaps to elements in imported MicroStation files (default). |

PDFOSNAP determines if osnap works with geometry in attached PDF files:

| DgnOsnap | Comments |
|---|---|
| 0 | Ignores MicroStation elements. |
| 1 | Snaps to elements in imported MicroStation files (default). |

OSNAPZ determines how osnaps work in 3D:

| OSnapZ | Comments |
|---|---|
| 0 | Uses the z coordinate of the current point (default). |
| 1 | Uses the value stored in the ELEVATION system variable. |

TEMPOVERRIDE determines whether temporary overrides are available:

| TempOverride | Comments |
|---|---|
| 0 | Does not allow temporary overrides (off). |
| 1 | Allows temporary overrides (on). |

OSOPTIONS determines how osnaps operate during dynamic UCS mode:

| OsOptions | Comments |
|---|---|
| 0 | Operates normally during dynamic UCS. |
| 1 | Ignores hatch objects. |
| 2 | Ignores geometry with negative Z values. |

APSTATE reports whether the aperture box cursor is on or off.

3. Select one or more object snap modes. You can turn on all osnap modes with the **Select All** button, and turn them all off with the **Clear All** button.
4. Not all object snap options are listed in this dialog box. Click the **Options** button to view more. AutoCAD displays the Drafting tab of the Options dialog box:

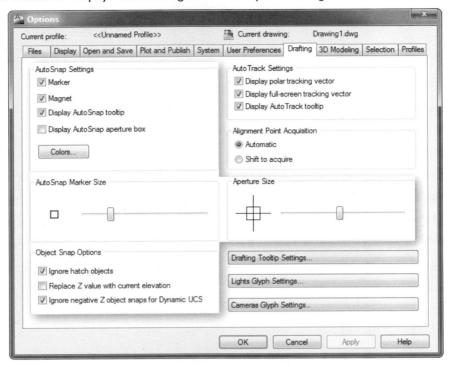

Here you select from the many options that affect how object snaps show up in the drawing. For example, if you changed AutoCAD's background color, you may want to change the marker color from a hard-to-see color to one of higher visibility. The options have the following meanings:

**Marker** — toggles the display of the osnap icons.

**Magnet** — toggles whether the cursor automatically moves to the osnap point.

**Display AutoSnap Tooltip** — toggles whether the tooltip appears, which labels the object snap by name.

**Display AutoSnap Aperture Box** — indicates AutoSnap mode is on, through a small square appearing at the center of the cursor.

**Colors** — selects any color for the marker.

**AutoSnap Marker Size** — makes the marker (icon) larger and smaller.

**Ignore Hatch Objects** — determines whether osnaps snap to the lines that make up hatch patterns.

**Replace Z values with current elevation** — determines whether osnaps use the z coordinate or the value of the ELEVATION system variable.

**Ignore negative Z object snaps for Dynamic UCS** — snaps only to positive z-axis locations when dynamic UCS is turned on.

**Aperture Size** — makes the object snap cursor larger and smaller.

5. Click **OK** to exit the dialog box.
6. AutoCAD returns you to the Drafting Settings dialog box. Click **OK** to exit this dialog box.

### -OSnap

The -OSNAP command displays a prompt at the command line. It lists the current osnap modes, and then prompts you to enter additional modes. You can enter a single osnap mode, such as **Int**, or several modes separated by commas, as shown below:

> Command: **-osnap**
> Current osnap modes: End,Mid,Cen
> Enter list of object snap modes: **int,qua**

You need only type the first three letters of each mode name. For example, "int" means **INTersection** and "qua" means **QUAdrant**.

You cannot selectively turn off osnap modes. Instead, you turn off other modes by entering new modes. In the example above, entering **int** and **qua** turns off **end**, **mid**, and **cen**.

### TUTORIAL: USING OBJECT SNAP MODES

In the following tutorial, you use osnap modes to draw a line from the end point of an arc to the center of a circle, precisely.

1. With the **CIRCLE** command, draw a circle; also draw an arc of any size with the **ARC** command. Their locations in the drawing are not important, except that they should be apart.
2. Start the **LINE** command.
3. When AutoCAD prompts you, do not pick a point. Instead, hold down the **CTRL** (or **SHIFT**) key, and then press the right mouse button to access the shortcut menu.
    Specify first point: *(Hold down CTRL key, and then press right mouse button.)*

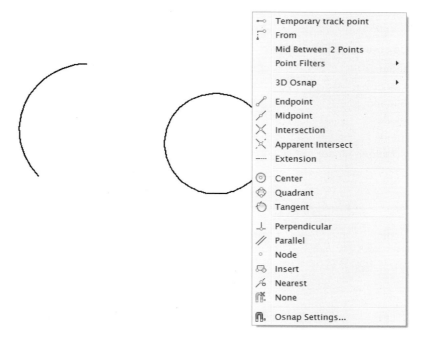

4. From the menu, choose **Endpoint**. Notice that the aperture square appears at the intersection of the cross hair cursor.

## TEMPORARY OVERRIDES

*Temporary overrides* temporarily turn on several object snap and drawing modes. For example, when you are drawing with all osnap modes off, you can hold down SHIFT+E to turn on the ENDpoint osnap mode temporarily.

(The override does not work in reverse, unfortunately: when ENDpoint is on, holding down SHIFT+E does not turn it off temporarily. The workaround is to press SHIFT+D to disable all osnaps temporarily.)

| Temporary Override | Keyboard Shortcut(s) | |
|---|---|---|
| Disable osnap enforcement | SHIFT+A | SHIFT+' |
| Enable osnap enforcement | SHIFT+S | SHIFT+; |
| Enable CENter osnap | SHIFT+C | SHIFT+, |
| Enable ENDpoint osnap | SHIFT+E | SHIFT+P |
| Enable MIDpoint osnap | SHIFT+V | SHIFT+M |
| Toggle ortho mode | SHIFT | ... |
| Toggle osnap tracking mode | SHIFT+Q | SHIFT+] |
| Toggle polar mode | SHIFT+X | SHIFT+. |
| Disable tracking and all osnap modes | SHIFT+D | SHIFT+L |

There are two shortcuts for each mode, one for left-handed people and the other for right-handed.

When you find that none of these temporary overrides works, then it could be that the TEMPOVERRIDE system variable has been turned off — or, the shortcuts were changed with the CUI command.

In addition to the shortcuts listed above, the following shortcuts toggle osnap and other modes:

| Toggles | Keyboard Shortcut(s) | |
|---|---|---|
| Toggle osnap mode | F3 | CTRL+F |
| Toggles ortho mode | F8 | CTRL+L |
| Toggle snap mode | F9 | CTRL+B |
| Toggle polar mode | F10 | CTRL+U |
| Toggle osnap tracking mode | F11 | ... |

5. Place the cursor over one end of the arc, and click. The line snaps precisely to the end of the arc.

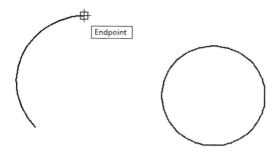

6. At the next prompt of the **LINE** command, type **cen**, and then press **ENTER**, as follows:
   Specify next point or [Undo]: **cen** *(Press **ENTER**.)*

7. Notice that AutoCAD prompts you with "of". It waits for you to choose the geometry:
   of *(Place the cursor over any part of the circle's circumference, and then click.)*

   Notice how the line snaps to the precise center of the circle.

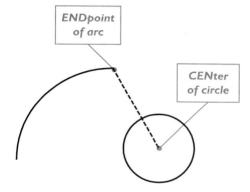

8. Continue the line, using the **Center** object snap on the arc. To find the arc's center, you move the cursor to one of these locations:
   * Move the cursor onto the arc, and the tracking line appears.
   * Or, move the cursor roughly to where you expect the arc's center to be located.

   Either way, the Center icon appears, and when you click, the line snaps to the center point of the arc.

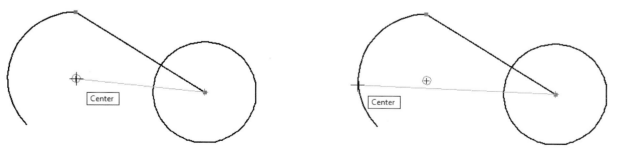

**Left:** *Finding the arc's center at its center.*
**Right:** *Finding the arc's center on the arc.*

9. Try other object snap modes on the line, arc, and circle. For example, use the Tangent object snap to construct lines tangent to circles and arcs.

 **OTRACK**

Otrack is short for "object snap tracking"; Autodesk also calls it AutoTrack™, complete with the trademark symbol. Otracking adds alignment paths to the object snap process.

Alignment paths are thin dotted lines that show the osnap's relationship to geometry, as illustrated later.

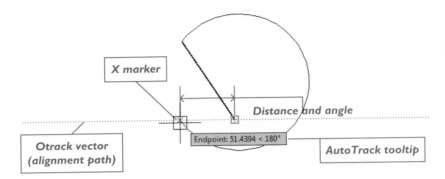

Think of otrack as a "super extension" osnap. At least one object snap mode must be turned on for otracking to work.

When any prompt asks you to 'Select objects:', move the cursor around the drawing to see geometric relationships. In the figure above, AutoTrack has placed a marker (the **x**) at the endpoint of the arc and in-line with the endpoint of the line. After a moment, a tooltip appears, reporting the object snap mode and the angle between the geometry.

### OTRACK OPTIONS

The Options dialog box's Drafting tab lets you control how object tracking looks on your screen. Check out the AutoTrack Settings and Alignment Point Acquisition sections:

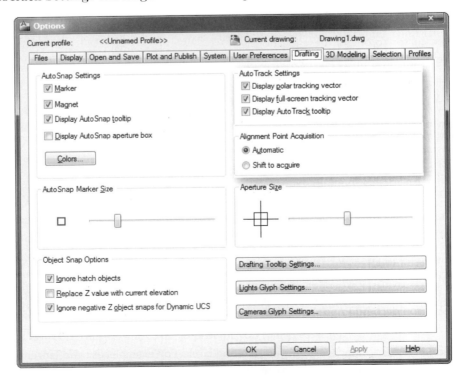

## AutoTrack Settings

**Display Polar Tracking Vector** displays tracking vectors (dotted lines) at polar angles (angles that divide into 90 degrees) defined by the Polar Tracking tab of the Drafting Settings dialog box.

**Display Full-Screen Tracking Vector** displays tracking vectors as infinite construction lines.

**Display AutoTrack Tooltip** toggles the display of AutoTrack and ortho tooltips.

## Alignment Point Acquisition

**Automatic** displays tracking vectors any time the aperture is over an object snap point.

**Shift to Acquire** displays tracking vectors only when you hold down the SHIFT key while the aperture is over an object snap point.

### TUTORIAL: TOGGLING OTRACK MODE

Like polar mode, there is no "OTRACK" command; instead, you use shortcuts, system variables, or dialog boxes to turn it on and off.

To turn on otrack mode, use one of these methods:

- From the status bar, click **OTRACK** or ⦦ (turned on when button looks blue).
- On the keyboard, press function key **F11**.
- In the Drafting Settings dialog box's Object Snap tab, turn on the **Object Snap Tracking On** option.
- Or change the value of the **TRACKPATH** system variable:

| TrackPath | Polar Tracking Vector | Object Snap Tracking Vector Is |
|---|---|---|
| 0 | Full screen. | Full screen. |
| 1 | Full screen. | Between cursor and alignment point. |
| 2 | Not displayed. | Full screen. |
| 3 | Not displayed. | Between cursor and alignment point. |

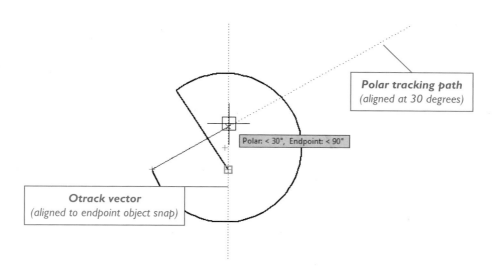

*Polar tracking path (aligned at 30 degrees)*

*Otrack vector (aligned to endpoint object snap)*

You can use otracking together with temporary tracking. At the 'Select objects:' prompt, enter **tt** (short for tracking) to show alignment paths relative to the temporary tracking point.

The technical editor notes that OTRACK and TT together eliminate the need for the rays, xlines, and other construction lines discussed next.

## RAY AND XLINE

The RAY and XLINE commands place construction lines in the drawing. Construction lines are useful for creating drawings, helping to reference things like center lines and offset lines. The construction lines, unfortunately, are plotted by AutoCAD, so you should place them on frozen or no-plot layers.

RAY draws "semi-infinite" lines, which have start points, but no endpoints. XLINE draws infinitely-long lines, which have no start or endpoints.

### TUTORIAL: DRAWING CONSTRUCTION LINES

1. To draw construction lines, start the **RAY** or **XLINE** command with one of these methods:
   - In the ribbon's 2D Home panel, choose the **Construction Line** button in the Draw panel.
   - Or, at the 'Command:' prompt, enter the **ray** or **xline** commands.
     
     Command: **xline** *(Press ENTER.)*
   - Alternatively, enter the **xl** alias at the 'Command:' prompt.

2. Specify two points that lie along the construction line:
   
   Specify a point or [Hor/Ver/Ang/Bisect/Offset]: *(Pick point 1.)*
   
   Specify through point: *(Pick point 2.)*

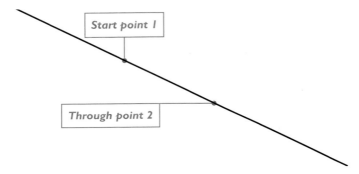

3. Because the **RAY** and **XLINE** commands automatically repeat, press **ENTER** or **ESC** to exit them:
   
   Specify through point: *(Press ENTER to exit command.)*

The RAY command differs from XLINE by first asking for the starting point:

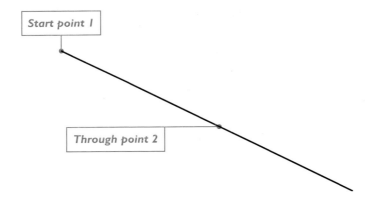

Command: **ray**
Specify start point: *(Pick point 1.)*
Specify through point: *(Pick point 2.)*
Specify through point: *(Press ENTER to exit command.)*

## DRAWING CONSTRUCTION LINES: ADDITIONAL METHODS

The XLINE command contains these options that provide additional methods for drawing construction lines.

- **Hor** draws horizontal construction lines:

    Command: **xline**

    Specify a point or [Hor/Ver/Ang/Bisect/Offset]: **h**

    Specify through point: *(Pick a point.)*

    Specify through point: *(Press ENTER to exit command.)*

- **Ver** draws vertical construction lines:

    Command: **xline**

    Specify a point or [Hor/Ver/Ang/Bisect/Offset]: **v**

    Specify through point: *(Pick a point.)*

    Specify through point: *(Press ENTER to exit command.)*

- **Ang** draws construction lines at a specified angle:

    Command: **xline**

    Specify a point or [Hor/Ver/Ang/Bisect/Offset]: **a**

    Enter angle of xline (0) or [Reference]: *(Enter an angle, or type* **r** *for the reference option.)*

    Specify through point: *(Pick a point.)*

    Specify through point: *(Pick another point.)*

    Specify through point: *(Press ENTER to exit command.)*

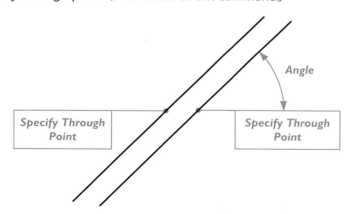

- **Bisect** draws construction lines that bisect an angle defined by a vertex and two endpoints:

    Command: **xline**

    Specify a point or [Hor/Ver/Ang/Bisect/Offset]: **b**

    Specify angle vertex point: *(Pick a point.)*

    Specify angle start point: *(Pick a point.)*

    Specify angle end point: *(Pick a point.)*

    Specify angle end point: *(Press ENTER to exit command.)*

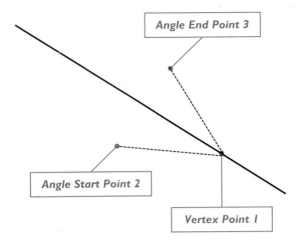

- **Offset** draws construction lines parallel to existing lines and construction lines:

    Command: **xline**

    Specify a point or [Hor/Ver/Ang/Bisect/Offset]: **o**

    Specify offset distance or [Through] <Through>: *(Enter a distance, or type* **t** *for the through option.)*

    Select a line object: *(Pick a line.)*

    Specify side to offset: *(Pick a point.)*

    Select a line object: *(Press* ENTER *to exit command.)*

### TUTORIAL: DRAWING WITH CONSTRUCTION LINES

In this tutorial, you draw the true surface of a rectangular inclined face. You use much of what you have learned so far in this chapter: grid, snap, ortho, object snaps, and construction lines.

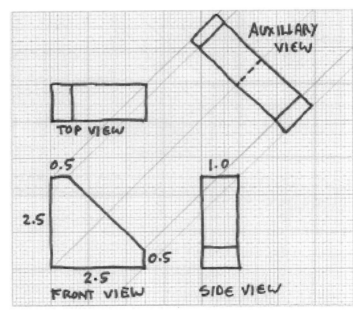

1. Start AutoCAD with a new drawing.
2. Change the following settings:

   Turn on **SNAP**; ensure the snap spacing is 0.5 units.

   Turn on **GRID**.

   Turn on **ORTHO**.

   Turn on **OSNAP**; ensure that object snap modes **Intersection** and **Perpendicular** are turned on.

3. With the LINE command, draw the multiview drawing shown below.

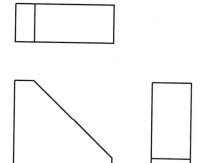

4. Use the SNAP command's **Rotate** option to rotate the snap to align with the sloping edge of the front view:

   Command: **snap**

   Specify snap spacing or [ON/OFF/Aspect/Rotate/Style/Type] <0.5000>: **r**

   Specify base point <0.0000,0.0000>: *(Pick point 1.)*

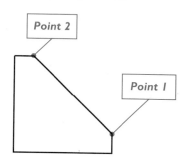

   Because you set the object snap modes earlier in Step 2, you should have no difficulty snapping to the ends of the sloped edge.

   Specify rotation angle <0>: *(Pick point 2.)*

   Angle adjusted to 315

5. Turn off **SNAP**.

   With the RAY command, draw the series of parallel construction lines shown below.

   Command: **ray**

   Specify start point: *(Pick point 1.)*

   Specify through point: *(Pick point 2.)*

   Specify through point: *(Press ESC.)*

   Repeat the command to draw four more construction lines. *Hint:* Press ESC to end a command that repeats itself; press the spacebar to repeat the command.

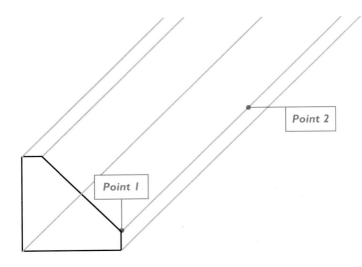

6. With the construction lines in place, you can now draw the auxiliary view with the **LINE** command. Remember that the object is 1" wide.

    *Hint*: Use the NEArest object snap to start drawing from the construction line.

    Command: **line**

    Specify first point: **nea**

    to *(Pick point 1.)*

    Specify next point or [Undo]: *(Pick point 2.)*

    Specify next point or [Undo]: *(Move cursor toward point 3, and then enter 1.)* **1**

    Specify next point or [Close/Undo]: *(Pick point 4.)*

    Specify next point or [Close/Undo]: **c**

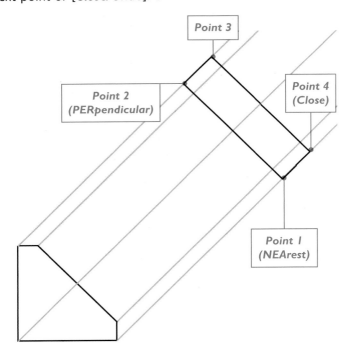

7. Draw the two edge lines, as follows:
    Command: **line**
    Specify first point: *(Pick point 1.)*
    Specify next point or [Undo]: *(Pick point 2.)*
    Specify next point or [Undo]: *(Press ESC.)*

    Command: *(Press ENTER to repeat the command.)*
    LINE Specify first point: *(Pick point 3.)*
    Specify next point or [Undo]: *(Pick point 4.)*
    Specify next point or [Undo]: *(Press ESC to end the command.)*

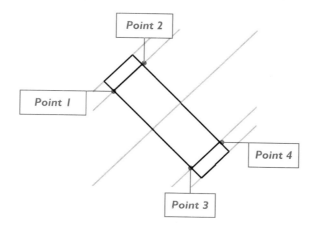

8. Delete the construction lines with the **ERASE** command.
9. Save the drawing as *auxview5.dwg*.

## REDRAW AND REGEN

The REDRAW and REGEN commands clean up and update the drawing. REDRAW cleans up the display by removing blip marks (if any) and blanked out areas. REGEN is short for "regeneration," and updates the display by recalculating all the vectors making up the drawing. During a regeneration, AutoCAD recalculates line endpoints, hatched areas, and so on. A redraw is always faster than a regen.

If you have more than one viewport displayed, REDRAW and REGEN operate on the current viewport only. Use the REDRAWALL and REGENALL commands to clean up and update all viewports.

### TUTORIAL: USING THE REDRAW COMMAND

1. To clean up the display, start the **REDRAW** command with one of these methods:
   - At the 'Command:' prompt, enter the **redraw** command.

        Command: **redraw** *(Press ENTER.)*

   - Alternatively, enter the **r** alias at the 'Command:' prompt.

   The screen may flicker as AutoCAD cleans up the drawing.

The commands associated with REDRAW work similarly:

| Command | Alias |
| --- | --- |
| RedrawAll | ra |
| Regen | re |
| RegenAll | rea |

### TRANSPARENT REDRAWS

A redraw can be executed while another command is active. This is called a transparent redraw. To perform this, enter an apostrophe ( ' ) before the command. For example, to perform a transparent redraw during the LINE command:

Command: **line**

Specify first point: *(Pick a point.)*

Specify next point or [Undo]: **'redraw**

Resuming LINE command.

Specify next point or [Undo]: *(Pick another point.)*

Specify next point or [Undo]: *(Press* ENTER.*)*

When the transparent redraw is completed, the previously-active command resumes. REGEN is not a transparent command.

## ZOOM

The ZOOM command enlarges and reduces the view of the drawing.

Most drawings are too large and too detailed to work with on a typical computer's 19" — or even 24" — screen. CAD operators routinely zoom in to small areas to work on details. When zoomed out, you see more of the drawing; when zoomed in, you see less of the drawing but more detail. Think of zoom as a magnifier and shrinker.

Let's consider this analogy: imagine that your drawing is the size of the wall in your room. The closer you walk to the wall-size drawing, the more detail you see, but the less you see of the entire drawing. When you move a great distance away from the wall, you see the entire drawing, but not much detail.

The wall does not change size, only your viewing distance. The ZOOM command works in the same way: you enlarge and reduce the drawing size on the screen, but the drawing itself does not change size (or scale).

Because zooming is performed often, AutoCAD has many ways to zoom.

### TUTORIAL: USING THE ZOOM COMMAND

1. To change the viewing size of the drawing, start the **ZOOM** command with one of these methods:
   - In the ribbon's 2D Home panel, choose a **Zoom** button in the Utilities panel.
   - On the navigation bar, click the  Zoom button.
   - Right-click anywhere in the drawing, and from the shortcut menu select **Zoom**.
   - Or, at the 'Command:' prompt, enter the **zoom** command.
   
     Command: **zoom** *(Press* ENTER.*)*
   - Alternatively, enter the **z** alias at the 'Command:' prompt.

2. The command displays its many options:

   Specify corner of window, enter a scale factor (nX or nXP), or
       [All/Center/Dynamic/Extents/ Previous/Scale/Window/Object] <real time>:

The prompt indicates there are 12 options, and there are more that are hidden: two are **Left** and **Vmax**, and the third is the mouse wheel.

Many of AutoCAD's commands have a single default option, but Autodesk managed to endow this one with three defaults! These are the three defaults:

> (1) Specify corner of window, (2) enter a scale factor (nX or nXP), or (3) <real time>:

Here's what they mean:

**Specify corner of window** — click a point on the screen, and AutoCAD executes the **Window** option.

**Enter a scale factor** — type a number, and AutoCAD executes the **Scale** option.

**<real time>** — press ENTER, and AutoCAD executes real-time zooms.

### ZOOM COMMAND: ADDITIONAL METHODS

The ZOOM command contains these options for zooming in and out. The options are shown at the command prompt.

| Zoom Option | Meaning |
| --- | --- |
| **Shows Entire Drawing (zoom out):** | |
| All | Shows the entire drawing, or the limits of the drawing, whichever is larger. |
| Extents | Shows everything in the drawing. |
| Vmax | Zooms out to the maximum without invoking a regen (hidden option). |
| **Shows Portion of Drawing (zoom in):** | |
| Center | Zooms about a center point. |
| Object | Zooms to the extents of the selected objects. |
| Window | Zooms in to a rectangular area specified by two points. |
| Left | Zooms relative to a lower left point (hidden option). |
| **Shows All or Portion of Drawing (zoom out or in):** | |
| real time | Zooms in real time with mouse movement. |
| Previous | Shows the previous view, whether a zoom or pan. |
| Scale | Zooms by absolute and relative factors in model space and paper space. |
| Dynamic | Displays a zoom/pan box for interactive zooming (obsolete). |

Despite the many options, I find I use only a few frequently — specifically **Extents**, **Window**, and **Previous**. Some CAD users find they don't need the ZOOM command at all: they use the mouse wheel to zoom in and out of the drawing transparently.

Let's look at the ZOOM command's options in alphabetical order.

### Zoom All

The **All** option displays the entire drawing on the screen. This typically displays the entire area of the limits and the extents of the drawing, including the area of the drawing outside the limits.

Note this difference between the **Extents** and **All** options: **Extents** displays the extents of the drawing, while **All** displays either the drawing extents or the drawing limits, depending on which is larger.

> Command: **zoom**
> Specify corner of window, enter a scale factor (nX or nXP), or
>     [All/Center/Dynamic/Extents/Previous/Scale/Window/Object] <real time>: **a**

Occasionally, ZOOM **All** has to regenerate the drawing twice. If this is necessary, AutoCAD displays the following message on the prompt line:

> * * Second regeneration caused by change in drawing extents.

When the limits are changed, the entire drawing area is not shown until the next ZOOM **All** is performed.

### Zoom Extents

The **Extents** option displays the drawing at its maximum size on the display screen. This results in the largest possible display, while showing the entire drawing.

    Command: **zoom**

    Specify corner of window, enter a scale factor (nX or nXP), or
        [All/Center/Dynamic/**Extents**/Previous/Scale/Window/Object] <real time>: **e**

 **Note** The fastest access to **ZOOM Extents** is through double-clicking the middle button or mouse wheel.

### Zoom Object

The **Object** option displays selected objects at their maximum size on the screen. This option is useful for inspecting one or more objects visually.

    Command: **zoom**

    Specify corner of window, enter a scale factor (nX or nXP), or
        [All/Center/Dynamic/Extents/Previous/Scale/Window/**Object**] <real time>: **o**

### Zoom Previous

The **Previous** option returns to the last zoom or pan you viewed. This option is useful if you need to move frequently between two areas. AutoCAD remembers the last ten view changes; simply use the **Z P** aliases several times in a row. Since the view coordinates are stored automatically, you do not need any special procedure to access them.

    Command: **zoom**

    Specify corner of window, enter a scale factor (nX or nXP), or
        [All/Center/Dynamic/Extents/**Previous**/Scale/Window/Object] <real time>: **p**

### Zoom Scale

The **Scale** option enlarges or reduces the entire drawing (original size) by a numerical factor. For example, entering **5** results in a zoom that increases the drawing to five times its normal size — AutoCAD zooms in. The zoom is centered on the screen's center point.

When an **x** follows the zoom factor, the zoom is computed relative to the current display. For example, entering **5x** makes the drawing five times larger than its current zoom factor.

Only positive values can be used for zoom scales. To zoom out (i.e., display the drawing smaller), use a decimal value smaller than 1. For example, **0.5** results in a view of the drawing that is one-half its normal size. (As of Release 14, ZOOM **All** and **Extents** leave a bit of room around the drawing, effectively doing a zoom 0.95x.)

    Command: **zoom**

    Specify corner of window, enter a scale factor (nX or nXP), or
        [All/Center/Dynamic/Extents/Previous/**Scale**/Window/Object] <real time>: **s**

    Enter a scale factor (nX or nXP): **5**

It is not necessary to enter the "s" for the **Scale** option. Enter a number after starting the command. Then AutoCAD assumes you want the scaled zoom.

The **XP** option scales the model viewport relative to the paper space viewport.

### VIEW TRANSITION

The ZOOM command can employ "view transition." When zooming in and out, AutoCAD executes an animation that makes objects appear to move closer and further away.

View transitions are useful for seeing where the view "ends up" in the drawing, particularly in 3D; the drawback is that zooms and pans take longer. Real-time zooms still occur in real time; they don't have the animation.

View transitions are controlled by these system variables:

VTENABLE specifies when view transitions take place (using bit codes):

| VtEnable | Meaning |
|---|---|
| 0 | View transitions turned off. |
| 1 | Turned on for the ZOOM and -PAN commands. |
| 2 | Turned on for view rotation. |
| 4 | Turned on for scripts. |

VTDURATION specifies the duration of view transitions. The default value is 750 milliseconds; range of values is 0 (no transition) to 5000 (very slow transition).

VTFPS specifies speed at which the view transition takes place. The range is 1 to 30fps (frames per second).

### GROUPED VIEW-CHANGE UNDO

Undoing a series of view changes, such as repeated pans, can be annoying. The value of 16 in the UNDOCTL system variable groups together all zooms and pans as a single action.

Alternatively, you can turn on the setting in the Options dialog box. From the Tools menu, select **Options**, and then choose the **User Preferences** tab.

## Zoom Window

The **Window** option uses a rectangular window to show the area, which you specify with two screen picks.

```
Command: zoom
Specify corner of window, enter a scale factor (nX or nXP), or
    [All/Center/Dynamic/Extents/Previous/Scale/Window/Object] <real time>: w
Specify first corner: (Pick point 1.)
Specify other corner: (Pick point 2.)
```

A box is shown around the area to be zoomed.

It is not necessary to enter the "w" for the **Window** option. Simply pick two points after starting the command; AutoCAD assumes you want a windowed zoom.

### Zoom Realtime

With today's computers and fast display drivers providing the horsepower, AutoCAD can zoom in real time. Real time means that the zoom changes continuously as you move the mouse. To enter real-time zoom mode:

> Command: **zoom**
> Specify corner of window, enter a scale factor (nX or nXP), or
> [All/Center/Dynamic/Extents/Previous/Scale/Window/Object] <real time>: *(Press ENTER.)*
> Press ESC or ENTER to exit, or right-click to display shortcut menu.

Once in real-time zoom mode, the cursor changes to a magnifying glass. To zoom, hold down the left mouse button, and then move the mouse up (toward the monitor) and down (away from the monitor). As you move the mouse up, the drawing becomes larger; as you move the mouse down, it becomes smaller.

While in real-time zoom mode, right-click the mouse to see the shortcut menu:

| Option | Meaning |
| --- | --- |
| Exit | Exits the ZOOM command. |
| Pan | Switches to real-time pan mode (discussed later in this chapter). |
| Zoom | Switches to real-time zoom mode (default). |
| 3D Orbit | Switches to interactive 3D rotation. |
| Zoom Window | Displays a cursor with small rectangle. Select two points, which AutoCAD displays as the new zoomed view. Remains in real-time zoom. |
| Zoom Original | Displays the drawing as it was when you began real-time zooming. |
| Zoom Previous | Displays the extents of the drawing's objects. |

To dismiss the shortcut menu without choosing one of its options, click anywhere outside the menu. When the drawing is the size you want, press ENTER or ESC to exit real-time zoom mode and the ZOOM command.

### Rarely Used Zoom Modes

**Zoom Center** — allows you to specify the center point of the zoom. You then indicate the magnification or height for the new view. If an "x" follows the magnification value, such as 5x, the zoom factor is relative to the current display.

**Zoom Dynamic** — performs dynamic zooming and panning, and toggles between zoom and pan modes. This option is a predecessor to real-time zoom and pan, and is now rarely used.

### Transparent Zoom

A transparent zoom can be executed while another command is active. Enter 'ZOOM at almost any prompt (notice the apostrophe prefix.) There are, however, restrictions:

You cannot perform a transparent zoom if a regeneration is required, which means you cannot zoom outside the generated area — see ZOOM **Dynamic**.

Transparent zooms also cannot be performed when certain commands are in progress. These include the VPOINT, PAN, VIEW, and the ZOOM command itself. Commands on the ribbon use transparent zooms automatically.

## PAN

The PAN command moves the view around the drawing.

Many times, you zoom into an area of the drawing to see more detail. You may want to "move" the screen a short distance, to continue working while remaining at the same zoom magnification. This sideways movement is called "panning."

A pan is similar to placing your eyes at a certain distance from a paper drawing, and then moving your head about the drawing. This would allow you to see all parts of the drawing at the same distance from your eyes.

### TUTORIAL: PANNING THE DRAWING

1. To move the view of the drawing, use the PAN command by one of these methods:
   - In the ribbon's 2D Home panel, choose the **Pan** button in the Utilities panel.
   - On the navigation bar, click the Pan button.
   - Or, at the 'Command:' prompt, enter the **pan** command.

     Command: **pan** (Press ENTER.)
   - Alternatively, enter the **p** alias at the 'Command:' prompt.

2. Hold down the left mouse button; notice the hand cursor.

   Move the mouse; notice that the drawing moves with the mouse.

   Right-click to see the same shortcut menu as displayed during real-time zoom.

3. To exit pan mode, press **ENTER** or **ESC**.

   Press ESC or ENTER to exit, or right-click to display shortcut menu. (Press ESC.)

### TRANSPARENT PANNING

A pan can be performed transparently while another command is in progress. To do this, press-drag the middle mouse button (or mouse wheel).

Alternatively, you can enter 'PAN at any prompt, except those expecting you to enter text. The same restrictions apply as for transparent zooms.

### SCROLL BARS

Like most other Windows applications, AutoCAD provides a pair of scroll bars that allow you to pan the drawing horizontally and vertically (though not diagonally). I find them more useful than other forms of panning; however, Autodesk turns them off by default.

(To turn on the scroll bars, enter the OPTIONS command, and then choose the **Display** tab. In the Window Elements section, click the **Display scroll bars in drawing window** option, and then click **OK**.)

There are three ways to use the scroll bars:

- Click the arrow at either end of the scroll bar. This pans the drawing in one-tenth increments of the viewport size.
- Click and drag on the scroll bar button. This pans the drawing interactively as you move the button. This is also the way to pan by a very small amount.
- Click anywhere on the scroll bar, except on the arrows and button. AutoCAD pans the drawing by 80 percent of the view.

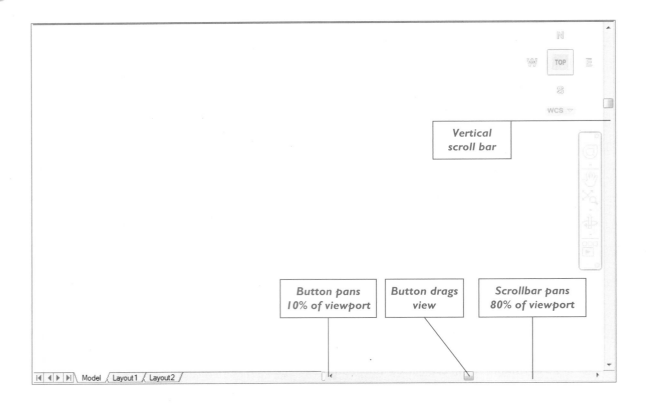

### -PAN

The **-PAN** command is the command-line alternative to panning. It prompts you to pick a pair of points, much like the LINE command:

> Command: **-pan**
>
> Specify base point or displacement: *(Pick a point.)*

As you move the cursor, AutoCAD draws a dragline.

> Specify second point: *(Pick another point.)*

After the second pick point, AutoCAD pans by drawing by the distance indicated with the two pick points.

The technical editor reports, "I never use any variants of PAN or ZOOM, except for ZOOM All. I just use the left mouse button to pan and the scroll wheel to zoom."

 **NAVIGATION BAR**

The navigation bar is a handy collection of buttons that let you move about the drawing visually. When on, the bar appears at the right edge of the drawing window, looking faded. When you move the cursor over the bar, it become legible. Several buttons that previously appeared on the status bar were relocated here in AutoCAD 2011.

From top to bottom, the navigation bar's button provides the following services:

**Steering Wheel** — activates the steering wheel for "walking" through drawings (equivalent to using the '**NAVSWHEEL** command); not covered by this book.

**Pan** — pans the drawing in realtime ('**PAN** command).

**Zoom** — zooms the drawing in real-time ('**ZOOM** command).

**3D Orbit** — rotates the drawing view in three dimensions ('**3DORBIT** command).

**Show Motion** — opens the show motion interface ('**NAVSMOTION** command); not covered by this book.

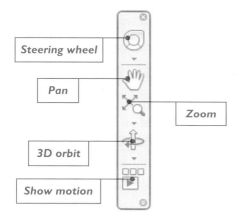

**NAVBAR** AND **NAVBARDISPLAY**

The NAVBAR command toggles the display of the navigation bar in the current viewport or space:

    Enter an option [ON/OFF] <ON>: *(Enter ON or OFF.)*

In contrast, the NAVBARDISPLAY system variable toggles the display of the navigation bar in *all* viewports and layouts.

The navigation bar contains shortcut menus that allow you to access additional commands and to customize it.

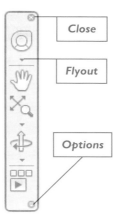

Click the ⊗ close button to close the navigation bar.

Click the ▼ flyout button to access command options for the steering wheel, zoom, and orbit functions.

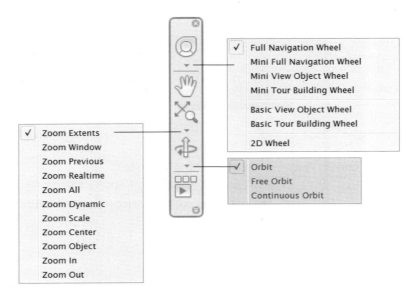

Click the ⊖ dropdown button to access options for the navigation bar.

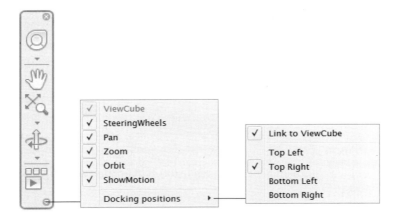

The initial shortcut menu toggles the display of the five buttons. (The **ViewCube** option is a puzzler, since it cannot be used; instead, use the OPTIONS command"s 3D Modeling tab's **Display ViewCube** option.)

The secondary shortcut menu positions the navigation bar:

**Link to View Cube** locates the navigation bar near the view cube. When this option is turned off, however, the navigation bar cannot be moved.

**Top Left**, etc. positions the navigation bar and view cube relative to the drawing window.

## CLEANSCREENON

The CLEANSCREENON command minimizes the user interface to maximize the drawing. It turns off the title bar, toolbars, and window edges.

 **Note** You can make the drawing area even larger by turning off the scroll bars and layout tabs. This is done through the **Display** tab of the Options dialog box. In addition, you can drag the Command Prompt away from the bottom of the screen area to make it float, and then make it transparent.

### TUTORIAL: TOGGLING THE DRAWING AREA

1. To make the drawing area as large as possible, start the **CLEANSCREENON** command:
   - On the keyboard, press **CTRL+0** (zero).
   - On the status bar, click the CleanScreen button.
   - At the 'Command:' prompt, enter the **cleanscreenon** command.

        Command: **cleanscreenon** *(Press ENTER.)*

2. To return to the "normal" screen, use the **CLEANSCREENOFF** command.
        Command: **cleanscreenoff** *(Press ENTER.)*

You can press CTRL+0 (zero) to toggle between maximized and regular views.

**VIEW**

The VIEW command stores and recalls views of the drawing by name. This lets you quickly move about a drawing. Think of it as combining the ZOOM and PAN commands, without needing to specify zoom ratios and pan directions.

You can save the current display as a view, or else window an area to define the view. Naturally, if you are saving the current view, you need to zoom and pan into that view first. Views can be stored relative to the current UCS (user-defined coordinate system), such as when using an architectural plan with an angled wing of the building or when working with 3D drawings.

You take advantage of named views outside of the VIEW command in two ways. When AutoCAD starts, you can specify the name of a view to be displayed when the drawing opens. And, when AutoCAD plots, you can specify a named view with the PLOT command.

Named views can be connected with layer names. That means you can access a view that automatically turns off certain layers for a more specific view. Views can also define visual styles and backgrounds.

The name of views can be changed with the RENAME command, and views can be deleted from drawings with the VIEW command's "hidden" **Delete** option.

### PREDEFINED VIEWS

AutoCAD provides several predefined views — six standard orthographic views and four standard isometric views. These are meant for viewing 3D drawings. The orthographic views show the top, bottom, left, right, front, and back of drawings, while the isometric views show the southwest, northwest, northeast, and southeast corners of drawings.

You can access these named views from the View tab's Views palette on the ribbon.

### TUTORIAL: STORING AND RECALLING NAMED VIEWS

1. To create named views, start the **VIEW** command with one of these methods:
   - In the ribbon's View tab, choose **Named Views**.
   - Or, at the 'Command:' prompt, enter the **view** command.

     Command: **view** *(Press* ENTER.*)*
   - Alternatively, enter the aliases **v** or **ddview** (the old name for this command) at the 'Command:' prompt.

   Notice the View dialog box.

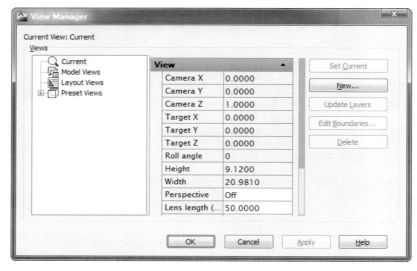

2. Click **New** to create a new named view. Notice the New View dialog box — which you can access directly with the **NEWVIEW** command.

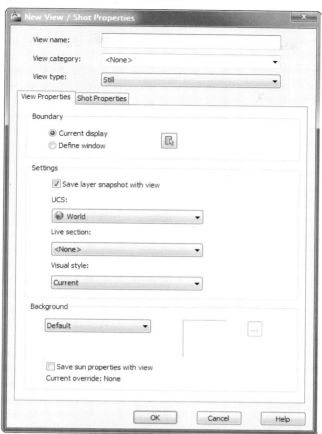

3. In **View Name**, enter a name for the view.

   Later, when it comes time to recall the view, you specify it by its name. That's why it makes sense to give views descriptive names, such as "Titleblock."

4. Determine the view's boundary:
   - **Current display** uses the current display as the named view.
   - **Define window** defines a view smaller than the current display through the **Define Window** button. AutoCAD dismisses the dialog boxes temporarily, and then prompts you to pick two points that define the rectangular window.

     Specify first corner: *(Pick a point.)*

     Specify opposite corner:  (Pick another point.)

5. Decide whether you want the visibility of layers changed with the view.

   When the **Save layer snapshot** option is turned on, AutoCAD remembers which layers were frozen, locked, turned off, and so on. When you later restore the view, the same layers are frozen, etc.

   You can ignore the other options — **UCS**, **View Category**, **Background**, and so on — because they are meaningful mainly for working with sheet sets and 3D drawings.

   (The Shot Properties tab is for the **NEWSHOT** command's show-motion animation feature, not covered by this book.)

6. Click **OK**. AutoCAD adds the view name to the list.

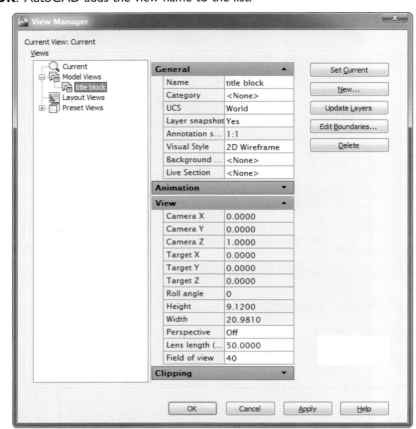

7. To recall the view, select its name (such as "Title block"), and then click **Set Current**.
   A small triangle points to the current view name.
8. Click **OK**, and notice how AutoCAD fills the screen with a closeup of the drawing's title lock (or whatever you windowed).

## TUTORIAL: RENAMING AND DELETING NAMED VIEWS

You can rename and delete named views through the shortcut menu. Right-click a view name:

To delete views: select a name, and then click **Delete**.

To rename views: under the General section, click the **Name** field, and then change the name.

The meaning of the items in this shortcut menu is as follows:

| Shortcut Menu | Comments |
| --- | --- |
| Set Current | Makes the named view current. |
| New | Creates new named views. |
| Update Layers | Changes layer visibility to the current setting. |
| Edit Boundaries | Shows the windowed view, with non-view areas in gray. |
| Delete | Removes the named view from the drawing. |

*Warning!* AutoCAD removes the view without warning.

## TUTORIAL: OPENING DRAWINGS WITH NAMED VIEWS

When you use the OPEN command, AutoCAD displays the Select File dialog box. One option you may have overlooked is **Select Initial View**, which is normally turned off. When turned on, AutoCAD presents the list of view names, from which you select one.

To open a drawing with an initial view:

1. Enter the **OPEN** command.
2. In the Select File dialog box, follow these steps:
    a. Select the *bias1.dwg* drawing file.
    b. Click **Select Initial View** to turn it on (a check mark shows).

    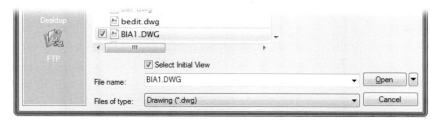

    c. Click **Open**.
3. AutoCAD starts to open the drawing, and then displays the Select Initial View dialog box. Select the "KYOTO" from the list, and then click **OK**.

    "M" means the named view was saved in model space.
    "P" means the view was saved in a layout.
4. AutoCAD finishes opening the drawing, and displays the view.

## TUTORIAL: STARTING AUTOCAD WITH NAMED VIEWS

AutoCAD originally ran on another operating system called DOS (short for "disk-based operating system upon which Windows was also designed.) In those days, it was common to start a program with command-line switches, which instructed the program how to start. These switches are still available for AutoCAD, and the /v switch specifies the named view to display when AutoCAD opens a drawing. The drawing, naturally, needs to have the named view; otherwise AutoCAD ignores /v, and instead shows the last saved view.

To use the /v switch, edit AutoCAD's command line, as follows:

1. On the Windows desktop, right-click the AutoCAD icon.
2. From the shortcut menu, select **Properties**.
3. In the Properties dialog box, select the **Shortcut** tab.
4. In the Target text box, add the path and file name of the drawing. An example is shown in boldface text, but may differ for your computer:

    "C:\AutoCAD 2011\acad.exe" **"c:\autocad 2011\sample\file name.dwg" /v titleblock**

5. Click **OK** to close the dialog box, and double-click the AutoCAD icon to see if it loads the drawing with the named view.

 **Notes** A space is required after the switch. Double quotes are needed when there are spaces in the file name.

You can have more than one shortcut icon on the Windows desktop. To make copies of icons, hold down the **CTRL** key, and then drag the icon. When you let go, Windows creates the copy, which you can then rename and edit.

### TUTORIAL: CHANGING WINDOWED VIEWS

AutoCAD makes it easy to change window views, by showing the current windowing.

1. Open the *bias1.dwg* file.
2. Enter the **VIEW** command, and then select the "KYOTO" view.
3. Click the **Edit Boundaries** button.

   Notice that the dialog box disappears. The area outside the window is shown in gray.

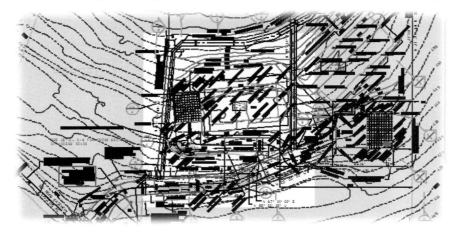

4. At the command prompt, AutoCAD asks you:

   Specify first corner: *(Pick a point.)*

   Specify opposite corner: *(Pick another point.)*

5. After you pick two points for the new window, AutoCAD shows the area in white (or black, depending on the background color of the drawing area), and asks again:

   Specify first corner (or press ENTER to accept): *(Press **ENTER**.)*

   Pick new points, or else press **ENTER** to accept the change.

   The dialog box returns.

6. Click **OK** to dismiss the dialog box.

## EXERCISES

1. Draw the following base plate from lines and circles.
   a. Set snap to 0.25 and grid to 1.0.
   b. Use the appropriate object snaps to assist your drafting.

   The dimensions need not be exact, but it may be helpful to know that the circles have a radius of 1.0.

   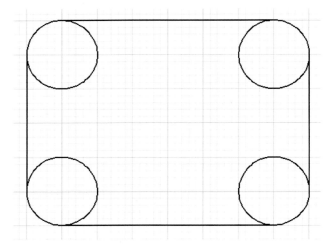

2. Draw the following hook plan using lines and arcs. The dimensions need not be exact, but it may be helpful to know that the photograph is full-size. The hook is 2.7" wide and the curved parts are 0.75" in diameter.

   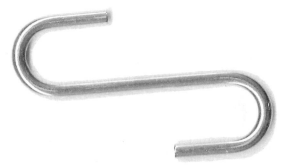

3. Draw the front, side, and top views of the bracket made of 1/4" sheet metal. The units are shown on the sketch.

   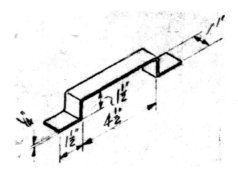

4. Draw an isosceles triangle with lines.
    a. Its angles are 45-45-90, and the two sides are 5 units.
    b. Use polar tracking set to 45 degrees and polar snap set to 5 units.

5. Draw the following hanger detail using.
    a. Set the snap to 0.25 and the grid to 1.0.
    b. Use the ZOOM All command to see the entire drawing area.
    d. Turn on ortho mode.
    d. Use the appropriate object snaps to assist your drafting.
   Follow the dimensions shown in the sketch.

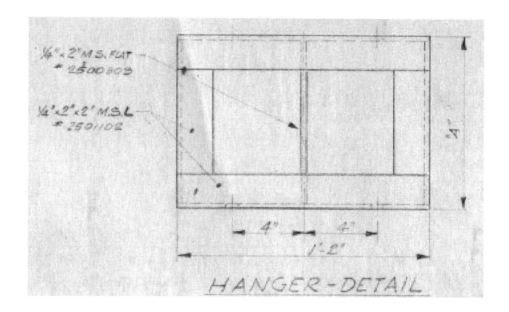

6. Draw the profile of the electrical switch spring shown below.
    a. Use polylines, switching between line- and arc-drawing mode, as necessary.
    b. Set the pline width to 0.1 units.
    c. Snap and grid are 1.0 units.

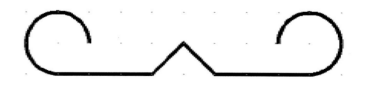

7. Use object snap tracking to assist you in creating the third view (shown in gray) of the industrial strength door wedge. The wedge is 9 units long, 2 units tall, and 3 units wide.

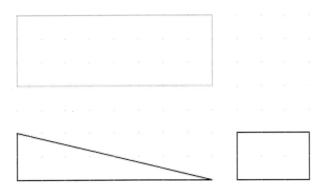

8. Use construction lines to assist you in creating the true view of the object shown below. The units are shown on the sketch.

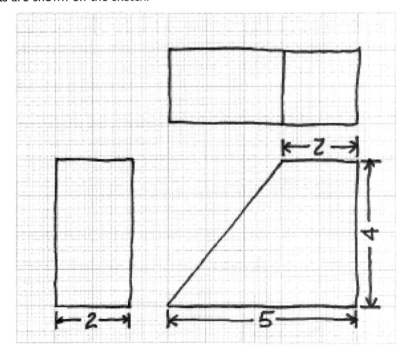

9. Using object snaps, place an arc tangent to the two circles.

   Then, draw the remaining lines, using the object snaps shown.

   The larger circle has a radius of 4 units, the smaller a radius of 2 units.

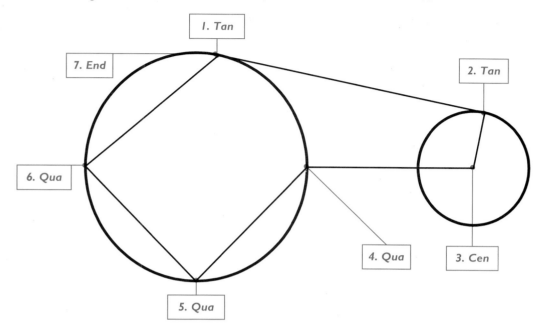

10. Draw the 3.7: wide by 3.5" tall diskette (89mm x 94mm).

    The figure below illustrates the diskette full size, 1:1; take your measurements directly from the photograph.

11. Draw the 3.9" wide by 1.75" tall cross brace for ceiling lamps. Take your measurements directly from the photograph, which is shown below at full-size, 1:1 scale.

12. Draw the 2.25" wide by 1.0" tall (57mm x 25mm) latch face for a security dead bolt lock. Take your measurements directly from the full-size photograph, below.

## CHAPTER REVIEW

1. What command allows you to enlarge and reduce the visual size of the drawing?
2. What is the purpose of the **PAN** command?
3. What is the **Realtime** option under the **ZOOM** command used for?
4. When a real-time zoom is in progress, can you override it? If so, how?
5. How do the **PAN** and **ZOOM** commands differ?
6. What does *toggle* mean?
7. What is a *transparent* command? How is this option invoked?
8. How do the **REDRAW** and **REGEN** command differ?
9. List four ways by which the grid can be toggled:
    a.
    b.
    c.
    d.
10. What is *temporary tracking* used for?
11. Name two ways to create *construction lines*:
    a.
    b.
12. What is the *aperture*?
13. What object snap mode would you choose to snap to a point object?
14. Can you snap to the midpoint of arcs?
15. When does AutoSnap come into effect?
16. Can different x and y spacings be given to the snap? Can it coincide with the grid spacing?
17. What advantages do rays and xlines have over grids?
18. What modes provide great accuracy for creating true horizontal and vertical lines?
19. What purpose does **APPint** mode serve?
20. Which command controls the extent of the grid?
21. Although the grid is not a part of the drawing, can it be plotted?
22. Where does a ray start? End?
23. Describe the function of the navigation bar.
24. In addition to the snap angle, the **Angle** option of the **SNAP** command also affects:
    a.
    b.
    c.
    d.
25. Which mode does **F8** toggle?
26. What is the difference between *ortho* and *polar* modes?
27. What are *tooltips*?

28. When would you use the **<** symbol in AutoCAD?
29. What is an *alignment path*?
30. Describe the function of *PolarSnap*.
31. Provide the meaning of the following abbreviations:

    osnap

    otrack

    appint

    tt
32. Define the following object snap abbreviations:

    int

    cen

    qua

    end
33. Which part of the object geometry do the following object snaps snap to?

    ins

    nod

    per

    tan
34. When would you use the **From** object snap?
35. Describe how the EXTension object snap operates.
36. Can AutoCAD snap to the intersection of two circles?

    If so, in how many places could the object snap take place?
37. Why does AutoCAD sometimes defer snapping to a tangent?
38. Which command is toggled by the following function keys?

    F3

    F9

    F10

    F11
39. What is the purpose of AutoSnap's magnet?
40. How do the **REDRAW** and **REDRAWALL** commands differ?
41. When might you turn off **REGENAUTO**?
42. Name the command that matches the alias:

    z

    sn

    xl

    ds
43. Which command maximizes the drawing area?
44. What is the function of the **DSVIEWER** command?
45. Name four commands that take advantage of *named views*.

    a.

    b.

    c.

    d.

46. How do you delete named views?
47. What happens when the lined grid is rotated?
48. Describe *view transitions*.
49. How would you turn off view transitions?
50. What keystrokes would you press to turn on MIDpoint object snap temporarily?

# CHAPTER 6
## Drawing with Efficiency

In addition to precision, another advantage of CAD over hand drafting is efficiency. "You should never have to draw anything twice," said John Walker, one of the founders of Autodesk, Inc. By drawing efficiently, you complete projects in less time — or more projects in the same time. In this chapter, you learn to use some of AutoCAD's commands for creating and reusing content, such as symbols (blocks) and hatch patterns:

**BLOCK** creates reusable components called "blocks."
**INSERT** places blocks in drawings.
**BASE** changes the base point of drawings.
**WBLOCK** exports blocks and drawings as *.dwg* drawing files.
**BEDIT** edits blocks.
**HATCH** places hatch patterns and solid fills; **GRADIENT** places gradient fills.
**HATCHEDIT** edits associative hatch patterns and fills.
**BOUNDARY** creates single regions from disparate areas.
**SELECT** selects objects by their location; **QSELECT** selects objects by their properties.
**DDSELECT** controls selection options.
**SELECTSIMILAR** selects all objects that share the same properties (new to AutoCAD 2011).
**ADDSELECTED** creates new objects based on properties of another (new to AutoCAD 2011).
**HIDEOBJECTS**, **ISOLATEOBJECTS**, and **UNISOLATEOBJECTS** hide, isolate, and show objects independently of layers (new to AutoCAD 2011).
**FILTER** selects objects based on their properties and location.
**DRAWORDER**, **TEXTTOFRONT**, and **HATCHTOBACK** (new to AutoCAD 2011) control the order in which objects are displayed.

---

**NEW TO AUTOCAD 2011** IN THIS CHAPTER
- **HATCH** and **GRADIENT** commands display the new Hatch Creation tab on the ribbon.
- Hatches and gradients can now be transparent and take on colors independently of the layer.
- **HATCHTOBACK** command displays hatch patterns behind all other objects.
- Selection cycling now displays a context menu.
- **SELECTSIMILAR** command selects all objects that share the same subset of properties.
- **ADDSELECTED** command creates new objects based on properties of another object.
- **HIDEOBJECTS**, **ISOLATEOBJECTS**, and **UNISOLATEOBJECTS** commands hide, isolate, and show objects independently of layers.

## BLOCK AND INSERT

The **BLOCK** command creates symbols that AutoCAD calls "blocks." Its complement is the **INSERT** command, which places blocks in drawings.

Using blocks in drawings gives you several distinct advantages. Entire libraries of blocks can be used over and over again for repetitive details. By using pre-drawn blocks, you reduce the amount of drafting required to complete drawings. For instance, AutoCAD includes more than a dozen libraries containing hundreds of blocks in the \DesignCenter and \Dynamic Blocks folders, found under \AutoCAD 2011\ Sample. Thousands of additional blocks are available online at Autodesk's seek.autodesk.com site.

**Left:** *Architectural blocks displayed by AutoCAD DesignCenter.*
**Right:** *Window blocks displayed by Autodesk Seek Web site.*

Blocks and *nested blocks* are excellent for building drawings from "pieces." (A nested block is a block placed within another block.) Several blocks require less space than several copies of the same objects, because AutoCAD stores only the information for the original block definition.

*Block of a telephone symbol made of a couple of dozen lines.*

**Note** It is more efficient to insert blocks than to copy the same objects over and over. The **COPY** command makes a complete copy each time you use it; in contrast, AutoCAD creates a single definition of a block, and then points to it when you make "copies" with the **INSERT** command. This reduces the drawing's file size and improves the display time.

While standard blocks are static, *dynamic blocks* can be modified interactively according to the designer's intentions. Dynamic blocks are easily stretched, rotated, mirrored, and aligned to other objects — if the blocks are defined that way. In addition, lookup tables allow one block to display itself in multiple forms.

Blocks can be used with *attributes*, which are text records that can be visible or invisible. (Attributes can only be attached to blocks.) The data from attributes can be exported to tables in drawings and to spreadsheet programs for further analysis. This is useful in facilities management, for example, where multiple occurrences of desks, chairs, and computers are found. Each is stored as a block, and has attributes associated with it, such as the person's name and telephone number.

AutoCAD stores blocks in the drawing in which they were made. You can share them with other drawings by using DesignCenter, or by exporting blocks as *.dwg* drawing files to disk with the **WBLOCK** command; AutoCAD can import any *.dwg* drawing file as a block.

## Editing the Content of Blocks

Because a block is a group of objects combined into a single object, you move, scale, copy, and erase it as though it were a single object. Indeed, blocks are so "tight" that it is difficult to edit their individual members. The three ways to edit the objects in blocks are as follows, in historical order:

**EXPLODE** command — breaks blocks into their individual parts; when you are done editing the parts, you can use the **BLOCK** command to recombine them.

**REFEDIT** command — "checks out" blocks, and then the **REFSET** command edits them by adding and removing objects; when you are done, the **REFCLOSE** command adds the changes to the block definitions.

**BEDIT** command — opens blocks in the Block Editor, where the objects can be edited; when you are done, the **BSAVE** command saves the changes to the original blocks. The **BSAVEAS** command saves the changes to new blocks, preserving the old ones.

### TUTORIAL: CREATING BLOCKS

There are three basic pieces of information AutoCAD needs before it creates blocks: (1) a name for the block, so that you can later identify it, (2) x, y, z coordinates for the insertion point, so that AutoCAD knows where to place the block, and (3) the objects that make up the block.

With that in mind, let's create a block with the BLOCK command. (Dynamic blocks are created separately with the BEDIT command.)

1. Start AutoCAD and then open the *block-006.dwg* file, the drawing of a louver detail.
2. Move the cursor over the lines. Notice that each one is highlighted, indicating the drawing is made of individual lines.
3. To turn this detail into a block, start the **BLOCK** command with one of these methods:
   - In the ribbon's 2D Home tab, choose the **Create** button from the Block panel.
   - Or, at the 'Command:' prompt, enter the **block** command:

     Command: **block** *(Press ENTER.)*

   - Alternatively, enter the aliases **b**, **bmod**, or **bmake** (one of the command's older names, short for "block make") at the keyboard.

   Notice the Block Definition dialog box:

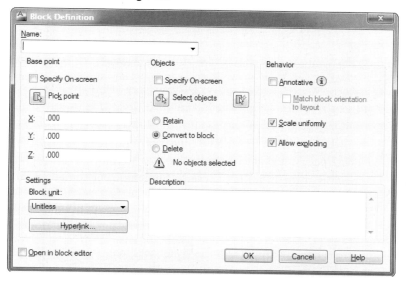

4. Select the objects that make up the block by following these steps:
   a. Choose the **Select Objects** button.
   b. Notice that the dialog box disappears and that AutoCAD prompts you:

   Select objects: *(Press* CTRL+A *to select all of the objects.)*
   Select objects: *(Press* ENTER *to return to the dialog box.)*

   You can use any method of object selection, such as Window or Fence, but in this case you press CTRL+A because you want all of the louver as part of the block.

   After you press ENTER to finish selecting objects, the dialog box reappears. Notice that AutoCAD reports the number of objects you selected (200, in this case), and that AutoCAD constructs an icon of the selected objects.

5. Blocks need names to identify them. Give this block a name by entering something like "Louver Detail" in the **Name** field.

   You can enter nearly anything, from a single letter to 255 characters. Not permitted are the following characters: < > / \ " ' : ; ? * | =. Spaces are permitted.

   **Notes** If you were to give the block a name that already exists in the current drawing, AutoCAD would warn you, saying: "Blockname is already defined. Do you want to redefine it?" Most times you respond **No**, and then change the name. Sometimes, however, you deliberately want to redefine an existing block, because perhaps you made an error in creating the block or you wish to replace it with an updated block. In this case, answer **Yes** to the question.

   If you need to see the names of blocks already defined in the drawing, click the down arrow in the Name field. AutoCAD displays the names of blocks in the current drawing.

   Block names that begin with * (asterisk), such as *X20, are created by AutoCAD. These are called *anonymous blocks*.

6. All blocks need a *base point*, the point where the block is later placed into the drawing. Usually, the base point is at the lower left corner of the block, but sometimes it may make more sense to locate it elsewhere for convenience.

In the figure below, the logical location for base point is in the center of the sprinkler head's. This makes it easy to insert and attach it to existing objects, such as lines representing the irrigation piping.

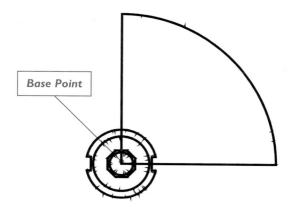

For this block, go with the traditional lower left corner. The base point is specified by coordinates in **X, Y**, and **Z** fields. Generally, you have no idea what the x, y, z coordinates should be, so you choose the **Pick Point** button, as follows:

   a.  Click the ▯ **Pick Point** button. Notice that the dialog box disappears temporarily.

   b.  AutoCAD prompts you:

           Specify insertion base point: *(Pick a point.)*

     I recommend using an object snap mode to make the pick accurate, such as ENDpoint for the end of a line or CENter for the center of a circle or arc.

   c.  After you pick the insertion point, notice that the dialog box reappears.

AutoCAD fills in the **X, Y**, and **Z** fields. Normally, the **Z** value is 0, unless you have set the elevation or selected a point on a 3D object.

**Note**  The **Specify On-screen** option allows you to leave the base point decision until after you click **OK** to dismiss the dialog box. At that point, AutoCAD will prompt you for the base point.

7.  You can ignore all of the dialog box's other settings for the purposes of this tutorial. Click **OK**, and AutoCAD creates the block — even though it's hard to tell it happened.

8. Pass the cursor over the newly-created block. Notice that the entire block is highlighted, which means that those 200 lines now make up a single entity.

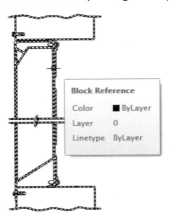

### Suggestions for Designing Blocks

When creating blocks, it is best to make them the size you intend for use. For example, office desks are commonly 24" x 36", so it makes sense to draw them that size.

Sometimes, however, you don't know what size will be used. For example, the block of a tree symbol could be inserted at 2' or 3' or 5' or whatever size you need. When you don't know how large the inserted block will be, draw the block at unit size. This means that the block fits inside a 1-unit square:

When it comes time to insert the block, the scale factor option sizes the block correctly and automatically.

Blocks should be drawn on layer 0 so that they insert on the current layer. If a block is created on a layer other than 0, say a layer named "Doors," then the block is be inserted on the current layer, but all constituent objects are placed on layer Doors.

#### CREATING BLOCKS: ADDITIONAL METHODS

The Block Definition dialog box contains a number of options that help you create blocks.

### Objects

The **Objects** section of the dialog box has these settings:

☐ **Specify On-screen** prompts you to select objects after you click the **OK** button to close the dialog box.

**Select** objects clears the dialog box, and then prompts you to select objects in the drawing. After pressing **ENTER**, the dialog box returns.

**Quick Select** displays the Quick Select dialog box, where you can select objects based on their properties; see the **QSELECT** command later in this chapter.

- **Retain** keeps the objects making up the block in the drawing as individual objects, as well as stores them in the new block definition.
- **Convert to Block** erases the objects, and replaces them with the block in the same position in the drawing (default).
- **Delete** erases the objects, while storing the block definition in the drawing. The block seems to disappear, which can freak out new AutoCAD users: "Wha' happened to my drawing!!??" You can insert the block with the **INSERT** command, or use the **OOPS** command to bring back the individual elements.

**Objects Selected** reports the number of objects selected for the block. You cannot create a block with no objects.

## Behavior

The **Behavior** section determines how the block reacts during and after insertion.

☐ **Annotative** gives the block annotative scaling, which means it is displayed only when the viewport scaling matches the block's annotative scale factor. (Clicking the **i** icon displays online help for this feature.)

☐ **Match Block Orientation to Layout** forces the block to remain "upright," even when rotated with the **ROTATE** command or through layout rotation for plotting. In the figure below, the boxed "Arbour" text is a block, with orientation turned on. When all objects in the drawing are rotated, the block remains in place.

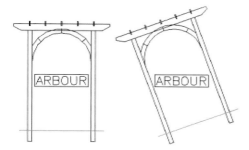

☐ **Scale Uniformly** fixes the scale factor ratio at 1:1:1. This means the block cannot be inserted with different scale factors in the x, y, and/or z directions. It's a good idea to turn on this option when you need the block to remain proportional.

☑ **Allow Exploding** allows the block to be reduced to its constituent parts with the **EXPLODE** command. Exploding causes block insertions to sever their link with block definitions, and to lose attribute data, if any. To prevent the block from being exploded, turn off this option.

## Settings

The **Settings** section sets settings.

**Block Unit** specifies units for the block. Usually, AutoCAD makes one drawing unit equal to the current unit. When you drag blocks from DesignCenter or *i-Drop*-enabled Web sites into drawings, AutoCAD automatically scales the block. (i-Drop is Autodesk technology that allows you to drag blocks from Web sites directly into drawings.)

Select a unit from the droplist:

| | | | |
|---|---|---|---|
| Inches | Nanometers | Feet | Microns |
| Miles | Decimeters | Millimeters | Decameters |
| Centimeters | Hectometers | Meters | Gigameters |
| Kilometers | Astronomical Units | Microinches | Light Years |
| Mils | Parsecs | Yards | Angstroms |
| Unitless | | | |

(1 parsec = 3.26 light years; 1 astronomical unit = average distance between Earth and Sun; 1 angstrom = typical size of an atom, or 1 ten-billionth of a meter; 1 microinch = 1 millionth of an inch.)

**Hyperlink** displays the Insert Hyperlink dialog box for attaching hyperlinks to blocks.

☐ **Open in Block Editor** opens blocks in the Block Editor after you click **OK**. This allows you to edit blocks further, or to turn them into a dynamic blocks.

## Description

The **Description** field is for text displayed by the INSERT and DESIGNCENTER commands.

**Description** provides room for you to include a long (or short) description of the block. This item is completely optional.

## INSERTING BLOCKS

After you define blocks with the **BLOCK** command or copy them from other drawings, you can place them with the **INSERT**, **DESIGNCENTER**, or **TOOLPALETTE** commands.

**INSERT** needs at least two pieces of information to place blocks: (1) the name, and (2) the insertion point. In addition, you can optionally specify the scale factors, normally 1.0; the rotation angle, usually 0 degrees; and other options. (**DESIGNCENTER** and **TOOLPALETTE** need only the insertion point.)

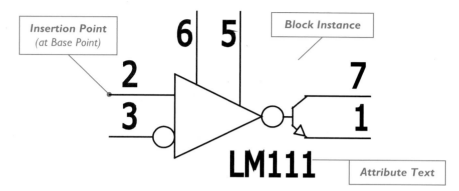

The insertion point is the x, y-coordinate in the drawing where the block is inserted, as illustrated below. More specifically, the block is inserted at its base point, which was defined early during the block's creation. The base point becomes the insertion point.

After blocks are inserted, you can use the INSertion object snap to snap to their insertion points.

### TUTORIAL: PLACING BLOCKS

1. Continuing from the last tutorial, place the block in the drawing. Start the **INSERT** command with one of these methods:
   - In the ribbon's 2D Home panel, choose the **Insert** button from the Block panel.
   - Or, at the 'Command:' prompt, enter the **insert** command:

     Command: **insert** *(Press* **ENTER.***)*

   - Alternatively, enter the aliases **i**, **ddinsert** (the command's old name), or **inserturl** at the keyboard.

   Notice the **Insert** dialog box.

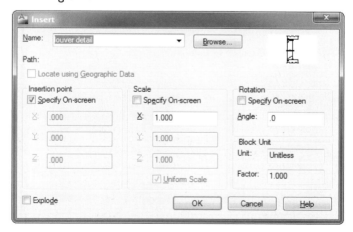

2. A drawing can contain many blocks, and so you need to select it by its name. Choose the name from the Name droplist.

(You can insert *any* drawing file as a block; click **Browse** to select the *.dwg* file name. Alternatively, you can use DesignCenter to search for and insert blocks and drawings.)

3. To specify the point at which the block is to be inserted point, you can enter the x, y, and z coordinates here in the dialog box — if you know them.

It is easier, though, to use the **Specify on Screen** option to pick a point later.

4. Click the **OK** button. Notice that AutoCAD prompts you:

Specify insertion point or [Scale/X/Y/Z/Rotate]: *(Pick a point, enter an option, or type the x, y, z coordinates.)*

After you specify the insertion point, AutoCAD places the block in the drawing. Technically, this is called "inserting an instance of the block definition."

 **Note** The inserted block resides on the layer that is current at the time the block is placed. Objects within the block retain the color, linetype, and so on of their original layer (i.e., the layer they were on before they were made part of the block). These properties are named "ByBlock."

There is one exception: if objects were on layer 0 when they were created, then they take on the properties of the layer on which they are inserted.

### INSERTING BLOCKS: ADDITIONAL METHODS

The Insert dialog box contains several options:

**Scale** changes the size of the block instance.

**Rotation** rotates the block instance.

**Explode** inserts the block with its original components, not as an instance.

In addition to the dialog box, the command-line version of the command lists a number of options, including some that are hidden from you:

| Insert Option | Meaning |
|---|---|
| Basepoint | Relocates the block's base point. |
| Scale | Specifies a single scale factor for the x, y, and z-directions. |
| X | Scales the block in the x-direction. |
| Y | Scales the block in the y-direction. |
| Z | Scales the block in the z-direction. |
| Rotate | Rotates the block counterclockwise. |

The following options are hidden:

| | |
|---|---|
| **PScale** | Sets the preview scale factor for the x, y, and z-directions. |
| **PX** | Sets the preview scale factor and insertion scale for the x-direction. |
| **PY** | Sets the preview scale factor and insertion scale for the y-direction. |
| **PZ** | Sets the preview scale factor and insertion scale for the z-direction. |
| **PRotate** | Sets the preview rotation angle and insertion scale factor. |

### Basepoint

The **Basepoint** option allows you to specify a new base point, the point at which the block is inserted. This can make it easier inserting blocks relative to other objects, such as placing a bathtub symbol into the corner of the bathroom. AutoCAD temporarily places the block in the drawing, and then prompts you:

Specify base point: *(Pick a point to relocate the base point.)*

Pick a point, and AutoCAD proceeds with the 'Insertion point:' prompt, allowing you to place the block in its intended location.

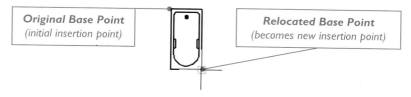

Changing the base point affects only this block insertion; it does not affect the block reference's base point. In other words, you need to reuse this option every time you insert the block and need a different base point.

### Scale

AutoCAD normally uses a unit scale factor for the inserted block. ("Unit" means that the scale factor is 1.0, so that the block is inserted at its original size.) There are times, however, when you might want the block to be a different size.

When the scale factor is more than 1.0, the block is made larger; when the scale factor is less than 1.0, the block is smaller.

The three scale factors — X, Y, and Z — can be the same or different, negative or positive. (When the **Scale Uniformly** option is turned on during block creation, only the X Scale factor is available.) When the scale factors are different from each other, the block is stretched; when negative, the block is mirrored, including its text; when positive, the block is inserted the right way around.

The illustration below shows the effect of some different scale factors on inserted blocks:

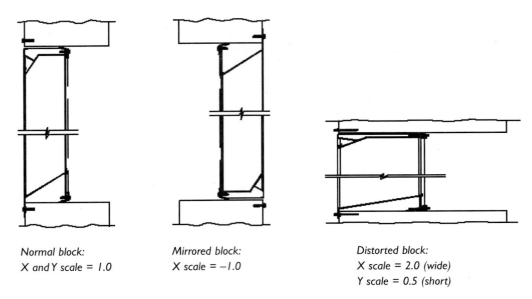

*Normal block:*
*X and Y scale = 1.0*

*Mirrored block:*
*X scale = –1.0*

*Distorted block:*
*X scale = 2.0 (wide)*
*Y scale = 0.5 (short)*

The **Specify On-screen** option determines whether you specify the scale factor in the dialog box or in the drawing:

- ☐ Specify the scale factor in the **X**, **Y**, and **Z** text entry boxes.
- ☑ Specify the factor in the drawing. After you click the Insert dialog box's **OK** button, Auto-CAD will prompt you:

    Enter X scale factor, specify opposite corner, or [Corner/XYZ] <1>: *(Press* **ENTER** *to accept the default factor, enter a different x-scale factor, pick a point, or specify an option.)*

    Enter Y scale factor <use X scale factor>: *(Press* **ENTER** *to make the y scale factor the same as x, or enter the y-scale factor.)*

AutoCAD presents a large number of options at the command line. Here's what they mean:

| Scale Option | Meaning |
| --- | --- |
| X scale factor | Scales the block in the x-direction. |
| Opposite corner | Specifies a rectangle that defines the x and y scale factors; the insertion point is the first corner. |
| Corner | Has the same effect as the above option. |
| XYZ | Scales the block independently in the x, y, and z-directions. |
| Y scale factor | Scales the block in the y-direction. |
| Use X scale factor | Makes the y scale factor equal to the x scale factor. |

The **Uniform Scale** option makes the y and z scale factors the same as the x factor.

### Rotation Angle

You can have AutoCAD rotate the block about its insertion point relative to the original orientation of the block. The **Specify On-screen** option determines when the specification takes place:

- ☐ Specify the rotation angle in the **Angle** text entry box.
- ☑ Specify the angle in the drawing. After you click **OK**, AutoCAD prompts you:

Specify rotation angle <0>: *(Enter the rotation angle, or pick two points to indicate the angle.)*

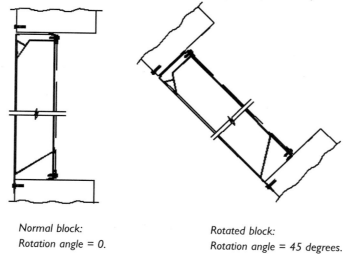

*Normal block:*
*Rotation angle = 0.*

*Rotated block:*
*Rotation angle = 45 degrees.*

If you choose to indicate the angle, move the cursor. AutoCAD ghosts a rubber band cursor between the previously-set insertion point and the cursor. Move the cursor until the rubber-banded line shows the desired angle, and then click. The distance between the two points is irrelevant. The angle, shown by rubber-banded line between the insertion point and the angle point, determines the angle of insertion.

### Explode

When you turn on the **Explode** option, AutoCAD inserts the block's constituent parts in the drawing — the lines, arcs, and text that make up the block definition.

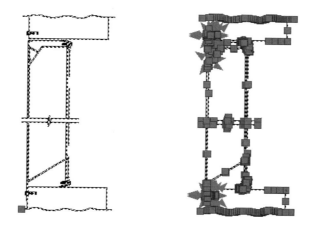

**Left:** *Block, with a single grip (at lower left corner).*
**Right:** *Exploded block, with grips for every constituent part.*

This option is not available when the **Allow Exploding** option was turned off during block creation. Note that AutoCAD limits you to specifying a single scale factor for exploded blocks, which applies equally to the x, y, and z directions.

## Preview

The Preview window shows you what the block looks like.

Icons in the preview window indicate special blocks:

## DESIGNCENTER AND TOOL PALETTES

The drawback to the **INSERT** command is that it provides a preview of just one block at a time. Often times, it is easier to pick out blocks visually from a larger group based on their look, rather than on their names. Other times, you may not be sure of the block's location: which drawing or folder is it stored in? For these reasons, you may find it easier to use Design Center or Tool palettes to insert blocks, which let you see groups of blocks, as well as use drag'n drop insertion.

This section describes how blocks work with Design Center and Tool palettes.

### Inserting Blocks with DesignCenter

DesignCenter displays all block definitions in the current drawing, as well as blocks in drawings stored on your computer, on the local network, and the Internet.

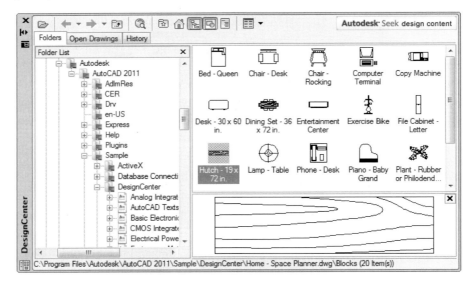

After you find the block you want, you simply drag it into your drawing. There are no prompts to answer, because AutoCAD determines the scale factor automatically, and determines the insertion point by where you let go of the mouse button. Object snaps help you place blocks accurately.

When you need more control, right-click the block's icon in DesignCenter, and then select **Insert Block** from the shortcut menu. AutoCAD displays prompts on the command line for setting insertion point, scale factor, and rotation angle options before the block is inserted in the drawing.

### Inserting Blocks with Tool Palettes

Whereas DesignCenter displays all blocks definitions in drawings automatically, the Tool palette is meant to contain the blocks you use the most; you add them manually. You add blocks to the palette by dragging them from the drawing or DesignCenter onto a palette. Once on the palette, the blocks become available to any drawing you open or start.

Once blocks are on the palette, you can drag them into drawings — just as with DesignCenter.

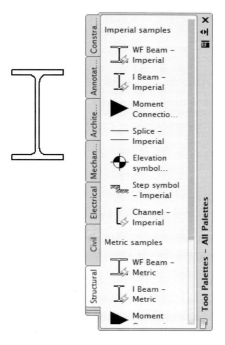

When you need to control placement of a block, right-click its icon, and then select **Properties** from the shortcut menu. The Tool Properties dialog box lets you change the scale factor, rotation angle, and so on. These changes are semi-permanent: they affect future insertions of the block until you again change the properties.

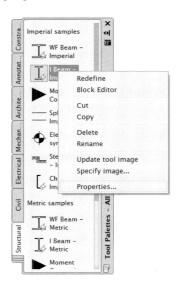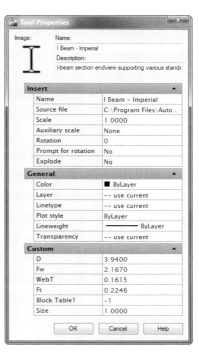

(A drawback to Tool palettes is that they display images of blocks that no longer exist. Dragging the icon of non-existent blocks results in nothing.)

**Note** You can drag .dwg files from Windows Explorer into drawings. AutoCAD treats them differently, depending on where you drag them.

**Drag files to the title bar** — AutoCAD opens the .dwg files as drawings.
**Drag files into the current viewport** — AutoCAD inserts the .dwg files as blocks.

### CREATING AND INSERTING BLOCKS: ADDITIONAL METHODS

AutoCAD has a couple of commands useful for working with blocks:

- **BASE** specifies the base point (origin) of the drawing.
- **WBLOCK** exports blocks as .dwg files on disk.

Let's look at each one.

### Base

The INSERT command can insert other drawings as blocks into the current drawing. The inserted drawing is normally inserted with coordinates 0,0 as its base point. If you wish to change the base point, use the BASE command on the drawing before inserting it.

Command: **base**
Specify base point <default>: *(Pick the point for the new base point.)*

Save the drawing to save the new base point, and then insert it in the other drawing.

### WBlock

The WBLOCK command is the complement of the INSERT command. Whereas INSERT inserts .dwg drawing files as blocks, WBLOCK saves all or part of the current drawing as another .dwg file on disk. This is an alternative to using the DesignCenter for sharing blocks.

- At the 'Command:' prompt, enter the **wblock** command.
- Alternatively, enter the **w** alias at the keyboard.

Command: **wblock** *(Press ENTER.)*

AutoCAD displays the Write Block dialog box, which looks similar to the BLOCK command's dialog box. The primary difference is in the Source area, where you choose what you want written to disk.

### Source

**Block** saves blocks as .dwg files; the blocks must exist in the current drawing.

**Entire Drawing** saves the whole drawing with an exception: unused layers, blocks, styles, and dimension styles are not saved. This is a quick way to clean up a drawing.

**Objects** saves parts of the drawing; you specify the base point, and then select the objects.

### Destination

In the **Destination** area, specify the file name and location (drive and folder.)

When you change the file name extension from .dwg to .dxf, AutoCAD saves the drawing or block in DXF format. DXF is short for "drawing interchange format." (With AutoCAD 2004, Autodesk removed WBLOCK's ability to save drawings in XML format — extended markup language.)

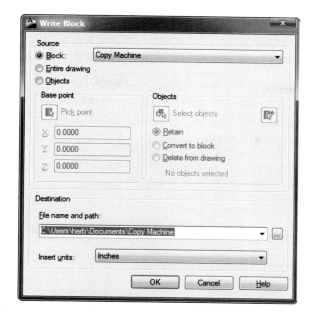

## REDEFINING BLOCKS

After you insert the same block many times in drawings, you may need to change the block. By redefining one block, you can change all inserted blocks of the same definition. This is an especially powerful feature: imagine being able to change 100 identical drawing parts in a single operation!

Here's how:

1. Explode a block that has been inserted. This action breaks the link to its block definition and reduces the block to its constituent parts.
2. Edit the exploded block.
3. Use the **BLOCK** command to convert the edited parts back into a block. Give this "new" block the same name. You may select the name from the **Name** droplist.
4. Click **OK**. AutoCAD displays the following dialog box:

5. Click **Yes**. Notice that all other blocks of the same name change, including those inserted with **MINSERT**.

If the block was dynamic, all of its dynamism is lost when exploded.

If the block is another drawing that was inserted whole, edit the original drawing, and then reinsert it as described above.

## BEDIT

The **BEDIT** command invokes the Block Editor environment for editing blocks, as well as making them *dynamic* (interactive).

The Block Editor is a separate editing environment. Many (but not all) AutoCAD commands are available for drawing and editing the blocks. In addition, the editor includes unique commands for constructing dynamic blocks. The Block Editor performs these functions:

- Creates new blocks.
- Edits existing blocks.
- Adds dynamic actions to blocks.
- Converts dimensional constraints to parameter constraints.

Inside the editor, there are several special commands for all block definitions (those operating only inside the Block Editor). Additional system variables affect the display of dynamic blocks only while in the Block Editor.

### TUTORIAL: EDITING BLOCKS WITH THE BLOCK EDITOR

1. From the CD, open *bedit.dwg*, a drawing that contains a block.
2. To edit the block with the Block Editor, invoke the **BEDIT** command by one of these methods:
   - In the ribbon's 2D Home panel, choose the **Block Editor** button from the Block panel.
   - Or, at the 'Command:' prompt, enter the **bedit** command:

     Command: **bedit** *(Press* ENTER.*)*
   - Alternatively, enter the **be** alias at the keyboard.

 **Note** There are two alternate, faster methods to start the Block Editor: selecting a block, and then entering the **BEDIT** command opens the block in the Block Editor. Even faster is double-clicking the block, which also opens it in the Block Editor.

3. (Click **Cancel** if AutoCAD asks, "Do you want to see how dynamic blocks are created?")
   Before entering the Block Editor, AutoCAD displays the Edit Block Definition dialog box.

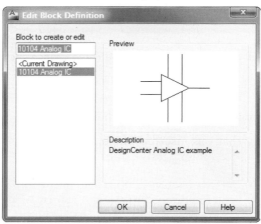

The dialog box provides you with three options:
- To edit an existing block, select its name from the list.
- To edit the entire drawing, select **<Current Drawing>**.
- To create a new block from scratch, enter a new name in the **Block to create or edit** field.

(When a ⚡ yellow lightning icon appears in the corner of the preview window, this indicates a dynamic block is selected.)

4. Select a block to be edited. For this tutorial, select "10104 Analog IC," and then click **OK**. Notice the Block Editor. It looks much like regular AutoCAD, but the background color is pale gray, and it includes a ribbon and palette specific to block editing. In addition, the UCS icon is shifted to the block's insertion point.

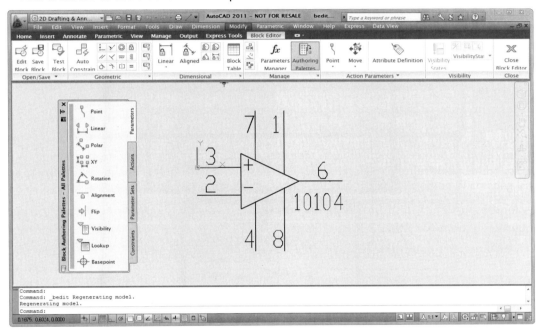

The toolbar contains some commands useful for block editing, as illustrated below; the remaining commands are specific to editing dynamic blocks, and are not discussed here.

5. Make some changes to the block. For instance, select the lines making up the triangle, and thicken them with the **Lineweights** droplist; use the STYLE command to change the font to Arial.
6. Use the BSAVE command to save the changes.
7. On the toolbar, click the **Close Block Editor** button. Back in AutoCAD's drawing editor, notice that previous insertions of the block take on the changes you made.

 HATCH

The HATCH command places hatch and fill patterns in drawings, while the HATCHEDIT command edits the patterns.

### ABOUT HATCHING

Sometimes it can be helpful to view objects as if they were cut apart, called a "section." Mechanical designers use sections to show interior details of parts; civil engineers detail profiles by showing sections along roadways and railways; architects use sections through entire structures to show how buildings are to be constructed.

Hatching is used to show solid parts of sections, and AutoCAD easily hatches section views: after drawing a section, apply hatch patterns to cut areas with the HATCH command.

The spacing between hatch lines should be relative to the scale of the section, while the pattern should be oriented at 45 degrees from the main lines of the cut area, whenever possible.

#### Full Sections
Full sections cut across the entire object, and are usually cut wherever the view needs to be clarified.

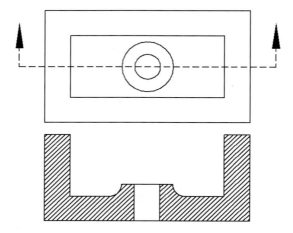

Parts that are "cut" are shown with crosshatching.

#### Revolved, Offset, and Removed Sections
It can be helpful to view rotated sections of objects, to show their cross section more clearly. Such sections are referred to as revolved sections.

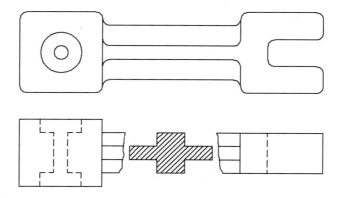

Offset sections are cut along an uneven line. Offset sections should be used carefully; change the cutting plane only to show essential elements.

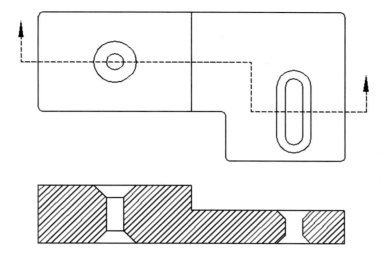

Removed sections are similar to revolved sections, except the section is not placed at the point where the section was cut.

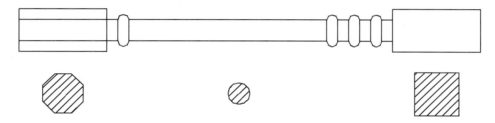

### How Hatching Works

The HATCH command generates a boundary around an area automatically, and then places an associative hatch pattern within the area. Associative hatch patterns automatically update themselves when you change their boundaries, which is very helpful. If you prefer that the patterns not update themselves, then place non-associative hatches.

Hatches differ from other objects in that the lines making up the pattern are treated as one object. When you select one line, the entire hatch is selected.

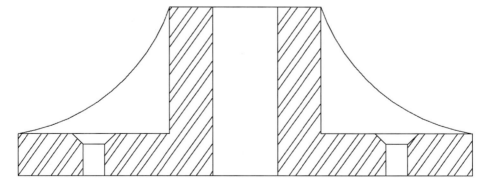

To place hatch patterns, AutoCAD needs to know at least these parameters: (1) the name of the hatch pattern, (2) the area to be hatched, and (3) if it is associative. Optionally, you can specify the scale, rotation angle, and style of boundary detection. You can also require that the boundary be retained, and you can create simple custom patterns.

## BASIC TUTORIAL: PLACING HATCH PATTERNS

1. Start AutoCAD, and then open the *hatching.dwg* file.
2. To place hatch patterns in the drawing, start the **HATCH** command with one of these methods:
   - In the ribbon's 2D Home tab, choose the **Hatch** button from the Draw panel.
   - Or, at the 'Command:' prompt, enter the **hatch** command:

        Command: **hatch** *(Press ENTER.)*

   - Alternatively, enter the aliases **h**, **bh**, or **bhatch** (one of its old names, short for "boundary hatch") at the keyboard.

    Notice that the ribbon changes to display the Hatch Creation tab.

3. AutoCAD prompts you:

        Pick internal point or [Select objects/seTtings]: *(Move cursor into area to be hatched.)*

   Move the cursor into the area to be hatched. Notice that AutoCAD previews the hatch pattern.

   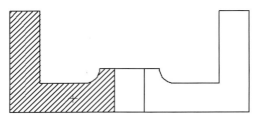

   (In AutoCAD 2010 and earlier, the Hatch and Gradient dialog box would appear. If it appears now, then turn on the **HPDLGMODE** system variable by setting it to 1.)

    **Note** You can enter **u** at the 'Pick internal point' prompt to unhatch an area.

4. Click to place the pattern, and then repeat in the other half of the drawing.

   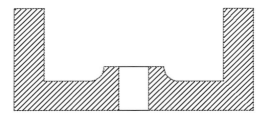

5. Move the cursor to the ribbon, and then choose a different pattern from the Pattern panel.

   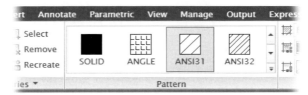

To see more patterns at once, click the ⏷ expand button.

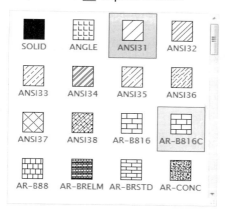

You can scroll through the list to see additional samples, as well as a variety of types of gradient fills. To fill areas will solid colors, choose SOLID.

6. Choose the AR-B816C sample, a pattern that represents brickwork. Notice that the hatching changes in the drawing...

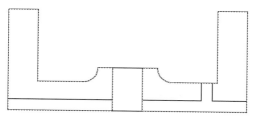

7. ...but the scale is too large. To adjust the scale, click the 🔲 hatch scale spinners, or enter a scale factor (found in the Properties panel). A scale of 0.03 ought to be just about right.

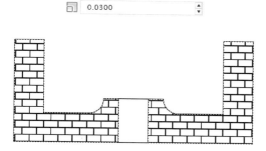

You can use other properties to change the angle and transparency of the pattern.

8. Brick is normally red, with gray being the cement binding the bricks together:

a. From the 🔲 Hatch Color droplist, choose gray.

b. From the ![icon] Background Color droplist, choose red.

Notice that the hatch pattern instantly changes colors.

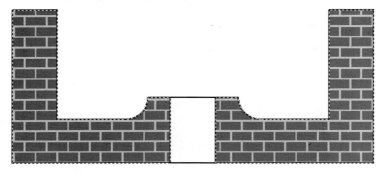

9. Press **ENTER** to end the **HATCH** command.

### PLACING HATCH PATTERNS: ADDITIONAL METHODS

The ribbon's Hatch Creation tab has panels that group related functions. The following sections go through each of the panels. Similar functions can be found in the Hatch and Gradient dialog box.

**Boundaries Panel**

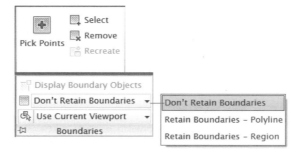

Hatches must be contained within *boundaries*; otherwise AutoCAD will not place the hatch. Boundaries are one or more objects that form closed polygons, and include objects like lines, polylines, circles, and arcs.

Before placing a hatch pattern, you identify the objects that form the closed boundary. If the boundary is not quite closed, you can use the **Gap Tolerance** option (found in the Options panel) to tell AutoCAD to hatch boundaries with gaps.

## HATCH AND GRADIENT DIALOG BOX

Prior to AutoCAD 2011, the **HATCH** and **GRADIENT** commands displayed the Hatch and Gradient dialog box. (When **HPDLGMODE** = 1, they will do so in the current release of AutoCAD as well.) This dialog box can also be reached by entering **T** at the **HATCH** command's 'Pick internal point or [Select objects/seTtings]' prompt.

The drawback to the dialog box is that it is not interactive, as is the ribbon's Hatch Creation tab.

Click the **Gradient** tab to view the options for placing gradient fills:

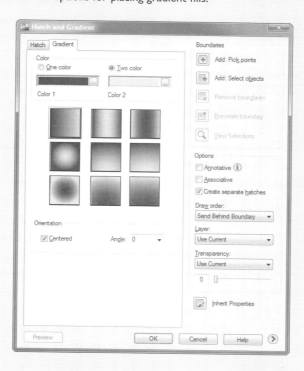

There are several ways by which you identify boundaries:

- Pick points within areas to be hatched, and then let AutoCAD find the boundaries.
- Select objects that make up the boundary.
- Remove islands from inside boundaries.

### Defining Boundaries by Picking Areas

The simplest way to define the boundary is to show AutoCAD by selecting a point within the area to be hatched at 'Pick internal point' prompt. Let's look at an example. The illustration shows a chair with areas that could be hatched: seat, back, arm rests.

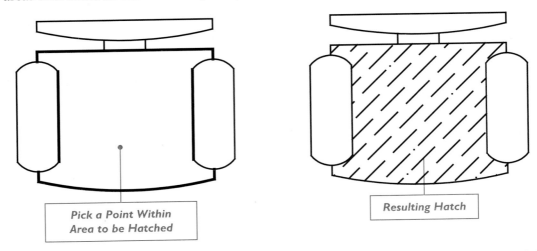

*Pick a Point Within Area to be Hatched*

*Resulting Hatch*

AutoCAD analyzes the area, looking for leakage. Unknown to you, AutoCAD then places a polyline as the boundary around the inside of the area. (You can place boundaries by themselves with the BOUND-ARY command, discussed later in this chapter.)

If you select a point in an area that is not contained, AutoCAD displays this warning:

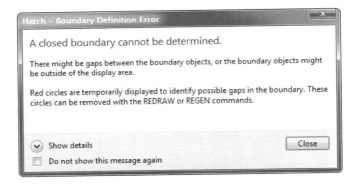

In addition, AutoCAD highlights the gaps with red circles.

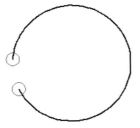

Click **Close** to close the dialog box, and then consider your options:

- Draw a line to close off the area, or use the **JOIN** command to close an arc.
- Define a larger maximum for the allowable gap (Options panel).

## Defining Boundaries by Selecting Objects

You can select the objects that will contain the hatch pattern. The illustration below shows a shape to be hatched that's been constructed with four lines.

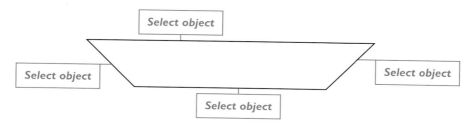

To place a hatch in the polygon, choose the **Select** button. AutoCAD prompts you to select the objects that will form the boundary. Place the pickbox over each object, and then click.

Select objects or [picK internal point/seTtings]: *(Pick one or more objects.)*

 **Note** Use the **Window** or **Crossing** option to select all the sides of the boundary in one operation.

As soon you are finished selecting objects, AutoCAD floods the area with the current hatch pattern.

## Removing Boundaries

Sometimes the area to be hatched is not as straightforward as the examples hatched above. Hatch areas commonly contain *islands*, other closed areas within the boundary — such as the two illustrated below.

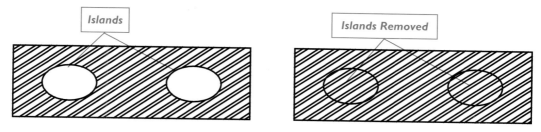

AutoCAD initially assumes you do not want islands hatched. If, however, you want patterns drawn right through islands, then choose the **Remove** button. AutoCAD prompts you for the boundaries (islands) to remove:

Select boundary to remove: *(Pick object.)*

## Recreating Boundaries

The **Recreate** button creates a polyline around a selected hatch pattern, and then associates the hatch with the boundary.

### Additional Boundary Options

Click the ▼ **Boundaries** label to see additional options for boundaries. The **Display Boundary Objects** option is available only when editing hatches; see the HATCHEDIT command.

### Boundary Retention

Boundaries are drawn as polylines or region objects. AutoCAD normally erases the boundary after the HATCH command is finished. To retain the polyline boundary in the drawing, choose **Retain Boundaries**.

There are two types of objects that AutoCAD uses for creating boundaries: polylines (for compatibility with older versions of AutoCAD) and regions.

### Selecting Boundary Sets

When you pick a point inside a boundary, AutoCAD analyzes all objects visible in the current viewport. You can, however, change the set of objects AutoCAD examines. In large drawings, reducing the set lets AutoCAD operate faster. Click the **Use Current Viewport** button for the following options:

> **Use Current Viewport** — examines all objects visible in the current viewport to create the boundary (default).
>
> **New** — prompts you to select objects from which to create the boundary set. AutoCAD includes only objects that can be hatched.
>
> **Existing Set** — creates the boundary from the objects selected with the **New** option.
> You must use the **New** option before the **Existing Set** option.

### Select a Hatch

The first step is to select the type of pattern. The Pattern droplist displays (a) hatch patterns defined by the *acad.pat* or *acadiso.pat* hatch pattern definition files, (b) gradient fills, (c) solid fills, and (d) user defined.

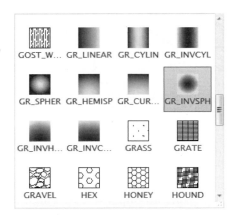

Each time you select a pattern, the preview in the drawing changes to match. You can freely select from among patterns, gradients, and the solid fill — unlike in earlier releases of AutoCAD.

### ISO Patterns

ISO hatches appear when you start a new drawing with the *acadiso.dwt* template drawing. ISO patterns permit a pen width to be assigned to the pattern. (See Properties, below.) "ISO" is short for the International Organization of Standards.

### User Defined Patterns

When you select **User** as the hatch type, you work with a single pattern consisting of simple, parallel lines. You specify three parameters: (a) angle, (b) scale, and (d) single or double.

The **Angle** option is the same as described earlier. Turning on the **Double** in the Properties panel causes AutoCAD to draw the pattern twice, the second time at 90 degrees to the first pattern, creating a crosshatch. The **Scale** option replaces Spacing of earlier releases, which specifies how far apart the parallel lines are drawn.

The illustration below shows an example of combining a linetype with a user-defined hatch pattern. To create this effect, follow these steps:

1. Use the **LINETYPE** command to load a linetype; below, I used the Dashed linetype.
2. Set the dashed linetype as current.
3. Enter **h** to start the **HATCH** command.
4. From the Patterns droplist, choose **User**.
5. In the Properties panel, change the **Angle** to 45 and turn on the **Double** option.

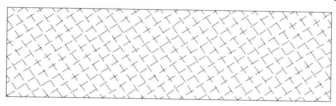

You may need to adjust the hatch pattern **Scale** setting, as well as change the linetype scale with LTSCALE.

## Hatch Properties

The Properties panel affects the look of the hatch patterns and gradient fills.

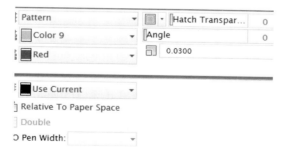

### Pattern

Although you generally select hatches, gradients, and solid fills from the **Pattern** panel, the ⌧ Pattern droplist replicates the droplist from the Hatch and Gradient dialog box.

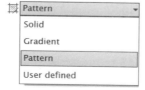

## Hatch and Background Color

Hatches and solid fills normally take on the color of the layer upon which they are placed, whereas gradients have colors you have to specify in the ribbon. As of AutoCAD 2011, you can also specify hatch and background colors for patterns.

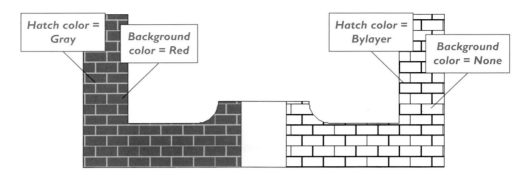

*Left: Hatch and background colors set to red and gray, respectively.*
*Right: Hatch and background colors set to their defaults.*

The **Hatch Color** droplist shows 'Use Current' as the default, meaning it uses the layer's color. To change the color of the hatching (lines making up the pattern), click the droplist and then choose another color.

The **Background Color** droplist shows 'None' as the default, meaning it uses the background color of the drawing area (or paper). To change the color of the background (areas between pattern lines), click the droplist and then choose another color.

## Hatch Transparency

Hatches, gradients, and solid fills can be faded. Since they can take up large areas, it can be useful to make them transparent. Drag the Hatch Transparency slider to vary the transparency.

Alternatively, you can have the hatch take on the same transparency as specified by its layer (ByLayer) or by the block it may be part of (ByBlock).

Transparency does not work under two conditions: (a) if Transparency is turned off on the status bar, and (b) if your computer's graphics board does not support the feature.

### Angle and Scale

Hatch patterns can be drawn at varying angles and scales. The sample swatches in the dialog box are shown at an angle of 0 degrees, even if the lines of the pattern are drawn at angles.

Set an angle by dragging the **Angle** slider, or else use the keyboard to enter the angle.

Scale multiplies the hatch pattern by a factor. For instance, changing the value in **Scale** to 2 doubles the spacing between lines; changing it 0.5 reduces the spacing to half-size. Scaling hatch patterns is similar to scaling text and linetypes: enter the inverse scale factor in the Scale box. For example, if the drawing is to be plotted at a scale of 1:100, then the hatch pattern should be applied at a scale of 100.

Click the spinner buttons to change the scale factor by 0.001 per click, or use the keyboard to enter a scale factor.

Better yet, click the **Annotative** button (found in Options panel) and the pattern scales itself.

 **Note** The OSNAPHATCH system variable performs the useful function of determining whether object snap modes will snap to hatch patterns and gradient fills. When set to **0**, the default, then all osnap modes ignore hatch and fill patterns; when set to **1**, osnap modes select component objects within the hatch.

 Layer

The Layer droplist lets you choose the layer upon which to place the hatch pattern. 'Use Current' places the hatching on the current layer. Click the droplist to choose from other layer names.

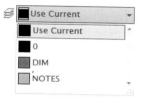

Should you wish to have the hatch on a new layer, then you'll need to use the LAYER command to create it before starting the hatching — or move it there afterwards with the Properties palette.

 **Note** Hatches can be handled more easily if they are put on their own layer. They can also be turned off and frozen to speed redraw time. Be sure that the layer linetype is continuous. Although the hatch pattern may contain dashed lines and dots, the linetype should be continuous to ensure a proper hatch.

### Relative to Paper Space

When you turn on the **Relative to Paper Space** option, AutoCAD scales hatch patterns appropriately for each layout, using the layout's scale factor. (This option is grayed out in model space.)

To change the thickness of the pattern lines, use lineweights.

### Double Hatching

The **Double Hatching** option is available only when you choose the User hatch pattern. It repeats the pattern at 90 degrees, as described earlier.

### ISO Pen Width

The **ISO pen width** option is available only when you select an ISO hatch pattern. It allows you to specify the width of the lines making up the ISO pattern.

### Origin Panel

The default origin of hatch patterns is 0,0 — coinciding with the origin of drawings. In many cases, that's fine, but sometimes you want adjacent hatches to line up with each other, or the pattern to begin at the edge of its boundary.

In the figure, the first hatch pattern has its origin at 0,0; notice that the brick pattern does not begin at the left or bottom edge. To correct this problem, the origin was moved to the lower left corner of the wall, and now the bricks line up correctly.

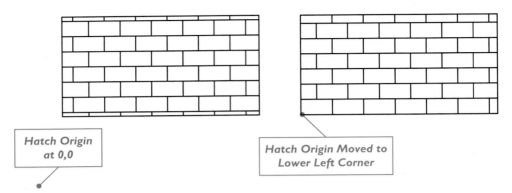

Hatch Origin at 0,0

Hatch Origin Moved to Lower Left Corner

The **Origin** panel allows you to relocate the origin, which is applied to all subsequent hatches — until you change the origin again. This is useful for ensuring brick patterns line up with the edges of walls. (Prior to AutoCAD 2006, the hatch origin was relocated using the SNAPBASE system variable.)

The **Set Origin** button specifies a different starting point for the hatch pattern. (The default is 0,0). AutoCAD prompts you:

> Specify origin point: *(Pick a point in the drawing to relocate the origin of the hatch pattern.)*

Additional Origin Options

Click the ▼ **Origin** bar to reveal additional options for setting the origin point of the hatch pattern.

Each hatch pattern has an *extents*, an imaginary rectangle that encompasses the hatch object. When the pattern is rectangular, then the extents match up; otherwise, they do not, as illustrated by the figure.

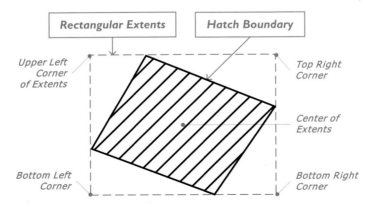

The Origin panel has these buttons that immediately set the origin to the corners of the extents. Select one of the buttons to set the origin to the lower left, lower right, upper left, upper right, or center of the extents, respecitvely.

The **Use Current Origin** button reads the value stored in the HPORIGIN system variable, which is 0,0 until it is changed by one of the earlier buttons.

The **Store as Default Origin** button stores the new hatch origin coordinates in the in the HPORIGIN system variable.

### Options Panel

The **Options** panel contains miscellaneous settings for hatch patterns. As it turns out, the ribbon contains most, but not all, options available for hatch patterns. For those you need the Hatch and Gradient dialog box, which you can access by clicking the dialog box button.

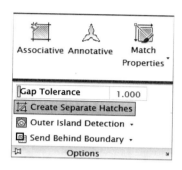

### Associative and Non-Associative Hatch Patterns

When AutoCAD draws hatch patterns, they are *associative*, which means they associate with their boundaries: change a boundary, and the hatch pattern updates by enlarging or shrinking itself to fit. AutoCAD draws the associative hatch pattern as a single object.

In some rare cases, you may want to place *non-associative* hatch patterns. This pattern is "dumb," not knowing its boundary, or its parameters. AutoCAD draws the non-associative hatch pattern as a collection of independent lines collected into an anonymous block.

The **Associative** button switches new hatch patterns between associative or non-associative. If you are not sure, keep the option set to Associative, because you can later use the explode command to change associative hatch patterns to non-associative; going the opposite direction is not possible.

### Annotative Scaling

The **Annotative** option turns on the annotative scaling property. When this property is on, the hatch pattern does not appear in model space unless its scale factor matches that of the viewport. See the chapter on text for more about annotative scaling.

### Matching Properties

Associative hatch patterns are self-aware. They know their parameters, such as pattern name, scale, and origin. The **Match Properties** button lets you copy hatch pattern properties, and in fact does exactly the same thing as the Match Properties button in the Home tab, but with an adjustment for hatch patterns. (This option is named "Inherit Properties" in the Hatch dialog box.) The button displays two options:

**Use Current Origin** — copies the properties of an existing hatch pattern, but uses the origin coordinates stored in the HPORIGIN system variable.

**Use Source Hatch Origin** — copies all properties of the selected hatch, including its origin.

You use this button to make a new hatch pattern look exactly like an existing one.

### Additional Options

Click the ▼ **Options** bar to reveal additional options for creating hatch patterns.

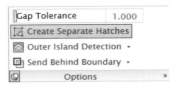

### Gap Tolerance

In the earlier days of AutoCAD, if a boundary had a gap, the hatch pattern could "leak" out, covering much of the rest of the drawing. Hatches are made of many lines, and displaying them was a processor-intensive operation. Leaky hatches could bog down the slow computers of the day. Hence, CAD operators were careful to ensure hatches would not leak.

Then Autodesk programmers wrote code that checked for leaks before applying the hatching. If it found a gap, AutoCAD would warn, "Valid hatch boundary not found." But CAD operators still had to fix gaps manually.

As of AutoCAD 2005, hatches could deal with gaps in the boundary as large as 5000 units. (Warning! Hatching to a gapped boundary leaves associative hatching turned off until you turn it back on again.) The **Gap Tolerance** slider lets you adjust the size of allowable gap/

### Create Separate Hatches

When you select two or more boundaries, AutoCAD can fill them with one hatch, or each with its own. The **Create Separate Hatches** button determines the number of patterns created when more than one boundary is selected: one for all, or one for each.

### Island Detection

When the area to be hatched contains other objects, the **Island Detection** droplist lets you choose which get hatched. *Island detection* solves the problem of how to hatch closed areas inside an outer boundary and text. The island detection options are as follows:

**Normal** — hatches inward from the outermost boundary, skips the next boundary, and hatches the next. In addition, text is not hatched. In the figure below, notice how the text is bordered by the hatch. With this style, an invisible window protects text from being obscured by hatching.

**Outer** — hatches only the outermost enclosed boundary. The hatch continues only until it reaches the first inner boundary, and continues no further. Text is not hatched.

**Ignore** — hatches all areas defined by the outer boundary, with no exceptions. This style even hatches through text.

**No** — turns off island detection.

The **Ignore** and **No** settings seem to operate identically, but there is a subtle difference between them: Ignore sets the HPISLANDDETECTION system variable to 2, while No sets a different system variable, HPIS-LANDDETECTIONMODE, to 0. In effect, No is a toggle that turns the other three modes on or off.

The figures below illustrate the island detection modes:

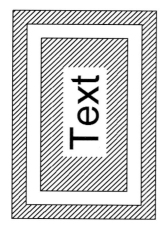 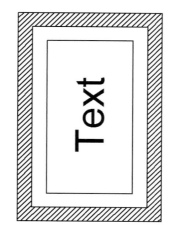 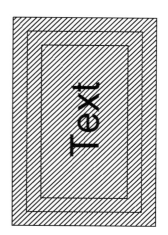

**Left:** *Islands and text hatched using Normal style.*
**Center:** *Outer island only hatched with Outer style.*
**Right:** *All islands and text hatched using Ignore style.*

## Draw Order

Hatch patterns and filled areas can overwhelm the drawing when they are displayed over top of other objects. This occurs when the hatches are placed later in the drawing process. The ▦ Draw Order droplist determines whether hatch patterns are displayed in front of or behind their boundary and other objects.

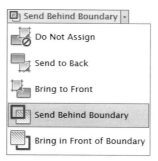

The best solution is to select **Send to Back**, which ensures the hatch patterns are visually underneath all other objects in the drawing. This option is particularly useful for areas filled with solid color or gradients. You can use this option, together with the DRAWORDER and TEXTTOFRONT commands (described later), to control the display order of all objects in drawings.

Separately, you can have the hatch patterns in front of or behind their boundaries. Bringing them in front of the boundary hides the boundary, while maintaining it in the drawing.

## Close

The ✕ **Close Hatch Creation** button exits the HATCH command.

**Notes** You can use the **TRIM** command to trim hatch and fill patterns. The process is similar to trimming other objects. (See Chapter 9, "Additional Editing Options.") Start the **TRIM** command, and then select an object as the cutting edge. At the "Select object to trim" prompt, pick the portion of the hatch you want trimmed (removed).

Conversely, the hatch pattern can be used to trim other objects. Trimmed hatches are associated with the new boundary.

**Left:** *Hatch pattern and circle — before trimming.*
**Center:** *The circle trimming the hatch.*
**Right:** *The hatch lines trimming the circle.*

## GRADIENT

The **GRADIENT** command places gradient fills.

*Gradients* are colors that change in intensity (more white or more black), or change from one color to another. Using gradients allows you to create 3D-like effects in 2D drawings, as the examples below illustrate.

**Left:** *Cylinder gradient pattern.*
Right: *Curved gradient pattern.*

The **GRADIENT** command displays the Hatch Creation tab on the ribbon. It looks identical to the one for hatch patterns, except for two controls specific to gradient fills: Centered and Tint.

AutoCAD supports two types of gradients, one- or two-color:

**One-color** — a single color gradually shifts from dark to light, such as dark blue to light blue.
**Two-color** — two colors gradually shift from one to the other, such as red to green.

The ribbon controls unique to gradients are as follows:

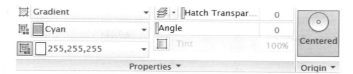

The **Tint** slider changes the strength of the dark-light range of single-color gradients. It ranges from black (0) to white (100), as illustrated below. At 50%, the color is uniform.

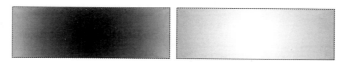

*Left:* Tint set to 0 (black).
*Right:* Tint set to 100 (white).

The **Centered** button toggles the gradient between being centered in the fill area, or starting at the left edge.

*Left:* Centered turned on...
*Right:* ...and turned off.

Other hatch controls take on a slightly different meaning when applied to gradients:

The Hatch and Background Color droplists of hatch patterns become Gradient Color 1 and 2 for gradients. The **Gradient Color 1** droplist is used together with the Tint slider to control one-color gradients.

The **Gradient Color 2** droplist determines the second color for two-color gradients. Clicking the button switches between one- and two-color gradients.

The **Angle** slider rotates the gradient inside the fill area.

*Left:* Angle at 0 degrees.
*Right:* Angle at 45 degrees.

# HATCHEDIT

The HATCHEDIT command changes the properties of hatch patterns and gradient fills. The command displays a dialog box that looks exactly the same as that of the HATCH command, but with some options grayed out, meaning they are unavailable.

Alternatively, double-click a hatch or gradient, and AutoCAD displays the Hatch Editor tab on the ribbon, and the Properties palette. (The Hatch Editor tab is identical to that of the Hatch Create tab.) The tab and palette allow you interactively to change the look of the patterns.

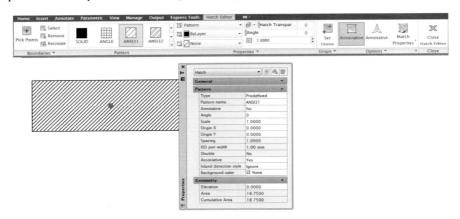

To make one hatch or gradient look like another one, use the **Match Properties** button. AutoCAD prompts you:

Select hatch object: *(Pick a hatch pattern.)*

Pick another pattern. AutoCAD copies its properties, name, colors, and rotation angle, and then applies it to the selected hatch or gradient.

## Direct Editing of Hatch Boundaries

The boundaries of hatch patterns can be stretched larger and smaller simply by dragging one of the grips (colored squares) found on the boundary, as illustrated below.

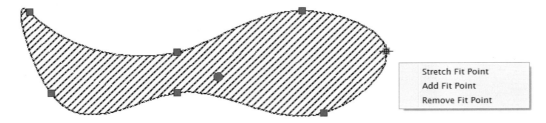

Pause the cursor over a grip, and the shortcut menu provides additional editing operations, such as adding and removing points.

## BOUNDARY

The BOUNDARY command creates single regions from disparate areas.

Earlier, you read how AutoCAD creates a temporary boundary out of a polyline or region to hold the hatching. To work with boundaries independent of hatch patterns, use the BOUNDARY command. The boundary is selected in the same manner, and is constructed as a polyline or region. The difference is that no hatch is placed within the boundary.

> **Note** The BOUNDARY command converts the outlines of objects to polylines; this allows you to select them more easily for editing.

You probably won't notice the new boundary in the drawing, because AutoCAD traces over the objects defining the boundary. To edit it, for example to move or color it, use the **Last** object selection mode, which selects the last-drawn object in the drawing.

### TUTORIAL: CREATING BOUNDARIES

1. To place boundaries in the drawing, use the **BOUNDARY** command with one of these methods:
   - In the ribbon's 2D Home panel, choose the **Boundary** button from the Draw panel.
   - At the 'Command:' prompt, enter the **boundary** command.

    Command: **boundary** (Press ENTER.)

   - Alternatively, enter the aliases **bo** or **bpoly** (the command's original name, short for "boundary polyline") at the keyboard.

   Notice the Boundary Creation dialog box.

2. Select the object you want used for the boundary: a polyline or a region:
   **Polyline** boundaries are more easily edited.
   **Region** boundaries can be analyzed for properties, such as the centroid (geometrically-weighted center), using the **MASSPROP** command.

3. Click **Pick Points** to pick a point within the area to be bounded. If the area is enclosed, AutoCAD creates the boundary, makes the following report, and exits the command:
    BOUNDARY created 1 polyline

    If AutoCAD finds the area is not closed, a dialog box complains in somewhat misleading terms, "Valid hatch boundary not found."

## SELECT

The SELECT command selects objects in drawings. You typically select objects before editing them. For example, you can select all objects that are inside an area, or all objects in the entire drawing.

Most commonly, objects are not selected with SELECT, but during other commands that present the 'Select objects:' prompt. For example, you can start the MOVE command, and then select the objects you wish to move. Both the SELECT command and the 'Select objects:' prompt have identical options.

More usefully, AutoCAD allows you to select objects without this command or the prompt. When no command is active, you can pick one or more objects: AutoCAD highlights them, and displays grips as shown on the previous page. (These small squares are known as "handles" in other software application; in AutoCAD, handles are unique identifiers assigned to objects.) You then edit the object by manipulating its grips; press ESC to exit this direct editing mode.

As the cursor passes over them, objects take on highlighting and thicken their lines. When you click the mouse button, the object is selected. AutoCAD identifies selected objects by displaying them with highlighting — as if they were drawn with a dashed line.

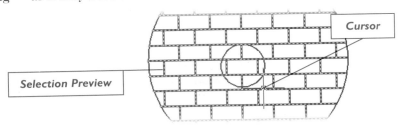

Selection preview is controlled by options found in the Selection tab of the Options dialog box, described later in this chapter.

### TUTORIAL: SELECTING OBJECTS BY LOCATION

1. To select objects in the drawing, use the SELECT command with one of these methods:
   - At the 'Command:' prompt, enter the **select** command.
   - Alternatively, enter a selection option at any "Select objects" prompt.

   Command: **select** *(Press ENTER.)*

2. Notice that AutoCAD prompts you to select objects:

   Select objects: *(Select an object.)*

   1 found Select objects:

   As you select objects, AutoCAD highlights them, and keeps a running tally of the number of objects selected, as in "1 found."

3. You can keep selecting objects until you press ENTER to exit the command:

   Select objects: *(Select more objects.)*

   3 found Select objects: *(Press ENTER to end object selection.)*

4. After pressing ENTER to exit the SELECT command, the highlighting disappears. The objects you selected are added to AutoCAD's *selection set*.

You may wonder, what good is the SELECT command? Not a lot, which is why many CAD operators never use it. You can, however, access selection sets made with the SELECT command through the **Previous** option at any 'Select objects:' prompt. In this way, the SELECT command is good for creating a selection set used later by other commands. The selection set is changed the next time an object is selected.

SELECT is one of the rare commands that does not list its options. To force it to do so, enter **?**, as follows:

```
Command: select
Select objects: ?
*Invalid selection* Expects a point or
Window/Last/Crossing/BOX/ALL/Fence/WPolygon/CPolygon/Group/Add/Remove/Multiple/Previous/Undo/AUto/SIngle/SUbobject/Object: (Enter an option.)
```

That probably seems like too many options for you! It is useful to know about all of them, but in practice you probably use just a half dozen — All, Previous, Last, Window, Crossing, and point are the ones I use most frequently, while the technical editor uses Auto the most. "Because," he says, "Auto is the most efficient, and is the default. Most of the other options are left over from older versions of AutoCAD."

| Select Option | Meaning |
|---|---|
| Expects a point | Selects one object under the cursor. |
| Window | Selects all objects fully within a rectangle defined by two points. |
| Last | Selects the most recently created visible object. |
| Crossing | Selects objects within, touching, and crossing a rectangle defined by two points. |
| BOX | Selects all objects within and/or crossing a rectangle specified by two points. If the selection rectangle is picked from:<br>• Right to left, performs Crossing selection.<br>• Left to right, Window selection. |
| All | Selects all objects in the drawing, except those residing on frozen and locked layers. |
| Fence | Selects all objects crossing a selection line. The fence line can cross itself. |
| WPolygon | Selects all objects completely within a selection polygon, which can be any shape but cannot cross itself; AutoCAD closes polygon. |
| CPolygon | Selects all objects within and crossing a selection polygon, which can be any shape but cannot cross itself; AutoCAD closes polygon. |
| Group | Selects all objects comprising a named group. |
| CLass | Selects object classes defined by add-on software, such as Autodesk Map. |
| Add | Switches selection mode to Add, after being in Remove mode. Objects selected by any means listed above are added to the selection set. |
| Remove | Switches selection mode to Remove. Objects selected by any means listed above are removed from the selection set. As an alternative, you can hold down the SHIFT key to remove objects from the selection set. |
| Multiple | Selects objects without highlighting them. Also selects two intersecting objects when the intersecting point is selected twice. |
| Previous | Adds objects that were previously selected to the selection set. The Previous selection set is ignored when you switch between model and paper space. |
| Undo | Removes the object most recently added to the selection set. |
| Auto | Selects objects by three methods. Picking a point in a blank area of the drawing starts Window or Crossing mode, depending on how you move the cursor:<br>• Moving right to left performs Crossing selection.<br>• Moving left to right performs Window selection.<br>• Picking an object selects it. |
| Single | Selects the first object(s) picked, and then does not repeat the "Select objects:" prompt. |
| SUbobject | Selects subobjects on 3D models: vertices, edges, and faces; equivalent to holding down the CTRL key during object selection. |
| Object | Exits subobject selection mode. |

## ALL (Select All Objects)

The **All** option selects all the objects in the drawing, except those on frozen and locked layers — most of the time. For some commands, such as COPYCLIP, the **All** option selects only those objects visible in the viewport; in model space, it selects all objects, visible or not.

> **Notes** As a shortcut, you can press the **CTRL+A** keys, which selects all objects in the drawing (subject to the restrictions noted above). To select nearly all objects in the drawing, first select all objects with the **All** option, and then use the **Remove** (explanation follows) option to "deselect" the exceptions.

## Pick (Select Single Object)

To select a single object, click on it with the cursor. There is no need to enter any options at the "Select objects:" prompt:

   Select objects: *(Pick an object.)*

In this mode, AutoCAD selects a single object. To select another object, pick again; to "unselect" the object, hold down the **SHIFT** key. To select from two or more overlapping objects, follow this procedure:

1. Hold down the **CTRL** key while selecting an object. If the object overlaps another, AutoCAD displays <Cycle on> at the 'Command:' prompt.
2. Release the **CTRL** key.
3. Click repeatedly (cycle) until AutoCAD selects the object you want.
4. Press the **ENTER** key to exit cycle mode and return to the 'Select objects:' prompt.

## Visual Selection Cycling

AutoCAD 2011 adds a new way to select overlapping objects. A context menu lists the names of overlapping objects. As you pass the cursor over each object name, the related object is highlighted in the drawing.

The context menu is controlled by the SELECTIONCYCLING system variable: (0) off, (1) on, but does not display the context menu, and (2) on, and displays context menu.

## W (Select Within a Window)

One of the most common forms of object selection is the **W** option (short for "window"). It places a rectangle around the objects to be selected. You define the rectangle by picking its opposite corners, as follows:

   Select objects: **w**
   Specify first corner: *(Pick point 1.)*
   Specify opposite corner: *(Pick point 2.)*

Objects entirely within the window rectangle are selected. If an object crosses the rectangle, it is not selected (see Crossing mode). Only objects currently visible on the screen are selected.

You can place the selection rectangle so that all parts of those objects you want to choose are contained in the windows, while those that you don't want are not entirely contained. With this method, you select objects in an area of your drawing where objects overlap.

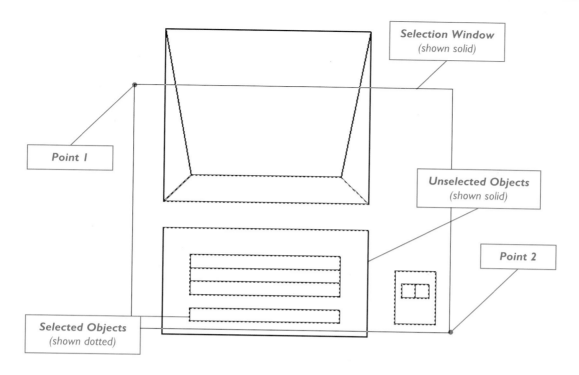

## C (Select with a Crossing Window)

The **C** option (short for "crossing") is similar to the **Window** option, with an important difference: objects crossing or touching the rectangle are included in the selection set — as well as those objects entirely within the rectangle. Define the rectangle by picking its opposite corners.

    Select objects: **c**

    Specify first corner: *(Pick point 1.)*

    Specify opposite corner: *(Pick point 2.)*

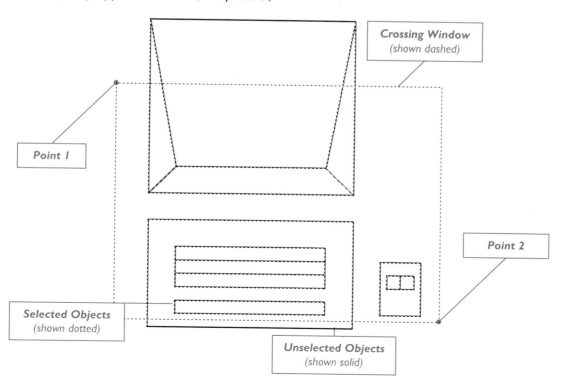

Notice that the crossing window is dashed and filled with a transparent green color. This distinguishes it from the Window rectangle, which is made of solid lines and blue color.

The color and opacity of area selections can be changed with the Options dialog box: choose the **Selection** tab, and then click the **Visual Effect Settings** button.

BOX (Select with a Box)

The **BOX** option allows you to use either the crossing or window rectangle to select objects. Define the rectangle by picking its opposite corners.

After you pick the first corner of the box, move the cursor either to the right or to the left. If you move to the right, the result is the **Window** selection: AutoCAD selects objects completely within the blue selection rectangle.

When you move to the left, the result is the crossing selection: AutoCAD selects objects within the rectangle, as well as those crossing it. In addition, the rectangle is drawn with dashed lines and filled with green color.

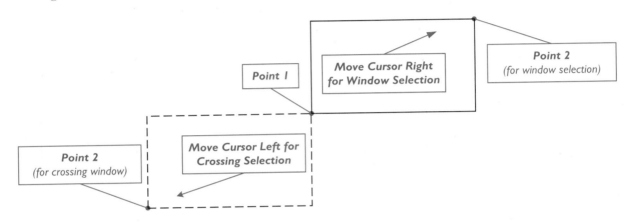

F (Select with a Fence)

The **F** option (short for "fence") uses a polyline to select objects. The fence is displayed as a dashed line; all objects crossing the fence are selected. You can construct as many fence segments as you wish, and can use the **Undo** option to undo a fence line segment.

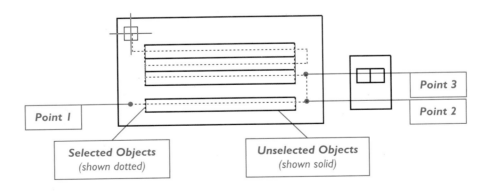

Select objects: **f**

First fence point: *(Pick point 1.)*

Specify endpoint of line or [Undo]: *(Pick point 2.)*

Specify endpoint of line or [Undo]: *(Pick point 3.)*

*et cetera*

### AU (Select with the Automatic Option)

The **AU** option (short for "automatic") combines the pick and **Box** options. It is the default for most selection operations. After you enter **AU** in response to the "Select objects" prompt, you select a point with the pickbox. If an object is found, the selection is made.

If an object is not found, the selection point becomes the first corner of the **Box** option. Move the box to the right for Window, or to the left for Crossing. The **AU** option is excellent for all users who wish to reduce the number of modifier selections.

### WP (Select with a Windowed Polygon)

The **WP** option (short for "windowed polygon") selects objects by placing a polygon window around them. The polygon window selects in the same manner as the **Window** option: all objects completely within the polygon are selected. Objects that cross or are outside the polygon are not selected.

The difference is that the **WP** option creates a multisided window, instead of a rectangle. Let's look at a sample command sequence:

> Select objects: **wp**
> First polygon point: *(Pick point 1.)*
> Specify endpoint of line or [Undo]: *(Pick point 2.)*
> Specify endpoint of line or [Undo]: *(Pick point 3.)*

After you enter the first polygon point, build the window by placing one or more endpoints. The polygon rubber-bands to the cursor intersection, always creating a closed polygon window filled with the color blue.

Undo the last point entered by typing **U** in response to the prompt. Pressing ENTER closes the polygon window and completes the process.

Note that the polygon window must not cross itself or rest directly on a polygon object. If it does, AutoCAD warns "Invalid point, polygon segments cannot intersect," and refuses to place the vertex.

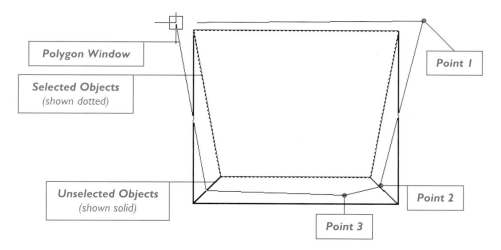

### CP (Select with a Crossing Polygon)

The **CP** option (short for "crossing polygon") works in the same manner as the **WP** option, except that the polygon functions in the same manner as a crossing window. All objects within or crossing the polygon are selected. The crossing polygon is displayed with dashed lines filled with green, similar to a crossing window.

### SI (Select a Single Object)

The **SI** option (short for "single") forces AutoCAD to issue a single "Select objects:" prompt. (All other selection modes repeat the "Select objects:" prompt until you press ENTER.) You can use other selection options in conjunction with **SI** mode, such as the Crossing selection shown in this example:

> Select objects: **si**
>
> Select objects: **c**
>
> Specify first corner: *(Pick point 1.)*
>
> Specify opposite corner: *(Pick point 2.)*
>
> *nnn* found

The Single option is useful for efficient single object selection in macros, because it does not require pressing ENTER to end the object selection process.

### SU and O (Select Subobjects and Objects)

The **SU** option (short for "subobjects") selects faces, edges, and vertices of 3D models. This is equivalent to holding down the CTRL key during object selection. The **O** option (short for "objects") exits subobject selection mode.

### M (Select Through the Multiple Option)

Each time you select an object, AutoCAD scans the entire drawing to find the object. If you are selecting an object in a drawing that contains a large number of objects, there can be a noticeable delay.

The **M** option (short for "multiple") forces AutoCAD to scan the drawing just once. This results in shorter selection times. Press ENTER to finish the object selection and begin the scan. Note that AutoCAD does not highlight the objects selected until you press ENTER .

> Select objects: **m**
>
> Select objects: *(Pick three times.)*
>
> 3 selected, 3 found

### P (Select the Previous Selection Set)

The **P** option (short for "previous") uses the previous selection set. This very useful option allows you to perform several editing commands on the same selection set, without reselecting the objects. In addition, you can add to and remove from the **Previous** selection set.

### L (Select the Last Object)

The **L** option (short for "last") selects the last object drawn still visible on the screen. When the command is repeated, and **Last** is used a second time, AutoCAD chooses the same last object and reports "1 found (1 duplicate), 1 total." This option is useful for immediately editing an object just drawn.

### G (Select the Group)

The **G** option (short for "group") adds the members of a named group to the selection set. (Groups are covered later in this chapter.) The option prompts:

> Enter group name: *(Enter a group name.)*

## Changing Selected Items

You can add and remove objects from the group of selected objects by using modifiers. Modifiers must be entered after you select at least one object, and before you press ENTER to end object selection.

### U (Undo the Selected Option)

The **U** option (short for "undo") removes the most recent addition to the set of selections. If the undo is repeated, you will step back through the selection set. This shortcut replaces the two-step process of using the **Remove** option, and then remembering the objects to pick.

### R (Remove Objects from the Selection Set)

The **R** option (short for "remove") removes objects from the selection set by any object selection method.

> Select objects: **r**
> Remove objects: *(Pick an object.)*
> 1 found, 1 removed, 2 total

**Notes** The **Remove** option is useful when you need to select a large number of objects with the exception of one or two objects located in the area. Select all the objects, then use the **R** option to remove the excess objects.

**SHIFT+Select** removes objects in **Add** mode, but adds objects in **Remove** mode.

### A (Add Objects to the Selection Set)

The **A** option (short for "add") adds objects to the set. **Add** is usually used after **Remove**. **Add** changes the prompt back to 'Select objects:' so you may add objects to the selection set. Hold down the SHIFT key to remove objects.

#### Canceling the Selection Process.

Pressing ESC at any time cancels the selection process and removes the selected objects from the selection set. The prompt line returns to the 'Command:' prompt.

Pressing ENTER ends the selection process, and continues with the editing command's other options.

### Selecting Objects During Commands

To use the selection options at any 'Select objects:' prompt, enter one of the abbreviations listed in the table on the earlier page (abbreviations shown in uppercase letters). Alternatively, just pick two points, because the **Auto** mode is the default.

For example, to move all the objects within the selection rectangle, do the following

> Command: **move**
> Select objects: **w**
> Specify first corner: *(Pick point 1.)*
> Specify opposite corner: *(Pick point 2.)*
> 5 found Select objects: *(Press ENTER to end object selection.)*
> Specify base point or displacement: *(Pick a point.)*
> Specify second point of displacement or <use first point as displacement>: *(Pick another point.)*

Notice that the "Select objects:" prompt repeats until you press ENTER. This means you can keep selecting (and deselecting) objects until you are satisfied with the selection set.

## QSELECT

The QSELECT command selects objects based on their properties, instead of on their location, as with the SELECT command. For example, use QSELECT to select all objects with hidden linetypes on a specific layer, or all circles with a radius larger than one inch.

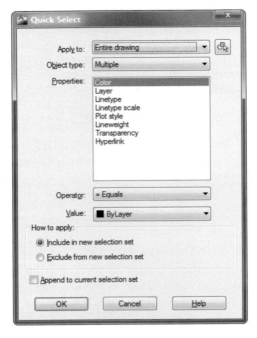

The dialog box is also accessible through the **Quick Select** button in several other commands, such as BLOCK and PROPERTIES.

### TUTORIAL: SELECTING OBJECTS BY PROPERTIES

1. To select objects by their common properties, start the QSELECT command with one of these methods:
   - In the ribbon's 2D Home tab, choose the **Quick Select** button from the Utilities panel.
   - Right-click, and from the shortcut menu choose **Quick Select**.
   - At the 'Command:' prompt, enter the **qselect** command:

      Command: **qselect** *(Press ENTER.)*

   Notice the Quick Select dialog box.

2. The selection can be made from the entire drawing or a subset of the drawing, called the "current selection." To select a subset, click the **Select Objects** button (next to the Apply To droplist). AutoCAD prompts you:

      Select objects: *(Select one or more objects.)*
      Select objects: *(Press ENTER to return to the dialog box.)*

   Notice that the Apply To droplist now has two options. Select one:
   - Entire Drawing.
   - Current Selection.

3. The **Object Type** droplist narrows down the selection to specific objects. Only the objects found in the drawing or the current selection are listed here. To include all objects, select "Multiple."

4. Narrow down your selection further by picking one of the items in the **Properties** list. The object you selected earlier affects the content of this list, which contains only properties specific to the object. You can select only one property from this list.
5. The **Operator** list further narrows the range AutoCAD searches for. Depending on the property you select, the choice of operator includes:

   | Operator | Meaning |
   | --- | --- |
   | = | Equals |
   | <> | Not Equal To |
   | > | Greater Than |
   | < | Less Than |
   | * | Wildcard Match (selects all). |

   Use the *Wildcard Match operator with text fields that can be edited, such as the names of blocks.
6. The **Value** field goes with the Operator list: if AutoCAD can determine values, it lists them here; if not, you type the value.
7. Under **How to Apply**, include or exclude objects matching these parameters from the existing selection set.
8. The **Append to Current Selection Set** check box determines whether these are added to (on) or replace (off) the current selection set.
9. Choose **OK** to exit the dialog box. AutoCAD highlights the objects selected, and reports the number selected:

   *nnn* item(s) selected.

### CONTROLLING THE SELECTION: ADDITIONAL METHODS

In addition to selection modes and direct selection, AutoCAD has these commands:

- **PICKBOX** changes the size of the square pick cursor.
- **DDSELECT** changes selection options; displays the Selection tab of the Options dialog box.
- **FILTER** selects objects based on their properties.
- **SELECTURL** highlights all objects that contain hyperlinks.

Let's look at how these commands work.

### Pickbox

When an editing command displays the "Select objects:" prompt, the cursor is supplemented by a small square called the "pickbox." When you place the pickbox over the object, and then click (press the left mouse button), AutoCAD scans the drawing, and selects the object under by the pickbox.

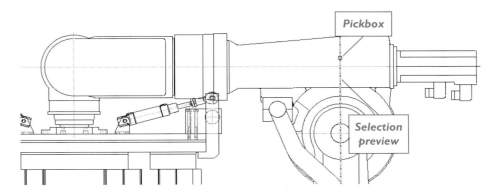

The pickbox can be made larger and smaller. The Selection tab of the Options dialog box allows you to change the pickbox size; or, you just might prefer to use the PICKBOX system variable:

Command: **pickbox**

Enter new value for PICKBOX <3>: *(Enter a value between 0 and 50.)*

The size of the pickbox is measured in pixels. A large pickbox forces AutoCAD to scan through more objects, which can take longer in complex drawings on slow computers.

## DdSelect

The DDSELECT command allows you to change selection modes. (The command is actually an alias for displaying the **Selection** tab of the Options dialog box.)

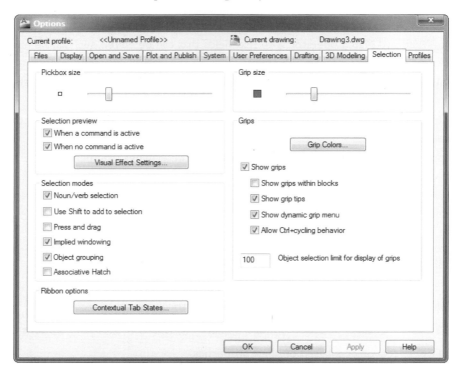

- Right-click, and from the shortcut menu choose **Options**.
- At the 'Command:' prompt, enter the **ddselect** command.

Command: **ddselect** *(Press ENTER.)*

Let's look at each option.

### Pickbox Size

The pickbox is the square cursor that appears at the 'Select objects:' prompt. Here, you can change its size from small to large.

### Selection Preview

**Selection preview** highlights objects as the pickbox cursor moves over them. You can turn the mode on and off.

**When a Command Is Active** means that selection preview works only when a command displays the 'Select objects:' prompt.

When **No Command Is Active** means that selection preview works any other time.

The **Visual Effect Settings** button displays a dialog box that controls the visual effects of selection previews and windowed selections. For selection previews, you can choose whether dashed lines, thickened lines, or both are displayed. For windowed and crossing selections, you can choose the color and opacity. (Autodesk documentation is incorrect in suggesting these affect the appearance of windowed areas only during selection preview.)

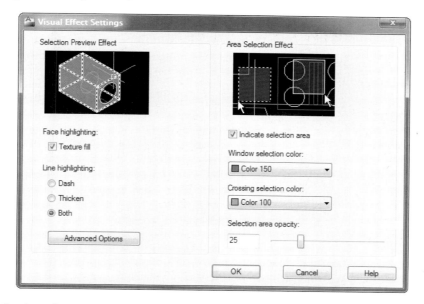

The **Advanced Options** button accesses further options, specifically which objects should be excluded from selection previewing.

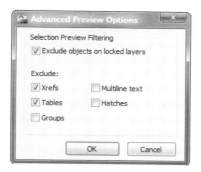

### Noun/Verb Selection

The "normal" way to edit drawings is to (1) first select the editing command, and then (2) select the objects to modify. Technically, this is choosing the verb (the action represented by the edit command), and then the noun (the object of the action represented by the selection set).

As an alternative, AutoCAD allows you to reverse the procedure: (1) first select the object(s) to edit, and then (2) choose the editing command. This is called "noun/verb selection."

When the **Noun/Verb Selection** option is turned on (as it is by default), AutoCAD places the pickbox at the intersection of the cross hairs. The presence of the pickbox indicates that you can select objects before entering one of these editing commands:

    ALIGN   LIST   ARRAY   MIRROR   BLOCK   MOVE   CHANGE   PROPERTIES   CHPROP

    ROTATE   COPY   SCALE   DVIEW   STRETCH   ERASE   WBLOCK   EXPLODE   BEDIT

Here is the command sequence to erase a single object, for example:

        Command: *(Select the object.)*

        Command: **erase** *(Press* ENTER.*)*

### Use Shift to Add to Selection

When selecting, you normally choose the object(s), and then press ENTER when you are finished. As you select each object, it is automatically added to the selection set. To remove objects from the selection set, hold down the SHIFT key while picking them.

As an alternative, turn on the **Use Shift to Add to Selection** option. Now you must hold down the SHIFT key to add objects to the selection set; this is similar to the method used by many other Windows programs. When you make more than one selection without holding the SHIFT key, the previous selections are removed from the selection set.

### Press and Drag

The traditional AutoCAD method for selecting objects through windowing is to (1) click one corner, (2) move the cursor to the other corner, and then (3) click the other corner.

As an alternative, turn on the **Press and Drag** option. You build a window by (1) clicking one corner, and then (2) dragging (by holding down the mouse button) the cursor to the other corner. Like the **Shift to Add** option, this is used by most other Windows programs.

### Implied Windowing

You previously learned how to use a selection window or a crossing window to select objects. All you have to do is enter either a **W** or a **C** to invoke the window mode.

The **Implied Windowing** option does this automatically. (The option is on by default.) You "imply" the window by clicking in an empty area of the drawing. When AutoCAD does not find an object within the area of the pickbox, it assumes that you want to use windowing. When you move the cursor to the right, Window selection mode is entered; move to the left, Crossing selection mode is entered.

Because the first point entered describes the first corner of the window, be sure to select a desirable position.

### Object Grouping

When the **Object Grouping** option is turned on, AutoCAD selects the entire group when you pick one object in the group. When off, just the picked object is selected. (More on groups follows in this chapter.)

As a shortcut, you can toggle object grouping mode with CTRL+SHIFT+A. Each time you press the key combination, AutoCAD comments:

Command: (Press CTRL+SHIFT+A.) <Group off>

"Group off" means objects are selected, while "Group on" means the entire group is selected:

Command: (Press CTRL+SHIFT+A.) <Group on>

Curiously, CTRL+H does the same thing. Technically, it toggles the PICKSTYLE system variable between **0** (individual objects selected from the group) and **1** (entire group selected). PICKSTYLE can also have the values of **2** and **3** for toggling the selection of associative hatch patterns.

### Associative Hatch

When the **Associative Hatch** option is turned on, AutoCAD selects the boundary object(s) when you pick an associative hatch pattern. When off, only the hatch is selected.

### Filter

The FILTER command displays a dialog box that selects objects based on their properties. You can save the resulting selection set by name, and then access it at the 'Select objects:' prompt.

- At the 'Command:' prompt, enter the **filter** command.
- Alternatively, enter the **f** alias at the keyboard.

    Command: **filter** *(Press* ENTER.*)*

- Because FILTER is a *transparent* command, it can be invoked during another command to filter the selection set:

    Command: **erase**

    Select objects: **'filter**

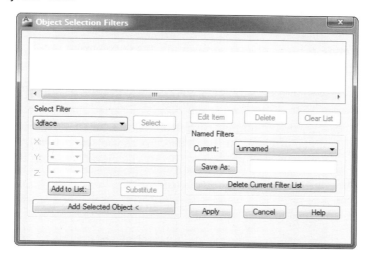

The selection set created by FILTER can be accessed via the **P** (previous) selection option. In the following tutorial, you erase all construction lines from a drawing:

1. Start the **ERASE** command:

    Command: **erase**

2. Then invoke the FILTER command transparently by prefixing the command with the quotation mark:

    Select objects: **'filter**

    Notice that AutoCAD displays the Object Selection Filters dialog box.

3. In the Select Filter droplist, select **Xline**.

    Xline is near the end of the list. Here's a quick way to get to any item in an alphabetical list: after clicking on the droplist, press **x** on the keyboard. You are taken to the first word starting with "x".

4. Click the **Add to List** button. Notice that AutoCAD adds this text to the "list" (the large white area at the top of the dialog box):

    Object           =Xline

5. Click **Apply**. Notice that the dialog box disappears, and that the FILTER command displays prompts on the command line:

    Applying filter to selection.

    This means that AutoCAD will apply the filter (search for all xlines) to the objects you now select.

6. Specify that the filter should apply to the entire drawing with the **All** option:
   >>Select objects: **all**
   9405 found 9401 were filtered out.

   The double angle bracket ( **>>** ) indicates AutoCAD is currently in a transparent command. (The **ERASE** command continues later.)

7. Return to the **ERASE** command by exiting the **FILTER** command. Press **ENTER**:
   >>Select objects: *(Press* ENTER *to exit the* FILTER *command.)*
   Exiting filtered selection.
   Resuming ERASE command.

   AutoCAD returns to the **ERASE** command, and picks up the four objects selected by the **FILTER** command:
   Select objects: 4 found

8. Press **ENTER** to erase the four objects and exit the **ERASE** command:
   Select objects: *(Press* ENTER *to exit the* ERASE *command.)*

   AutoCAD uses the selection set created by the **FILTER** command to erase the xlines.

In addition to filtering objects, this command also filters x, y, z coordinates, such as the center point of circles. Other options become available in the dialog box as you select them. For example, when you select **Elevation**, the **X** text entry box allows you to specify the elevation using the following operators:

| Operator | Meaning |
|---|---|
| < | Less than. |
| <= | Less than or equal to. |
| = | Equal to. |
| != | Not equal to. |
| > | Greater than. |
| >= | Greater than or equal to. |
| * | All values. |

Specifying elevation > 100, for example, means that all objects with an elevation greater than 100 will be added to the filtered selection set.

In addition, you can group filter sequences using these operators:

| Operator | Meaning |
|---|---|
| **Begin OR with **End OR | Include any of these items. |
| **Begin AND with **End AND | Include all of these items. |
| **Begin NOT with **End NOT | Include none of these items. |
| **Begin XOR with **End XOR | Include none of these items if one item is found. |

You can save filter definitions by name to *.nfl* (short for "named filter") files on disk for use in other drawings or editing sessions.

**SELECTSIMILAR**

The **SELECTSIMILAR** command adds similar objects to the selection set based on their properties. This is a simpler version of the **QSELECT** command.

For example, select a circle when Name and Layer settings are turned on. AutoCAD will then also select all other circles on the same layer.

    Command: **selectsimilar**

    Select objects or [SEttings]: *(Select one or more objects, or type* **SE** *for settings dialog box.)*

    Select objects or [SEttings]: *(Select additional objects, or press* **ENTER** *to exit.)*

At the 'Select objects' prompt, you can enter any of the selection modes from the **SELECT** command, such as **w** for Window or **f** for Fence selection mode.

The **SEttings** option displays the Select Similar Settings dialog box:

Toggle the items whose properties should determine the selection set. Most are straightforward, but the meaning of these two might be puzzling:

    **Object Style** — refers to styles, such as text styles, dimension styles, and table styles.

    **Name** — refers to object types, such as lines or circles.

When Object Style is on, this command restricts its selection to objects whose styles match the selected one. Select a piece of text, and only text that has the same style is selected.

When Name is on, this command restricts its selection to objects that match the selected one. Select a line, for instance, and only other lines are selected. When off, any object matching the properties is selected.

### TUTORIAL: SELECTING SIMILAR OBJECTS

In this tutorial, you use the **SELECTSIMILAR** command to select all other objects in a drawing that match one.

1. Start AutoCAD, and then open the *redline-2.dwg* file.
2. Enter the **SELECTSIMILAR** command using one of the following methods:
   - Select an object, right-click, and then choose **Select Similar** from the shortcut menu.
   - At the 'Command:' prompt, enter **SELECTSIMILAR**:

    Command: **selectsimilar**

3. If you entered the command at the ,'Command:' prompt then AutoCAD prompts you. Choose a line of text, as follows:

    Select objects or [SEttings]: *(Select a line of text.)*

    Select objects or [SEttings]: *(Press* **ENTER**.*)*

    If you selected the command from the shortcut menu, there is no prompt; AutoCAD immediately selects all similar objects.

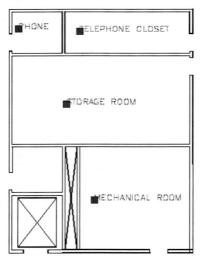

    Notice that all lines of text are selected.
4. You can now change their properties and perform other editing tasks with them. For this tutorial, change the text to red: from the ribbons Color droplist, choose **Red**.
5. Cancel the selection by pressing **ESC**.

Keep this drawing open for the next tutorial.

 **ADDSELECTED**

The **ADDSELECTED** command creates new objects based on the object type and properties of the selected object. For example, select a line, and this command launches the **LINE** command, giving the new line the same properties.

### TUTORIAL: DRAWING SIMILAR OBJECTS

In this tutorial, you use the **ADDSELECTED** command to create a new piece of text based on the properties of existing text. Continue with the *redline-2.dwg* from the previous tutorial.

1. Enter the **ADDSELECTED** command using one of the following methods:
    - Select a piece of text, right-click, and then choose **Add Selected** from the shortcut menu.
    - At the 'Command:' prompt, enter **ADDSELECTED**:

    Command: **addselected**

2. Notice that AutoCAD prompts you to select an object. Choose a piece of text:

    Select object: *(Choose a line of text.)*

    If you selected the command from the shortcut menu, there is no prompt. AutoCAD immediately proceeds to the **TEXT** command.

3. Notice that AutoCAD launches the **TEXT** command; follow its prompts:
   _text
   Current text style: "STANDARD"  Text height: 0'-10"  Annotative: No
   Specify start point of text or [Justify/Style]: *(Pick a point inside the storage room.)*
   Specify rotation angle of text <E>: *(Press* ENTER.*)*
4. Enter 'STORAGE' at the prompt, and then press **ENTER** to end the command:
   Enter text: **STORAGE** *(Press* ENTER.*)*

Notice that AutoCAD places the text in red with the Standard style. (Keep the drawing open for the next tutorial.)

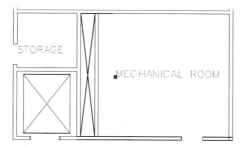

## HIDEOBJECTS AND ISOLATEOBJECTS

The **HIDEOBJECTS** command hides selected objects independently of the layer settings.

    Command: **hideobjects**
    Select objects: *(Choose one or more objects.)*
    Select objects: *(Press* ENTER *to end command.)*

The **ISOLATEOBJECTS** command is the opposite: it hides unselected objects, displaying only those that you select (ie, isolate).

    Command: **isolateobjects**
    Select objects: *(Choose one or more objects.)*
    Select objects: *(Press* ENTER *to end command.)*

You use these commands to hide objects quickly, without needing the Layers dialog box or palette. When objects are hidden or isolated, a  light bulb icon on the status bar changes from yellow to red. Left or right-click the icon to access this shortcut menu:

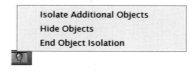

The options operate as follows:

    **Isolate Additional Objects** — executes the **ISOLATEOBJECTS** command.
    **Hide Objects** — executes the **HIDEOBJECTS** command.
    **End Object Isolation** — executes the **UNISOLATEOBJECTS** command.

**UnisolateObjects**

The **UNISOLATEOBJECTS** command displays objects hidden by the **HIDEOBJECTS** and **ISOLATEOBJECTS** commands:

> Command: **unisolateobjects**
>
> *n* object(s) unisolated.

Instead of entering this command, you can click the light bulb icon on the status bar.

### TUTORIAL: HIDING AND ISOLATING OBJECTS

In this tutorial, you use the commands described above to hide, isolate, and unisolate objects in the *redline-2.dwg* drawing from the previous tutorial.

1. Enter the **HIDEOBJECTS** command using one of the following methods:
   - Select a piece of text, right-click, choose **Isolate** from the shortcut menu, and then **Hide Objects**.
   - At the 'Command:' prompt, enter **HIDEOBJECTS**:

   > Command: **hideobjects**

2. Notice that AutoCAD prompts you to select objects. Choose a piece of text:

   Select objects: *(Choose one or more lines of text.)*

   Select objects: *(Press* ENTER.*)*

   If you selected the command from the shortcut menu, there is no prompt; AutoCAD immediately hides the selected objects.

3. Notice that the selected text disappears from view.

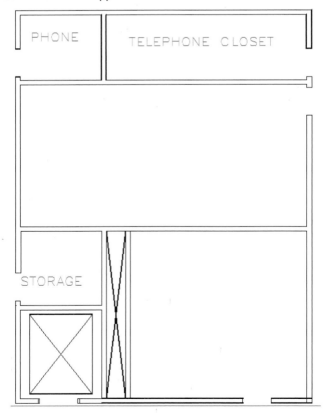

On the status bar the yellow light bulb turns red. This alerts you that one or more objects are hidden; there is no other indication.

## Isolating Objects

4. Enter the **ISOLATEOBJECTS** command using one of the following methods:
   - Select a remaining piece of text, right-click, choose **Isolate** from the shortcut menu, and then **Isolate Objects**.
   - At the 'Command:' prompt, enter **ISOLATEOBJECTS**:

     Command: **isolateobjects**

5. Notice that AutoCAD prompts you to select objects. Choose additional pieces of text:

   Select objects: *(Choose one or more lines of text.)*

   Select objects: *(Press ENTER.)*

   If you selected the command from the shortcut menu, there is no prompt; AutoCAD immediately isolates the selected objects.

6. Notice that all objects disappear from view, except for the text you selected.

## Unisolating Objects

7. Enter the **UNISOLATEOBJECTS** command using one of the following methods:
   - Right-click, choose **Isolate** from the shortcut menu, and then **End Object Isolation**.
   - At the 'Command:' prompt, enter **UNISOLATEOBJECTS**:

     Command: **unisolateobjects**

8. Notice that AutoCAD immediately makes all objects visible. In addition, the red light bulb turns yellow.

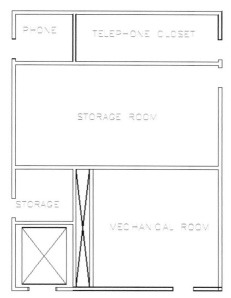

# DRAWORDER

The **DRAWORDER** command determines the order in which overlapping objects are displayed. In the figure below, the text "Using AutoCAD" is obscured by the rectangle. By changing the display order — the order in which AutoCAD redraws objects — you can make the text visible on top of the inappropriately placed rectangle.

**Left:** *Text above the rectangle.*
**Right:** *Text below the rectangle.*

### TUTORIAL: SPECIFYING DRAWORDER

1. To specify the display order of overlapping objects, start the **DRAWORDER** command with one of these methods:
   - In the ribbon's 2D Home tab, choose the **Bring to Front** button from the Modify panel.
   - At the 'Command:' prompt, enter the **draworder** command.

     Command: **draworder** *(Press* ENTER.*)*
   - Alternatively, enter the **dr** alias at the keyboard.

2. Select one or more objects:
   Select objects: *(Select one or more objects.)*
   1 found Select objects: *(Press* ENTER *to end object selection.)*

3. Specify the display order for the selected objects(s)
   Enter object ordering option [Above object/Under object/Front/Back] <Back>: *(Press* ENTER.*)*
   Regenerating model.

### CHANGING THE DRAW ORDER: A DDITIONAL METHODS

The **DRAWORDER** command has four options for controlling the visual overlapping of objects. The first two listed below are useful when three or more objects overlap.

- **Above object** places the object visually on top of other selected objects.
- **Under object** places the object under other selected objects.
- **Front** places the object above all other objects.
- **Back** places the object below all other objects.

AutoCAD provides additional controls over draw order for specific objects.

## HpDrawOrder

The **HPDRAWORDER** system variable determines whether hatch and fill patterns are above or below their boundaries and other objects. See the **HATCH** command earlier in this chapter.

## HatchToBack

The HATCHTOBACK command immediately places all hatch patterns below all other objects.

> Command: **hatchtoback**
> 
> *n* hatch object(s) sent to back.

This command also applies to gradient and solid fills.

## TextToFront

The TEXTTOFRONT command determines whether text and/or dimensions are always placed on top (or in front) of overlapping objects in the drawing.

> Command: **texttofront**
> 
> Bring to front [Text/Dimensions/Both] <Both>: *(Press* ENTER.*)*

This command ensures text is always legible, not obscured by other objects. It applies to all text and dimensions in the drawing. It does not, however, work when model space objects lie over text in a layout, and vice versa.

These commands are not modes; if hatches, gradients, text and dimensions are obscured by later objects, you must run HATCHTOBACK and TEXTTOFRONT again.

## EXERCISES

1. In this exercise, you use some of the options of the **SELECT** command to erase portions of a valve housing drawing.

   a. Open the *edit1.dwg* drawing.

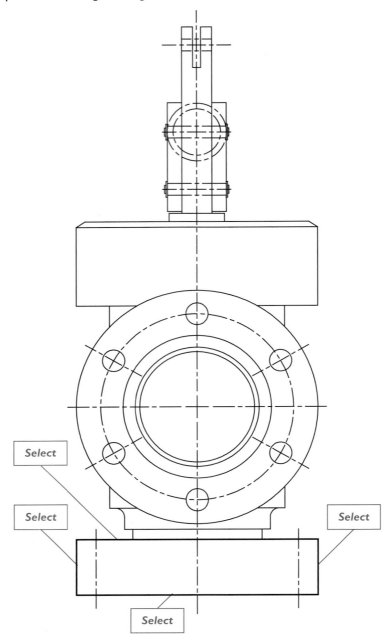

   b. Use the **ERASE** command to delete some of the objects.
   c. You should now see a pickbox on the screen. Place the pickbox over the bottom line of the part, as shown in the illustration, and click. The line should be highlighted.
   d. One by one, select the remaining lines, as shown in the figure.
   e. Finally, press **ENTER**. The four lines are removed from the drawing.

2. Repeat the above exercise on the same drawing with the **Window** object selection mode.
    a. But first, use the **u** command to undo the erasure.
    b. Select **ERASE** again, and enter **w** in response to the "Select objects:" prompt.
    c. Consult the figure for the points referenced in the following command sequence.
        Command: **erase**
        Select objects: **w**
        Specify first corner: *(Select point 1.)*
        Specify other corner: *(Select point 2.)*

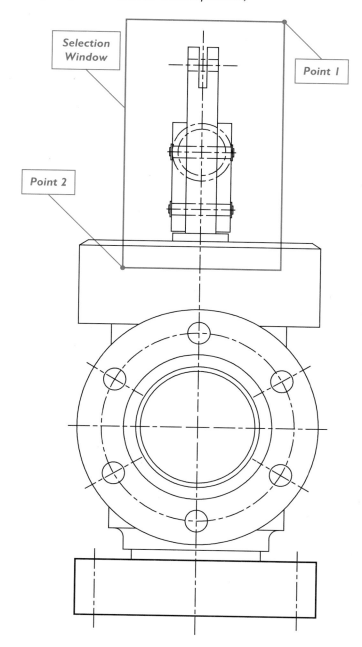

    d. All items within the window are selected and highlighted. Notice that objects that extend outside the window area (but are not wholly contained therein) are not selected.
   Select objects: *(Press* **ENTER**.*)*
    e. Press **ENTER** and the selected objects are deleted.

3. Repeat the above exercise on the same drawing with the **Crossing** window object selection mode.
   a. Again, first undo the erasure with the **u** command.
   b. Select **ERASE** again, entering **C** (for "crossing") as the option. Refer to the following command sequence and the figure.

   Command: **erase**
   Select objects: **c**
   Select first corner: *(Select point 1.)*
   Select other corner: *(Select point 2.)*

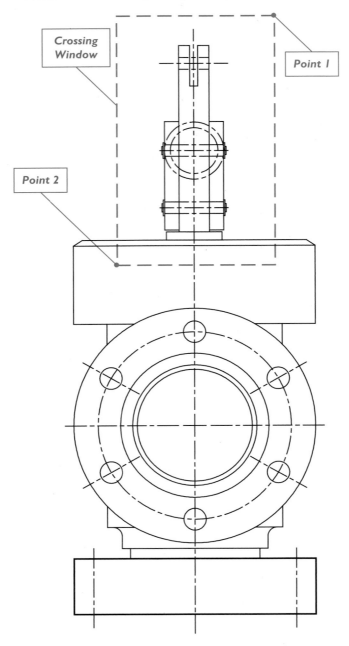

Notice that all the objects within, crossing, and touching the window are selected (as noted by the highlighting).

4. Continuing from the above exercise, remove objects from the selection set.
   a. After you placed the crossing window in the previous exercise, AutoCAD asked you to select more objects. This time, enter **R** for remove. The command sequence continues:

   Select objects: **r**

   Remove objects: *(Select one of the horizontal lines.)*

   1 found, 1 removed Remove objects: *(Select the other horizontal line.)*

   1 found, 1 removed Remove objects: *(Press* ENTER.*)*

   The objects you removed from the object selection set are not erased.
   b. Exit the drawing, discarding the changes you made so that the edits are not recorded.

5. In this exercise, you create a block from objects within a drawing, and then insert the block in the drawing several times.
   a. Open the drawing named *office.dwg*.
   b. Window the entire desk, and then start the **BLOCK** command.
   c. Name the block "SDESK", and then pick an insertion point.

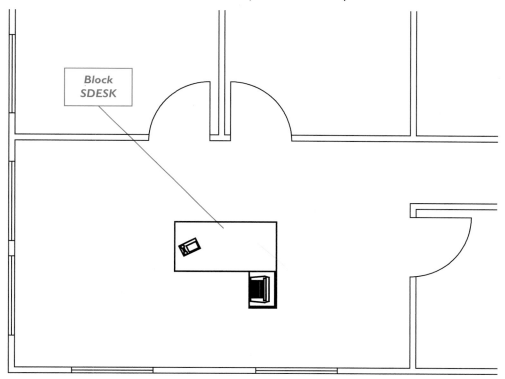

   d. Click **OK**. The block is now stored with the drawing, and can be used as many times as you require.
   e. With the **INSERT** command, place the SDESK block in the drawing three times, using these scales and rotation angles:

   Scale = 1.0    Angle = 0 degrees
   Scale = 1.5    Angle = 90 degrees
   Scale = 0.75   Angle = 45 degrees

6. In this exercise, use DesignCenter to place blocks in a drawing.
    a. Open the site plan drawing named *insert.dwg*. The landscape items are drawn for you and stored as blocks in the drawing.
    b. Use **INSERT** to view blocks, and then insert them in the drawing to create a landscaping design of your own.

7. In this exercise, you draw an electronic part called a "diode," and then insert it as an array of blocks.
    a. Using the **PLINE** command, construct the semi-conductor symbol shown below.

    b. With the **BLOCK** command, select the diode, and name it "DIODE"; use the left end of the diode as the insertion point.
    c. You have now created the block DIODE in the drawing. Use the **INSERT** command, followed by the **ARRAY** command to place a 4x6 array of diodes.
    d. You may need to use the **ZOOM All** command to see the completed array.

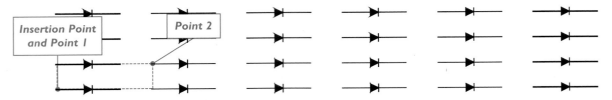

8. In this exercise, you hatch portions of a drawing of several hot air balloons.
    a. Open the drawing named *solids.dwg*.
    b. Use the **HATCH** command to place the solid hatch pattern in a variety of colors in some of the balloon areas.
    c. Now try placing a variety of hatch patterns in the same drawing.

9.  In this exercise, you specify the area to be hatched with a pick point. The figure shows an object containing three areas that could be hatched. You need to hatch the square and circle, but not the triangle.

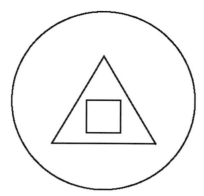

  a.  Draw the circle, triangle, and square using the **CIRCLE** and **POLYGON** commands.
  b.  Enter the **HATCH** command.
  c.  In the **Pattern** panel, select **ANSI31**.
  d.  Click the **Options** panel, and then choose **Normal Island Detection**. .
  e.  Hover the cursor the circle, but outside the triangle.
  f.  Your hatch should look similar to the figure below. If not, change the hatch options and/or your pick points. When the preview hatch looks right, press **ENTER** to apply the pattern.

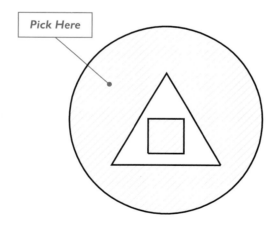

*(If the Hatch and Gradient dialog box appears, then follow these instructions:*
  c.  In the **Pattern** list box, select **ANSI31**.
  d.  Select the **Advanced** tab, and then ensure that **Normal** appears in the **Style** list box

and that the **Island Detection** method has **Flood** selected.

e. Select the **Pick Points<** button, and then click inside the circle, but outside the triangle.

f. Right-click to return to the dialog box. Choose the **Preview** button. Your hatch should look similar to the figure above. If not, change the hatch options and/or your pick points. When the preview hatch looks right, choose **OK** to apply the hatch pattern.)

10. In this exercise, you use hatch styles to control the behavior of hatch patterns.

    a. Open the *hatch.dwg* drawing file.

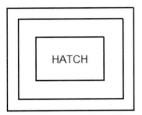
NORMAL

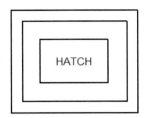
OUTERMOST

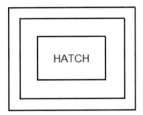
IGNORE

    b. Start the **HATCH** command, and set the following options:

       Pattern                 ANSI31
       Scale                   2
       Angle                   0
       Island Detection Style  Normal
       Select objects          W  (or pick a point between the two outer rectangles)

    c. Place the window around the object, and then press **ENTER**. Your drawing should look similar to the following illustration:

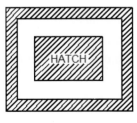
NORMAL

11. Repeat the above exercise on the same drawing, but use the **Outer** style. The drawing should look similar to the following illustration:

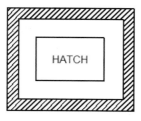

OUTERMOST

12. Repeat the above exercise on the same drawing, but use the **Ignore** style of island detection. Notice that the hatch has ignored the boundaries and the text. Your drawing should look similar to the following illustration:

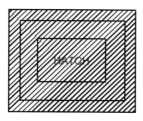

IGNORE

## CHAPTER REVIEW

1. Describe the purpose of DesignCenter.
2. What are two ways to insert a block from DesignCenter into a drawing?
   a.
   b.
3. Describe two ways to change blocks to a new definition.
4. Which key do you hold down to select more than one item?
5. Explain the effect of turning off the OSNAPHATCH system variable.
6. Can hatch patterns be trimmed?
   Can hatch patterns be used as trim boundaries?
7. How would you force text and dimensions to display in front of other overlapping objects in drawings?
8. Name three uses for the Block Editor:
   a.
   b.
   c.
9. Explain what the HATCH command does.
10. If you did not want the solid filled areas in your drawing to plot, what could you do?
11. What are *hatch boundaries*?
12. What are *blocks*?
13. What is the *base point* of a block?
14. When placed in drawings, how do blocks handle their layer definitions?
15. What is a *nested* block?
16. How would you create a separate drawing file from an existing block?
17. How do you place blocks in drawings?
18. How can you place blocks in other drawings?
19. When you place blocks in the drawings, what is the *insertion point*?
20. Can you place one AutoCAD drawing in another AutoCAD drawing?
21. Name two advantages of using blocks.
22. What commands are combined to create the **MINSERT** command?
23. How would you change the origin point of a hatch?
24. When do hatch patterns obscure text?
25. When an editing command is invoked, must a selection option, such as SIngle or Window, be entered before an object is selected?
26. When choosing a set of objects to edit, other than by using the **U** command, how can you alter an incorrect selection without starting over?
27. If you wish to edit an item that was not drawn last, but was the last item selected, would the Last selection option allow you to select the item?
28. How do you increase and decrease the size of the pickbox?
29. When entering BOX in response to the "Select objects:" prompt, how are you then allowed to choose objects?
30. What makes it evident that an item has been selected?

31. When a group of items is selected with the Window option, does an item become a part of the selection set as long as it is partially inside the window?
32. In the object selection process, how is the BOX option different from the AUtomatic option?
33. What is the purpose of the tool palette?
34. How do you add blocks from a tool palette to the drawing?
35. Describe the purpose of the BASE command.
36. When might you use the Hatch and Gradient dialog box's Draw Order option?
37. What kind of hatches can the HATCHEDIT command not change?
38. Can the HATCHEDIT command copy the properties of one hatch pattern and apply them to another?
39. What is the difference between the SELECT and QSELECT commands?
40. Can you change the size of the *pick* cursor?
41. What is the purpose of the HATCH command's Gap Tolerance option?
42. What is the purpose of the DRAWORDER command?
43. Explain how *auxiliary scale* differs from the normal scale factor.

# III

# Editing Drawings

# CHAPTER 7
## Properties and Layers

As you create drawings, you sometimes need to set the properties of objects, such as color, layer assignment, or linetype pattern. In this chapter, you learn how to modify the properties of any object using the following commands:

**MATCHPROP** matches the properties of one object to other objects.

**COLOR** changes the colors of objects.

**TRANSPARENCY** changes the transparency of objects (new to AutoCAD 2011).

**LWEIGHT** changes the display width of objects.

**LINETYPE** changes the patterns of objects.

**LTSCALE** and **CELTSCALE** change the scale factor of linetypes.

**LAYER** displays the Layer Properties Manager palette.

**LAYERSTATE** accesses the Layer States Manager dialog box.

**SETBYLAYER** resets layer properties in overridden viewports.

**AI_MOLC** and **CLAYER** set the current layer.

**LAYERP** and **LAYERPMODE** set the layers to their previous state.

**LAYISO** isolates selected layers by fading out all other layers.

**PROPERTIES**, **QUICKPROPERTIES**, and **ROLLOVER TIPS** report and/or change properties of selected objects.

---

**NEW TO AUTOCAD 2011** IN THIS CHAPTER
- Object transparency is a new property.

# MATCHPROP

The MATCHPROP command "reads" the properties of one object, and then applies them to other objects.

At times, you may want a group of objects to match the properties of another object. For example, you may accidentally place several doors on the Landscape layer, instead of the Door layer. This command lets you make this change. Or, you may realize that some lines drawn with hidden linetype should be in another linetype — whose name you don't recall.

You could use the Properties toolbar to change layers, colors, linetype, and so on, but it doesn't work well if you are not sure of the exact layer and linetype name. Complex drawings have hundreds of layer names, many of which look similar. One sample drawing provided with AutoCAD has the following layer names: ARCC, ARCCLR, ARCDIMR, ARCDSHR, ARCG, ARCM, ARCR, ARCRMNG, ARCRMNR, and ARCTXTG. And that's just the first ten! Similarly, linetypes can look confusingly alike. (Layers, linetypes, and other properties are discussed later in this chapter.)

The MATCHPROP command solves those problems. It lets you copy the properties from one object to a selection set of objects. You must, however, be sure to select objects in the correct order: (1) select the single object whose properties to copy, and then (2) select the object(s) to take on those properties.

| Property | Objects Affected |
|---|---|
| Color | All objects, except OLE objects. |
| Dimension | Dimensions, leaders, and tolerances; also paints dimension styles. |
| Hatch | Hatches; also paints the hatch pattern. |
| Layer | All objects, except OLE objects. |
| Linetype | All except attributes, hatches, multiline text, OLE objects, points, & viewports. |
| Linetype Scale | All objects except attributes, hatches, multiline text, OLE objects, points, and viewports. |
| Lineweight | All objects. |
| Material | All objects. |
| Multileader | Multileaders. |
| Plot Style | All OLE objects; unavailable when PSTYLEPOLICY is 1. |
| Polyline | Polylines; also paints width and linetype generation, but not fit/smooth, variable width, or elevation. |
| Shadow display | All objects. |
| Thickness | Arcs, attributes, circles, lines, points, 2D polylines, regions, text, and traces. |
| Text | Text and multiline text only; also paints the text style. |
| Transparency | All objects. |
| Viewport | Viewports; also paints on/off, display locking, standard or custom scale, shade plot, snap, grid, as well as UCS icon visibility and location settings, but not clipping, UCS-per-viewport, or layer freeze/thaw state. |
| Visual style | All objects. |

### TUTORIAL: MATCHING PROPERTIES

1. Open the *match properties.dwg* file, the drawing of several multi-colored objects — and an equal number of monochrome objects.
2. To match the properties of one object to others, start the **MATCHPROP** command with one of these methods:
   - In the Home tab's Clipboard panel, choose **Match Properties**.
   - At the 'Command:' prompt, enter the **matchprop** command:

     Command: **matchprop** *(Press* ENTER.*)*
   - Alternatively, enter the aliases **ma** or **painter** (the old name in AutoCAD LT) at the 'Command:' prompt.

3. Notice that AutoCAD prompts you at the command line:
   Select source object: *(Pick a single object.)*

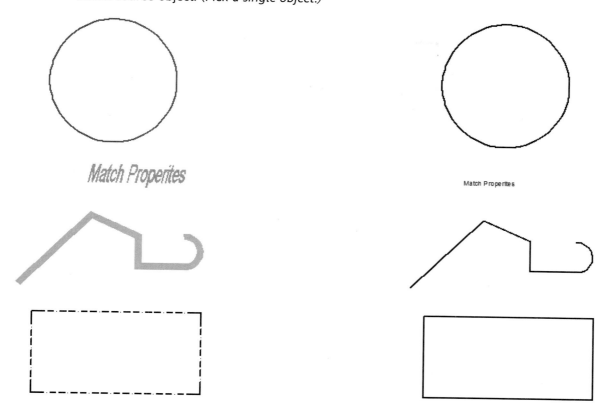

Notice that the cursor turns into a paintbrush.

On the command bar, AutoCAD lists the *active settings,* properties to be "picked up" by the paintbrush.

   Current active settings:  Color Layer Ltype Ltscale Lineweight Transparency Thickness
         PlotStyle Dim Text Hatch Polyline Viewport Table Material Shadow display Multileader

4. Select the objects to which the properties should be applied.
   Select destination object(s) or [Settings]: *(Pick one or more objects, or enter* **S***.)*

5. You can continue selecting objects using windows, fences, single picks, and other selection modes. When done, press **ENTER** to exit the command.
   Select destination object(s) or [Settings]: *(Press* **ENTER***.)*

## Settings

At the "Select destination object(s) or [Settings]:" prompt, enter **S** to display the Property Settings dialog box.

The properties listed by this dialog box are more extensive than those listed by the Properties toolbar. Not all properties, however, work with all objects. For example, it makes no sense to match the hatch properties of text objects, since text cannot be hatched.

By default, all properties are on, meaning they will all be copied.

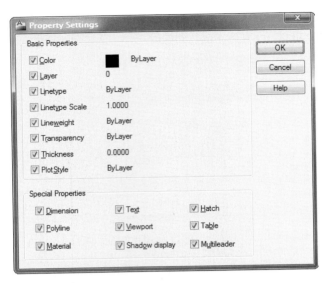

When you turn off properties, AutoCAD remembers this the next time you use MATCHPROP, which is why it lists "Current active settings."

 COLOR

Colors typically identify objects to segregate them visually.

In AutoCAD, the default color of objects is white or dark gray, depending on the background color of the viewport. If the background is dark, then objects are displayed in white; if light, then AutoCAD automatically switches to black objects. (In drawings begun with the *acad3d.dwt* template file, the default color is pale gray-blue.)

You don't have to stick with one of black, white, or pale blue. You can set colors before you draw objects, or change their colors afterwards:

**Presetting colors.** For objects you plan to draw, the best method is to specify colors through the **Color** option of the LAYER command; less preferred is to set the working color through the COLOR command.

**Changing colors.** For objects already in the drawing, the *fastest* method is to use the Color Control droplist on the Object Properties panel; alternatives are to use the **PROPERTIES** or **CHPROP** command. The *best* method, however, is to move the objects to the appropriate layer, so that they take on the color of the layer.

It is important to understand that the COLOR command overrides the colors set by the LAYER command. Here is the difference between them: LAYER sets the color for objects drawn on each layer; COLOR sets the color for subsequently drawn objects. Thus, it is possible for a layer to contain objects of different colors, regardless of the color set for the layer.

### TUTORIAL: CHANGING COLORS

1. To change colors, start the **COLOR** command with one of these methods:
   - In the ribbon's Home tab, click the Color Control droplist on the Properties panel, and then choose **Select Color**.
   - Or, at the 'Command:' prompt, enter the **color** command:

      Command: **color** *(Press* ENTER.*)*

- Alternatively, enter the aliases **col**, **colour** (the British spelling), or **ddcolor** (an old name) at the 'Command:' prompt.

2. In all cases, AutoCAD displays the Select Color dialog box.

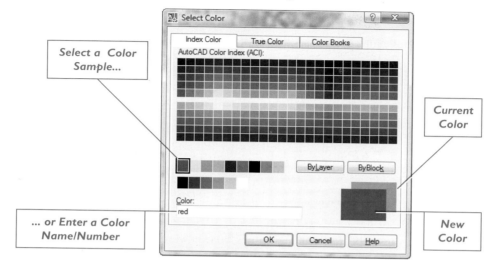

3. Select a color sample (one of the colored squares), or enter a color name/number.

4. Click **OK**. All the objects you now draw take on the new color — until you change color again.

**Note** The Select Color dialog box is used by more than just the **COLOR** command; it is also accessed by any other command that sets colors, such as **LAYER**, **OPTIONS**, and **PROPERTIES**.

(*History*: Some drafters were upset when Autodesk introduced the COLOR command with AutoCAD v2.5. They felt that layer settings should always determine the colors of objects. Like this book's technical editor, these drafters feel it is wrong to override the layer's color setting with the COLOR command. Indeed, overriding colors is a poor method of working with CAD, which you may come across in your place of work. AutoCAD 2008 partially solved the problem by adding the SETBYLAYER command, which forces objects to take the color, linetype, lineweight, and plot style of their assigned layers.)

### UNDERSTANDING AUTOCAD'S COLOR SYSTEM

AutoCAD lets you specify colors by a number of methods:

- Select a color swatch from the dialog box.
- Specify a number between 1 to 255.
- Enter an RGB (red, green, blue) value, or an HSL (hue, saturation, luminance) value.
- Select a color book name and number.
- Type a color name such as "blue" (valid for the first seven colors only).

By the way, *hue* refers to differences in a color, while *shade* refers to differences in gray.

(*History*: In early releases of AutoCAD, color was used to control pens in plotters. The color red, for example, could be assigned to pen #1, which in turn could be a red-colored pen or a black pen with a width of 0.1". Today in AutoCAD, you can control the plotter by color or with styles. More details follow in Chapter 20, "Plotting Drawings.")

### Color Numbers and Names

Of the 16.7 million colors available on computers, AutoCAD identifies the first seven by name and number, and the first 255 by number. These color numbers are called the "AutoCAD Color Index," or ACI for short. Beyond that, colors are identified by RGB value (amounts of red, green, and blue), HSL (hue, saturation, and luminance), or by color book number.

The first seven color names and their associated ACI numbers are listed below in numerical order:

| ACI | Name | Abbreviation* |
|---|---|---|
| 1 | Red | R |
| 2 | Yellow | Y |
| 3 | Green | G |
| 4 | Cyan | C |
| 5 | Blue | B |
| 6 | Magenta | M |
| 7 | White (black) | W |

* These abbreviations are used at the command line during commands like -COLOR, CHANGE, and -LAYER.

You may be wondering about my reference to the first 255 colors? Before AutoCAD 2004, the program worked with just 255 colors, because early graphics boards were typically limited to displaying 1, 4, 16, or 256 colors. These are the ones you see in the Select Color dialog box's Index Color tab. (Colors 250 through 255 are shades of gray.) Every CAD package adopts a different order for colors, and so Autodesk identified AutoCAD's as the ACI — AutoCAD Color Index.

### Background Color 0

Color 0 is a special ACI color number that you never use in drawings. It identifies the background color of the drawing area (a.k.a viewport color). The background can be any color, but typically is white or black; AutoCAD 2011 uses a very dark blue (RGB = 33,40,48).

If you want to change the background color, use the dialog box accessed through the **A| Options | Display | Colors** menu sequence. This dialog box lets you change the colors of the background of the model layout, layout viewports, plot preview, block editor, and 3D projections views, as described in Chapter 3, "Setting Up Drawings."

### Color 7

Because black lines are nearly invisible against the dark gray background, AutoCAD automatically changes color 7 (white) to the opposite of the background color (color #0) — here is a time in life when white is black, and black is white. Officially, however, color 7 is white, because AutoCAD's original background color was black. For this reason, you may encounter color 7 being referred to as "white," even though it appears black on your screen, as well as black when plotted on paper.

### Colors ByLayer, ByBlock, and ByEntity

In addition to the special colors numbers 0 and 7, AutoCAD has three more special "colors": the oft-seen ByLayer, less-commonly seen ByBlock, and rarely-seen ByEntity. These solve the problem of what to name colors when they are controlled by layers, blocks, and entities (the old term for objects), respectively.

### ByLayer Color #256

**ByLayer** means that the color of objects is determined by the layer's color. Choosing ByLayer causes subsequently drawn objects to inherit the layer's color. For example, if a layer is set to red, then all objects drawn on that layer are colored red automatically. (Layers are discussed later in this chapter.) Color ByLayer is also known as color #256.

### ByBlock Color #0

**ByBlock** means that objects in blocks take on the color assigned to the block. Choosing ByBlock causes objects to be drawn in white until they are turned into a block. When the block is inserted, the objects inherit the color of the block insertion. This color is also known as entity color #0.

### ByEntity Color #257

**ByEntity** means that each object carries its own color, rather than one assigned by the layer or block. ByEntity is only used by system variables, such as OBSCUREDCOLOR and INTERSECTIONCOLOR, although programmers can use it in their software programs. This color has a value of 257.

### True Color

AutoCAD now displays millions of colors. Technically, computer software works with three sets of 8-bit colors, which translates into 16.7 million hues (a.k.a. 24-bit color). In addition, another eight bits are used by today's graphics boards to display transparent objects; this is known as 32-bit color. Drawings support transparency of objects as of AutoCAD 2011.

It gets frustrating selecting a specific color out of millions of choices, and so AutoCAD provides several methods to help you make your selection. These methods are through RGB, HSL, and color books.

### RGB

RGB is short for "red, green, blue," the most common system for selecting colors. It specifies three sets of numbers that range from 0 to 255. Each number represents the amount of red, green, or blue, ranging from black (0) to full color (255).

As you may recall from elementary school physics, all colors of light are represented by varying amounts of red, green, and blue. Yellow is made from mixing red and green, for instance, and orange is made from full red (255) and half green (127), and no blue (0).

The hue of blue used in this book is represented by R=38, G=133, and B=187. Black consists of no colors (0,0,0), while white is all colors (255,255,255).

To specify a color in AutoCAD with the RGB system, follow these steps:

1. In the Select Color dialog box, select the **True Color** tab.
2. Under Color Model, select **RGB**.
3. Drag the sliders for Red, Green, and Blue.

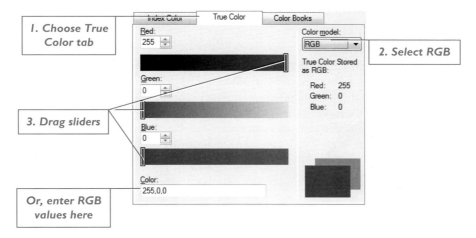

Alternatively, enter an R,G,B color triplet, (such as 120,222,217) in the **Color** text box.

4. Click **OK**.

## HSL

HSL is short for "hue, saturation, luminance." It is a second system for specifying colors, which varies the *hue* (color), *saturation* (amount of color ranging from gray to full color), and *luminance* (brightness of color ranging from darkened to bright).

Hue numbers range from 0 to 360, starting at 0 = red. The colors progress through the colors of the rainbow: yellow is 60, green is 120, cyan is 180, blue is 240, violet is 300, and then red again at 360.

Saturation and luminance each range from 0 to 100 percent, where 0 is none (black) and 100 is full (white). The HSL value for the blue used in this book is H=202, S=66, and L=44.

To specify a color through HSL:

1. In the Select Color dialog box, select the **True Color** tab.
2. Under Color Model, select **HSL**.

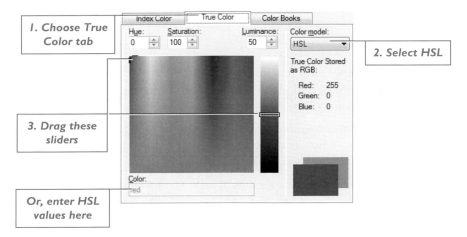

3. Move the cross slider around to select the hue and saturation. Move the horizontal slider up and down to choose luminance.

   Alternatively, In the Color text box, enter a color triplet, such as 120,222,217.
4. Click **OK**.

 **Note** AutoCAD includes a drawing showing its colors in *color wheels*. In the *\autocad 2011\samples* folder, open *colorwh.dwg*. On the left, looking like a smoothly shaded donut, is the true color wheel made of 35,840 tiny tiles. (The coloring makes it appear like a 3D torus.)

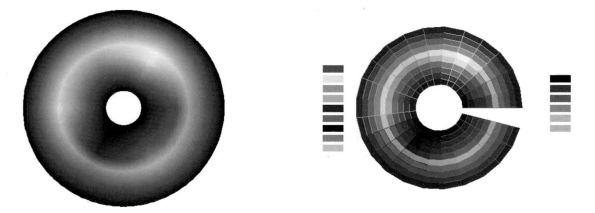

On the right is the original 256-color ACI color wheel. It is flanked on the left by the seven named colors and on the right by six shades of gray.

RGB and HSL let you set precise colors in your drawings. For example, you can use the color of a client's logo.

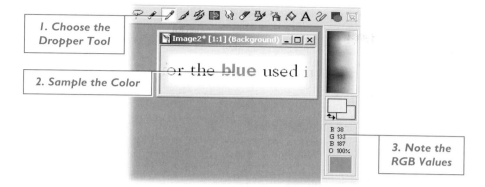

Here's how I do it: get a sample of the color, either by grabbing a screen or by scanning a page. Open the sample in a raster editor like PaintShop Pro. Use the color dropper tool to sample the color; note the RGB values, and then enter the three numbers into AutoCAD's Select Color dialog box.

Color Books

AutoCAD provides a third naming system used by the publishing and design industries, called "color books." Colors are seen differently by individuals, and so a standardized system for specifying a select number of colors was needed. Color books consist of swatch names and numbers.

For example, the name of the blue color used in this book is Pantone 2925. When I specify this name, the identical hue of blue shows up in my desktop publishing software, in the printed book, and in AutoCAD.

AutoCAD supports three of these systems: the American Pantone, the European RAL, and the Japanese DIC systems.

**The Pantone Matching System** (www.pantone.com) was designed in 1963 to specify colors for the graphic arts, textiles, and plastics industries. Today, designers typically work with Pantone's fan-format book of standardized colors. This system is used primarily in North America.

**The RAL color system** (www.ral.de) was designed back in 1927 to standardize colors by limiting the number of color gradations, first to just 30 and now to over 1,600. RAL (Reichs Ausschuß für Lieferbedingungen – German for "Imperial Committee for Supply Conditions") is administered by the German Institute for Quality Assurance and Labeling. This system is used primarily in Europe.

**The DIC Color Guide** (www.dic.co.jp, Japanese language only) is the Japanese standard for specifying colors. DIC is the abbreviation for the ink company that came up with the system, Dainippon Ink and Chemicals.

Here's how to specify colors by Color Books in AutoCAD:

1. In the Select Color dialog box, choose the **Color Books** tab.
2. Under Color Books, select a DIC, Pantone, or RAL book name.
3. Move the slider up and down to choose a color group, and then pick a specific color name.
    As an alternative, in the Color text box, enter a color book specification.
4. Click **OK**.

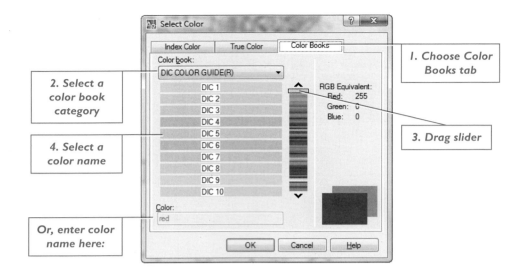

### CHANGING COLORS: ADDITIONAL METHODS

AutoCAD provides another way to change colors through the Color Control droplist in the Properties panel of the ribbon. The control affects color by two methods, depending on the order in which you use them:

> **Set the working color** (subsequent objects taking on the color) — select a color from the droplist.
>
> **Change** the color of objects(s) — first select the objects, and then a color from the droplist.

### TUTORIAL: SETTING AND CHANGING COLORS

In the following tutorial, we look at both methods of changing colors.

1. Start AutoCAD with a new drawing, and then draw a few lines. Notice that they are colored white or black — depending on the background color of your screen.
2. From the ribbon's Properties panel, click on the **Color Control** droplist. Notice the list a of basic colors: ByBlock, ByLayer, and the first seven colors, red through white.
3. Select **Red**.
4. Draw a few more lines. Notice that they are colored red.
5. Now change the red lines to blue: select the red lines by picking them. They become highlighted and show grips (squares).
6. From the Color Control droplist, select **Blue**. Notice that the lines turn blue.
7. Press ESC to "unselect" the lines (remove the highlighting and grips).

## TRANSPARENCYDISPLAY

The **TRANSPARENCYDISPLAY** system variable toggles the transparency of objects. (There is a **TRANSPARENCY** command, but it is for turning on transparency for grayscale images attached to drawings.)

Transparency is used to make some objects look dimmer than others. This can make drawings less confusing to view. Transparency can be plotted, but the feature is turned off by default, because plotting with transparency can take much longer as AutoCAD must rasterize all objects in the drawing.

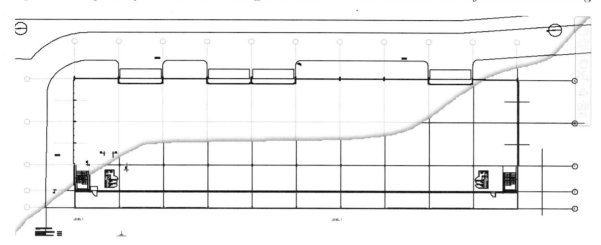

*Upper left:* Transparency of grid lines turned on.
*Lower right:* Transparency turned off.

You can apply transparency to individual objects and blocks (**CETRANSPARENCY** system variable), as well as control it through layers (**LAYER** command and **SETBYLAYERMODE** system variable), and through the Transparency control in the ribbon's Properties panel:

**ByLayer** applies transparency to all objects assigned to the layer.

**ByBlock** applies transparency to all objects making up the block.

**0** turns off transparency; objects are opaque.

**90** makes objects the most transparent.

It turns out that you cannot use any of the methods listed above retroactively to change the transparency of objects; instead, you have to select the objects, and then change the setting with the **PROPERTIES** command.

**Note**  If no change in transparency occurs, then it is likely that it is turned off. To turn it on, click the (or TR) **Transparency** button on the status bar. Another reason may be that transparency level is set to 0.

The transparency of hatch patterns (**HPTRANSPARENCY** system variable), locked layers (**LAYLCK** command), external references, and plots (**PLOTTRANSPARENCYOVERRIDE** system variable) can be controlled independently.

 **LWEIGHT**

The **LWEIGHT** command (short for "lineweight") changes the visual width of lines and other objects.

(*History*: In old releases of AutoCAD, only polylines and traces could have a width (weight); all other objects were drawn one pixel wide, as thin as the display. Widths were assigned at plot time based on the object's color.)

Using a variety of weights (heavy and thin lines) helps make drawings clearer. Weights can be assigned to layers or to individual objects. In the figure below, the upper part of the drawing has lineweights turned off; in the lower part, lineweights are turned on.

Lineweights range from 0.05 mm (0.002") to 2.11 mm (0.083"). AutoCAD displays the values in millimeters or inches. The lineweight of 0 is compatible with earlier versions of AutoCAD, and displays as one pixel in model space. It is plotted at the thinnest width of which the plotter is capable.

When on, the weight appears when both displayed and plotted. The display is constant, in pixels. This means that as you zoom out, the lines appear to get wide — WYSINQWYG — "what you see is not quite what you get," in the words of the technical editor.

There is an exception: objects with lineweights under a certain value are displayed one pixel wide in model space. The default is 0.025mm but can be adjusted by the **Display Scale** slider in the Linetype Settings dialog box, described below.

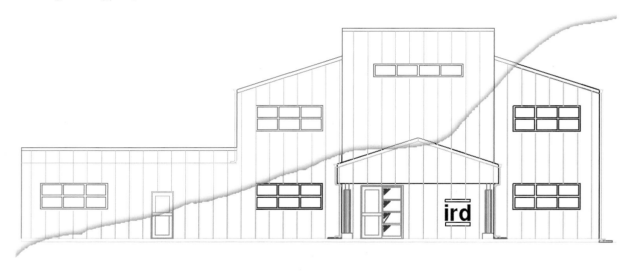

**Upper-left:** Lineweights turned off.
**Lower-right:** Lineweights turned on.

At plot time, objects with weight are plotted at the exact same width. Objects copied to the Clipboard (through CTRL+C) retain their lineweight data when pasted (CTRL+V) back into AutoCAD, but may lose their weight when pasted into other Windows applications.

Some drafters prefer to define lineweight through layers, as with colors; they do not approve of overriding the layer's lineweight ByLayer setting.

AutoCAD includes three named lineweights:

**ByLayer** — objects take on the lineweight defined by the layer.
**ByBlock** — objects take on their block's lineweight.
**Default** — displays the default value of 0.25 mm, or another default value.

**Note** Lineweights are meant as a visual aid. Do not use lineweights to represent the actual width of objects. If printed circuit board traces are 0.05 inches wide, use polylines with width of 0.05 inches; don't draw lines with a weight of 0.05 inches.

## TUTORIAL: CHANGING LINEWEIGHTS

1. In new drawings, lineweights are not displayed. To turn on their display, choose the ⊞ (or **LWT**) **Lineweight** button on the status bar. Lineweights are on when the button looks blue.
2. To change the lineweights of objects, start the **LWEIGHT** command with one of these methods:
    - In the ribbon's Home tab, click the **Lineweight** droplist in the Properties panel.

### AUTOCAD'S DEFAULT LINEWEIGHTS

The table below shows the default lineweight values used by AutoCAD and their equivalent values for industry standards. There are 25.4 mm per inch; 72.72 points per inch.

| Millimeters | Inches | Points | Pen Size | ISO | DIN | JIS | ANSI |
|---|---|---|---|---|---|---|---|
| 0.50 | 0.002 | | | | | | |
| 0.90 | 0.003 | $1/4$ pt | | | | | |
| 0.13 | 0.005 | | | | ✓ | | |
| 0.15 | 0.006 | | | | | | |
| 0.18 | 0.007 | $1/2$ pt | 0000 | ✓ | ✓ | ✓ | |
| 0.20 | 0.008 | | | | | | |
| 0.25 | 0.010 | $3/4$ pt | 000 | ✓ | ✓ | ✓ | |
| 0.30 | 0.012 | | 00 | | | | 2H or H |
| 0.35 | 0.014 | 1 pt | 0 | ✓ | ✓ | ✓ | |
| 0.40 | 0.016 | | | | | | |
| 0.50 | 0.020 | | 1 | ✓ | ✓ | ✓ | |
| 0.53 | 0.021 | $1 1/2$ pt | | | | | |
| 0.60 | 0.024 | | 2 | | | | H, F, or B |
| 0.70 | 0.028 | $2 1/4$ pt | $2 1/2$ | ✓ | ✓ | ✓ | |
| 0.80 | 0.031 | | 3 | | | | |
| 0.90 | 0.035 | | | | | | |
| 1.00 | 0.039 | | $3 1/2$ | ✓ | ✓ | ✓ | |
| 1.06 | 0.042 | 3 pt | | | | | |
| 1.20 | 0.047 | | 4 | | | | |
| 1.40 | 0.056 | | | ✓ | ✓ | ✓ | |
| 1.58 | 0.062 | $4 1/4$ pt | | | | | |
| 2.00 | 0.078 | | | ✓ | ✓ | | |
| 2.11 | 0.083 | 6 pt | | | | | |

- On the status line, right-click ✚, and then choose **Settings** from the shortcut menu.
- At the 'Command:' prompt, enter the **lweight** command:

    Command: **lweight** *(Press ENTER.)*

- Alternatively, enter the aliases **lw** or **lineweight** at the 'Command:' prompt.

Notice that AutoCAD displays the Lineweight Settings dialog box.

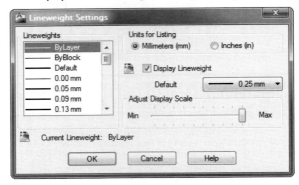

Under Lineweights is the list of weights available in AutoCAD. (This is the same list displayed by the Lineweights droplist in the Properties panel.) You cannot customize lineweights.)

3. Select a lineweight; this becomes the default until you change the lineweight again. Other items of interest in the dialog box include:

    **Units for Listing** — selects between inches and millimeters (the default).

    **Display Lineweight** check box — functions identically to the **LWT** button on the status bar.

    **Default** list box — specifies the lineweight with which all objects are drawn, unless overridden by layer.

    **Adjust Display Scale** slider — controls the display scale of lineweights in model space. This slider is beneficial if your computer displays AutoCAD on a high-resolution monitor, which tends to make lines look thinner. Experiment with adjusting the lineweight scale to see if you get a better display with different lineweights. As you move the slider, notice that the widths of lines in the Lineweights list change in clusters of 4-5 weights, and not uniformly.

4. Click **OK** to dismiss the dialog box.

### Lineweight Control

As an alternative to using commands and dialog boxes, you can change lineweights directly through the Lineweight Control droplist. As with color, this control affects lineweights by two methods, depending on the order in which you use it:

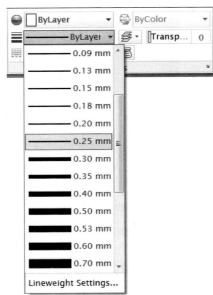

**Set** the working lineweight (subsequent objects taking on the selected lineweight): first select a lineweight from the droplist, and then draw.

**Change** the lineweight: first select the objects, and then a lineweight from the droplist.

# LINETYPE

The **LINETYPE** command sets the patterns of lines.

Linetypes are used to identify objects in 2D drawings. For example, solid lines represent the edges of objects; AutoCAD refers to solid lines as "continuous lines." Dashed lines represent hidden lines. The figure shows an object containing edges that are hidden in some views. These edges are defined with the Hidden linetype.

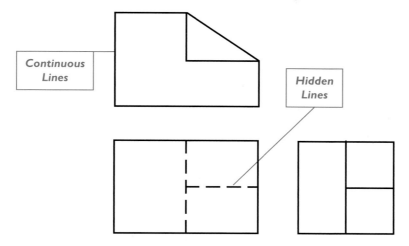

AutoCAD provides many linetypes, and you can create custom ones. AutoCAD employs three classes of linetype:

**Simple** — consist of line segments, dots, and gaps; these are the most commonly-used linetypes.

**ISO** — resemble simple linetypes, except that they conform to standards set by the International Organization of Standards, and can have pen widths assigned to them.

**Complex** — use dashes and spaces like simple linetypes, but add characters and shapes, such as HW (hot water pipes) or squiggles (to indicate insulation).

AutoCAD stores linetype definitions in *acad.lin* and *acadiso.lin*, and the shapes for complex linetypes in *ltypeshp.shp* — all found in the *C:\Users\<login>\AppData\Roaming\Autodesk\AutoCAD 2011\ R18.1\enu\Support* folder. You can create your own linetypes, as well.

AutoCAD 2011 added the upright parameter to the definition of complex linetypes. "U" means that shapes and text are displayed upright, no matter the angle of the object.

## TUTORIAL: CHANGING LINETYPES

Using linetypes in drawings takes two steps: (1) loading them into the drawing, and (2) selecting them for use. Applying them by layer is the preferred method.

1. To change the linetype, start the **LINETYPE** command with one of these methods:
    - In the ribbon's tab, click the **Linetype** droplist on the Properties panel, and then choose **Other**.
    - At the 'Command:' prompt, enter the **linetype** command:

        Command: **linetype** *(Press ENTER.)*
    - Alternatively, enter the aliases **lt**, **ltype**, or **ddltype** (the old name) at the 'Command:' prompt.

Notice that AutoCAD displays the Linetype Manager dialog box:

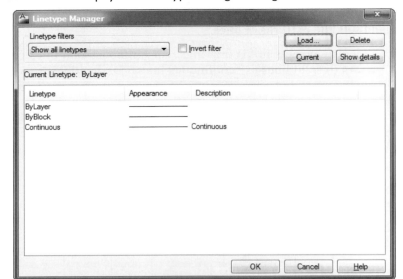

The dialog box lists the linetypes loaded into the current drawing. New drawings have three — Continuous and the same two special names that appear in colors and lineweights:

**ByLayer** — assigned by layers.

**ByBlock** — assigned by blocks.

3. To load additional linetypes, choose the **Load** button.

   AutoCAD displays the Load or Reload Linetypes dialog box. (Unless you know you have linetype definitions stored in a file other than *acad.lin*, you can ignore the File button.)

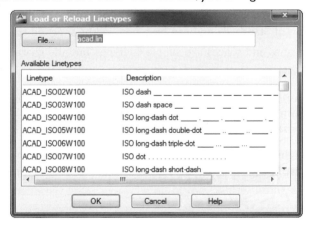

4. I find it easier to load all linetypes at once, and then later remove any that remain unused. To do so, right-click on any linetype, and then choose **Select All**. AutoCAD highlights all linetypes.
5. Choose **OK**, and AutoCAD loads all linetypes.
6. Back in the Linetype Manager dialog box, select a linetype, and then click **Current**.
7. Choose **OK**.

From now on, all objects are drawn in this linetype — until you change it again.

| Linetype | Description |
| --- | --- |
| ACAD_ISO02W100 | ISO dash __ __ __ __ __ __ __ __ __ |
| ACAD_ISO03W100 | ISO dash space __  __  __  __  __  __ |
| ACAD_ISO04W100 | ISO long-dash dot ____ . ____ . ____ . ____ . _ |
| ACAD_ISO05W100 | ISO long-dash double-dot ____ .. ____ .. ____ . |
| ACAD_ISO06W100 | ISO long-dash triple-dot ____ ... ____ ... ____ |
| ACAD_ISO07W100 | ISO dot . . . . . . . . . . . . . . . . . . . . |
| ACAD_ISO08W100 | ISO long-dash short-dash ____ __ ____ __ ____ _ |
| ACAD_ISO09W100 | ISO long-dash double-short-dash ____ __ __ ____ |
| ACAD_ISO10W100 | ISO dash dot __ . __ . __ . __ . __ . __ . |
| ACAD_ISO11W100 | ISO double-dash dot __ __ . __ __ . __ __ . |
| ACAD_ISO12W100 | ISO dash double-dot __ . . __ . . __ . . |
| ACAD_ISO13W100 | ISO double-dash double-dot __ __ . . __ __ . |
| ACAD_ISO14W100 | ISO dash triple-dot __ . . . __ . . . __ . . . |
| ACAD_ISO15W100 | ISO double-dash triple-dot __ __ . . . __ __ . |
| BATTING | Batting SSSSSSSSSSSSSSSSSSSSSSSSSSSSSSSSSSSSSS |
| BORDER | Border __ __ . __ __ . __ __ . __ __ . __ __ |
| BORDER2 | Border (.5x) _ _ . _ _ . _ _ . _ _ . _ _ . |
| BORDERX2 | Border (2x) ____  ____ . ____  ____ . ____ |
| CENTER | Center ____ _ ____ _ ____ _ ____ _ ____ |
| CENTER2 | Center (.5x) __ _ __ _ __ _ __ _ __ _ __ |
| CENTERX2 | Center (2x) _____ __ _____ __ _____ |
| DASHDOT | Dash dot __ . __ . __ . __ . __ . __ . __ |
| DASHDOT2 | Dash dot (.5x) _._._._._._._._._._._._ |
| DASHDOTX2 | Dash dot (2x) ____ . ____ . ____ . ____ |
| DASHED | Dashed __ __ __ __ __ __ __ __ __ __ __ __ |
| DASHED2 | Dashed (.5x) _ _ _ _ _ _ _ _ _ _ _ _ _ _ _ |
| DASHEDX2 | Dashed (2x) ____  ____  ____  ____  ____ |
| DIVIDE | Divide ____ . . ____ . . ____ . . ____ |
| DIVIDE2 | Divide (.5x) __._._ __._._ __._._ __._ |
| DIVIDEX2 | Divide (2x) _____ . . _____ . . |
| DOT | Dot . . . . . . . . . . . . . . . . . . . . |
| DOT2 | Dot (.5x) . . . . . . . . . . . . . . . . . . |
| DOTX2 | Dot (2x) . . . . . . . . . . . |
| FENCELINE1 | Fenceline circle ----0-----0----0-----0----0--- |
| FENCELINE2 | Fenceline square ----[]-----[]----[]-----[]---- |
| GAS_LINE | Gas line ----GAS----GAS----GAS----GAS----GAS--- |
| HIDDEN | Hidden __ __ __ __ __ __ __ __ __ __ __ __ __ |
| HIDDEN2 | Hidden (.5x) _ _ _ _ _ _ _ _ _ _ _ _ _ _ _ _ |
| HIDDENX2 | Hidden (2x) ____ ____ ____ ____ ____ ____ |
| HOT_WATER_SUPPLY | Hot water supply ---- HW ---- HW ---- HW ---- |
| PHANTOM | Phantom _____ __ __ ____ __ __ ____ |
| PHANTOM2 | Phantom (.5x) ___ _ _ ___ _ _ ___ _ _ ___ |
| PHANTOMX2 | Phantom (2x) _____  ____  ____  _ |
| TRACKS | Tracks -|-|-|-|-|-|-|-|-|-|-|-|-|-|- |
| ZIGZAG | Zig zag /\/\/\/\/\/\/\/\/\/\/\/\/\/\/\/\/\ |

## Global, Local, and Annotative Linetype Scales

AutoCAD lets you specify linetype scales in three ways:

**Globally** — used for all objects in the drawing. The global linetype scale is set by the **LTSCALE** command, described later.

**Locally** — used for individual objects. Although it is poor drafting practice to use more than one linetype scale in drawings, local scaling is sometimes needed for complex linetypes, such as Batting and Hot Water Supply.

**Annotatively** — used for scale-specific viewports.

### Local (Object) Linetype Scaling

To change the linetype scale of individual objects, use either the **PROPERTIES** command (as described later in this chapter) or the **CELTSCALE** system variable (short for "current entity linetype scale").

There is a difference between the two: the **PROPERTIES** command changes the linetype scale of selected objects that already exist in the drawing, while the **CELTSCALE** system variable changes the scale of objects you plan to draw.

**CELTSCALE** can be tricky to use, because it sets the linetype scale relative to the value set by the **LTSCALE** command setting: when **LTSCALE** is set to 3.0 and **CELTSCALE** is set to 0.75, objects will be drawn with a linetype scale of 2.25 (3.0 x 0.75 = 2.25).

Both global and object linetype scaling are available as features "hidden" in the Linetype Manager dialog box. Click the **Details** button to reveal them.

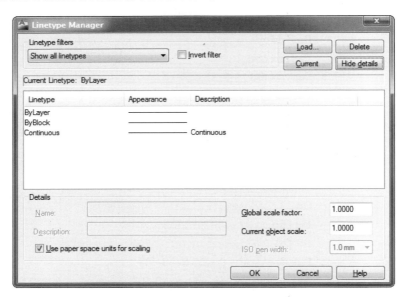

**Global scale factor** — sets the linetype scale for all objects in the drawing. This scale factor takes effect with the next drawing regeneration.

**Current object scale** — sets the linetype scale for subsequently-created objects.

**ISO pen width** — selects from one of the ISO's pen widths, such as 1.00 mm. This option is grayed out when non-ISO linetypes are selected. To activate, double-click an ISO linetype.

**Use paper space units for scaling** — scales linetypes identically in paper space and model space through the **PSLTSCALE** system variable.

Annotative Scaling

Annotative scaling makes linetypes appear at the correct size in model space viewports. This includes the Model tab and viewports in Layout tabs. When a scale factor is assigned a viewport (or in Model tab), then the scale factor for all linetypes is adjusted automatically by AutoCAD.

The MSLTSCALE system variable (short for "model space linetype scale") toggles annotative scaling of linetypes. When MSLTSCALE is set to 1, linetypes are scaled to match the current viewport scale; this is the default setting. (The variable is set to 0 in drawings opened from AutoCAD 2007 and earlier.)

To use MSLTSCALE correctly, you have to set all linetype scale system variables to 1:

>   CELTSCALE = 1
>   LTSCALE = 1 (or another size correct for plotting)
>   MSLTSCALE = 1
>   PSLTSCALE = 1

The changes due to annotation scaling appear with the next regeneration (REGEN command).

## LINETYPE SCALING (LTSCALE)

Because linetypes are constructed of line segments and dashes, the pattern can look too small or too large for drawings. How do you tell when the scale is wrong? When the pattern lines do not look continuous.

Like hatch patterns, linetypes need to be set to an appropriate scale; fortunately, the same scale factor applies to both hatch patterns and linetypes. When you figure out the correct factor for one, you can use it for the other, as well as for text and dimensions.

The scale of linetypes is adjusted by the LTSCALE (short for "linetype scale") command:

>   Command: **ltscale**
>   Enter new linetype scale factor <1.0000>: *(Enter a value.)*
>   Regenerating model.

Numbers larger than 1.0 result in longer line segments, while numbers smaller than 1.0 create shorter line segments.

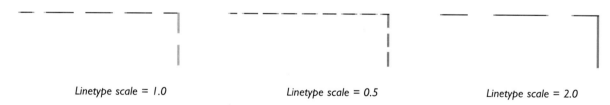

Linetype scale = 1.0          Linetype scale = 0.5          Linetype scale = 2.0

To display the new linetype scale correctly, AutoCAD automatically performs a regeneration. If a regeneration does not occur after you change the linetype scale, force a regeneration with the REGEN command.

If, after changing the scale, the linetypes still appear continuous, change LTSCALE to yet a different value. It is possible for the linetype scale to be too large or small to display. If lines take longer to draw, it is likely that the scale is too small.

General rule: the linetype scale is the inverse of the plot scale. For example, when the plot scale is 1/8" = 1' (1:96), the linetype scale factor is 96. The exception is when your drawing uses annotative scale factor, in which case AutoCAD 2008 (and later) uses the viewport scale for linetypes when MSLTSCALE = 1 (the default setting).

### Linetype Control

As with color and lineweights, you can change linetypes directly through the ribbon's **Linetype Control** droplist. The control affects linetypes by two methods, depending on the order in which you use it:

**Set the working lineweight** (objects subsequently taking on the selected linetype) — select a linetype name from the droplist.

**Change the linetype of existing objects(s)** — first select the objects, and then a linetype name from the droplist.

**Note** If you (or someone else) creates custom linetypes, AutoCAD has two ways to share them. One is to exchange a copy of the *.lin* file that stores the custom definitions. The other is to use DesignCenter to drag a copy of the linetype from one drawing to another.

## PROPERTIES

The **PROPERTIES** command displays a palette for changing (almost) all properties of objects.

The Properties palette displays different sets of properties depending on the object(s) you select: text, mtext, 3D face, 3D solid, multiline, arc, point, attribute definition, polyline, block insertion, ray, body, region, circle, shape, dimension, mleader, 2D solid, ellipse, spline, external reference, hatch, tolerance, image, trace, leader, viewport, line, xline, and more.

That's because each object has a somewhat different set of modifiable properties. For example, you can modify the endpoints of a line through x, y, and z coordinates; you cannot, however, modify the endpoint of an arc; instead, you have to alter its start and end angle.

### TUTORIAL: CHANGING ALL PROPERTIES

1. To change all the properties of objects, start the **PROPERTIES** command with one of these methods:
   - Double-click an object.
   - At the keyboard, press **CTRL+1**.
   - At the 'Command:' prompt, enter the **properties** command:

     Command: **properties** *(Press ENTER.)*

   - Alternatively, enter the aliases **pr**, **props**, **mo** (short for "modify"), **ch**, (short for "change"), **ddchprop**, or **ddmodify** (its old name) at the 'Command:' prompt.

   Notice that AutoCAD displays the Properties palette. (See figure, next page.)

Notice that there are two colors of field: white and gray. White means you can modify the property, while gray means you cannot.

The effect of changes you make in this window depends on whether or not objects are selected — much like changing settings in the Properties panel.

When no objects are selected (as shown below), changing the properties affects objects drawn from now on.

When one or more objects are selected, changing the properties affects the selected objects immediately.

You can collapse unneeded sections by clicking the double button, which makes the palette smaller.

Notice the section named "General" located in the upper part of the window. The General section includes these modifiable properties:

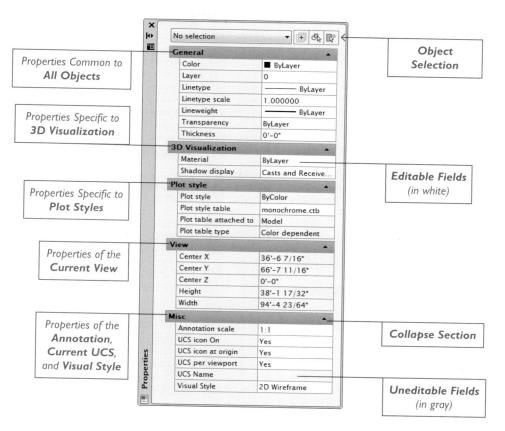

- **Color** — selects the color square to change colors.
- **Layer** — selects a layer name.
- **Linetype** — selects a linetype name.
- **Linetype Scale** — specifies a different scale factor.
- **Lineweight** — selects a preset lineweight.

2. To inspect the properties of one or more objects, pick them in the drawing. You do not dismiss the Properties palette: simply move the cursor into the drawing, and then pick the objects. (To select all objects in the drawing, press **CTRL+A**.)

Notice that the palette changes its content to reflect the properties of the selected object(s), such as the circle at left and the text at right:

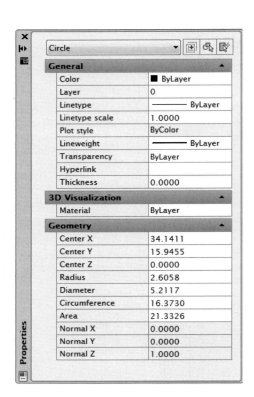
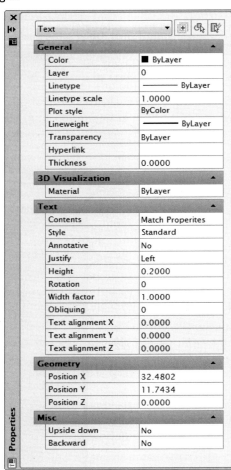

*Left:* Properties for a circle.
*Right:* Properties for text.

As an alternative to picking, choose one of the buttons at the top of the palette:

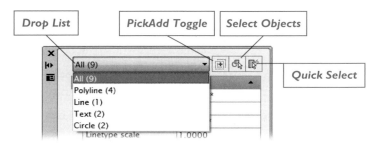

The droplist shows the variety (and, in parentheses, the number) of objects.

- **Quick Select** displays the Quick Select dialog box, described in Chapter 7, "Drawing with Efficiency."
- **Select Objects** causes AutoCAD to prompt you (unnecessarily):
    Select objects: *(Select one or more objects.)*
    Select objects: *(Press* ENTER *to end object selection.)*

**Toggle value of PICKADD sysvar** changes how you select objects:

Plus sign means you select additional objects by picking them.

Number 1 means you select additional objects by holding down the **SHIFT** key.

You can modify any property whose value has a white background.

3. In the Geometry section, click a field, such as **Height**.
4. Change the value to a different number. Notice that the geometry changes in the drawing.

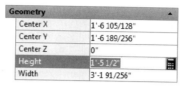

As alternatives:
- Click the button found on the toolbar. AutoCAD prompts you:

    Pick a point in the drawing: *(Move the cursor, and then pick a point.)*

    Use object snaps to make accurate picks.
- Click the calculator icon to bring up the Quick Calculator palette.

5. When done with this palette, you can click the **x** or enter the **PROPERTIESCLOSE** command.

**Note** You can easily display the Properties window by double-clicking (almost) any object in the drawing. Exceptions include double-clicking text, which displays the text editing window.

## CHANGING PROPERTIES: ADDITIONAL METHODS

AutoCAD provides other ways to change properties:

- **QUICKPROPERTIES** command changes properties in a floating palette near the cursor.
- Rollover tips display object properties in a tooltip.

The **QUICKPROPERTIES** command toggles the display of a properties palette that floats near the cursor.

Command: **quickproperties**

<Quick properties on.>

Enter the command a second time to turn off Quick Properties. Alternatively, press **CTRL+SHIFT+P** or click the (or QP) **Quick Properties** button on the status bar.

### TUTORIAL: USING QUICK PROPERTIES

To use this palette, follow these steps:

1. Open *map.dwg*.
2. Ensure Quick Properties is turned on at the status bar.

3. Choose an object in the drawing. Notice that the palette appears near the cursor.

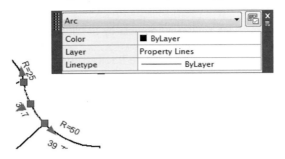

4. Move the cursor over the palette. Notice that it expands to show additional properties.

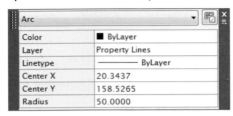

5. As in the Properties palette, you can change the values of properties. Click on any white field to select another value for the related property. For instance, I use Quick Properties often to change the color of a selected object, as follows:
   a. Next to Color, click **ByLayer**. Notice the droplist.
   b. Choose a color from the droplist. Notice that the object changes color.

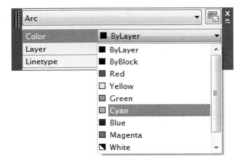

6. Normally, the palette floats near the cursor. You can change this and other settings by clicking the 📇 **Options** button (located under the **X**):

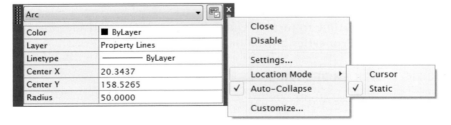

**Close** — closes the palette, but will reappear the next time you select an object.

**Disable** — closes the palette and will no longer display for the current object (arcs, in this case). It will, however, be displayed for other objects.

**Settings** — sets display options, several of which are duplicated by this menu. This option displays the Quick Properties tab of the Drafting Settings dialog box.

**Location Mode** — determines the location of the palette, either Cursor (floating near the cursor) or Static (at a fixed position). I find I prefer Static; otherwise, the palette tends to obscures the drawing right where I am working.

**Auto-Collapse** — determines whether the palette becomes smaller when the cursor leaves its territory. I prefer this option turned off, since the collapsing and expanding action of the palette becomes irritating.

**Customize** (or ▣ ) — lets you choose the individual properties for every object type displayed by this palette. It displays the Customize User Interface dialog box.

## QuickProperties Settings

The Quick Properties tab of the DSETTINGS command provides the following options for controlling the placement of the Quick Properties palette:

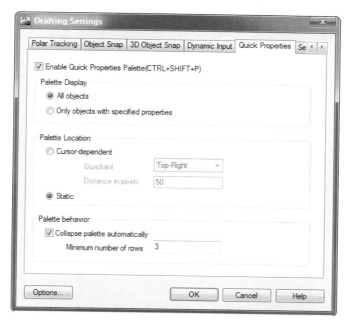

**Quick Properties On** — toggles the display of the palette.

**Display Per Object Type** — determines if the palette is always displayed, or just for specified objects.

**Location Modes** — specifies where the palette appears: in a fixed position (called "Float" by Autodesk), or near the cursor. When it is near the cursor, you have the option of specifying the distance away from the cursor, as well as which cursor quadrant it appears in.

**Size Settings** — controls the default size of auto-collapsed palettes.

## Rollover Tips

Rollover tips display object properties in a tooltip as the cursor passes over objects. To use rollover tips, follow these steps:

1. Ensure rollover tips are turned on:
   a. Enter the **OPTIONS** command.
   b. Click the **Display** tab.
   c. In the Windows Elements section, check that the **Rollover Tips** option is turned on.

d. Click **OK** to exit the dialog box.
2. Move the cursor over an object. Notice the tooltip that appears.

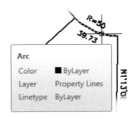

Rollover tips differ from Quick Properties in the following ways. You do not select the object; just roll the cursor over objects. Rollover tips cannot edit object properties.

The content of the rollover tip is customizable through the CUI command's Rollover Tooltips mode. There you can choose among dozens of properties for a hundred object types. In the case of arcs, you have 28 properties to choose from, ranging from the useful (Arc Length) to the obscure (Normal Y).

# LAYER

The **LAYER** command applies common sets of properties to objects assigned to layers.

Nearly all CAD drawings are constructed using layers that can be turned on and off, and changed. Layers go back to the days of traditional drafting, which sometimes involved a method of drawing called "overlay drafting." Drafters overlaid sheets of semitransparent drafting media so that the drawings showed through each other. Multiple sheets could be printed together, resulting in prints that showed the complete design.

The bottom sheet was typically referred to as the "base drawing." Each additional sheet showed different items. For instance, when a drafter prepared a set of floor plans, he typically drafted separate drawings showing the items to be removed, plumbing, electrical, and so on. Often, the floor plan is the base drawing, with each discipline, such as electrical and plumbing, placed on overlay sheets.

AutoCAD's layers are like sheets of glass stacked on top of one another. You place different aspects of the drawing on the sheets of glass (layers), yet are able to see through all the layers. The work appears as though it were one drawing.

AutoCAD goes two steps further. You can turn layers on or off, so that they are either visible or invisible, as well as change their properties. The properties of a layer affect all objects assigned to the layer, but can be overridden with other commands, such as **COLOR**, **LINETYPE**, and **PROPERTIES** (described earlier this chapter). In turn, these overrides can be overridden.

Before adding objects to a layer, you switch to the appropriate layer. Anything you now draw is placed on this layer.

The working layer is called the "current layer." AutoCAD can have only one layer current at a time. The current layer cannot be frozen. (Frozen layers are invisible and cannot be edited.) A layer named "0" exists in all new drawings; it has some special properties, and cannot be renamed or removed.

**Note** The **LAYER** command displays layer information in a palette. Like other palettes, it is meant to be open all the time. As you change layer properties in the palette, they are reflected immediately in the drawing.

If you prefer to use the layer dialog box, enter the **CLASSICLAYER** command. Alternatively, if you want the **LAYER** command to show the dialog box, set the new **LAYERDLGMODE** system variable to 1. (You can still access the layer palette with the **LAYERPALETTE** command.)

## TUTORIAL: CREATING LAYERS

1. To create layers, start the **LAYER** command with one of these methods:
   - In the ribbon's 2D Home tab, click the **Layer Properties** button on the Layers panel.
   - Or, at the 'Command:' prompt, enter the **layer** command:

     Command: **layer** *(Press* ENTER.*)*

   - Alternatively, enter the aliases **la** or **ddlmodes** (an old name for this command) at the 'Command:' prompt.

   Notice that AutoCAD displays the Layer Properties Manager palette.

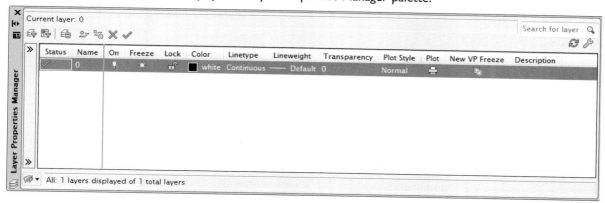

2. Click the  **New Layer** icon. (Alternatively, press ALT+N.) Notice that AutoCAD creates a new layer named "Layer1."

   To the right of the name are icons indicating that the layer is turned on, thawed (not frozen), unlocked, and colored white/black; has continuous linetype, and default lineweight; and will be plotted/printed. (More about these settings later.)

3. To change the new layer's name, edit the "Layer1" text, type a new name — anything up to 255 characters long.

4. To make the layer current, click the **Set Current** button. (Alternatively, press ALT+C.)

   Notice that the name immediately appears on the ribbon's Layer droplist. (This behavior is different from the Layer dialog box in AutoCAD 2008 and earlier.)

**Notes** AutoCAD creates some layers on its own. Two layers you must become acquainted with are **0** and Defpoints. Every new drawing contains the layer named 0 (zero). This layer cannot be removed or renamed. Layer 0 has a special property for creating blocks, of which you learn elsewhere in this book.

The first time you draw dimensions, AutoCAD creates a layer named Defpoints (short for "definition points"). This layer contains data that AutoCAD needs to keep its dimensions associative. The layer cannot be removed. You can rename it, but then AutoCAD creates a new Defpoints layer with the next dimension. AutoCAD does not plot anything you draw on this layer, accidentally or otherwise. Students sometimes become frustrated to find that part of their drawing won't plot, because they accidentally drew on layer Defpoints but had overridden colors and linetypes, so everything looked correct. (Use the **No Print** toggle to make *any* layer non-printing.)

A third layer created by AutoCAD is AShade, which holds data regarding renderings.

### Shortcut Keystrokes

The following shortcut keystrokes operate only while the Layer Properties Manager palette is displayed:

| Keystroke | Meaning |
|---|---|
| ALT+N | Creates new layer. |
| ALT+D | Deletes selected layer(s). |
| ALT+C | Makes the selected layer as current. |
| ALT+P | Creates new property filter. |
| ALT+G | Creates new group filter. |
| ALT+S | Displays layer states manager. |

## Working with Layers

The Layer Properties Manager palette provides you with a fair degree of control over layers.

### Status

The **Status** column uses icons to report the status of layers:

| Status | Comment |
|---|---|
| ✓ | Layer is current. |
| ▱ | Layer is in use (has at least one object). |
| ▱ | Layer is empty (can be deleted, unless part of an xref). |

### Turning Layers On and Off

When layers are on, their objects are seen and can be edited; the status is indicated by the yellow light bulb icon in the On column. When off, the objects on that layer are not seen; the light bulb icon turns blue-gray.

To turn one or more layers on or off, first highlight the layer(s) by selecting their name(s), and then choose the light bulb icon. The "on" column immediately reflects the change.

On and Off were used in the earliest versions of AutoCAD to toggle the display of layers, and are rarely used anymore; Freeze and Thaw are now preferred.

### Freezing and Thawing Layers

Frozen layers cannot be seen or edited; their status is shown by the snowflake icon in the Freeze column. Thawed layers can be seen and edited, just like "on" layers; the symbol for thawed layers is the sun icon.

To freeze or thaw one or more layers, highlight the target layer(s), and then choose the sun or snowflake icon. It is more efficient to freeze layers than to turn them off. Objects on frozen layers are ignored by AutoCAD during regenerations and other computationally-intensive operations.

### Locking and Unlocking Layers

In most cases, objects on locked layers can be seen, but not edited. You can still use these objects for osnaps, trimming, and extending.

Locked layers show a padlock icon in the Lock column. Unlocked layers are like "on" layers: their objects are seen and can be edited; an unlocked layer shows an open padlock.

In drawings, locked layers are faded. When you attempt to select objects on locked layers, AutoCAD shows a padlock icon near the cursor, as shown below.

To unlock layers, select the target layer(s), and then choose the 🔒 padlock icon(s) in the Lock column.

### Setting Layer Colors

To set the color for objects on layers, first select the layer names(s), and then click on the color square. AutoCAD displays the Select Color dialog box. Choose a color, and then click **OK**.

From now on, all objects on the layer are displayed in that color, unless overridden by the COLOR command or by a block's properties.

### Setting Layer Linetypes

When you select a linetype for a layer, it affects all the objects on that layer. First select the layer names(s) for which you wish to set a linetype, and then choose the linetype. AutoCAD displays the Select Linetype dialog box. If necessary, load linetypes. Select a linetype and choose **OK**.

### Setting Layer Lineweights

You can specify the lineweights for objects residing on layers.

1. Select the layer name(s) for which you wish to set a lineweight.
2. Choose the lineweight in the layer palette, usually "Default." This action causes the Lineweight dialog box to appear.

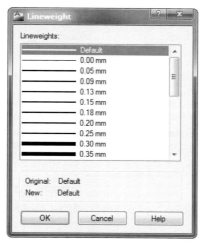

3. Select a lineweight, and then choose **OK**.

## Setting Transparency

Transparency determines the "see-through-ness" of objects. It ranges from 0 (fully opaque) to 90 (nearly fully transparent). To change the transparency:

1. Select the layer names(s) for which to set the transparency.
2. Click the transparency value, typically "0." Notice the Layer Transparency dialog box.

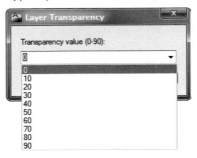

3. Enter a transparency level, or else choose a value from the droplist.
4. Click **OK**.

## Setting the Layer Plot Style

Plot styles determine how all objects residing on a layer are plotted. To change the plot style:

1. Select the layer names(s) for which to set a plot style.
2. Click the plot style name, typically "Solid." This causes the Select Plot Style dialog box to appear.

3. Choose plot style from those available in the dialog box, and then click **OK**.

Plot styles are available only when the feature is turned on, as described in Chapter 20; otherwise, the plot style names are shown in gray and cannot be changed.

 Setting Layer Print Toggles

You can specify that some layers print, while others do not. Under the Plot column, choose the printer icon for the layer(s) you don't want to print or plot. To allow the layer to print, simply choose the icon.

## CONTROLLING LAYER FILTERS

Layer filters can slow down AutoCAD, so Autodesk added system variables that help speed up the display of the layers palette and dialog box.

The **SHOWLAYERUSAGE** system variable toggles the display of icons. (The icons indicate whether layers have objects or are empty.) Turn off this system variable (set it to **0**) to improve the speed of the dialog box and palette.

When drawings have more than 99 filters *and* the number of filters exceeds the number of layers, the **LAYERFILTERALERT** system variable determines when excess filters can be deleted, thereby improving performance:

| *LayerFilterAlert* | *Meaning* |
|---|---|
| 0 | Filters are never deleted. |
| 1 | All filters are deleted without warning the next time the layer dialog box is opened. |
| 2 | All filters are optionally deleted; prompts 'Do you want to delete all layer filters now?' the next time the Layer dialog box is opened. |
| 3 | Filters are selectively deleted; displays dialog box for selecting filters to delete the next time the drawing is opened. |

**Notes** To select *all* layers, right-click any layer name to display a shortcut menu. Choose **Select All** to highlight all layer names. Choose **Clear All** to deselect all layers.

To change the width of the columns in the layer dialog box, grab the black bar separating column tiles and drag left or right.

### Description

The **Description** column permits a descriptive sentence for each layer.

### Layer Controls in Layouts

In layout mode (paper space), more columns are added to the layer dialog box. To see them, switch to layout mode by selecting any layer tab. In the Layer Properties Manager palette, scroll the layer listing all the way to the right.

AutoCAD can work with tiled or floating viewports. You can freeze layers in floating viewports — something you cannot do in model view — to display different sets of layers in different viewports.

AutoCAD allows you to do the following to layers in floating viewports: freeze the active viewport; freeze new viewports; and override the color, linetype, lineweight, and plot style in specific viewports.

###  Layout Freeze Modes

The **New VP Freeze** column freezes specified layers when a new viewport is created. Select the layer name(s), and then choose the icon in the New VP Freeze column. When the icon has a shining sun, the layer is thawed; when the icon has a snowflake, the layer is frozen.

The **VP Freeze** column automatically freezes the specified layers in the current viewport. Select the layer name(s), and then choose the icon in the VP Freeze column (formerly named "Current VP Freeze"). As always, when the icon has a shining sun, the layer is thawed; when the icon has a snowflake, the layer is frozen. When the icon is gray, however, AutoCAD cannot display floating viewports, because the drawing is not in layout mode.

### Layout Property Modes

The **VP Color** column allows you to override the layer color for the current viewport. You use it like this:

1. In model space, draw a rectangle.
2. Switch to a layout tab, and then create two viewports. Double-click a viewport to make it current.
3. Start the **LAYER** command.
4. In the layer palette, change **VP Color** from "White" to "Red."

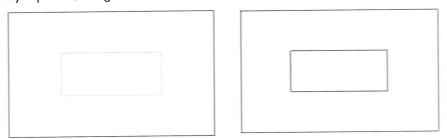

Notice that the rectangle is red in one viewport (shown at left, below), but remains black in the other (at right).

The same trick can be played using linetypes, lineweights, and plot styles. Each viewport can display the drawing in a different manner. The viewport is said to have its layer properties "overridden."

## Resetting Overridden Layout Properties

To return the viewport properties to the layers' default settings, use the SETBYLAYER command. Technically, this command resets the properties of objects to ByLayer. The command applies only to viewports in layouts.

### Selecting Layers

Some types of drawings may have many, many layer names. In theory, AutoCAD drawings can have an unlimited number of layers; in practice, drawings may contain thousands. Working your way through long lists of layers can become tedious.

For example, open the *8th floor.dwg* file. Start the LAYER command, and notice the text on the status line of the dialog box:

> All: 225 layers displayed of 225 total layers.

Scroll through the list of layers to appreciate its length. Two hundred and twenty-five is a lot of layers! The reason is that this drawing includes six other drawings, known as "external references." The original drawing has 27 layers; the xrefs contribute the additional 198 layers.

To help out, AutoCAD allows you to sort and to shorten the list of names using several techniques:

- By sorting names and properties alphabetically in columns.
- By searching for layer names.
- By inverting layer listings.
- By using group and property filters.
- By deleting unused layers.

### Sorting by Columns

The names of the columns — **Name**, **On**, **Freeze**, **Lock**, and so on — are actually buttons. For instance, click once on the **Name** column button, and the column sorts layer names in alphabetical order: 0 - 9 followed by A - Z. See figure at left, below.

**Left:** *Layer names in alphabetical order.*
**Right:** *Layer names sorted in reverse alphabetical order.*

Click a second time, and the column sorts in reverse order: Z - A followed by 9 - 0. The figure above at right illustrates layer names sorted in reverse order starting with Z.

To sort layers by color number, click the **Color** column button. The first color is red. Choose it a second time to reverse the order of color numbers, starting with the highest color number used by drawing.

Similarly, clicking the **On**, **Freeze**, and other column titles sorts the layers by state. Click once, and the column sorts by light bulb, sun, or open padlock. Click a second time, and the column reverses the sorting by dim bulb, snowflake, or closed padlock.

**Notes** You can change the order of the columns by dragging their headers around. For instance, if you want Color next to On, simply drag the Color header next to On.

The Status and Name columns are "frozen"; as you drag the scroll bar, they remain in view, while other columns slide past.

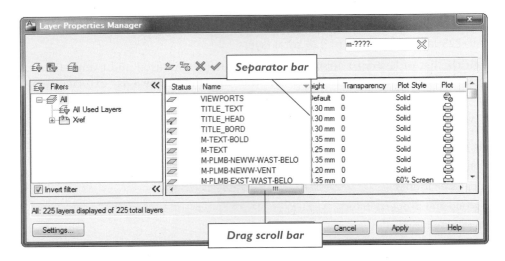

You can hide the display of columns. Right-click any column header, and then select the column name you want hidden. (The check marks indicate the headers that are displayed.)

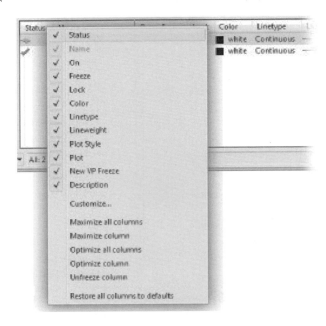

## Searching for Layer Names

The **Search for Layer** field reduces the list of layer names displayed by the dialog box.

Continuing with the *8th floor.dwg* sample drawing:

1. Click in the **Search for Layer** field. Notice that an asterisk appears ( * ). This is a "wild card" character. It lists all layer names.

2. Let's look for all layers starting with the letters TITLE and ending with any characters. The filter would read:
    title*

   This wildcard lists layers named TITLE_BORD, TITLE_HEAD, and TITLE_TEXT. (It would not list VIEW_A, VIW21, or 3RD_VIEW, because the names don't match TITLE*.)

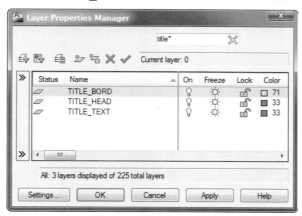

3. Another wild card is the question mark ( ? ), which matches single characters. Use it to find all layers that start with "m-" and have four characters before the next dash, like so:
    m-????-*

   The results is all mechanical layers (m) that also have TEXT, PLMB, and so on in their names.

4. Now erase the asterisk, leaving the **m-????-** on its own. Notice that no layer names appear, because none consists of just those letters.

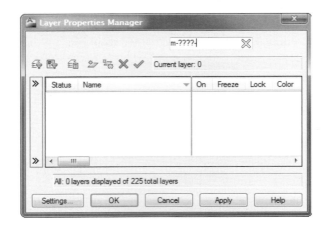

5. Use the **Invert Filter** option to display all layers, *except* those that contain **m-????-**: Follow these steps:
   a. Click the ≫ **Filter Tree** button. Notice the flyout.
   b. Click the ☑ Invert filter **Insert Filter** check box.

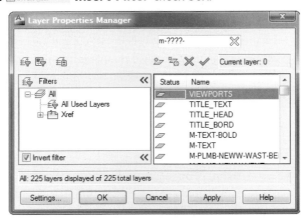

*Warning!* When the **Invert Filter** is applied to **All**, the layer list is empty.

### Creating Group Filters

When drawings have large numbers of layers, it's useful to organize them into groups and subgroups. For example, you can create a group of all layers to do with electrical or with title block information.

AutoCAD includes three groups: All layers, All Used Layers, and Xref (layers from externally-referenced drawings). These work with the Invert Filter option. For example, choose the **Xref** group and then click **Invert Filter**: this forces AutoCAD to display only layers in the base drawing and none from xrefs.

Layer groups are controlled by the tree view at the left of the layer palette. It shows groups of layers in a hierarchy. The topmost group of the hierarchy is always **All**, and holds the names of all layers. Select it, and the list view (the right half of the dialog box) lists the names of all layers.

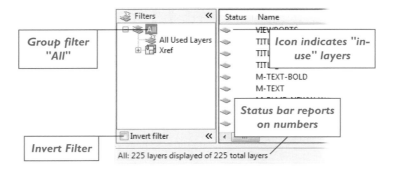

On the status line, AutoCAD reports the number of layers displayed by the filter.

Underneath the All group is the **All Used Layers** group, which contains all layers that have at least one object. When you click it, you see the list of all used layers in the list view. Click **Invert Filter** to see only the names of empty layers, those that contain no objects.

When drawings contains xrefs (short for "externally referenced drawings"), the tree view also includes a group named **Xref**. Click it to see the names of all layers found in xrefs. Below the **Xref** group are the names of the xref drawings. Each is a subgroup that contains the names of layers found in that particular drawing.

Notice that the status line now reports that just 26 layer names are displayed.

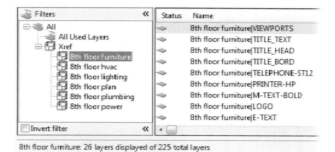

Group filters are groups of layers that you drag from the list view into the group filter name. Group filters are static; they change only when you add and subtract layers to and from the group.

 **Notes** To view all layers *not* in xrefs, select the **Xref** group, and then turn on the **Invert Filter** option.

Although layers may appear to be empty, sometimes they contain objects from block definitions.

### TUTORIAL: CREATING GROUP FILTERS

In this tutorial, you create a group filter that lists all text-related layers. Follow these steps:

1. Start AutoCAD with the *8th floor.dwg* drawing.
2. Enter the **LAYER** command to open the Layer Properties Manager dialog box.
3. Click the  **New Group Filter** button. (Alternatively, press **ALT+G**.)

    Notice that AutoCAD creates an empty filter with the generic name of "Group Filter 1."

4. Change the name to something more meaningful: **Text**.

5. Now it's time to populate the group filter with all layer names that contain "text":
   a. Click the **All** group filter to display all layer names.
   b. In the Search for layer 🔍 search bar, enter "*text*" to select all layer names containing "text."

   c. Select all the layer names. (Choose the first name, and then press **CTRL+A**.)

   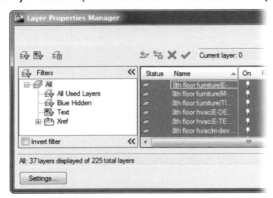

6. Drag the selected name over to the Filters pane, and deposit them on the **Text** name.

   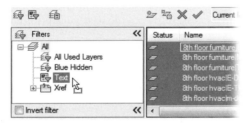

   *Don't worry!* When you drag layer names around this dialog box, you do not move or lose them; AutoCAD copies them.

7. Select the **Text** filter. Notice that the list view shows a shortened list of layer names.

   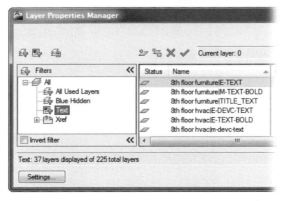

8. Select the **All** filter to see all names again.

## TUTORIAL: CREATING PROPERTY FILTERS

AutoCAD also supports *property filters*. These reduce the number of layer names displayed, based on their properties — color, linetype, frozen status, and so on. For example, you could have a group that holds layers with blue color and Hidden linetype. (Note that group filters handle layer names only.)

Property filters are dynamic; as you create (or delete) layers of the same properties, AutoCAD automatically adds and removes them from the filter group. There is no need to drag layer names into property filters, as you had to with group filters.

In this tutorial, you create a group filter that lists all layers that are colored blue and have the Hidden linetype. Follow these steps.

1. Start AutoCAD with the *8th floor.dwg* drawing.
2. Enter the **LAYER** command. Notice the Layer Properties Manager palette.
3. Click the  **New Property Filter** button. (Alternatively, press **ALT+P**.) Notice that AutoCAD displays the Layer Properties Filter dialog box.

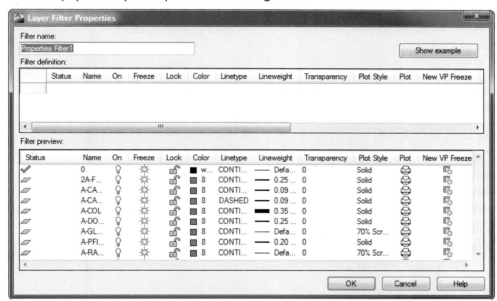

In the upper half, you define the filter(s); the lower half displays the names of layers and their properties.

4. In the **Filter Name** field, replace the generic name "Properties Filter 1" with:

   Filter Name:    **Blue Hidden**

5. Under each heading (Status, Name, On, ...) you enter restrictions that shorten the list of layers eligible for the filter. In this tutorial, this filter is supposed to display just those layers colored blue:

   a. Under **Color**, click the blank area. Notice a small button that appears.

b. Click the  button. Notice the Select Color dialog box.
c. Select the blue color (index color 5), and then click **OK**.

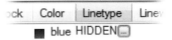

Notice that the list of layer names shortens to just those with color blue.

6. Apply a further restriction: just those layers with the Hidden linetype. Repeat the steps of #5, but this time select the **Hidden** linetype when the Select Linetype dialog box appears.

In the lower half of the dialog box, the list of layers is again shortened to the one blue layer with Hidden linetype.

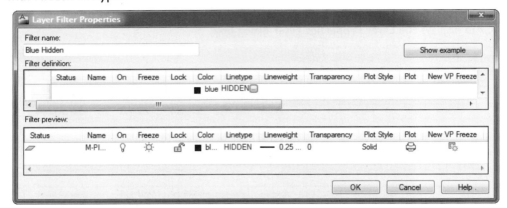

7. Click **OK**.
8. In the Layer Properties Manager, select the **Blue Hidden** property filter, and only one layer appears.

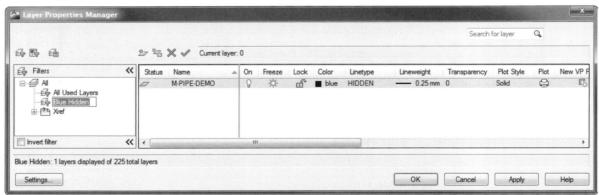

9. Create a new layer with color blue and linetype Hidden. Notice that it is automatically added to the property filter.

**Notes** To edit a property filter, double-click its name in the tree view. This causes the Layer Properties Filter dialog box to appear.

Group and Property filters apply only to the names of layers appearing in the layer palette. They have no effect on the drawing.

## Deleting Layers

AutoCAD allows you to delete empty layers, those with no objects.

But you cannot delete layers 0 and DefPoints, as well as externally-referenced layers — even when these are empty. Some seemingly empty layers cannot be erased, because they are part of an unused block definition.

Select one or more layer names, and then choose the **Delete Layer** button. (Alternatively, press ALT+D.) If the layer cannot be erased, AutoCAD displays the following dialog box to explain why not:

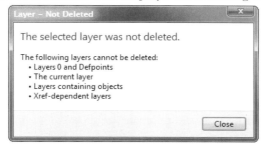

If the layer is empty, then it is erased immediately.

(In the layer dialog box, the layer is not erased right away. Rather, an  icon in the Status column marks the layer for deletion; this allows you to change your mind. The layer is deleted when you click **Apply** or **OK**.)

## Layer States Manager

The **Layer States Manager** saves the current *state* of layers, and then restores it at a later time. The state of layers includes names and properties, such as whether they are thawed or frozen, as well as their colors and linetypes.

Once you save a layer state by name, you can edit the state, and export it for sharing with others.

### TUTORIAL: CREATING AND APPLYING LAYER STATES

1. To create layer states, start the **LAYERSTATE** command with one of these methods:
   - In the ribbon's 2D Home tab, click the **Layer States Manager** button on the Layers panel.
   - Or, at the 'Command:' prompt, enter the **layerstate** command:

     Command: **layerstate** *(Press ENTER.)*
   - You can also access the dialog box through the Layer Properties Manager palette.

Notice that AutoCAD displays the Layer States Manager dialog box.

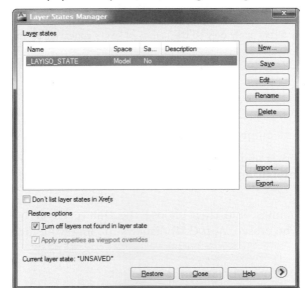

2. To create a new layer state, click **New**. Enter a name and description, and then click **OK**.

3. Back in the Layer States Manager dialog box, you determine the properties to be saved. Click the ⊙ **More** button to expand the dialog box, and then choose from the list of properties.

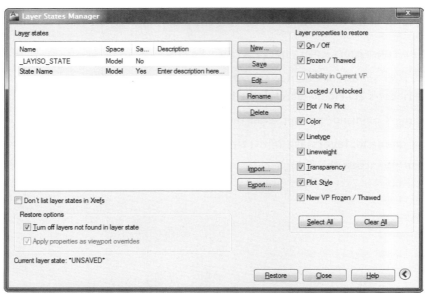

4. To save the layer state to disk, click the **Export** button. Provide a name and folder for the *.lay* file, and then click **Save**.
5. The easy way to apply a layer state is to select its name from the ribbon's Layers panel.

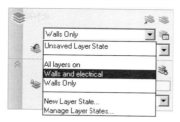

To import a layer state, follow these steps:

1. Click the **Import** button.
2. Select the folder and name of a file, and then click **Open**.

Duplicate layer states names are not imported. As of AutoCAD 2008, you can import layer states from *.dwg* (drawing), *.dws* (drawing standards), and *.dwt* (drawing template) files — as well as from *.las* (layer state) files.

### CONTROLLING LAYERS: ADDITIONAL METHODS

AutoCAD provides several other methods to control layers:

- **Layers** panel controls layers on the ribbon.
- **CLAYER** system variable makes a layer current quickly.
- **LAYMCUR** command makes the selected object's layer current (formerly **AI_MOLC**).
- **LAYERP** command restores the previous layer state.
- **Layer Tools** manipulate layers through individual commands.
- **LAYISO** command isolates selected layers by freezing or locking all other layers.

Let's look at each of these.

### Layers on the Ribbon

The ribbon's Layer panel select the current layer without using the LAYER command. (The layout of this panel changed in AutoCAD 2010.)

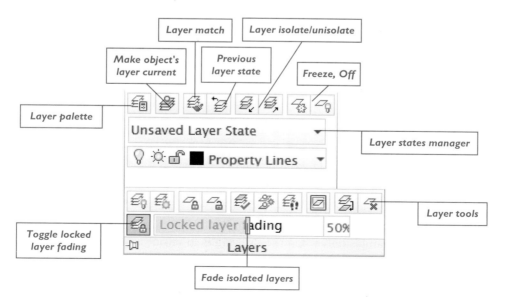

### Changing the Current Layer

To change the current layer, click the down arrow shown below. It reveals a list of layer names, along with icons signifying their status. (The actual names of layers vary according to the drawing currently open.)

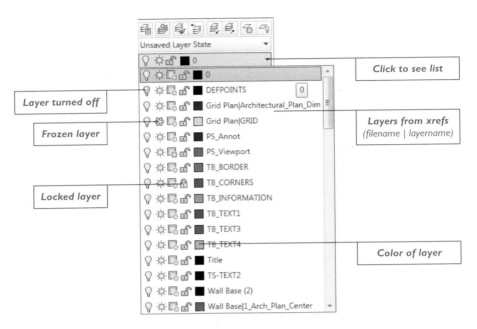

Select the name of a layer to make it current. It's that simple.

There is, unfortunately, one catch. You cannot make current any layer whose name is shown with a snowflake, because it is frozen. (You cannot see or work with frozen layers).

### Changing the Layer's Status

The drop-down box lists a quintet of icons beside each layer name. The colored square icon represents the color assigned to this layer. Each of the other four icons has states:

**Light bulb** on or off — layer is on (default) or off.

**Sun** or **snowflake** — layer is thawed (default) or frozen.

**Sun** or **snowflake on square** — layer is thawed (default) or frozen in the current viewport; this icon changes only when the drawing is in paper space (layout mode).

**Padlock** — open or closed: layer is unlocked (default) or locked.

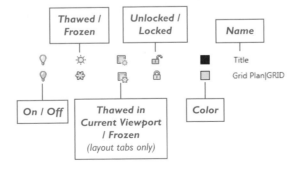

You change the status of each of the four icons simply by clicking it. Note that the thawed/frozen icon in current viewport icon changes only when the drawing is in layout mode (paper space). Clicking the color square displays the Select Color dialog box, allowing you to change the layer color.

**LayMCur**

An easy way to switch to another layer is with the LAYMCUR command (short for "Layer Make Current"; formerly the AI_MOLC command). Suppose you are interested in switching to the layer holding a green dotted line whose name you're not sure about. You use the feature as follows:

1. Choose the **Make Object's Layer Current** button. AutoCAD prompts you:
   Select object whose layer will become current: *(Pick an object.)*
2. Select the green dotted line. AutoCAD reports:
   *layername* is now the current layer.

Check the ribbon, and you will see that layer *layername* is now current.

### CLayer

Because I prefer the keyboard to palettes and dialog boxes, I find that the CLAYER system variable (short for "current layer") is the fastest way to switch between layers. It works like this:

Command: **clayer**

New value for CLAYER <default>: *(Enter the name of the layer, and then press* ENTER.*)*

AutoCAD immediately makes that layer current.

 **LayerP** and **LayerPMode**

The LAYERP command (short for "layer previous") undoes changes to layer settings, much like the ZOOM **Previous** command restores the previous view.

Command: **layerp**

Restored previous layer status.

The command, however, cannot undo the changes to renamed, deleted, purged, and newly-created layers. (Purging is described in a later chapter.) When you rename a layer and change its properties, only the properties are changed back, not the name. A related command is LAYERPMODE (short for "layer previous mode"). It toggles whether AutoCAD tracks changes to layers.

Command: **layerpmode**

Enter LAYERP mode <ON>: *(Enter* **ON** *or* **OFF**.*)*

When turned off, the LAYERP command does not work, and so AutoCAD reminds you:

Layer-Previous is disabled. Use LAYERPMODE to turn it on.

### Layer Tools

The layer tools were formerly part of Express Tools, a collection of commands not supported by Autodesk. In AutoCAD 2007, Autodesk changed the status of layer tools to fully-supported. Many of them now appear as buttons on the ribbon's Layer panel. Click the **Layers** button to access them.

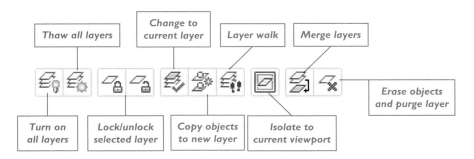

The layer tools are:

**LAYCUR** (layer current) — changes the layer of selected objects to that of the current layer.

**LAYMCH** (layer match) — changes the layers of selected objects to that of a selected object.

**LAYMCUR** (layer make current) — makes the selected object's layer current, just like the old **AI_MOLC** command.

**LAYMRG** (layer merge) — moves objects to another layer, and then removes the layer of the moved objects.

**LAYWALK** (layer walk) — displays objects on selected layers.

**LAYVPI** (layer viewport isolate) — isolates the selected object's layer in the current viewport by freezing its layer in all other viewports. This command works only in paper space with two or more viewports.

**LAYISO** (layer isolate) — turns off all layers except those holding selected objects (detailed below). **LAYUNISO** turns on layers that were turned off with the last **LAYISO** command.

**LAYDEL** (layer delete) — erases all objects from the specified layer, and then purges the layer from the drawing. The current layer cannot be deleted. If you make a mistake, you can use the **U** command later to restore the layer name and its objects.

**LAYFRZ** (layer freeze) — freezes the layers of the selected objects. **LAYTHW** thaws all layers.

**LAYLCK** (layer lock) — locks the layer of the selected object. **LAYULK** unlocks the layer of a selected object.

**LAYOFF** (layer off) — turns off the layer of the selected object. **LAYON** turns on all layers, except frozen layers.

## LAYISO

The LAYISO command (short for "layer isolation") isolates one or more layers, causing the objects on all other layers to fade away — hence isolating the selected ones. The other layers are either frozen (made invisible) or faded — your choices. When faded, layers are locked, not frozen; this lets you see but not edit them.

The LAYLOCKFADECTL system variable determines the amount of fading. You may find it easier to use the Locked Layer Fading slider on the ribbon's Layer panel. The maximum fade amount is 90% (not quite entirely faded); the default is 50%.

The related LAYUNISO command returns layers to normal. Let's take at look at how this works.

1. Open *map.dwg*.
2. Start the **LAYISO** command:
   - In the ribbon's 2D Home tab, click the **Layer Isolate** button in the Layers panel.
   - At the 'Command:' prompt, enter the **layiso** command:

     Command: **layiso**
     Current setting: Lock layers, Fade=50

3. Type "s" to specify the **Settings** options:
   Select objects on the layer(s) to be isolated or [Settings]: **s**
   Enter setting for layers not isolated [Off/Lock and fade] <Lock and fade>: *(Press* ENTER.*)*

   The Settings options have the following meaning:

   **Off** freezes layers.

   **Lock and fade** locks layers and specifies the percentage to fade layers, from 0% to 90%.

4. I find that the maximum of 90 works most effectively. Press ENTER at the 'Enter setting for layers not isolated' prompt, and AutoCAD prompts you as follows:
   Enter fade value (0-90) <50>: **90**

5. Select one or more objects on the layers you want to *isolate* (display fully, not fade). For this tutorial, select the black property lines; this will fade the blue text and the dimensions.
   Select objects on the layer(s) to be isolated or [Settings]: *(Pick black property lines.)*

   Note that picking one object also selects its layer, and hence automatically selects all other objects on the layer. You need pick just one property line to choose them all.

6. Press ENTER to exit object selection.
   Select objects on the layer(s) to be isolated or [Settings]: *(Press ENTER.)*
   Layer Property Lines has been isolated.

   AutoCAD reports the name of the layer(s) isolated. Notice how the objects on the other (unselected) layers are isolated (faded), as illustrated below:

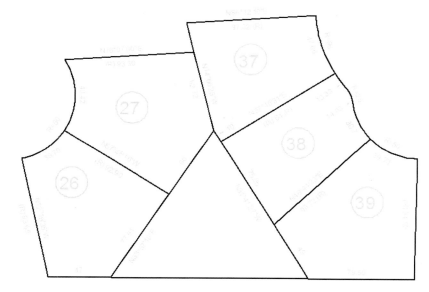

You may find it much easier to use the ribbon's Layers panel to control layer isolation; for example, the Locked Layer Fading slider lets you interactively control the amount of fading displayed by locked layers.

Follow the four steps illustrated below to use the panel's layer isolation controls.

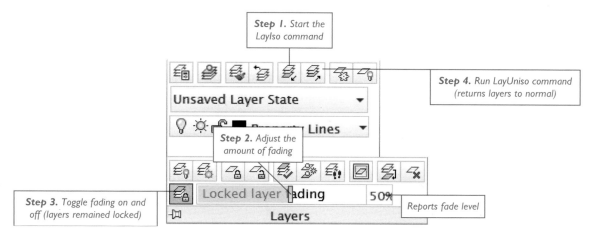

## EXERCISES

1. In this exercise, you change the colors of objects.
   a. Open AutoCAD to start a new drawing.
   b. Use the **LAYER** command to set the color of layer "0" to cyan (light blue).
   c. Draw some lines. Are they drawn in color?
   d. Use the **COLOR** command to set the current color to yellow.
   e. Draw three circles. Did the **COLOR** command override the layer's color setting?
   f. Next, set the current color to red.
   g. Draw three boxes. Do they appear in the correct color?
   h. Start the **PROPERTIES** command, and then select all the objects on the screen.
   i. Select blue as the color. Did all the objects change to blue?

2. In this exercise, you create a layer, and then freeze and thaw it.
   a. Create a layer named "Mylayer."
   b. Set "Mylayer" as the current layer. Do you see the layer name in the ribbon at the top of the screen?
   c. Set the layer color to green, and then draw some objects on the layer.
   d. Using the Layer droplist, freeze layer "Mylayer", and then set the current layer to "0." Do the objects you drew disappear?
   e. Thaw "Mylayer". Do the objects reappear?

3. In this exercise, you create several layers, and then change their properties.
   With the **LAYER** command, create seven layers using these names:
   
   Landscape
   Roadways
   Hydro
   ToBeRemoved
   StormSewer
   CableTV
   Building

   Using the layer palette:
   a. Change the color of layer "Landscape" from white to green.
   b. Change the lineweight of layer "Roadways" from Default to 0.083" (2.11mm).
   c. Change the linetype of layer "Hydro" from Continuous to Gas_line.
   d. Change the "ToBeRemoved" layer from white to red, and from Continuous to Dashed linetype.
   e. Change the linetype of the "CableTV" layer from Continuous to Hidden, and change the lineweight to 0.020" (0.51mm).
   f. Make the "CableTV" layer current by choosing the **Current** button.

g. Draw some lines. Do they appear in hidden linetype? If the lines do not look thick, ensure the **LWT** button is depressed on the status line.

h. Use the Layer droplist to change to the "ToBeRemoved" layer, and then draw some lines. Do they appear red and dashed?

i. Again, use the Layer droplist to change to the "Building" layer, and then draw some lines. Do they appear black and continuous?

4. In this exercise, you copy the properties from one object to all others in the drawing.

   a. Continue with the drawing you created in the previous exercise.

   b. Start the **MATCHPROP** command.

   > Select source object: *(Select one of the red, dashed lines drawn on the "CableTV" layer.)*
   >
   > Select destination object(s) or [Settings]: **all**
   >
   > Select destination object(s) or [Settings]: *(Press* ENTER.*)*

   Do all the objects become red and dashed?

## CHAPTER REVIEW

1. What do hidden lines show in drawings?
2. Name two ways linetypes can be applied to objects in AutoCAD:
    a.
    b.
3. How do you control the length of individual segments in dashed lines?
4. How do you load all linetypes into a drawing?
5. What is the difference between *global* and *local* linetype scaling?
6. Is it possible to save the current state of layers?
   If so, how?
7. Describe five properties that the **PROPERTIES** command changes.
8. What is meant when the color of objects is changed to **ByLayer**?
9. Why would you use the **PROPERTIES** command instead of the Properties panel on the ribbon?
10. Can the Properties panel change objects?
    If so, how?
11. What is the **MATCHPROP** command used for?
12. Why would a CAD drafter use layers?
13. What layer option do you use to create new layers?
14. How do you turn on frozen layers?
15. What is the difference between *locking* layers and *freezing* layers?
16. How do you obtain a listing of all layers in drawings?
17. Can you have objects of more than one color on the same layer?
    Explain.
18. What do the following layer symbols mean?
    Snowflake
    Open lock
    Light bulb glowing
    Printer
    Colored square
19. Name a benefit to using lineweights.
20. Is it possible to add custom linetypes to drawings?
    Custom lineweights?
21. What color is designated by R?
    What is color 7?
    What is a *color book*?
22. Describe what is meant by *RGB* in terms of colors.
23. Should you use lineweights to represent objects with width?
24. What is the purpose of the **LWT** button on the status bar?
25. Describe the steps to changing the lineweight of circles using the Properties palette:
    a.
    b.

26. Explain how these three classes of linetype differ:
    Simple
    ISO
    Complex
27. Before using linetypes in new drawings, what must you do first?
28. What is the name of the linetype used to show hidden edges?
29. **LTSCALE** is set to 0.5 and **CELTSCALE** is set to 5.0. At what scale is the next linetype drawn?
30. Under what condition can layers be deleted?
    Can layer 0 be deleted?
31. Describe two ways to manage long lists of layer names:
    a.
    b.
32. When drawings are in model space, why do the **Current VP Freeze** and **New VP Freeze** columns not appear in the Layer palette?
33. What is the purpose of the **Show all used layers** filter in the layers palette?
34. Can you assign lineweights to layers?
    Assign linetypes?
    Assign hatch patterns?
35. When linetypes are assigned to layers, are all objects on those layers displayed with that linetype?
    If not, why not?
36. Explain the purpose of the following filters:
    Group
    Properties
37. What is the purpose of the **CLAYER** system variable?
38. Which keystroke shortcut displays the Properties palette?
39. Describe how the **PICKADD** system variable affects object selection.
40. Can the Properties palette be used to change the endpoints of line segments?

# CHAPTER 8
## Constructing Objects

Drawing the same objects over and over was a tedious part of hand drafting. In Chapter 6, "Drawing with Efficiency," you learned to insert blocks to place parts quickly in drawings. In some circumstances, however, blocks are not the best method.

This chapter introduces other means of making copies — using mirrored copies, parallel offsets, and arrays of copies. After completing this chapter, you will be able to employ the following AutoCAD commands for constructing objects from existing objects:

**COPY** makes one or more identical copies of objects.
**MIRROR** makes mirrored copies.
**MIRRTEXT** determines whether text is mirrored.
**OFFSET** makes parallel copies.
**MEASURE** and **DIVIDE** place copies of points and blocks along objects.
**ARRAY** constructs linear, rectangular, and polar copies.
**FILLET** and **CHAMFER** create rounded and angled corners.
**JOIN** joins similar objects into one.
**REVERSE** reverses the direction of objects.
**REVCLOUD** creates revision clouds.
**MARKUP** inserts marked-up .dwf files from Design Review.

---

**NEW TO AUTOCAD 2011** IN THIS CHAPTER
- **JOIN** now joins more kinds of objects.
- **MIRRHATCH** determines whether hatches are mirrored by the **MIRROR** command.

## COPY

The COPY command makes one or more copies of objects.

The copies are identical to the original; all that changes is the location in the drawing. This command is meant for use within drawings; to copy objects between drawings, use the COPYCLIP command.

To use the COPY command, you need to tell AutoCAD three things: (1) the objects to be copied, (2) the point from which the copying takes place, and (3) the location to place the copies. The most recent displacement is the default value the next time the COPY command is used — but only if the **Displacement** option is used. The COPY command repeats until you press ESC.

### TUTORIAL: MAKING COPIES

1. To copy one or more objects, start the **COPY** command:
   - In the ribbon's 2D Home tab, choose the **Copy** button from the Modify panel.
   - At the 'Command:' prompt, enter the **copy** command:

     Command: **copy** *(Press ENTER.)*
   - Alternatively, enter the aliases **co** or **cp** at the 'Command:' prompt.
   - You can also use **CTRL+C** and **CTRL+V** to copy and paste objects in the current drawing, to other drawings, and to documents in other applications.

2. In all cases, AutoCAD prompts you to select the objects you want copied:
   Select objects: *(Pick one or more objects.)*
   Select objects: *(Press ENTER to end object selection.)*

3. Identify the point from which the displacement is measured:
   Specify base point or [Displacement] <Displacement>: *(Pick point 1.)*

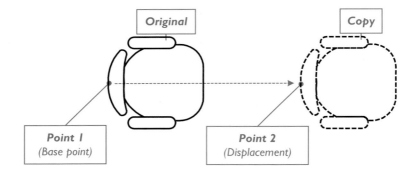

4. Identify the location for the copied object:
   Specify second point of displacement or <use first point as displacement>: *(Pick point 2.)*

5. The command repeats the last prompt so that you can place additional copies. Pick additional points, or press **ESC** to end the command.
   Specify second point or [Exit/Undo] <Exit>: *(Press ESC.)*

Students often have difficulty understanding the concept of displacement. This is the distance from the original object (known as the base point) to the location of the copied object. (See the detailed discussion below.)

You can choose any point you like as the base point, but some points make more sense than others. Examples include the lower-right corner of rectangles, the center of circles, and the 0,0 origin of drawings. In addition, it is often handy to use object snaps, such as INTersection or INSertion, to pick base points precisely.

To copy objects vertically or horizontally, hold down the SHIFT key to turn on ortho mode. (If ortho mode is already on, SHIFT turns it off temporarily.)

AutoCAD has several ways to determine the displacement, which you learn about in the next tutorial.

### MAKING COPIES: ADDITIONAL METHODS

This command also has the following options:

- **Displacement** option specifies the displacement distance.
- **COPYMODE** system variable determines whether the command repeats.

## Displacement

The **Displacement** option displays the displacement from the prior use of the COPY command:

Specify base point or [Displacement] <Displacement>: **d**

Specify displacement <1.0000, 2.0000, 0.0000>: *(Press ENTER, or enter a new displacement.)*

Whether you press ENTER or enter coordinates, AutoCAD displaces the selected objects, and then exits the COPY command.

## More on Displacements

When AutoCAD prompts, 'Specify base point or [Displacement]:' this is a hint of what is to come.

Specify base point means you can enter x, y coordinates (such as 4,5) or pick a point in the drawing. Both methods can be used as the base point, but the x, y coordinates could be interpreted differently by AutoCAD. Displacement is a hint that the prompt to follow will select one of the two methods, depending on your next action.

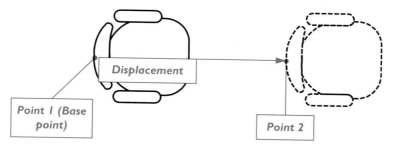

Specify second point of displacement means that you should enter another x, y coordinate or pick another point. Both actions place the copy at a distance that AutoCAD calculates from the two sets of x, y coordinates or two pick points. You can, of course, mix and match coordinate entry and pick points.

It's not clear from the prompt, but AutoCAD wants you to press ENTER at <use first point as displacement>. AutoCAD interprets the coordinates you entered at the earlier prompt (such as 4,5) as relative distances. In this case, there is no need to use the @ prefix. AutoCAD places the copy 4 units right and 5 units up from the original.

This option can, unfortunately, have unexpected results, because, as this book's technical editor Bill Fane once remarked, "Pressing ENTER sometimes makes the copied objects end up near Hawaii — not a problem," he adds, "if you live in Maui." This problem occurs when you use the mouse to pick the first point, and then just press ENTER for the second point; AutoCAD interprets the second point as 0,0. He recommends using the ZOOM Extents command to find the "missing" copy.

1. Start the **COPY** command, and then select objects to copy:

   Command: **copy**

   Select objects: *(Select one or more objects.)*

   Select objects: *(Press* ENTER.*)*

2. Enter coordinates for the base point:

   Specify base point or [Displacement] <Displacement>: **4,5**

3. Press ENTER to place the copy by a relative distance and end the command:

   Specify second point of displacement or <use first point as displacement>: *(Press* ENTER *to interpret the first point as relative coordinates.)*

   Specify second point or [Exit/Undo] <Exit>: *(Press* ENTER *to exit the command.)*

 **Notes** You may enter x, y coordinates for 2D displacement, or x, y, z coordinates for 3D displacement. In addition, you can use direct distance entry to specify the displacement.

As an alternative to the **COPY** command, you can use the right-click copy/move/insert technique.

Or, use the **Copy** option of grips editing:

\*\* STRETCH \*\*

Specify stretch point or [Base point/Copy/Undo/eXit]: **c**

Become familiar with AutoCAD's many object selection modes, because they are crucial for working efficiently with the commands in this (and the next) chapter. You may wish to review the **SELECT** command in Chapter 6, "Drawing with Efficiency."

##  MIRROR

The **MIRROR** command makes copies that are mirrored.

The command saves time when you are drawing symmetrical objects. Draw a half, or a quarter, of a group of objects, and then construct the others by mirroring them. You have the option to retain or delete the originals, as well as to decide whether text should be mirrored or not.

Other drawing programs use the phrases "flip horizontal" and "flip vertical" in place of mirror. As alternatives to this command, you can use the **Mirror** option found in grips editing, or insert blocks with a negative x- or y-scale factor (which mirrors the block).

To use the **MIRROR** command, you need to tell AutoCAD three things: (1) the objects to be mirrored, (2) the line about which the mirroring takes place, and (3) whether the source objects should be erased.

## TUTORIAL: MAKING MIRRORED COPIES

1. To make mirrored copies of one or more objects, start the **MIRROR** command:
   - In the ribbons's 2D Home tab, choose the **Mirror** button from the Modify panel.
   - Or, at the 'Command:' prompt, enter the **mirror** command:

       Command: **mirror** (Press ENTER.)

   - Alternatively, enter the **mi** alias at the 'Command:' prompt.

2. In all cases, AutoCAD prompts you to select the objects you want mirrored:

   Select objects: (Pick one or more objects.)

   Select objects: (Press ENTER to end object selection.)

3. Identify the points that define the mirror "line":

   Specify first point of mirror line: (Pick point 1.)

   Specify second point of mirror line: (Pick point 2.)

   The *mirror line* is the line about which the objects are mirrored. It need not be an actual line; two points will do.

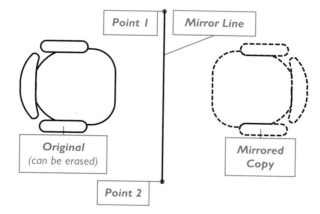

4. Decide whether you want the original objects(s) erased:

   Delete source objects? [Yes/No] <N>: (Enter **Y** or **N**.)

**Notes** To make the mirror line absolutely horizontal or vertical, hold down the **SHIFT** key to turn on ortho mode temporarily.

Polar mode is handy for making mirrored copies at specific angles.

## MAKING MIRRORED COPIES: ADDITIONAL METHOD

Whether text and hatches are mirrored is determined by the **MIRRORTEXT** and **MIRRHATCH** system variables, respsectively.

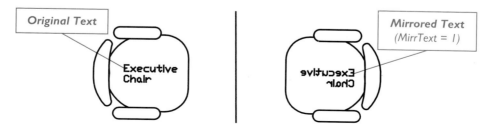

Sometimes objects contain text and hatches. The dilemma is whether to mirror them, which makes text read backwards and hatches take on a different meaning. The MIRRTEXT system variable determines how mirroring affects text, as follows:

Command: **mirrtext**

New value for MIRRTEXT <0>: *(Enter **1** or **0**.)*

The 0 and 1 have the following meaning:

| MirrText | Comment |
|---|---|
| 0 | Text is not mirrored (default). |
| 1 | Text is mirrored (default in AutoCAD 2004 and earlier). |

The same values apply to the MIRRHATCH system variable:

| MirrHatch | Comment |
|---|---|
| 0 | Hatches and gradients are not mirrored (default). |
| 1 | Hatches and gradients are mirrored (default in AutoCAD 2010 and earlier). |

##  OFFSET

The OFFSET command makes parallel copies.

This command constructs copies that are parallel to the source object; AutoCAD limits you to making offset copies of one object at a time. When the objects have curves or are closed, the copies become larger or smaller, depending on whether they are on the inside or outside of the source.

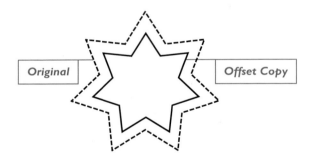

To create offset copies, AutoCAD needs to know three pieces of information, and in this order: (1) the offset distance, (2) the objects to offset, and (3) the side on which to place the offset copies.

It may seem counter-intuitive first to specify the distance and then to select the object, but "that's the way the Mercedes bends," as the driver said after his automobile accident. (*Pun credit*: technical editor.)

AutoCAD offsets lines, arcs, circles, ellipses, elliptical arcs, polylines (2D only), splines, rays, and xlines. Sometimes, unexpected results happen, as illustrated below. The dashed lines are copied offset from the original polyline (shown as the heavy line).

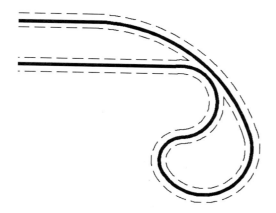

*The thick line indicating the original polyline; the dashed lines, the offset copies.*

As an alternative to this command, use the MLINE command to place parallel lines.

**TUTORIAL: MAKING OFFSET COPIES**

1. To make offset copies of an object, start the **OFFSET** command:
   - In the ribbons's 2D Home tab, choose the **Offset** button from the Modify panel.
   - Or, at the 'Command:' prompt, enter the **offset** command:

     Command: **offset** *(Press* ENTER.*)*
   - Alternatively, enter the **o** alias at the 'Command:' prompt.

2. In all cases, AutoCAD displays the current settings:

   Current settings: Erase source=No   Layer=Source   OFFSETGAPTYPE=0

   And then prompts you for the offset distance:

   Specify offset distance or [Through/Erase/Layer] <Through>: *(Enter a distance.)*

3. Select the object to offset:

   Select object to offset or [Exit/Undo] <Exit>: *(Select one object.)*

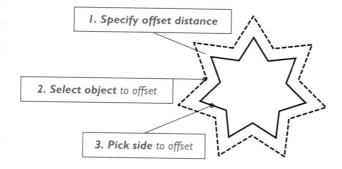

4. Pick the side on which the offset should be placed:

   Specify point on side to offset or [Exit/Multiple/Undo] <Exit>: *(Pick point 3.)*

5. Press **ENTER** to exit the command:

   Specify point on side to offset or [Exit/Multiple/Undo] <Exit>: *(Press* ENTER.*)*

The "Select object to offset:" prompt repeats to allow you to offset as often as you wish, but just one object at a time. Press ENTER to terminate the command.

For the offset distance, you can pick two points on the screen, or enter a number representing the distance. If you get the message, "That object is not parallel with the UCS," this means that the direction of the object's Z axis is not parallel to the current user coordinate system.

### MAKING OFFSET COPIES: ADDITIONAL METHODS

The OFFSET command's **Through** option lets you specify a point through which the copy is offset. In addition, two system variables let you preset parameters.

- **Through** option combines the distance and side options.
- **Erase** option erases the source object.
- **Layer** option specifies the destination layer.
- **OFFSETGAPTYPE** system variable determines how polyline gaps are handled.

Let's look at each.

### TUTORIAL: THROUGH

The **Through** option constructs the offset copy "through" a point. In effect, it combines the "Specify offset distance" and "Side to offset" options.

1. Draw a circle.
2. Start the OFFSET command, and then enter the **Through** option.

   Command: **offset**

   Specify offset distance or [Through]: **t**

3. Select the circle, and then pick the point through which the copy should be offset:

   Select object to offset or <exit>: *(Pick the circle.)*

   Specify through point: *(Pick point 1.)*

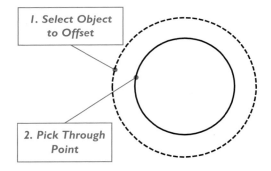

4. Continue making offset copies, or press ENTER to end the command.

   Select object to offset or <exit>: *(Press ENTER.)*

### Erase

The **Erase** option erases the source object:

   Erase source object after offsetting? [Yes/No] <No>: *(Type **Y** or **N**.)*

"Yes" erases the source object, while "No" retains it. AutoCAD remembers this setting until you change it. (There is no system variable associated with it.)

### Layer

The **Layer** option specifies the destination layer:

   Enter layer option for offset objects [Current/Source] <Source>: *(Type **C** or **S**.)*

"Current" places offset objects on the current layer, while "Source" places them on the same layer as the source object.

## OffsetGapType

Offsetting polylines can be tricky. For this reason, AutoCAD includes the OFFSETGAPTYPE system variable to help you decide how potential gaps between polyline segments should be handled. The choices are extending lines, creating fillets (arcs), or creating bevels (chamfers).

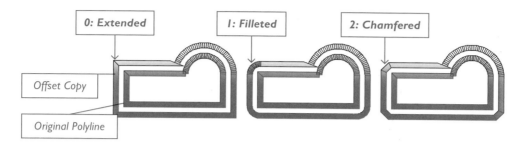

Command: **offsetgaptype**
Enter new value for OFFSETGAPTYPE <0>: *(Enter a number between 0 and 2.)*

| OffsetGapType | Comment |
|---|---|
| 0 | Gaps filled with extended line segments (default). |
| 1 | Gaps filled with filleted segments, creating arcs. |
| 2 | Gaps filled with chamfered line segments, creating beveled edges. |

## DIVIDE AND MEASURE

The DIVIDE and MEASURE commands place copies of points and blocks along objects.

DIVIDE divides an object into an equal number of parts, while MEASURE works with a specific distance — number of segments versus length of segment. Both commands place either blocks or points along the object.

(The MEASURE command has nothing to do with measuring distances or lengths; for those, use the DIST and LIST commands.)

Like the OFFSET command, dividing and measuring work with just one object at a time. Use the cursor to select a single object, because you cannot use Window, Crossing, or Last selection modes. In addition, AutoCAD is limited to working with lines, arcs, circles, splines, and polylines; picking a different object results in the complaint:

> Cannot divide that object. * Invalid*

AutoCAD does not place a point or block at the start or end of open objects.

 **Note** After using the DIVIDE and MEASURE commands to place points on an object, you can snap to the points with the **NODe** object snap.

To use the DIVIDE and MEASURE commands, you need to tell AutoCAD three things: (1) the object to be marked, (2) the number of markers, and (3) whether the markers are points or blocks.

## TUTORIAL: DIVIDING OBJECTS

1. To see the effect of the **DIVIDE** command better, first change visibility of points.
   Command: **pdmode**
   Enter new value for PDMODE <0>: **4**

2. To divide an object into equal parts, start the **DIVIDE** command:
   - At the 'Command:' prompt, enter the **divide** command:
   
     Command: **divide** *(Press* ENTER.*)*
   
   - Alternatively, enter the **div** alias at the 'Command:' prompt.

3. In all cases, AutoCAD prompts you to select the single object to divide:
   Select object to divide: *(Select one object.)*

4. Specify the number of divisions:
   Enter the number of segments or [Block]: *(Enter a number between 2 and 32767.)*

The **DIVIDE** command results in evenly spaced points, and with one fewer point than you would expect: divide a line by six, and AutoCAD places five points, creating six divisions. (Count the number of points and divisions on the line illustrated below.) The points and blocks are independent of the line; you can move and erase them at will.

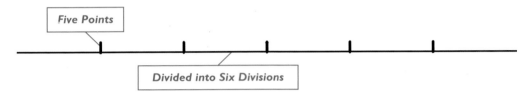

The **MEASURE** command operates similarly to the **DIVIDE** command, the difference being that you specify the length of segment along which to space the points.

## TUTORIAL: "MEASURING" OBJECTS

1. To place points along an object at specific distances, start the **MEASURE** command:
   - At the 'Command:' prompt, enter the **measure** command.
   
     Command: **measure** *(Press* ENTER.*)*
   
   - Alternatively, enter the **me** alias at the 'Command:' prompt.

2. In all cases, AutoCAD prompts you to select the single object to divide:
   Select object to measure: *(Select one object.)*

3. Specify the number of divisions:
   Specify the length of segment or [Block]: *(Enter number higher than 1.)*

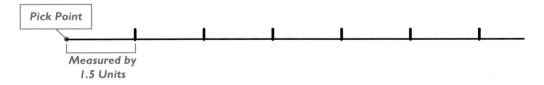

The **MEASURE** command needs a starting point, which depends on the object. As the figure above illustrates, measurement evenly distributes points as does division; there is usually a section left over at the end.

For open objects — lines, arcs, splines and open polylines — the starting point is the endpoint closest to your pick point; for closed polylines, it's the point where you began drawing the polyline. For circles, it is at the current snap angle, which is usually 0 (at the circle's 3 o'clock point); the measurement is made in the counterclockwise direction.

### DIVIDING AND MEASURING: ADDITIONAL METHODS

Both commands' **Block** option lets you specify a block to place along the object, in place of a point. And, since points tend to be invisible, the PDMODE and PDSIZE system variables (accessed by DDPTYPE) are useful.

- **Block** option places blocks along the object.
- **DDPTYPE** command changes the look of the points.

Let's look at each.

### Block

The **Block** option places blocks (symbols) along the object. The block must already exist in the drawing. The option operates identically for both commands, and so both commands have the same problem in that they don't provide you with a list of block names; entering ? does not help. If you are unsure of the block's name or even of its existence in the drawing, use the DesignCenter to help you.

1. Start either command. Enter the **b** option, and then the name of a block:
   Enter the number of segments or [Block]: **b**
   Enter name of block to insert: *(Type name.)*
2. Decide whether you want the block aligned with the object (Y) or at its own orientation (N):
   Align block with object? [Yes/No] <Y>: *(Enter **Y** or **N**.)*
3. Continue with the command.

"Yes" means the inserted block turns with the divided object, such as arc, circle, or spline. "No" means the block is always oriented in the same direction.

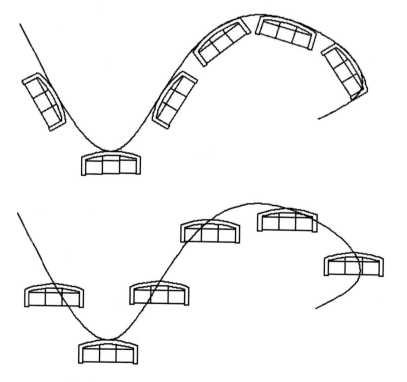

*Blocks aligned with a spline (top) and unaligned (bottom).*

If you are unsure of the block's name or even of its existence in the drawing, use the DesignCenter to help you.

### DdPType

Recall from Chapter 4, "Drawing with Basic Objects," that the DDPTYPE command changes the look and size of points. Points are normally invisible (for all intents and purposes), so it may be useful to change their size. The figure below illustrates the before-and-after difference.

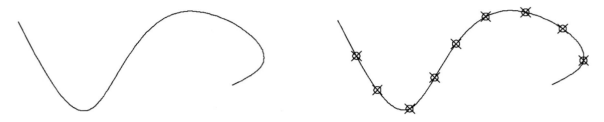

*Left: Spline divided by points of type 0 (dots)...*
*Right: ...and of type 35 (circle-x).*

## ARRAY

The ARRAY command makes evenly-spaced copies in linear, rectangular, and round patterns.

There are times when you want to place multiple copies of object in patterns, like rows of seats in a movie theater, or columns of parking spaces at a shopping center. If you were using traditional drafting techniques, you would draw each one separately — tediously.

Already in the chapter, you saw how the COPY command (with its **Multiple** option) and the DIVIDE and MEASURE commands can place many copies in drawings. The ARRAY command, however, proves superior for making copies in precise rows, columns, matrices, circles, and semicircles. After you array an object, each copy can be edited separately — unlike the similar MINSERT command. For placing many copies in random places, however, COPY is better.

The largest number of rows and columns you can enter is 32,767; the smallest number is 1. Because a 32767 x 32767-array creates just over a billion elements, AutoCAD limits the total number to 100,000 — otherwise your computer system would overload. (The seemingly arbitrary value of 32,767 comes from one less than $2^{15}$).

You can change the upper limit to another value between 100 and 10,000,000 with the MAXARRAY system registry variable. At the command prompt, enter "MaxArray" exactly as shown to set the limit to 1000:

Command: **(setenv "MaxArray" "1000")**
"1000"

To create arrays, AutoCAD needs to know (1) the type of array, rectangular or polar, (2) the object(s) you plan to array, which must already exist in the drawing, and (3) the parameters of the array.

## TUTORIAL: LINEAR AND RECTANGULAR ARRAYS

A linear array copies objects in the horizontal direction (in a row or the x direction) or the vertical direction (in a column or the y direction). A rectangular array copies the objects in a rectangular pattern made up of rows and columns.

1. To make an array of copies, start the **ARRAY** command:
   - In the ribbons's 2D Home tab, choose the **Array** button from the Modify panel.
   - At the 'Command:' prompt, enter the **array** command:

     Command: **array** *(Press* ENTER.*)*

   - Alternatively, enter the **ar** alias at the 'Command:' prompt.

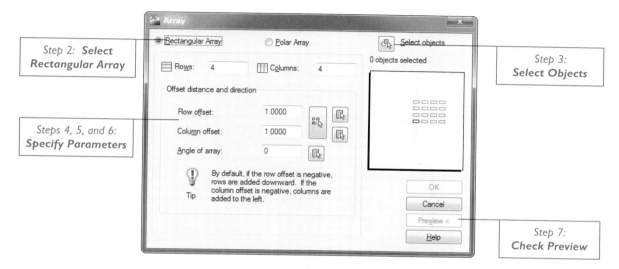

2. In all cases, AutoCAD displays the Array dialog box:

   Notice that the dialog box has two radio buttons (the round buttons at the top of the dialog box) that determine the type of array:
   - **Rectangular** — displays options for creating linear and rectangular arrays.
   - **Polar** — displays options for creating polar (circular and semicircular) arrays.

3. Choose **Select Objects** to select the object(s) to array. The dialog box disappears, and this prompt appears:

   Select objects: *(Select one or more objects.)*
   Select objects: *(Press* ENTER *to return to the dialog box.)*

**Notes** You can avoid the Select Objects step by selecting the objects *before* starting the **ARRAY** command.

Sometimes it's difficult to distinguish between *rows* and *columns*. Rows go side to side, while columns go up and down. Still puzzled? Look at the preview window, or click the **Preview** button for a sneak peak.

4. In the **Rows** and **Columns** text boxes, specify the number of copies to make in rows and columns. When you enter a 1 for both, AutoCAD complains,

   Only one element; nothing to do.

   ...because a 1x1 array consists of the original object only.

5. The **Row Offset** and **Column Offset** options measure the distance between the elements of the array. You can enter a specific distance, or click the adjacent buttons to select the distances in the drawing:

   Specify the distance between rows (or columns): *(Specify a distance.)*

   Enter a distance, or pick two points that specify the distance between row or column elements in the array. After picking the second point, the dialog box returns.

   The **Pick Both Offsets** button clears the dialog box, and AutoCAD prompts you at the command line:

   Specify unit cell: *(Pick a point.)*
   Other corner: *(Pick another point.)*

   Pick two points, creating a rectangle that specifies the row and column distance between elements in the array.

   After you pick the second point, the dialog box returns.

    **Notes** AutoCAD normally creates rows to the right, and columns *upwards*. To have AutoCAD draw the array elements in the other direction, enter negative values for the row and column offsets.

   If you use the mouse to pick two points and the second point is below or to the left of the first, then AutoCAD automatically sets negative values.

   To save file space in drawings with arrays, first turn the objects to be arrayed into a block, and then array the block.

6. **Angle of array** tilts the rectangular array at an angle; note that the object itself does not tilt, but the elements of the array are staggered by the angle. Entering an angle of 180 degrees draws the array downward. (To select the angle in the drawing, click the **Pick Angle of Array** button.)

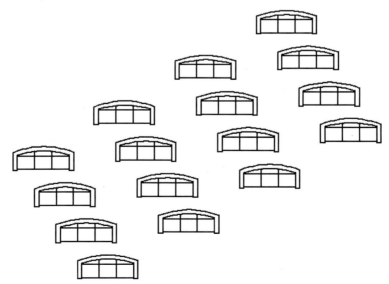

*A 4x4 rectangular array at an angle of 30 degrees.*

7. Click **Preview** to see what the array will look like. (As of AutoCAD 2009, the dialog box is replaced by the command prompt.)

   Pick or press Esc to return to dialog box or <right-click to accept array>: *(Move the mouse, or enter an option.)*

   **Drag the mouse** to pan the drawing, useful for when the array extends beyond the border of the viewport.

   **Twirl the roller wheel** to zoom in and out.

   **Click the drawing** (or press **ESC**) to return to the dialog box.

   **Right-click** to accept the array, and end the command.

## TUTORIAL: POLAR AND SEMICIRCULAR ARRAYS

Polar arrays arrange objects in circular patterns. To construct a polar array, you must define the angle between the items (from center to center, not actually between) and either the number of items or degrees to fill. You have the option of rotating (or not rotating) each object as it is arrayed.

1. Start the **ARRAY** command.
2. Click the **Polar Array** radio button. Now let's take a look at the options for creating a round array.

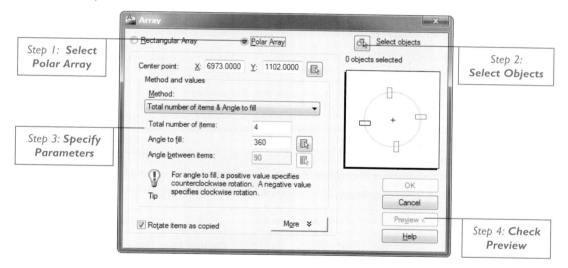

3. Select the **Center point** of the polar array; think of it as being the same as the center of a circle. AutoCAD automatically picks a point for itself using a method I have never figured out. (The technical editor agrees: "It's totally random!") But you can change it: enter new x, y-values or click the **Pick Center Point** button.

4. To create a polar array, you must provide data for any two of the following options. The Method droplist lets you pick the pair of options from the three possible choices:

   **Total Number of Items** — specifies the number of elements in the array. AutoCAD draws them to fit the polar route.

   **Angle to Fill** — changes the angle. By default, AutoCAD creates a 360-degree polar array, but you can create an arc array, instead. For example, specifying 270 degrees gives you three-quarters of a polar array.

   **Angle Between Items** — specifies the angle between the elements of the array.

5. Decide whether you want AutoCAD to **Rotate Items as Copied**. When turned on (the check mark appearing), AutoCAD makes sure the items "face the center."

**Left:** *A polar array with rotated objects.*
**Right:** *Polar array with unrotated objects.*

6. Click the **More** button to see more options for polar arrays:

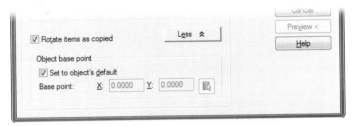

The **Object base point** option is tricky to grok, but essentially you get to pick where AutoCAD begins to measure the distance from the object to the center of the polar circle. AutoCAD uses the following default values, which change depending on the object being arrayed:

| Object | Default Base Point |
|---|---|
| Lines, polylines, donuts, 3D polylines, rays, splines | Starting point. |
| Arcs, circles, ellipses | Center point. |
| Polygons, rectangles | First corner drawn. |
| Xlines | Midpoint. |
| Blocks, mtext, text | Insertion point. |
| Regions | Grip point. |

7. Click **Select objects** to pick one or more objects to array. Click **Preview** to see what the array will look like.

# FILLET AND CHAMFER

The FILLET and CHAMFER commands create rounded and angled corners, respectively.

The FILLET command connects two lines or polylines with a perfect intersection, or with an arc of specified radius. Fillets can also connect two circles, two arcs, a line and a circle, a line and an arc, or a circle and an arc.

The two objects need not touch to be filleted, including parallel lines. This allows you to intersect two non-touching lines. In the case of parallel lines, the shorter line is extended to match the longer one, and then a 180-degree arc is drawn between their ends.

To use the CHAMFER and FILLET commands, you need to tell AutoCAD two things: (1) the size of chamfer or fillet, and (2) the two objects to be chamfered or filleted.

Holding down the SHIFT key temporarily changes the fillet and chamfer distances to zero.

### TUTORIAL: FILLETING OBJECTS

1. To fillet a pair of objects, start the FILLET command:
   - In the ribbons's 2D Home tab, choose the **Fillet** button from the Modify panel.
   - At the 'Command:' prompt, enter the **fillet** command:

     Command: **fillet** *(Press* ENTER.*)*

   - Alternatively, enter the **f** alias at the 'Command:' prompt.

2. In all cases, AutoCAD first displays the current fillet settings, and then asks you to select two objects:

   Current settings: Mode = TRIM, Radius = 0.0000
   Select first object or [Undo/Polyline/Radius/Trim/Multiple]: *(Pick object 1.)*
   Select second object or shift-select to apply corner: *(Pick object 2.)*

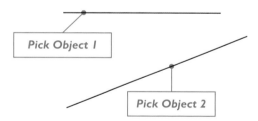

3. And the command is done! Because the radius was set to 0 (the fresh-off-the-distribution-CD default value), AutoCAD creates a clean intersection, without the arc you might have been expecting. (Notice that AutoCAD extends the two lines so that they intersect.)

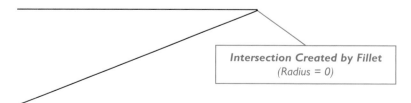

## DIFFERENT FILLET RESULTS FOR DIFFERENT OBJECTS

Depending on the objects involved, the fillet differs.

### Intersecting Lines

Two intersecting lines are trimmed back (or extended, as necessary), so that an arc fits between them. A zero-radius fillet connects two lines with a perfect intersection. The picked segments are trimmed off.

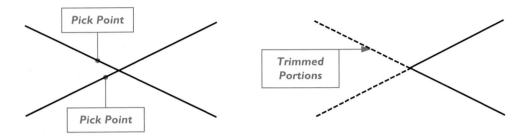

### Parallel Lines

Two parallel lines are filleted with a radius equal to their offset distance.

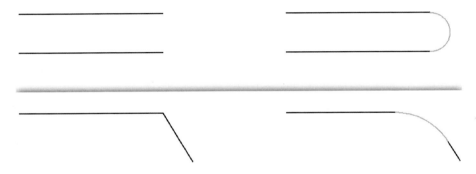

*Left:* Parallel and intersecting lines prior to filleting.
*Right:* Lines after filleting (fillet arcs shown in gray).

### Polylines

You can fillet an entire polyline in one operation when you select the **P** option (short for "polyline"). The fillet radius is placed at all vertices of the polyline. If arcs exist at any intersections, they are changed to the new fillet radius. Note that the fillet is applied to the continuous polyline.

*Left:* Original polyline.
*Right:* Filleted with the Polyline option (fillets shown in gray).

If the fillet radius is too large for a line or polyline segment, AutoCAD does not apply the fillet, making the following observation:

    3 were too short

When a line and a polyline are filleted together, all three objects (the line, the fillet arc, and the polyline) are converted to a single polyline.

## Arcs and Circles

When you fillet Lines, arcs, and circles, there are often several possible fillet combinations, depending on *where* you select the objects. AutoCAD attempts to fillet the end point closest to your pick point.

The figures illustrate several combinations between a line and arc. Observe the different ways AutoCAD places the fillet based on the pick point.

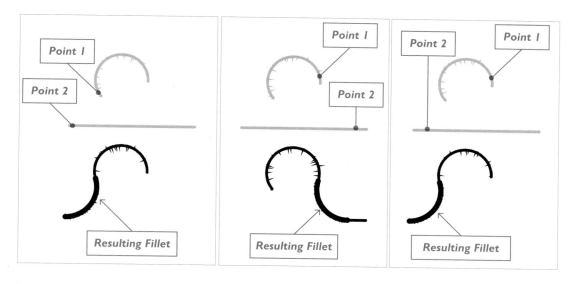

The original line and arc shown in gray; the resulting fillet shown by the heavy arcs.

The result of filleting two circles also depends on the location of the pick points used to select the circles. The figure illustrates the three possible combinations, each using different pick points.

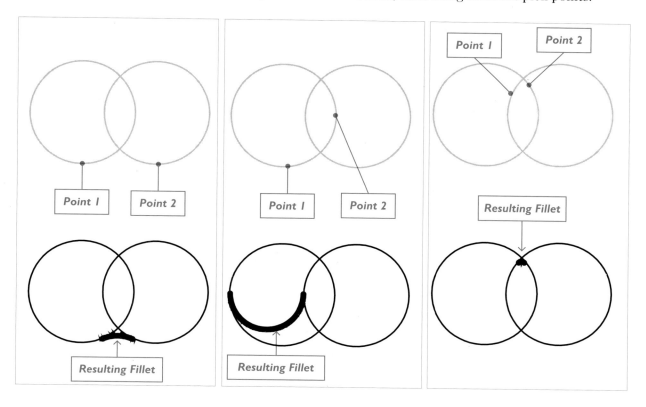

The original pre-fillet circles shown in gray; the resulting fillet shown by the heavy arcs.

**Notes** When you select two objects for filleting, and get an undesirable result, use the **Undo** option to undo the fillet. Try to respecify points closer to the endpoints you want to fillet.

If you have several filleted corners to draw, construct your intersections at right angles, and then fillet each later. This allows you to continue the **LINE** command without interruption, and requires fewer commands.

Changing an arc radius by fillet is cleaner and easier than erasing the old arc and cutting in a new one. Let AutoCAD do the work for you! **FILLET** remembers the last-used radius.

You can "clean up" line intersections by setting the fillet radius to zero and filleting the intersections. Shift-select the second object to set the radius to 0 temporarily and get sharp corners.

### CONSTRUCTING FILLETS: ADDITIONAL METHODS

The command has several options for special cases:

- **Undo** undoes the last fillet operation.
- **Polyline** treats polylines differently.
- **Radius** changes the radius of three-dimensional arrays.
- **Trim** determines what happens to the leftover bits.
- **Multiple** continues the command to fillet additional objects.

Let's look at each.

### Undo

The **Undo** option undoes the last fillet operation, without requiring you to exit the command to access the U or UNDO commands.

### Polyline

The **Polyline** option fillets all vertices of a single polyline. When you fillet a polyline without using this option, AutoCAD expects you to place a fillet between two adjacent segments. At the "Select first object" prompt, select a single polyline.

Select 2D polyline: *(Select a single polyline.)*

### Radius

The **Radius** option determines the radius of the fillet arc. When set to 0, the command ensures that the two lines match precisely; an arc is not created.

Specify fillet radius <1.0000>: *(Enter a radius.)*

If the fillet radius is too large for a line, AutoCAD complains:

Radius is too large *Invalid*

In that case, use a radius smaller than the shortest line. Press SHIFT to override temporarily the radius to 0.0.

### Trim

The **Trim** option determines what happens to the trimmed bits. When on (the default), the command trims away the selected edges, up to the fillet arc's endpoint. When off, no trim occurs.

Enter Trim mode option [Trim/No trim] <Trim>: *(Enter **T** or **N**.)*

### Multiple

The **Multiple** option (formerly **mUltiple**) repeats the command until you press ESC:

Select first object or [Undo/Polyline/Radius/Trim/Multiple]: *(Press ESC to exit command.)*

## CHAMFER

The CHAMFER command trims segments from the ends of two lines or polylines, and then draws a straight line or polyline segment between them. The distance to be trimmed from each segment can be different or the same. The two objects do not have to intersect, but they must be capable of intersecting if there were extended. Unlike fillets, parallel lines cannot be chamfered.

Chamfering only works with line segments, such as lines, 2D polylines, and traces. It does not work with arc segments, such as arcs, circles, and ellipses.

### TUTORIAL: CHAMFERING OBJECTS

1. To chamfer a pair of objects, start the **CHAMFER** command:
   - In the ribbons's 2D Home tab, choose the **Chamfer** button from the Modify panel.
   - At the 'Command:' prompt, enter the **chamfer** command:

     Command: **chamfer** *(Press ENTER.)*
   - Alternatively, enter the **cha** alias at the 'Command:' prompt.

2. In all cases, AutoCAD first displays the current chamfer settings:
   (TRIM mode) Current chamfer Dist1 = 0.0000, Dist2 = 0.0000

3. Before selecting objects, change the chamfer distance from the current setting of 0:
   Select first line or [Undo/Polyline/Distance/Angle/Trim/mEthod/Multiple]: **d**
   Specify first chamfer distance <0.0000>: *(Enter a value, such as .5.)*

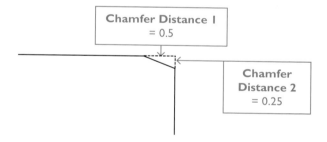

4. Chamfers work with two distances, which can be different. Enter the second distance:
   Specify second chamfer distance <0.5000>: *(Enter another value, such as .25.)*

5. Now pick the two lines to chamfer:
   Select first line or [Undo/Polyline/Distance/Angle/Trim/mEthod/Multiple]: *(Pick line 1.)*
   Select second line or shift-select to apply corner: *(Pick line 2.)*

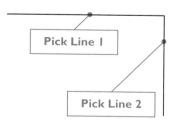

AutoCAD creates a chamfer at the intersection.

You can specify the amount to be trimmed by entering either a numerical value, or by showing AutoCAD the distance using two points on the screen.

### CONSTRUCTING CHAMFERS: ADDITIONAL METHODS

The command has several other options for special cases:

- **Undo** undoes the last chamfer operation.
- **Polyline** chamfers a single polyline.
- **Angle** specifies the chamfer angle.
- **Trim** determines whether end pieces are saved.
- **mEthod** specifies the chamfer method.
- **Multiple** continues the command to fillet additional objects.

Let's look at each.

### Undo

The **Undo** option undoes the last chamfer operation, without requiring you to exit the command to access the U or UNDO commands.

### Polyline

Like filleting, chamfering a polyline is different from chamfering a pair of lines. If you had used the PLINE command's **Close** option to finish the polyline, AutoCAD chamfers all corners of the polyline; if not, the final vertex is not chamfered.

> Select 2D polyline: *(Pick a polyline.)*

If the polyline contains arcs, they are not chamfered. If some parts of the polyline do not chamfer, it could be that the segments are too short or parallel to each other. In that case, AutoCAD warns: "*n* were too short."

### Angle

As an alternative to specifying a chamfer by two distances, you can specify a distance and an angle.

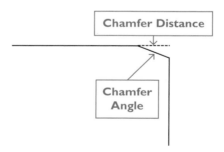

> Specify chamfer length on the first line <1.0000>: *(Enter the distance.)*
> Specify chamfer angle from the first line <0>: *(Enter the angle.)*

### Trim

Normally, the CHAMFER command erases the line segments not needed after the chamfer. The **Trim** option, however, determines whether you keep the excess lines.

> Enter Trim mode option [Trim/No trim] <Trim>: *(Enter **T** or **N**.)*

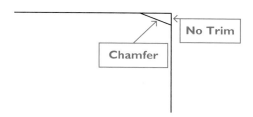

### mEthod

The default method specifies two distances. If you prefer the distance-angle method, the **mEthod** option (formerly **Method**) lets you change the default:

> Enter trim method [Distance/Angle] <Angle>: *(Type **D** or **A**.)*

### Multiple

The **Multiple** option (formerly **mUltiple**) repeats the command until you press ESC:

> Select first object or [Polyline/Radius/Trim/mUltiple]: *(Press ESC to exit command.)*

##  JOIN

The JOIN command joins similar objects into one.

This command is useful for closing arcs, turning them into circles. As of AutoCAD 2011, some objects no longer need to be coplanar. The following objects can be joined, subject to conditions specified by AutoCAD's prompt:

| Source Object | Prompt |
| --- | --- |
| Line | Select lines to join to source. |
| Polyline segment | Select objects to join to source. |
| 3D polyline | Select any open curves to join to source. |
| Arc | Select arcs to join to source or [cLose]. |
| Polyline arc | Select objects to join to source. |
| Elliptical arc | Select elliptical arcs to join to source or [cLose]. |
| Spline, Helix | Select any open curves to join to source. |

AutoCAD does not let you join a more complex object to a simpler one. Joining a spline to a polyline, for example, results in the following error message: "0 segments added to polyline." You have to do the reverse: join simpler objects to complex ones.

**Colinear lines** join into a single line, and can include lines with overlaps and gaps between them. (*Colinear* means the lines are lined up in a row like cars in a train on a straight track.)

**Arcs** and **elliptical arcs** join into a single arc counterclockwise from the source arc. They can have overlaps and gaps between them, but they must lie in the same imaginary circle. Arcs dimensioned with DIMARC are converted into circles disassociated from their dimensions.

**Polylines**, **lines**, and **arcs** join into a single polyline; they *cannot* have gaps between them. When the first object selected is a polyline, this works like the PEDIT Join command, although PEDIT has the added benefit of a fuzz factor that joins segments with gaps.

**Splines** and **helices** join into a single objects; they cannot have gaps between them. Splines and helices can be joined to each other.

**3D polylines** join to other open objects, such as lines, elliptical arcs, helices, and splines. The 3D polyline must be selected first, because the other objects are converted to 3D polylines.

The source object determines the properties of the joined objects.

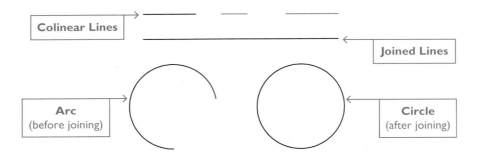

To use the JOIN command, you need to tell AutoCAD two things: (1) the source object, and (2) the other objects to join; when arcs and elliptical arcs are selected, there is also the choice of whether you want the open objects closed.

### TUTORIAL: JOINING OBJECTS

1. To join two or more objects, start the **JOIN** command:
   - In the ribbons's 2D Home tab, choose the **Join** button from the Modify panel.
   - At the 'Command:' prompt, enter the **join** command:

     Command: **join** *(Press* ENTER.*)*
   - Alternatively, enter the **j** alias.

2. AutoCAD prompts you to select the source object, which can be a line, arc, elliptical arc, polyline, or spline:
   Select source object: *(Select one object.)*

3. AutoCAD prompts you to select the objects to join. The prompt varies, depending on the source object selected; the following prompt is for arcs:
   Select arcs to join to source or [cLose]: *(Select one or more objects.)*

4. AutoCAD repeats the prompt until you press **ENTER** to exit the command:
   Select arcs to join to source or [cLose]: *(Press* ENTER.*)*

### JOINING OBJECTS: ADDITIONAL METHODS

The JOIN command has one option: **cLose** closes arcs and elliptical arcs.

   Command: **join** *(Press* ENTER.*)*
   Select source object: *(Select one arc or ellipse.)*
   Select arcs to join to source or [cLose]: **l**
   Arc converted to a circle.

 **Note** Use the **BREAK** command to convert circles into arcs; use the **JOIN** command to convert arcs into circles.

 **REVERSE**

The REVERSE command reverses the direction of lines, polylines, splines, and helices.

**TUTORIAL: REVERSING OBJECTS**

1. Open the *Reverse.dwg* file, a drawing containing lines with linetypes.

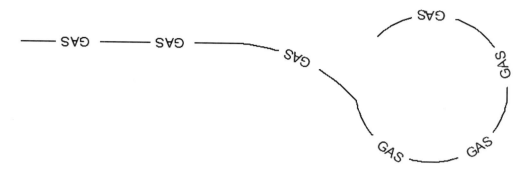

2. To see the effect of reversing directions of objects, start the **REVERSE** command:
   - In the ribbons's 2D Home tab, choose the **Reverse** button from the Modify panel.
   - At the 'Command:' prompt, enter the **reverse** command:

   Command: **reverse** *(Press ENTER.)*

3. AutoCAD prompts you to select one or more objects, which can be lines, polylines, splines, and/or helixes:

   Select objects: *(Select the lines.)*

4. AutoCAD repeats the prompt until you press ENTER to exit the command:

   Select objects: *(Press ENTER.)*

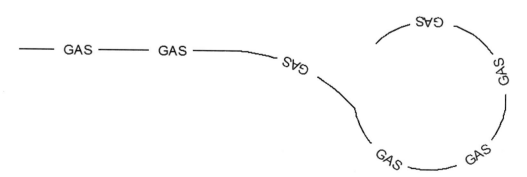

Notice that the linetype text has switched orientation, indicating the lines are reversed.

- As of AutoCAD 2011, linetype definitions can now use the "U" (upright) parameter to forces text and shapes always to be displayed "upright." This reduces the need for the REVERSE command.

 **REVCLOUD**

The REVCLOUD command creates revision clouds.

Revision clouds are often used to highlight areas in drawings that require attention, such as a revision or a potential error. Revision clouds are sometimes called "markups" or "redlines," because they were often drawn with red pencils to stand out in drawings.

AutoCAD's REVCLOUD command creates revision clouds, or converts other objects into revision clouds — specifically circles, ellipses, closed polylines, and closed splines. Because you cannot invoke transparent zooms and pans (other than through the mouse wheel) during the command, ensure you can see the entire area before starting.

 **Note** Before starting REVCLOUD, switch to a layer set to red. That makes it easier to turn on and off the display of the revision cloud, and makes it the traditional color of red to boot.

To use the REVCLOUD command, you need to tell AutoCAD two things: (1) the starting point, and (2) the cloud path.

### TUTORIAL: REDLINING OBJECTS

In this tutorial, you mark up a drawing using the revcloud command.

1. Start AutoCAD, and then open the *17_20.dwg* drawing file.
2. Revision clouds are traditionally made in red. Use the **LAYER** command to create a new layer named "Revisions" and set to color **Red**.
3. To markup the drawing, start the **REVCLOUD** command:
   - In the ribbons's 2D Home tab, choose the **Revision Cloud** button from the Draw panel.
   - At the 'Command:' prompt, enter the **revcloud** command.

   Command: **revcloud** *(Press ENTER.)*

4. AutoCAD displays the current settings, and then prompts you to start drawing the revision cloud:
   Minimum arc length: 0.5000   Maximum arc length: 0.5000   Style: Normal
   Specify start point or [Arc length/Object/Style] <Object>: *(Pick a point.)*

5. Move your cursor. Notice that AutoCAD automatically creates the cloud pattern.
   As an alternative, you can define the size of arcs by picking points.
   Guide crosshairs along cloud path...

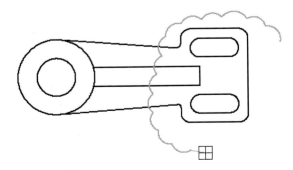

6. When the cursor is close to the start point, AutoCAD automatically closes the cloud:
    Revision cloud finished.

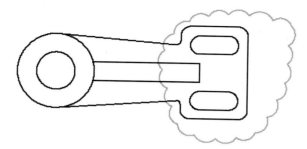

After placing the revision cloud, you can edit it like other objects in the drawing — through copying, grips editing, erasing, and so on.

### REDLINING OBJECTS: ADDITIONAL METHODS

To help you in redlining, AutoCAD has these options:

- **Arc length** defines the size of the arcs making up the clouds.
- **Object** converts existing objects into revision clouds.
- **Style** switches between simple and calligraphic arcs.

Let's look at each.

### Arc length

The **Arc length** option determines the size of the arcs making up the clouds, which are scale-dependent, like linetypes and hatch patterns. REVCLOUD saves the arc length as a factor of the DIMSCALE system variable, so that clouds drawn after the dimension scale changes still look the right size when the scale factor changes.

When you specify different minimum and maximum lengths, AutoCAD draws clouds with random-size arcs. AutoCAD limits the maximum arc length to three times the minimum length.

Specify minimum length of arc <0.5000>: *(Enter a value.)*
Specify maximum length of arc <0.5000>: *(Enter a value.)*

### Object

The **Object** option lets you convert existing objects into revision clouds.

Select object: *(Select one object.)*
Reverse direction [Yes/No] <No>: *(Enter Y or N.)*
Revision cloud finished.

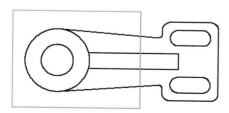 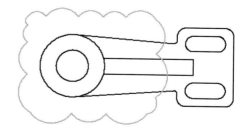

**Left**: The original rectangle object.
**Right**: Object converted to revision cloud.

You can convert closed polylines, circles, ellipses, and closed splines to revision clouds. After applying this option, what happens to the original object? That depends on the setting of the DELOBJ system variable. If set to 1 (the default), the original is erased; if set to 0, the original object stays in the drawing.

Command: **delobj**
Enter new value for DELOBJ <1>: *(Enter **1** or **0**.)*

### Style

The **Style** option switches between simple and calligraphic arcs. Calligraphic arcs, added in AutoCAD 2005, have a variable width, as illustrated below.

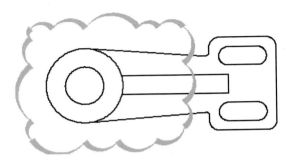

To switch between normal and calligraphic arcs, use the **Style** option, as follows:

Specify start point or [Arc length/Object/Style] <Object>: **s**
Select arc style [Normal/Calligraphy] <Normal>: **c**
Arc style = Calligraphy

To convert calligraphic arcs to normal arcs, use the PEDIT command's **Width** option.

##  MARKUP

The MARKUP command imports marked-up *.dwf* files from Design Review (formerly DWF Composer; DWF is short for "design Web format"). It displays the Markup Set Manager, which places and edits redline DWF markups in drawings.

Despite its name, MARKUP does not mark up drawings! Its purpose is to *manage* marked up drawings; the markup process takes place outside of AutoCAD, using Autodesk's free Design Review software.

This command opens only *.dwf* files that contain markup data; it won't open any other kind of markup file, not even unmarked-up DWF data, or work with drawings marked up with AutoCAD's older RMLIN command.

(*History*: MARKUP replaces the RMLIN command [short for "Red Markup Line In"], which imported XML-format *.rml* markup files created by Autodesk's old Volo View drawing viewer software. MARKUP replaced RMLIN in AutoCAD 2006, and DesignReview replaced Volo View.)

### Design Review

To use the MARKUP command, you must have access to Design Review, the free viewing, redlining, and printing software from Autodesk. (You can download a copy from www.autodesk.com/designreview, following registration.) Design Review acts as a publisher, merging files from many sources — such as raster images, word processing documents, spreadsheets, and drawings from other CAD packages — into a single, multi-page document. (To convert non-AutoCAD documents into DWF format, download and install the free DWF Writer printer driver. More information is available at www.autodesk.com/dwf.)

The markup cycle works like this:

### In AutoCAD

1. Create and edit drawings.
2. Export drawings in DWF format using the **EXPORTDWF** command. (Alternatively, you can use any of the **PLOT**, **PUBLISH**, **AUTOPUBLISH**, **EXPORTDWFX**, or **3DDWF** commands See Chapter 20 for details.)
3. The *.dwf* file appears in Design Review, which should open automatically. If not, then start the program by clicking its icon.

### In Design Review

4. Open the *.dwf* file in Design Review.
5. Mark up the drawings using the markup tools found on the ribbon

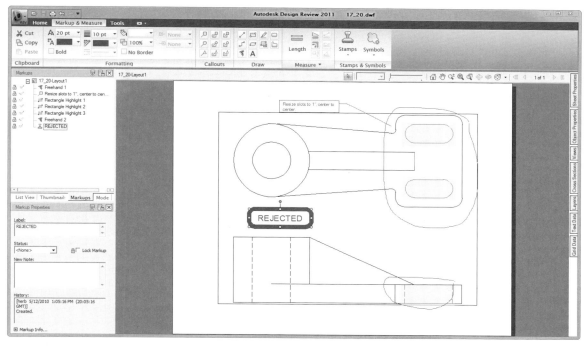

6. Save the markups with the **File | Save** command.

### Back in AutoCAD

7. Enter the **MARKUP** command.
8. In the Markup Set Manager palette, choose **Open**, and then select the marked-up *.dwf* file.

9. To view the markups, double-click the markup name in the Markup Set Manager window, such as "17-21.dwf" shown in the figure below.

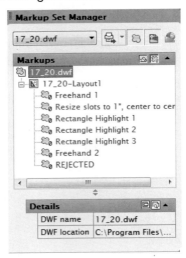

10. You can now edit the drawing and/or the markups in AutoCAD.

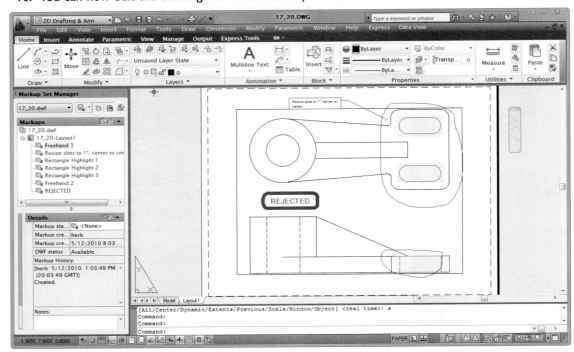

## Markup Options

Many of the MARKUP command's options are "hidden" in shortcut menus. Depending on where you right-click in the Markup Set Manager palette, you get menus with differing sets of commands. The most-commonly used shortcut menu is found by right-clicking a markup sheet (indicated by the red revision cloud icon).

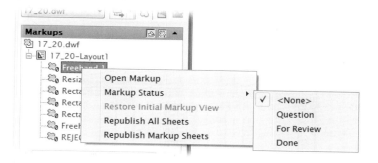

The menu's options are as follows:

| Command | Comment |
| --- | --- |
| Open Markup | Finds and opens the original *.dwg* drawing file, and makes active the layout associated with the markup; the *.dwf* file is also opened, but is hidden until you press ALT+4. |
| Markup Status | Allows you to change the status of markups:<br>　　**<None>** has no status.<br>　　**Question** indicates additional information is needed.<br>　　**For Review** indicates changes should be reviewed by another.<br>　　**Done** indicates the markup is implemented. |
| Restore Initial Markup View | Restores the original markup view. |
| Republish All Sheets | Reoutputs all sheets in DWF format with the changes made. |
| Republish Markup Sheets | Overwrites the previous *.dwf* file with changes made to the drawing and the markup status. |

Marked-up *.dwf* files can also be opened in AutoCAD with the OPENDWFMARKUP command.

The MARKUPCLOSE command closes the Markup Set Manager window.

# EXERCISES

1. Open the *edit3.dwg* drawing file, an architectural elevation.

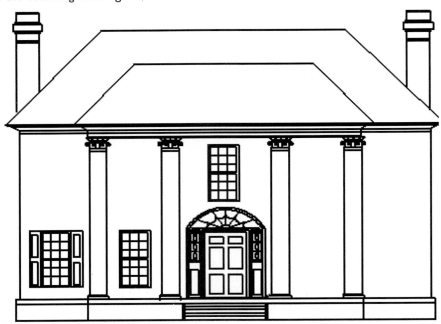

   Use the **COPY** command to copy the windows from the left side to the right side.
   Then, copy all the windows on the lower level (including those you just copied) to the upper level. When copying, use the object selection options you think will work best.

2. Open the *edit4.dwg* drawing file, a landscaping plan.

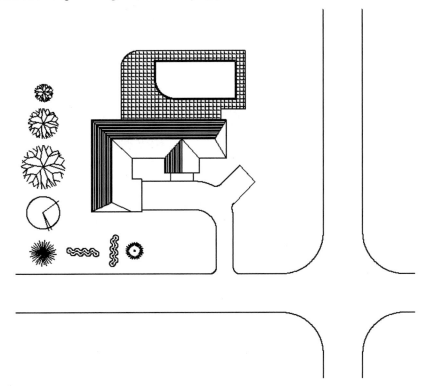

Use the **COPY** command to copy the landscaping blocks (trees, shrubbery, and so on) and to create a landscape scheme.

3. Connect two lines with fillets of varying radius.
   First, draw lines similar to those in the illustration.

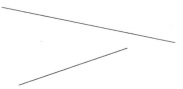

   a. Connect the two lines with a fillet of radius of 0.15.
   b. Use the **SHIFT** key to set the fillet radius to zero, and then apply the fillet to the two lines again. Do the two lines now connect in a perfect intersection?

4. Open the *edit6.dwg* drawing file, a practice drawing.
   Fillet the objects to achieve the results shown on right side of the figure.

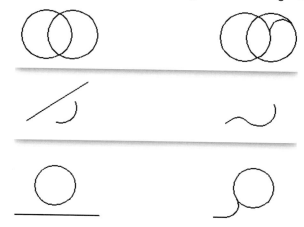

5. Suppose that you have to add another circle to your drawing that is identical to an existing circle.
   Draw a circle, and then use the **COPY** command to place a second circle.
   Is the second circle an exact copy of the first?

6. In this exercise, you create a rectangular array.
   a. Start a new drawing.
   b. As a visual aid, turn on the grid with a value of one.
   c. Draw a circle with the center point located at 1,1 and a radius of one.
   d. Select the circle, and then start the **ARRAY** command.

e. In the dialog box, enter the following options:

| | |
|---|---|
| Type of array | **Rectangular** |
| Rows | 3 |
| Columns | 5 |
| Row offset | 2 |
| Column offset | 2 |

Your array should look like the following illustration:

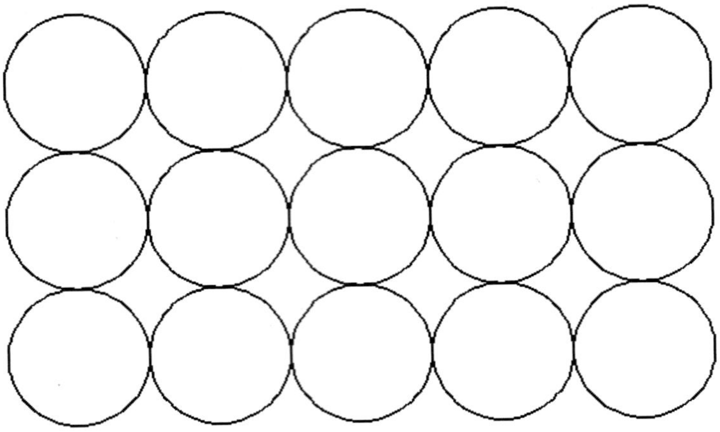

Save the drawing with the name *array.dwg*.

7. Start another drawing to create a polar array.
   a. Draw a square with sides of one unit each.
   b. Start the **ARRAY** command, and enter these options:

   | | |
   |---|---|
   | Select objects | *Select the square* |
   | Type of array | **Polar** |
   | Center point | 2,2 |
   | Total number of items | 4 |
   | Angle to fill | 270 |
   | Rotate items as copied | **Off** |

   The polar array should look like the one in the figure. Save the drawing with the name *parray.dwg*.

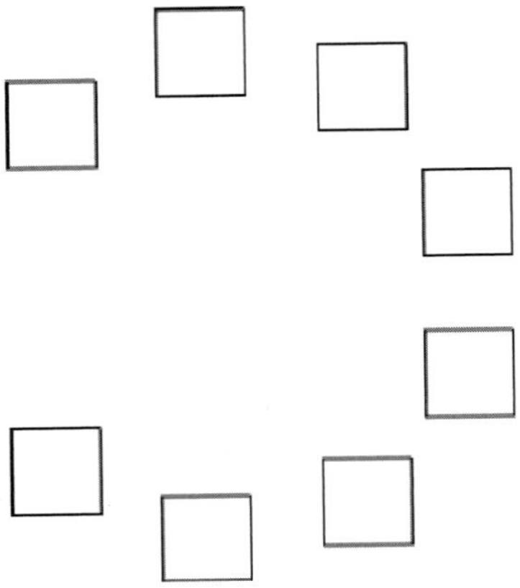

8. Open the *mirror1.dwg* drawing file, a floor plan.

   Suppose that you are designing a house, and you want to reverse the layout of the bathroom.

   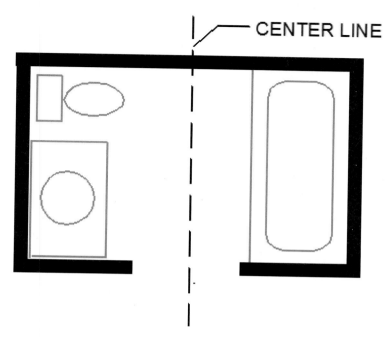

   Reverse the room using the **MIRROR** command. *Hint:* Turn on ortho mode (**F8**), and delete the source objects.

9. In this exercise, the **PDMODE** system variable allows you to see points placed by the **DIVIDE** command.
   a. Start a new drawing.
   b. Draw a circle: center the circle in the viewport, and use a radius of 3.
   c. Start the **DIVIDE** command, and use 8 for the number of segments.
   d. Do you see any difference to the circle?
   e. Set the **PDMODE** system variable to 34.
   f. If necessary, use the **REGEN** command to make the new point style visible.

10. Draw a symbol similar to the one in the figure.
    a. Which two commands make it easy to draw the symbol?

    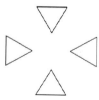

    b. Turn the symbol into a block, and name it "Symbol."
    c. Draw a circle on the screen, and then start the **DIVIDE** command. Use 8 segments, along with the **Block** option, with blocks aligned with the object.
    d. Repeat the exercise, but this time don't rotate the block.

11. Let's "measure" an object.
    a. Draw a horizontal line 6 units in length.
    b. Set the point mode to 34 to make the points visible.
    c. With the **MEASURE** command, specify a segment length of 1.0.

12. In this exercise, you use the **ELLIPSE** command to draw a can.
    a. Draw one ellipse with the **ELLIPSE** command. Don't worry about the size.
    b. Next, use the **COPY** command to copy the ellipse to create the top of the can. *Hint:* Use ortho mode to align the copy perfectly with the original ellipse.
    c. Now use the **LINE** command to draw lines between the two ellipses, as shown in the figure. *Hint:* Use QUADrant object snap to capture the outer quadrants of the ellipses.

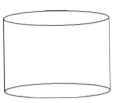

13. In this exercise, you use the **OFFSET** command to help draw a cityscape plan.
    a. Open the *offset.dwg* drawing file.

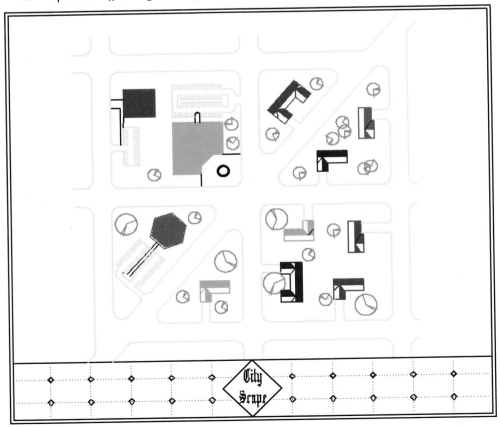

    b. The edges of the streets are drawn with polylines. Use the **OFFSET** command's **Through** option to offset the curbs by 6".

14. Draw the object shown in the figure below. Use the **CHAMFER** command to bevel all its corners.

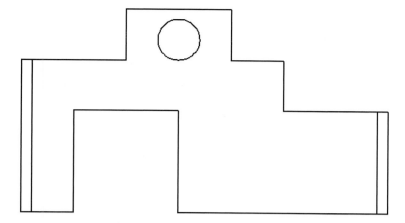

15. Open the *edit5.dwg* drawing, a piping diagram. Use the **REVCLOUD** command to place revision clouds around the two unfinished areas.

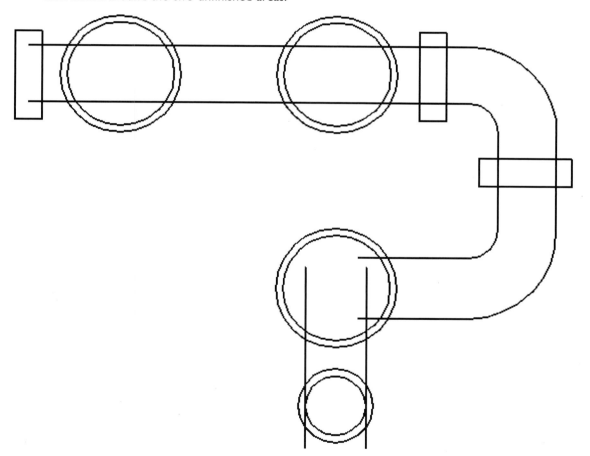

16. Draw the profiles of the aluminum extrusions diagrammed and dimensioned below. Apply fillets, where required, and hatch with a solid fill. All dimensions are in inches.

   a. Angle.

   | A | B | C | D |
   |---|---|---|---|
   | 4.000 | 1.000 | 0.155 | 0.125 |

   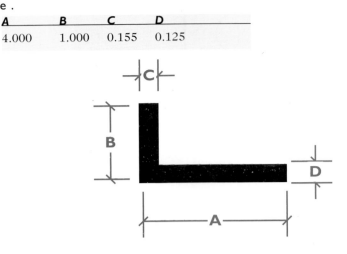

   b. Square tube.

   | A | T |
   |---|---|
   | 4.000 | 0.145 |

   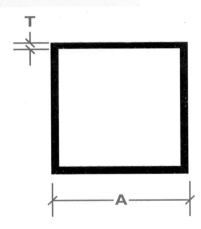

   c. Rectangular tube with rounded corners.

   | A | B | T | R1 | R2 |
   |---|---|---|----|----|
   | 2.250 | 1.750 | 0.125 | 0.125 | 0.125 |

   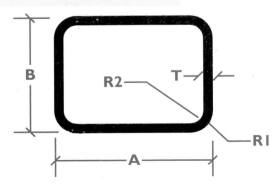

d. Rectangular bar with rounded corners.

| A | B | R |
|---|---|---|
| 3.250 | 0.375 | 0.030 |

e. Tee.

| A | B | T1 | T2 | R |
|---|---|---|---|---|
| 2.000 | 2.000 | 0.125 | 0.125 | 0.015 |

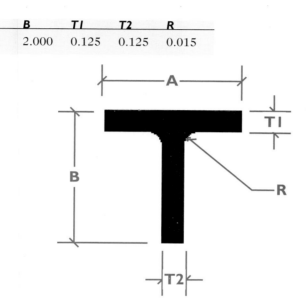

f. Channel with rounded corners.

| A | B | C | D | R |
|---|---|---|---|---|
| 3.000 | 1.500 | 0.375 | 0.375 | 0.375 |

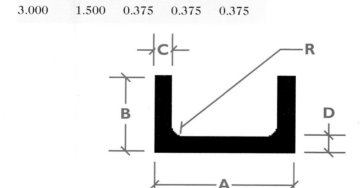

17. Draw the outline and fold lines of the diecut cardboard packaging that holds a tube of toothpaste. Cutlines are shown in black; fold lines in white. All dimensions are in mm. Print the drawing, and then cut and fold along the lines to recreate the cardboard box.

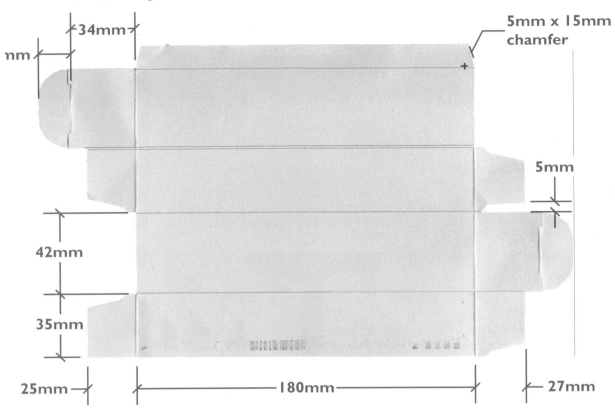

18. Open the *join.dwg* drawing.

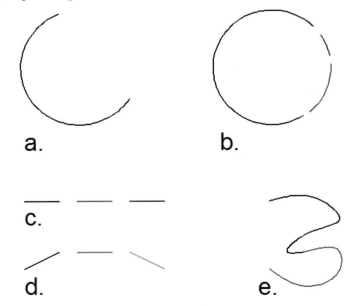

a.       b.       c.       d.       e.

Use the **JOIN** command to perform the following tasks:
- a. Turn the arc into a circle.
- b. Join the three arcs into one circle.
- c. Join the three lines into one line.
- d. Join the three lines into one line.
- e. Join the two splines into one spline.

Which one of the above tasks did not work? Why?

What color did each joined entity become? Why?

## CHAPTER REVIEW

1. What two changes can a fillet make to an intersection?
2. When making multiple copies with the **COPY** command, is each copy relative to the first base point entered, or to the point of the last copy made?
3. Can objects of different types be filleted (such as a line and an arc), or must they be alike?
4. What are the three types of arrays?
5. Describe the function of the **MIRROR** command.
6. Which system variable controls whether text is mirrored?
7. What is the *mirror line*?
8. What objects can you use with the **MEASURE** command?
9. How many objects can be offset at one time?
10. Describe the purpose of the **Through** option of the **OFFSET** command?
11. Draw something using the **OFFSET** command's **Through** option.
12. What is the purpose of the **CHAMFER** command?
13. What is the procedure for setting **CHAMFER** distances?
14. Describe the purpose of the **ARRAY** command.
15. Describe a task for which the **REVERSE** command is useful.
16. Can the **ARRAY** command draw arrays at an angle?
17. Under what condition does the **ARRAY** dialog box's **OK** button stay grayed (and unavailable)?
18. To construct an array in the -x direction, you enter a negative value in which option?
19. What could happen to your computer if you construct too large an array?
20. Define *displacement*, as used by the **COPY** command.
21. Which mode helps you copy objects precisely horizontally and vertically?
22. What are the aliases for the following commands:

    **COPY**

    **MIRROR**

    **OFFSET**

    **FILLET**
23. What problem can occur when the **OFFSET** command is applied to curved objects?
24. How does the **OFFSETGAPTYPE** system variable affect polylines?
25. Does the **MEASURE** command measure the lengths of objects?
26. What two kinds of objects does the **DIVIDE** command place along objects?
27. When you specify 5 segments for the **DIVIDE** command, how many objects does it place?
28. Can parallel lines be filleted?
29. Describe the purpose of the **FILLET** and **CHAMFER** commands' **Trim** options.
30. What is the difference between a fillet and a chamfer?
31. Can all vertices of a polyline be chamfered at once?
    If so, how?
32. Describe the purpose of the **REVCLOUD** command.
33. What is another name for "redlines"?
34. How does the **DELOBJ** system variable affect the **Object** option of the **REVCLOUD** command?

35. Explain the purpose of the **MARKUP** command.
36. Describe the role of the Design Review (DWF Composer) software.
37. How are DWF files created?
38. What command converts arcs into circles?
39. Which key do you hold down to make a zero-radius fillet? A zero-distance chamfer?
40. What is the purpose of the **Undo** option of the **COPY** command?
41. What are *colinear* lines?

# CHAPTER 9
## Additional Editing Methods

In previous chapters, you learned numerous methods of creating and changing objects. In this chapter, you learn AutoCAD's other editing commands, including some that may already be familiar to you from grips editing.

This chapter summarizes the remaining editing commands used in two-dimensional drafting:

> **ERASE** deletes objects from drawings, and **OOPS** brings them back.
>
> **BREAK** removes portions of objects.
>
> **TRIM, EXTEND,** and **LENGTHEN** change the length of open objects.
>
> **STRETCH** makes portions of objects larger and smaller.
>
> **MOVE** moves objects in the drawing.
>
> **ROTATE** rotates objects.
>
> **SCALE** changes the size of objects.
>
> **CHANGE** changes the size, properties, and other characteristics of objects.
>
> **EXPLODE** and **XPLODE** reduce complex objects into their simplest forms.
>
> **PEDIT** edits polylines.
>
> **Grips Editing** and **Direct Distance Entry** edit objects directly.
>
> **TRACKING** and **M2P** assist with direct drawing.

---

**NEW TO AUTOCAD 2011** IN THIS CHAPTER

- Individual vertices of polylines can now be edited by holding down the **CTRL** key.

---

**ERASE**

The ERASE command deletes objects from drawings. As an alternative to this command, you can select one or more objects, and press the DEL (delete) key.

**TUTORIAL: ERASING OBJECTS**

1. From the CD, open the *erase.dwg* drawing file.

2. To erase one or more objects, start the ERASE command:
   - In the ribbons's 2D Home tab, pick the **Erase** button in the Modify panel.
   - At the 'Command:' prompt, enter the **erase** command:

     Command: **erase** *(Press ENTER.)*
   - Alternatively, enter the **e** alias at the 'Command:' prompt.

3. In all cases, AutoCAD prompts you to select the objects you want erased:
   Select objects: *(Pick one or more objects.)*
   Select objects: *(Press ENTER to end object selection.)*

   Notice that the selected objects disappear from the drawing.

At the "Select objects:" prompt, the crosshair cursor changes to the tiny, square pickbox for selecting one object at a time (the point method of selection).

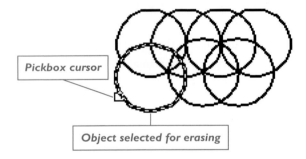

If you wish to use another object selection option, such as Window or Crossing, enter the option name at the "Select objects:" prompt. Following erasure, it may be necessary to use the REDRAW command to refresh the screen.

**RECOVERING ERASED OBJECTS**

AutoCAD has two commands for restoring erased objects:
- **OOPS** returns the last erased object(s).
- **U** undoes the last operation, including erasure.

Let's look at each.

## Oops

The OOPS command restores the objects that were last erased from the drawing.

Command: **oops**

Notice that the erased object(s) are returned to the drawing.

This command also works in conjunction with the BLOCK and -BLOCK commands, if the original objects were erased in making the block.

## U

If you find yourself unable to use OOPS to return erased objects, remember that you can also reverse the deletion with the U command.

**Notes** To erase the object you just drew, enter ERASE with the **Last** option. A peculiarity is that AutoCAD only erases the last-drawn object if it is visible on the screen.

Similarly, to erase the previous selection set, use ERASE **Previous**. You can erase the entire drawing with the ERASE **All** command. To do the same using keyboard shortcuts, press CTRL+A and then DEL.

## BREAK

The BREAK command partially and fully erases objects, as well as cracks them.

To "break" an object, select two points on it. The portion of the object between the two points is erased. This command removes a portion of almost any object, and shortens open objects — depending on the place you pick. The following objects can be broken: arcs, circles, ellipses, elliptical arcs, lines, polylines, rays, splines, and xlines.

BREAK can remove an object entirely, like the ERASE command. Here's how: for the first point, snap to one end of a line, and then snap to the other endpoint for the second point. The entire line disappears from the drawing. Use U to recover, if necessary.

BREAK is useful for "cracking" objects: click the same point twice, and BREAK creates two segments that touch. BREAK operates on just one object at a time.

### TUTORIAL: BREAKING OBJECTS

1. To remove a portion of an object, start the **BREAK** command:
   - In the ribbons's 2D Home tab, pick the **Break** button in the Modify panel.
   - At the 'Command:' prompt, enter the **break** command:

     Command: **break** *(Press ENTER.)*

   - Alternatively, enter the **b** alias at the 'Command:' prompt.

2. In all cases, AutoCAD prompts you to select the single object you want broken:
   Select object: *(Pick one object at point 1.)*

   The point you pick is crucial, because it becomes the first point of the break.

3. Pick a second point, the other end of the break:
   Specify second break point or [First point]: *(Pick point 2.)*

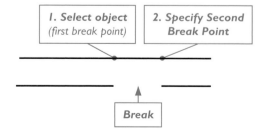

**Note** AutoCAD displays the 🔒 padlock icon when you attempt to select objects on locked layers. Objects on locked layers can be seen but not edited. To edit these objects, you must first unlock their layer(s).

### BREAKING OBJECTS: ADDITIONAL METHOD

By default, the first point you select on the object becomes the first break point. To redefine the first break point, enter **F** in response to the prompt. Select another point to be the first break point.

Redefining the first point is useful when the drawing is crowded, or when the break occurs at an intersection where pointing to the object at the first break point might select the wrong object.

- **First** defines the first pick point.

Let's look at this option.

 **First**

The **First** option allows you to redefine the first pick point, providing greater control over the segment being broken out.

> Command: **break**
> Select object: *(Pick object.)*
> Specify second break point or [First point]: **f**
> Specify first break point: *(Pick point 1.)*
> Specify second break point: *(Pick point 2.)*

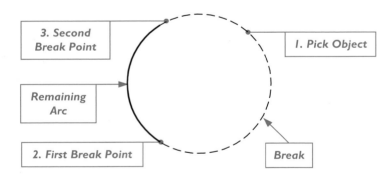

 **Note** The first tip I ever learned from technical editor Bill Fane was that sometimes it is easier to create an arc by drawing a circle, and then using the **BREAK** command to remove the unwanted portion.

The BREAK command affects objects in different ways:

| Object | Break Action |
|---|---|
| Arcs, Lines | Removes portion of line or arc between the pick points. When one point is on the line, but the other point is off the end of the line, the line is "trimmed back" to the first break point. |
| Circles | Breaks into an arc. Unwanted piece is determined by going counterclockwise from the first point to the second point. |
| Ellipses | Breaks into an elliptical arc in a manner like circles. |
| Splines | Reacts in the same manner as lines and arcs. |
| Traces | Reacts in the same manner as lines, except that the new endpoints are trimmed square. |
| Polylines | Cuts wide polylines squarely, as with traces. Breaking closed polylines creates open polylines. |
| Viewports | Does not brake. |
| Xlines | Breaks into rays. |

This command cannot break blocks, dimensions, multilines, or regions. For dimensions, use the DIM-BREAK command.

## TRIM, EXTEND, AND LENGTHEN

The TRIM, EXTEND, and LENGTHEN commands change the length of open objects.

The TRIM command shortens objects by defining other objects as cutting edges; any portion of the object beyond the cutting line is "cut off."

AutoCAD needs to know two pieces of information for trimming: (1) the object(s) to be used as the cutting edges, and (2) the portion of the object to be removed.

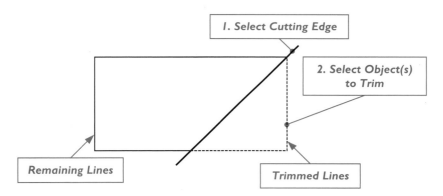

Cutting edges are any combination of arcs, circles, ellipses, lines, viewports, rays, regions, splines, text, xlines, hatches, 2D and 3D polylines. Polylines with nonzero width trim to the center line of the polyline.

A quick method is to select all objects as cutting edges: at the 'Select objects <select all>:' prompt, press ENTER, and AutoCAD selects all objects in the drawing, including those not in the current view. A previously undocumented method of quickly selecting many objects to trim is the **Fence** selection mode. Each part that crosses the fence line is picked and trimmed.

### TUTORIAL: SHORTENING OBJECTS

1. To shorten one or more objects, start the **TRIM** command:
   - In the ribbons's 2D Home tab, pick the **Trim** button in the Modify panel.
   - At the 'Command:' prompt, enter the **trim** command:

     Command: **trim** *(Press ENTER.)*

   - Alternatively, enter the **tr** alias at the 'Command:' prompt.

2. In all cases, AutoCAD reports the current trim settings, and then prompts you to select cutting edges:

   Current settings: Projection=UCS Edge=None

   Select cutting edges ...

   Select objects or <select all>: *(Pick one or more objects as cutting edges, or press ENTER to select all objects in the drawing.)*

   Select objects: *(Press ENTER to end object selection.)*

3. Select the object to be trimmed. *Warning!* AutoCAD trims the portion you select.

   Select object to trim or shift-select to extend or

   [Fence/Crossing/Project/Edge/eRase/Undo]: *(Pick objects to trim, or enter an option.)*

4. AutoCAD repeatedly prompts you to select additional objects to trim. When done trimming, press **ENTER**:

   Select object to trim or shift-select to extend or

   [Fence/Crossing/Project/Edge/eRase/Undo]: *(Press ENTER to exit.)*

When you select objects that cannot serve as cutting edges, such as blocks and dimensions, AutoCAD displays the message:

   No edges selected.

If an object cannot be trimmed, AutoCAD displays:

   Cannot TRIM this object

If the object to be trimmed does not intersect a cutting edge, AutoCAD complains:

   Object does not intersect an edge.

### Trimming Closed Objects

To be trimmed, closed objects, such as circles and polygons, must be intersected at least twice by the cutting edge, such as by a line drawn through two points on a circle's circumference. If only one cutting edge intersects the closed object, AutoCAD complains:

   Circle must intersect twice.

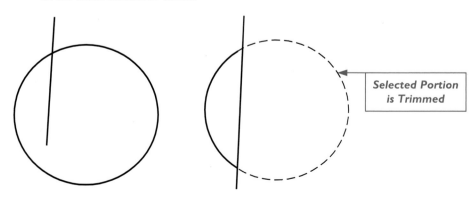

## Trimming Polylines

Polylines are trimmed at the intersection of the center line of the polyline and the cutting edge. The trim is a square edge. Therefore, if the cutting edge intersects a polyline of nonzero width at an angle, the square-edged end may protrude beyond the cutting edge.

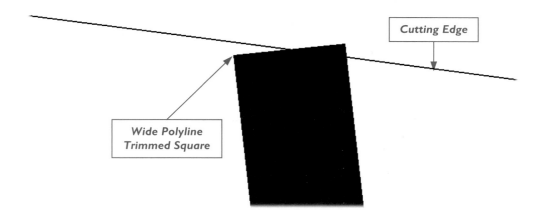

 **Notes** The **TRIM** and **EXTEND** commands can switch roles:

To *extend* an object during the **TRIM** command, hold down the **SHIFT** key while selecting the object. Similarly, to *trim* an object during the **EXTEND** command, hold down the **SHIFT** key while selecting the object.

Hatch patterns can be cutting edges or the objects being trimmed. In other words, the **TRIM** command can trim hatch and fill patterns, and use them to trim other objects. (The **EXTEND** and **LENGTHEN** commands have no effect on hatches.)

### TRIMMING OBJECTS: ADDITIONAL METHODS

The **TRIM** command has several options:

- **Select all** selects all objects as cutting edges.
- **Fence** and **Crossing** select objects to be trimmed.
- **Project** determines trimming in 3D space.
- **Edge** toggles actual and implied cutting edges in 3D space.
- **eRase** erases objects.
- **Undo** undoes the last trim.
- **SHIFT-select** extends the object to the cutting line, instead of trimming.

Let's look at each option.

### Select All

The **Select all** option selects all objects in the drawing as cutting edges — at least those not on frozen layers. When you press **ENTER** at the 'Select objects or <select all>:' prompt, AutoCAD selects all objects, and then stops repeating the 'Select objects:' prompt.

### Fence and Crossing

The **Fence** and **Crossing** options select objects to be trimmed. Their inclusion in the prompt may be puzzling to you: can't any selection mode be used? No. For some peculiar reason, the original **TRIM** command allowed you only to pick objects singly for trimming.

The **Fence** option prompts:

> Specify first fence point: *(Pick a point.)*
>
> Specify next fence point or [Undo]: *(Pick another point.)*

And continues repeating the prompt until you press ENTER. You can back up along the fence path by entering the **Undo** option.

Using the **Crossing** option can be ambiguous; Autodesk notes that AutoCAD follows the rectangular crossing window clockwise from the first pick point. The **Crossing** option prompts:

> Specify first corner: *(Pick a point.)*
>
> Specify opposite corner: *(Pick another point.)*

The crossing window must cross objects; otherwise nothing is trimmed. When you pick an object at the 'Specify first corner:' prompt, AutoCAD abandons **Crossing** mode, and selects the object instead.

### Project

The **Project** option is used in 3D drafting to specify the projection AutoCAD uses when it trims objects.

> Enter a projection option [None/Ucs/View]: *(Enter an option.)*

The sub-options are as follows:

> **None** — specifies no projection; AutoCAD trims only objects that actually intersect with cutting edges.
>
> **UCS** — projects the objects onto the x, y-plane of the current UCS (user-defined coordinate system). AutoCAD trims objects, even if they do not intersect with the cutting edge in 3D space—in effect, flattening 3D objects onto the 2D plane.
>
> **View** — specifies a projection along the current view direction. AutoCAD trims objects that look as if they should be trimmed from your viewpoint, even if they don't physically intersect.

### Edge

The **Edge** option is also used in 3D drafting, and determines whether objects are trimmed at an actual cutting edge (objects crossing physically), or at an implied cutting edge (objects appearing to cross):

> Enter an implied edge extension mode [Extend/No extend]: *(Enter an option.)*

The options are as follows:

> **Extend** — projects the cutting edge so that it intersects the object(s).
>
> **No extend** — trims objects at cutting edges that physically intersect; cutting edges are not extended.

### eRase

The **eRase** option erases objects, just like the ERASE command itself. It is a convenient way of removing whole objects that don't require trimming.

### Undo

The **Undo** option undoes the last trim. This allows you to undo trim errors without leaving the command.

### Extend (SHIFT-select)

When you hold down the SHIFT key at the "Select object to trim or shift-select to extend" prompt, AutoCAD extends the object to the cutting line, instead of trimming — in effect, reversing the TRIM command. The complementary option is available in the EXTEND command.

## EXTEND

The EXTEND command extends objects in a drawing to meet a boundary object. It functions very much like the TRIM command. Instead of a cutting edge, this command uses boundary objects — the boundary to which selected objects are extended.

AutoCAD needs to know two things: (1) the object(s) to be used as boundary objects, and (2) the object to extend.

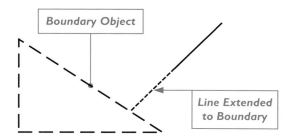

Boundary objects can be lines, arcs, circles, polylines, ellipses, splines, rays, xlines, regions, blocks, and viewports (in paper space). If several boundary edges are selected, the objects are extended to the first boundary encountered. If none can be reached, AutoCAD complains:

> No edges selected.

Generally, closed objects cannot be extended, specifically text, splines, 3D solids, xlines, mutlilines, ellipses, donuts, hatches, revclouds, regions, 3D surfaces, and points. (Dimensions can be extended.) If the object cannot be extended, AutoCAD displays:

> Cannot EXTEND this object.

Take care! You must point close to the end of the object to extend.

### TUTORIAL: EXTENDING OBJECTS

1. To extend one or more objects, start the **EXTEND** command:
   - In the ribbons's 2D Home tab, pick the **Extend** button in the Modify panel.
   - At the 'Command:' prompt, enter the **extend** command:

     Command: **extend** *(Press ENTER.)*

   - Alternatively, enter the **ex** alias at the 'Command:' prompt.

2. In all cases, AutoCAD reports the current extend settings, and then prompts you to select boundary edges:

   Current settings: Projection=UCS Edge=None

   Select boundary edges ...

   Select objects or <select all>: *(Pick one or more objects as boundaries.)*

   Select objects: *(Press ENTER to end object selection.)*

3. Select the object to be extended. AutoCAD extends the end that you select.

   Select object to extend or shift-select to trim or [Fence/Crossing/Project/Edge/Undo]: *(Pick objects to be extended.)*

4. AutoCAD repeatedly prompts you to select additional objects to extend. When done extending, press **ENTER**:

   Select object to extend or shift-select to trim or [Fence/Crossing/Project/Edge/Undo]: *(Press ENTER to exit command.)*

### Extending Polylines

Polylines of nonzero width are extended until the centerline meets the boundary object; similarly, objects extend to the centerline of the polylines. Only open polylines can be extended. If you attempt to extend a closed polyline, AutoCAD complains:

> Cannot extend a closed polyline.

When wide polylines and the boundary intersect at an angle, a portion of the square end of the polyline may protrude over the boundary. Extending tapered polylines adjusts the length of the segments that taper — the taper extends over the longer length.

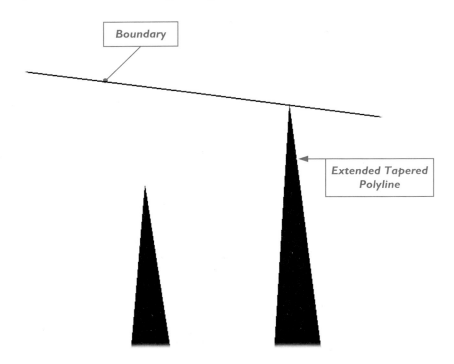

*A tapered polyline extended to a boundary, its taper extended as well.*

### EXTEND OPTIONS

The **EXTEND** command's options are identical to those of **TRIM**'s.

### LENGTHEN

The **LENGTHEN** command is a faster version of **TRIM** and **EXTEND**: it changes the length of open objects, making them longer or shorter; it does not work with closed objects. **LENGTHEN** is faster, because you do not need cutting edges or boundaries: you just point anywhere in the drawing for the change to occur, or else specify a percentage change numerically.

This command works with lines, arcs, open polylines, elliptical arcs, and open splines.

#### TUTORIAL: LENGTHENING OBJECTS

1. To change the length of an open object, start the **LENGTHEN** command:
   - In the ribbons's 2D Home tab, pick the **Lengthen** button in the Modify panel.
   - At the 'Command:' prompt, enter the **lengthen** command:

     Command: **lengthen** *(Press* ENTER.*)*

   - Alternatively, enter the **len** alias at the 'Command:' prompt.

2. Notice that AutoCAD prompts you to select an object. AutoCAD reports the length of the object:

   Select an object or [DElta/Percent/Total/DYnamic]: *(Select one open object.)*
   Current length: 24.6278

3. AutoCAD repeats the prompt. It wants you to select an option:

   Select an object or [DElta/Percent/Total/DYnamic]: *(Enter an option, such as **dy**.)*

4. Curiously, AutoCAD has forgotten which object you selected; you need to select it a second time — or, you can select a different object:

   Select an object to change or [Undo]: *(Select one open object.)*

5. As you move the cursor, notice that AutoCAD ghosts in the lengthened line. When satisfied with the new length, click.

   Specify new end point: *(Move the cursor, and then click.)*

6. AutoCAD repeats the prompt. When done lengthening, press **ENTER**:

   Select an object to change or [Undo]: *(Press **ENTER** to exit the command.)*

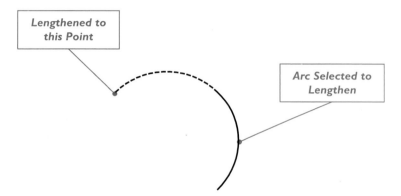

**Note** When you select an object at the initial "Select an object:" prompt, AutoCAD reports on its length, as follows: *Current length: 10.2580*. For an arc, AutoCAD also reports its angle:

Current length: 8.4353, included angle: 192.

When you select a closed object, such as a circle or closed polyline, AutoCAD complains, "This object has no length definition." You cannot lengthen dimensions, hatch patterns, and splines.

### LENGTHENING OBJECTS: ADDITIONAL METHODS

The LENGTHEN command changes the length of open objects by four methods:

- **Delta** — changes by an incremental amount.
- **Percent** — changes by a percentage.
- **Total** — changes to the total amount.
- **DYnamic** — changes by cursor movement.

Let's look at each option.

### Delta

The **DElta** option changes the length by adding the indicated amount to the object. Enter a negative number to shorten the object. As an alternative to typing a value, you can indicate the amount by picking two points anywhere in the drawing.

> Command: **lengthen**
> Select an object or [DElta/Percent/Total/DYnamic]: **de**
> Enter delta length or [Angle] <0.0000>: *(Enter a number, or pick two points on screen.)*
> Select an object to change or [Undo]: *(Select object.)*
> Select an object to change or [Undo]: *(Press ESC to exit the command.)*

Note that this command lengthens the open object at the end you select.

The **Angle** option changes the angle of a selected arc or polyarc. The command lengthens the end you select:

> Enter delta angle <0>: *(Type an angle, or pick two points.)*
> Select an object to change or [Undo]: *(Select the arc.)*
> Select an object to change or [Undo]: *(Press ESC to exit the command.)*

The arc is lengthened by the angle you specify. When you select two points, AutoCAD uses the angle of the rubber-band line, not the length of the line.

The angle you specify must be small enough for the arc not to total 360 degrees. If the angle is too large, AutoCAD curtly informs you: "Invalid angle."

### Percent

The **Percent** option changes the length by a percentage of the object. For example, entering **25** shortens a line to 25 percent of its original length, while entering **200** doubles its length.

> Select an object or [DElta/Percent/Total/DYnamic]: **p**
> Enter percentage length <100.0000>: *(Enter a percentage, such as **25**.)*
> Select an object to change or [Undo]: *(Select the object.)*
> Select an object to change or [Undo]: *(Press ESC to exit the command.)*

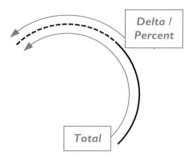

### Total

The **Total** option changes the length of a line by an absolute length. For example, a value of **5** changes the line to a length of 5.0 units, no matter its existing length.

> Select an object or [DElta/Percent/Total/DYnamic]: **t**
> Specify total length or [Angle] <1.0000>: *(Enter a value, such as **5**.)*
> Select an object to change or [Undo]: *(Select the object.)*
> Select an object to change or [Undo]: *(Press ESC to exit the command.)*

This option changes arcs to an absolute length. The **Angle** option changes the length of an arc to the included specified angle.

### Dynamic

The **DYnamic** option visually changes the length of open objects. Notice that this option reverses the order: (1) first you select the object to lengthen, and (2) then you specify the length.

>Select an object or [DElta/Percent/Total/DYnamic]: **dy**
>
>Select an object to change or [Undo]: *(Select the object.)*
>
>Specify new end point: *(Pick a point.)*
>
>Select an object to change or [Undo]: *(Press ESC to exit the command.)*

After you select the object, the length of the object changes as you move the cursor. You can use object snap modes to make the dynamic lengthening more accurate.

## STRETCH

The **STRETCH** command makes portions of objects larger and smaller, and moves selected objects, while retaining their connections to other objects. As you will see, it is one of the most useful editing commands in AutoCAD — a distant second only to the **UNDO** command. The command works with lines, arcs, elliptical arcs, solids, traces, rays, splines, and polylines.

### STRETCHING — THE RULES

You must understand the rules associated with the **STRETCH** command to execute it properly:

*Rule 1: Crossing Selection.* You can use any object selection mode, but at least one must be **Crossing** or **CPolygon**. If you do not use a crossing selection, AutoCAD complains, "You must select a crossing or polygon window to stretch." The command is no longer restricted to using the last window specified; instead, **STRETCH** now remembers all the objects selected by multiple **Crossing** selection rectangles. The window must include at least one vertex or endpoint of the object.

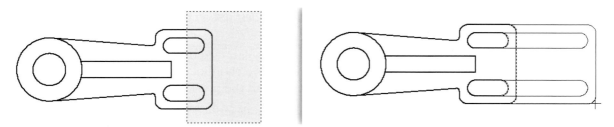

*Objects crossing the window being stretched.*

*Rule 2: Move vs. Stretch.* If the object is completely inside the selection, it will be moved rather than stretched. Objects — specifically arcs, elliptical arcs, lines, polyline segments, 2D solids, rays, traces, and splines — entirely within the selection window are moved as with the **MOVE** command. The endpoints within the selection window are moved, while those outside remain fixed.

Arcs are stretched like lines, except that the arc's center, start, and endpoints are adjusted so the distance from the midpoint of the chord to the arc is constant. You can stretch either end of the polyline, and move its vertices. Spline-curve and fit-curve polylines can be stretched if the crossing window includes the original vertex (even though it is invisible). When tangent-arc polyline segments are stretched, they lose tangency.

This command does not stretch text. (The technical editor suggests, "Use TEXT to place 'The Truth' in the drawing, and then try to stretch it. You cannot. This proves that you cannot stretch the truth.")

*Rule 3: Definition Points.* Some objects, such as circles and blocks, cannot be stretched; they are either moved or left alone by the **STRETCH** command, depending on their definition point. If this point lies inside the selection window, the object is moved; if outside, it is not affected.

| Object | Definition Point |
|---|---|
| Point | Center of the point. |
| Circle | Center point of the circle. |
| Block | Insertion point. |
| Text | Insertion point of the text line. |

### TUTORIAL: STRETCHING OBJECTS

1. To stretch one or more objects, start the **STRETCH** command:
   - In the ribbons's 2D Home tab, pick the **Stretch** button in the Modify panel.
   - At the 'Command:' prompt, enter the **stretch** command:

     Command: **stretch**
   - Alternatively, enter the **s** alias at the 'Command:' prompt.

2. In all cases, AutoCAD prompts you to select the objects using a crossing selection mode.

   If you do not enter a selection option, such as **Crossing**, AutoCAD defaults to **AUtomatic**.

   Select objects to stretch by crossing-window or crossing-polygon...

   Select objects: **c**

   Specify first corner: *(Pick a point.)*

   Specify opposite corner: *(Pick another point.)*

   Select objects: *(Press ENTER to end object selection.)*

3. Identify the points that measure the displacement:

   Specify base point or Displacement <Displacement>: *(Pick a point, using an object snap mode if necessary.)*

   Specify second point of displacement: *(Pick the destination point.)*

For the **Displacement** option, you can enter Cartesian, polar, cylindrical, or spherical coordinates — without the @ prefix, because AutoCAD assumes relative coordinates. The next time you use the STRETCH command, enter the **Displacement** option to recall the previous displacement distance.

An alternative to the STRETCH command is grips editing, as described later.

**MOVE**

The MOVE command moves objects in the drawing.

It is similar to the COPY command in that AutoCAD needs to know two things: (1) the objects to move, and (2) the displacement — the distance to move. (See Chapter 8, "Constructing Objects.")

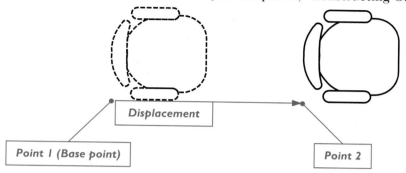

### TUTORIAL: MOVING OBJECTS

1. To move one or more objects, start the MOVE command:
    - In the ribbons's 2D Home tab, pick the **Move** button in the Modify panel.
    - At the 'Command:' prompt, enter the **move** command:

        Command: **move** *(Press ENTER.)*
    - Alternatively, enter the **m** alias at the 'Command:' prompt.

2. In all cases, AutoCAD prompts you to select the objects to move:

    Select objects: *(Select one or more objects.)*

    Select objects: *(Press ENTER to end object selection.)*

3. Identify the point from which the displacement is measured:

    Specify base point or Displacement <Displacement>: *(Pick point 1.)*

4. And pick the second point:

    Specify second point of displacement or <use first point as displacement>: *(Pick point 1.)*

The first point need not be on the object to be moved; using a corner point or another convenient point of reference on the object makes the displacement easier to visualize. I find that object snap and ortho mode help make the move more precise.

The next time you use the MOVE command, select the **Displacement** option to recall the previous displacement distance.

An alternative to the MOVE command is grips editing, as described in Chapter 10. The GTDEFAULT system variable determines whether the MOVE or MOVE3D command is activated when entering "move" in 3D views; the same holds for ROTATE and ROTATE3D.

### Command-less Moving

You can move objects in AutoCAD without using any commands. Here's how:

1. Select an object. Notice the blue grips.
2. Grab the object away from the grips by holding down the **left** mouse button.
3. Wait.

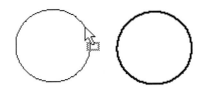

4. After a second or two, the move cursor appears. You can now drag the object in the drawing, effectively moving it.

Alternatively, you can move objects by dragging specific grips, such as the center grip on circles arcs, and the insertion point of blocks and text.

 ROTATE

The ROTATE command rotates objects about a base point. (The base point is the point about which selected objects rotate.)

You can specify the rotation angle three ways: by entering an angle, by dragging the angle, or by choosing a reference angle. (The similar sounding REVOLVE command creates 3D solid models.)

### TUTORIAL: ROTATING OBJECTS

1. To turn one or more objects, start the ROTATE command:
   * In the ribbons's 2D Home tab, pick the **Rotate** button in the Modify panel.
   * At the 'Command:' prompt, enter the **rotate** command:

     Command: **rotate** *(Press ENTER.)*
   * Alternatively, enter the **ro** alias at the 'Command:' prompt.

2. In all cases, AutoCAD indicates the current rotation settings, and then prompts you to select the objects you want rotated:

   Current positive angle in UCS:  ANGDIR=counterclockwise  ANGBASE=0

   Select objects: *(Pick one or more objects.)*

   Select objects: *(Press ENTER to end object selection.)*

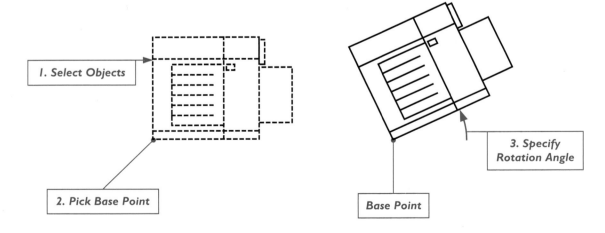

3. Identify the point about which to rotate the objects:

    Specify base point: *(Pick a point, such as the lower left corner.)*

    The base point need not be on the object, but can be anywhere in the drawing.

4. Specify the angle of rotation:

    Specify rotation angle or [Copy/Reference] <90>: *(Enter an angle, such as **45**.)*

You can specify a simple angle, or change one angle to another. For example, if an object is currently oriented at 58 degrees, and you wish to rotate the object to 26 degrees, rotate the object by the difference of –32 degrees. Positive angles rotate counterclockwise; negative angles, clockwise.

### ROTATING OBJECTS: ADDITIONAL METHODS

The ROTATE command has alternative options for specifying the angle:

- **Reference** option rotates relative to another angle.
- **Copy** option rotates a copy of the object.
- **Dragging** option rotates the object in real-time.

Let's look at each.

### Reference

The **Reference** option aligns the object to another. You don't need to know the angle of the other object, but object snaps are most useful. In this example, we want Desk B to be at the same slant as Desk A.

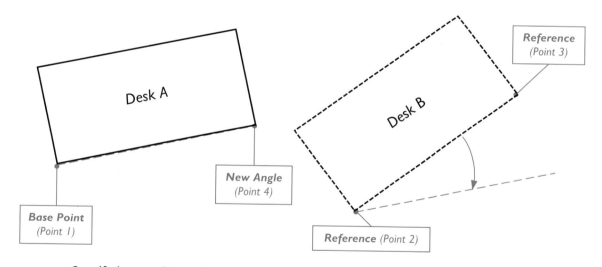

Specify base point: **end**

of *(Pick point 1.)*

Specify rotation angle or [Reference]: **r**

Specify the reference angle <0>: **end**

of *(Pick point 2.)*

Second point: : **end**

of *(Pick point 3.)*

Specify the new angle: **end**

of *(Pick point 4.)*

## Copy

The **Copy** option rotates a copy of the object, leaving the original in place. When you enter **c** at the 'Specify rotation angle or [Copy/Reference] <90>:' prompt, AutoCAD reports:

Rotating a copy of the selected objects.

The option is temporary. The next time you use the ROTATE command, you have to reenter the **Copy** option.

## Dragging

To rotate the objects in real time, move the cursor in response to the "Specify rotation angle or [Reference]:" prompt. Keep an eye on the status line as it reports the x, y, z coordinates and angle. Click to fix the object at the location shown on the screen.

## SCALE

The SCALE command changes the size of objects. Contrary to what the name suggests, this command does not change the scale of objects in the drawing — but their size. The x, y, and z directions of the objects are changed equally.

AutoCAD needs to know three things: (1) the objects to be resized, (2) the base point from which the objects are resized, and (3) the amount to resize.

### TUTORIAL: CHANGING SIZE

1. To resize one or more objects, start the **SCALE** command:
   - In the ribbons's 2D Home tab, pick the **Scale** button in the Modify panel.
   - At the 'Command:' prompt, enter the **scale** command:

     Command: **scale** *(Press ENTER.)*

   - Alternatively, enter the **sc** alias at the 'Command:' prompt.

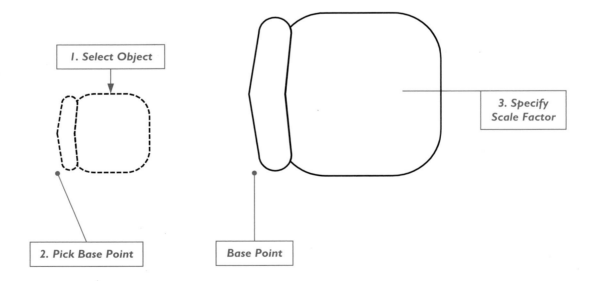

2. In all cases, AutoCAD prompts you to select the objects you want resized:
   Select objects: *(Pick one or more objects.)*
   Select objects: *(Press ENTER to end object selection.)*

3. Identify the point about which the objects are rotated:

    Specify base point: *(Pick a point, such as the lower left corner.)*

    The base point can be located anywhere in the drawing. This point remains stationary, and the object is resized from that point.

4. Specify the resize factor:

    Specify scale factor or [Copy/Reference] <1.0000>: *(Enter a factor, such as **2** or **0.5**.)*

Entering a decimal factor makes objects smaller: a factor of 0.25 makes an object 25 percent of the original size. Factors larger than 1.0 increase the size of the objects. For example, entering 2.0 makes the object twice the original size in the direction of all three axes.

Instead of entering the scale factor, you can specify the size by dragging the cursor or by referencing a known length, and then entering a new length.

### CHANGING SIZE: ADDITIONAL METHODS

The SCALE command has two options:

- **Copy** rotates a copy of the objects.
- **Reference** changes the size relative to other distances.

Let's look at them.

### Copy

The **Copy** option scales a copy of the objects, leaving the original in place. When you enter **c** at the 'Specify scale factor or [Copy/Reference] <1.0>:' prompt, AutoCAD reports:

    Scaling a copy of the selected objects.

The option is temporary. The next time you use the SCALE command, you have to reenter the **Copy** option.

### Reference

The **Reference** option adjusts objects to a "correct" size. This is particularly useful when you scan a drawing, and then bring the raster image into AutoCAD with the IMAGE command. To trace over the scanned image, you first have to change the image to the right size: you achieve this with the **Reference** option. All you need is the true length of just one line in the scanned image.

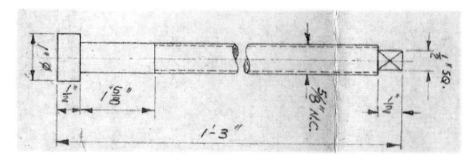

1. Use the **IMAGEATTACH** command to place the *part6.tif* raster file in a new drawing. Use the following parameters:

    | | |
    |---|---|
    | Insertion point | **0,0** |
    | Scale | **1.0** |
    | Rotation | **0** |

    If necessary, use the **ZOOM Extents** command to see the image.

2. Start the SCALE command, and then select the raster image by clicking on the black rectangle that surrounds it:

   Command: **scale**

   Select objects: *(Pick the edge of the raster image.)*

   Select objects: *(Press ENTER to end object selection.)*

3. For the base point, select the lower left corner of the raster image. Since this coincides with its insertion point, you can enter 0,0:

   Specify base point: **0,0**

4. At the next prompt, specify the **Reference** option:

   Specify scale factor or [Copy/Reference] <1.0>: **r**

5. Select the end of the longest known length in the raster image, such as the 1'-3" dimension:

   Specify reference length <1>: *(Pick one end of the raster line.)*

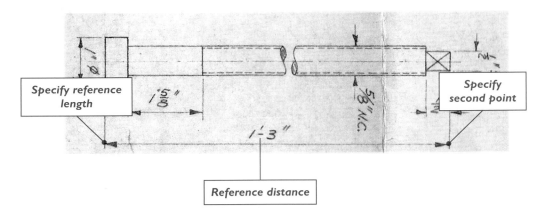

6. Select the other end of the raster line; this allows AutoCAD to measure the distance of the line:

   Specify second point: *(Pick other end of raster line.)*

7. For the new length, enter the value of the known dimension, 15" in this case.

   Specify new length: **15**

Notice that AutoCAD resizes the scanned image, making it the correct size.

If the image is skewed, you can use the ROTATE command in the same way, employing its **Reference** option to specify that a line be 0 degrees. If the scanned image is warped, you need to correct it with "rubber sheeting," a feature not available in AutoCAD.

The technical editor provides this workaround for changing the aspect ratio of images: convert it to a block, and then insert it with differing x and y scale factors; explode the block, and the image maintains its new aspect ratio. He's also found that IntelliCAD-based CAD systems allow unequal scaling of images, which AutoCAD honors when opening the *.dwg* file.

The SCALE command is also useful for converting an Imperial drawing to metric, and vice versa.

## CHANGE

The CHANGE command changes the size, properties, and other characteristic of objects. It is similar to, but has more capabilities than, CHPROP, but PROPERTIES is the easiest and most versatile of all. (See Chapter 7.)

### TUTORIAL: CHANGING OBJECTS

1. To change one or more objects, start the **CHANGE** command:
   - At the 'Command:' prompt, enter the **change** command:

     Command: **change** *(Press ENTER.)*

   - Alternatively, enter the **-ch** alias at the 'Command:' prompt.

2. Notice that AutoCAD prompts you to select the complex objects you want to change:

   Select objects: *(Pick one or more objects.)*

   Select objects: *(Press ENTER to end object selection.)*

3. The **Properties** options are somewhat different from those of the CHPROP command:

   Specify change point or [Properties]: *(Pick a point.)*

The CHANGE command's **Elev** option changes the elevation of objects, an option missing from CHPROP.

Although the Properties palette is generally handier, CHANGE offers this advantage: the **Change Point** option reacts differently depending on the object selected, as described next.

### CHANGING OBJECTS: ADDITIONAL METHODS

Here's how the CHANGE command's **Change Point** option affects different objects:

- **Lines** change endpoints.
- **Circles** change radii.
- **Blocks** change insertion points and rotation angles.
- **Text** changes properties.
- **Attribute text** changes properties.

### Lines

The **Change Point** option moves the endpoints of lines closest to the pick point.

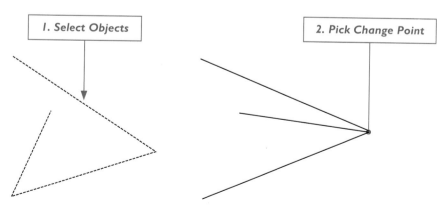

When ortho mode is on, AutoCAD makes the lines parallel to either the x or y axis, depending on which is closest. This is a quick way to straighten out a bunch of crooked lines, and is not available in any other AutoCAD command.

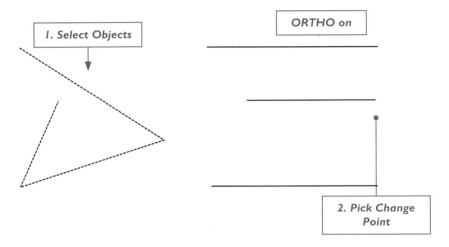

### Circles

The **Change Point** option changes the circle's radius. When more than one circle is selected, AutoCAD repeats the prompt for each circle:

Specify new circle radius <no change>: *(Enter a new radius, or press* ENTER.*)*

Press ENTER to keep the circle's radius.

### Blocks

The **Change Point** option changes the location and rotation of blocks.

Specify new block insertion point: *(Pick a new insertion point, or press* ENTER.*)*

Specify new block rotation angle <current>: *(Enter a new angle, or press* ENTER.*)*

Press ENTER to keep each option in place.

### Text

The **Change Point** option changes the position of the text, as well as the text's other properties. (Until the PROPERTIES command's predecessor, DDMODIFY, came along, this was the only way to change text properties in a drawing.)

Specify new text insertion point <no change>: *(Pick a new insertion point, or press* ENTER.*)*

Enter new text style <current>: *(Enter a different text style, or press* ENTER.*)*

Specify new height <current>: *(Enter a new height, or press* ENTER.*)*

Specify new rotation angle <current>: *(Enter a new angle, or press* ENTER.*)*

Enter new text <current>: *(Enter a new line of text, or press* ENTER.*)*

Press ENTER to keep each option as is. If you select more than one line of text, AutoCAD repeats the prompts for the next one.

### Attribute Definitions

The **Change Point** option changes the text and properties of attributes that are not part of a block.

Specify new text insertion point: *(Pick a new insertion point, or press* ENTER.*)*

Enter new text style <current>: *(Enter a different text style, or press* ENTER.*)*

Specify new height <current>: *(Enter a new height, or press* ENTER.*)*

Specify new rotation angle <current>: *(Enter a new angle, or press ENTER.)*
Enter new text <current>: *(Enter new text, or press ENTER.)*
Enter new tag <current>: *(Enter a new tag, or press ENTER.)*
Enter new prompt <current>: *(Enter a new prompt, or press ENTER.)*
Enter new default value <current>: *(Enter a new default value, or press ENTER.)*

Press ENTER to keep each option in place.

## EXPLODE AND XPLODE

The **EXPLODE** and **XPLODE** commands reduce compound objects to their simplest forms.

The commands "break down" blocks, polylines, and other compound objects into basic lines, arcs, and text. Some objects must be exploded several times before they are finally reduced to vector primitives.

### TUTORIAL: EXPLODING COMPOUND OBJECTS

1. To explode one or more compound objects, start the **EXPLODE** command:
   - In the ribbons's 2D Home tab, pick the **Explode** button in the Modify panel.
   - At the 'Command:' prompt, enter the **explode** command:

     Command: **explode** *(Press ENTER.)*

   - Alternatively, enter the **x** alias at the 'Command:' prompt.

2. In all cases, AutoCAD prompts you to select the compound objects you want to explode:
   Select objects: *(Pick one or more objects.)*
   Select objects: *(Press ENTER to end object selection.)*

The objects are exploded, but you might not see any difference. Different compound objects react differently to being exploded.

**Arcs** do not explode, unless they are part of a nonuniformly scaled block. When the block is exploded, arcs become elliptical arcs.

**Circles** do not explode, unless they are part of a nonuniformly scaled block. When the block is exploded, circles becomes ellipses.

**Blocks** explode into the lines and other objects originally used to define them. Nested blocks must be exploded a second time. Attribute text is deleted; attribute definitions are redisplayed.

**Xrefs** and **blocks** inserted with the **MINSERT** command cannot be exploded.

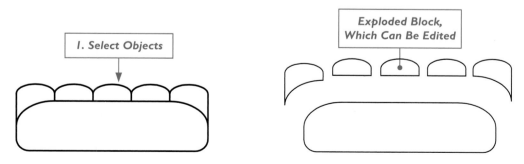

**Left:** *Block before being exploded.*
**Right:** *After exploding, with parts pulled apart for clarity.*

**Polylines** explode into lines and arcs. Width and tangent information is discarded; the resulting lines and arcs follow the center line of the old polyline. If the exploded polyline has segments of width, AutoCAD displays the message:

> Exploding this polyline has lost width information.
> The UNDO command will restore it.

The new lines and arcs are placed on the same layer as the polyline, and inherit the same color.

**Left:** *Polyline with variable width before being exploded.*
**Right:** *Polyline after being exploded, with lines and arcs pulled apart for clarity.*

**Three-dimensional polylines** explode into line segments. Linetypes are retained.

**Associative dimensions** explode into basic objects (lines, polyline arrowheads, and text), which are placed on the same layer as the original dimensions, and inherit that layer's properties.

**Constraints** explode into nothingness: when constrained objects are exploded, the constraints are erased.

**Leaders** explode into lines and multiline text; the arrowheads explode into 2D solids. Depending on how the leader was constructed, resulting objects could also include splines, block inserts of arrowheads and annotation blocks, and tolerance objects. More than one explode may be required.

**Multiline text** explodes into single-line text.

**Multilines** explode into lines and arcs.

**Single-vertex polyface meshes** explode into point objects. Two-vertex meshes explode into lines. Three-vertex meshes, into 3D faces.

**Tables** explode into lines and mtext.

**3D solids** explode into bodies, which can be further exploded into regions. Regions, in turn, explode into lines, arcs, and circles.

**Legacy mesh objects** explode into 3D faces; **3D mesh objects** also explode into 3D faces.

**Surfaces** explode into lines, arcs, circles, and polylines.

 **Note** When a block is inserted with different scale factors in the x, y, and/or z directions, it is called a "nonuniformly scaled block." When exploded, the results may be different from what you expect. Arcs, for example, are converted to elliptical arcs to maintain their distortion.

Sometimes, nonuniformly scaled blocks contain objects that cannot be exploded, such as bodies, 3D solids, and regions. In that case, AutoCAD places them in an *anonymous* block with the **\*E** prefix in the block's name.

## XPLODE

The XPLODE command controls what happens when compound objects are exploded.

### TUTORIAL: CONTROLLED EXPLOSIONS

1. To explode compound objects with control, start the **XPLODE** command:
   - At the 'Command:' prompt, enter the **xplode** command.

     Command: **xplode** *(Press ENTER.)*

2. AutoCAD prompts you to select the compound objects you want to explode:
   Select objects to XPlode.

   Select objects: *(Pick one or more objects.)*

   Select objects: *(Press ENTER to end object selection.)*

3. Enter an option:
   Enter an option

   [All/Color/LAyer/LType/LWeight/Inherit from parent block/Explode] <Explode>: *(Enter an option.)*

4. If you had selected more than one object, an additional prompt appears:
   Enter an option [Individually/Globally] <Globally>: *(Enter I or G.)*

The **Global** option applies the previous options (color, layer, and so on) to all selected objects. The **Individually** option applies the previous options to each object, one at a time.

### All

The **All** option prompts you to specify the color, linetype, lineweight, and layer on which the exploded objects should land. The prompts are the same as those for the four following options:

### Color

The **Color** option sets the color of exploded objects:

[Red/Yellow/Green/Cyan/Blue/Magenta/White/BYLayer/BYBlock/Truecolor/Colorbook] <BYLAYER>: *(Enter a color name or number.)*

### Layer

The **Layer** option sets the layer of the exploded objects:

Enter new layer name for exploded objects <current>: *(Enter the name of an existing layer.)*

### LType

The **LType** option sets the linetype of the exploded objects:

Enter new linetype name for exploded objects <BYLAYER>: *(Enter the name of a linetype.)*

### Inherit from Parent Block

The **Inherit from parent block** option sets the color, linetype, lineweight, and layer of the exploded objects to that of the originating block.

### Explode

The **Explode** option reduces compound objects in the same manner as the EXPLODE command.

**PEDIT**

The PEDIT command (short for "polyline edit") edits polylines.

Recall that the DONUT, RECTANGLE, and POLYGON commands create 2D polylines, so this command edits them. This command also edits 3D polylines drawn with the 3DPOLY command, and 3D polyfaces drawn with AutoCAD's 3D surfacing commands. As of AutoCAD 2011, polylines and splines can be converted back and forth.

While this command is useful for certain methods of polyline editing, you may find it easier to edit them directly in other cases. Both methods are discussed in this section. As a reminder, here are the parts of a polyline:

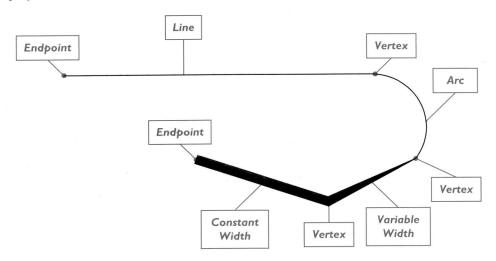

### TUTORIAL: STARTING THE PEDIT COMMAND

1. To edit one or more polylines, start the PEDIT command:
   - In the ribbon's 2D Home tab, pick the **Edit Polyline** button in the Modify panel.
   - At the 'Command:' prompt, enter the **pedit** command:
   
     Command: **pedit** *(Press ENTER.)*
   - Alternatively, enter the **pe** alias at the 'Command:' prompt.

2. Notice that AutoCAD prompts you to select the polyline you want edited:
   Select polyline or [Multiple]: *(Pick a polyline.)*

3. Many options are listed:
   Enter an option [Close/Join/Width/Edit vertex/Fit/Spline/Decurve/Ltype gen/Reverse/Undo]:

Of the options listed by the prompt, only some are usefully employed through the PEDIT command. These are listed below, and then discussed in greater detail on the following pages. Other options, especially **Edit vertex**, are better done through direct editing. In some cases, you might even prefer to use the Properties palette to edit polylines.

| PEdit Option | Alternative | Comment |
| --- | --- | --- |
| Close or Open | Direct editing | The Close and Open options add and remove the closing segment only; direct editing removes *any* segment. |
| Join | JOIN | JOIN command is easier to use. |
| Width | PROPERTIES | The Width option applies to the entire polyline; the Properties palette applies variable widths to segments. |
| Edit vertex | Direct editing | The Edit vertex option is painful to use. |
| Fit | ... | Coverts polyline segments to arcs. |
| Spline | ... | Converts polylines to splines. |
| Decurve | ... | Converts curved segments to lines. |
| Ltype gen | ... | Ensures linetypes are continuous through vertices. |
| Reverse | REVERSE | REVERSE command is easier to use. |
| Undo | UNDO | The Undo option applies to the last action inside the PEDIT command; UNDO command applies to everything. |

In the following sections, I'll describe almost all of the PEDIT command's options, leaving out **Edit vertex**, because it is more easily handled by direct editing.

## Close/Open

The **Close** option closes open polylines. It connects the last point to the first point of the polyline. This is like the PLINE command's **Close** option, except it allows you to close the polyline afterwards.

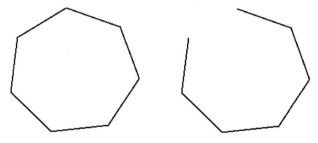

**Left:** *Closed polyline (heptagon).*
**Right:** *Open polyline.*

When the polyline is closed, the **Open** option replaces **Close**. It opens closed polylines by removing the last segment added by the **Close** option, as well as by other commands, such as POLYGON and the PLINE command's **Close** option.

*Direct editing option:* Hold down the CTRL key, and then select the segment you wish to remove. Press the DEL key.

## Join

The **Join** option converts and connects polylines and non-polyline objects (lines and arcs only) to the original polyline. AutoCAD prompts:

Select objects: *(Select one or more lines, arcs, and polylines.)*

The objects you select become part of the original polyline; if the polyline was curve-fitted, it is first decurved.

AutoCAD determines which arcs and lines share common endpoints with the original polyline, and then merges them into that polyline. To join successfully, the objects must touch, or be within a fuzz factor distance, as described later in this chapter. If endpoints are too far away to join, use the FILLET or CHANGE command to extend them to a perfect match.

*Alternative commands:* Consider the JOIN command, and see the **Multiple** and **Jointype** options discussed later in this chapter.

### Width

The **Width** option changes the width of the entire polyline. The width is applied uniformly to all segments of the polyline, unlike the PLINE command's **Width** variable option.

**Left**: Polyline with varying widths.
**Right**: Polyline after applying the **Width** option.

AutoCAD prompts you:

Specify new width for all segments: *(Enter a value.)*

You can type the width at the keyboard, or else pick two points in the drawing.

*Alternative command:* The **PROPERTIES** command applies constant or variable widths to the entire polyline or selected segments, as illustrated by the palette below.

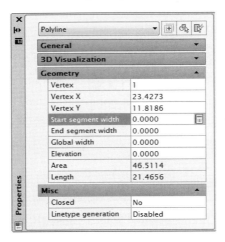

To adjust a single segment, you cannot hold down the CTRL key; instead, you need to click the spinner buttons in the **Vertex** field, as shown below. AutoCAD marks the current vertex with an **X**.

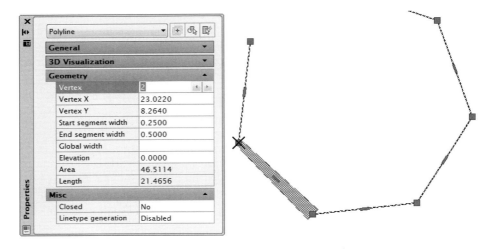

### Edit vertex

The suboptions of the **Edit vertex** option edit the vertices of a polyline. (Recall that the vertex is the connection between polyline segments, as well as between the two endpoints of open polylines.) The name, "edit vertex," is somewhat misleading, because it also edits the segments between pairs of vertices.

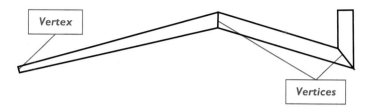

To keep track of which vertex is being edited, AutoCAD displays a marker in the shape of an X. Press **N** and **P** to move the marker to the next and previous vertex. A second marker, an arrow, is displayed when you work with the **Tangent** option.

> Command: **pedit**
> Select polyline or [Multiple]: *(Pick a polyline.)*
> Enter an option [Close/Join/Width/Edit vertex/Fit/Spline/Decurve/Ltype gen/Undo]: **e**
> [Next/Previous/Break/Insert/Move/Regen/Straighten/Tangent/Width/eXit]: *(Enter an option.)*

 **Note** Quite frankly, the **Edit vertex** options are a pain to use. In many cases, it much easier to edit the polyline using grips. Use these options only to insert and delete vertices. Use the Properties palette to set the widths of individual segments.

### Fit

The **Fit** option constructs smooth curves from the vertices in the polyline. The curve consists of arcs joined at vertices, as illustrated by the figure. Notice how each segment has at least one arc associated with it. AutoCAD inserts extra vertices and arcs, where necessary.

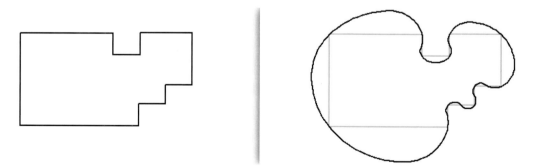

*Left: Original polyline.*
*Right: Polyline after applying the **Fit** curve option; the original polyline shown overlaid by gray lines.*

### Spline

The **Spline** option converts the polyline into a true spline object.

(*History*: Prior to AutoCAD 2010, the option turned polylines into an approximated spline, using the vertices as the control points to approximate a B-spline curve; if the original polyline contained arcs, they were converted to straight segments.)

The **Decurve** option and the SPLINEDIT command reverse the process, converting splines into polylines.

### Decurve

The **Decurve** option negates the effect of the **Fit curve** and **Spline** options by straightening all segments of the polyline, as well as by removing extra vertices that may have been added.

### Ltype gen

The **Ltype gen** option (short for "Linetype generation") controls the generation of linetypes through the vertices of the polyline.

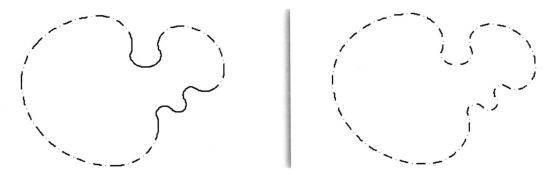

*Left: Polyline with **Ltype gen** turned off, part of the polyline lacking the line pattern.*
*Right: Polyline after turning on the option, all of the polyline patterned.*

Normally, AutoCAD starts and stops the pattern of dashes, dots, and gaps at each vertex. Turning on this option forces AutoCAD to treat the entire polyline as a single segment, resulting in a uniform linetype pattern along the entire length of the polyline.

Enter polyline linetype generation option [ON/OFF] <Off>: *(Enter **ON** or **OFF**.)*

In most cases, you probably want this option turned on.

 **Note** Linetype generation does not work on polylines with tapered segments.

### Reverse

The **Reverse** option operates in the same way as the REVERSE command described in the previous chapter: it reverses the direction of polylines.

### Undo

The **Undo** option reverses the last PEDIT operation without requiring you to exit the command to access the U command.

### EDITING POLYLINES: ADDITIONAL METHODS

Of the PEDIT command's many options, some are not obvious:

- **Select** option selects a line, arc, or spline.
- **Multiple** option selects more than one polyline.
- **Jointype** option determines how multiple objects are joined into a single polyline.

Let's look at each.

### Select

The "Select polyline" prompt misleads, because you can also select non-polyline objects — specifically lines, arcs, or splines. In this case, AutoCAD notices, and ask you:

> Command: **pedit**
> Select polyline or [Multiple]: *(Pick a line, arc, or spline.)*
> Object selected is not a polyline.
> Do you want to turn it into one? <Y>: *(Enter **Y** or **N**.)*

When you respond with "Y", the object is converted to a polyline. When you select an object other than a line, arc, or spline, the conversion fails, and AutoCAD complains:

> Object selected is not a polyline.

### Mutiple

You can select more than one polyline by entering the **Multiple** option, as follows:

> Command: **pedit**
> Select polyline or [Multiple]: **m**
> Select objects: *(Pick one or more polylines.)*
> Select objects: *(Press ENTER to end polyline selection.)*

### Jointype

The **Join** option normally merges lines, arcs, and polylines into a single polyline, but they all need to be touching. When you use the **Multiple** option at the start of the PEDIT command, however, they need not touch! AutoCAD displays the following prompt:

> Join Type = Extend
> Enter fuzz distance or [Jointype]<0.0000>: *(Enter a distance, or type **J**.)*

The *fuzz distance* determines how far apart the objects can be, and still be joined to the original polyline.

The **Jointype** option specifies how distant objects should join:

> Enter a vertex editing option
> Enter join type [Extend/Add/Both] <Extend>: *(Enter an option.)*

**Extend** — option joins the selected objects by extending and trimming them to fit.
**Add** — option joins the selected objects by bridging them with a straight segment.
**Both** — option joins the selected objects by extending or trimming; if that is not possible, AutoCAD adds the straight segment.

When you attempt to join two non-touching polylines, AutoCAD complains:

> 0 segments added to polyline

No matter which option you pick, fit-curved and splined polylines lose their curvature, and are converted back to the original frame.

## CONTROLLING POLYLINE EDITING

AutoCAD provides a number of system variables that control how polyline editing takes place:

- **PEDITACCEPT** toggles suppression of the **PEDIT** command's 'Object selected is not a polyline' prompt.
- **PLINECONVERTMODE** determines how splines are converted to polylines.
- **SPLINESEGS** specifies how many line segments approximate spline-fitted polylines.
- **SPLINETYPE** specifies the type of spline created by the Spline option.

### PEditAccept

The **PEDITACCEPT** system variable toggles the display of the **PEDIT** command's 'Object selected is not a polyline' prompt.

| SplineType | Meaning |
|---|---|
| 0 | Prompt is displayed (default) |
| 1 | Prompt is not displayed. |

### PLINECONVERTMODE

The **PLINECONVERTMODE** system variable determines how splines are converted to polylines with the **PEDIT** command's **Spline** option.

| PLineConvertMode | Meaning |
|---|---|
| 0 | Line segments (default) |
| 1 | Arc segments |

### SPLINESEGS

The **SPLINESEGS** system variable specifies the number of line segments to approximate spline-fitted polylines created by the **PEDIT** command's **Spline** option.

| SplineSegs | Meaning |
|---|---|
| 8 | Default value |
| 32767 | Maximum value |

Enter a positive number for splines, a negative number for fit-curves.

### SPLINETYPE

The **SPLINETYPE** system variable specifies the type of curve created by the **PEDIT** command's **Spline** option.

| SplineType | Meaning |
|---|---|
| 5 | Quadratic B-spline |
| 6 | Cubic B-spline (default) |

# GRIPS

Grips allow direct editing of objects — without commands.

In traditional AutoCAD usage, you enter an editing command and then select the objects to edit. The reverse is also possible: first select the objects, and then edit them directly without entering commands.

Grips are small, colored squares that indicate where the object can be edited. For example, lines have a grip at either end, plus another at their midpoint; circles have grips at the center point and at the four quadrant points.

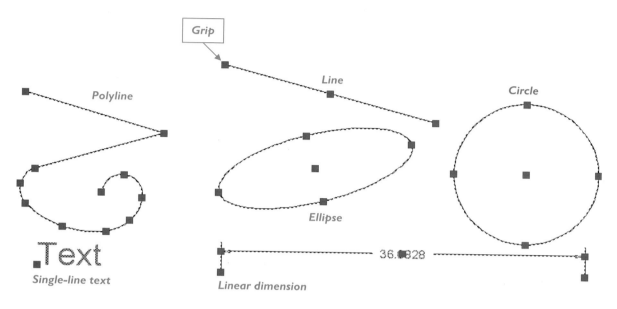

Grips editing can be quick. For example, a quick way to erase objects is to (a) select them with the cursor, and then (b) hit the DEL key.

**Note** To select all objects in the drawing, press **CTRL+A** (short for "all") at the command prompt:

Command: *(Press* **CTRL+A***.)*

AutoCAD selects everything in the drawing, except those on frozen and locked layers.

## DIRECT EDITING WITH GRIPS

Once an object is selected, you can perform two editing operations with it, in most cases:

**Move** — drag the center grip to move the object.

**Stretch** — drag one of the other grips to make the object larger or smaller.

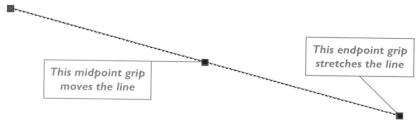

To move objects using grips, follow these steps:

1. Click on the object to select it. Notice the highlighting and the grips.
2. Click on the center grip. Notice that it turns red.

3. Move the mouse; don't drag! (To move orthogonally, hold down the **SHIFT** key.)
4. Click to place the moved object in its new location.
5. When done, press the **ESC** key to unselect the object.

The same actions apply to stretching objects, except that you select one of the other grips.

## Grip Color and Size

As a visual aid, standard grips are assigned colors and names:

| Grip Color | Name | Meaning |
|---|---|---|
| Blue | Cold | Grip is not selected. |
| Cyan | Dynamic | Grip's actions are defined by the Block Editor. |
| Pink | Hover | Cursor is positioned over grip. |
| Red | Hot | Grip is selected, and object can be edited. |

### TUTORIAL: EDITING WITH GRIPS

Although you can move and stretch objects directly through grips, AutoCAD provides other commands at the command prompt. After objects are selected, a prompt appears at the command line or dynamic prompt.

These are Stretch, Move, Rotate, Scale, Copy, and Mirror. (They also appear in the shortcut menu, should you right-click at this point.) In addition to these six, you can use other AutoCAD commands with grips. Examples include ERASE, COPYCLIP, PROPERTIES, MOVE, SCALE and STRETCH. Some editing commands, such as TRIM and OFFSET, cannot be used with grips.

1. Select an object in the drawing. The object is highlighted.

Notice that one or more blue grips are placed on the object (sometimes called "cold"). You can select as many objects as you wish by selecting more than once. If you cannot select more than one object, hold down the **SHIFT** key on the keyboard when selecting the grips.

2. Move the cursor over a blue grip. As the cursor gets close, notice that it jumps to the grip, and that the grip changes color to orange. The grip is said to be to "warm" as you "hover."

3. Click the orange grip. Notice that the grip box changes color to red to denote selection (called "hot"). The hot grip is the base point from which some editing actions, such as stretch and rotate, take place.

On the command line or dynamic prompt, AutoCAD prompts you, as follows:
** STRETCH **

Specify stretch point or [Base point/Copy/Undo/eXit]:

4. You can now take one of the following actions:
   - Drag the grip to stretch (or move) the object, depending on the grip you selected.

   - Or, press the spacebar to see the other editing options: Move, Rotate, Scale, and Mirror. These are described next.
   - Or, right-click a hot grip to see these editing options in a shortcut menu.

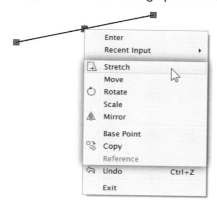

5. Select a command, such as Move or Rotate, and then follow the prompts at the command line (or dynamic prompt).
6. When done with grips editing, press ESC to clear the hot grip and selected objects.

## GRIP EDITING MODES

When you select a single grip (so that it turns red), AutoCAD displays editing options on the command line or dynamic prompt. The options cycle as you press the spacebar or ENTER. (Right-click to see the shortcut menu listing the same options.)

** STRETCH **

Specify stretch point or [Base point/Copy/Undo/ eXit]: *(Press spacebar.)*

** MOVE **

Specify move point or [Base point/Copy/Undo/ eXit]: *(Press spacebar.)*

** ROTATE **

Specify rotation angle or [Base point/Copy/Undo/Reference/eXit]: *(Press spacebar.)*

** SCALE **

Specify scale factor or [Base point/Copy/Undo/Reference/eXit]: *(Press spacebar.)*

** MIRROR **

Specify second point or [Base point/Copy/Undo/ eXit]: *(Press spacebar.)*

Press the spacebar until the desired edit option is listed, and then proceed with the command. For example, to resize the object, press the spacebar until **SCALE** appears, and then enter a scale factor at the "Specify scale factor:" prompt. The selected grip point becomes the *base point* for editing.

The commands and their options are covered in detail later in this section.

**Note** It is possible to work with two hot grips at a time, notes the technical editor. For instance, draw two lines, and then select both. While holding down the **SHIFT** key, select two (or more) grips. Release the **SHIFT** key, and then drag one of the hot grips. As you do, the other line also moves along.

## GRIP EDITING OPTIONS

Let's look at each grip editing option.

### Stretch Mode

**Stretch** functions like the STRETCH command, except that AutoCAD uses the hot grip to determine the stretch results.

> ** STRETCH **
>
> Specify stretch point or [Base point/Copy/Undo/eXit]: *(Pick a point, or enter an option.)*

**Stretch.** When you select the end grip of a line or arc, or quad grip of a circle, the object is stretched. Lines and arcs "stretch" longer or shorter, while circles change their diameter. The figure illustrates stretching these objects. (For clarity, the hot grips are shown in black, and the cold ones in white.)

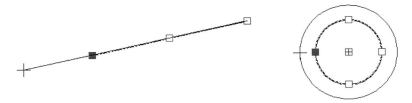

The line is lengthened by the stretching action. The circle is not stretched into an ellipse, as you might think from the word "stretch;" instead, its diameter is being changed.

**Move.** When you select the midpoint grip of a line, or arc, or the center of a circle, the object is moved, but not stretched. In the figure, the line is being moved, because the grip is at the center of the line; similarly, the entire circle is being moved.

### Move Mode

**Move** moves the selected objects. All selected objects are moved, relative to the hot grip. It matters not which grip you choose, unlike with Stretch mode.

> ** MOVE **
>
> Specify move point or [Base point/Copy/Undo/eXit]: *(Pick a point, or enter an option.)*

Objects are moved by the distance between the base point and the cursor location. To specify the distance and direction of the move, you can either pick a point in the drawing, or enter absolute or relative coordinates.

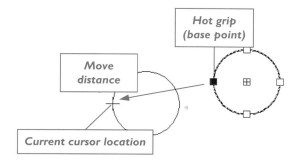

For example, enter relative coordinates that specify a distance of 50 units at an angle of 45 degrees from the x-axis:

Specify move point or [Base point/Copy/Undo/eXit]: **@50<45**

## Rotate Mode

The **Rotate** option rotates the selected objects around the base point. The selected objects are rotated, relative to the hot grip.

** ROTATE **

Specify rotation angle or [Base point/Copy/Undo/ Reference/eXit]: *(Pick a point, or enter an option.)*

Rotation occurs around the hot grip, unless you use the **Base point** option to relocate it. When rotating objects, you can dynamically set the rotation angle by moving the cross hairs, or by specifying a rotation in degrees.

## Scale Mode

If you wish to resize the selected objects, cycle through the mode list until **Scale** is shown on the command line. The selected objects are resized relative to the hot grip.

** SCALE **

Specify scale factor or [Base point/Copy/Undo/ Reference/eXit]: *(Pick a point, or enter an option.)*

The hot grip serves as the base point for resizing. Dynamically resize the objects by moving the cross hairs away from the base grip. You can also resize the selected objects by entering a scale factor.

## Mirror Mode

**Mirror** mode mirrors the selected object(s) relative to the hot grip. This option differs from the MIRROR command in that the objects are moved — not copied! — about the mirror line. The mirror line is created by the cursor, with the first point being established by the hot grip (unless you change it with the **Base point** option). AutoCAD prompts you for the second point of the mirror line:

\*\* MIRROR \*\*

Specify second point or [Base point/Copy/Undo/ eXit]: *(Pick a point, or enter an option.)*

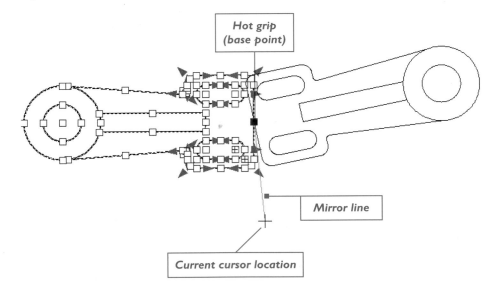

### Grip Editing Options

Each grip editing option has a set of common suboptions.

#### Base point

Normally, the hot grip is the base point for the editing operation. To change it, enter **B**, and AutoCAD prompts you:

Specify base point: *(Pick a point.)*

The point you pick becomes the new base point.

#### Copy

Enter **C** to copy the selected object, leaving the original intact. The prompt changes to:

\*\* STRETCH (multiple) \*\*

Specify stretch point or [Base point/Copy/Undo/eXit]: *(Pick a point.)*

The selected object is copied to the point you pick. The prompt repeats itself so that you can make multiple copies, as indicated by the word "(multiple)." Press ENTER or ESC to exit the command.

#### Undo

Enter **U** to undo the last operation.

#### Reference

Enter **R** to provide a reference for the Scale and Rotate modes. For Scale mode, AutoCAD prompts:

Specify reference length <1.0000>: *(Enter a length, or pick two points.)*

Sometimes the Reference option can be difficult to understand. Here is an example. Suppose you have an object that is 6 units in length, but you wish to resize it to 24 units. Select the **Reference** option, and then enter original length, as follows:

Specify scale factor or [Base point/Copy/Undo/Reference/eXit]: **r**

Reference length <1.0000>: **6**

AutoCAD then prompts for the new length.

<New length>/Base point/Copy/Undo/Reference/eXit: **24**

AutoCAD calculates the scale factor, and then resizes the object.

For **Rotate** mode, AutoCAD prompts:

Specify reference angle <0>: *(Enter an angle, or pick two points.)*

The reference angle is the angle at which the object is currently rotated. AutoCAD next prompts for the "New angle." The new angle is the angle you want to rotate the object.

eXit

Enter **X** or press ESC to exit grips editing.

## Other Grip Editing Commands

You can also use other commands when objects are highlighted with grips. Here is a sample:

ERASE deletes the gripped objects from the drawing. As an alternative, press the DEL key. Caution: selecting all objects and then pressing DEL erases the entire drawing! Use U to recover.

COPYCLIP copies the gripped objects to the Clipboard; CUTCLIP cuts the objects. As alternatives, you can press CTRL+C and CTRL+X, respectively.

PROPERTIES displays the Properties palette for the gripped objects.

MATCHPROP copies the properties of one gripped object; if more than one object is gripped, AutoCAD complains, "Only one entity can be selected as source object."

ARRAY creates rectangular and polar arrays of the gripped objects.

BLOCK creates a block of gripped objects.

EXPLODE explodes gripped objects; if they cannot be exploded, AutoCAD complains, "1 was not able to be exploded."

Grips on Arcs

Arcs have additional grips that are triangular in shape and perform specific editing tasks, as illustrated by the figure below.

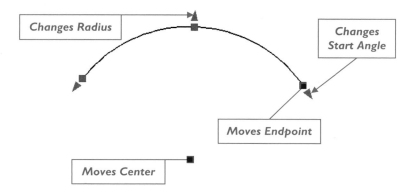

**Square grips** — move the arc; move its endpoints.

**Triangular grips** — change the angle of the start and endpoints; changes the radius of the arc.

### Grips on Polyline

Polylines have additional grips that edit vertices, and line and arc segments, as illustrated by the figure below.

You can select the entire polyline, or just one of its segments. To choose one segment: (a) hold down the CTRL key, and then (b) pick the segment you wish to edit.

To edit the polyline or segment: (a) hover the cursor over a grip (don't select it!), and then (b) choose an option from the shortcut menu that appears.

> **Square grips** — move the vertex; add and remove vertices.
>
> **Rectangular grips** — move the segment; convert segments between lines and arcs; add vertex.

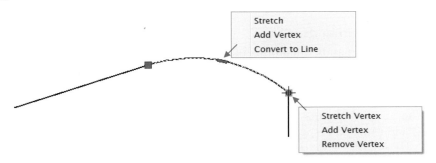

When you choose **Add Vertex**, AutoCAD displays the ✢ add icon. The vertex is added at the location of the cursor: (a) move the cursor along the segment, and then (b) click to position the new vertex.

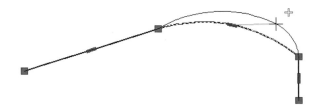

The **Remove Vertex** option removes the current vertex, and then makes the two adjoining segments into one. If they were curved, the new segment becomes a straight line.

When you choose **Convert to Arc**, AutoCAD displays the ⌒ arc icon. The radius of the arc is determined by the location of the cursor: (a) move the cursor, changing the radius, and then (b) click to specify the radius.

This direct editing method does not, unfortunately, allow you to specify the arc radius precisely by entering a radius at the keyboard.

### Grips on Hatches

The boundaries of hatch patterns and gradients sport exactly the same grips as do polylines. The hatches themselves have an additional round grip, located at their centers.

Select the hatch pattern, and then pause the cursor over the round grip. Notice the shortcut menu that appears:

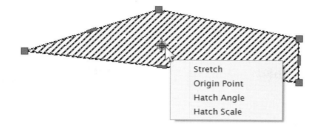

**Stretch** — moves the hatch pattern, and does not stretch it.

**Origin Point** — moves the hatch's origin, useful for making patterns line up with geometric features; pick a point in the drawing, or enter x, y coordinates at the prompt:

Specify origin point: *(Enter x, y coordinates.)*

**Hatch Angle** — changes the angle of the pattern; move the cursor, or enter an angle at the prompt:

Specify hatch angle: *(Enter an angle.)*

**Hatch Scale** — changes the size of the pattern; move the cursor, or enter a scale factor at the prompt:

Specify hatch scale: *(Enter a scale factor.)*

## RIGHT-CLICK MOVE/COPY/INSERT

A simpler alternative to grips editing is right-click move/copy/insert. It is meant as a faster method to copy and move objects, as well as to insert them as blocks. (This action is not documented by Autodesk, so I don't know its official name.)

It works like this:

1. Select one or more objects.

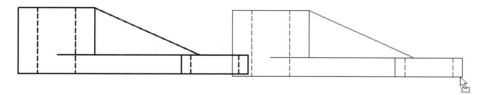

2. Holding down the *right* mouse button, drag the object. (*Important!* Drag the object, do not drag a grip.) Notice the copy icon at the cursor.

3. Let go of the mouse button. AutoCAD displays a small shortcut menu with these options:

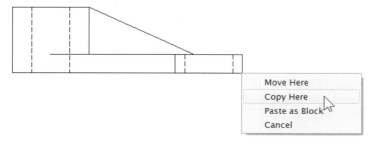

| Option | Meaning |
|---|---|
| Move Here | Moves the selected objects. |
| Copy Here | Copies the selected objects, and locates them. |
| Paste As Block | Turns the objects into a block, and then inserts them. |
| Cancel | Cancels the operation; the objects remain as they were. |

4. Choose an option from the shortcut menu.

When you choose the **Paste As Block** option, AutoCAD names the block automatically with a name like "A$C39FB63B0". Use the RENAME command to change the block's name to something more meaningful.

## DIRECT DISTANCE ENTRY

Direct distance entry allows "pen up" movement during drawing and editing commands. (Pen up means that the cursor changes its position without drawing or editing objects.)

Direct distance entry is an alternative to entering polar or relative coordinates that rely on a distance and an angle. To show the angle, you move the mouse; then you type the distance. With ortho or polar modes turned on, direct distance entry is an efficient way to draw lines.

Direct distance entry can be used any time a command prompts you to specify a point, such as "Specify next point:".

### TUTORIAL: DIRECT DISTANCE ENTRY

1. To draw a line 10 units long using direct distance entry, enter a drawing or editing command:
   Command: **line**

2. Before you can use direct distance entry, you must pick an initial point:
   Specify first point: *(Pick a point.)*

3. At the next prompt, move the mouse to indicate the angle, and then enter the distance at the keyboard:
   Specify next point: *(Move the mouse, and you see a rubber band line. At the keyboard, type* **10** *and press* ENTER.*)*

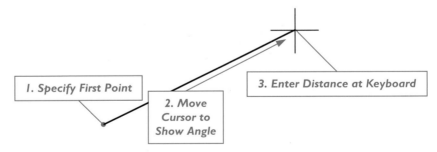

4. End the command:
   Specify next point: *(Press* ENTER.*)*

AutoCAD draws a line segment ten units long in the direction you move the mouse. You can use direct distance entry for drawing polylines, arcs, multilines, and most other objects. Direct distance entry does not make sense with some drawing commands, like CIRCLE and DONUT, because they want a radius.

Additionally, you can use direct distance entry with editing commands, such as MOVE, STRETCH, and COPY.

##  TRACKING

The **Tracking** modifier allows "pen up" movement during drawing and editing commands.

Whereas direct distance entry lets you draw relative distances, tracking lets you move relative distances within a command. Tracking is not a command, but a command option; you can use tracking only within another command.

Upon entering tracking mode, AutoCAD automatically switches to ortho mode, and then prompts for the "First tracking point:". (If polar mode is on, AutoCAD switches it off and turns on ortho mode.) Once you exit tracking mode, AutoCAD changes ortho and polar back to their original states.

Some commands keep prompting you for additional points, such as the LINE and PLINE commands. With these, you can go in and out of tracking mode as often as you like.

### TUTORIAL: TRACKING

1. To employ tracking, enter a drawing or editing command:
   Command: **line**

2. At a "Specify first point:" or "Specify next point:" prompt, enter **tracking**, **track**, or **tk**.
   Specify first point: **tk**

3. AutoCAD enters tracking mode. Notice that ortho mode is turned on. (Look at the OR-THO button on the status bar.)
   Move the mouse a distance, and then click.
   First tracking point: *(Move the mouse and click.)*

4. This time, move the mouse, and then enter a distance.
   Next point (Press Enter to end tracking): *(Move the mouse in another direction, and then enter a distance, such as **2**.)*

5. Press **ENTER** to exit tracking mode, and resume drawing with the **LINE** command.
   If ortho mode was off when you started the line command, AutoCAD switches it off automatically upon exiting tracking mode.
   Next point (Press Enter to end tracking): *(Press **ENTER**.)*
   To point: **2,3**
   To point: **10,5**

6. Switch back to tracking mode, and enter distances as absolute x, y coordinates:
   To point: **tk**
   First tracking point: **5,10**

7. Exit tracking mode, draw another line, and exit the **LINE** command.
   Next point (Press Enter to end tracking): *(Press **ENTER**.)*
   To point: **3,2**
   To point: *(Press **ENTER**.)*

If you want tracking to move in angles other than 90 degrees, use the **Rotate** option of the SNAP command to change the angle. Since the SNAP command is transparent, you can change the tracking angle in the middle of tracking, as follows:

   Command: **line**
   Specify first point: *(Pick a point.)*
   Specify next point: **tk**
   First tracking point: *(Move cursor and pick a point.)*

Next point (Press Enter to end tracking): **'snap**
>>Specify snap spacing or [ON/OFF/Aspect/Rotate/Style/Type] <0.5000>: **r**
>>Specify base point <0.0000,0.0000>: *(Press* ENTER *to accept default.)*
>>Specify rotation angle <0>: **45**

Resuming LINE command.
Next point (Press Enter to end tracking): *(Move cursor and pick a point.)*
Next point (Press Enter to end tracking): *(Press* ENTER *to end tracking.)*
Specify next point: *(Pick a point.)*
Specify next point: *(Press* ENTER *to exit command)*

**Note** Tracking normally assumes you want to switch direction each time you use it. For example, if you first move north (or south), AutoCAD assumes you next want to move east (or west).

It is not easy, however, to back up or track forward in the same direction. For example, if you track north and then want to track north some more, you find the cursor wanting to move east or west, but not north or south. To make the tracking cursor continue in the same direction, take it back to its most recent starting point. Then move in the direction you want.

### OSNAP WITH OBJECT TRACKING

Technical editor Bill Fane points out that it is possible to employ tracking without using the TK option. AutoCAD can acquire coordinates from other objects through object tracking. Follow these steps:

1. Turn on object tracking: On the status bar, click the  (or OTRACK) **Object Tracking** button.
2. Turn on ENDpoint object snap: right-click the ▭ (or OSNAP) **Object Snap** button on the status bar, and then choose ✎ **Endpoint** from the shortcut menu.
3. Draw a line at an angle, similar to the one illustrated below.

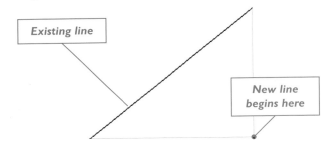

4. You want to draw a second line that starts at the point illustrated above. Enter the **LINE** command:
    Command: **line**

5. At the prompt, move the cursor over the lower endpoint of the existing line. Wait until the tag "Endpoint" appears.

   Specify first point: *(Move cursor over endpoint, but do not click.)*

   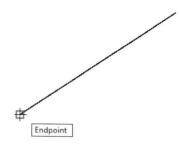

   AutoCAD has acquired the coordinates of the endpoint. It indicates this through the small plus marker found at the end of the line.

6. Move the cursor to the line's other endpoint, and pause. Again, the "Endpoint" tag appears after a moment.

   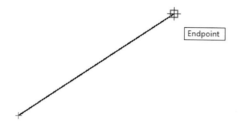

   AutoCAD has acquired the coordinates of the second endpoint.

7. Move the cursor to where the new line is to begin, at the "intersection" of the two endpoints. You'll know you've arrived when the dashed object tracking lines appear, as illustrated below. An X marks the intersection.

   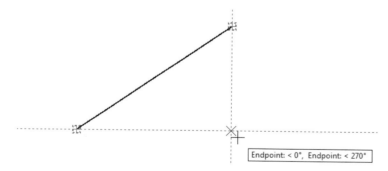

8. Now click to fix the first point:

   Specify first point: *(Pick the point.)*

9. You can continue using this technique to find other points through object tracking, such as the "intersection" of the midpoint and endpoint.

This technique is not limited to lines, but can be used with any objects, such as placing a circle in the center of a hexagon.

This method also works with editing commands, such as MOVE and COPY. Object tracking pretty much eliminates the need for construction lines when editing orthographic views.

## M2P

The **M2P** modifier finds the midpoint between two picked points — hence "m2p." (Alternatively, you can enter "mtp.")

M2P sounds like an object snap, but is not. (Nor it is an early version of music file.) It is entered during any prompt that asks you to specify a point. You can employ object snaps during the M2P process.

### TUTORIAL: M2P

1. To benefit from **M2P**, enter a drawing or editing command:
   Command: **line**

2. At a "Specify first point:" or "Specify next point:" prompt, enter **m2p** or **mtp**.
   Specify first point: **m2p**

3. AutoCAD prompts you for the first "mid" point. I recommend using object snaps to help place the point accurately. In this example, the CENter object snap is specified:
   First point of mid: **cen**
   of *(Pick point 1.)*

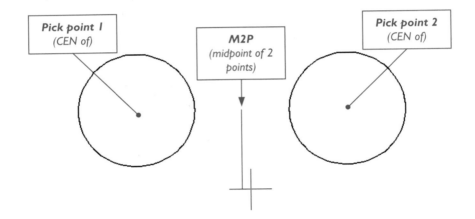

4. AutoCAD prompts you for the second point:
   Second point of mid: **cen**
   of *(Pick point 2.)*

5. The **LINE** command carries on. AutoCAD draws the line starting at the midpoint of the two pick points.
   Specify next point: *(Continue with the command.)*

## EXERCISES

1. In this exercise, you erase objects from a drawing, and then bring them back.
    a. Open the edit1.dwg drawing file, a valve housing.
    b. Use the **ERASE** command to delete some of the objects.
    c. Next, issue the **OOPS** command. Did the objects return?

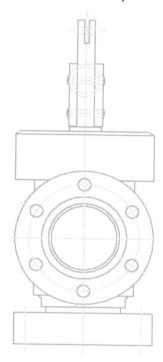

    d. Repeat the **ERASE** command with the **All** option. Did everything disappear?
    e. This time, use the **U** command to bring back the drawing.

2. In this exercise, you assemble a jigsaw puzzle.
    a. Open the *edit2.dwg* file, a drawing of puzzle pieces.
    b. Use the **MOVE** command to move the pieces into position, leaving a small space between them. The most effective method is to move the pieces roughly into position, then zoom in, and finely position the pieces using object snap.
    c. You may also want to use grips editing to move pieces into place.

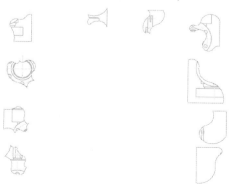

3. In this exercise, you practice breaking and trimming lines.

   a. Open the *edit5.dwg* drawing file, a piping diagram.

   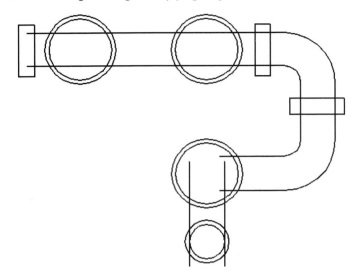

   b. Use the **BREAK** command to break each of the objects in the drawing, achieving the result shown.

   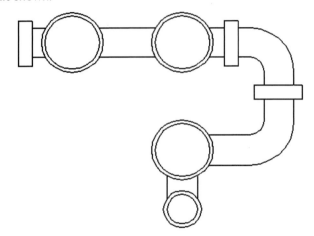

   c. Quit the drawing — don't save your work!
   d. Open *edit5.dwg* again; this time use the **TRIM** command to produce the same result as shown above. Here's a timesaving tip: at the "Select cutting edges" prompt, just press **ENTER** to select all objects in the drawing.
   e. You may need to use the **ERASE** command to clean up.
   f. Which command did you find easier — **BREAK** or **TRIM**?

4. In this exercise, you work with several editing commands.

   a. Open the *edit7.dwg* drawing file, a piece of sheet metal.

   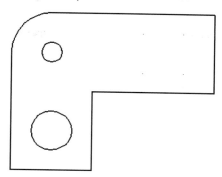

   b. Suppose that you are instructed to adjust the design by using the **ERASE** command to erase the four points and the lower circle.

   c. Your boss changes his mind. Put back the four points you just erased.

   d. After erasing the lower circle, your boss decides the remaining circle should be moved down by 2.0 units. Use CENter object snap, ortho mode, and direct distance entry to move it accurately.

   e. After reviewing the drawing, your boss's boss feels the sheet metal needs two holes. Add another circle, identical to the remaining circle. Use grips editing to copy the circle.

   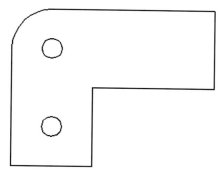

   f. You learn from the product design department that the sheet metal is too large. A notch needs to be taken out. Turn on snap mode, and set the snap spacing to 0.25. Use the **BREAK** and **LINE** commands to draw the notch.

   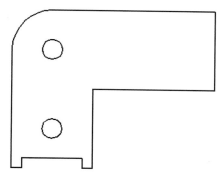

g. The product safety department reviews the drawing, and feels the sheet metal may be hazardous to children aged three years and younger. You are asked to round off two corners. Add a fillet of 0.5 to each of the two right corners.

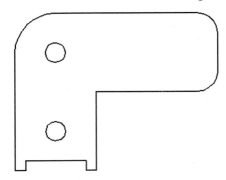

h. At last, all departments are satisfied. While the scenario in this exercise may seem silly, you will probably experience such changes to your drawings in the workplace. Save the drawing.

5. In this exercise, you trim objects.
   a. Draw four intersecting lines about 6 units long and 2 units apart, as illustrated by the figure. Which command quickly makes the second, parallel line?

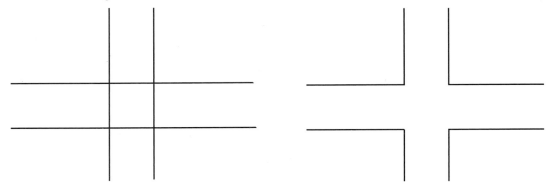

*Left:* Four lines before trimming.   *Right:* After trimming.

   b. Use the **TRIM** command to trim the intersections. Remember to use the trick of responding **all** to the "Select cutting edges" prompt.

6. In this exercise, you extend objects.
   a. Draw two vertical parallel lines 6 units long and 6 units apart.
   b. Between them, draw one horizontal line 4 units long, as illustrated by the figure.

c. Extend the horizontal line to the vertical lines with the **EXTEND** command.

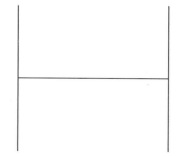

d. To extend the other end (remembering to choose both vertical lines as boundaries), respond to the repeating 'Select object to extend' prompt by selecting a point at the other end of the horizontal line.

7. In this exercise, you rotate objects.
   a. Draw the arrow symbol illustrated by the figure.
   b. Using the **ROTATE** command, turn the arrow by 45 degrees about the center of the arrow's base. Which object snap mode helps you find the midpoint of a line?
   c. Using grips editing, turn the arrow a further 90 degrees.

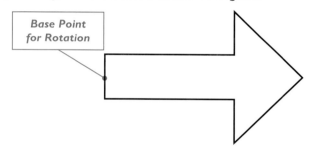

   d. Draw a cam similar to the one illustrated by the figure.
   e. Use Window object selection to rotate the entire cam by -45 degrees.

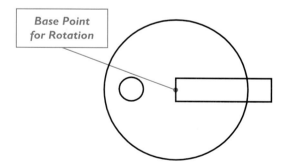

8. In this exercise, you resize objects.
   a. With the **POLYGON** command, draw the triangle illustrated by the figure. The edge is 2 units long.
   b. Use the **SCALE** command to double the size of the triangle.

   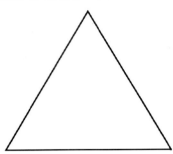

   c. Repeat the command, but this time use the **Reference** option again to double the size of the triangle.
      Specify scale factor or [Reference]: **r**
      Specify reference length <1>: **2**
      Specify second point: **4**

   Did AutoCAD resize all the lines making up the triangle?

9. In this exercise, you "stretch" a window to move it within the wall — without editing the wall.
   a. Open the *stretch1.dwg* file, a drawing of a wall cross section.

   b. Use the **STRETCH** command to move the window along the wall. Which selection mode must you use?

   Did the wall stretch?

10. Open the *stretch2.dwg* file, a drawing of a pencil.

    Use the **STRETCH** command to make the pencil shorter.

11. In this exercise, you lengthen and shorten objects with a different command.
    a. Draw an arc, using these parameters:
       Start point    **2,5**
       Second point   **0,3**
       Endpoint       **2,1.5**
    b. Start the **LENGTHEN** command, and select the arc. What is its length and included angle?
    c. Use the **DElta** option to add 2 units to the length.
    d. Use the **Total** option to change the arc to 5 inches. Did it become longer or shorter?
    e. Use the **Percent** option to change the arc to 100%. Did the arc change?

12. In this exercise, you change objects.
    a. Ensure ortho mode is turned off.
    b. Draw four lines, randomly, as illustrated by the figure.

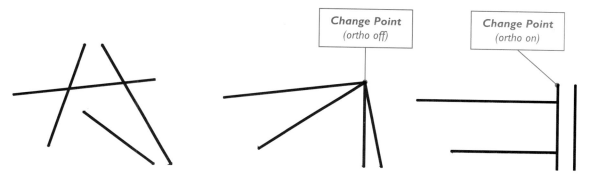

    c. Use the **CHANGE** command to select the lines, and then use the **Change Point** option to give the lines a common endpoint.
    d. Now use the **CHANGE** command's **Properties** option to change the lines to red.

13. In this exercise, you explode objects.
    a. Open the *edit2.dwg* file, the drawing of the puzzle.
    b. Select one of the puzzle pieces. What does the single grip tell you?
    c. With the piece still selected, use the **EXPLODE** command. Does the piece look different?
    d. Select the piece again. Has the number of grips changed? What does this tell you?
    e. Use the **EXPLODE** command a second time. What message does AutoCAD give you?
    f. Select a different, unexploded piece.
    g. Start the **XPLODE** command, and specify a color of red. What happens to the puzzle piece?

14. In this exercise, you edit polylines.
    a. Open the *offset.dwg* file, the drawing of the small town.
    b. Select any of the polylines that define the edges of streets.
    c. Use the **PEDIT** command's **Width** option to change the width to 0.1 units.
    d. Exit the **PEDIT** command, and then start the **PROPERTIES** command.
    e. Click the **Quick Select** button, and select *all* polylines by entering these parameters:

    | | |
    |---|---|
    | Apply to | **Entire drawing** |
    | Object type | **Polyline** |
    | Properties | **Layer** |
    | Operator | **=** |
    | Value | **0** |

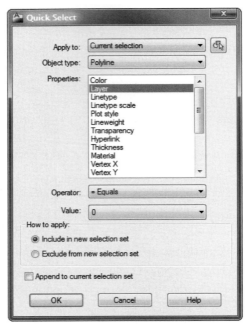

    f. After you click **OK**, notice that AutoCAD highlights all polylines.

g. In the Geometry section of the Properties palette, change the value of Global Width from *VARIES* to **0.1** units.

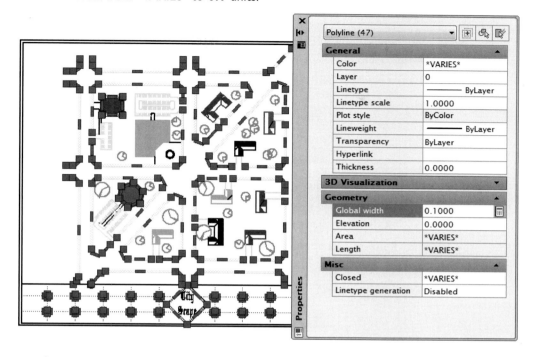

h. Press **ESC** to remove the grips and highlighting. The drawing should have fatter road lines, like the figure below.

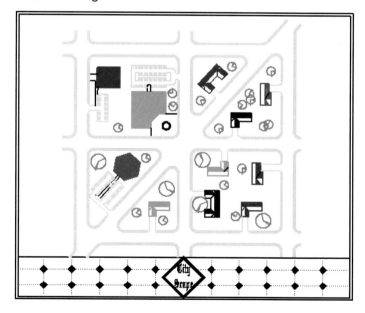

15. In this exercise, use grips to edit objects in the drawing.
    a.  Open the *17_35.dwg* file, a drawing of an angle.
    b.  Select a circle, and then click on the center grip.
    c.  Drag the grip to another place in the drawing. Does the circle move?

    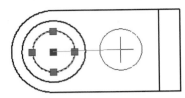

    d.  Now click one of the four quadrant grips.
    e.  Drag the grip. Does the circle become larger?

    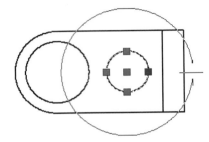

    f.  Press **ESC** to remove the grips.
    g.  Select the entire drawing, and then click on any grip.
    h.  Press the spacebar to see ** MOVE ** on the command line.
    i.  Drag the grip. Do all parts of the drawing move?
    j.  Press the spacebar again to see ** ROTATE ** on the command line.
    k.  Drag the grip. Does the drawing rotate around the grip?
    l.  Enter **b** to access the base point option:
        Specify base point: *(Pick another point.)*
    m.  Move the cursor. Does the drawing now rotate about the new point?
    n.  Press the spacebar again to see ** SCALE ** on the command line.
    o.  Drag the grip. Does the drawing change its size?
    p.  Press the spacebar again to see ** MIRROR ** on the command line.
    q.  Drag the grip. Do you see a mirrored copy?

16. Use the **LINE** command with direct distance entry to draw the object defined by the following relative coordinates:
    Point 1: 0,0
    Point 2: @3,0
    Point 3: @0,1
    Point 4: @−2,0
    Point 5: @0,2
    Point 6: @−1,0
    Point 7: 0,0

17. Use the **PLINE** command with direct distance entry to draw the object defined by the following relative coordinates:

    Point 1: 0,0
    Point 2: @4<0
    Point 3: @4<90
    Point 4: @4<180
    Point 4: @4<270

18. Use direct distance entry and tracking to draw the following fuse link. Each square represents one unit.

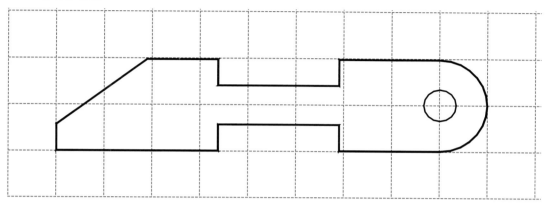

19. Use direct distance entry and tracking to draw the following cylinder. Each square represents 2 units.

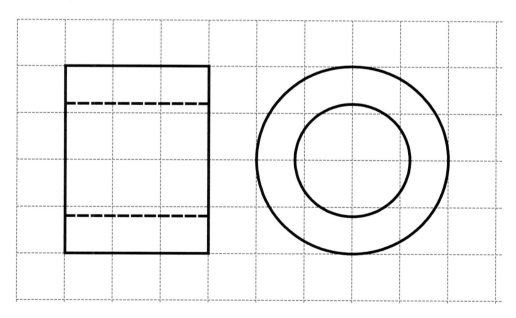

20. Use direct distance entry and tracking to draw the following baseplate. Each square represents a half-unit.

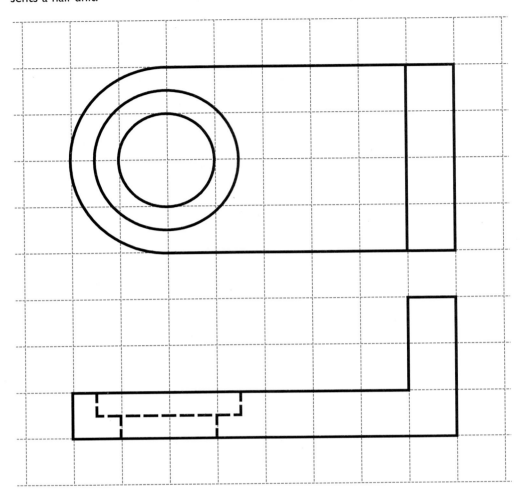

21. Use direct distance entry and tracking to draw the following wrench. Each square represents one unit.

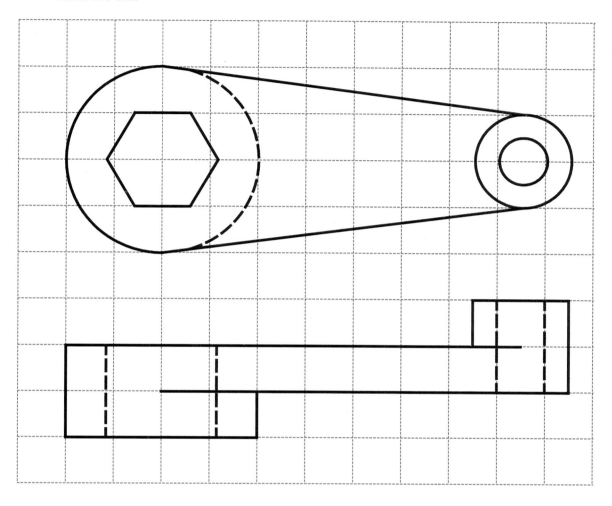

## CHAPTER REVIEW

1. What is the difference between hot and cold grips?
2. When moving objects, must the base point be on the selected objects?
3. What objects are affected by the **BREAK** command?
4. In breaking objects, what happens if you do not enter **F** for selection of the first point?
5. Which commands restore objects just erased?
6. What happens when you explode blocks?
7. Describe what happens when you explode polylines.
8. What are *cutting edges*?
9. At what point are polylines trimmed?
10. What is the special requirement before circles can be trimmed?
11. Name the border line used by the **EXTEND** command.
12. Describe three ways to rotate objects.
    a.
    b.
    c.
13. Can objects be resized differently in the x and y axes with the **SCALE** command?
14. What value changes objects to half their original size?
15. Describe the purpose of the **Copy** option of the **ROTATE** and **SCALE** commands.
16. Name the command that explicitly edits polylines?
17. Which parts of the polyline are the *vertices*?
18. How do you edit polyline vertices?
19. What types of vertex editing can you perform?
20. List the aliases for the following commands:

    **TRIM**

    **BREAK**

    **EXTEND**

    **PEDIT**
21. When can the **OOPS** command be used?
22. Is it possible to erase all objects in the drawing?
23. Can the **BREAK** command shorten objects?
24. What does a circles become after you apply the **BREAK** command?

    What does an ellipse become?

    An xline?
25. Describe how the **LENGTHEN** command differs from the **TRIM** and **EXTEND** commands.
26. Can the **TRIM** command extend open objects?

    If so, how?
27. Can the **EXTEND** command extend closed objects?

    If so, how?
28. Do blocks and dimensions work as cutting edges?

    Do hatches?
29. In what kind of drafting are the **TRIM** command's **Project** and **Edges** options used?

30. List the two things AutoCAD needs to know before extending objects:
    a.
    b.
31. Do blocks work as boundary edges?
    Do hatches?
32. Can circles be extended?
33. Is it acceptable to select all objects as cutting edges?
34. Can arcs be lengthened with the **LENGTHEN** command?
    Shortened?
35. Describe how the **LENGTHEN** command's **Total** and **Delta** options differ.
36. When does the **STRETCH** command only move objects?
37. Can objects be moved using grips editing?
38. What does the **ROTATE** command rotate objects about?
39. Where does 0 degrees point, by default?
    Can the direction be changed?
    If so, how?
40. In which direction does AutoCAD measure negative angles, by default?
41. What three pieces of information does AutoCAD need to know before resizing objects?
    a.
    b.
    c.
42. Describe how the **CHANGE** command's **Change Point** option changes lines with
    Ortho mode turned off:
    Ortho mode turned on:
43. What is the difference between the **EXPLODE** and **XPLODE** commands?
44. How are arcs in nonuniformly scaled blocks exploded?
45. Under what condition must blocks be exploded more than once?
46. What are polylines exploded into?
47. Can you specify the color of exploded objects?
    If so, how?
48. Describe how AutoCAD reacts when you select lines and arcs with the **PEDIT** command?
49. Can lines and arcs that don't touch be turned into a single polyline?
    If so, how?
50. During polyline vertex editing, which keystrokes move the x marker from vertex to vertex?
51. How does the **Straighten** option affect a polyline?
52. What is the difference between the **Fit** and **Spline** options of the **PEDIT** command?
53. Are the splines created by **PEDIT** true splines?
54. Explain why the **Ltype gen** option is important?
57. What does the padlock icon indicate?
58. What are the default colors for the following grips:
    Hot:
    Cold:
    Hover:

59. Name three editing operations you can perform with grips.
60. How do you deselect gripped objects?
61. What is the primary difference between *tracking* and *direct distance entry*?
62. What is the primary purpose of grips editing?
63. Where are grips located on lines?

    On circles?

    On blocks?
64. Using grips, can more than one object be stretched at a time?
65. Describe the purpose of the **Base point** option in grips editing.
66. What happens when you drag a circle by its center point grip?

    What happens when you drag a circle by its quad grip?
67. Can direct distance entry be used with editing commands?
68. What is the alias for tracking?
69. Name the mode that AutoCAD automatically turns on when you enter tracking mode?
70. Other than the standard six grips editing commands, name three other editing commands that work with gripped objects:

    a.

    b.

    c.
71. Can you use the BLOCK command with gripped objects?
72. How would you find the midpoint between two other points?

# IV

# Text and Dimensions

# CHAPTER 10

## Placing and Editing Text

Text is used in many areas of drawings: title blocks that identify drawings, callouts that identify parts, bills of material that list parts, and paragraphs of text that warn and explain. Traditionally, placing text by hand in a paper drawing was laborious; drafters drew each letter individually. In contrast, placing text in an AutoCAD drawing requires little effort. This chapter shows how to place text in drawings, and then how to change the text and its properties:

**TEXT** places text in drawings, a line at a time.
**MTEXT** places paragraphs of formatted text.
**TEXTTOFRONT** displays text in front of other objects.
**QLEADER** and **MLEADER** place callouts in drawings.
**TEXTEDIT** edits text.
**STYLE** define named text styles based on fonts.
**MLEADERSTYLE** defines named multiline leader styles.
**SPELL** checks the drawing for unfamiliar words.
**FIND** searches and optionally replaces text.
**SCALETEXT** and **JUSTIFYTEXT** change the size and justification of text.
**Annotation** property displays correctly-scaled text in model space.

---

**NEW TO AUTOCAD 2011** IN THIS CHAPTER
- **STYLE** command now identifies the names of missing fonts. (**OPEN** command now lets you ignore missing fonts.)
- **MTEXT** command now includes the Stack function in the Text Editor tab.

## TEXT IN DRAWINGS

The figure below illustrates many uses of text in drawings. Can you spot the general notes, title block, section view numbers, leaders, and dimensions? Methods for creating some of these are described in this and following chapters.

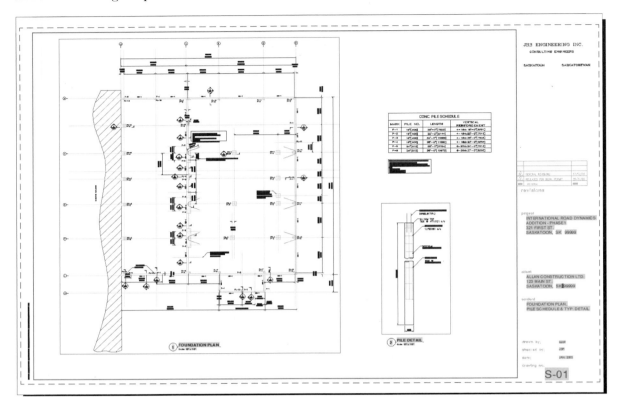

In the days of hand drafting, neophyte drafters learned how to print letters and numbers neatly first, and only then how to draw geometry. Today, AutoCAD does the lettering for us, but the same rule applies: clear text is crucial to understanding drawings.

The use of text in drafting is governed by standards. Many North American industries use the American National Standards Institute (ANSI) style of letters and numbers. European companies use ISO lettering (International Organization for Standardization), while other companies and countries set standards of their own. (See en.wikipedia.org/wiki/CAD_standards for examples of drafting standards.)

Here are some basic guidelines for sizing text in drawings:

**Headings** are 3/16-inch (0.1875") or 5mm high.

**Note text** is 1/8-inch (0.125") or 3.5mm high.

**Text** is left-justified, where each line of text aligned at its left edge.

In Imperial drawings based on *acad.dwt*, the default text height is 0.2". Metric drawings based on *acadiso.dwg* use 2.5mm as the default.

You can calculate the size of the text manually, or let AutoCAD do it automatically through annotative scaling.

## FONTS

In drawings, the look of text is based on fonts, and AutoCAD is versatile at displaying fonts. They can be stretched, compressed, obliqued (slanted), and mirrored (reversed); in some cases, the text can even be placed vertically. You can apply colors, lineweights, and plot styles to text — though not linetypes.

The default font in every new AutoCAD drawing is the common Arial font. Other clear fonts often used by drafters are called "Simplex" and "RomanS."

# Arial font
## Simplex font

**Left:** *AutoCAD's default style, the Arial font.*
**Right:** *AutoCAD's Simplex font preferred by many drafters.*

Fonts are the "design" of text letters. Caslon, for example, is the name of the font used for the text in this paragraph, while Gill Sans is used for the headings and tutorial text. Arial, Times Roman, and `Courier` are among the most commonly-used fonts in documents.

Fonts are defined by files, and AutoCAD supports several formats. There is AutoCAD's own SHX format (stored in *.shx* files) that Autodesk developed many years ago. But today it is common to use TrueType fonts (*.ttf* files) in drawings; these were designed by Apple and are found on almost all computers. A third format, PFB (*.pfb* files), was designed by Adobe and is typically used for high-end typesetting; it can be used in AutoCAD, but only after conversion to SHX format with the **CONVERT** command.

AutoCAD has two primary commands for placing text. **TEXT** places lines of text in drawings, and is like the minimalist Notepad text editor. **MTEXT** places paragraphs of text and is more akin to a full-featured word processing application.

## TEXT

The **TEXT** command places lines of text in drawings.

This command is useful for placing many bits of text all over a drawing quickly, with minimal worry about formatting.

### TUTORIAL: PLACING LINES OF TEXT

In this tutorial, you add some words to a new drawing.

1. Start AutoCAD with a new drawing with the *acad.dwt* template.
2. To place lines of text in drawings, start the **TEXT** command:
   - In the ribbon's Home tab, pick the **Single Line Text** button in the Annotate panel.
   - At the 'Command:' prompt, enter the **TEXT** command:

     Command: **text** *(Press* **ENTER**.*)*

   - Alternatively, enter the **dt** or **dtext** (the old name for this command) aliases at the 'Command:' prompt.

3. In all cases, AutoCAD reports the current text style, height, and annotation scale settings, and then prompts you for the location of the start point:

   Current text style: "Standard" Text height: 0.2000 Annotative: No

Specify start point of text or [Justify/Style]: *(Pick a point anywhere in the drawing, or specify x, y coordinates.)*

4. Specify the height of the text and its rotation angle:
   Specify height <0.2000>: **0.75**

   Specify rotation angle of text <0>: *(Press* **ENTER** *to accept the default.)*

   In most cases, the rotation angle will be 0 degrees; sometimes, however, you may need to place text at an angle, such as at 45 or 90 degrees.

5. Enter a line of text, and then press **ENTER**. (This command is unusual in that it has no prompt on the command line.)
   **Using** *(Press* **ENTER**.*)*

   As you type text, notice that AutoCAD displays the I-beam cursor, called that because it looks like the letter I. The cursor has a pair of short horizontal lines that represent the top of uppercase letters and the bottom of descenders. (The *descender* is the part of lowercase letters that hangs below the *baseline* — for instance, the letters g, q, and y have descenders.)

6. After you press **ENTER**, notice that the cursor jumps to the start of the next line.
7. Enter additional lines of text. Notice that AutoCAD places them under the previous line.
   **AutoCAD** *(Press* **ENTER**.*)*

   If you make a mistake, press **BACKSPACE** to erase one character at a time; alternatively, highlight a group of characters to change or erase them.

8. To exit the command, press **ENTER** again, without entering any text:
   *(Press* **ENTER** *to exit the command.)*

*Careful!* When you end the TEXT command by pressing ESC, the text is erased; so, remember to press ENTER to end the command and preserve the text.

The text you enter is surrounded by a gray bounding box, which shows the extent of the text.

During the **TEXT** command, you'll find that ribbon selections, command options, and other functions are "locked out." The cursor turns into the international symbol for no: ⊘. Only keyboard entry is permitted during this command.

The next time you use the **TEXT** command, the last text string you entered is *highlighted* (shown in a pattern of dots):

# Using
# AutoCAD

AutoCAD does this for a good reason: to help you place additional lines of text. When you press **ENTER** at the 'Specify start point of text:' prompt, the I-beam cursor is placed on the next line after the previous text string — as though you had never exited the command:

Command: *(Press* **ENTER** *to repeat the* **TEXT** *command.)*

TEXT Current text style: "Standard" Text height: 2.0000

Specify start point of text or [Justify/Style]: *(Press* **ENTER** *to continue text below the last line.)*

AutoCAD skips the height and rotation prompts, because it assumes you want them to be the same as before.

# Using
# AutoCAD

# Using
# AutoCAD
# for text

*Pressing* **ENTER** *at the 'Specify start point of text or [Justify/Style]' prompt continues text below the text entered earlier.*

### PLACING TEXT ALL OVER DRAWINGS

The **TEXT** command is useful for placing text quickly in many places on the drawing: simply click the cursor in another location, and then continue typing. This trick is handy for filling out title sheets and other text intensive jobs.

AutoCAD calls these pockets of text "blocks" (not to be confused with blocks made of objects.) During the **TEXT** command, you can move between text blocks by pressing the **TAB** key. Press **SHIFT+TAB** key to return to a previous text block; press **TAB** again to jump ahead to the next text block.

To split a line of text in two, use the cursor arrow keys (or the mouse) to locate the cursor at the break, and then press **ENTER**.

### PLACING TEXT: ADDITIONAL METHODS

The **TEXT** command allows you to control the justification and style of the text, as listed by the options below.

- **Justify** specifies a text justification mode.
- **Style** selects a predefined text style.
- **Height** specifies the height of the text.

- **Rotation angle** specifies the angle at which the line of text is rotated.
- **Control Codes** add formatting and special characters.
- **DTEXTED** system variable controls the look and feel of the command.

Let's look at each.

## Justify

When you enter text, AutoCAD left justifies it automatically; in most cases, this is exactly what you want. (In AutoCAD, the starting point of text is called the "insertion point.") For a different text alignment, then enter the **Justify** option. Other common justifications include center-aligned (often used in title blocks) and right-justified, used for text placed to the left of leader lines. These justification modes are illustrated below.

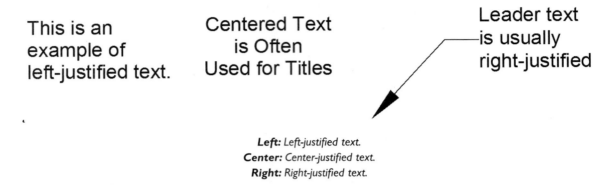

**Left:** Left-justified text.
**Center:** Center-justified text.
**Right:** Right-justified text.

The default justification is left-justified, which means that all lines of text start at the left edge. To see the list of all text justification modes, respond with a **J** (short for "justify") at the 'Specify start point of text or [Justify/Style]:' prompt, like this:

Specify start point of text or [Justify/Style]: **j**

As it lists the names and abbreviations of justification modes, AutoCAD asks you to enter an option:

Enter an option [Align/Fit/Center/Middle/Right/TL/TC/TR/ML/MC/MR/BL/BC/BR]:

The list of justification modes is long, because text can be justified (or aligned) both vertically and horizontally. AutoCAD uses one- and two-letter abbreviations to designate each alignment option, such as A for align and TL for top-left. The one-letter justification modes are most commonly used.

(*History*: The one-letter options were present in the original AutoCAD 25 years ago; the two-letter options were added a few years after.)

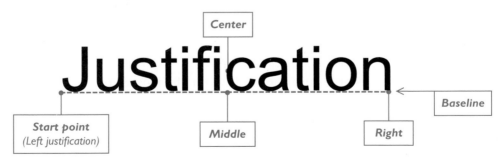

| Justify | One-letter Options |
|---|---|
| Start point | Left-justifies text (default). |
| A | **Align** fits text between two points, adjusting the height appropriately. |
| C | **Center** centers text on the baseline. |
| F | **Fit** fits text between two point at a specific height. |
| M | **Middle** centers text vertically and horizontally; equivalent to MC. |
| R | **Right** aligns text at its right edge. |

When it comes to the two-letter options, the first letter describes the vertical alignment; the second, the horizontal alignment. For example, **TL** is top-left. The figure illustrates the alignment modes listed in the table. Note that **B** refers to *bottom*, not baseline.

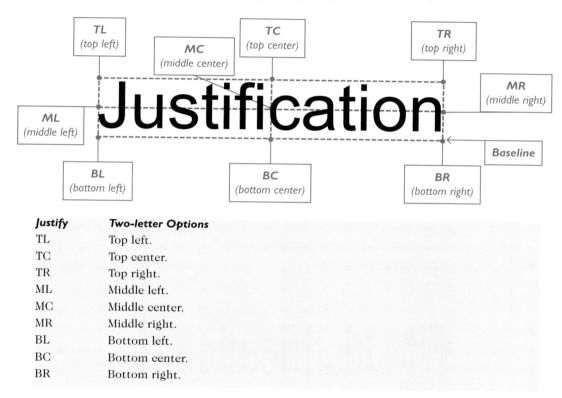

| Justify | Two-letter Options |
|---|---|
| TL | Top left. |
| TC | Top center. |
| TR | Top right. |
| ML | Middle left. |
| MC | Middle center. |
| MR | Middle right. |
| BL | Bottom left. |
| BC | Bottom center. |
| BR | Bottom right. |

 **Note** If you know the alignment you want, you can use this shortcut: simply enter the one- or two-letter abbreviation for the alignment. (It is not necessary to enter the **J** option.) For example, to right-justify text:

Specify start point of text or [Justify/Style]: **r**

Align

The **Align** option requires you to select two points, and then it fits the text between them. AutoCAD automatically adjusts the text height so that the baseline of the text fits perfectly between the two pick points. Note that the two points can be placed at any angle in relation to each other.

Command: **text**
Current text style: "Standard"  Text height:  0.7500  Annotative:  No
Specify start point of text or [Justify/Style]: **a**
Specify first endpoint of text baseline: *(Pick point 1.)*
Specify second endpoint of text baseline: *(Pick point 2.)*
*(Type text.)*
*(Press* ENTER *to exit command.)*

There are no "height" and "rotation angle" prompts, because the pick points determine the angle of the text. AutoCAD determines the height. As you type text, AutoCAD automatically sizes the text, making it smaller as you add more characters.

When you pick two points in reverse order (right to left), AutoCAD draws the text upside down.

Fit

The **Fit** option is similar to Aligned in that it prompts you for two points between which to place the text. But it differs in that it also prompts for the text height. AutoCAD fits the text between the two points, and draws the text at the height you specified.

*Examples of text fitted between two points, but assigned different heights.*

Specify start point of text or [Justify/Style]: **f**
Specify first endpoint of text baseline: *(Pick point 1.)*
Specify second endpoint of text baseline: *(Pick point 2.)*
Specify height <0.2>: *(Enter a height.)*
*(Type text.)*
*(Press* ENTER *to exit command.)*

The text is squashed or stretched horizontally to fit between the two points, as illustrated above. Fitted text is often used in constrained areas, for example when labeling small closets. ("Or," suggests, the technical editor, "You could come out of the closet and put the label in the hall.")

When you pick the second point to the left of the first, AutoCAD draws the text upside down.

## Style

The **Style** option specifies the name of a predefined text style. Styles must be created before you can use them with the TEXT command. Exceptions are the Standard and Annotative styles, which are included in all new drawings.

Styles are covered in detail later in this chapter; for now, it is sufficient to know that they define many parameters of the text, such as the font it uses, and whether it is shown boldface or has a predefined height.

**Note** The **Style** option (of the TEXT command) and the **STYLE** command do not perform the same tasks:

- **Style** option *selects* the style to use with text.
- **STYLE** command *creates* and modifies text styles.

## Height

The **Height** option determines the height of the text, which can be any distance — as tall as the orbit of Pluto, if need be. (The technical editor corrects the author by noting that text can in fact be taller than the orbit of Pluto. The tallest text can be as tall as $9.49 \times 10^{94}$ units — considerably taller than the known universe. "Tower of Babel-size font," remarks the copy editor.) The default is 0.2" in Imperial drawings and 2.5mm in metric ones.

Specify height <0.2>: *(Enter a height such as* **17.5**, *or press* **ENTER**.*)*

Once you change the height, it becomes the new default, until you change it again.

There are two conditions under which the 'Specify height' prompt doesn't show up: one is when the height is predefined by the style; the other is during the Align justification.

### Determining the Height of Text

Like hatch patterns and linetypes, text must be scaled appropriately for drawings. If you were to sketch a picture of your house on a piece of paper, you would draw it small enough to fit the paper. (There might be a sheet of paper big enough to draw your house full size, but it would be hard to find!) Because scale is so important for text, a review may be in order.

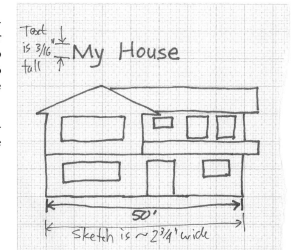

In a sketch of my 50-foot house, I drew it 2 3/4" wide. My drawing of the house is 218 times smaller than its actual size. The drawing scale of my sketch is 1:218.

Here is how I arrived at 1:218:

1. Convert feet and inches to the common unit of inches. That makes 50' equal to 600", .
2. Divide the drawing width by the real-world wide:

$$\frac{2.75"}{600"} = 1:218$$

The 1:218 is the scale factor.

In AutoCAD, however, scaling is done in reverse. Instead of drawing the house smaller, I draw it full size in the drawing editor. The 50-foot house is drawn 50 feet long in AutoCAD, at a scale of 1:1.

The text I wrote on the paper is full size, which just happens to measure 3/16" tall in my sketch. The text cannot be drawn at 1:1, because otherwise its 3/16"-height would be too, too small. Text must to be 218 times larger. When I place the text in the house drawing, I specify a height of 40" (3/16" x 218). This may seem much too tall (nearly four feet high!), but trust me that when it is plotted, it will look exactly right.

The figure below illustrates text placed at several heights — some too small, some just right, and some too large. (The copy editor calls this the "Goldilocks Factor.") Whether it is right or wrong depends on the plotted size.

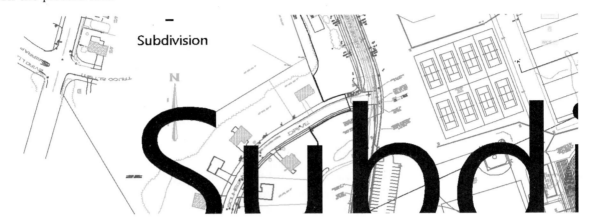

*Text ranging in height from 3" (too small), to 26" (just right) and 349" (too tall).*

## Annotative Scaling

There are two approaches to solving the text height problem. One is to work out the text size and then enter it when the **TEXT** command prompts you, as described above.

The other way is to use *annotative scaling*, in which AutoCAD determines the text height automatically. You tell it the plotted height (a.k.a. the paper height), and then AutoCAD sizes the text based on the viewport's scale factor.

Annotative scaling is available only when it is turned on with the **STYLE** or **PROPERTY** commands. Anything affected by annotative scaling has the 🔺 icon near it. (The icon represents the end view of an engineer's scale ruler.) All new drawings contain the Annotative text style, although Standard continues to be the default.

When annotative scaling is turned on, the **TEXT** command changes the wording of the height prompt by adding the word "paper":

Specify paper height <0.2000>: **0.125**

You enter the height at which you wish the text to appear plotted on *paper*. Thus, text $1/8$" (0.125 units) tall is plotted $1/8$" tall, no matter the scale of the viewport. This eliminates the need to figure out the scale factor. But there are some nuances involved in annotative scaling, as detailed at the end of this chapter.

## Rotation Angle

Lines of text can be placed at any angle in drawings. You specify the angle when you are prompted:

Specify rotation angle of text <0>: *(Enter an angle, or pick two points.)*

Press **ENTER** to accept the default angle, 0 degrees in this case. The text is rotated about the insertion point, whose location varies depending on the justification mode.

Once the text angle is set, the angle remains the default until you change it. The figure illustrates text placed at several angles.

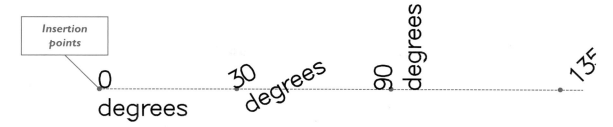

The angle is affected by the ANGBASE and ANGDIR system variables, as are other angles in AutoCAD.

## Control Codes

AutoCAD allows you to add *metacharacters* (also known as "control codes) to text. Metacharacters mean something other than themselves. (As a professor of English literature, the copy editor notes that all text means something other than itself. We will sidestep the philosophical discussion over what is meta and what isn't.) Control codes specify underlined text and symbols, such as the degree and plus/minus.

While the **TEXT** command is running, you can type codes that start with two percent characters: %%. The following table lists the available codes and their meaning:

| Control Code | Meaning | Sample |
|---|---|---|
| %%c | Diameter symbol | Ø |
| %%d | Degree symbol | ° |
| %%o | Start and stop overline | Text |
| %%% or % | Percent symbol | % |
| %%p | Plus-minus symbol | ± |
| %%u | Start and stop underlining | Text |
| %%nnn | Any ASCII character *nnn* | %%126 = ~ |

Let's look at an example with the following text string:

Command: **text**

Current text style: "Standard"  Text height: .02  Annotative: No

Specify start point of text or [Justify/Style]: *(Pick a point.)*

Specify height <0.02>: *(Press ENTER.)*

Specify rotation angle of text <0>: *(Press ENTER.)*

**If the piece is fired at 400%%dF for %%utwenty%%u hours,**
**it will achieve %%p95%%% strength.**

*(Press ENTER.)*

Codes are initially displayed instead of the effect; once you begin typing the next text, the code disappears, replaced by the formatting:

> If the piece is fired at 400°F for <u>twenty</u> hours,
> it will achieve ±95% strength.

Notice that you turn on and off the underscore (underlining) by typing %%u two times — once to turn on, and once to turn it off. If you forget to turn it off, AutoCAD does it for you at the end of the line.

You can "overlap" symbols, for instance, using the degree symbol between the underscore symbols. To get **Fire at 400°F.**, enter this:

%%uFire at 400%%dF%%u.

## ASCII Characters

ASCII characters refer to the ASCII (short for "American Standard Code for Information Interchange") character set, which assigns number codes to symbols.

To use a symbol in the text, enter two percent signs (%%) followed by the ASCII character code. For example, to place a tilde (~) in your text, enter %%126.

## DTextEd

Over the years, Autodesk has changed the way the TEXT command operates in AutoCAD, and so it added the DTEXTED system variable to control the changes. (The sysvar is named after the old DTEXT command, short for "dynamic text editor.") Changing the value of the system variable changes the way the TEXT command works, as follows:

> **2** — Displays no prompt for entering text. You can click to continue text elsewhere in the drawing. Default action of **TEXT** since AutoCAD 2007.

> **1** — displays the 'Text:' prompt at the command line; you can click elsewhere in the drawing to start new text strings. Simulates the action of **TEXT** in AutoCAD 2005 and earlier.

> **0** — like DTextEd=2, but you cannot click to continue text elsewhere. In this way, it simulates the action of **TEXT** in AutoCAD 2006.

I recommend you leave this system variable set to 2.

## TEXT SHORTCUT MENU

The **TEXT** command sports a shortcut menu with useful options. To see the menu, right-click text during the command. (This works when **DTEXTED** is set to 0 or 2; when set to 1, the shortcut menu doesn't appear.)

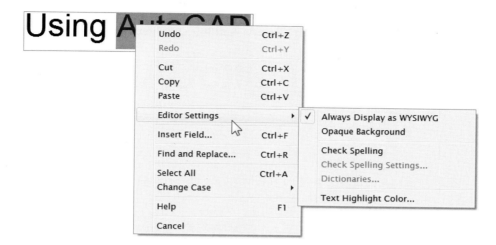

### Editor Settings

The significant options are described next.

#### Always Display as WYSIWYG

The **Always Display as WYSIWYG** option edits text in-place. (WYSIWYG is short for "what you see is what you get.") When this option is on, and text is at some unusual angle, the text editor edits it at the same angle, as illustrated below, left.

When off, text is rotated to be clearly legible while it is being edited, as illustrated below, at right. AutoCAD colors the background to alert you that the text will not be placed the way you see it.

**Left:** *Display as WYSIWYG turned on...*
**Right:** *...and turned off.*

#### Opaque Background

The **Opaque Background** option fills the bounding box with gray (default = off). The color helps distinguish the text input area, and appears only while editing; when you exit the command, the color disappears.

**Left:** *No background color.*
**Right:** *Opaque background..*

(This option is not available when **Always Display as WYSIWYG** is turned off, since AutoCAD automatically turns on the colored background.)

### Check Spelling

The **Check Spelling** option toggles real-time spell checking. As you enter or edit text, AutoCAD checks the spelling of each word against its list of correctly-spelled words. When it finds a word missing from the list, AutoCAD underlines the word with a red dashed line, as shown below.

You can backspace to correct the spelling of the word, or else right-click to see a list of alternative spellings.

Choose **More Suggestions** for a sub-menu filled with additional words. Or, choose **Add to Dictionary** when you want AutoCAD to recognize the word in the future.

### Check Spelling Options and Dictionaries

The **Check Spelling Options** items display the Check Spelling Settings dialog box, while the **Dictionaries** option displays the Dictionaries dialog box. Both are the same as found in the Spell dialog box, described later in this chapter.

### Text Highlight Color

The **Text Highlight Color** option specifies the color of highlighted text (default = bluejeans blue, which I find too dark). Use this option to change the color from the Select Color dialog box. In the figure below, "AutoCAD" is shown with the highlight color.

### Insert Field

The **Insert Field** option displays the Field dialog box. This allows you to add automatic text, as described fully in Chapter 17, "Tables and Fields." Examples of field text include the current date and time, the file name of the drawing, and the plot scale.

Field text is identified by a dark gray background, such as the date shown below.

## Using AutoCAD on April 10

When the field text shows "----", this means the text hasn't been determined yet, such as the name of a drawing that is not yet saved. When "####" appears, it means values have not yet been assigned, such as with a sheet set.

### Update Field

After the field is placed in the text, the shortcut menu shows two additional options: Update Field and Convert Field to Text.

The **Update Field** option forces AutoCAD to update the value of fields. This option updates fields now, instead of later. Normally, AutoCAD waits to update fields until certain events occur, such as when a file is opened, saved, plotted or regenerated; or when the **ETRANSMIT** command is used.

### Convert Field to Text

The **Convert Field to Text** option does exactly that: the field is converted to static text. This is useful when you no longer want AutoCAD to update fields.

### Find and Replace

The **Find and Replace** option displays a simple dialog box for finding (and optionally replacing) text.

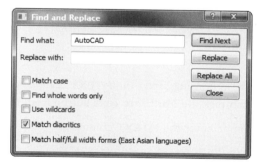

I am not sure of the utility of this option, because the **TEXT** command is meant for small amounts of text, and so finding a specific word is not a problem. A better method is to use the **FIND** command, which searches the entire drawing.

### Select All

The **Select All** option selects all the text in the bounding box.

### Change Case

The **Change Case** option changes text to all UPPERCASE or all lowercase. The change applies to selected text only, and not to field text.

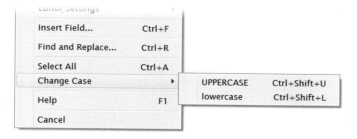

# A MTEXT

The MTEXT command places paragraphs of text in drawings, and provides many formatting options.

While TEXT places text one line at a time, MTEXT (short for "multiline text") is best for adding multiple paragraphs of text. Like a desktop publishing program, it fits the text within invisible boundaries that let you fit the block inof text to specific areas, or to change margin widths easily. (The boundary is not printed or plotted.)

MTEXT permits many text enhancements, such as varying heights and colors, adding bullet points, and stacking fractions. The figure below illustrates paragraph text:

**NOTES:**

**GENERAL**

1.) PROTECT AND SAFEGUARD EXISTING BUILDING AND SERVICES WHICH MAY AFFECTED BY THIS WORK. REPORT ANY UNFORESEEN CONDITIONS TO THE GENERAL CONTRACTOR AND CONSULTANT BEFORE PROCEEDING.
2.) ALL DIMENSIONS TO EXISTING TO BE VERIFIED ON SITE.
3.) GIVE TOP OF FOUNDATION, A TROWEL FINISH. THE FOUNDATION MUST BE SQUARE, LEVEL, AND SMOOTH.

**CONCRETE PILES**

1.) ALL PILES ARE TO BE CENTERED WITHIN 50MM OF LOCATION SHOWN ON   PLAN AND SHALL NOT BE OUT OF PLUMB BY MORE THAN 2% OF PILE LENGTH.
2.) PROTECT AND SAFEGUARD EXISTING BUILDING AND SERVICES WHICH MAY BE AFFECTED BY THIS WORK.
3.) CONSOLIDATE TOP 3000 OF DRILLED CAST IN PLACE CONCRETE PILES WITH MECHANICAL VIBRATOR.

## TUTORIAL: PLACING PARAGRAPH TEXT

In this tutorial, you learn how to start the MTEXT command.

1. To place paragraphs of text in drawings, start the MTEXT command:
   - In the ribbon's Home tab, pick the **Multiline Text** button from the Annotate panel.
   - At the 'Command:' prompt, enter the **MTEXT** command:

     Command: **mtext** *(Press ENTER.)*
   - Alternatively, enter the aliases **t** or **mt** at the 'Command:' prompt.

2. In all cases, AutoCAD reports the current settings, and then prompts you to pick the two corners of the bounding box :

   Current text style: "Standard"  Text height: 0.2  Annotative: No

   Specify first corner: *(Pick point 1.)*

   Specify opposite corner or [Height/Justify/Line spacing/Rotation/Style/Width/Columns]: *(Drag bounding box to point 2.)*

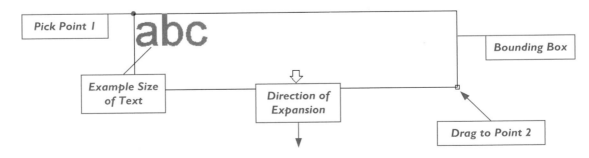

Placing and Editing Text    **497**

The bounding box specifies the space in which the text should fit. Don't worry about the box being exactly the right size; you can always make it bigger or smaller later. In any case, AutoCAD automatically makes it longer as text is added; the direction that the bounding box expands is indicated by the direction of the arrow icon you see in the box.

(You can use the **MTJIGSTRING** system variable to change the preview text, shown as "abc" above.)

3. When the text editor appears, it consist of two parts: (a) a tab on the ribbon for formatting text, and (b) a floating tab bar for setting margins and tabs.

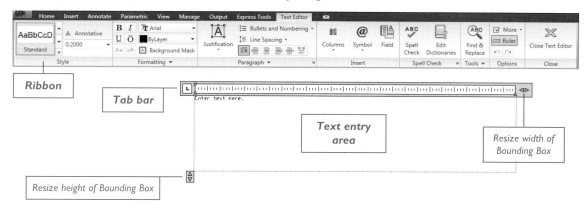

4. Type text into the text entry area.
5. To format the text, you must *highlight* it first, by dragging the cursor over the text you want formatted. Then, from the ribbon toolbar, select a format option.
6. When done, click **OK**. Notice that AutoCAD places the text in the drawing. To exit the mtext editor without saving changes, press **ESC**.

 **Note** To insert the Euro symbol, turn on **NUMLOCK**, hold down the **ALT** key, and then type **0128** on your keyboard's numeric keypad. The € (Euro currency) symbol has been included with all AutoCAD fonts since Release 2000.

## RIBBON MTEXT TAB

The MText tab of the ribbon allows you to apply styles to text; more importantly, it lets you override styles with other properties, such as font and color. We look below at each panel found in the tab.

### Style Panel

Select a style name from the Style droplist. Only text styles defined in the drawing are listed here; you create additional styles with the **STYLE** command. The style applies to all text in the bounding box; you cannot selectively apply styles. You can, however, override styles with font, text height, and other properties provided in the Text Formatting toolbar.

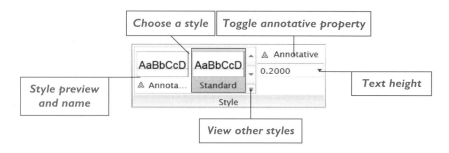

### Annotative Property

Click the ⚙ **Annotative** button to change the font scaling to annotative. You learn more about this option near the end of this chapter.

### Text Height

If you want some (or all) of the text at different heights, highlight the text, and then select a height from the Text Height dropbox.

If the height you need isn't on the list, simply type the measurement into the droplist, and then press **ENTER**. AutoCAD adds the height to the list.

## Formatting Panel

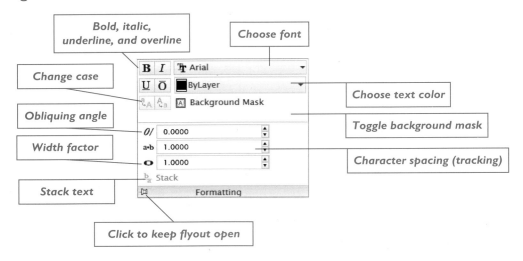

### Bold, Italic, Underline, and Overline

To **boldface** text, highlight the text, and then choose the **B** **Bold** button. Choose the button a second time to "unboldface" the text. Pressing **CTRL+B** performs the same function.

When the **B** on the button looks gray, the current font cannot be boldfaced — usually a problem with certain AutoCAD fonts.

To italicize text, highlight the text, and then choose the *I* **Italic** button. (As an alternative, press **CTRL+I**.) When the **I** button is gray, the text cannot be italicized.

To underline text, highlight the text, and then choose the **U** **Underline** button. All fonts can be underlined; **CTRL+U** is the keystroke shortcut.

To overline text, highlight the text, and then choose the **Ō** **Overline** button. All fonts can be overlined; **CTRL+O** is the keystroke shortcut.

### UPPER and lower Case

The **Uppercase** and **Lowercase** buttons change selected text to all UPPERCASE or all lowercase. There are a couple of keyboard shortcuts that bypass the menu. Select text, and then press the following keys:

**CTRL+SHIFT+U** converts selected text to all UPPERCASE.

**CTRL+SHIFT+L** converts selected text to all lowercase.

## Font

Select a font name from the Font droplist. The fonts listed are those installed on your computer. The list probably includes all TrueType fonts included with Windows, as well as fonts installed by AutoCAD and other applications. TrueType fonts have a tiny **T** logo in front of them; AutoCAD fonts are prefixed by a tiny Autodesk **A** logo. (Printer fonts, PostScript fonts, and other fonts are not listed, because AutoCAD does not support them directly.)

You can use one or more fonts in one block of mtext, but remember to highlight the text before selecting the name of a font:

> To highlight *some* text, click and drag it with the cursor.
>
> To select *all* text, press **CTRL+A**. Alternatively, right-click the text and choose **Select All** from the shortcut menu.

## Color

Highlight a portion of the text, and then select a color from the ■ **Color** list box. You can choose Bylayer, Byblock, the seven basic AutoCAD colors, and the other 16.7 million colors and color books.

## Background Mask

The **Background Mask** option places a colored rectangle behind paragraphs of text. The size of the background mask is the same as the mtext window, which may be larger or smaller than the area of the text. You can resize the mask with grips editing.

To create the background mask, follow these steps:

1. Click the **Background Mask** button on the ribbon.
2. In the dialog box, turn on the **Use Background Mask** option.
3. Select the fill color, and then click **OK**.

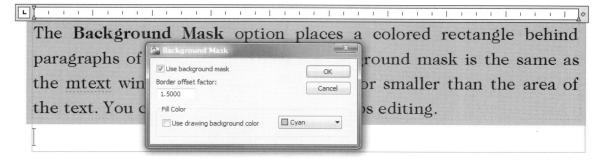

If you turn on the **Use Drawing Background Color** option, the background color is (usually) white, masking the other drawing entities behind it. You may need to use the TEXTTOFRONT command to display text on top of other objects.

**Note** To create white text on a colored background, select a True Color as follows:
1. In the Text Formatting toolbar's Color droplist, select **Select Color**.
2. In the Select Color dialog box, click the **True Color** tab.
3. Select any color.
4. Drag the **Luminance** slider to the top, turning the text color to white.
5. Click **OK**. The text is white.

### Oblique Angle, Tracking, and Width Factor

The *0/* **Oblique Angle** option changes the slant of selected text, almost like applying italics, except that text can be slanted forward and backwards by a specific amount (up to 85 degrees).

The *a-b* **Tracking** option changes the distance between characters. By reducing the tracking, you can squeeze more text into a line.

The *o* **Width Factor** option makes selected characters wider and narrower. Again, by making characters narrower, you can squeeze more text into a line. (The **STYLE** command illustrates these properties later in this chapter.)

### Stack

The **Stack** button stacks fractions by changing side-by-side text, like 11/32, to show the 11 over the 32. (This button is not new to AutoCAD 2011, but is new to the Text Editor tab.) Using this feature takes these steps:

1. Type the fraction, such as 11/32.
2. Highlight the entire fraction.
3. Click the **Stack** button.

AutoCAD replaces the numbers with a stacked fraction. To "unstack" stacked fractions, repeat steps 2 and 3.

This button is not limited to stacking numbers: it stacks any combination of text, numbers, and symbols — whatever is in front of the slash goes on top; whatever is behind goes underneath.

Also, you are not limited to working with slash symbols. The carat ( ^ ) and the hash or pound mark ( # ) indicate different kinds of stacking:

**Slash** ( / ) stacks with a horizontal bar.      15/16      $\frac{15}{16}$

**Carat** ( ^ ) stacks without a bar (tolerance style).  12^12   $\genfrac{}{}{0pt}{}{12}{12}$

**Pound** ( # ) stacks with a diagonal bar.   12#12   $^{12}/_{12}$

When you right-click stacked text, a new option is added to the shortcut menu that appears: **Stack Properties**. It displays the dialog box illustrated below.

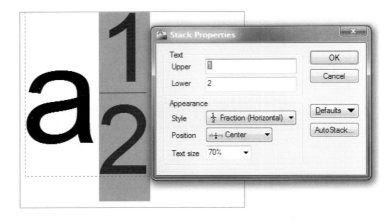

**Upper** and **Lower** fields specify the fraction text. This lets you change the values.

**Style** droplist specifies the three slash styles noted above: horizontal, diagonal, and none.

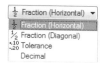

**Position** droplist specifies the position of the fraction relative to the baseline of text on either side of the fraction: top, center, or bottom.

**Text size** droplist specifies the size of fraction text relative to regular text. The range is from 50% (half as tall) to 100% (same size); the default is 70%.

**Defaults** button saves the current settings as the default, or restores previously saved defaults.

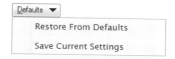

**AutoStack** button displays the AutoStack Options dialog box.

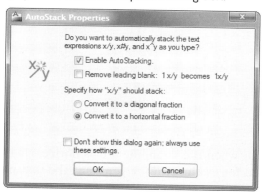

**Enable AutoStacking** option toggles the automatic stacking of numbers on either side of the /, #, and ^ characters.

**Remove Leading Blank** removes blanks between whole numbers and stacked fractions. This option is available only when AutoStacking is turned on.

**Convert It to a Diagonal Fraction** creates stacked numbers with the diagonal slash, regardless of the stack character (/, #, and ^).

**Convert It to a Horizontal Fraction** creates stacked numbers with the diagonal slash, regardless of the stack character.

**Note** When drawings with diagonal stacked fractions are opened in AutoCAD Release 14 or earlier, they are converted to horizontal fractions, but are restored when opened in AutoCAD 2000 or later.

## Paragraph Panel

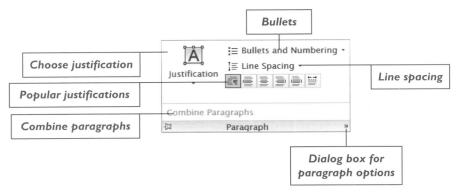

### Justification

The **Justification** button changes the justification of the entire mtext block relative to the bounding box. You can choose from left, center, and right; and from top, middle, and bottom.

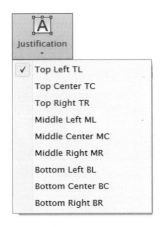

Vertical justification works only when the bounding box is noticeably longer than the paragraphs of text. You would select **Bottom**, for example, when you want the text block to rest at the bottom of the bounding box.

### Popular Justifications

The **Horizontal Justification** buttons change the justification of the text relative to the sides of the bounding box. You can choose from Left, Center, Right, Justify, and Distribute.

Both **Justify** and **Distribute** force the left and right margins to be even. The difference between the two is how the last line in a paragraph is handled: Justify left justifies it, while Distribute forces the last line to fit between the left and right margins, as illustrated below.

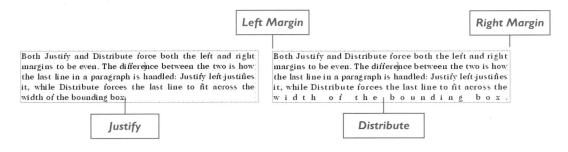

## Bullets

The ⋮≡ **Bullets** button specifies the format of bullets: letters, numbers, symbols, or none.

AutoCAD automatically numbers (or letters) and indents the text for you. Each paragraph (text that ends with a hard return) is bulleted separately.

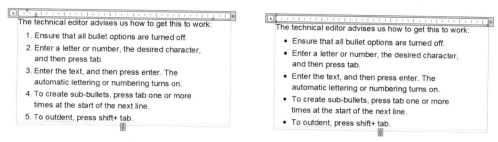

**Left:** *Automatically numbered paragraphs.*
**Right:** *Bulleted paragraphs.*

Click the **Bullets** button for this list of options:

**Off** — removes bulleting from the mtext.

**Numbered** — applies number bullets to paragraphs of text; numbers are Roman numerals only.

**Lettered** — applies letter bullets to paragraphs of text. Once it reaches Z, it continues with AA, BB, and so on. Letters can be all uppercase or all lowercase.

**Bulleted** — places round bullets, like this one •.

**Start** — starts letter bullets from A; starts numbered bullets from 1.

**Continue** — adds selected paragraphs to the previous list.

**Allow Auto-list** — applies bullet formatting as you type.

**Use Tab Deliminter Only** — restricts automatic bullet creation to when the **TAB** key is pressed.

**Allows Bullets and Lists** — applies bullets to all text objects that look like lists (namely those that begin with one or more letters or numbers or a symbol, followed by punctuation, a space created by **TAB**, and that end with **ENTER** or **SHIFT+ENTER**). When off, this option removes bullet formatting and turns off all other options.

To change the bullet style, click the **Options** button (looks like a downward pointing arrow), and then select **Bullets and Lists** from the menu.

You can also add punctuation after letter and number bullets: periods ( . ), commas ( , ), close parentheses ( ) ), close angle brackets ( > ), close square brackets ( ] ), and close curly braces ( } ). The technical editor advises us how to get this to work:

1. Ensure that all bullet options are turned off.
2. Enter a letter or number, the desired character, and then press **TAB**.

3. Enter the text, and then press **ENTER**. The automatic lettering or numbering turns on.
4. To create sub-bullets, press **TAB** one or more times at the start of the next line.
5. To outdent, press **SHIFT+ TAB**.

### Line Spacing

The **Line Spacing** buttons change the spacing between lines. Select the text, and then choose a spacing. This increases the spacing.

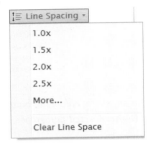

### Combine Paragraphs

The **Combine Paragraphs** option groups selected paragraphs into a single paragraph. You need to select at least two paragraphs for this option to operate.

### Paragraph Options

The **Paragraph** option (named **Indents and Tabs** in earlier releases of AutoCAD) displays a dialog box that controls all aspects of paragraphs.

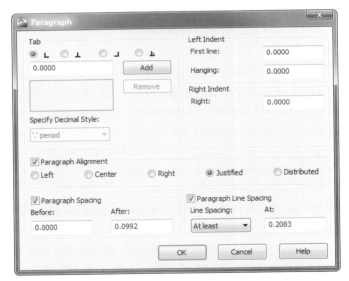

**Tab** section lets you enter exact tab stop positions. Tab stops line up columns of data.

**Left Indent** and **Right Indent** sections allow you to enter precise distances to indent the first line and the paragraph (i.e. every line).

**Paragraph Alignment** section selects the type of paragraph justification.

**Paragraph Spacing** section sets the distance between paragraphs, defined as a distance before and after the paragraph.

**Paragraph Line Spacing** section sets the spacing between lines, despite the word "paragraph" in the name.

### Insert Panel

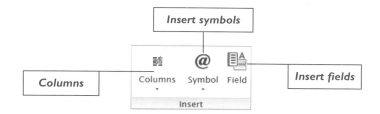

### Columns

The **Columns** button sets the number of columns. It splits the mtext block into two or more columns. This is useful when you need to fit the text into an area that is wider than it is deep.

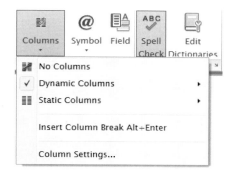

You can choose from the dynamic or static columns, or else force the start of a new column with ALT+ENTER.

**Dynamic** — AutoCAD creates columns automatically as text is added to the bounding box.

**Static** — you tell AutoCAD the number of columns you want, 2 or more.

**Columns Settings** — opens a dialog box that defines all column options in one place.

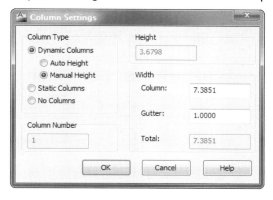

### Insert Symbol

The **@ Symbol** button displays a menu of symbols commonly used in drawings. Click **Other** to display the Windows Character Map dialog box, which provides access to all symbols. (Ironically, "@" is not one of the available symbols; it is available through the keyboard.)

Behind the symbol names are Unicode numbers for each character, such as \U+2248. Some of the symbols can be used for Greek characters; for example, **Ohm** is the same as **omega**.

One symbol is invisible: the "non-breaking space." When you place it between two words, it prevents AutoCAD from using that space to wrap the sentence. For instance, if your text includes the phrase "one inch," you might want a non-breaking space between the "one" and "inch" to keep the two words together.

**Other** opens the Windows Character Map dialog box, which contains all the characters available for the font.

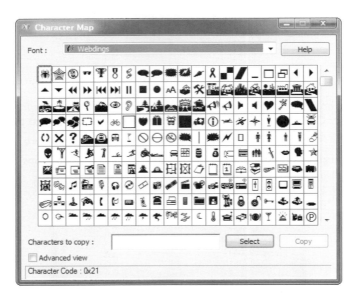

Inserting a symbol in this way requires this rather awkward process:

1. Choose the font name.
2. Select the symbol.
3. Click **Select**. If you wish, choose additional symbols, clicking **Select** after you choose each one.
4. Click **Copy**.
5. Clicking **x** to close the dialog box.
6. Right-click in the mtext editor where you want the symbol places.
7. Select **Paste**.

### Insert Fields

The **Insert Field** button displays the Field dialog box. See the FIELD command elsewhere in this chapter.

## Spell Check Panel

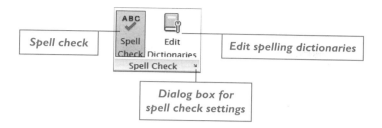

### Spell Check

The **Spell Check** button displays the Spell Check dialog box; see the **SPELL** command later in this chapter.

### Edit Dictionaries

The **Edit Dictionaries** button displays the Dictionaries dialog box, described later.

### Spell Check Settings

The **Spell Check Settings** button displays the Check Spelling Settings dialog box, also described later.

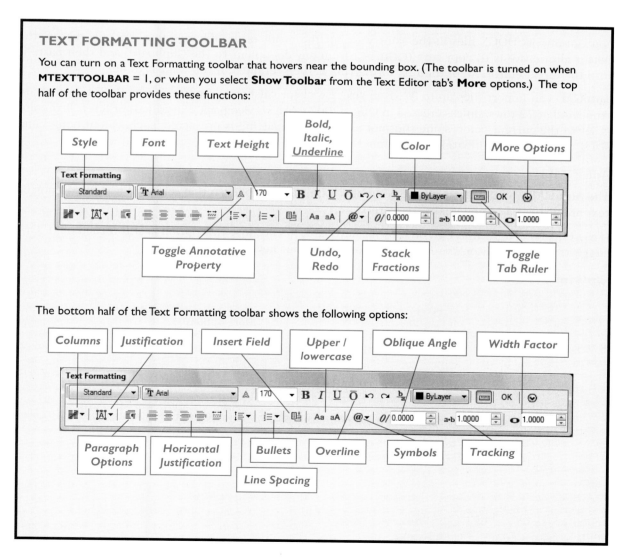

**TEXT FORMATTING TOOLBAR**

You can turn on a Text Formatting toolbar that hovers near the bounding box. (The toolbar is turned on when **MTEXTTOOLBAR** = 1, or when you select **Show Toolbar** from the Text Editor tab's **More** options.) The top half of the toolbar provides these functions:

The bottom half of the Text Formatting toolbar shows the following options:

### Tools Panel

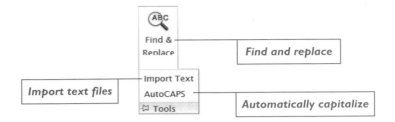

#### Find and Replace

The **Find and Replace** button displays the Find dialog box; see the FIND command later in this chapter.

#### Import Text

The **Import Text** button displays the Select File dialog box, from which you can select document files that were previously saved in plain text (ASCII format) or RTF (rich text format). RTF files are imported with formatting intact.

The Import Text option will also import text from documents saved in popular word processing formats, such as Word (DOC). To do so, enter *.doc into the dialog box's **File Name** field. These documents are imported without formatting and may display curious character strings, such as ÐÏà¡±áÿÿÿÿÿÿÿ — these are internal formatting codes used by the documents.

You can import DOCX files in the same way, but the result is all gibberish text. For this reason, the better alternative is to copy text (CTRL+C) from any document to the Clipboard, and then paste it into the mtext editor. All of the text formatting is then retained, if you want it to be.

AutoCAD can import files up to 32KB in size. Excel spreadsheets imported through RTF format are truncated at 72 rows, unless created in Office 2002 (with service pack 2 installed) or later. Text color is set to the current color; some (but not all) other formatting is preserved, such as font names and sizes, though not justification or columns.

#### AutoCAPS

The **AutoCAPS** toggle turns on the computer's CapsLock mode, so that all typed text and imported text are displayed in uppercase characters.

If the CapsLock light won't go off on your keyboard, it's because this option is still turned on. ("Or," suggests the copy editor, "you spilled Gatorade on your keyboard.")

### Options Panel

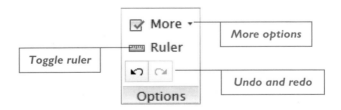

### More Options

The ☑ **More** button displays the following menus. (This option was formerly known as "Options.")

### Character Set

The **Character Set** option changes the character set to match local language requirements. Not all fonts support all character sets.

### Remove Formatting

The **Remove Formatting** option removes bold, italic, and underline formatting from the selected text. The submenu offers these options:

**Remove Paragraph Formatting** — removes formatting from paragraphs, such as justification.

**Remove All Formatting** — removes all format overrides. This option returns the font to the current style.

### Editor Settings

The **Editor Settings** item controls the display of elements in the mtext editor, and is similar to the one displayed by the TEXT command:

**Always Display as WYSIWYG** — edits text parallel to the viewing plane, even when at an angle.

**Show Toolbar** — toggles the display of the entire Text Formatting toolbar.

**Opaque Background** — toggles the gray background of the text entry area.

**Text Highlight Color** — selects a color for highlighted text.

With all options turned off, you have a bare bones text editor that looks very similar to that of the TEXT command:

### Toggle Ruler

Click the ▭ **Ruler** button to hide and display the tab ruler, located atop the boundary box.

### Undo and Redo

To reverse the last operation, choose the ↶ **Undo** button. Or press CTRL+Z.

Choose the ↷ **Redo** button to "undo" the undo. Or press CTRL+Y.

**Close Panel**

Clicking the ✖ **Close Text Editor** button exits the MTEXT command; you can edit the text later by double-clicking it.

## MTEXT SHORTCUT MENU

The limited space provided by the ribbon means there isn't enough room to show all available mtext operations. Right-clicking the text entry area displays a shortcut menu with additional options. The content of the menu varies according to the text selected, but typically looks like this:

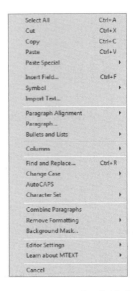

The first four options on the shortcut menu are standard: **Select All** (CTRL+A), **Cut** (CTRL+X), **Copy** (CTRL+C), and **Paste** (CTRL+V).

### Paste Special

Before you can paste, you have to copy some text to the Clipboard, either text from the drawing or from another document. The regular Paste option pastes text from the Clipboard with all formatting intact. For instance, if the text contains several fonts and colors, they are preserved when pasted as mtext.

The **Paste Special** item provides ways to paste text with different ways of keeping or not keeping the formatting. Its options can be confusing to sort out, not least because Autodesk's documentation lacks explanations for them. Here are the variations on Paste Special:

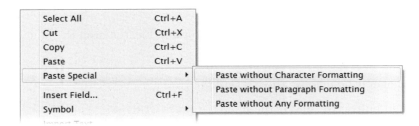

**Paste without Character Formatting** — pastes the text and the applies the current AutoCAD style (removes source formatting), although tabs and indents are preserved.

**Paste without Paragraph Formatting** — pastes text, keeps original character formatting (such as font and boldface), but removes paragraph formatting, such as indents and centered justification.

**Paste without Any Formatting** — pastes text with the current style. (Frankly, I have not figured out how this one differs from pasting without character formatting.)

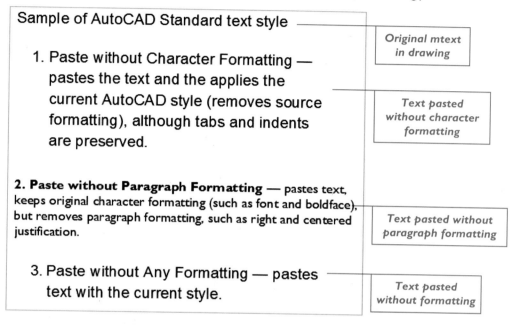

### MTEXT TEXT ENTRY AREA

The text entry area is a rectangle that represents the mtext bounding box. The box is topped by a tab bar that sets tabs, indents, and margins. A different indent and tab setting can be applied to each paragraph.

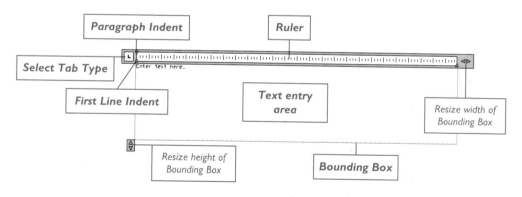

### First Line Indent

The first line indent shows how far in the first line of each paragraph is indented. You move its arrowhead along the tab bar by dragging it back and forth with the cursor.

You can also create hanging indents: move this indent farther left than the paragraph indent. Hanging indents are created automatically when you apply bullets and lists.

### Paragraph Indent

The paragraph indent shows how far the entire paragraph should be indented. Only the left indent can be modified; to change the right indent, drag the boundary box margin.

### Tabs

Click the tab bar to set tabs; existing tabs can be dragged back and forth along the tab bar. AutoCAD supports the commonly used left tabs (illustrated below), as well as right, center, and tabs for lining up text by decimals.

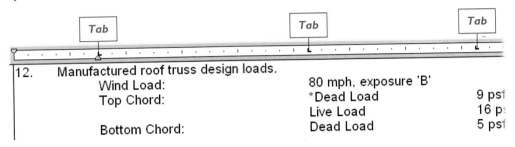

Change the tab type by clicking the **L** icon at the left end of the tab bar.

Right-clicking the tab bar reveals a shortcut menu:

**Paragraph** displays the Paragraph dialog box that you read about earlier.

### Set Mtext Width and Set Mtext Height

The **Set Mtext Width** and **Set Mtext Height** options display identical-looking dialog boxes that specify a precise width or height for the bounding box.

As a less-precise alternative, you can just drag the right edge and bottom edges of the bounding box, making it wider and longer.

## PLACING PARAGRAPH TEXT: ADDITIONAL METHODS

When the MTEXT command starts up, it displays a prompt line with many options. Each time you respond to an option (except the **Width** option), AutoCAD redisplays the following MTEXT prompt — until you pick the opposite corner:

Specify opposite corner or [Height/Justify/Line spacing/Rotation/Style/Width/Columns]:

**Height** specifies the height of text.
**Justify** specifies the justification of text in the bounding box.
**Line spacing** specifies the spacing between lines of text.
**Rotation** rotates the bounding box.
**Style** selects the text style.
**Width** specifies the width of the bounding box.
**Columns** specifies the number of columns.

Let's look at each.

### Height

The **Height** option changes the height of the text used by the mtext editor. It prompts you as follows:

Specify height <0.2000>: *(Type a number or indicate a height.)*

### Justify

The **Justify** option selects the justification and positioning of text within the bounding box. AutoCAD prompts:

Enter justification [TL/TC/TR/ML/MC/MR/BL/BC/BR] <TL>: *(Enter an option, or press ENTER.)*

The justification options are the same as for the MTEXT command, except that in this case, there are two areas where justification applies: (1) text and (2) flow.

Text justification is left, center, or right, relative to the left and right boundaries of the rectangle. Flow justification positions the block of text top, middle, and bottom relative to the top and bottom boundaries of the rectangle.

### Linespacing

The **Linespacing** option changes the spacing between lines of text, sometimes called the "interline spacing" or "leading." The option has two methods for specifying the spacing: **At least** and **Exactly**. The **At least** option specifies the minimum distance between lines, while **Exactly** specifies the precise distance between lines of text.

Enter line spacing type [At least/Exactly] <At least>: **a**

Enter line spacing factor or distance <1x>: *(Enter a value, including the **x** suffix.)*

The value you enter is a multiplier of the standard line spacing distance, which is defined in the font.

### Rotation

The **Rotation** option specifies the rotation angle of the bounding box.

Specify rotation angle <default>: *(Enter an angle, or pick two points.)*

This option rotates the entire block of text. For example, enter **90** degrees to place text sideways at the edge of a drawing.

 **Note** When you use the mouse to show the rotation angle, AutoCAD calculates the angle as follows:

The *start* of the angle is the positive x-axis.
The *end* of the angle is the line anchored by the "Specify first corner:" pick; the other corner is now in line with the indicated angle.

### Style

The **Style** option selects a text style to use for the multiline text.

> Enter style name (or ?) <Standard>: *(Enter a name, or press* **?** *for a list of style names.)*

When you enter **?** in response to this prompt, AutoCAD displays the names of text styles defined in the drawing. You create new text styles with the **STYLE** command (described later in this chapter) or with a style borrowed from another drawing using the DesignCenter.

### Width

The **Width** option specifies the width of the bounding box.

> Specify width: *(Enter a value, or pick a point to show the width.)*

A width of 0 (zero) has special meaning. AutoCAD draws the multiline text as one long line (no word wrap), as if there were no bounding box at all.

### Columns

The **Columns** option specifies the number of columns.

### TEXTTOFRONT

The **TEXTTOFRONT** command brings text to the front of the display order. It works with text and dimensions. (Use the **HATCHTOBACK** to move back patterns behind everything else, and the **DRAWORDER** command for all other objects.)

You can force AutoCAD to display all text over top of all other objects in the drawing, like this:

> Command: **texttofront**
>
> Bring to front [Text/Dimensions/Both] <Both>: **t**
>
> *n* object(s) brought to front.

This command is much more efficient than the **DRAWORDER** command, because it selects only text (and/or dimensions), instead of any object.

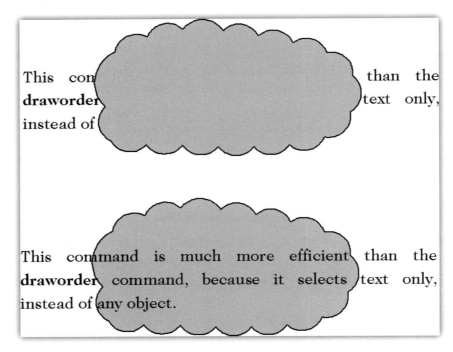

**Top**: *Text behind another object.*
**Above**: *Text brought to the front.*

**STYLE**

The **STYLE** command defines named text styles, which are based on fonts.

A *style* is a collection of properties applied to the font of your choice, such as slanting the text and specifying the height. Styles are often used in word processing and desktop publishing, and work the same way in AutoCAD. The idea is to preset most text properties so that you don't need to do it each time you start the **TEXT** or **MTEXT** command. (Some properties, such as color and layer, cannot be set by styles.)

Text styles are also used by attributes, fields, leaders, and dimensions. Styles are stored and accessed by names of your choice of up to 256 characters. Here is some of the most important information stored by styles (not all fonts support all properties):

> **Font file** associates a *.ttf* (TrueType) or *.shx* (AutoCAD) file with the style. PostScript *.pfb* font files can be used after conversion to *.shx* with the **COMPILE** command.
>
> **Font style** selects regular, boldface, or italicized text.
>
> **Height** specifies the height of the text. Normally, this is set to 0, which means that the **TEXT** command prompts you for the height. Entering a height here means the command doesn't pester you later.
>
> **Annotative** toggles annotative scaling.
>
> **Orientation** makes text independent of drawing and layout rotation.

You met a number of these properties earlier through the Text Editor tab. When you change a style, all text assigned to that style also changes.

### TUTORIAL: DEFINING TEXT STYLES

1. To define one or more text styles, start the **STYLE** command:
   - In the ribbon's Annotate tab, choose **Manage Text Styles** in the Text panel's style name droplist.
   - At the 'Command:' prompt, enter the **STYLE** command.

     Command: **style** *(Press* **ENTER**.*)*
   - Alternatively, enter the aliases **st** or **ddstyle** (the old name) at the 'Command:' prompt.

   Notice that AutoCAD displays the dialog box:

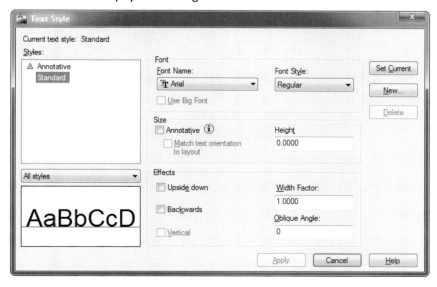

2. The **Styles** list provides the names of styles already in the drawing. Use this list to change the properties of existing styles. (Options that do not apply to a font or style are grayed out.)
3. To create a new style, click **New**. Notice that AutoCAD displays the New Text Style dialog box. This is where you name the style.

Change the default name "style1" to anything else — up to 255 characters long — and then choose **OK**. Notice that AutoCAD adds the name to the Styles list.

 **Notes** To rename a style, right-click its name, and then choose **Rename** from the shortcut menu. You can rename all styles except Standard.

The **Delete** option erases selected styles from drawings, but only if they are unused. If a style is used by any amount of text, AutoCAD complains, "Style is in use, can't be deleted." The Standard style can never be deleted.

4. From the **Font Name** droplist, select a font. There are many font files available for Auto-CAD, which comes with an large selection: specialty fonts for map making (symbols instead of letters), cursive writing, and so on. The font file should be appropriate to the application:
    - Most mechanical applications use the ANSI type lettering, often called "Leroy." The Leroy lettering machines were made by K&E, and permitted precise lettering on hand-drawn plans. The Simplex font (also known as "RomanS," short for Roman, single stroke) closely approximates this style.
    - Architectural drawings typically use "hand-lettered" fonts, such as City Blueprint.
    - Engineering applications often use the Simplex font, since it closely resembles a traditional font constructed with a lettering template.

Multistroke SHX fonts use a code in which "S" denotes single stroke fonts, "D" duplex or double stroke, "C" complex (also double stroke), and "T" triple stroke.

As of AutoCAD 2011, font names display a  warning icon when AutoCAD cannot find the font file.

5. Select any of the following options, if available (not grayed out):

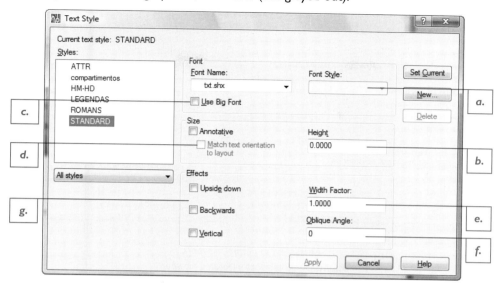

a. The Font Style droplist selects Regular, **Boldface**, *Italic*, or ***Bold Italic***. Not all font styles are available, because some fonts cannot handle these effects.

b. The **Height** text box specifies the height of the text, which is measured from the baseline to the top of uppercase letters. There are three ways to specify the height:

- Determine the scale of the drawing before deciding the text height, as described earlier in this chapter.
- Enter a value of **0**. This gives you the flexibility to change the height as the text is being placed. This applies particularly if you are not sure of the scale until the end of the drawing process.
- Turn on the **Annotative** option, and then specify the true (paper) height.

c. The **Use Big Font** option refers to a class of AutoCAD SHX fonts that handle more than 256 characters, primarily for Chinese and other languages with thousands of characters. TrueType fonts don't need this option.

d. **Annotative** scales the text automatically according to the layout scale, as described later. When Annotative is turned on, then the following option becomes available:

**Match Text Orientation to Layout** ensures the text is always displayed "correctly," no matter how the rest of the drawing is rotated. The name of this option is somewhat misleading, because it works even when the layout is not rotated. In the figure below, the entire drawing (including the text) is rotated 90 degrees. The text keeps its orientation, which makes it easier to read.

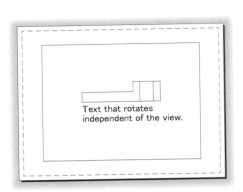
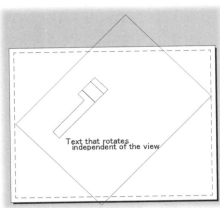

**Left:** *Text in normal viewport.*
**Right:** *Unrotated text in rotated viewport.*

e. The **Width Factor** determines the width of characters. The width of 1.0 is "standard." A decimal value, such as 0.85, draws text narrower by 15%, which is useful for condensed text. Values greater than 1 expand text.

## 2.0 Width Factor
### 0.5 width factor

f. **Oblique Angle** determines the slant of characters. A zero obliquing angle draws text that is "straight up." Positive angles, such as 15 degrees, draw text that slants forward, while negative angles draw text that slants backward. Valid values are between –85 and 85 degrees.

*15° Obliqued Text*
*-15° Obliqued Text*

g. **Upside Down** draws text upside down.

## ʇxǝʇ uʍop ǝpᴉsdՈ

**Backwards** means the text is drawn in reverse. Backwards text is useful for plotting drawings on the back side of clear media, or for mold drawings used by the casting and injection-molding industries.

## txeT sdrawkcaB

**Vertical** draws text vertically, which is *not* the same as rotating text by 90 degrees. Text is drawn so that each letter is vertical and the text string itself is vertical. TrueType and some AutoCAD fonts cannot be drawn vertically.

**Note** Obliquing, underscoring, or overscoring should not be used with vertical text, since the result will look incorrect.

6. Click **Set Current** to make the style current; this means any text you draw now takes on this style.
7. Click **OK** to exit the dialog box.

### USING TEXT STYLES

Styles modify the properties of a font, such as its height, width, and slant. One font can be used by many styles, but each style can specify only one font.

The difference between fonts and styles is sometimes confusing. You cannot use fonts directly in AutoCAD; fonts can only be used through styles. (The exception is in the mtext editor, but even there fonts only override styles.)

1. To use a font, you must create a style with the **STYLE** command.
2. Then, in the **TEXT** and **MTEXT** commands, you specify the style to use for the text you place.

Each line or paragraph of text is assigned a style; you may change the style with the **PROPERTIES** command. When the style changes, the look of the text also changes. To change text globally, change the style with the **STYLE** command.

Often, drafting offices create a standard selection of styles, each of which is used for a particular type of text in drawing. For example, one style is used for notes, another for titles, and others for the different parts of the title block. Using styles ensures consistency across drawings, no matter the drafter.

### Styles Panel

An alternative method to applying styles is using the ribbon's Home | Annotate panel.

1. Select one or more lines of text.
2. Pick a style name from the style droplist:

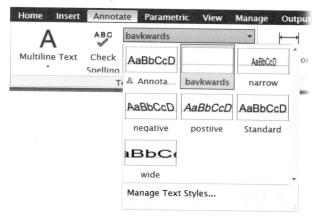

Notice that AutoCAD changes the look of the text to match the style.

### DEFINING STYLES: ADDITIONAL METHODS

Several system variables affect the look and quality of TrueType fonts.

- **FONTALT** specifies font substitution.
- **TEXTFILL** toggles the fill of TrueType fonts for plots.
- **TEXTQLTY** adjusts the quality of TrueType fonts for plots.

## FontAlt

The **FONTALT** system variable specifies the name of the font to use when other fonts cannot be found.

Often, different computers have different collections of installed fonts. When you receive a drawing from another AutoCAD drafter, the drawing may use fonts not found on your computer. So that text doesn't go missing, AutoCAD substitutes the *simplex.shx* font for missing fonts. (If the Simplex font cannot be found, AutoCAD displays the Alternate Font dialog box, so that you can select another font file.)

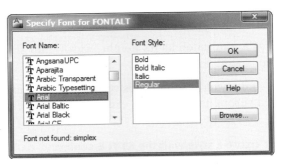

Command: **fontalt**

Enter new value for FONTALT, or . for none <"simplex">: *(Enter the name of a TrueType or AutoCAD font.)*

The technical editor notes that problem fonts can be caused by the setting of AutoCAD's **FONTALT** system variable, the contents of the *acad.fmp* file, and/or font substitutions made by Windows.

## TextFill

The **TEXTFILL** system variable toggles the fill of TrueType fonts for plots. When on (the default), fonts are filled; when off, only the outline of the font is plotted. This system variable affects only the plotting of fonts, not the display of fonts in drawings.

Command: **textfill**

Enter new value for TEXTFILL <1>: **0**

## TextQlty

The **TEXTQLTY** system variable specifies the resolution of TrueType font outlines (technically, the tessellation fineness) when plotted: 0 means the text is not smoothed; 100 means maximum smoothness. The difference is seen only when drawings are plotted; this system variable does not affect the display.

Command: **textqlty**

Enter new value for TEXTQLTYL <50>: **100**

## QLEADER

The **QLEADER** command places callouts (a.k.a leaders) in drawings. The figure below illustrates examples of leaders (at left) and multileaders (at right) in a drawing.

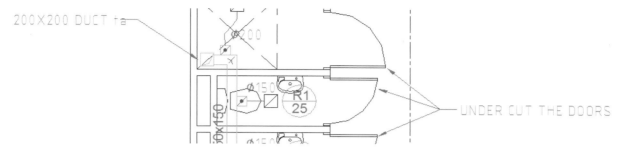

AutoCAD has several commands for creating leaders. The differences among them are as follows:

- **QLEADER** (short for "quick leader") — includes a Settings dialog box for fashioning the leader's properties. This is the command you'll want to use most of the time.

- **MLEADER** (short for "multiline leader") — groups multiple leader lines and block annotation. Mleaders can have preset styles created through the **MLEADERSTYLE** command, and can collect multiple leaders into one. Unlike leaders and qleaders, mleaders are limited to mtext and blocks; they do not include tolerances or selected objects.

- **LEADER** — enters all options at the command line. This is an earlier method of creating leaders, and has been left in AutoCAD for compatibility reasons.

- **DIM: LEADER** — also enters options at the command line, the original method.

### TUTORIAL: PLACING LEADERS

In the following tutorial, you draw a rectangle, and then specify its size through a callout, placed as a leader.

1. Open a new drawing in AutoCAD with the *acad.dwt* template.
2. Use the **RECTANGLE** command to draw a rectangle of any size.
3. To place the callout in the drawing, start the **QLEADER** command:
   - At the 'Command:' prompt, enter the **QLEADER** command:

     Command: **qleader** *(Press ENTER.)*
   - Alternatively, enter the **le** alias at the 'Command:' prompt.
4. AutoCAD prompts you for the *first point*, which is the tip of the leader's arrowhead:

   Specify first leader point, or [Settings] <Settings>: *(Pick a point near the rectangle, shown as point 1 in the following figure.)*

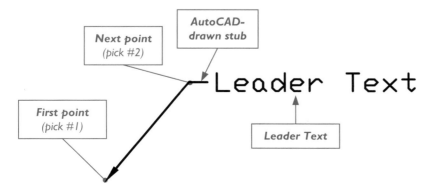

5. AutoCAD prompts you to pick the *next point*.
   Specify next point: *(Pick point 2.)*

6. From now on, each of your picks indicates the location of the next vertex in the leader line — until you press **ENTER** to stop drawing it. (Later, AutoCAD will add the short horizontal stub automatically.)
   Specify next point: *(Press **ENTER** to end drawing the leader line.)*

7. Enter **0** for unconstrained width. (If the width of the text must be constrained, enter the width here.)
   Specify text width <0.0000>: *(Press **ENTER**.)*

8. Enter the text, and then press **ENTER**. For this tutorial, enter the size of the rectangle, such as 24"x36".
   Enter first line of annotation text <Mtext>: *(Enter text, such as **24"x36"**.)*

9. Press **ENTER** twice in a row to exit the command.
   Enter next line of annotation text: *(Press **ENTER** twice to exit command.)*

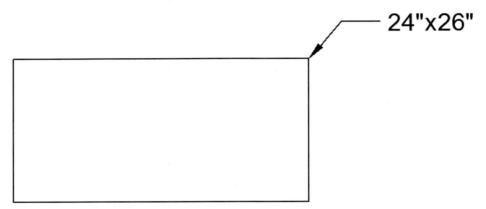

If the leader's size overwhelms the drawing, use the **DIMSCALE** system variable to change the size prior to placing the leader. To make the leader text and arrowhead smaller, use values between 0 and 1; for larger leaders, use values above 1.

If you use object snap to touch the leader's start point to an object, the leader is associated with the object. This means that when you move the object, the arrowhead moves with the object, although the leader text stays in place. (Associativity is available when associative dimensioning is turned on through the **DIMASSOC** system variable — which is the case in most drawings.)

Many of the leader's properties, such as line color and arrowhead scale, are determined by system variables related to dimensioning, all of which start with **DIM**. To modify the properties, use the **DIMSTYLE** command, as described in the next chapter.

You can use grips editing to modify the leader's position. To edit the leader text, double-click the text (not the leader line), and AutoCAD displays the familiar mtext editor.

### PLACING LEADERS: ADDITIONAL METHODS

The **Settings** option displays a dialog box for specifying the look and operation of "q"-leaders.

Command: **qleader** *(Press **ENTER**.)*

Specify first leader point, or [Settings] <Settings>: *(Type **S**.)*

Notice the dialog box.

## Annotation Type

The options you select under **Annotation Type** affect the prompt displayed by the QLEADER command.

> **MText** (default) — displays prompts for creating leader text from mtext:
> Specify text width <0.0000>: *(Enter a width, or press ENTER.)*
> Enter first line of annotation text <Mtext>: *(Enter text.)*
> Enter next line of annotation text: *(Press ENTER to end the command.)*

> **Copy an Object** — prompts you to select another line of text already in the drawing. This can be multiline text (including other leader text), single-line text, a tolerance object, or a block.
> Select an object to copy: *(Select one object.)*
>
> **Tolerance** — displays the Tolerance dialog box when the command requires annotation; see Chapter 16, "Geometric Dimensioning and Tolerancing."
>
> **Block Reference** — prompts you to insert a block in place of the leader text:
> Enter block name or [?]: *(Enter the name of a block)*
> Specify insertion point or [Scale/X/Y/Z/Rotate/PScale/PX/PY/PZ/PRotate]: *(Pick a point.)*
> Enter X scale factor, specify opposite corner, or [Corner/XYZ] <1>: *(Enter a scale factor.)*
> Enter Y scale factor <use X scale factor>: *(Press ENTER.)*
> Specify rotation angle <0>: *(Enter an angle.)*
>
> **None** — removes extra prompts; draws the standard arrowhead and leader line.

## MText Options

The **MText Options** become available only when the **MText** option was selected (see above).

> **Prompt for Width** (default = on) — toggles the display of prompts for the width of the mtext leader text:
> Specify text width <0.0000>: *(Enter a width, or press ENTER.)*
>
> **Always Left Justify** (default = off) — toggles left-justification of mtext, regardless of leader location. AutoCAD normally adjusts the text justification based on the leader line's orientation. When this option is turned on, multiple lines of text are left-justified when the leader angles left, as illustrated by the figure below. When this option is turned off, multiple lines of text are right-justified.
>
> **Frame Text** (default = off) — toggles the addition of a rectangular frame around the mtext annotation.

Annotation Reuse

The options under **Annotation Reuse** determine whether the leader text or other annotation is reused by subsequent QLEADER commands.

**None** (default) — causes AutoCAD to prompt you for an annotation, as described above.

**Reuse Next** — uses the next annotation you create for subsequent leaders. There is no prompt for specifying mtext, blocks, tolerances, or copying objects.

**Reuse Current** — reuses the current annotation for future leaders.

Leader Line

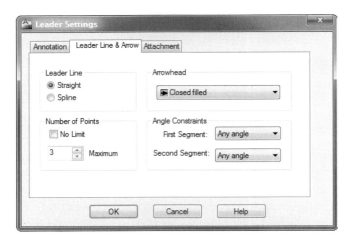

**Straight** — draws the leader line using straight line segments.

**Spline** — draws the leader from a spline object, resulting in a curved leader.

Number of Points

**No Limit** — command prompts you to "Specify next point:" until you get tired and press **ENTER**.

**Maximum** — command stops prompting you after a fixed number, such as 3. You can always stop earlier by pressing **ENTER**. You must set the number to one more than the number of leader segments you want to create: 3 means that you are prompted for two segments; recall that AutoCAD draws the third segment automatically. The minimum maximum is 2; maximum is 999!

## Arrowhead

Select one of the arrowheads shown by the droplist:

**User Arrow** — lists blocks in the drawing, one of which can be selected as the leader's arrowhead.

## Angle Constraints

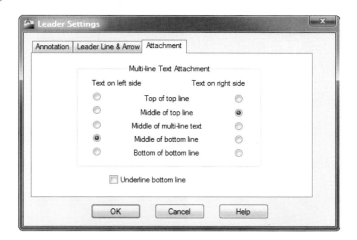

**First Segment** — specifies the angle of constraint for the first segment. "Any angle" means the segment is not constrained; otherwise, select from Horizontal (0 degrees), 90, 45, 30, or 15 degrees.

**Second Segment** — specifies the angle-of-constraint for the second leader segment.

## Multiline Text Attachment

The **Multiline Text Attachment** options are available only when the **Mtext** option is selected on the Annotation tab. These options align the text with the leader line, and can be set differently for left- and right-justified text.

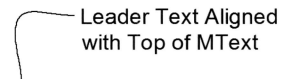 

**Top of Top Line** — aligns the leader line with the top of the top mtext line.
**Middle of Top Line** — aligns the leader line with the middle of the top mtext line.
**Middle of Multiline Text** — aligns the leader line with the middle of the mtext.
**Middle of Bottom Line** — aligns the leader line with the middle of the bottom mtext line.
**Bottom of Bottom Line** — aligns the leader line with the bottom of the bottom mtext line.

If none of the above options works for you, you may have selected the wrong side: select an option for the correct side.

**Underline Bottom Line** — attaches the leader line to the bottom of the mtext, and underlines the last mtext line.

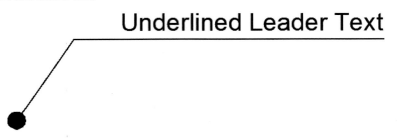

 **MLEADER**

The **MLEADER** command draws leaders with multiple lines, among other effects, such as the ones illustrated below.

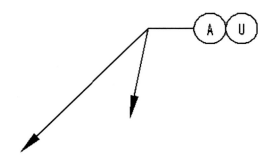

The initial mleader can be drawn just like a qleader, and then you can add more leader lines, combine annotation from multiple leaders into a single leader, and so on. The **MLEADEREDIT** command adds and removes leader lines; the **MLEADERCOLLECT** command collects annotations. The **MLEADERSTYLE** command determines the default look of mleaders.

Multiline leaders can have text or blocks, but not tolerance symbols. By default, the blocks are the bubbles illustrated above, but can also be any user-defined block. In addition, mleaders can have multiple blocks per leader line.

When the drawing contains a number of mleaders near each other, the **MLEADERALIGN** command lines them up. Mleaders support annotative scaling, so that they appear at the correct size in model space.

**TUTORIAL: DRAWING MULTILINE LEADERS WITH TEXT (MLEADER)**

1. To draw leaders with multiple leader lines, start the **MLEADER** command:
   • In the ribbon's Home tab, choose the **Multileader** button in the Annotate panel. Or in the Multileaders panel of the Annotative tab.

- At the 'Command:' prompt, enter the **multileader** command:

    Command: **mleader** *(Press* ENTER.*)*

- Alternatively, enter the **mld** alias at the 'Command:' prompt.

2. In all cases, AutoCAD prompts you to specify the start point of the leader:

    Specify leader arrowhead location or [leader Landing first/Content first/Options] <Options>: *(Pick point 1.)*

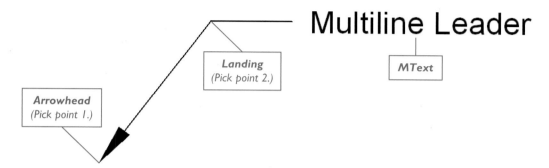

3. The prompts that follow vary, depending on the mleader style. The default style, "Standard," next prompts you to pick the end of the leader line, which also happens to be the start of the landing line:

    Specify leader landing location: *(Pick point 2.)*

4. As soon as you pick the landing location, the mtext editor opens — either in the floating toolbar or on the ribbon, depending on the setting of the **MTEXTTOOLBAR** system variable (described earlier in this chapter).

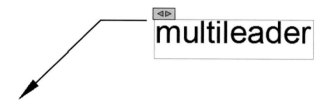

5. Enter text for the leader, and then click the mtext editor's **OK** button to exit the command.

### MLeader Command Options

This command has the following options:

    Specify leader arrowhead location or [leader Landing first/Content first/Options] <Options>: *(Enter an option.)*

**Specify leader arrowhead location** — specifies the endpoint of the arrowhead; the rest of the leader is drawn from there.

**leader Landing first** — draws the landing first, followed by the arrowhead location.

**Content first** — draws the "content" (mtext or blocks) first, followed by the arrowhead location.

**Options** — leads you to a further set of options. Quite frankly, it is easier to set these options with the **MLEADERSTYLE** command, but for the sake of completeness I include them here:

    Enter an option [Leader type/leader lAnding/Content type/Maxpoints/First angle/Second angle/eXit options] <eXit options>:

**Leader type** — draws the leader from straight lines, a spline, or draws none at all.

**leader lAnding** — specifies the length of the landing, or no landing at all.

**Content type** — specifies mtext, a block, or no content. If a block, then specifies its name. After the **MLEADER** command has been used at least once in a drawing, then these bubble blocks become available:

| | | |
|---|---|---|
| _DetailCallout | _TagBox | _TagCircle |
| _TagSlot | _TagTriangle | _TagHexagon |

**Maxpoints** — specifies the maximum number of leader vertices; default and minimum = 2.

**First angle** — constrains the angle of the first leader segment.

**Second angle** — constrains the angle of the second leader segment; default = 0 degrees (horizontal).

### TUTORIAL: ADDING LEADERS TO MLEADERS (MLEADEREDIT)

Once one multiline leader is in the drawing, you can add additional leader lines to it with the **MLEADEREDIT** command. This command adds and removes leader lines; it works with one mleader object at a time.

1. To add leaders to mleaders, start the **MLEADEREDIT** command:
   - In the ribbon's Home tab, choose the **Add Leader** or **Remove Leader** button from the Annotate panel.
   - At the 'Command:' prompt, enter the **MULTILEADEREDIT** command:

     Command: **mleaderedit** (Press ENTER.)
   - Alternatively, enter the **mle** alias at the 'Command:' prompt.

2. In all cases, AutoCAD prompts you to select an mleader:

   Select a multileader: (Pick an mleader.)

3. Specify whether you want to add or remove leader lines; for this tutorial, press **ENTER** to take the default, Add:

   Select an option [Add leader/Remove leader] <Add leader>: (Press ENTER.)

4. Indicate where you want the arrowhead located; AutoCAD then positions the leader line:

   Specify leader arrowhead location: (Pick a point.)

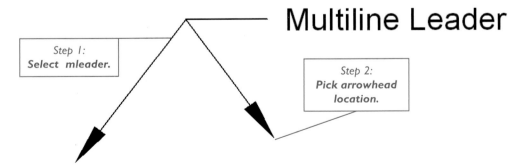

 **Note** The second leader line can end up on the other side of the text, depending on how you position the cursor.

5. Add additional leader lines, or press **ENTER** to exit the command:
   Specify leader arrowhead location: *(Press* **ENTER**.*)*

To remove leader lines, restart this command and then select the **Remove** option. Pick the leaders to be removed, as follows:

   Command: **mleaderedit**
   Select a multileader: *(Pick a multileader.)*
   Select an option [Add leader/Remove leader] <Add leader>: **r**
   Specify leaders to remove: *(Pick one or more leader lines.)*
   Specify leaders to remove: *(Press* **ENTER** *to exit the command.)*

**Notes** The **Remove** option allows you to remove all leaders from mleaders, leaving just the mtext or block. Since it is still seen as an mleader object, you can add leader lines to it again.

The Multileader toolbar and panel don't use the **MLEADEREDIT** command to add and remove leaders. Instead, they use a pair of commands undocumented by Autodesk:

   **AIMLEADEREDITADD** — adds leaders to mleaders.
   **AIMLEADEREDITREMOVE** — removes leaders from mleaders.

The advantage of this pair of commands is that they work more quickly, because they exclude the "Select an option [Add leader/Remove leader]" prompt.

## TUTORIAL: DRAWING MLEADERS WITH BUBBLES (  MLEADERSTYLE)

Multiline leaders can show bubbles in place of text. To do so, you first need to set up a new mleader style. This is done with the MLEADERSTYLE command.

1. **To create a new mleader style, start the MLEADERSTYLE command:**
   - In the ribbon's Annotate tab, choose the **Multileader Style** arrow in the Multileader panel.
   - At the 'Command:' prompt, enter the **mleaderstyle** command:
      Command: **mleaderstyle** *(Press* **ENTER**.*)*
   - Alternatively, enter the **mls** alias at the 'Command:' prompt.

2. In all cases, AutoCAD displays the Multiline Leader Style dialog box. To create the new style, click **New**.

3. Enter a name for the style. For this tutorial, enter "Round Bubble," and then click **Continue**.

4. Since we want to change the content of the mleader, choose the **Content** tab.
5. From the Multileader Type droplist, select **Block**.

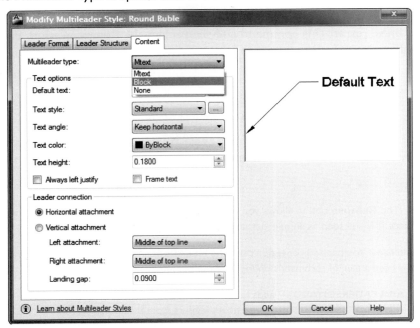

6. Notice that the dialog box changes to show options related to blocks. From the Source Block droplist, select **Circle**.

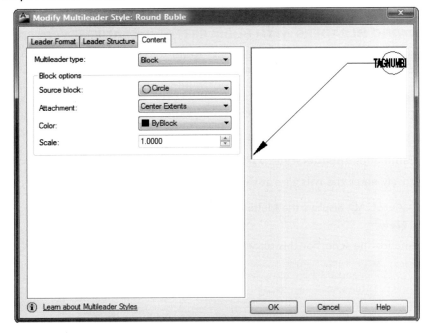

7. Click **OK** to exit the first dialog box.
8. Back in the original dialog box, click **Set Current** to make this new style current.
9. Click **Close** to exit the dialog box.

10. Use the **MLEADER** command to place multiline leaders with the new style.

    Command: **mleader**

    Specify leader arrowhead location or [leader Landing first/Content first/Options] <Options>: *(Pick a point.)*

    Specify leader landing location: *(Pick another point.)*

    Enter attribute values

    Enter tag number <TAGNUMBER>: **1**

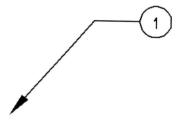

**Notes** The other style options are very similar to those of the Settings dialog box found in the **QLEADER** command. The **CMLEADERSTYLE** system variable stores the name of the current mleader style name.

You can easily create styles from existing mleaders that you've modified:
1. Select the mleader, and then right click.
2. From the shortcut menu, select **Multileader Style**, and then **Save As New Multileader Style**.
3. Notice the dialog box. Enter a name, and then click **OK**.

To use the new style, select it from the droplist in the Multiline toolbar or panel. Styles created this way remember the mtext, and will prompt you:

Overwrite default text [Yes/No] <No>: *(Type **Y** or **N**.)*

## TUTORIAL: COLLECTING BUBBLES INTO ONE MLEADER (  MLEADERCOLLECT)

Multiline leaders can have two or more bubbles each. To have multiple bubbles, you first create a number of mleaders with a single bubble each, and then use the **MLEADERCOLLECT** command to combine them. This is commonly done when the leaders all point to the same feature, as illustrated below.

1. Draw two or more bubble mleaders following the steps described above.

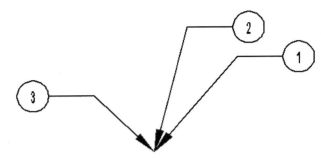

2. To collect multiple multiline leaders into one, start the **MLEADERCOLLECT** command:
   - From the Multileader toolbar, choose the **Collect Multileaders** button.
   - In the ribbon's Annotate tab, choose the **Collect** button in the Multileader panel.
   - At the 'Command:' prompt, enter the **MLEADERCOLLECT** command:

       Command: **mleadercollect** *(Press ENTER.)*
   - Alternatively, enter the **mlc** alias at the 'Command:' prompt.

3. In all cases, AutoCAD prompts you to select two or more mleaders:

    Select multileaders: **all**

    Select multileaders: *(Press ENTER to exit object selection.)*

4. AutoCAD asks where you would like to position the collected mleader:

    Specify collected multileader location or [Vertical/Horizontal/Wrap] <Horizontal>: *(Pick a point, or enter an option.)*

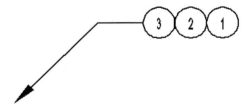

### MLeaderCollect Options

The **MLEADERCOLLECT**'s options include:

   Specify collected multileader location or [Vertical/Horizontal/Wrap] <Horizontal>:

   **Vertical** — stacks the bubbles vertically.

   **Horizontal** — aligns the bubbles horizontally, as illustrated above.

   **Wrap** — wraps the bubbles; AutoCAD asks you for the number of bubbles to wrap around, and how far to wrap them:

   Specify wrap width or [Number]: *(Enter a distance, or the number of bubbles.)*

You can always do the stacking or wrapping yourself, because the added bubbles (#2 and #1, in the figure above) are not part of the leader with #3. You can drag the bubbles to new locations, as required.

**Notes** AutoCAD calls numbers inside the bubbles "tags," but they actually are attributes. To edit them, double-click the bubble (which is a block). In the Edit Attributes dialog box, you can change the tag number to any other number or text.

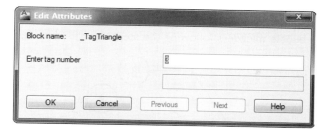

You can change the bubble type through the **PROPERTIES** command. In the Block section, click the **Source block** droplist, and then choose another style of bubble, as illustrated below.

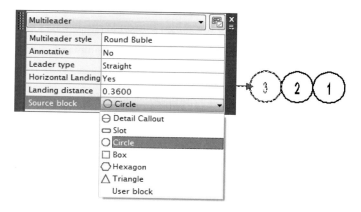

The predefined blocks (a.k.a bubbles) are illustrated below.

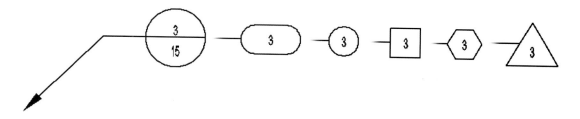

**Left to right:** *Callout, slot, circle, box, hexagon, and triangle.*

## TUTORIAL: ALIGNING MLEADERS (MLEADERALIGN)

When numerous mleaders are in the same area of the drawing, they can look messy. The **MLEADERALIGN** command lines them up nicely.

1. Draw two or more bubble mleaders following the steps described earlier.

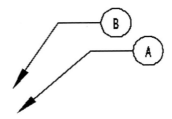

2. To align multiple multiline leaders, start the **MLEADERALIGN** command:
   - In the ribbon's Annotate tab, choose the **Align** button in the Leaders panel.
   - At the 'Command:' prompt, enter the **MLEADERALIGN** command:

     Command: **mleaderalign** *(Press ENTER.)*
   - Alternatively, enter the **mla** alias at the 'Command:' prompt.

3. In all cases, AutoCAD prompts you to select two or more mleaders:

   Select multileaders: **all**

   Select multileaders: *(Press ENTER to exit object selection.)*

4. AutoCAD asks you to pick one mleader; the others will align with it.

   Current mode: Use current spacing

   Select multileader to align to or [Options]: *(Pick one multiline leader. In this tutorial, I've picked A.)*

5. Pick a point to define the alignment. This is like drawing an xline:

   Specify direction: *(Pick a point.)*

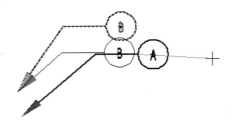

The screen can get messy as AutoCAD shows you the mleaders in their original and new positions. As you move the cursor, the bubbles of the other mleaders align themselves between the cursor and the base mleader.

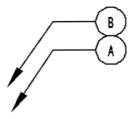

Turn on ortho mode to align mleaders vertically or horizontally.

## MLeaderAlign Options

This command has an **Options** option:

> Enter an option [Distribute/make leader segments Parallel/specify Spacing/Use current spacing] <Use current spacing>: *(Enter an option.)*

**Distribute** — distributes the mleader content (bubbles or mtext) between two points. The distance and angle between the two points determine the spacing and angle of the content.

> Specify first point or [Options]: *(Pick a point.)*
>
> Specify second point: *(Pick another point.)*

**Make leader segments Parallel** — makes the last leader segment of each mleader parallel. AutoCAD prompts you to select the one mleader to which the other leader lines are made parallel.

**Left:** Two separate multilines.
**Right:** B made parallel with A.

**Specify Spacing** — allows you to specify the spacing between each leader.

> Specify spacing <0.000000>: *(Enter a number, such as 0.5.)*
>
> Select multileader to align to or [Options]: *(Pick one mleader.)*
>
> Specify direction: *(Pick a point to indicate the angle.)*

**Use current spacing** — keeps the spacing of the leaders (default).

# TEXTEDIT

The **TEXTEDIT** command edits text. (The name was changed in AutoCAD 2010 from **DDEDIT**, short for "dynamic dialog editor," *dynamic dialog* being an old reference to dialog boxes.)

This command edits all AutoCAD text, whether single-line text, mtext, dimensions, fields, or attributes. For each type of text, it displays a different user interface.

## TUTORIAL: EDITING TEXT

In the following tutorial, you edit text with the **TEXTEDIT** command.

1. Open the *textedit.dwg* file. It contains a sample of each type of text found in AutoCAD.

   | Single-line text | ATTRIBUTE_DEFINITION | Thursday, April 29, 2010 |
   |---|---|---|
   | Multi-line paragraph text | Attribute | ⊢—— 3.0000 ——⊣ |

2. To edit text in drawings, start the **TEXTEDIT** command:
   - Double-click the text; this action bypasses the 'Select annotation to edit' prompt.
   - At the 'Command:' prompt, enter the **textedit** command:

     Command: **textedit** *(Press ENTER.)*

   - Alternatively, enter the **ed** alias at the 'Command:' prompt.

3. In all cases, AutoCAD prompts you to select the text:

   Select an annotation object or [Undo]: *(Select another line of text, or press ESC to exit the command.)*

   AutoCAD provides different text editors, depending on the text you selected:
   - The in-place editor for single-line text placed with the **TEXT** command, as well as for text in fields, tables, leaders, and dimensions:

   **Left:** *Single line text...*
   **Right:** *...and dimension edited in-place*

   - The Text Formatting toolbar or ribbon for multi-line text placed with the **MTEXT** command:

   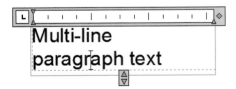

- The Edit Attribute Definition dialog box for attribute definitions placed with the **ATTDEF** command, as well as for text in mleader bubbles.

- The Enhanced Attribute Editor for attributes inserted with blocks.

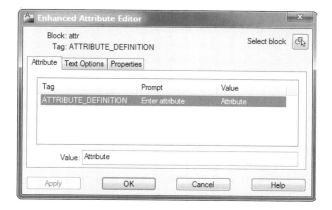

**Notes** As an alternative to entering the **TEXTEDIT** command, you can double-click single-line and multiline text, and attribute definitions. AutoCAD automatically displays the appropriate editor.

Double-clicking leaders and dimension text displays the Properties window, which can also be used to edit text.

Attributes placed with the **INSERT** command are edited with a separate command, **ATTEDIT**.

### Single-line Text Editor

When you select single-line text placed by TEXT, AutoCAD displays the inplace editor. Each line of text is edited independently of the others. See the TEXT command at the beginning of this chapter.

You can edit a group of text blocks by holding down the ALT key, and then selecting each text block. When done editing, press ENTER. If you want to discard the editing changes, press ESC.

### Multiline and Leader Text Editor

When you select leaders, dimensions, or text created by the MTEXT command, AutoCAD displays the Multiline Text tab (or Text Formatting toolbar) — identical to the one displayed by MTEXT.

Make your changes, and choose the **Close Text Edtor** button. AutoCAD returns to the prompt:

> Select an annotation object or [Undo]: *(Select another paragraph of text, or press* ESC *to exit the command.)*

### Attribute Text Editor

When you select attribute definitions created by the ATTDEF command, AutoCAD displays the Edit Attribute Definition dialog box. (Attributes are not discussed in this book.) Change the tag, prompt, and default text, and then choose the **OK** button.

### Enhanced Attribute Editor

When you select a block containing attributes, AutoCAD displays the Enhanced Attribute Editor dialog box. This is the same dialog box displayed by the **EATTEDIT** command. Change the attributes, and then choose the **OK** button

#### EDITING TEXT: ADDITIONAL METHODS

An alternative to the **TEXTEDIT** command is the **PROPERTIES** command, which displays a palette that allows you to edit text as well as to change many text properties — color, layer, insertion point, the background mask, and so on. The only property that the Properties palette cannot change is the font, which must be changed with the **STYLE** command.

To access this palette, select the text, and then right-click. From the shortcut menu, select **Properties**. AutoCAD displays the Properties palette.

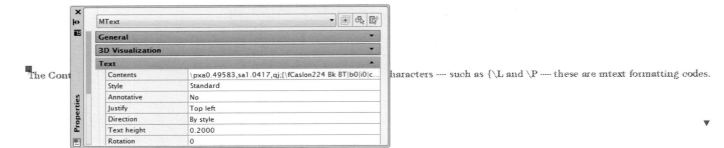

The **Contents** area displays the text of the text object. When you see strange characters — such as {\L and \P — these are mtext formatting codes.

If the text is paragraph text, choose the **[...]** button, which causes AutoCAD to open the familiar mtext editor.

## SPELL

The **SPELL** command checks the drawing for words unfamiliar to AutoCAD.

### TUTORIAL: CHECKING SPELLING

1. To check the spelling of text, start the **SPELL** command:
   - In the ribbon's Annotate tab, click the **Check Spelling** button in the Text panel.
   - At the 'Command:' prompt, enter the **SPELL** command:

     `Command: **spell** (Press ENTER.)

   - Alternatively, enter the **sp** alias at the 'Command:' prompt.

   In all cases, AutoCAD displays the Check Spelling dialog box.

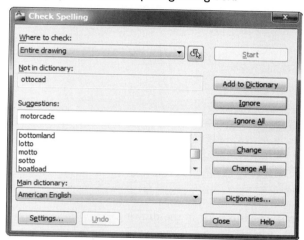

2. Click **Start** to begin checking the spelling of words.
3. Confirm the spelling of words AutoCAD does not recognize.
4. When done, AutoCAD reports: "Spelling check complete." Click **OK** to exit the dialog box.

This command works with words placed or imported by the **TEXT**, **-TEXT**, **MTEXT**, **-MTEXT**, **LEADER**, **QLEADER**, **MLEADER**, and **ATTDEF** commands. When it comes to attributes, values are spell checked, but not tags. Because '**SPELL** can be run transparently, you can use it while other commands are active.

# FIND

The FIND command searches for, and optionally, replaces text.

**TUTORIAL: FINDING TEXT**

1. To find and replace text, start the **FIND** command:
   - In the ribbon's Annotate tab, enter text in the **Find Text** field of the Text panel.
   - At the 'Command:' prompt, enter the **FIND** command.

     Command: **find** *(Press ENTER.)*

2. In all cases, AutoCAD displays the Find and Replace dialog box.

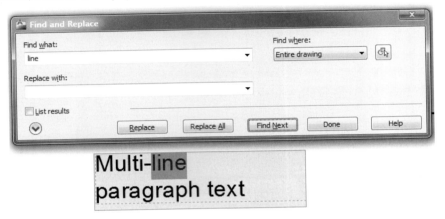

3. Enter a word or phrase to search for in the drawing.
4. To narrow the search, click **More**.

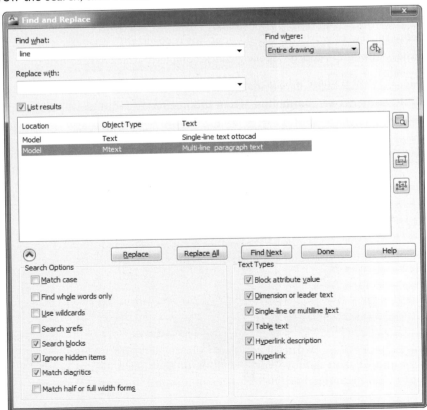

The expanded dialog box lets you narrow the search to specific kinds of words, such as those found only in dimensions or in table text. Field text is considered a form of mtext.

5. Optionally, enter a replacement word or string.
6. Click **Find Next** to find the first instance of the phrase.
   If necessary, click **Replace** or **Replace All**.
7. When done, click **Close**.

The **U** command reverses changes made by this command.

## SCALETEXT AND JUSTIFYTEXT

The SCALETEXT and JUSTIFYTEXT commands change the size and justification of text.

After placing text in drawings, you may find it necessary to change the height of the text, or its justification.

AutoCAD provides the SCALETEXT command to change the height (scale) of text. It makes selected text larger or smaller. The advantage of this command over PROPERTIES or SCALE is that it resizes multiple text objects without changing their location; the PROPERTIES command may not do it accurately, while the SCALE command scales everything, including location, relative to the base point.

The JUSTIFYTEXT command changes the justification of text. It can, for example, change left-justified text to right-justified. The style of justification of mtext can be changed with the mtext editor through the MTEDIT command.

### TUTORIAL: RESIZING TEXT

1. To resize text, start the **SCALETEXT** command:
   - At the 'Command:' prompt, enter the **SCALETEXT** command.

     Command: **scaletext** *(Press ENTER.)*

2. AutoCAD prompts you to select text.

     Select objects: *(Pick one or more lines of text.)*

     Select objects *(Press ENTER to end object selection.)*

3. Pick a base point, or select a justification point:

     Enter a base point option for scaling [Existing/Align/Fit/Center/Middle/Right/TL/TC/TR/ML/MC/MR/BL/BC/BR] <Existing>: *(Enter an option, or press ENTER.)*

4. Specify the new height:

     Specify new height or [Match object/Scale factor] <3/16">: *(Enter a height, or an option.)*

**Left:** *Original text 0.2" tall; the square indicating the base point.*
**Right:** *Same text scaled to be 0.3" tall.*

When AutoCAD prompts you to select objects, you can choose a mix of text and non-text objects; AutoCAD filters out the non-text objects automatically.

The SCALETEXT command changes the height of selected text relative to a *base point*. The base point can be the existing insertion point or one of the many text justification points. By default, the base point is the existing insertion point.

**Match Object** — matches the height to that of another text object:
>Select a text object with the desired height: *(Pick another text object.)*
>Height=3/16"

**Scale Factor** — scales the text by a factor. A factor larger than 1 enlarges the text, while a value under 1 reduces it.
>Specify scale factor or [Reference]: *(Enter a factor, such as **2**.)*

**Reference** — makes text identical to that of the **SCALE** command: the text is scaled relative to another size.

### TUTORIAL: REJUSTIFYING TEXT

1. To change the justification of text, start the **JUSTIFYTEXT** command:
   - At the 'Command:' prompt, enter the **JUSTIFYTEXT** command.

   >Command: **justifytext** *(Press ENTER.)*

2. AutoCAD prompts you to select text.
   >Select objects: *(Pick one or more lines of text.)*
   >Select objects *(Press ENTER to end object selection.)*

3. AutoCAD prompts you to select a justification option; the option you select will override the justification of the existing text.
   >Enter a justification option [Existing/Align/Fit/Center/Middle/Right/ TL/TC/TR/ML/MC/MR/ BL/BC/BR] <Existing>: *(Press ENTER.)*

The same rules apply to the **JUSTIFYTEXT** and **SCALETEXT** commands regarding object selection. The U command can undo any damage.

##  ANNOTATION SCALING

Annotation scaling ensures only correctly-scaled objects appear in Model tab and model space viewports.

*Annotation* refers to objects that annotate drawings. (Annotation means "explanatory text.") In AutoCAD, this includes the following objects:

| Single-line text | Mtext | Field text |
| --- | --- | --- |
| Attribute text | Dimensions | Tolerances |
| Multiline leaders | Hatch patterns | Linetypes |

As described earlier in this chapter, text needs to be scaled larger to appear at the correct size when the drawing is plotted smaller to fit the paper or vice-versa. Annotation scaling is a system devised by Autodesk to solve the problem semi-automatically. It works like this:

>A layout viewport has its scale factor set to **1:100**. Draw text with an annotative style. AutoCAD automatically assigns the text the inverse scale factor of **100:1**.

There is no "Annotation" command; instead, annotation is a property integrated into some commands, styles, and user interface elements. Let's work through a tutorial, and then examine the additional options.

> ### WHERE ANNOTATIVE SCALING HAPPENS
>
> Annotative scaling is scattered about AutoCAD in many nooks and crannies. It is a property that can be set in styles and commands, and changed after the fact with these commands and system variables. Annotative styles are indicated in lists and on screen by the triangle icon:
>
> **Style Commands**
>     **DIMSTYLE** — affects dimensions and tolerances.
>     **MSTYLELEADER** — affects multiline leaders.
>     **STYLE** — affects single-line text, mtext, and fields.
>
> **Drawing Commands**
>     **ATTDEF** and **BATTMAN** — create and edit attributes.
>     **INSERT** — reports annotatively-scaled blocks and dynamic blocks.
>     **HATCH** and **HATCHEDIT** — draw and edit hatches.
>     **TEXT**, **MTEXT**, and **TEXTEDIT** — draw and edit text, mtext, and fields.
>     **MSLTSCALE** system variable — scales all linetypes annotatively.
>
> **PROPERTIES Command**
> The **PROPERTIES** command changes the annotative property for single-line text, mtext, fields, dimensions, hatches, attributes, multileaders, leaders, qleaders, and tolerances. Typical annotation-related properties are shown below for text.
>
> | Text | |
> |---|---|
> | Contents | Annotative Text |
> | Style | Annotative |
> | Annotative | Yes |
> | Annotative scale | 1:50 |
> | Justify | Left |
> | Paper text height | 0" |
> | Model text height | 10" |
> | Match orientation to layout | No |

**Note** AutoCAD identifies objects with annotative scaling through the triangular icon. As you pass the cursor over objects, they highlight, and those with annotative scaling display the icon near the cursor.

(*History:* The icon represents the end view of a triangular scale ruler used by architects and engineers. Being triangular in cross-section, it provided 12 scale factors, two along each of the six edges. The technical editor notes that these rulers are excellent for firing elastic bands at coworkers.)

### TUTORIAL: AUTOMATICALLY SCALING TEXT

1. In AutoCAD, open the *annotative.dwg* drawing file.
2. Switch to layout mode by clicking the **Layout1** tab or button.
3. Enter model space by double-clicking inside the viewport border.

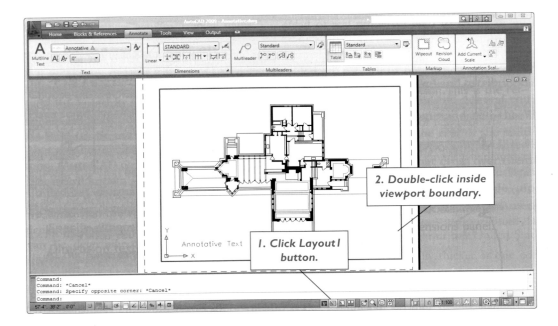

Notice the tools for working with annotative scaling. These can appear in two places, depending on how AutoCAD is set up: one possible location is on the drawing bar (above the command bar); the other is on the status bar (below the command bar).

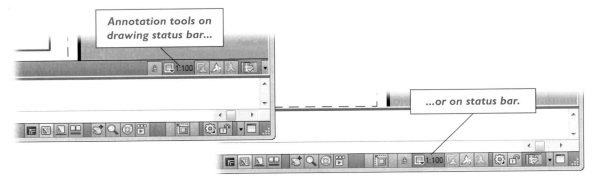

(The location is switched through the **OPTION** command's **Display Drawing Status Bar** option in the Display tab.)

4. The drawing is not correctly scaled in the viewport; it should be scaled to meet two criteria: (1) the entire drawing must fit the viewport, and (2) the drawing must be at a standard scale factor. To select an appropriate scale factor, follow these steps:
    a. Use the **ZOOM Extents** command to fit the drawing to the viewport.
    b. Notice the scale factor next to **VP Scale**: it reads 0.010998 or something similar.
    c. Click **VP Scale**, and then select the nearest standard scale factor of **1:100**. Notice that the drawing reacts by zooming slightly smaller, and that the Annotation Scale factor matches.
5. To test the annotation feature, place some text with the **TEXT** command. Follow these steps:
    a. Switch to Model tab.

**Notes** The **VP Scale** button displays a long list of scale factors. You can edit this list with the **SCALELISTEDIT** command, removing and adding scale factors. This single command affects the scale factors available to viewport scaling, plot scaling, and annotation scaling.

To reduce the list further, set the **HIDEXREFSCALES** system variable to 1, which hides scale factors imported from attached xref files.

As of AutoCAD 2011, scale factors can be edited in the Options dialog box. Choose the User Preferences tab, and then click the **Default Scale List** button.

## ANNOTATION BAR

There is no command for controlling annotation scaling, because it is a property. You control it through the "annotation bar" (illustrated below), or through system variables, properties, some editing and style commands, and the Properties palette.

In Model tab, these annotation controls are displayed on the status bar:

In Layout tab, only two controls are displayed when in paper space:

When in a layout's model space, the following controls are shown:

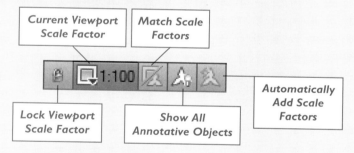

**Lock Viewport Scale Factor** — toggles locking of the viewport scale; also locks the annotation scale. This option is also available through the viewport's properties.
**Current Viewport Scale Factor** — reports the scale of the current viewport; click the arrow to select another scale factor.
**Match Scale Factors** — makes the viewport and annotative scale match.
**Show All Annotative Objects** — toggles the display of annotative objects:
   **On** — all objects are displayed, regardless of their annotative scale.
   **Off** — only objects whose annotative scale matches the viewport scale are displayed.

**Automatically Add Scale Factors** — toggles addition of annotative scales to objects:
   **On** — when the viewport scale changes, the new scale factor is added to annotative objects automatically.
   **Off** — objects keep their annotative scale factors fixed.

The technical editor explains why: "If you zoom and pan while entering text in a layout tab, you mess up the viewport's scale. When you re-enter model space, the annotations are 'missing,' because their scale no longer matches the viewport scale. For this reason, it is best to enter annotative text in Model tab."

   b. From the Dashboard's Text panel or the Styles toolbar, select the **Annotative** text style.

   c. Start the **TEXT** command.

   Command: **text**

   d. Notice that the prompt reports that the annotative property is turned on:

   Current text style: "Annotative"  Text height: 0'-0"  Annotative: Yes
   Specify start point of text or [Justify/Style]: *(Pick a point in the viewport.)*

   e. Notice that the prompt asks for "paper height": this is the height of the text as you want it to appear when plotted. In this case, enter a height of 0.2", which is close to the ideal of 3/16":

   Specify paper height <0'-0">: **0.2"**
   Specify rotation angle of text <0>: *(Press ENTER.)*

   **Annotative Text** *(Press ENTER twice.)*

Notice that the text appears at the correct size in the viewport. As illustrated below, the floorplan is 70 feet wide, yet the text is legible despite it being just 0.0017 feet tall (0.2").

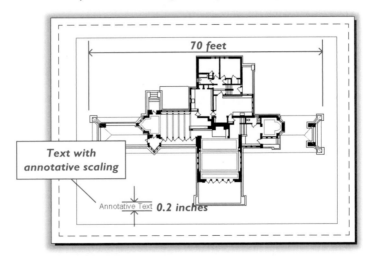

6. To see why this is possible, use the **LIST** command, and then select the text. I've highlighted the important parts below:

   | | |
   |---|---|
   | TEXT | Layer: "0" |
   | Space: | Model space |
   | Handle = | 5d5 |
   | Style = | "Annotative" |
   | Annotative: | Yes |
   | **Annotative scale:** | **1:100** |
   | Typeface = | Arial |
   | start point, | X=  9'-0"  Y= -14'-6"  Z=  0'-0" |
   | **paper text height** | **0'-0"** |
   | **model text height** | **1'-8"** |
   | text | Annotative Text |

The text contains the annotative scale property of 1:100, and then applies it to text in model space ("model text height"). We entered a height of 0.2". Behind the scenes, AutoCAD automatically inverted the scale to draw the text at 0.2" x 100:1 = 20" (same as 1'-8" in feet-inches).

The **paper text height** is set to 0'0", even though you entered 0.2". That's because AutoCAD reverse-calculates the height from the annotative scale factor multiplied by the model text height. The paper text height changes automatically when the associated annotative scale changes.

7. To see the effect of annotative scaling, change the scale of the viewport. From the status bar, click **VP Scale** and select **1:50**. Notice that the text disappears. (If necessary, pan the drawing to where the text is/was located.)

8. One of the status bar buttons lets you see annotative-scaled objects, no matter what the viewport scaling is. Click the **Annotation Visibility** button. (This button toggles the **ANNOALLVISIBLE** system variable.) Notice that the text reappears and is twice as large.

9. When you work with several viewports scales, it can be a pain to assign annotative scales to a large numbers of objects. To have AutoCAD automatically assign the scale factors, follow these steps:

    a. Click the **Automatically Add Scales** button. (This button toggles the **ANNOAUTOSCALE** system variable.)
    
    b. Change **VP Scale** to something else, such as **1:40**. Notice that the text does not become larger (as you might expect from step 7), but returns to its former size.
    
    c. Change **VP Scale** to **1:50**. Again, the text is the "correct" size. No matter which viewport scale you now select, the annotatively-scaled text appears at constant size.

**Notes** When objects have more than one annotative scale, the selection cursor shows a double triangle icon, as illustrated below

Autodesk warns against objects having too many annotative scale factors, saying that this could slow down your computer. The **OBJECTSCALE** command lets you add and remove scale factors from objects.

## ANNOTATION SCALING: ADDITIONAL METHODS

While no single command controls annotative scales, these system variables and commands fine tune the system:

- **OBJECTSCALE** command edits annotative scales associated with objects.
- **SELECTIONANNODISPLAY** and **XFADECTL** system variables dim the display of annotative objects whose scale differs from the viewport scale.
- **ANNOUPDATE** command updates objects to match a recently-changed style.
- **ANNORESET** command forces the scale and visibility to match the current settings.
- **ANNOTATIVEDWG**, **DIMANNO**, and **MSLTSCALE** system variables control the annotative properties of drawings inserted as blocks, dimensions, and linetypes, respectively.
- **SAVEFIDELITY** system variable determines whether drawings are saved with visual fidelity to earlier releases of AutoCAD.

Let's look at each.

**ObjectScale**

The **OBJECTSCALE** command edits annotative scales associated with objects through a dialog box.

> Command: **objectscale**
> Select annotative objects: *(Pick one or more objects.)*
> Select annotative objects: *(Press ENTER to end object selection.)*

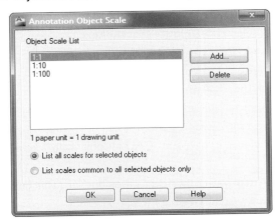

The dialog box lists the scale factor(s) assigned to the selected object(s).

> **Add** — adds scale factors to the objects, in a roundabout manner; displays the following dialog box:

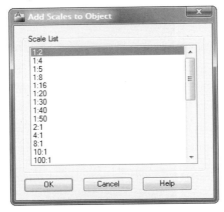

This dialog box copies the list maintained by the **SCALELISTEDIT** command, which also handles scale factors for viewport and plot scaling; if you don't see the scale factor you need, then you need to use the **SCALELISTEDIT** command to add it, as detailed in Chapter 20 on plotting.

**Delete** deletes scale factors. You can select one or more for deletion, but you cannot delete all of them, because at least one annotative scale must remain. (To remove the last remaining annotative scale, use the **PROPERTIES** command to turn off the object's Annotative property.)

The following options determine which scale factors are listed by the Annotation Scale Factor dialog box:

- ⊙ **List All Scales For Selected Objects** — displays all scales factors from all the objects you selected.
- ○ **List Scales Common to All Selected Objects Only** — displays only the scale factors that the selected objects share.

## SelectionAnnoDisplay and XFadeCtl

The **SELECTIONANNODISPLAY** system variable toggles the display of annotative objects whose scale differs from the viewport scale.

> Command: **selectionannodisplay**
> Enter new value for SELECTIONANNODISPLAY <1>: *(Type 1 or 0.)*

**0** — Only the annotative objects that match the viewport scale are displayed.

**1** — All annotative objects are displayed, with non-current ones faded (default).

At left in the figure below, SELECTIONANNODISPLAY is on. The annotative text is selected and displays its multiple-scale representations; those not at the current viewport scale are faded (light gray).

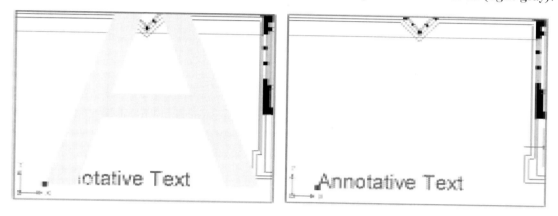

*Left:* SELECTIONANNODISPLAY = 1, XFADECTL = 90.
*Right:* SELECTIONANNODISPLAY = 0. *(Value of* XFADECTL *doesn't matter.)*

At right, SELECTIONANNODISPLAY is off, and so only the representation that matches the viewport scale is displayed.

The level of dimming is controlled by a second system variable, XFADECTL (short for "xref fade control"). The level is a percentage between 0 (off) and 90 (maximum dimming):

> Command: **xfadectl**
> Enter new value for XFADECTL <50>: *(Enter a value between 0 and 90.)*

 **Notes** When annotative text has multiple representations, you can move each independently of the others, as illustrated below. To do this, set a viewport scale that matches one representation, and then move it; select the next viewport scale, and so on.

While you can move multiple representations independently of each other, you cannot change properties, such as color, independently.

The **DRAWORDER** command has no effect on changing the display order when multiple representations overlap, because they all represent the same text, albeit at different sizes.

## AnnoUpdate

The **ANNOUPDATE** command updates objects to match a recently-changed style.

> Command: **annoupdate**
> n found n were updated

Use this command to update annotative objects when the related styles are changed, such as text and dimension styles.

## AnnoReset

The **ANNORESET** command forces the location of scale representations to match the current settings.

> Command: **annoreset**
> Reset alternate scale representations to current position
> Select objects: *(Pick one or more objects.)*
> Select objects: *(Press* ENTER *to end object selection.)*

The technical editor also notes that this command does not reset annotatively-scaled objects to their original locations, but instead moves them to match the current location of the one used to trigger the change.

## AnnotativeDwg

The **ANNOTATIVEDWG** system variable controls the annotative property of drawings inserted as blocks into other drawings. It is read-only when the drawing contains at least one annotative object; otherwise, you can change the value between 0 and 1.

> Command: **annotativedwg**
> ANNOTATIVEDWG = 0 (read only)

- **0** — Behaves nonannotatively.
- **1** — Behaves annotatively.

## DimAnno

The **DIMANNO** system variable reports whether the current dimension style has the annotative property.

> Command: **dimanno**
> DIMANNO = 1 (read only)

- **0** — dimension style is not annotative.
- **1** — dimension style is annotative; use the **DIMSTYLE** command to change the annotative property, which affects the size of dimension text and arrows. See Chapter 15, "Editing Dimensions."

## MsLtScale

The **MSLTSCALE** system variable toggles the annotative property of linetypes in model space (short for "model space linetype scale").

> Command: **msltscale**
> Enter new value for MSLTSCALE <0>: *(Type 1 or 0.)*

- **0** — linetypes are not scaled by the annotation scale, the default for drawings imported from AutoCAD 2007 and earlier.
- **1** — linetypes are scaled by the annotation scale, the default for drawings created in AutoCAD 2008 and later.

This system variable does not come into effect until after the next drawing regeneration. (Use the **REGENALL** command, if necessary.) To use this system variable effectively, all other linetype scale-related system variables should also be set to 1:

LtScale = 1 or a scale correct for plotting.
CeLtScale = 1
PsLtScale = 1

### SaveFidelity

The **SAVEFIDELITY** system variable determines whether drawings with multiple annotatively-scaled objects are saved with visual fidelity. Earlier releases of AutoCAD cannot handle these objects, and so visual fidelity ensures that they are saved to separate layers

Command: **savefidelity**

Enter new value for SAVEFIDELITY <1>: *(Type 1 or 0.)*

**0** — Drawings save only annotative objects at the current scale, and discard the others.
**1** — Drawings save each scaled representation of annotative objects on separate layers (default).

The problem with multiple representations of annotatively-scaled text is that they display correctly only in AutoCAD 2008 and later. This system variable determines what to do when drawings are opened in earlier releases of AutoCAD.

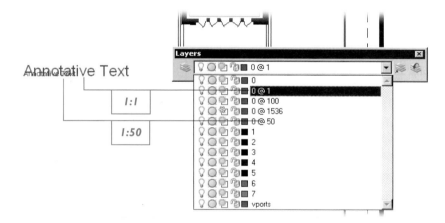

When set to 1 (the default), drawings contain extra layers whose names are marked with a scale factor, such as "0 @ 50," as illustrated above. This layer contains all annotative objects found on layer 0 bearing the scale factor of 1:50. The layer is frozen when first opened in AutoCAD 2007 (or earlier), and can be thawed to view the text. Everything returns to normal when re-opened in 2008 or later.

To speed up the display of complex drawings, turn off **SAVEFIDELITY** until saving the file for an earlier release, and then turn it off again afterwards.

## EXERCISES

1. In this exercise, you practice placing text.

   Start AutoCAD with a new drawing. With the **TEXT** command, place following lines of text:
   
   THIS IS THE FIRST LINE OF TEXT
   
   THIS IS THE SECOND & CENTERED LINE OF TEXT

2. In this tutorial, you place text centered on a point.

   The text "Part A" is to be centered both vertically and horizontally on a selected point. Use the **TEXT** command to place the text. Which justification mode did you use?

3. In this exercise, you place text with a variety of alignments.

   The figure illustrates several text strings. The placement point is marked with a solid dot. Use the **TEXT** command to place the text as shown.

   Kitchen   Radius
             AutoCAD
   Drill-thru   .Title
   Capacitor   Isometric View
               Section A-A

4. In this exercise, you place the following line of text at a variety of angles:
   
   This text is rotated
   
   Using the **TEXT** command, place the following line of text at the following angles:
   a. 0 degrees.
   b. 45 degrees.
   c. 90 degrees.
   d. 135 degrees.
   e. 180 degrees.
   
   Use the same base point for each line.

5. In this exercise, use control codes to construct the following text string in the drawing.
   
   The story entitled <u>MY LIFE</u> is ±50% true.

6. In this exercise, you practice using mtext. Start a new drawing, and use the **MTEXT** command to place several lines of text in the drawing, such as your name and address.
   
   Your First and Last Name
   
   1234 First Avenue
   
   Anytown, BC
   
   V8C 1T2 Canada
   
   Click the **OK** button to exit the mtext editor.

7. Continuing from the previous exercise, double-click the mtext. Does the mtext editor reappear?

   Make the following changes to parts of the text:
   - Font: **Times New Roman**
   - Color: **Red**

   Click the **OK** button to exit the mtext editor. Did the changes come into effect?

8. Continuing from the previous exercise, turn on the **QTEXT** command.
   a. Did the text change its look?
   b. If no, which command did you forget to use?

9. In this exercise, you import text from another file.
   a. If necessary, create a brief text file in Notepad, the text editor included with every version of Windows. Save the file in *.txt* format.
   b. Import the text file into your drawing with the **MTEXT** command.

10. Create five different styles that represent very different text appearances, such as wide text, slanted, and so on.
    a. Which command creates styles?
    b. Place examples of each style in the drawing.

11. Using only the **MTEXT** command, design your own business card with at least two fonts, three font variations, and two colors.

    Standard business cards are rectangles that measure 3.5" wide by 2" tall. Do not design a logo — just position the text. Include at least your name, address, and Internet information.
    a. Which command is useful for drawing rectangles?
    b. How many business cards can you fit onto a vertical A-size sheet?

    (The technical editor quotes Lynn Allen: "You know you've been working with AutoCAD too long when you use it to design wedding invitations.")

12. Use a combination of text and drawing commands to design a logo for an engineering office. The logos often include variations on the owners' names.

13. In this and the next exercise, you practice drawing leaders.
    a. Draw a rectangle.
    b. Attach a leader to each of the four corners.
    c. Use the following text for each leader:
       - North East Corner
       - North West Corner
       - South East Corner
       - South West Corner

14. Continuing from the previous exercise,:
    a. Place a splined leader pointing to the center of the rectangle.
    b. Change the arrowhead to a dot.
    c. Use the following annotation:
       Parcel of Land.

15. Continuing from the previous exercise, double-click the annotation of the splined leader, and change the text to:
    Disputed Parcel of Land.
    Not to be Subdivided.

16. In this and the following exercise, you correct and change text in drawings.
    a. Open the *spelling.dwg* drawing file, a drawing that contains text with spelling errors.

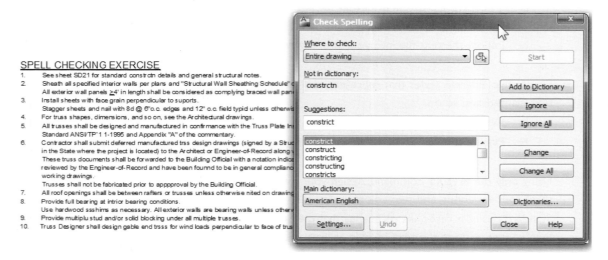

    b. Correct the spelling.
    c. How many mistakes does AutoCAD's **SPELL** command find?
    d. How many *real* spelling mistakes are there?

17. Continuing with the same drawing, find and replace the following phrases:

    | Find | Replace With |
    |---|---|
    | sheet SD1 | sheet AA-01 |
    | Building Official | local Building Official |

    Which command did you use to find and replace words in drawings?

18. Open the *insert.dwg* drawing file, a house plan.

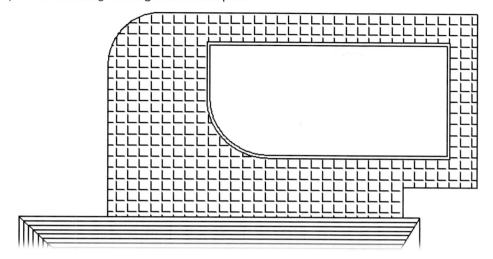

Use the **MTEXT** command to place the following note over a white rectangle on the patio area at the rear of the house. Use the background mask feature.
   Interlocking Belgium Bricks

19. Continuing with the same drawing, use the **SCALETEXT** command to reduce the size of the text you added in the previous exercise.

20. Continuing with the previous exercise, use the **JUSTIFYTEXT** command to change the justification to **Center**. How does the text change?

21. In this exercise, we see the effect of using the following buttons on annotative text:

    From left to right, these are:

    **Annotation Scale** (**CANNOSCALE** system variable) — changes the annotation scale.

    **Annotation Visibility** (**ANNOALLVISIBLE** system variable) — toggles visibility of text at all annotation scales.

    **Add Annotation Scales Automatically** (**ANNOAUTOSCALE** system variable) — toggles addition of scales to objects as annotation scale changes.

    a. Start a new drawing, and then turn off all of the settings, as listed below. You can use the controls on the status bar, or change the values of the system variables.

    | | | | |
    |---|---|---|---|
    | Annotation Scale | 1:1 | | (**CANNOSCALE** = 1:1). |
    | Annotation Visibility | | off | (**ANNOALLVISIBLE** = 0) |
    | Add Annotation Scales Automatically | | off | (**ANNOAUTOSCALE** = -4) |

### Creating an Annotative Text Style

b. Create a new text style named "Anno" with the following properties:

    Annotative ☑ (on)
    Paper Text Height **0.25**
    Font Name **Arial**

Set the style current, and then close the dialog box.

c. With the **TEXT** command, place the following text in the drawing in model space:
    Annotative Text = 0.25

d. With the **RECTANGLE** command, draw a rectangle that encompasses the extent of the text, as illustrated below. This rectangle will help you see the size and location of the original text as the annotation scale changes.

| Annotative Text = 0.25 |

e. Change **Annotation Scale** to 1:2.
What happens to the text?
Why did the text change in that way?

### Controlling Annotation Visibility

f. Click the **Annotation Visibility** button (or set **ANNOALLVISIBLE** = 1).
Why does the text appear?

g. Change **Annotation Scale** to 1:4,
Click the **Add Annotation Scales Automatically** button (or set **ANNOAUTOSCALE** = 4).
Why does the text change its look?

h. Change **Annotation Scale** back to 1:2, and then select the text.
What do you see on the screen?

### Resetting and Updating Annotative Text

i. With the text still selected, use its grip to move it a few inches away.
What happens to the other representations of the text?

j. With the text still selected, use the **ANNORESET** command.
To see the effect of the command, select the text.
What happens to the other representations of the text?

k. Return to the **STYLE** command, and change the **Paper text height** to **0.5**.
Click **Apply** and then **Close**.
What happens to the text?

l. Select the text, and then enter the **ANNOUPDATE** command.
What happens to the text?

m. Select the text.
What happens to the other representations?

*(Keep this drawing open for the following exercise.)*

22. In this exercise, you create *oriented* text. Continue with the drawing from the previous exercise.

    a. Start the **STYLE** command, and then turn on the **Match Text Orientation to Layout** option for the "Anno" text style.
       Click **Apply** and then **Close**.
    b. Use the **ANNOUPDATE** command to update the style of the text in the drawing.
    c. Use the **ROTATE** command to rotate all objects in the drawing by 90 degrees.
       What happens to the annotative text?

23. Use the **MLEADER** command to create the following leaders:

    a. 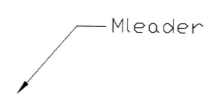     b.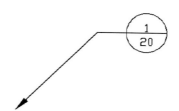

    c. Draw the following multi-bubble mleader.
       Which command did you use to create it?

    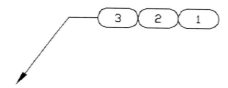

    d. Open the *mleader.dwg* file.
       Add one more leader to make it look like the following figure.
       Which command did you use to add the leader?

## CHAPTER REVIEW

1. Name three ways you can see a list of the text styles stored in the drawing.
    a.
    b.
    c.
2. What are *fonts*?
   How do styles differ from fonts?
3. What four text properties are altered by the **STYLE** command?
    a.
    b.
    c.
    d.
4. There are several methods of placing text. List three of them, with a brief explanation of each:
    a.
    b.
    c.
5. When an underscore or overscore is added to text, is the entire string altered, or can words be treated separately?
   Explain.
6. When AutoCAD asks for a text height, is a numerical entry required?
   Explain.
7. List five alignment modes available through the **TEXT** command.
    a.
    b.
    c.
    d.
    e.
8. Can text be rotated at any angle?
   How does the oblique angle differ from the text angle?
9. Does annotative scaling apply to hatch patterns and dimensions?
10. How can the text height be altered each time you enter text without redefining the style?
11. How can you compress and expand the text width?
12. Which formats of text can AutoCAD import into drawings?
13. What height should note text be in drawings?
14. Name four areas where text is used in drawings:
    a.
    b.
    c.
    d.
15. Can AutoCAD use *any* TrueType font found on your computer?
16. What happens when you open a drawing that contains a font not found on your computer?

17. Which command is better for placing lines of text at many locations over the drawing: **TEXT** or **MTEXT**?
18. List the command associated with each alias:

    dt

    t

    ed

    st
19. Can you access the menu while entering text?
20. What do the three horizontal lines of the I-beam cursor represent?

    a.

    b.

    c.
21. Which one of the following letters has a descender?

    a

    B

    d

    j

    L
22. Explain the meaning of the following justification codes:

    TL

    MC

    BR

    TR
23. Is BR justification the same as the **Right** justification? Explain.
24. Describe the one difference between **Align** and **Fit** justification modes.
25. What height should text be for a drawing scaled at:

    1:100

    1" = 50'

    1:1

    1" = 6"

    Show your work for each calculation.
26. Why is it wrong to use 1/8"-high text in an A-size drawing of a house?
27. When might you place rotated text in drawings?
28. Explain the meaning of the following control codes:

    %%u

    %%%

    %%d

    %%c
29. If you fail to turn off underlining, what happens on the next line of text?
30. What does *string* mean?
31. If the Euro symbol is not on your keyboard, how would you enter it?

32. Can you use more than one font with the **TEXT** command?

    With the **MTEXT** command?

33. Explain the meaning of the following mtext editor keyboard shortcuts:

    **CTRL+B**

    **CTRL+I**

    **CTRL+A**

34. How does AutoCAD stack text on either side of these characters?

    /

    ^

    #

35. Can different colors be applied to text placed with the **TEXT** command?

    With the **MTEXT** command?

36. Explain the meaning of the following zeros:

    Text Height = 0

    Mtext Width = 0

37. Can only numbers be used for stacked text?

38. Is it possible to use a text editor or word processor other than AutoCAD's built-in mtext editor?

    If so, how?

39. List the three primary parts of a leader:

    a.

    b.

    c.

40. What is the quickest way to edit the text of a leader?

    To edit the position of leader line?

41. Can leaders use only one kind of arrowhead?

42. Name three kinds of leader annotation:

    a.

    b.

    c.

43. What does a splined leader line look like?

44. Can leaders be attached to objects?

45. List three ways to edit text:

    a.

    b.

    c.

46. Is a different command needed to edit text created by the **TEXT** and **MTEXT** commands?

47. Does a different editor display for text created by the **TEXT** and **MTEXT** commands?

48. Can you have more than one text style in drawings?

49. What happens when you apply a different style to a line of text?

50. What happens to text when you make changes to a style?

51. Which font most closely approximates the Leroy lettering guides?

52. Can TrueType fonts have the vertical style?

53. Name one advantage to using styles in drawings.

54. What is the purpose of the **TEXTFILL** and **TEXTQLTY** system variables? When do they come into effect?
55. Can the **SPELL** command be used for text placed by commands other than **TEXT** and **MTEXT**?
56. Explain a benefit of the **MTEXT** command's background mask.
57. What happens to orientated text when it is rotated?
58. Describe the purpose of the **SCALETEXT** command. And the **JUSTIFYTEXT** command.
59. How would you change the size of text in a drawing being converted from imperial to metric units?
60. How do you apply annotative scaling to text?
61. How does the **MLEADER** command differ from the **QLEADER** command?
62. Define "content" of multiline leaders.
63. Label the parts of the multi-line leader illustrated below:

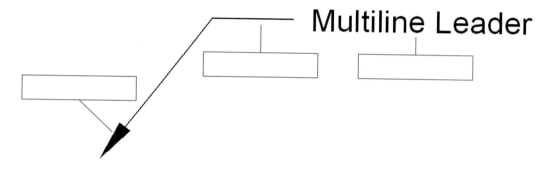

64. Describe how annotative text differs from normal text.
65. When would you use the **OBJECTSCALE** command?

# CHAPTER 11

## Placing Dimensions

Dimensions play an important role in defining the size, location, angle, and other attributes of objects in drawings. Dimensions show the length and width of objects, the diameter of holes, and the angles of sloped parts.

This chapter describes how to place dimensions in the drawing using these commands:

**DIMLINEAR** places horizontal, vertical, and rotated dimensions.

**DIMALIGNED** places dimensions aligned with objects.

**DIMBASELINE** and **DIMCONTINUE** add baseline and continuous dimensions.

**QDIM** generates a variety of continuous dimensions.

**DIMRADIUS** and **DIMDIAMETER** place radial and diameter dimensions.

**DIMJOGGED** draws radial or linear dimensions with jogged leader or dimension lines.

**DIMARC** measures the lengths of arcs.

**DIMCENTER** places center marks on arcs and circles.

**DIMANGULAR** places angular dimensions.

## INTRODUCTION TO DIMENSIONS

Dimensions are used in drawings to eliminate scale rulers and protractors to work out lengths and angles; dimensions placed with AutoCAD are far more accurate than any measurement with a ruler.

Some disciplines, such as architecture, dimension only the most important parts of the drawing and not every detail. Other disciplines, such as mechanical, completely dimension their drawings. Either way, dimensions are there to inform, and so should not interfere with the drawing.

The figure below illustrates a small portion of a sample architectural drawing provided with AutoCAD.

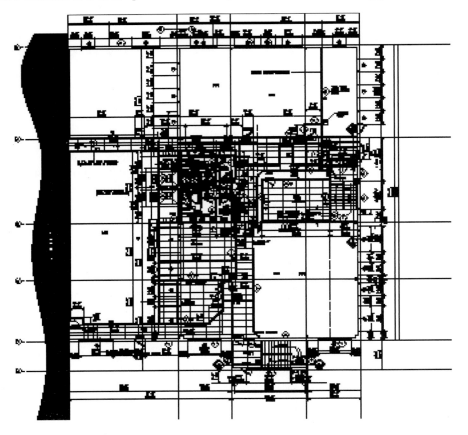

### THE PARTS OF DIMENSIONS

Linear dimensions consist typically of a dimension line with an arrowhead at each end, a pair of extension lines that extend to the part being dimensioned, and some text indicating the measured distance or other information.

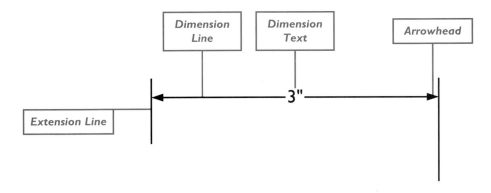

## Dimension Lines

The dimension line indicates the distance or angle being measured, as illustrated by the figure above. It is a simple line with arrowheads or "ticks" at each end. The arrowheads point at the extension lines.

When there is too little space between the extension lines, the dimension line is placed outside the extension lines (as illustrated below) or, in some cases, not drawn when there is too little room.

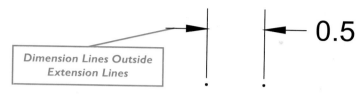

When angles are measured, the dimension line is an arc, instead of a straight line.

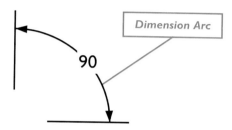

To report the distance or angle, text is placed on or near the dimension line. AutoCAD automatically creates a gap in the dimension line to make room for the text.

Normally, dimension lines take on the properties of their layer. AutoCAD allows you to override the layer setting by specifying the lineweight, color, and linetype for the dimension and extension lines separately through *dimension styles*.

## Extension Lines

Nearly all dimensioning commands begin by prompting you to pick two points or an object: these become the starting points for the extension lines. Extension lines (sometimes called "witness lines") indicate the geometric feature being dimensioned. Usually, extension lines are perpendicular to the dimension line, but need not be, such as in isometric dimensioning.

There are usually two extension lines, one at either end of the dimension line. Sometimes, the first or second extension line is not drawn, such as when it overlaps with another dimension; in rare cases, neither is drawn.

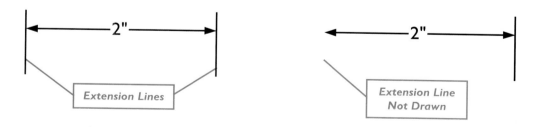

**Left:** Dimension with both extension lines.
**Right:** Dimension with one extension line left undrawn.

AutoCAD normally draws the extension lines long enough to span the distance from the origin to just beyond the dimension line — but AutoCAD can also fix the length of extension lines.

AutoCAD draws the extension lines a short distance beyond the dimension line. This is standard drafting practice. The default distance is 0.18 units, but can be changed in the Dimension Styles Manager dialog box.

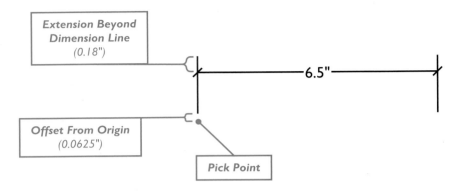

Similarly, AutoCAD offsets the extension lines from the pick points by a distance of 0.0625 units, again to mimic standard drafting practice. The offset distance is named "offset from origin," and can also be changed in the Dimension Style Manager dialog box.

## Arrowheads

Arrowheads are placed at either end of the dimension line, pointing to the extension lines. If there is insufficient room between the extension lines, then the arrowheads are placed outside. (The Dimension Style Manager dialog box lets you specify whether arrowheads or text should first be moved outside, or forced to fit inside the extension lines.)

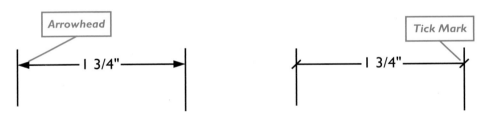

**Left:** Arrowheads and...
**Right:** ...tick marks.

The arrowheads can be replaced with tick marks and other symbols that AutoCAD includes; in addition, you can define your own. The figure at right illustrates all the arrowhead styles included with AutoCAD; to create custom arrowheads, you first define them as blocks.

The length of the arrowhead varies, depending on the scale of the drawing. Arrowheads that are too small or large prove difficult to read and can distract. Generally, arrowheads are 1/8 inch in length when used in small drawings, and 3/16 inch in larger drawings.

Arrowheads are usually drawn with a 1:3 aspect ratio so that the arrow is three times longer than it is wide. Other arrowheads, such as tick marks and circles, are 1:1.

## Dimension Text

The purpose of the dimension text is to report the distance, angle, or other information between the extension lines. Sometimes, the text reports other information, such as tolerances and alternate units of measurement. For example, some drawings require dimensions to show both metric and imperial units.

Typically you let AutoCAD measure the distance and determine the text, which it adds automatically; you can, however, override it and enter your own text with the mtext editor.

Dimension text appears in or near the dimension line, depending on whether there is enough room between the extension lines, or on the drafting standard. When there is too little room between the extension lines, the text is placed outside, or even some distance away, and in some cases referenced with a leader line. The Japanese JIS standard requires the text be placed on top of the dimension line, whereas in most other cases it is usually placed within the dimension line.

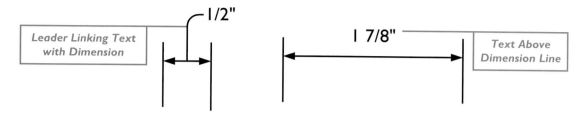

When the dimension text is *horizontal and centered*, the text is said by AutoCAD to be in the "home" position. Dimension text can be centered, left justified, or right justified on the dimension line. The text can be horizontal, vertical, or rotated.

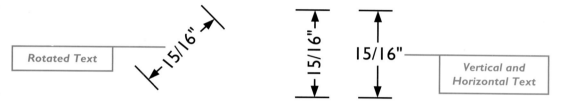

Feet and inches are separated by a dash, as in **5'-4"**. If there are no inches, the zero is usually included: **6'-0"**. AutoCAD allows you to choose whether you want the 0 displayed for feet and/or inches. If dimensions are stipulated strictly in inches, such as **72"**, you should use the inch mark to avoid confusion with feet and other units; in mechanical design, inch marks are not used.

Fractions are given either as common fractions, such as **1/2, 3/4,** and **9/16,** or as decimal fractions, such as **0.50** and **0.75**. In CAD, some text fonts lack the ability to stack fractions, such as $^1/_2$ and $^3/_4$. For this reason, the dash is used to separate inches from fractions and to avoid confusion, such as **3-1/2"**.

Dimension text indirectly follows the text style set by the STYLE command and the numerical format and precision set by the UNITS command. The style, units, and precision of dimensions are determined by the DIMSTYLE command. You can specify the style name, color of text, fill (background) color, and fixed height, as well as decide whether to surround the text with a box.

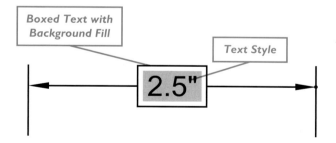

### Annotative Scaling

Like text and leaders, dimension can be annotatively scaled. This ensures that they appear at the correct size at specific model space scale factors, and eliminates the need to calculate a scale factor to make dimensions plot correctly.

## Tolerance Text

Dimension tolerances are plus and minus amounts appended to the dimension text. AutoCAD can add the tolerances automatically, but typically you specify the plus and minus amounts, which can be equal ("symmetrical") or unequal ("deviation").

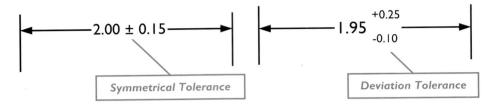

If tolerances are symmetrical, they are drawn with the plus/minus symbol. If deviation, they are drawn one above the other: above for the plus amount, below for the minus.

## Limits Text

Instead of showing dimension tolerances, the tolerance can be applied to the text itself — added to and subtracted from the dimension text. This is called "limits." The example below is a measurement of 4.00 units, with a tolerance of ±0.15.

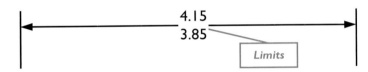

## Alternate Units

"Alternate units" show two forms of measurement, such as English and metric, on the same dimension line. The second set is usually shown in square brackets.

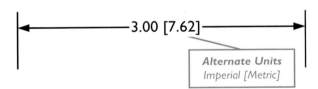

When alternate units are turned on in the dimension style, AutoCAD adds them automatically. When turned off, you can have AutoCAD calculate and add alternate units by entering a pair of square brackets ([]) when editing the dimension text.

### DIMENSIONING OBJECTS

AutoCAD places dimensions between any two points in a drawing. In addition, AutoCAD has a direct dimensioning mode, where it dimensions specific objects automatically: lines, polyline segments, arcs, polyline arcs, circles, vertices, and single points.

#### Lines and Polyline Segments

AutoCAD can dimension the lengths of lines and polyline segments with a single pick. The dimension is drawn completely, consisting of the dimension and extension lines, arrowheads, and text. This is accomplished typically with the DIMLINEAR and DIMALIGNED commands, as detailed later.

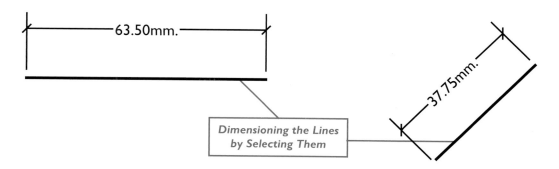

### Arcs, Circles, and Polyarcs

Arcs are dimensioned by their radius or length. When dimensioning the radius, place the leader line at an angle to avoid horizontal and vertical placements. The dimension line for the length of an arc follows the curve of the arc. The arc symbol designates arc length dimensions.

The letter "R," designating radius, prefixes the dimension text, such as **R2.125**. If a circle is dimensioned by its diameter, the dimension text is prefixed by the diameter symbol, such as Ø4.25.

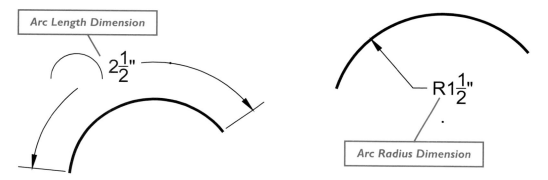

Objects constructed of several arcs are dimensioned in two stages: (1) locating the center of the arcs with horizontal or vertical dimensions, and (2) showing their radii with radius dimensioning. AutoCAD dimensions arcs and polyline arcs with a single pick. The resulting dimension depends on the command you use:

The DIMLINEAR and DIMALIGNED commands place dimension and extension lines, arrowheads, and text. The DIMRADIUS and DIMDIAMETER commands place dimension lines or leaders, arrowheads, and text.

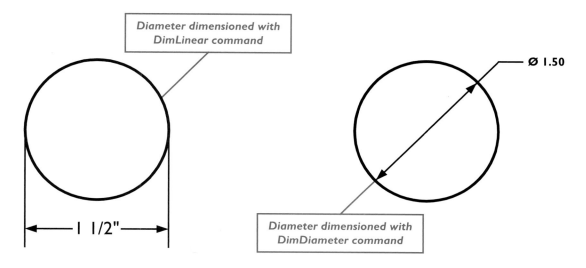

Arcs are usually dimensioned with radii, while circles are usually dimensioned with diameters.

## Wedges and Cylinders

Wedges are dimensioned in two views using three distances: length, width, and height.

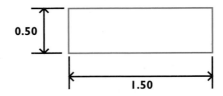

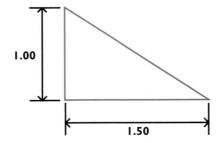

Cylinders are dimensioned for diameter and height. The diameter is typically dimensioned in the circular view. If a drill-through is dimensioned, it is described by a diameter leader.

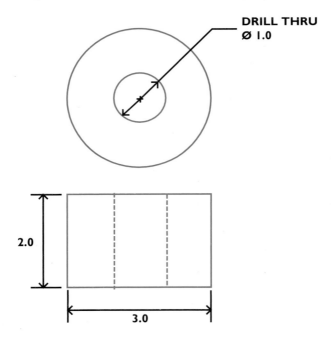

## Cones and Pyramids

Cones are dimensioned at the diameter and the height. Some conical shapes, such as truncated cones, require two diameter dimensions.

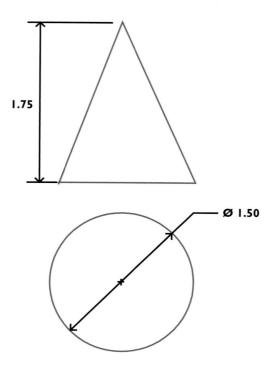

Pyramids are dimensioned like cones.

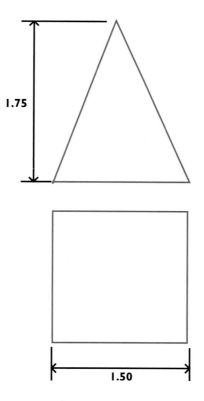

## Holes

Holes are created by drilling, reaming, boring, punching, or coring materials. You should dimension holes in drawings with notes that give the diameter, operation, and the number of holes, if more than one. The operation describes such types of holes as counter-bored, reamed, and countersunk.

Whenever possible, point the dimension leader to the hole in the circular view. Holes made up of several diameters can be dimensioned in their section.

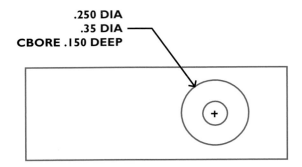

Drill sizes are designated as decimals or by a letter indicating the hole size. See www.gearhob.com/eng/design/drill_eng.htm.

## Vertices

AutoCAD measures the angle of lines, arcs, and polyline arcs with two picks using the DIMANGULAR command.

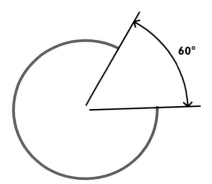

## Single Points

To "dimension" a single point, you typically use the LEADER, QLEADER, MLEADER, or DIMORDINATE commands. Leaders are typically pieces of text that point at a spot in the drawing through lines. Ordinate dimensions measure the x and y coordinates of objects relative to a base point.

**Left:** Ordinate dimension indicating x, y position.
**Right:** Leader indicating a note.

For drawing leaders, see Chapter 10, "Placing and Editing Text."

## Chamfers and Tapers

A chamfer is an angled surface applied to an edge. Use a leader to dimension chamfers of 45 degrees, with the leader text designating the angle and one (or two) linear distances.

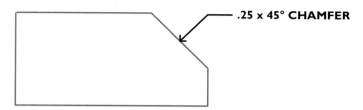

If the chamfer is not 45 degrees, dimensions showing the angle and the linear distances describe the part.

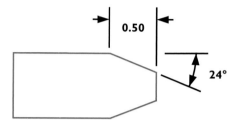

A taper can be described as the surface of a cone frustum (cone with the top sliced off). Dimension tapers by giving any three of these parameters: start diameter, end diameter, rate of taper, and/or the length.

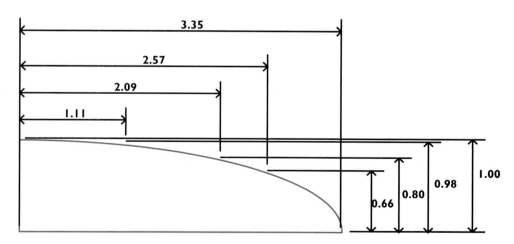

## DIMENSION MODES

AutoCAD constructs dimensioning semiautomatically. All you needs to provide it is some basic information, such as what to dimension, or where to start and end the dimensions. Then it constructs the dimension for you, measures the distances and angles, and then draws all the elements of the dimension.

AutoCAD's dimensions are *associative*. That means they automatically update when associated objects are moved and stretched. This is a very powerful feature and a great time-saver. Dimensions, however, do not need to be associative; AutoCAD can work with several levels of associativity:

**Fully associative** — standard mode where AutoCAD attaches the ends of the extension lines to the geometry of the object. Move the object, and the dimension moves with it; stretch the object, and the dimension text updates automatically. This is the preferred method.

**Partly associative** — Autodesk's first attempt at associativity. Extension lines are attached to *dimension points* placed on layer Defpoints (short for "definition points"). When stretching or moving objects, you must include the defpoints in the selection set; otherwise, the dimension does not update correctly.

**Non-associative** — dimensions are not attached to objects in any way.

**Trans-spatial** — dimensions created in paper space are attached to objects in model space; changing the model updates the dimensions in paper space.

**Dimensional constraints** — define distances within and between objects parametrically. In this case, the dimension drives the distance, either with a value (number) or a formula parameter. Dimensional constraints are covered in Chapter 13, "Applying Dimensional Constraints."

Dimensions are placed as if they were blocks: the lines, arrows, and text act as a single object. You can use the EXPLODE command to break apart dimensions.

**Note** When AutoCAD reports "Non-associative dimension created," this means that it was not able to attach the dimension to an object.

## Dimension Variables and Styles

Dimension variables determine how the dimensions are drawn; they are system variables specific to dimensions, sometimes called "dimvars" for short. Some dimvars store values, such as the color of the dimension line and the look of the arrowhead; others are toggles that turn values on and off, such as whether the first or second extension line should be displayed.

Just as styles determine the look of text, dimension styles determine the look of dimensions. Dimension styles ("dimstyles" for short) are created and modified with the DIMSTYLE command, which collects all the settings of dimension variables. Dimstyles are discussed in Chapter 12, "Editing Dimensions," and dimvars are listed in Appendix D, "Summary of Dimensional Variables."

## Dimension Standards

There are many ways to draw dimensions, because different industries and countries define standards differently. AutoCAD includes the ISO-25 dimensioning standard with its *acadiso.dwt* template drawing. (Other standards that were included in previous releases of AutoCAD, such as DIN, JIS, and Gb, are no longer part of AutoCAD.)

Most standards organizations make the dimension standards available to members at a cost (for example at www.buildingsmartalliance.org/ncs, — that's .org, not .com), while some are freely available at Web sites (for example at www.cadinfo.net/editorial/archdim1.htm).

## Object Snaps

Object snap modes are very helpful in placing dimensions accurately. For example, the ENDpoint and INTersection osnaps are useful for capturing the ends and intersections of lines, while the CENter and TANgent osnaps help with dimensions of arcs and circles.

# DIMLINEAR

The DIMLINEAR command places horizontal, vertical, and rotated dimensions.

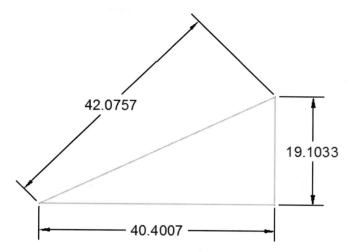

*Examples of rotated, horizontal, and vertical dimensions. The rotated dimension was placed at 45 degrees.*

You specify the location of the extension lines, and AutoCAD determines whether to draw a horizontal or vertical dimension automatically — based on where you place the dimension line.

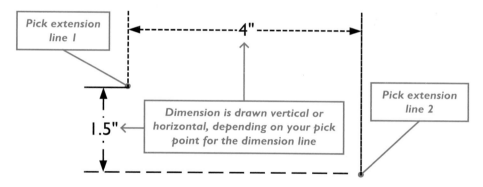

This command also applies dimensioning directly to objects.

## TUTORIAL: PLACING DIMENSIONS HORIZONTALLY

In this tutorial, you place horizontal dimensions on a rectangle. (If you find it difficult to draw dimensions horizontally with DIMLINEAR, you can use the undocumented DIMHORIZONTAL command, which draws only horizontal dimensions.)

Horizontal dimensions are drawn with just a couple of picks. To locate the extension lines, you select an object (line, arc, circle, or polyline) or pick two points (1 and 2, in the figure below). Another pick point (3) locates the dimension line, which also determines the location of the text and arrowheads, both of which are drawn automatically by AutoCAD.

1. Start AutoCAD with a new drawing using the *acad.dwt* template.
   Ensure that ENDpoint object snap is turned on.

2. Draw a 4x3 rectangle with the **RECTANGLE** command.

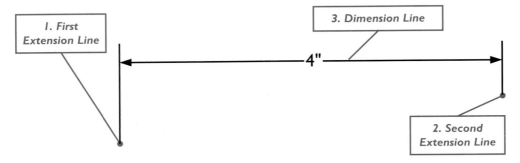

3. To dimension the width of the rectangle (horizontally), start the **DIMLINEAR** command:
   - In the ribbon's Annotation tab, choose the **Linear** button in the Dimensions panel.
   - At the 'Command:' prompt, enter the **DIMLINEAR** command.

     Command: **dimlinear** *(Press ENTER.)*

   - Alternatively, enter the aliases **dli** or **dimlin** at the 'Command:' prompt.

4. At the prompt, pick the lower left corner of the rectangle:
   Specify first extension line origin or <select object>: *(Pick point 1 to locate first extension line.)*

   This point specifies the origin of the first extension line.

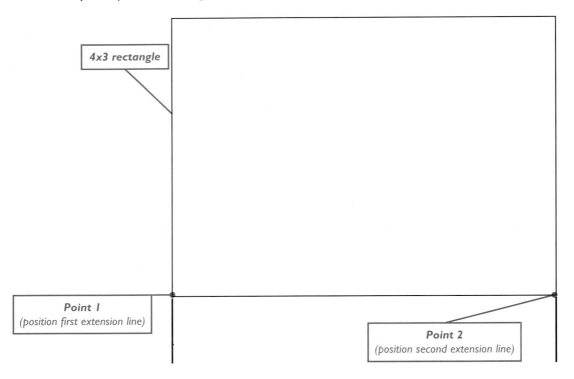

5. Pick the other lower corner of the rectangle:
   Specify second extension line origin: *(Pick point 2 to locate second extension line.)*

   Point #2 specifies the origin of the second extension line, and measures the distance reported by the dimension.

6. Pick a third point to located the dimension line:

    Specify dimension line location or [Mtext/Text/Angle]: *(Pick point 3 to place the dimension line.)*

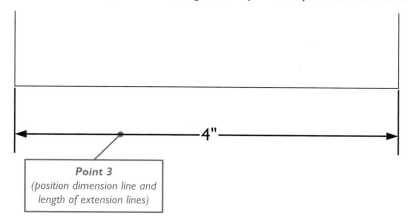

Point #3 specifies the distance of the dimension line away from the object; this point also determines the length of the two extension lines.

Notice that AutoCAD reports the length measurement at the command prompt, and draws the horizontal dimension.

Dimension text = 4"

## TUTORIAL: PLACING DIMENSIONS VERTICALLY AND ROTATED

The DIMLINEAR command also places dimensions vertically and at an angle (rotated).

To place dimensions vertically, pick the two points so that they are roughly vertical.

To place a rotated dimension, enter the DIMLINEAR command and then choose the **Rotated** option before locating the dimension line, as follows:

    Command: **dimlinear**
    Specify first extension line origin or <select object>: *(Pick a point.)*
    Specify second extension line origin: *(Pick another point.)*

Type **r** to specify the **Rotated** option:

    Specify dimension line location or
    [Mtext/Text/Angle/Horizontal/Vertical/Rotated]: **r**

AutoCAD asks you for the angle:

    Specify angle of dimension line <0>: **45**

And then place the dimension line:

    Specify dimension line location or
    [Mtext/Text/Angle/Horizontal/Vertical/Rotated]: *(Pick a point.)*

If you find the process frustrating, you can use the DIMVERTICAL and DIMROTATED commands to force AutoCAD to place the dimensions correctly.

### Dimensioning Objects

The DIMLINEAR command lets you dimension objects directly. This means that you don't need to pick points for the first and second extension line; AutoCAD finds the points for you.

This allows speedy, repeated dimensioning of drawings:

a. Press spacebar to repeat the command.
b. Press spacebar to choose the **Select Object** option.
c. Pick the object.
d. Pick the location of the dimension line.
e. Repeat: spacebar – spacebar – pick – pick.

#### TUTORIAL: DIMENSIONING OBJECTS

In this tutorial, you use the dimlinear command's object option to dimension a variety of geometry directly.

1. Open the *Dimension Objects.dwg* file, a drawing of several geometric shapes.

2. Start the **DIMLINEAR** command, and then press **ENTER** to dimension objects, as follows:
   Command: **dimlinear** *(Press ENTER.)*
   Specify first extension line origin or <select object>: *(Press ENTER.)*

3. AutoCAD prompts you to pick an object. Choose the line:
   Select object to dimension: *(Pick the line.)*

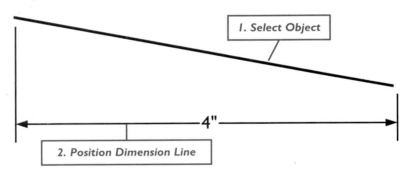

4. AutoCAD asks you to pick a location for the dimension line:
   Specify dimension line location or
   [Mtext/Text/Angle/Horizontal/Vertical/Rotated]: *(Position the dimension line.)*

   And then AutoCAD reports the length of the line in the command bar:
   Dimension text = 4.000

5. Continue the tutorial by dimensioning the circle:
   a. Press **ENTER** to repeat the **DIMLINEAR** command:
      Command: *(Press ENTER.)*
   b. Press **ENTER** to select an object:
      Specify first extension line origin or <select object>: *(Press ENTER.)*
   c. Pick the circle:
      Select object to dimension: *(Pick the circle.)*
   d. Locate the dimension line:
      Specify dimension line location or
      [Mtext/Text/Angle/Horizontal/Vertical/Rotated]: *(Position the dimension line.)*

6. Repeat to dimension the arc and the sides of the lines.

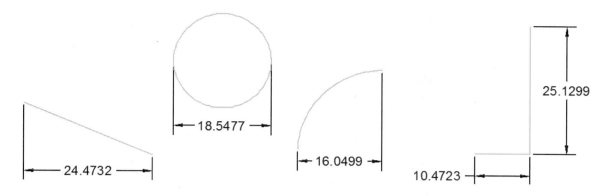

## LINEAR DIMENSIONS: ADDITIONAL METHODS

The DIMHORIZONTAL command has options to change the dimension text.

- **MText** displays the mtext editor.
- **Text** prompts you to edit the dimension text.
- **Angle** rotates the dimension text.
- **Horizontal** forces the dimension line horizontally.
- **Vertical** forces the dimension line vertically.
- **Rotated** rotates the dimension line.
- **Angle** rotates the dimension text.

Let's look at each option.

### MText

The **MText** option displays the mtext editor, allowing you to edit the dimension text. The editor should be familiar to you from Chapter 10, "Placing and Editing Text."

The mtext editor highlights the default dimension text. (Prior to AutoCAD 2006, the double angle bracket **<>** was used as a shorthand notation for the default dimension text.)

Here are some examples of how you can work with the default text:

- Replace default text with other text, and AutoCAD shows the new text on the dimension line.

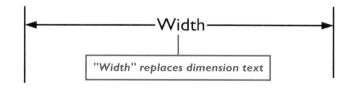

- Add text on either side of the default text, and AutoCAD shows the added text.

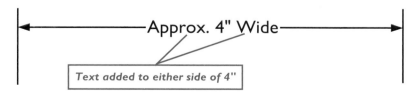

- Replace the default text with a space character, and AutoCAD erases the dimension text.

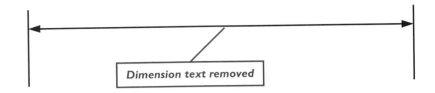

### Special Symbols

To add special characters, use the mtext editor's **Symbols** submenu. Alternatively, you can enter metacharacters to show these common symbols:

° (degrees) with %%d

± (plus or minus) with %%p

Ø (diameter) with %%c

### Text

The **Text** option prompts you to edit the dimension text at the command prompt.

Enter dimension text <4.0000>: **The distance is <> inches**

Include the double angle brackets ( <> ) to specify the location of AutoCAD's measured distance with the text you add. The result is: **The distance is 4.0000 inches**.

### Angle

The **Angle** option rotates the dimension text about its center point.

Specify angle of dimension text: *(Enter an angle, such as* **45**, *or use the mouse to show the angle.)*

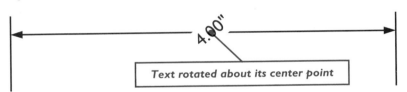

### Horizontal, Vertical and Rotated

The **Horizontal** option forces the dimension line to the horizontal, no matter where you locate the dimension line. This makes it operate like DIMHORIZONTAL.

Similarly, the **Vertical** option forces the dimension line to the vertical, just like the DIMVERTICAL command.

The **Rotated** option prompts you for an angle, and then draws the dimension line at that angle (measured from the positive x-axis), adjusting the lengths of the extension lines as required.

## DIMALIGNED

The DIMALIGNED command places dimensions aligned to two points.

This command is like the DIMLINEAR command, except that the pick points for the extension lines determine the angle of the dimension line. DIMALIGNED reports the true lengths of objects.

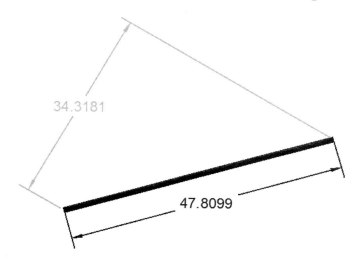

*In gray:* Rotated dimension drawn at 60 degrees with the DIMLINEAR command's Rotated option.
*In black:* Aligned dimension drawn with the DIMALIGNED command.

(Don't confuse rotated dimensions with aligned ones. In a rotated dimension, you specify the angle of the dimension line, distorting the lengths of the extension lines; in aligned dimensions, AutoCAD aligns the dimension line with the extension lines or the object being dimensioned. For the figure above, I specified an angle of 60 degrees for the rotated dimension.)

When you employ direct dimensioning, press ENTER and then select an object; DIMALIGNED reacts differently, depending on the object you select:

**Lines, Arcs, and Polylines** — endpoints of lines, arcs, and polylines determine the angle of the dimension line; AutoCAD measures the length of the line and polyline segments and the chord length of arcs.

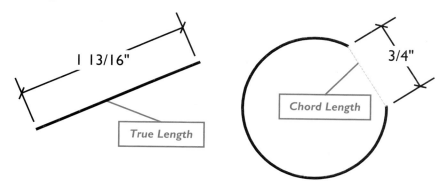

**Circles** — a point on the circle determines the angle of the dimension line; AutoCAD measures the circle's diameter.

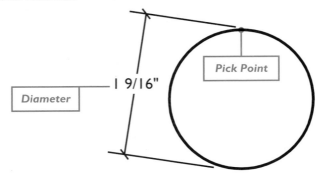

### TUTORIAL: PLACING ALIGNED DIMENSIONS

1. In AutoCAD, draw a line at an angle.
2. To place a dimension aligned to the line's angle, start the **DIMALIGNED** command:
   - In the ribbon's Annotation tab, choose the **Aligned** button in the Dimensions panel.
   - At the 'Command:' prompt, enter the **dimaligned** command.

     Command: **dimaligned** *(Press* ENTER.*)*
   - Alternatively, enter the aliases **dal** or **dimali** at the 'Command:' prompt.
3. In all cases, AutoCAD prompts you to select the object you want dimensioned:
   Specify first extension line origin or <select object>: *(Press* ENTER.*)*
   Select object to dimension: *(Pick one object.)*
4. Pick a point to place the dimension line:
   Specify dimension line location or
   [Mtext/Text/Angle]: *(Pick a point to place the dimension line.)*
   Dimension text = 10.0

This command's **Extension line**, **Mtext**, **Text**, and **Angle** options operate identically to those of the DIMLINEAR command.

# DIMBASELINE AND DIMCONTINUE

The DIMBASELINE command (also called "parallel dimensioning") continues dimensions from baselines, while the DIMCONTINUE command (also called "chain dimensioning") continues dimensions from the previous dimension.

Baseline dimensions tend to stack over one another, while continuous dimensions are usually located next to each other. The stack distance is 0.38, but can be changed. Continuous dimensions share common extension lines.

Both commands continue from the last placed dimension, which can be linear, angular, or ordinate. (These commands do not work with leaders, radial, or diameter dimensions.) You cannot use either command until at least one other dimension has been placed in the drawing. When no dimensions were created during the current drafting session, AutoCAD prompts you to select a dimension as the base.

### TUTORIAL: PLACING ADDITIONAL DIMENSIONS FROM A BASELINE

1. To place baseline dimensions, start the **DIMBASELINE** command:
    * In the ribbon's Annotation tab, choose the **Baseline** button in the Dimensions panel.
    * At the 'Command:' prompt, enter the **dimbaseline** command.

        Command: **dimbaseline** *(Press* ENTER.*)*
    * Alternatively, enter the aliases **dba** or **dimbase** at the 'Command:' prompt.

2. If necessary, AutoCAD prompts you to select a dimension to use as the base:
    Select base dimension: *(Pick a linear, angular, or ordinate dimension.)*

3. AutoCAD prompts you to pick the endpoint for the next dimension:
    Specify a second extension line origin or [Undo/Select] <Select>: *(Pick a point.)*
    Dimension text = 10.0

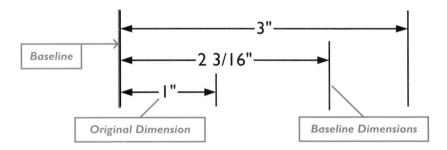

The baseline dimension continues from the first extension line of the previous dimension.

4. AutoCAD repeats the prompt for the next dimension:
    Specify a second extension line origin or [Undo/Select] <Select>: *(Pick another point.)*

5. Press **ESC** to exit the command:
    Specify a second extension line origin or [Undo/Select] <Select>: *(Press* ESC.*)*

## TUTORIAL: CONTINUING DIMENSIONS

The DIMCONTINUE command is identical to DIMBASELINE, except that it continues dimensions by placing new ones next to each other.

1. To continue dimensioning, start the **DIMCONTINUE** command:
   - In the ribbon's Annotation tab, choose the **Continue** button in the Dimensions panel.
   - At the 'Command:' prompt, enter the **dimcontinue** command.

     Command: **dimcontinue** *(Press* ENTER.*)*
   - Alternatively, enter the aliases **dco** or **dimcont** at the 'Command:' prompt.

2. If necessary, AutoCAD prompts you to select a dimension to use as the base:
   Select base dimension: *(Pick a linear, angular, or ordinate dimension.)*

3. AutoCAD prompts you to pick the endpoint for the next dimension:
   Specify a second extension line origin or [Undo/Select] <Select>: *(Pick a point.)*
   Dimension text = 10.0

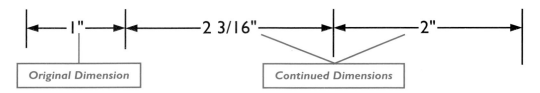

   Notice that the dimension continues from the second extension line of the previous dimension.

4. AutoCAD repeats the prompt for the next dimension:
   Specify a second extension line origin or [Undo/Select] <Select>: *(Pick another point.)*

5. Press ESC to exit the command:
   Specify a second extension line origin or [Undo/Select] <Select>: *(Press* ESC.*)*

The **Select** option selects another dimension from which to continue. AutoCAD prompts:

   Select continued dimension: (Pick an extension line.)

By picking an extension line, you determine the direction in which the dimension continues — to the left or to the right — assuming the dimension is horizontal.

## QDIM

The QDIM command places baseline and continuous style dimensions in one fell swoop.

It is a more powerful version of the DIMCONTINUE and DIMBASELINE commands: it places continued radial dimensions, applies the continued dimensions all at once, and more. Most importantly, it can dimension more than one object at a time. On the negative side, it is more complex than the two previous commands, and it cannot dimension blocks.

The figure below illustrates the continuous radius and continuous linear dimensions generated by this command. "While this is how AutoCAD dimensions radii," notes the technical editor, "it is poor practice. Dimensions should always be outside the parts."

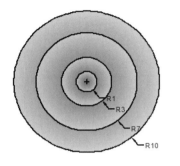
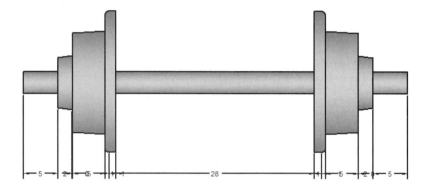

Perhaps the most powerful aspect of QDIM is its ability to change the type of dimension. For instance, you might first dimension objects in staggered mode, but some time later want to change to ordinate mode. Making the change is this simple: restart the command, select the same objects, and then specify ordinate mode. AutoCAD changes the staggered dimensions to ordinate dimensions.

### TUTORIAL: QUICK CONTINUOUS DIMENSIONS

1. To place continuous-style dimensions quickly, start the **QDIM** command:
   - In the ribbon's Annotation tab, choose the **Quick Dimension** button in the Dimensions panel.
   - At the 'Command:' prompt, enter the **qdim** command.

     Command: **qdim** *(Press ENTER.)*

2. In all cases, AutoCAD reports the dimensioning priority, and then prompts you to select the geometry to dimension:

   Associative dimension priority = Endpoint

   Select geometry to dimension: *(Select one or more objects.)*

   Select geometry to dimension: *(Press ENTER to end object selection.)*

   Unlike other dimensioning commands, you may select more than one object.

3. Pick a point to define the position of the dimension line:

   Specify dimension line position, or

   [Continuous/Staggered/Baseline/Ordinate/Radius/Diameter/datumPoint/Edit/seTtings] <Radius>: *(Pick a point.)*

   Depending on the objects you select, AutoCAD guesses at how to dimension them.

## QUICK DIMENSIONS: ADDITIONAL METHODS

The QDIM command boasts a host of options to force the type of continuous dimension and specify settings:

- **Continuous** forces continued dimensions.
- **Staggered** places staggered dimensions.
- **Baseline** places baseline dimensions.
- **Ordinate** places ordinate dimensions.
- **Radius** places radial dimensions.
- **Diameter** places diameter dimensions.
- **datumPoint** specifies the start point for baseline and ordinate dimensions.
- **Edit** edits continuous dimensions.
- **seTtings** selects the object snap priority.

Let's look at all the options.

### Continuous

The **Continuous** option forces continued dimensions, like those created by DIMCONTINUE.

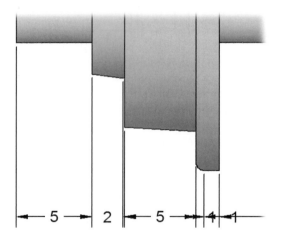

### Staggered

The **Staggered** option forces staggered dimensions, which are like stacked continuous dimensions. This option, however, requires an even number of dimensionable objects. As illustrated in the figure, the odd-numbered segment, unfortunately, is left out and is not dimensioned.

## Baseline

The **Baseline** option forces baseline dimensions, like those created by the DIMBASELINE command. The figure below shows the dimensions starting at the left end of the flywheel; you can change the start point with the **Datum point** option, as described later.

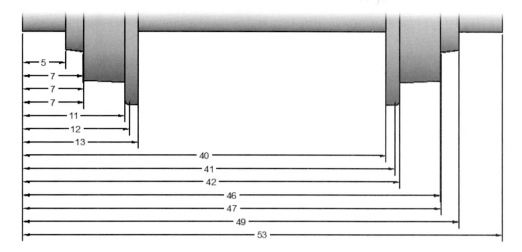

## Ordinate

The **Ordinate** option forces ordinate dimensions, like those created by the DIMORDINATE command, with two differences. DIMORDINATE prevents ordinate dimensions from ending on top of each other, while QDIM does not. Also, DIMORDINATE measures relative to the UCS, while this command measures relative to a datum point.

The figure illustrates x-ordinate dimensions starting with a value of 5.0, which is the distance from the origin (0,0). You can change the origin point with the **Datum point** option.

To create y-ordinate dimensions, move the cursor to the left or right side of the object during the "Specify dimension line position" prompt.

## Radius

The **Radius** option forces radius dimensions, and works only with circles and arcs. Selecting other circular objects causes AutoCAD to complain, "No arcs or circles are currently selected." The radial dimensions are placed at the point you pick at the "Specify dimension line position" prompt.

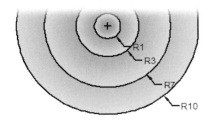

### Diameter

The **Diameter** option forces diameter dimensions.

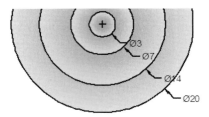

It operates identically to the **Radius** option. In both cases, it's a good idea to avoid placing dimensions horizontally, because they tend to bunch up. Because placing dimensions on parts is poor practice, you should manually move the radial and diameter dimensions. This is easily done with grips editing.

### datumPoint

The **datumPoint** option prompts you to select a new base point from which to measure baseline and ordinate dimensions.

   Select new datum point: *(Pick a point.)*

### Edit

The **Edit** option enters editing mode, and prompts you to add and remove dimension markers. These markers show up as an **x** at the endpoints of lines, and the center points of circles and arcs.

   Indicate dimension point to remove, or [Add/eXit] <eXit>: *(Enter an option.)*

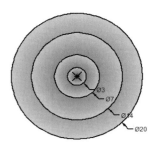
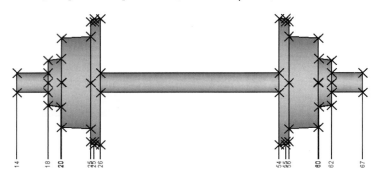

**Remove** removes markers from objects. To remove, pick a marker with the cursor. AutoCAD reports, "One dimension point removed."

**Add** adds markers to the drawing. To add, pick a point anywhere in the drawing; it need not be on an object. AutoCAD reports, "One dimension point added."

**eXit** exits **Edit** mode, and returns to the "Specify dimension line position" prompt.

### seTtings

The **seTtings** options doesn't set much, except whether endpoints or intersections have priority, and even then I am not convinced either setting makes any difference. AutoCAD displays the following prompt:

   Associative dimension priority [Endpoint/Intersection]: *(Enter E or I.)*

(The technical editor agrees: "You're right, it doesn't. AutoCAD always finds the endpoints.")

 **DIMRADIUS** AND **DIMDIAMETER**

The DIMRADIUS command places dimensions that measure the radius of circular objects, while the DIMDIAMETER command places dimensions that measure diameters.

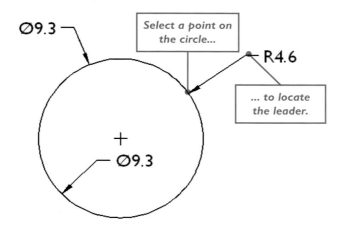

These commands dimension circles, arcs, and polyarcs (arcs that are parts of polylines). They do not dimension other curved objects, such as ellipses, splines, and regions.

The dimensions created by these two commands are more like leaders than the two-extension-line dimensions you have seen so far in this chapter. They don't prompt for a point, but immediately ask you to select a circle or an arc.

The second prompt asks you to "Specify dimension line location." Although this sounds like a single item (the dimension line), the point you pick determines several things at once, which can be difficult for new users:

- The angle of the leader line.
- The length of the leader line.
- The location of the dimension text.
- The placement of the center mark.
- The extent of an extension line.

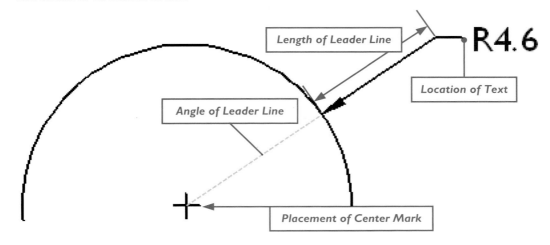

## TUTORIAL: PLACING RADIAL DIMENSIONS

1. To dimension a radius, start the **DIMRADIUS** command:
   - In the ribbon's Annotation tab, choose the **Radius** button in the Dimensions panel.
   - At the 'Command:' prompt, enter the **dimradius** command.

     Command: **dimradius** *(Press* ENTER.*)*
   - Alternatively, enter the aliases **dra** or **dimrad** at the 'Command:' prompt.

2. In all cases, AutoCAD prompts you to select the arc or circle you want dimensioned:
   Select arc or circle: *(Pick one arc, circle, or polyarc.)*
   Dimension text = 10.0

3. Pick a point to place the dimension leader:
   Specify dimension line location or [Mtext/Text/Angle]: *(Pick a point to position the dimension leader.)*

When you position the dimension beyond the end of the arc, AutoCAD adds an extension line automatically, as illustrated below:

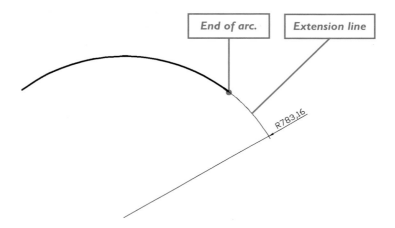

The **DIMDIAMETER** command operates identically:

Command: **dimdiameter**

Select arc or circle: *(Pick one arc, circle, or polyarc.)*

Dimension text = 10.0

Specify dimension line location or [Mtext/Text/Angle]: *(Pick a point to position the dimension.)*

The **Mtext**, **Text**, and **Angle** options are the same as those described earlier in this chapter under the **DIMHORIZONTAL** command.

### DIMJOGGED

The DIMJOGGED command draws radial dimensions with a jogged leader line (also called "foreshortened" radius dimensions).

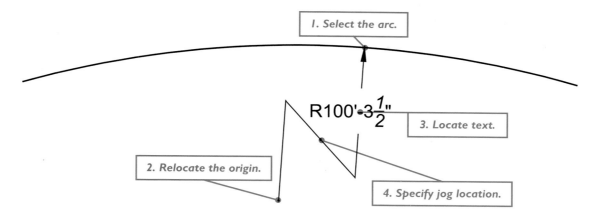

This command has more options, because AutoCAD also needs to know (1) where to relocate the dimension origin, and (2) where to place the jog.

The relocated origin replaces the center point, even though it should always point to the center. The jog location determines the middle point of the jog; the angle of the jog (called the "transverse angle") is determined by the dimension style. The segment containing the text always points to the arc's center point.

#### TUTORIAL: PLACING JOGGED RADIAL DIMENSIONS

1. To dimension a radius with a jogged leader, start the **DIMJOGGED** command:
   - At the 'Command:' prompt, enter the **dimjogged** command.

     Command: **dimjogged** *(Press ENTER.)*

   - Alternatively, enter the aliases **jog** or **djo** at the 'Command:' prompt.

2. In all cases, AutoCAD prompts you to select the arc or circle you want dimensioned:
   Select arc or circle: *(Pick one arc, circle, or polyarc.)*

3. Pick a point to relocate the origin point closer to the arc:
   Specify center location override: *(Pick a point.)*
   Dimension text = 101'-3 1/2"

4. Indicate the center of the dimension text:
   Specify dimension line location or [Mtext/Text/Angle]: *(Pick a point or enter an option.)*

5. Locate the jog's center:
   Specify jog location: *(Pick a point.)*

 **DIMARC**

The DIMARC command measures the lengths of arcs.

This command places dimensions along the circumference of arcs. The dimension can be along the full length of the arc, or just a portion.

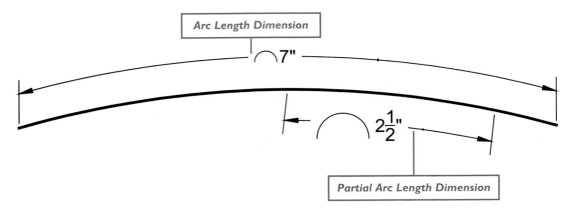

### TUTORIAL: PLACING ARC LENGTH DIMENSIONS

1. To dimension the length of an arc, start the **DIMARC** command:
   - At the 'Command:' prompt, enter the **dimarc** command.

     Command: **dimarc** *(Press* ENTER.*)*

   - Alternatively, enter the **dar** alias at the 'Command:' prompt.

2. In all cases, AutoCAD prompts you to select an arc or polyline arc:
   Select arc or polyline arc segment: *(Pick one arc or polyarc.)*

3. Indicate the position of the dimension text:
   Specify arc length dimension location, or [Mtext/Text/Angle/Partial/Leader]: *(Pick a point or enter an option.)*

### ARC LENGTH DIMENSIONS: ADDITIONAL METHODS

The DIMARC command has three options you are already familiar with — **Mtext**, **Text**, and **Angle**. Options unique to arc lengths are as follows:

- **Partial** measures parts of arc lengths.
- **Leader** toggles the addition of leaders.

The figure below shows the effect of the partial and leader options.

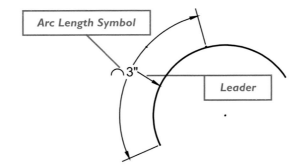

### Partial

The **Partial** option dimensions just part of the length of the arc. AutoCAD prompts you to pick two points along the arc between which the dimension is placed:

> Specify first point for arc length dimension: *(Pick a point on the arc.)*
> Specify second point for arc length dimension: *(Pick another point.)*

If ENDpoint osnap is on, this command will snap to the endpoints of the arc, negating the attempt at partial dimensioning.

The results differ, depending on where you place the partial dimension. If the dimension is between extension lines, the extensions lines are parallel to each other; when the dimension is outside, the extension lines are radial to the arc's center point.

### Leader

The **Leader** option toggles (turns on and off) the addition of a leader to arcs greater than 90 degrees. (This option does not appear when the arc is less is 90 degrees.) The leader points from the dimension text to the arc.

After entering the **Leader** option, the prompt changes to **No leader**:

> Specify arc length dimension location, or [Mtext/Text/Angle/Partial/Leader]: **l**
> Specify arc length dimension location, or [Mtext/Text/Angle/Partial/No leader]: *(Enter option.)*

## DIMANGULAR

The DIMANGULAR command dimensions angles, arcs, circles, and lines.

Unlike the other dimensioning commands in this chapter, this dimensions angles only — not lengths or diameters. The command requires a vertex and some means of determining the angle. The result depends on the object selected:

**Two Lines** — the vertex is the real or apparent intersection of the two lines; the angle is measured between the two lines.

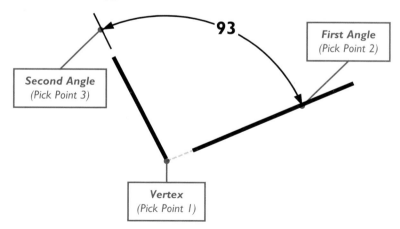

**Arc** — the vertex is the arc's center point; the angle is measured between two endpoints.
**Circle** — the vertex is the circle's center point; the angle is measured between the "Select circle" pick and the next pick point.
**Three Points** — the vertex is at the first pick point; the angle is measured between the next two pick points.

For intersecting lines, DIMANGULAR can draw four possible angles, each a supplementary pair adding up to 180 degrees. In the figure below, the supplementary pairs are shown in the same color.

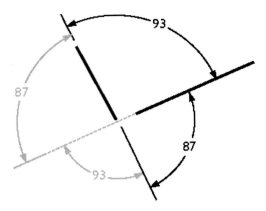

In the case of arcs and circles, two angles are possible: the minor arc (shown in black, below) and the major arc (shown in gray). Together they add up to 360 degrees.

### TUTORIAL: PLACING ANGULAR DIMENSIONS

1. To dimension an angle, start the **DIMANGULAR** command:
   - In the ribbon's Annotation tab, choose the **Angular** button in the Dimensions panel.
   - At the 'Command:' prompt, enter the **dimangular** command.

     Command: **dimangular** *(Press ENTER.)*
   - Alternatively, enter the aliases **dan** or **dimang** at the 'Command:' prompt.

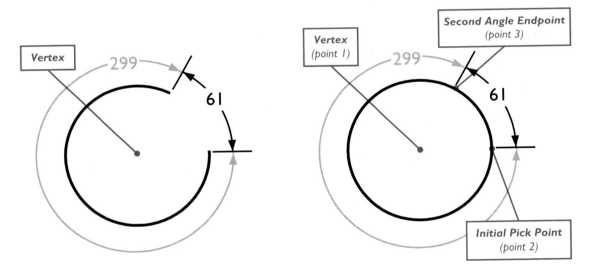

2. In all cases, AutoCAD prompts you to select the object you want dimensioned:
   Select arc, circle, line, or <specify vertex>: *(Press ENTER for the vertex option)*

3. Pick three points to specify the angle:
   Specify angle vertex: *(Pick a vertex at point 1.)*
   Specify first angle endpoint: *(Pick point 2.)*
   Specify second angle endpoint: *(Pick point 3.)*
   Non-associative dimension created.

4. Pick a point to locate the dimension:

    Specify dimension arc line location or [Mtext/Text/Angle/Quadrant]: *(Pick a point to position the dimension.)*

    Dimension text = 90

The **Mtext**, **Text**, and **Angle** options are the same as those described earlier in this chapter under the DIMHORIZONTAL command. The unique option is Quadrant.

### Quadrant

The **Quadrant** option forces the angular dimension to reside in one of the four quadrants:

   Specify quadrant: *(Drag the dimension around the arc.)*

Pick a quadrant for the dimension. When you drag the dimension beyond the end of the arc, AutoCAD adds an extension, just as with the DIMRADIUS command.

 DIMCENTER

The DIMCENTER command places center marks and lines on arcs and circles.

### TUTORIAL: PLACING CENTER MARKS

1. To mark the center of arcs and circles, start the **DIMCENTER** command:
   - In the ribbon's Annotation tab, choose the **Center Mark** button in the Dimensions panel.
   - At the 'Command:' prompt, enter the **dimcenter** command.

       Command: **dimcenter** *(Press ENTER.)*

   - Alternatively, enter the **dce** alias at the 'Command:' prompt.

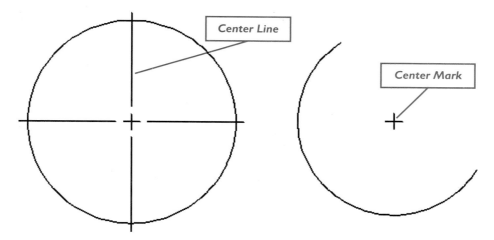

2. In all cases, AutoCAD prompts you to select the arc or circle you want dimensioned:
    Select arc or circle: *(Pick one arc, circle, or polyarc.)*

    Notice that AutoCAD places the center mark.

### DimCen

The DIMCEN system variable defines the size and look of the center mark. This variable affects center marks created by DIMRADIUS and DIMDIAMETER, although their center marks are drawn only when the dimension is placed outside the circle or arc.

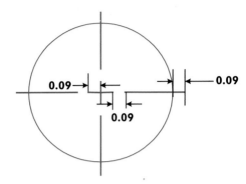

The default value is 0.09, which means that DIMCENTER draws the center mark lines 0.09 units long.

| DimCen | Meaning |
|---|---|
| 0 | Draws neither center lines nor center marks. |
| <0 | Draws center lines and center marks. |
| >0 | Draws center marks only. |

When the value is negative, such as -0.09, AutoCAD leaves a gap of 0.09 units between the mark and the line, as well as extends the line 0.09 units beyond the circumference of the circle or arc.

To draw center lines as well, change the value of DIMCEN to -0.09 (the negative value), as follows:

Command: **dimcen**

Enter new value for DIMCEN <0.0900>: **-0.09**

Normally, this value is set in the dimension style using the DIMSTYLE command.

## TUTORIAL: DIMENSIONING

In the following exercise, you dimension the fuse link below with some of the commands learned in this chapter. As you work through the tutorial, refer to the numbered points in the figure below.

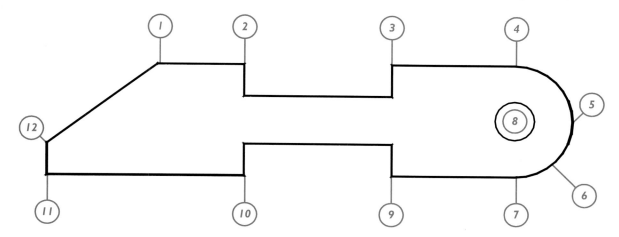

1. Open the *dimen.dwg* file, a drawing of a fuse link.

   This drawing has layers, dimension styles, and object snaps preset for you.

   Check the Layers toolbar to ensure layer "Dimensions" is current.

2. Start by placing a linear dimension along the top of the fuse link.

   From the Dimension menu, select **Linear**.

   Press **ENTER** to use the **Select Object** option, and then select the line between points 1 and 2.

   Locate the dimension line roughly 0.5 units above.

3. Continue dimensioning with the **DIMCONTINUE** command.

   Use INTersection object snap at points 3 and 4.

   Complete the dimension at point 5; do not dimension the arc at this time.

   Remember to press **ESC** to exit the command.

   Do the new dimension lines continue at the same height as the first dimensions?

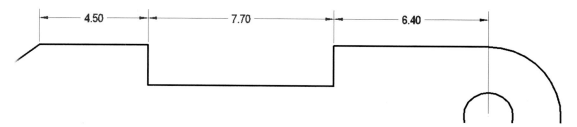

4. Now start a linear dimension along the bottom of the fuse link with the **DIMLINEAR** command.

   Place extension lines at points 11 and 10.

   Does INTersection object snap help you?

5. Switch to baseline dimensioning. From the Dimension menu, choose **Baseline**.

   Pick points 9 and 7 as the second extension line origins.

   Complete the command by picking point 7; do not dimension the arc yet.

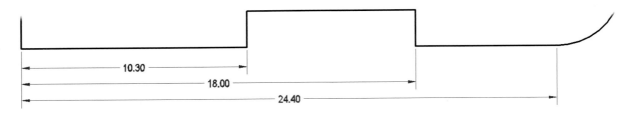

6. Place a vertical dimension by entering the **DIMLIN** alias.

   Choose point 11 as the first extension line origin, and point 12 as the second.

   If you make a mistake, use the **UNDO** command to eliminate the last dimension.

7. Dimension the angle at the left side with the **DIMALIGNED** command.

   Use the **Select Object** option to select the angled line between points 12 and 1.

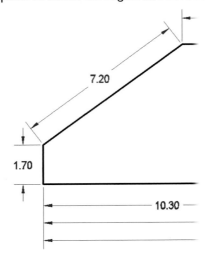

8. Select **Diameter** from the **Dimension** menu.

   Dimension the circle at point 8.

   Make sure the leader line extends outside the fuse link.

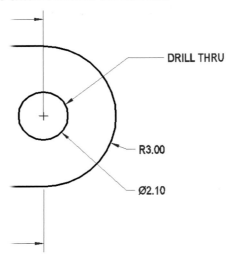

9. Dimension the arc with **DIMRADIUS**.

   Pick point 6, and then place the dimension leader outside the fuse link.

10. Construct a leader by entering the **QLeader** command. (See Chapter 10, "Placing and Editing Text," for more about leaders.)

    Start the leader at the circle (point 8), and end it outside the object.

    For the leader text, enter "Drill Thru."

    Remember to press **ENTER** to exit the **QLEADER** command.

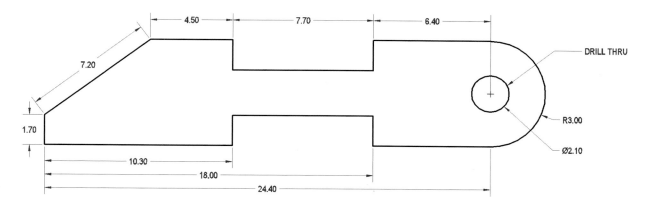

11. Save the completed drawing as *dimen.dwg*.

## TUTORIAL: ANNOTATIVE DIMENSIONING

Your boss has emailed you *Anno-Dim.dwg*, requesting that you dimension it. In his note, he remarks that the drawing will be needed in detail sheets with two different scale factors, one at 1:8 scale and the other at 1:16.

In this tutorial, you use annotative scaling to dimension the drawing.

1. Open the *Anno-Dim.dwg* file.
2. On the status bar, change the **Annotation Scale** to 1:8, the first value your boss specified.

3. You start the **DIMLINEAR** command:

   Command: **dimlinear**

   Specify first extension line origin or <select object>: *(Press ENTER.)*

   Select object to dimension: *(Select a line.)*

   Specify dimension line location or [Mtext/Text/Angle/Horizontal/Vertical/Rotated]: *(Pick a point.)*

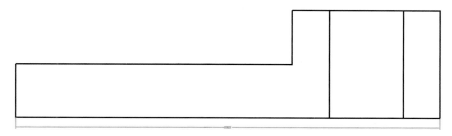

... but find that the dimension elements (shown in gray, above) are too small. Of course, you should be using annotative scaling.

4. You note that the drawing file lacks an annotative dimension style, and so you create one.
   a. In the ribbon's Annotate panel, click **Dimension Style** button
   b. Click **New**, and then name it "Annotative."

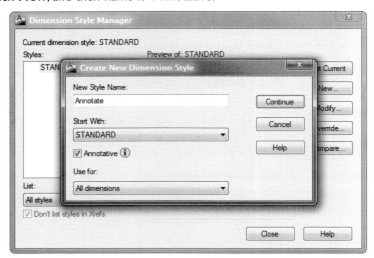

   c. Ensure that the **Annotative** option is turned on, and then click **Continue**.
   d. You don't need to change anything else, so click **OK**, **Set Current**, and **Close**.

The new annotative dimension style is now the current one.

5. Apply the **DIMLINEAR** command again. That's better!

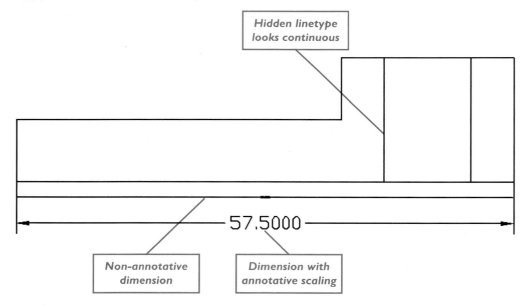

Notice that annotative scaling takes over the duty of the **DIMSCALE** system variable.

6. Erase the first linear dimension, the one that's scaled too small.

## Changing Linetype Scaling

The two vertical lines should be showing hidden lines, but they look continuous. The Properties palette confirms that they have the Hidden linetype, but are scaled incorrectly.

7. If necessary, change the value of **MSLTSCALE** to 1:

    Command: **msltscale**

    Enter new value for MSLTSCALE <0>: **1**

8. Use the **REGEN** command to update the drawing.

    Command: **regen**

    Notice how annotative scaling takes care of scaling linetypes correctly.

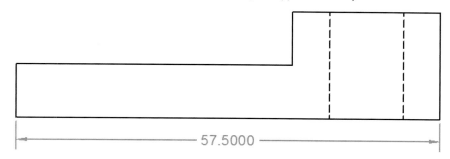

9. Add the other dimensions required by the drawing.

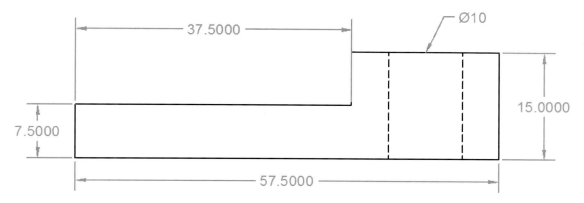

## Changing the Annotation Scale

The other scale your boss wanted on this drawing was 1:16.

10. Change **Annotation Scale** to 1:16.

11. Re-run the **DIMLINEAR** command.

Notice that the dimension text and arrowheads are twice as large — annotative scaling takes care of that.

12. But, oh no! You can see both scaled versions of the linear dimension. The quick way to deal with that is to turn off **ANNOALLVISIBLE**: click the **Annotation Visibility** button on the status bar.

Now you see just the dimensions drawn at the current annotation scale. That's better.

13. Finish dimensioning the drawing, and then save it.

An alternative approach is to apply one set of dimensions at one annotative scale. Once applied, select all, and then use the Properties palette to add additional annotative scale factors. The technical editor suggests that this is the better practice, because it does not duplicate dimensions; it add another display configuration to the existing one.

*Caution:* CAD programs "compatible" with AutoCAD cannot read annotative dimensions. Most strip them off without warning.

## EXERCISES

1. Start a new drawing.
    a. With the **POLYGON** command, draw a square approximately one-third the size of the screen.
    b. Use the **DIMLINEAR** command along with object snap INTersection to capture the lower left and lower right corners of the square.
    c. Place the dimension line below the bottom line of the square.

2. Continuing from the previous exercise, repeat the command for the upper line of the square, with this change:
    a. Instead of picking the corners, dimension the line by selecting it.
    b. Is the dimension text the same for the lower and upper lines?

3. Continuing:
    a. Dimension one of the vertical sides of the box.
    b. Is the vertical dimension the same as the horizontal dimension?

4. Next, dimension the remaining vertical line.
    a. At the 'Specify dimension line location or [Mtext/Text/Angle/Horizontal/Vertical/Rotated]' prompt, enter m.
    b. Does the mtext editor appear (in a dialog box or on the ribbon)?
    c. Type your name.
    d. Choose **OK** when finished.
    e. What appears along the dimension line?
    f. Save the drawing as *square1-4.dwg*.

5. Start a new drawing.
    a. Draw a scalene triangle.
    b. Use the **DIMALIGNED** command to dimension the three angled sides.
    c. Does AutoCAD draw the dimension lines parallel to the sides?

6. Draw two lines that are not parallel, similar to those shown below.
   Place an angular dimension on the two lines.

   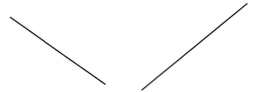

   Repeat the command to draw the supplemental angle.

7. Draw an arc and a circle.
   a. Dimension the minor and major arcs with the **DIMANGULAR** command.
   b. Save the drawing as *dimen5-7.dwg*.

8. Start a new drawing.
   a. Draw a diagonal line, similar to the ones in exercise #6.
   b. Use the following commands to dimension the line three times: horizontally, vertically, and rotated.
      **DIMHORIZONTAL**
      **DIMVERTICAL**
      **DIMROTATED**

9. Undo the three dimensions you drew in the previous exercise.
   a. Use the **DIMLINEAR** command to draw the same three orientations of dimension.
   b. Which commands did you find easier to use?
   c. Save the drawing as *linear8-9.dwg*.

10. Start a new drawing.
    a. Draw an arc and a circle on the screen.
    b. Dimension each with the **DIMDIAMETER** command.
    c. Practice placing diameter dimensions inside and outside the circle, and at different locations around the arc.

11. Repeat exercise #9 with the **DIMRAD** command. Save the drawing as *diarad10-11.dwg*.

12. Open the *qdim.dwg* drawing of a flywheel.
    a. Dimension the two views with the **QDIM** command.
    b. When complete, save the drawing as *flywheel12.dwg*.

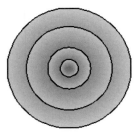
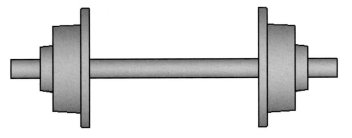

13. Open the *15-47.dwg* of a gasket.
    a. Dimension the drawing with the **DIMRAD** and **DIMDIA** commands.
    b. When complete, save the drawing by the same name.

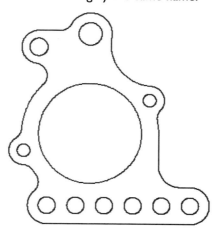

14. Open the following drawings, and then dimension the gaskets.
    a. Save as *5-43.dwg*.

b. Save as *15-44.dwg*.

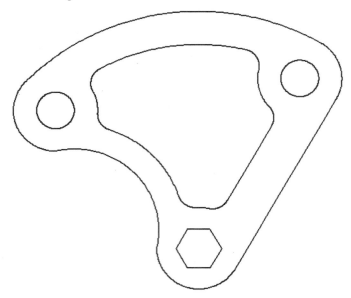

c. Save as *15-45.dwg*.

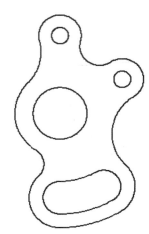

d. Save as *15-46.dwg*.

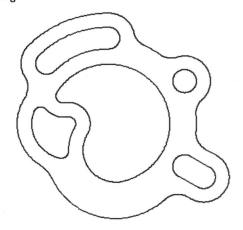

15. Open the following drawings. Dimension each drawing, and then save your work.
    a. *17_19.dwg:* Attachment for a three-point hitch.

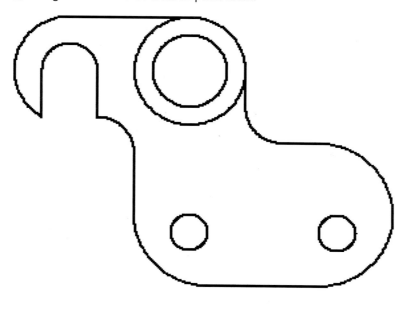

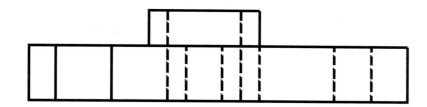

    b. *17_20.dwg:* Flag pole stabilizer.

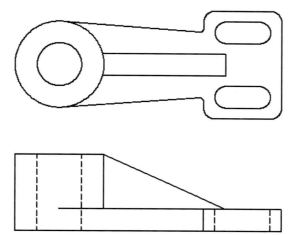

c. *17_21.dwg:* Reciprocating linkage.

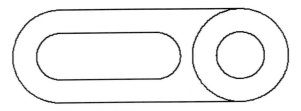
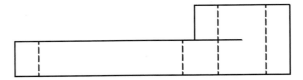

d. *7_22.dwg:* Socket linkage.

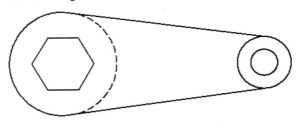
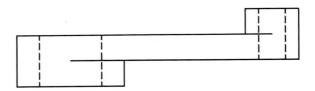

e. *17_23.dwg:* Attachment plate.

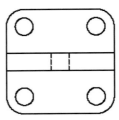

f. *17_24.dwg*: Hitch yoke.

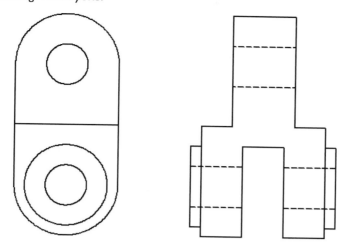

g. *17_25.dwg*: Hitch linkage.

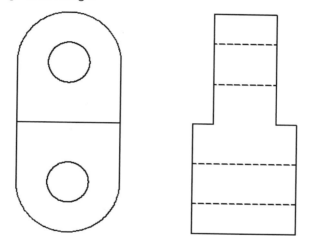

h. *17_26.dwg*: Axle support.

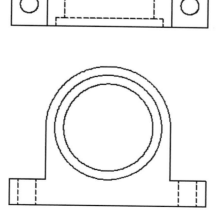

i. *17_27.dwg*: Cone of silence. (Technical editor comments: "Yes! Maxwell Smart lives!" Copy editor comments: "Sorry about that, Chief.")

j. *17_28.dwg*: Bushing.

k. *17_29.dwg*: Wedge.

l. *17_30.dwg:* Concrete footing.

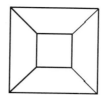

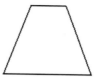

m. *17_31.dwg:* Heavy-duty axle support.

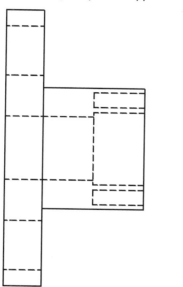 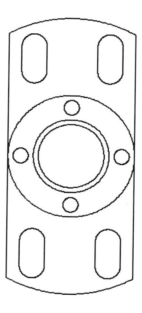

n. *14_32.dwg:* Paintbrush handle.

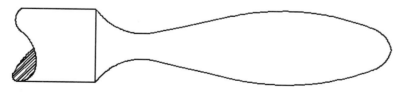

o. *17_33.dwg:* Spacer.

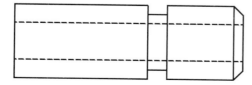

p. *17_34.dwg:* Base plate.

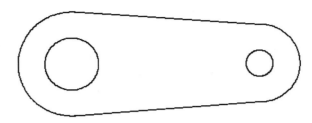

q. *17_35.dwg:* Angled base plate. Use the QDIM command.

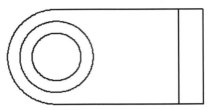
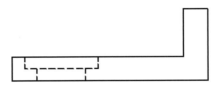

## CHAPTER REVIEW

1. A dimension measures _____.
2. Should extension lines touch the objects they reference?
3. Where should dimensions be placed, whenever possible?
4. Leaders indicate _____.
5. How many possible angles can **DIMANGULAR** draw on:
    Circles?
    Pair of lines?
6. Should you attempt to dimension every detail?
7. Under what conditions are dimension lines not drawn?
8. What is another name for a "witness line"?
9. How are the endpoints of extension lines determined?
10. Must a dimension always have two extension lines?
11. Sketch and label the parts of a measurement dimension. Include the following parts:
    Dimension line
    Arrowheads
    Extension lines
    Text

12. Sketch and label the parts of a radial dimension. Include the following parts:
    Leader line
    Arrowhead
    Text
    Radius Symbol

13. Sketch and label the parts of an extension line:
    Extension line
    Offset
    Extension beyond dimension line

14. List the types of linear dimensions that the DIMLINEAR command draws.
    a.
    b.
    c.
15. Explain the function of the DIMDIAMETER command?
16. What is the purpose of the QDIM command?
17. Sketch an example of a baseline dimension.

18. Must arrowheads always be arrows?
19. What is the aspect ratio of the standard arrowhead?
20. List three places where dimension text can be placed.
21. Must dimension text always be horizontal?
22. What angle does the DIMANGULAR command measure for the following objects:
    Circle
    Two Lines
    Arc
23. Explain why *direct dimensioning* is beneficial.
    Can any object be dimensioned directly?
24. Which dimensioning commands do you use most often with arcs and circles?
25. What is the purpose of the DIMANGULAR command?
26. Describe the meaning of "fully associative dimensioning."
27. Are all dimensions associative?
    Give an example.
28. How is the look of dimensions controlled?
29. Where in AutoCAD might you find dimension standards?
30. Name three object snaps useful for dimensioning:
    a.
    b.
    c.
31. Explain the purpose of the center mark.

    Describe the difference between a center mark and a center line.

    How do you instruct AutoCAD to draw one or the other?

32. Explain the meaning of the following options available in many dimensioning commands:

    Angle

    Mtext

    Text

33. What is the **<>** symbol meant for?
34. A dimension reads 2.5. Write out the text string that changes the dimension to read:
    a. "Adjust to 2.5m"

    b. "Turn by 2.5° increments."

    c. "Temperature range is 2.5°±0.5°"

35. Which part of an arc does the **DIMALIGNED** command measure?
    Of a circle?
    Of a line?
36. How do these commands differ?
    **DIMBASELINE**
    **DIMCONTINUE**
37. Can you apply the **DIMCONTINUE** command to a diameter dimension?
38. Write out the command for each alias:
    dra
    dco
    dli
    dimali
39. Can you dimension ellipses and splines with the **DIMRADIUS** command?
40. The **DIMJOGGED** command is best suited for which kinds of arcs and circles?
41. When would you use the **DIMARC** command?

# CHAPTER 12
## Editing Dimensions

You often need to change existing dimensions. This chapter describes how to edit the position and properties of dimensions. After completing this chapter, you will have an understanding of these commands:

**AIDIMFLIPARROW** flips the direction of arrowheads.
**DIMJOGLINE** adds jogs to dimension lines.
**DIMBREAK** adds breaks to dimension and extension lines.
**DIMSPACE** spaces groups of dimensions evenly.
**DIMEDIT** slants extension lines.
**DIMTEDIT** relocates dimension text.
**DDEDIT** changes dimension text with the mtext editor.
**DIMINSPECT** adds and removes inspection text.
**AIDIM** and **AI_DIM** prefix a group of undocumented dimension editing commands.
**DIMSTYLE** creates and modifies dimension styles.
**DIMOVERRIDE** applies changed dimvars to selected dimensions.
**TEXTTOFRONT** changes the display order of dimensions.
**Grips** edit dimensions and leaders directly.

 **AIDIMFLIPARROW**

The AIDIMFLIPARROW command flips the direction of arrowheads.

This command flips the direction of selected arrowheads by 180 degrees, which is useful for touching up dimensions when arrowheads would look better placed outside the extension lines, or vice versa. It is not documented by Autodesk, but appears in the shortcut menu when a dimension is selected.

### TUTORIAL: FLIPPING ARROWHEADS

1. Draw a linear dimension with the **DIMLINEAR** command.

2. To change the direction of an arrowhead, start the **AIDIMFLIPARROW** command:
   - Select the dimension, right-click, and then choose **Flip Arrow** from the shortcut menu.
   - At the 'Command:' prompt, enter the **aidimfliparrow** command.

   Command: **aidimfliparrow** *(Press* ENTER.*)*

3. AutoCAD prompts you to select objects, but you really can choose only one arrowhead at a time.

   Select objects: *(Select one or more arrowheads.)*

   AutoCAD flips the arrowhead, and draws a short dimension line.

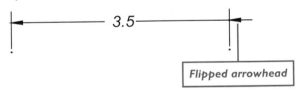

   Select objects: *(Press* ENTER *to end object selection.)*

When you select a dimension instead of an arrowhead, AutoCAD flips the arrowhead nearest to your pick point.

 **DIMJOGLINE**

The **DIMJOGLINE** command adds jogs to linear dimension lines.

Very wide drawings, such as road cross-sections and machinery shafts, use jogged dimension lines to show overall distance. This command adds and removes the jogs; you use the **PROPERTIES** command to specify the size of the jog, and can relocate the jog using grips editing.

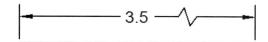

The **DIMJOGLINE** command works only with linear dimensions. You cannot give a dimension line more than one jog; to add jogs to radial extension lines, use the **DIMJOGGED** command, as described in the previous chapter.

### TUTORIAL: JOGGING DIMENSION LINES

1. To add a jog to dimension lines, start the **DIMJOGLINE** command:
    * In the ribbon's Annotate tab, choose the **Jog Line** button in the Dimensions panel.
    * At the 'Command:' prompt, enter the **dimjogline** command.

        Command: **dimjogline** *(Press ENTER.)*
    * Alternatively, enter the **djl** alias at the 'Command:' prompt.
2. In all cases, AutoCAD prompts you to select a dimension:
    Select dimension to add jog or [Remove]: *(Pick one dimension.)*
3. Specify the location of the jog:
    Specify jog location (or press ENTER): *(Pick a point.)*

When you press ENTER at the last prompt, AutoCAD places the jog halfway between the first extension line and the dimension text.

To change the size of the jog, use the Properties palette: in the Lines & Arrows section, change the value of **Jog Height Factor**.

To move the jog, use grips editing (as illustrated below), or else reuse this command and pick another location for the jog. Since dimension lines can have only one jog, specifying a "second" one relocates the jog.

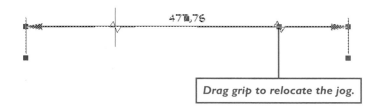

The **Remove** option removes the jog from the dimension line.

## DIMBREAK

The DIMBREAK command breaks extension and dimension lines where they cross other objects.

Breaking lines can make dimensions easier to read. AutoCAD remembers the locations of the breaks, so that when you move a dimension, the breaks move appropriately.

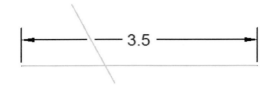

### TUTORIAL: BREAKING DIMENSION AND EXTENSION LINES

1. To break lines in dimensions, start the DIMBREAK command:
   - In the ribbon's Annotate tab, choose the **Break** button in the Dimensions panel.
   - At the 'Command:' prompt, enter the **dimbreak** command.

   Command: **dimbreak** *(Press ENTER.)*

2. In all cases, AutoCAD prompts you to select one or more dimensions:
   Select a dimension or [Multiple]: *(Select a dimension.)*

3. Choose an option; I recommend the **Auto** option, which lets AutoCAD handle the breaking job:
   Select object to break dimension or [Auto/Restore/Manual] <Auto>: **a**

   Notice that dimension and/or extension lines are broken.

To place breaks manually, enter "M" for the **Manual** option, which operates like the BREAK command:

Specify first break point: *(Pick a point.)*

Specify second break point: *(Pick another point.)*

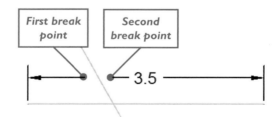

This option allows you to place the break anywhere, and not just where lines overlap. This is one of the few times in AutoCAD where something works better when object snap is turned off.

To break all dimensions in the drawing at once, use the **Multiple** option, as follows:

Select a dimension or [Multiple]: **m**

Select dimensions: **all**

Select dimensions: *(Press ENTER to exit object selection.)*

Enter an option [Break/Restore] <Break>: *(Press ENTER to accept the default.)*

When you stretch or move dimensions, the breaks follow. Should the dimensions no longer cross, the breaks heal, but reappear when you move the dimensions back. To remove breaks, use the **Restore** option. You can restore dimensions one by one, or all of them at once.

 **DIMSPACE**

The DIMSPACE command evenly spaces groups of dimensions.

The purpose of this command is to give a neater look to drawings with many dimensions.

**TUTORIAL: EVENLY SPACING DIMENSIONS**

1. To learn how to space dimensions evenly, draw several linear dimensions, as illustrated below.

2. Start the **DIMSPACE** command:
    - In the ribbon's Annotate tab, choose the **Adjust Space** button in the Dimensions panel.
    - At the 'Command:' prompt, enter the **dimspace** command.

        Command: **dimspace** *(Press ENTER.)*

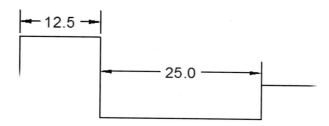

3. AutoCAD prompts you to select the base dimension. This is the dimension from which the others will be spaced evenly:
    Select base dimension: *(Pick one dimension.)*

4. Select the other dimensions to be spaced:
    Select dimensions to space: *(Pick one or more dimensions.)*
    Select dimensions to space: *(Press ENTER to exit object selection.)*

5. Enter a number for the size of the spacing between dimensions; or, do it the easy way, and press **ENTER** to let AutoCAD work out the spacing:
    Enter value or [Auto] <Auto>: *(Press ENTER.)*

    Notice that the dimensions have the same spacing vertically.

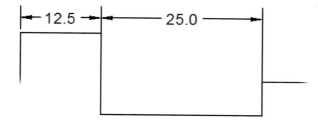

The **Auto** option determines the spacing distance by multiplying the dimension text height; Autodesk says the factor is 2x, but in the figure above, the factor works out to 5x. This system sometimes does not work well when DIMSCALE is set to extreme values.

You can enter a value of 0, which reduces the spacing to zero, making the dimensions look like overlapping continuous dimensions.

##  DIMEDIT, DIMTEDIT AND DDEDIT

The DIMEDIT and DIMTEDIT commands edit extension lines and dimension text.

The DIMEDIT command obliques extension lines so that they are no longer at right angles to the dimension line. *Obliqued* (slanted) extension lines are required by isometric drawings, and are also used in difficult dimensioning situations, such as dimensioning along shallow curves.

(Obliquing isn't applied to leaders, such as diameter, radius, and ordinates, because their angles can be changed with grips editing.)

The DIMTEDIT command changes the location of dimension text relative to the dimension line by rotating, relocating, and justifying (left, center, or right) it.

After these changes, either command can be used to "home" the text by sending it back to its default location: centered on the dimension line. One command's name is short for "dimension edit," while the other is short for "dimension text edit," yet both edit dimension text. The toolbar icons are unhelpful, DIMTEDIT showing a pencil, implying the dimension itself can be edited when it cannot.

The two commands share text editing tasks, making the differences between them confusing. The table below compares and contrasts the abilities of each command:

| *Editing Operation* | **DimEdit** | **DimTEdit** |
|---|---|---|
| **Extension Lines:** | | |
| Applies obliquing angle | ☑ | ... |
| | | |
| **Dimension Text:** | | |
| Rotates text | ☑ | ☑ |
| Moves text to default location (home) | ☑ | ☑ |
| Edits text | ☑ | ... |
| Drags text to new location | ... | ☑ |
| Centers text on dimension line | ... | ☑ |
| Right- and left-justifies text on dimline | ... | ☑ |

It would make more sense if DIMTEDIT handled only the editing of dimension text, while DIMEDIT handled the non-text parts of dimensions, such as extension lines, as well as functions ignored by AutoCAD, such as relocating and removing arrowheads. But they don't, and so we carry on.

Here's how to keep them straight:

**DimEdit** *obliques* extension lines. (Just ignore its other capabilities.)
**DimTEdit** changes the *position* of dimension text.
**DdEdit** *edits* dimension text with the mtext editor.

## TUTORIAL: OBLIQUING EXTENSION LINES

1. To change the angle of extension lines, start the **DIMEDIT** command:
   - In the ribbon's Annotate tab, choose the **Oblique** button in the Dimensions panel.
   - At the 'Command:' prompt, enter the **dimedit** command.
   
     Command: **dimedit** (Press ENTER.)
   - Alternatively, enter the aliases **ded** or **dimed** at the 'Command:' prompt.

2. In all cases, AutoCAD lists the options available:
   
   Enter type of dimension editing [Home/New/Rotate/Oblique] <Home>: **o**
   
   Enter **o** for the oblique option:

3. Select the dimensions whose extensions lines to slant. It is easiest to enter **all**, because AutoCAD will filter out non-dimension objects for you:
   
   Select objects: **all**
   
   Select objects: *(Press ENTER to end object selection.)*

4. Specify an angle of slant between 0 and 360 degrees:
   - To slant extension lines for isometric drawings, enter **30**.
     
     Enter obliquing angle (press ENTER for none): **30**
   - To slant in the other direction, enter negative angles, such as **-30**.
   - To keep the extension lines at their current angle (no change), press **ENTER**.
   - To return obliqued extension lines to perpendicular, enter **0** or **90**.
   - Or, pick two points to show AutoCAD the angle.
   
   AutoCAD obliques the extension lines of selected dimensions, as illustrated below:

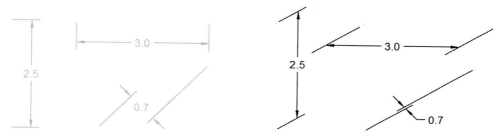

**Left:** Dimensions with normal, perpendicular extension lines.
**Right:** Same dimensions with extension lines obliqued by 30 degrees.

**Note** Be careful with obliquing angles that are very close to the angle of the dimension (1.0 degrees for horizontal dimensions, 89.0 degrees for vertical dimensions, and so on). They stretch the dimension into an unrecognizable jagged line. You can always use the **U** command to reverse undesirable obliquing.

## TUTORIAL: REPOSITIONING DIMENSION TEXT

1. To change the position of dimension text, start the **DIMTEDIT** command:
   - In the ribbon's Annotate tab, choose one of the **Justify** buttons in the Dimensions panel.
   - At the 'Command:' prompt, enter the **dimtedit** command.
     
     Command: **dimtedit** (Press ENTER.)
   - Alternatively, enter the **dimted** alias at the 'Command:' prompt.

2. AutoCAD prompts you to select one dimension:

    Select dimension: *(Select one dimension.)*

3. Enter an option to relocate the text:

    Specify new location for dimension text or [Left/Right/Center/Home/Angle]: *(Move cursor, or enter an option.)*

    - **Specify new location** relocates the text along the dimension line dynamically, and at the same time moves the dimension line perpendicular to the extension lines.

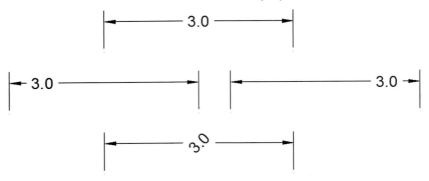

**Top:** *Text at home, centered position.*
**Middle:** *Text in left and right positions.*
**Above:** *Dimension text at a 45-degree angle.*

**Left** — moves the text to the left end of the dimension line.

**Right** — moves the text to the right end of the dimension line.

**Center** — centers the text on the dimension line.

**Home** — returns the text to the position defined by the dimension style.

**Angle** — rotates text about its center point; AutoCAD prompts you:

    Specify angle for dimension text: *(Enter an angle, such as **30**.)*

The Home position is defined by the dimension text, and normally it is centered on the dimension line. The JIS standard, however, centers text just above the dimension line.

Angled text is used with isometric dimensions.

### TUTORIAL: EDITING DIMENSION TEXT

1. To edit the value of the dimension text, start the **DDEDIT** command:
    - At the 'Command:' prompt, enter the **ddedit** command.

        Command: **ddedit** *(Press* ENTER.*)*

    - Alternatively, enter the **ed** alias at the 'Command:' prompt.

2. In all cases, AutoCAD prompts you to select the text to edit. Select a single dimension:

    Select an annotation object or [Undo]: *(Select a dimension or leader text.)*

    Notice that AutoCAD opens the Text Editor tab with the dimension text highlighted in purple.

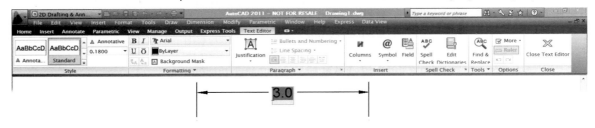

3. Recall from the previous chapter that you can edit dimension text, as follows:
   - Erase the text from the dimension by backspacing.
   - Change the value of the dimension text.
   - Add text before and after to create prefixes and suffixes.
   - Modify the look of the text (font, color, and so on) with the Text Editor tab.

4. When done editing the text, click **Close Text Editor** in the Text Editor tab.

5. AutoCAD repeats the prompt, and you can immediately select another dimension — or any other text object.

6. When done editing, press **ENTER**.

   Select an annotation object or [Undo]: *(Press* ENTER *to exit command.)*

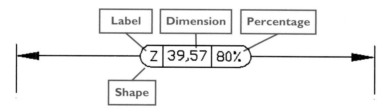

To apply a background mask to the dimension text, use the **Fill** option of the Properties palette.

## DIMINSPECT

The **DIMINSPECT** command adds and edits inspection-style text to dimensions.

Inspection dimensions are used in quality assurance (QA) testing to ensure that the manufactured part has precisely the correct dimensions. For example, if the percentage is shown as 80%, then 80% of the manufactured parts must meet this dimension to pass QA.

This command modifies existing dimension text, turning it into three parts, as illustrated below:

**Label** — identifies the dimension in tables and notes. You can enter any text for the label.

**Percentage** — specifies the inspection rate. You can enter any value, such as 80%, or a fraction, such as 23/74; in fact, it need not be a percentage at all: "Quite a few."

**Shape** — surrounds the dimension to make it stand out. AutoCAD provides oval and diamond shapes, or none at all, as illustrated above.

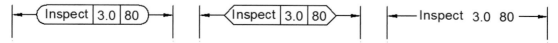

You select the type and value of inspection through a dialog box; the -DIMINSPECT command performs the same function at the command line.

## TUTORIAL: ADDING INSPECTION TEXT TO DIMENSIONS

1. To add inspection text to dimensions, draw a linear dimension similar to the one illustrated below.

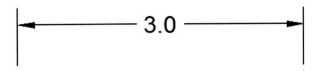

2. To add inspection text, start the **DIMINSPECT** command:
   - In the ribbons's Annotate tab, choose the **Inspect** button from the Dimensions panel.
   - At the 'Command:' prompt, enter the **diminspect** command.

      Command: **diminspect** *(Press* ENTER.*)*

3. In all cases, AutoCAD displays the Inspection Dimension dialog box:

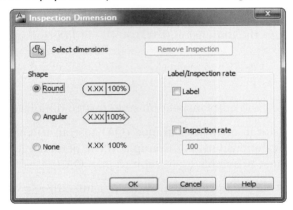

4. Click the **Select Dimensions** button.
5. Notice that AutoCAD prompts you at the command line to select dimensions:
   Select dimensions: *(Select one or more dimensions.)*
   Select dimensions: *(Press* ENTER *to return to the dialog box.)*

6. Back in the dialog box, you now specify the look of the inspection codes. For this tutorial, select the following options:

   | | |
   |---|---|
   | Shape | **Round** |
   | Label | **Inspect** |
   | Inspection Rate | **80%** |

7. Click **OK**. Notice that the dimension text change to look like this:

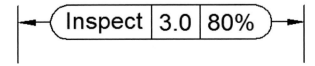

To change the values and/or shape of inspection dimensions, rerun the DIMINSPECT command, and then select the dimension. Alternatively, you can use the Misc section of the Properties palette to edit inspection dimensions.

**Note** You can use this command to create dimensions interactively with prefixes and suffixes, such as the one illustrated below. Set **Inspection Label** to "Length = " and **Inspection Rate** to " Inches." Alternatively, you can use the Properties palette to add and edit prefixes and suffixes.

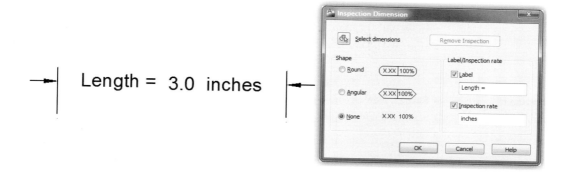

## AIDIM AND AI_DIM

AutoCAD has a group of undocumented commands that edit dimension text. These commands are meant for use in menu and toolbar macros, but are also handy for making quick fixes to mucked-up dimensions. This group of special commands all start with AI, and all are shortcuts to options within "official" commands and dimension variables:

| Special Command | Shortcut For | Comment |
| --- | --- | --- |
| AIDIMPREC | DIMDEC | Changes precision of fractions and decimals selectively. |
| AIDIMSTYLE | DIMSTYLE | Saves and applies up to six dimension styles. |
| AIDIMTEXTMOVE | DIMTMOVE | Relocates text with optional leader. |
| AI_DIM_TEXTABOVE | DIMTAD 3 | Moves text above dimension line for JIS compliance. |
| AI_DIM_TEXTCENTER | DIMTEDIT C | Centers dimension text. |
| AI_DIM_TEXTHOME | DIMTEDIT H | Returns text to home position. |

(The "AI" prefix indicates custom commands written by Autodesk Incorporated. Why do some commands start with **AI** and some with **AI_**? It seems that those starting with **AI** have multiple options, while those with **AI_** perform a single operation.)

You find all these commands in shortcut menus: select any dimension, and then right-click.

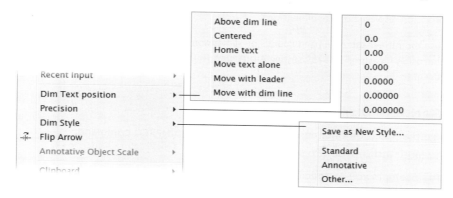

Here is how the commands relate to the shortcut menu:

| Shortcut Menu | Submenu | Related Command | Command Option |
|---|---|---|---|
| Dim Text Position | ▸ Above Dim Line | AI_DIM_TEXTABOVE | ... |
| | ▸ Centered | AI_DIM_TEXTCENTER | ... |
| | ▸ Home Text | AI_DIM_TEXTHOME | ... |
| | ▸ Move Text Alone | AIDIMTEXTMOVE | 0 |
| | ▸ Move With Leader | AIDIMTEXTMOVE | 1 |
| | ▸ Move With Dim Line | AIDIMTEXTMOVE | 2 |
| Precision | ... | AIDIMPREC | 0 – 6 |
| Dim Style | ▸ Save as New Style | AIDIMSTYLE | Save |
| | ▸ Standard, *etc.* | AIDIMSTYLE | 1 – 6 |
| | ▸ Other... | AIDIMSTYLE | Other |
| Flip Arrow | ... | AIDIMFLIPARROW | ... |

## AIDIMPREC

The **AIDIMPREC** command changes the precision of fractional and decimal dimension text and angles of selected dimensions. (It does not work leaders.) To affect a permanent change, change the dimstyle.

This command is far more convenient than its official alternative, which involves changing the value of the DIMDEC system variable and then using the -DIMSTYLE command's **Apply** option. AIDIMPREC rounds decimal text and angles to the nearest decimal place and fractional text to the nearest fraction, as shown by the table below:

| AiDimPrec Value | Nearest Decimal Units | Nearest Fractional Units |
|---|---|---|
| 0 | 0 | 1" |
| 1 | 0.0 | 1/2" |
| 2 | 0.00 | 1/4" |
| 3 | 0.000 | 1/8" |
| 4 | 0.0000 | 1/16" |
| 5 | 0.00000 | 1/32" |
| 6 | 0.000000 | 1/64" |

**Notes** The change in precision is not permanent; AutoCAD remembers the actual measurements and angles — which is why it is called "display *precision*." You can reapply the command to increase and decrease the precision displayed by the dimensions.

Because the **AIDIMPREC** command rounds off dimension text, it can create false measurements. For instance, if a dimension measures 3.4375", setting **AIDIMPREC** to 0 rounds down to 3". Similarly, a measurement of 2.5" rounds up to 3".

### TUTORIAL: CHANGING DISPLAY PRECISION

1. To change the precision of dimension text, start the **AIDIMPREC** command:
   Command: **aidimprec** *(Press* ENTER.*)*

2. AutoCAD prompts you to specify a precision:
   Enter option [0/1/2/3/4/5/6] <4>: *(Enter the number of decimal places.)*

3. Select one or more dimension objects. Entering **all** selects everything in the drawing; AutoCAD helpfully filters out non-dimension objects.
   Select objects: *(Enter* **all** *or pick one or more dimensions.)*

Select objects: *(Press ENTER to end object selection.)*

Notice that AutoCAD changes the display precision of the selected dimensions.

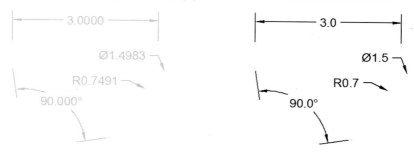

**Left:** *Dimension text displayed to four decimal places.*
**Right:** *Precision rounded to a single decimal place with the **AIDIMPREC** command.*

 **AIDIMTEXTMOVE**

The AIDIMTEXTMOVE command relocates text, either with or without moving the dimension line, and optionally adds a leader.

This command is more convenient than its official alternative: changing the DIMTMOVE system variable, followed by the DIMOVERRIDE command. AIDIMTEXTMOVE has the same options as DIMTMOVE:

| AiDimTextMove | DimTMove | Comment |
|---|---|---|
| 0 | 0 | Moves text within dimension line; moves dimension line. |
| 1 | 1 | Adds a leader to the moved text. |
| 2 (Default) | 2 | Moves text without leader line; moves the dimension line. |

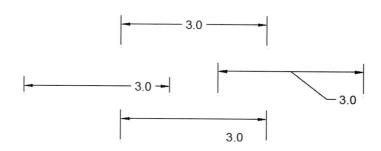

**Top:** *Text at home position.*
**Middle:** *Text moved with dimension line (at left) and with leader (at right).*
**Above:** *Text moved independently (without dimension line or leader).*

### TUTORIAL: MOVING DIMENSION TEXT

1. To move the dimension text, start the **AIDIMTEXTMOVE** command:
   Command: **aidimtextmove** *(Press ENTER.)*

2. AutoCAD prompts you to specify an option:
   Enter option [0/1/2] <2>: *(Enter an option.)*

3. Select one dimension. Although the command appears to allow you to select more than one dimension, it operates on the first-selected one only.
   Select objects: *(Pick one dimension.)*

Select objects: *(Press ENTER to end object selection.)*

4. Move the cursor to relocate the dimension text.

### Quick Dimension Text Moves

The following trio of commands quickly relocates dimension text relative to the dimension line: above, centered, or in the dimension line. These commands operate on just one dimension at a time, unfortunately.

| Command | DimTAD | Comment |
|---|---|---|
| AI_DIM_TEXTCENTER | 0 | Centers text vertically. |
| AI_DIM_TEXTHOME | 2 | Centers text horizontally and vertically. |
| AI_DIM_TEXTABOVE | 3 | Moves text above dimension line (JIS standard). |

(When DIMTAD is 1, AutoCAD moves text above the dimension line — when the dimension line is horizontal and text inside the extension lines isn't being forced to be horizontal.)

### Ai_Dim_TextCenter

The AI_DIM_TEXTCENTER command centers text along the dimension line. If the text is above or below the dimension line, AutoCAD keeps the vertical position, but centers the text horizontally. Select the dimension, right-click, and then choose **Dim Text Position | Centered** from the shortcut menu.

Command: **ai_dim_textcenter**
Select objects: *(Pick one dimension.)*
Select objects: *(Press ENTER to end object selection.)*

### Ai_Dim_TextHome

The AI_DIM_TEXTHOME command centers the dimension text in the home position.

Select a dimension, right-click, and then choose **Dim Text Position | Home Text** from the shortcut menu.

Command: **ai_dim_texthome**
Select objects: *(Pick one dimension.)*
Select objects: *(Press ENTER to end object selection.)*

### Ai_Dim_TextAbove

The AI_DIM_TEXTABOVE command moves text above the dimension line, which is helpful for making drawings JIS compliant.

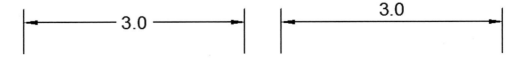

Select a dimension, right-click, and then choose **Dim Text Position | Above Dim Line** from the shortcut menu.

Command: **ai_dim_textabove**
Select objects: *(Pick one dimension.)*
Select objects: *(Press ENTER to end object selection.)*

## DIMENSION VARIABLES AND STYLES

How dimensions look and act depend on the settings of *dimension variables* (called "dimvars" for short). The variables control many properties of dimensions, such as the placement of dimension text within or above the dimension line, the use of arrowheads or tick marks, the color of the extension lines, and the overall scale. (Technically, dimvars are no different from system variables.)

The names of all dimvars begin with "dim," which makes them easy to identify, such as DIMSCALE for dimension scale and DIMTXT for the text style. Other names, however, can seem strange and are hard to decode. Who could guess that DIMTOFL means "draw the dimension line between the extension lines, even when the text is forced outside due to lack of space." I suppose "tofl" could be short for "text outside forced line." Or how about DIMTAD — "Text Adjusted relative to the Dimension line," perhaps? (The technical editor suggests DIMWIT as the name for a new dimension variable, short for "Weird Indecipherable Terminology.")

Like system variables, dimvars can be toggled (turned on and off), hold numeric values, or report on the status of dimensions. The advantage to dimvars is that they control almost every aspect of dimensions; the disadvantage is that there are so many of them (around 80) that it becomes difficult to remember them all; two-thirds alone are dedicated to formatting dimension text.

That's why AutoCAD has dimensions styles. Like text styles, these remember dimvar settings and easily control the look of dimensions. When you create or change a dimensions style with the DIMSTYLE command, AutoCAD changes the values of dimvars.

Creating dimension styles ("dimstyles" for short) is similar to creating text styles: you specify options, and then save them by name. All new AutoCAD drawings based on the *acad.dwt* template drawing have two dimension styles, named "Standard" and "Annotative."

### CONTROLLING DIMVARS

AutoCAD's dimension variables can be set through several methods, such as by using the Dimension Style Manager dialog box or the Properties palette, or by entering dimvar names at the 'Command:' prompt.

For example, to set an overall scale factor for dimensions, it can be faster to enter the DIMSCALE at the command prompt than to hunt it down in the multi-tabbed Dimension Style Manager dialog box:

    Command: **dimscale**
    New value for DIMSCALE <1.0>: *(Enter a scale factor, such as* **2.54**.*)*

The scale is not retroactive, however, and applies only to dimensions drawn subsequently. To change the scale of existing dimensions, you need to use the DIMSTYLE dialog box's **Override** option.

There are two dimvars not set by the Dimension Style Manager dialog box: DIMSHO (toggles whether dimensions are updated while dragged) and DIMASO (determines whether dimensions are associative or non-associative). Both have a value of 1, which means they are turned on. You can turn them off, but there isn't any good reason to do so.

(*History*: Both of these dimvars exist only for compatibility with old versions of AutoCAD. When computers were slow, generating highlighted images of dragged dimensions slowed the computer even more; this is no longer an issue.)

### SOURCES OF DIMSTYLES

When you create new drawings based on template drawings, they contain specific dimstyles. AutoCAD includes a *.dwt* template file that meets the international metric standard: *acad_iso.dwg* contains a dimstyle named "ISO," which conforms to the dimensioning standards of the International Organization for Standardization.

## Dimstyles from Other Sources

You can copy dimension styles from other drawings using a number of methods.

### DesignCenter

To copy dimstyles using DesignCenter, follow these steps:

1. Open the DesignCenter by pressing CTRL+2.
2. In DesignCenter's tree view, navigate to the drawing from which you wish to copy the dimstyle.
3. Open the **Dimstyles** item, and then choose a dimstyle.
4. Drag the dimstyle into the drawing. (Alternatively, you can right-click the dimstyle name, and then select **Add Dimstyle(s)** from the cursor menu.)

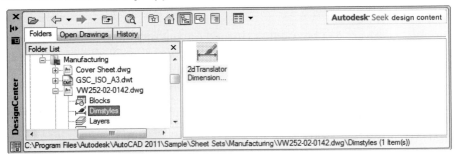

To check that the dimstyle was added to the drawing, follow these steps:

1. In the ribbon, choose the Annotate tab.
2. In the Dimensions panel, click the **Dimension Style** droplist. The name of the added dimstyle should appear on the list.

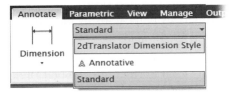

### Express Tools

Express Tools include the DIMEX and DIMIM commands for exporting and importing files that contain dimension styles. Follow these steps:

1. From the **Express Tools** tab, click the **DimStyle Export Style** or **Import Style** button on the Dimension panel.
2. Notice that AutoCAD displays the related dialog box; both are shown below.

**Left:** *Dialog box for importing dimension styles.*
**Right:** *Dialog box for exporting dimension styles.*

The two commands work with .dim files, which hold information about text styles in a DXF-like format.

### Binding External References

When working with externally-referenced drawings, you can use the XBIND command to *bind* (add in) dimension styles found in attached drawings. You cannot, however, set an externally-referenced dimstyle as current.

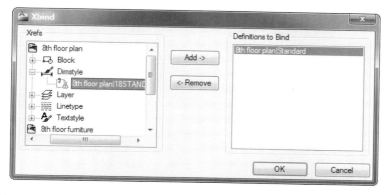

See appendix D, "Summary of Dimension Variables," for the complete list of dimension variables.

## DIMSTYLE

The DIMSTYLE command displays a dialog box for creating and modifying named dimension styles.

The tabbed dialog boxes of the Dimension Style Manager control almost all dimension variables, which in turn affect the look of dimensions in your drawings. A preview window provides a graphical overview of the current dimstyle.

New drawings contains at least two dimension styles, either "Standard" in Imperial drawings or "ISO-25" in metric drawings, and "Annotative" in all drawings. You can create as many dimstyles as you require, each with its own name — one for every drawing, if need be.

The dialog box is called a "manager," because it manages dimension styles in the drawing. With it, you create new styles, override and modify existing styles, compare differences between two styles, and delete styles.

### WORKING WITH DIMENSION STYLES

To create and change dimension styles, start the **DIMSTYLE** command:
- In the ribbon's Annotate tab, choose the **Dimension Style** button in the Dimensions panel.
- At the 'Command:' prompt, enter the **dimstyle** command.

    Command: **dimstyle** (*Press* ENTER.)

- Alternatively, enter the aliases **d**, **dst**, **dimsty**, or **ddim** (the command's old name) at the 'Command:' prompt.

In all cases, AutoCAD displays the Dimension Style Manager dialog box.

On the right, five buttons perform these important functions, discussed in detail below:

**Set Current** — sets the selected dimstyle as current. (After you exit the dialog box, newly created dimensions follow this style; existing ones are unchanged.)

**New** — creates new dimstyles. Usually, you rename an existing style, and then modify its parameters.

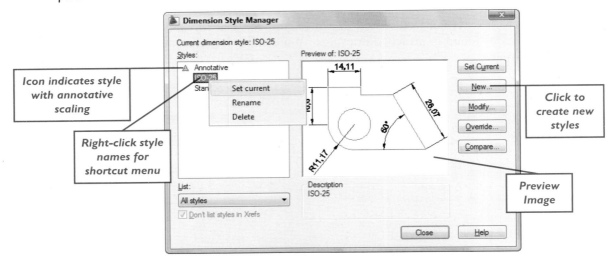

**Modify** — modifies parameters. After you exit the dialog box, all existing dimensions created with this style are retroactively and globally updated. (You tend to use this button more often than **Override**.)

**Override** — overrides the settings of the current dimstyle, affecting only dimensions created after the override is applied. (Externally-referenced dimstyles cannot be modified or overridden.)

**Compare** — contrasts the differences between two dimension styles.

 **Notes** The standard method of creating a new style is to copy an existing one, make changes, and then save the result. For this reason, AutoCAD displays "Copy of Standard" (or whatever the current style name is) in the **New Style Name** text field.

**Start With** text field lets you select which dimstyle to copy — provided the drawing contains two or more dimstyles. This allows you to start with an existing dimstyle (such as the ISO style stored in the template drawings) and modify it according to your needs.

**Annotative** determines whether the style uses annotative scaling.

**Use for** droplist lets you apply changes to all, or for a limited group of, dimensions:

    All dimensions
    Linear dimensions
    Angular dimensions
    Ordinate dimensions
    Radius dimensions
    Diameter dimensions
    Leaders and tolerance

## TUTORIAL: NEW — CREATING NEW DIMSTYLES

Follow these steps to create a new dimension style:

1. Enter the **DDim** command, and then click the **New** button. Notice the Create New Dimension Style dialog box.

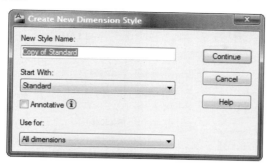

2. AutoCAD copies the current style, naming it "Copy of [dimstyle]." Enter a new name of up to 255 characters long.
3. Click **Continue**. Notice that AutoCAD displays the New Dimension Style dialog box, which is identical to the Modify version of the dialog box detailed below.

## MODIFY — MODIFYING DIMSTYLES

The **Modify** button leads to the Modify Dimension Style dialog box, which contains most of the options that affect dimensions. Here you modify the values to create and change dimstyles. Because there are so-o-o many options affecting dimensions, the dialog box segregates them into a series of tabs.

The following sections provide an overview of the contents of each tab:

**Lines** — specifies the properties of dimension and extension lines.

**Symbols and Arrows** — specifies the properties of arrowheads, sizes of center marks and arc length symbols, and angle of jogs.

**Text** — specifies the format, placement, and alignment of dimension text.

**Fit** — specifies the placement of dimension text, overall scaling of dimensions, and toggling of annotation.

**Primary Units** — specifies the format of units for primary linear and angular dimensions, and zero suppression.

**Alternate Units** — specifies the format of alternate (secondary) units.

**Tolerances** — specifies the format of dimension tolerance text.

**LINES TAB**

The Lines tab specifies the properties of dimension and extension lines.

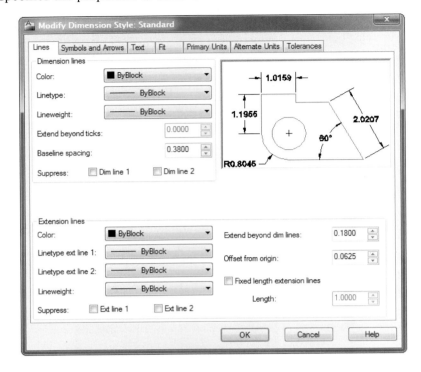

### Dimension Lines

The Dimension Lines section specifies the properties of dimension lines.

#### Color

The **Color** droplist specifies the color of dimension lines. The default color is Byblock, because dimensions are considered blocks. "ByBlock" means that dimension *lines* take on the color of the entire dimension. (In contrast, the default color of the dimensions themselves is Bylayer, meaning the dimension takes on the color defined by its layer.)

The color of the dimension line is stored in the dimension variable named DIMCLRD (short for "DIMension CoLoR Dimension line").

 **Notes** Although you can choose the color from this droplist, it is best to place dimensions on layers that specify the color of the dimension.

You can see the effect of changing variables through the Preview box: select the color blue from the Color droplist. Notice in the Preview box that the dimension line color changes from black to blue.

#### Linetype

To assign linetypes to dimension lines, click the **Linetype** droplist (stored in DIMLTYPE). The preview window updates to show the new linetype.

You cannot change the linetype if any dimstyle override is active. As with colors, it is better to specify the lineweight via the layer, rather than to change the lineweight here.

#### Lineweight

To change the lineweight of the dimension line, choose the droplist next to Lineweight and select a predefined width. The dimension line lineweight is stored in dimension variable DIMLWD.

As with colors and linetypes, it is better to specify the lineweight via the layer, than to change the lineweight here.

### Extend Beyond Ticks

The **Extend beyond ticks** option sets the distance that extension lines extend beyond the dimension line (stored in DIMDLE). The default is 0.0. This option applies only to the Architectural Tick arrowhead, an option that's set in the Symbols and Arrows tab.

### Baseline Spacing

The **Baseline spacing** option determines the distance between dimension lines when they are automatically stacked by the DIMBASELINE and QDIM commands. The distance is stored in dimvar DIMDLI.

### Suppress

AutoCAD can suppress the display of either or both dimension lines. To do so, select the **Suppress** check boxes to suppress **Dim Line 1** (stored in dimvar DIMSD1) or **Dim Line 2** (dimvar DIMSD2). In addition, the options suppress the display of associated arrowheads.

**Dim Line 1** does not refer to either the left or right half of the dimension line — the halves separated by the dimension text. Instead, the first half of dimension line drawn is determined by the initial pick point that places the first extension line.

When the dimension text does not cut the dimension line in two, these options have no effect on the dimension line.

## Extension Lines

The **Extension Lines** section specifies the properties of extension lines.

### Color, Linetype, and Lineweight

The **Color**, **Linetype**, and **Lineweight** options are similar to those of dimension lines. You can specify the color (stored in dimvar DIMCLRE) and lineweight (DIMLWE) of extension lines. In the case of linetypes, you can specify a different pattern for each extension line (DIMLTEX1 and DIMLTEX2).

### Suppress

The **Suppress** option hides the first (DIMSE1), the second (DIMSE2), or both extension lines. "Ext line 1" is the first extension line drawn, based on the points picked during the dimensioning command.

(*History*: Overlapping extension lines were an issue for pen plotters. When multiple extension lines were drawn on top of each other with pens, the paper eventually wore through.)

### Additional Options

The **Extend beyond dim lines** option specifies the distance the extension line protrudes beyond the dimension line (stored in DIMEXE).

The **Offset from origin** option sets the distance the extension line begins away from the pick point of the object being dimensioned (DIMEXO).

The **Fixed length extension lines** option specifies a fixed length for all extension lines (DIMFXLON). The default value is 0.18 units (DIMFXL).

## SYMBOLS AND ARROWS

The Symbols and Arrows tab specifies the properties of extension lines.

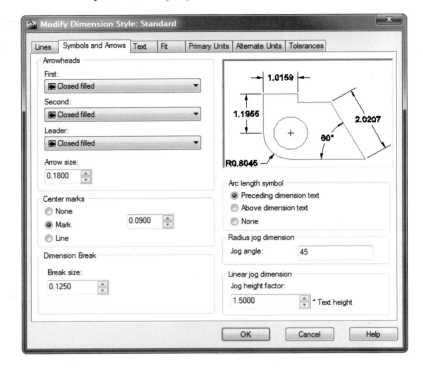

## Arrowheads

The Arrowheads section defines the type and size of arrowheads used by dimensions and leaders.

### First, Second, and Leader

The droplists let you choose from 20 predefined arrowhead styles (stored in DIMBLK). In addition to the standard arrowhead, AutoCAD includes arrowheads, such as the tick, the dot, open arrowhead, open dot, right angle arrowhead, or no head at all; see figure at the side.

When you select an arrowhead from the **First** droplist, AutoCAD changes the **Second** and **Leader** droplists to the same arrowhead. You can specify a different arrowhead for each end of the dimension lines (DIMBLK1 and DIMBLK2) and for the leader dimensions (DIMLDRBLK). When you specify a different one for each end, AutoCAD changes dimvar DIMSAH to 1.

Alternatively, select **User Arrow** to choose a custom-made arrowhead. The customized arrowhead is defined by a named block with the BLOCK command. (See the tutorial on the following pages.) The custom arrowhead's name is stored in DIMBLK.

Although radius and diameter dimensions consist of leaders, they use the arrowhead defined by **Second**. The arrowhead defined by **Leader** affects leaders placed by the LEADER, QLEADER, and MLEADER commands. (Ordinate dimensions do not use arrowheads.)

- Closed filled
- Closed blank
- Closed
- Dot
- Architectural tick
- Oblique
- Open
- Origin indicator
- Origin indicator 2
- Right angle
- Open 30
- Dot small
- Dot blank
- Dot small blank
- Box
- Box filled
- Datum triangle
- Datum triangle filled
- Integral
- None

### Arrow Size

The **Arrow size** option specifies the distance from the left to right end. For custom arrowheads, AutoCAD scales the unit block to size (0.18 units, by default). The value is stored in DIMASZ.

## TUTORIAL: CREATING CUSTOM ARROWHEADS

In this tutorial, you draw a custom arrowhead, and then use it with dimensions.

1. Start AutoCAD with a new drawing.
2. Set the grid to 1.0, or use xlines to demarcate a one inch square, because the arrowhead must fit within a 1"-square.

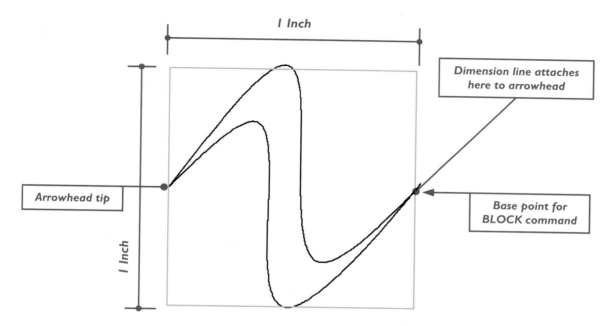

3. Draw the arrowhead with the tip pointing west. (The arrowhead illustrated above was drawn with the **SPLINE** and **OFFSET** commands.) The arrowhead attaches to the dimension line on the East. (See figure.)
4. Use the **BLOCK** command to turn the arrowhead into a block, and then give the block a meaningful name, such as "Spiral." *Important!* For the Base Point, pick the East end of the arrowhead, where it attaches to the dimension line, as illustrated above.
5. To use the new arrowhead with dimensions, follow these steps:
   a. Start the **DIMSTYLE** command, and then click **Modify**.
   b. Select the **Arrows and Symbols** tab.
   c. Under Arrowheads, in the First droplist, select **User Arrow**. Notice the Select Custom Arrow Block dialog box, which lists the names of all blocks defined in the drawing. (You can use any block found in the drawing as an arrowhead. Because it was not designed for use as an arrowhead, however, it would probably not work well.)

6. Select the name of the block you created earlier, and then click **OK**. Notice that AutoCAD automatically fills in the **Second** and **Leader** droplists with the same name. The preview image shows the custom arrowheads.
7. Click **OK** to exit the Symbols and Arrows tab.
8. Choose the **Set Current** button, and then **Close** to close the dialog box.
9. Place some dimensions with the custom arrowhead.

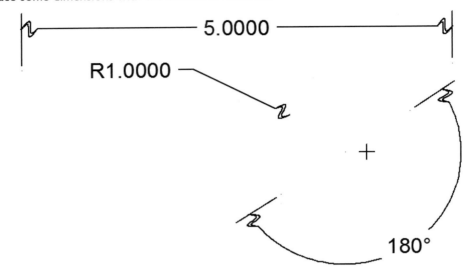

If the arrowhead is too small, increase its size with the DIMSTYLE command's **Arrow size** option.

### Center Marks for Circles

When you mark the centers of circles and arcs with the DIMCENTER command, AutoCAD determines the type of center mark by the value stored in the DIMCEN dimvar: center mark, center mark with extending lines, or no mark at all. Center marks are also placed by the DIMRADIUS and DIMDIAMETER commands.

The value of DIMCEN determines the length of the marks, the width of the gap, and the distance the lines extend beyond the circle. The default value is 0.09 units. When DIMCEN is

| | |
|---|---|
| -0.09 | Center lines and center marks are drawn. |
| 0 | No center mark or lines drawn. |
| 0.09 | Center marks drawn. |

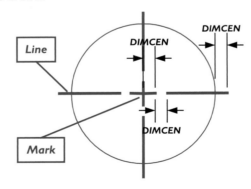

In the Dimension Styles dialog box, select a radio button:

**None** — places no center marks in circles and arcs.
**Mark** — places the center mark.
**Line** — places the center mark and lines.

**Size** specifies the size of the mark, gap, and line extension.

## Dimension Break

**Break Size** specifies the size of the gap created by the DIMBREAK command.

## Arc Length Symbol

The **Arc Length Symbol** option locates the arc length symbol (DIMARCSYM). It looks like an upside down "U." (The symbol cannot be customized.)

**Preceding Dimension Text** — places the symbol before the dimension text.

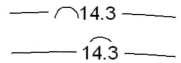

**Above Dimension Text** — places the symbol above the dimension text.

**None** — suppresses the display of the symbol.

## Radius Dimension Jog

The **Radius Dimension Jog** section specifies the angle used by jogged dimensions (DIMJOGANG). The default is 45 degrees.

Use the **Jog Angle** option to change the angle. Autodesk does not document the limits, but it turns out that the angle must be between 5 and 90 degrees. Enter an angle outside those limits, and AutoCAD does not allow you to exit the tab — until you enter a correct value.

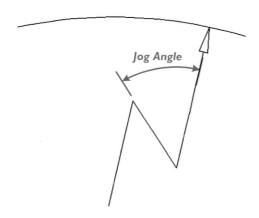

## Linear Jog Dimension

**Jog Height Factor** specifies the size of the jog created by the DIMJOGLINE command. The value is a factor (multiplier) of the text height, the default being 1.5x.

**TEXT**

The **Text** tab specifies the format, placement, and alignment of dimension text. There are additional settings that affect dimension text, and they are found in the Primary Units tab.

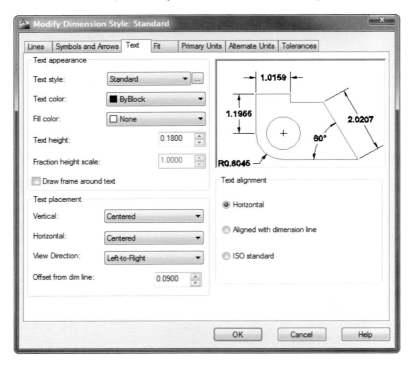

**Text Appearance**

Dimension text is treated independently of other text in the drawing: it uses the style defined by this dialog box, rather than styles defined by the STYLE command.

*Text Style*

The **Text Style** droplist selects a style previously defined by the STYLE command. The default is "Standard." Choose the [...] button to display the Text Style dialog box, which then lets you create and modify text styles. The dimension text style name is stored in dimvar DIMTXSTY.

*Text Color*

The **Text Color** droplist specifies the color of the dimension text (DIMCLRT). To choose a color not listed, pick **Select Color** at the bottom of the droplist.

*Fill Color*

The **Fill Color** option determines the color of a rectangle placed behind the dimension text (DIMTFILL and DIMTFILLCLR). One of the colors is called "Background," which takes on the drawing background color defined by the OPTIONS command. (From the application menu, select **Options**, and then choose the **Display** tab, and finally click the **Colors** button. Note that the color for the model and layout backgrounds can be chosen separately.)

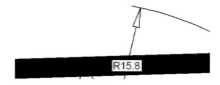

As an alternative, you can use the TEXTTOFRONT command to make dimension text stand out from other objects.

### Text Height

The **Text Height** option operates like the height setting in text styles: 0 means the same height as defined in the text style. Any other height entered here overrides the height defined by the style (DIMTXT).

### Fraction Height Scale

The **Fraction Height Scale** option specifies the size of fractions and tolerances as a multiplier of the dimension text (DIMTFAC). When set to 1 (default), for instance, fractions and tolerances are the same height as the dimension text; when set to 0.5, fractions and tolerances are half the size of dimension text.

### Draw Frame Around Text

The **Draw frame around text** check box adds a rectangle around the dimension text (DIMGAP < 0). The distance between the box and the dimension text is specified elsewhere in the tab with the **Offset from dim line** option.

## Text Placement

The **Text Placement** section specifies how the dimension text is placed relative to the dimension line.

### Vertical

The **Vertical** droplist specifies the vertical placement of dimension text (DIMTAD):

- **Centered** — centers dimension text on the dimension line between extension lines (default).
- **Above** — places text above the dimension line.
- **Outside** — places text on the side farthest from the extension line pick points.
- **JIS** — places text above the dimension line, as per the Japanese Industrial Standards.
- **Below** — places text under the dimension line.

### Horizontal

The **Horizontal** droplist specifies the horizontal placement of dimension text (DIMJUST).

- **Centered** — centers text between the extension lines and on the dimension line (default).
- **At Ext Line 1** — left-justifies text against first extension line.
- **At Ext Line 2** — right-justifies text against the second extension line.
- **Over Ext Line 1** — positions text vertically over the first extension line.
- **Over Ext Line 2** — positions text vertically over the second extension line.

### View Direction

The **View Direction** droplist specifies the reading direction of dimension text (DIMTXTDIRECTION).

- **Left-to-Right** — displays horizontal text left to right, and vertical text bottom to top.
- **Right-to-Left** — displays horizontal text right to left, and vertical text top to bottom.

### Offset From Dim Line

The **Offset from dim line** option specifies the gap between dimension text and the dimension line, as well as the space between the text and the frame (DIMGAP).

## Text Alignment

The **Text Alignment** option determines the orientation of dimension text when inside and outside the extension lines (DIMTIH and DIMTOH).

- **Horizontal** — forces text to be horizontal, whether or not it fits inside the extension lines.
- **Aligned with dimension line** — forces text to align with the dimension line.
- **ISO Standard** — aligns text with the dimension line when it fits inside extension lines; when outside extension lines, it is drawn horizontally.

## FIT

The **Fit** tab determines the placement of dimension text, arrowheads, leader lines, and dimension lines.

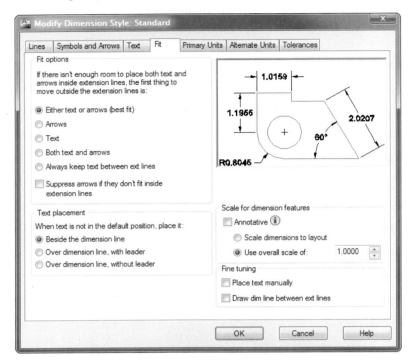

### Fit Options

The **Fit Options** section controls where text and arrowheads are placed when the distance between extension lines is too narrow (**DIMATFIT**).

**Either the text or the arrows, whichever fits best** — places the elements where there is room.

**Arrows** — fits only the arrowheads between extension lines, when space is available.

**Text** — fits text between extension lines, but places arrowheads outside when space is lacking.

**Both text and arrows** — forces both text and arrows outside the extension lines.

**Always keep text between ext lines** — keeps text always between the extension lines (**DIMTIX**).

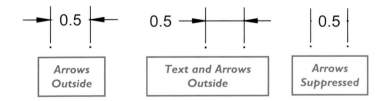

**Suppress arrows if they don't fit inside the extension lines** — draws no arrowheads when there is too little space between the extension lines (**DIMSOXD**).

### Text Placement

When AutoCAD cannot place the dimension text normally, you have these options (DIMTMOVE):

> **Beside the dimension line** — places text beside the dimension line.
>
> **Over the dimension line, with a leader** — draws a leader line between the dimension line and the text when there isn't enough room for the text.
>
> **Over the dimension line, without a leader** — draws no leader.

### Scale for Dimension Features

The **Scale for Dimension Features** section controls the size of text and arrowheads independently of the rest of the drawing, usually using the plot scale. The length of the dimension and extension lines is unaffected. (To change the weight of these lines, use the **Lineweight** option in the Lines and Arrows tab.)

> **Annotative** — toggles use of the annotative scaling property (DIMANNO). When on, the following two options become unavailable:
>
> **Use overall scale of** — sets the scale factor for text and arrowheads (DMSCALE). Entering a value of 2, for example, doubles the size of arrowheads and text.
>
> **Scale dimensions to layout** — scales dimensions to factors determined from the scale factor between model space and layouts. AutoCAD preforms the following actions:
>
> > In model tab, DIMSCALE IS set to 1.
> >
> > In all layout tabs, DIMSCALE is set to 0.
> >
> > In model space viewports of layout tabs, DIMSCALE is computed from the viewport's scale factor.
> >
> > In the paper space of layouts, DIMSCALE is computed from the current value of the ZOOM XP command.

### Fine Tuning

The **Fine Tuning** section presents a couple of miscellaneous items:

**Place text manually when dimensioning** determines whether dimension commands prompt you for the position of the dimension text, allowing you to change the text position, if necessary (DIMUPT). AutoCAD ignores the horizontal justification settings.

**Always draw dim line between ext lines** forces AutoCAD always to draw the dimension lines inside the extension lines (DIMTOFL). The arrowheads and text are placed outside, if there isn't enough room. This option also forces leaders to be drawn to the circle's center point, even when the leader would not normally be drawn.

## PRIMARY UNITS

The **Primary Units** tab sets the format of primary dimension units, as well as the prefix and suffix of dimension text. *Primary units* refers to regular dimension text, as opposed to alternate units shown in square brackets and tolerance text.

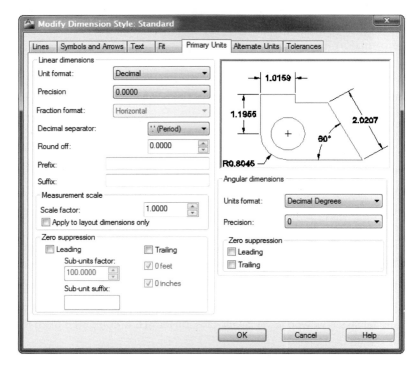

### Linear Dimensions

The **Linear Dimensions** section specifies the units for dimensions, such as feet or meters. Whereas the UNITS command controls the settings of units for *all* numbers, angles, and dimensions, this Primary Units tab overrides the settings for dimensions.

#### Units Format

The **Units format** droplist selects the format (**DIMLUIT**):

- **Scientific** — displays exponents.
- **Decimal** — used for metric dimensions (default).
- **Architectural** — displays feet, inches, and fractions.
- **Engineering** — displays feet, inches, and decimal fractions.
- **Fractional** — displays inches and fractions.
- **Windows Desktop** — displays units specified by the Windows operating system.

The Windows units are changed by the Numbers tab of the Regional Options dialog box, as found in the Windows Control Panel.

#### Precision

The **Precision** droplist selects the number of decimal places or fractional divisions to be displayed by dimensions. The range is from 0 to 8 decimal places, and from 0" to 1/256". The measured values are unaffected by this setting, which changes only the display.

### Fraction Format

The **Fraction format** droplist specifies how fractions are stacked (DIMFRAC):

> Horizontally
> Diagonally
> Not stacked

This option is available only when you select a units format with fractions, such as Architectural; otherwise, it is grayed out.

### Decimal Separator

The **Decimal separator** droplist specifies the decimal separator (DIMDSEP):

> Period (.)
> Comma (,)
> Space ( )

This option is available only when you select **Decimal** as the units format. This option is used typically in countries that use the comma ( , ) to separate units from decimals.

### Round Off

The **Round off** option specifies how to round off decimals and fractions of numbers; it does not apply to angles (DIMRND).

Setting this option to 0.33, for example, rounds the dimension text to the nearest 0.33. The measured values are unaffected by this setting, which changes only the display. This option is different from **Precision**, which truncates the display of decimal places.

### Prefix and Suffix

The **Prefix** and **Suffix** options provide room for alphanumeric values added in front of and behind the dimension text.

For example, to prefix every dimension with the word "Verify," enter that in the **Prefix** box. To suffix every dimension with "(TYPICAL)," enter that in the **Suffix** box. The DIMPOST dimvar stores both values using this format: "Verify<>(TYPICAL)".

### Measurement Scale

The **Measurement Scale** section specifies the scale factor applied to the *value* of the dimension text; to change the *size* of text and arrows, specify the overall scale factor in the Fit tab.

### Scale Factor

The **Scale Factor** option sets a scale factor that multiplies linear dimensions (DIMLFAC). By entering 25.4, for example, imperial dimensions (inches) are converted to millimeters (metric units).

The factor is not applied to angles. Alternate and tolerance values have their own scale factor settings.

### Apply to Layout Dimensions Only

The **Apply to layout dimensions only** option applies the linear scale factor only to dimensions created in layouts (paper space). When turned on, the scale factor is stored as a negative value in DIMLFAC.

### Zero Suppression

The **Zero Suppression** section determines whether zeros are displayed in dimension text.

### Leading and Trailing

The **Leading** and **Trailing** options specify whether leading or trailing zeros, and zero feet or inches, are suppressed (DIMZIN). When both options are turned on, **0.2500** becomes **.25**, for example, and **1.00** becomes **1**. Note that zero suppression overrides the setting of the **Precision** option.

### Sub-units

The following options operate only when zero-suppression is turned on, and with certain formats of units:

The **Sub-units Factor** option specifies the conversion factor, such as 100 for cm-to-m, or 12 for in-to-ft.

The **Sub-unit Suffix** option specifies the subunit, such as centimeters for meters, or inches for feet.

### 0 Feet and 0 Inches

The **0 feet** and **0 inches** options determine whether zero feet and zero inches are displayed. When both options are turned on, **0'–1/2"** becomes **1/2"** and **34'–0"** becomes **34'**. These feet and inches options become available only when a fraction unit format is selected, such as Architectural.

## Angular Dimensions

The **Angular Dimension** section specifies the display format for dimensions placed by the DIMANGULAR command. This allows angles to be formatted independently of linear dimensions.

### Units Format

The **Units format** droplist specifies the format of angular dimensions (DIMAUNIT) — curiously, the Surveyor's units are missing:

> **Decimal degrees** — displayed as DdD.dddd (default).
> **Degrees/Minutes/Seconds** — displayed as DD.MMSSdd.
> **Grads** — displayed as DDg.
> **Radians** — displayed as DDr.

### Precision

The **Precision** droplist specifies the number of decimal places or fractions (DIMADEC). This affects the display of the angle and not its measured value.

### Zero Suppression

The **Zero Suppression** section suppresses leading and trailing zeros of angles (DIMAZIN). As an example of suppressing leading zeros, the measured angle of **56°07"** is displayed as **56°7"**, and of suppressing trailing zeros, **56°00'** is displayed as **56°**.

## ALTERNATE UNITS

The **Alternate Units** tab selects the format and scale of alternate measurement units. The options are similar to the Primary Units tab; the following descriptions list only the differences.

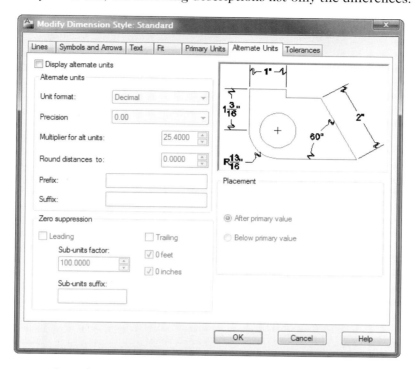

AutoCAD allows you to place dimensions with double units: a primary unit plus a second or alternate unit. [The alternate units appear in square brackets.] This is particularly useful for drawings that must show imperial and metric units. (Angles cannot have alternate units in AutoCAD.)

When you select options in this tab, the values are stored in this set of dimvars: DIMALTD specifies the decimal places for alternate units. DIMALTRND specifies the rounding of alternate units. DIMALTU specifies the format of alternate units, except for angular dimensions. DIMALTZ specifies zero suppression for alternate units.

The **Display alternate units** check box turns on alternate units (DIMALT); when off, all options are grayed out, meaning they are unavailable.

### Alternate Units

The **Alternate Units** section specifies the look of the alternate text; options are identical to that of the Primary Units tab.

### Zero Suppression

The **Zero Suppression** section determines whether leading and trailing zeros are suppressed; the options are identical to that of the Primary Units tab.

### Placement

The **Placement** section determines where the alternate units are placed (DIMAPOST):

> **After primary value** — places the alternate units behind the primary units.
> **Below primary value** — places the alternate units below the primary units

## TOLERANCES

The **Tolerances** tab controls the format of dimension text tolerances. It does not format tolerance symbols placed with the TOLERANCE command; they have no format options.

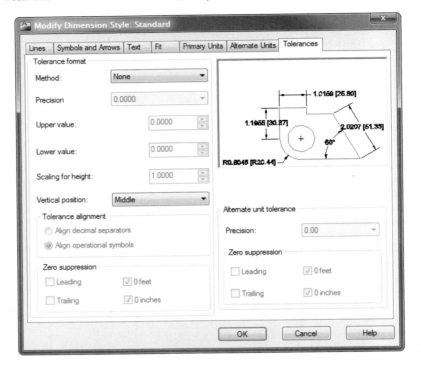

### Tolerance Format

The **Tolerance Format** section determines the style and look of tolerances.

Method

The **Method** droplist offers five styles of tolerance text (DIMTOL and DIMLIM):

**None** — suppresses the display of tolerance text (default), and grays out all options in this tab, except **Vertical position**, curiously enough.

**Symmetrical** — adds a single plus/minus tolerance, such as the 1/8" illustrated below (at left).

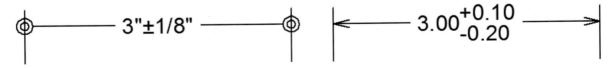

*Left: Symmetrical tolerance.*
*Right: Deviation tolerate.*

**Deviation** — adds a plus and minus tolerance, such as the +0.1 and -0.2 illustrated above (right).

**Limits** — places two dimensions, such as the 3.0 and 2.4 illustrated below (at right). AutoCAD arrived at these values by adding 0.5 to 3 (= 3.5), and subtracting 0.5 (= 2.5).

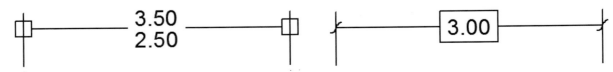

*Left: Limits tolerance.*
*Right: Basic tolerance.*

**Basic** — draws a box around the dimension text. (The distance between text and box is stored in **DIMGAP**.) No tolerances are displayed, as illustrated above, at left.

Dimvar DIMTOL appends the tolerances to the dimension text; DIMLIM displays dimension text as limits. Turning on DIMLIM turns off DIMTOL.

### Precision

The **Precision** droplist determines the number of decimal places displayed by both tolerance values. This allows tolerances to have a precision independent of the primary values.

### Upper Value and Lower Value

The **Upper Value** and **Lower Value** options specify the upper or plus value (DIMTP), and the lower or minus tolerance value (DIMTM).

### Scaling for Height

The **Scaling for height** option determines the size of the tolerance text relative to the main dimension text (DIMTFAC). A value of 1 indicates each character of tolerance (or limits) text is the same size as the primary dimension text. Changing the value to **0.5**, for example, reduces the tolerance text height by half.

### Vertical Position

The **Vertical position** droplist determines the relative placement of tolerance text (DIMTOLJ), whether it is aligned to the top, middle, or bottom of the primary dimension text.

### Tolerance Alignment

The **Tolerance Alignment** options align the deviation and limits tolerance text by decimal points or prefix symbols, such as + and -.

### Zero Suppression

The **Zero Suppression** options are identical to those found in the Primary Units tab.

### Alternate Unit Tolerance

The **Alternate Unit Tolerance** option specifies the number of decimal places for the tolerance values in the alternate units of a dimension (DIMALTTD). This option is available only when alternate units are turned on in the Alternate Units tab. Combining tolerances with alternate units leads to cluttered dimension text.

## SETTING THE CURRENT DIMSTYLE

To set a dimension style as active, select the dimstyle name from the list under Styles, and then choose the **Set Current** button. (The current dimension style name is stored in dimvar DIMSTYLE.)

Alternatively, you can use the ribbon's Dimensions panel or the Styles toolbar, which provide droplists of dimension and text styles in the drawing. To make a dimstyle current, select it from the droplist.

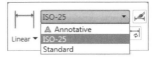

### Renaming and Deleting Dimstyles

It is not immediately apparent how to rename or delete dimension styles in the Dimension Style Manager. The commands are "hidden" in shortcut menus: select a dimension style name in the **Styles** list. Right-click to display the cursor menu, and then choose **Rename** or **Delete**.

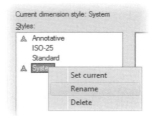

Not every style can be erased or renamed. If a style is in use, it cannot be erased. The "Standard" dimstyle can be renamed, but not deleted. Externally-referenced dimstyles cannot be renamed or deleted. Alternatively, you can use the RENAME command to rename dimstyles, and the PURGE command to remove unused ones.

## MODIFYING AND OVERRIDING DIMSTYLES

After you create one or more dimension styles for drawings, you might need to change the style. Perhaps the client wants dimension lines thicker than the extension lines, or the units changed to metric, or perhaps some extension lines removed.

Some changes are global, which means they apply to every dimension in the drawing, such as the thicker dimension lines. Other changes are local, which means they apply to selected dimensions, such as extension lines to be removed.

Some changes are retroactive, which means they apply to all dimensions already drawn. Other changes are proactive, which means they apply to dimensions not yet drawn. Here is how AutoCAD handles these four cases.

> **To apply global changes retroactively** — modify the dimension style with the **DIMSTYLE** command's **Modify** button. AutoCAD retroactively changes *all* dimensions in the drawing.
>
> **To apply global changes proactively** — modify the dimension style with the **DIMSTYLE** command's **Override** button, which creates "sub-styles." The *next* dimensions you draw reflect the change in dimension style.
>
> **To apply different dimension styles to selected dimensions (local changes)** — use the -DIMSTYLE command's **Apply** option to *selected* dimensions, and then apply the dimstyle to them, a local modification.
>
> **To change dimension variable settings of selected dimensions (local changes)** — use the **DIMOVERRIDE** command to specify the name of a dimension variable and then provide the new value. It prompts you to select the dimensions to which the local override should be applied.

**Note** The difference between *modifying* and *overriding* dimension styles is as follows:

**Modify** — modifies retroactively, and globally updates all existing dimensions created using that style.
**Override** — changes only dimensions created after the override is applied,

## TUTORIAL: GLOBAL RETROACTIVE DIMSTYLE CHANGES

In the following tutorial, you learn how to change the style of all dimensions in the drawing at once.

1. Start a new drawing with AutoCAD, and then draw some dimensions.
2. To change the style of *all* dimensions in the drawing, start the **DIMSTYLE** command:
3. In the dialog box, click **Modify**.

4. Make some changes to the dimension styles. For instance, choose the Lines tab, and then change the dimension line color to red.

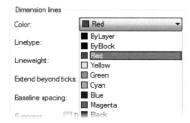

Notice that the preview window shows the dimension lines changed to red.

5. Click **OK**, and then **Close**.

Notice that all dimensions in the drawing now have red dimension lines.

## TUTORIAL: GLOBAL TEMPORARY DIMSTYLE CHANGES

In this, you override the dimension style temporarily, so that new dimensions take on the changed style; existing dimensions remain unaffected.

1. Continue with the drawing from the previous tutorial.
2. Start the **DIMSTYLE** command, and then click the **Override** button.
3. In the Lines tab, change the dimension line color to blue.

4. Click **OK**.

Notice that the Standard dimstyle has a sub-style named "<style override>." It is highlighted, meaning that it is current.

Also notice that in the Description area, the change to the blue dimension lines is listed: "Standard+Dim line color = 5 (blue)". (After exiting this dialog box, any dimension you draw from now on has dimension lines that are colored blue.)

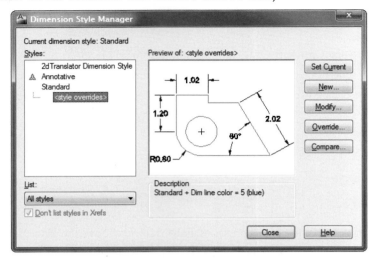

6. Click **Close**. Notice that none of the existing dimensions in the drawing changes.
7. Draw a linear dimension with the **DIMLINEAR** command. Notice that its dimension line is blue.

The names of overrides, unfortunately, are not listed by the droplist by the ribbon's Annotate panel.

**Note** You can turn overrides into permanent dimension styles. Here's how:

To turn overrides into *new* dimstyles, right-click "<style override>" (the AutoCAD-generated name), and then select **Rename** from the shortcut menu. Change the name to something meaningful, like "Blue Dimlines."

To make an override the current dimstyle, right-click "<style override>", and then select **Save to current style** from the shortcut menu. Notice that the override disappears, and that the current style shows the changes, such as blue dimension lines. Click **Close**, and all dimensions change to blue dimlines. (This is equivalent to using the **Modify** button in the first place.)

## STYLE OVERRIDES, DIMOVERRIDE, AND PROPERTIES

Overriding individual dimensions is making "local" changes.

The Dim Style droplist on the Dashboard's Dimension panel applies dimension styles to selected dimensions, as does the droplist on the Styles toolbar. The DIMOVERRIDE command applies changed dimension variables to selected dimensions.

### TUTORIAL: LOCAL DIMSTYLE CHANGES

1. For this tutorial, create a drawing with at least two dimension styles and several dimensions.
2. To apply a dimstyle to certain dimensions, select a few dimensions in the drawing. Notice that they are highlighted.
3. From the ribbon's Dimensions panel, click the **Dimension Style** droplist. Notice the list of dimension style names.

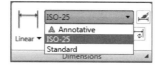

4. Select a different style name. Notice that the selected dimensions change their style.
5. Press ESC to remove the highlighting from dimensions.

### TUTORIAL: LOCAL DIMVAR OVERRIDES

In this tutorial, you change the color of one dimension's extension line to yellow.

1. For this tutorial, continue with the drawing from the last tutorial, the one with at least two dimensions styles.
2. To change a dimvar for selected dimensions, start the **DIMOVERRIDE** command.
   * In the ribbon's Annotate tab, click the **Override** button in the Dimensions panel.
   * At the 'Command:' prompt, enter the **dimoverride** command.

   Command: **dimoverride** (Press ENTER.)

   * Alternatively, enter the aliases **dov** or **dimover** at the 'Command:' prompt.
3. At the prompt, enter the name of a dimvar, such as **DIMCLRE**, which sets the color of extension lines:

   Enter dimension variable name to override or [Clear overrides]: **dimclre**
4. Enter a new value for the dimvar, such as yellow:

   Enter new value for dimension variable <BYBLOCK>: **yellow**
5. The prompt repeats so that you can override other dimvars. Press ENTER to continue and select the dimensions to override:

   Enter dimension variable name to override: (Press ENTER to continue.)
6. Select the dimensions to change:

   Select objects: (Select a dimension.)

   Select objects: (Press ENTER to end object selection.)

   Notice that the extension lines change to yellow.

**Note** You can override dimvars *during* dimensioning commands. In the following example of the DIMALIGNED command, the dimension line color is changed to green using the DIMCLRD dimvar:

> Command: **dimaligned**
> Specify first extension line origin or <select object>: **dimclrd**
> Enter new value for dimension variable <byblock>: **green**
> Specify first extension line origin <select object>: *(Continue with the DIMALIGNED command.)*

New dimension lines are now green, until you override again.

### Clearing Local Overrides

If you change your mind, you can clear local overrides. Start the DIMOVERRIDE command, and then enter the **Clear overrides** option:

> Command: **dimoverride**
> Enter dimension variable name to override or [Clear overrides]: **c**
> Select objects: *(Select the dimensions to return to normal.)*
> Select objects: *(Press ENTER to end object selection.)*

The dimension returns to its previous look, because the override is canceled.

## PROPERTIES

You can use the Properties palette to change many properties of dimensions locally. Open the palette with the PROPERTIES command, and then select the dimensions you wish to change.

## DRAWORDER FOR DIMENSIONS

You can force AutoCAD to display dimensions in front of all other objects in the drawing. To do this, use the TEXTTOFRONT command, as follows:

> Command: **texttofront**
> Bring to front [Text/Dimensions/Both] <Both>: **d**
> *n* object(s) brought to front.

This command is more efficient than the DRAWORDER command, because it selects all dimensions for you.

Because TEXTTOFRONT is a command, and not a setting, you may have to reuse this command as the drawing progresses.

## GRIPS EDITING

Grips allow you to edit dimensions and leaders directly — without needing to enter the names and options of dimension editing commands.

The grips on associative dimensions change the location of the dimension line, text, extension lines, and so on. The tricky part is knowing which grip performs what action; the grips themselves provide no clues, unfortunately. Thus, this section describes their actions.

## TUTORIAL: EDITING DIMENSIONS THROUGH OBJECTS

1. Draw a line, and then dimension it using the **Select Object** option, as follows:
   Command: **dimlinear**
   Specify first extension line origin or <select object>: *(Press ENTER.)*
   Select object to dimension: *(Pick the line.)*
   Specify dimension line location or [Mtext...]: *(Pick a point to position the dimension line.)*
   Dimension text = 2.00
2. Select the line — not the dimension! Notice the blue grips.
3. Select one of the end grips; it turns red.
4. Stretch the *object*, longer or shorter. Notice that the associated dimension updates itself, as illustrated by the figure.

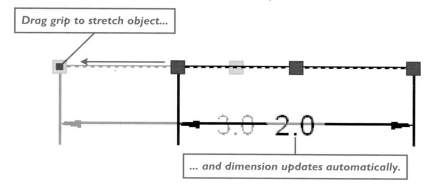

5. Select middle grip to move the line. Notice that the dimension tags along.
6. Press ESC to end grips editing.

### GRIPS EDITING OPTIONS

Grips perform six editing operations: stretch, copy, move, rotate, scale, and mirror. Here is how each one affects dimensions:

**Stretch** — changes the position of dimension parts, such as extension lines or text, while keeping the remainder of the dimension in place. The dimension remains partially associated with its object. When stretching the dimension line, the dimension text is updated; other parts of the dimension react differently to stretching, as detailed later.

**Copy** — copies the dimension, subject to the peculiarities of the selected grip.

**Move** — moves the entire dimension, but disassociates it from its object.

**Rotate** — rotates the entire dimension about the selected grip, but disassociates it from its object.

**Scale** — resizes the dimension, using the selected grip as the base point. The dimension remains associated with its object; dimension text is updated to reflect the new size. As with copying, scaling is subject to the peculiarities of the selected grip.

**Mirror** — does not mirror dimensions; instead, this option rotates them about the selected grip. Text is not mirrored, ignoring the setting of the **MIRRTEXT** system variable. The "mirrored" dimension is disassociated from its object.

# EXERCISES

1. In the following exercises, you modify the Standard dimension style.
    a. Start AutoCAD with a new drawing.
    b. Draw one each of the following dimensions — linear, aligned, diameter, and leader — similar to the figure below.

    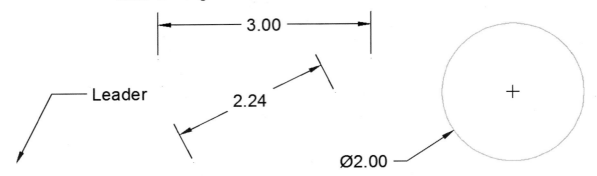

    Which commands did you need to draw each dimension?
    Linear dimension:
    Aligned dimension:
    Diameter dimension:
    Leader:

    c. Open the Dimension Style Manager dialog box. Which dimstyles are listed there?
    d. Modify the dimstyle: change the color of extension lines to red. When you exit the dialog box, what happens?
    e. Return to the Dimension Style Manager dialog box.
    f. Override the dimstyle: change the color of text to blue.
    g. When you exit the dialog box, what happens?
    h. Draw another linear dimension. Is it different from the others?
    i. Return to the Dimension Style Manager dialog box.
    j. Convert the override into a dimstyle called "Blue Text."
    k. When you exit the dialog box, what happens?
    l. Use the ribbon's Annotation tab to apply the Blue Text dimstyle to the aligned dimension. Does it change?
    m. Repeat, applying the dimstyle to the leader. Does it change?

2. Continuing with the drawing from above, create a new dimension style with the following settings:

   **Extension line**
   - Color — 8

   **Text**
   - Font — **Times New Roman**
   - Color — **155**
   - Vertical Placement — **JIS**

   **Arrowheads**
   - Both — **Architectural Tick**

   *Hint:* change the text style. Save the style by the name of "JayGray." Apply the style to all dimensions in the drawing. Save the drawing as *jaygray2.dwg*.

3. Continuing with the drawing from the previous exercise, use grips editing on the different grips of each dimension. Do not save the drawing.

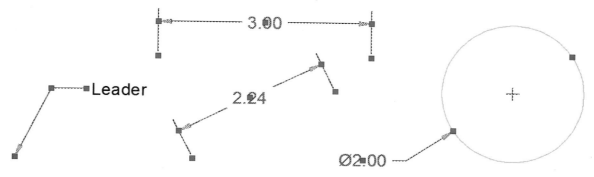

4. Start a new drawing. Open the DesignCenter, and then import dimension styles from the following drawing: *Architectural - Annotation Scaling and Multileaders.dwg*.

5. From AutoCAD's \sample folder, open the *Architectural - Annotation Scaling and Multileaders.dwg* file, a drawing of a set of stairs.
   a. Select a dimension, and then edit its text.
   b. Do the same to a leader.

6. Start a new drawing, and then draw a linear dimension.
   a. Add a jog to the dimension line. Which command did you use to add the jog?
   b. Draw a second linear dimension that crosses the first one.
   c. Add a gap where the two dimensions cross. Which command did you use to create the gap?

7. Open *structure.dwg* file, the drawing of an arbor. Space the linear dimensions evenly. Which command did you use to space the dimensions?

## CHAPTER REVIEW

1. Which command is used to create, edit, and delete dimension styles?
2. What is the difference between *modifying* and *overriding* dimensions?

    Modifying:

    Overriding:
3. Can the color and lineweight of dimension lines be set independently of extension lines?
4. Can extension lines be suppressed independently of each other?
5. Is it better to use dimension styles or layers to set the color and lineweight of dimension lines?
6. Can you define your own arrowhead for use by dimensions?
7. When the distance between extension lines is too tight, where does AutoCAD place the dimension text?
8. Describe some situations where you might need to edit dimensions.
9. What are *alternative units*?
10. When you move an object, do associative dimensions move also?
11. Describe how the following grips editing commands affect dimensions:

    Stretch:

    Copy:

    Move:

    Rotate:

    Scale:

    Mirror:
12. What is the purpose of the **AIDIMFLIPARROW** command?
13. Can extension lines have fixed lengths?
14. Explain the meaning of the following abbreviations:

    dimvar:

    dimstyle:

    dimlin:
15. What is the "T" in **DIMTEDIT** short for?
16. Describe the function of the **DIMEDIT** command's **Oblique** option.

    When is it useful?
17. How do you edit the wording of the dimension text?

    The properties of dimension text?

    The position of dimension text?
18. Briefly explain the differences among these three similar-sounding dimension-editing commands:

    **DDEDIT**:

    **DIMEDIT**:

    **DDIMTEDIT**:
19. If you make a mistake editing a dimension, which command returns the original dimension?
20. Can dimensions be scaled independently of the drawing?
21. What is the danger in rounding dimension text?
22. When AutoCAD rounds off dimension values, is the change permanent?

23. Round the following numbers to the nearest unit:

    2.5

    1.01

    4.923

    10.00

    6.489

24. Where does the JIS dimension standard expect text to be placed?
25. Describe the purpose of *dimension variables*.
26. Does every drawing have a dimension style?
27. How do you change the scale of dimensions relative to the rest of the drawing?
28. List two sources of dimension styles:

    a.

    b.

29. Why might you want to edit dimensions in the following manner:

    a. Create gaps in extension lines?

    b. Create jogs in the dimension line?

    c. Space dimensions evenly?

30. Briefly describe the purpose of the following dimvars. (For help, look up the tables in appendix D.)

    **DIMCLRE**:

    **DIMSE1**:

    **DIMJUST**:

    **DIMCLRT**:

    **DIMFIT**:

31. Find the name of the dimvar that does the following task. (For help, look up the tables in appendix D.)

    Specifies the size of the center mark:

    Enables alternate units:

    Specifies the length of the arrowhead:

    Specifies linear dimension units:

    Controls suppression of zeros:

32. Why does the Standard dimension style set **DIMALTF** (alternate scale factor) to 25.4?

    And why does the ISO-25 dimstyle set the same dimvar to 0.04?

33. What is the effect of annotative scaling on dimensions?
34. What is the command that corresponds to the alias?

    ddim:

    dimted:

    ded:

    dov:

    d:

35. Can you rename dimension styles?

    Purge dimension styles?

36. Can a drawing have more than one dimension style?

    Can a dimension have two different arrowheads?
37. Can you create your own arrowheads for dimensions?
38. Under what conditions do the arrowheads and text appear outside the extension lines?
39. What is the purpose of *zero suppression*?

    Why might you want to use zero suppression?
40. When are alternative units used?

    How are alternative units shown?
41. Briefly describe the sequence to change the style of all dimensions in the drawing retroactively.
42. Briefly describe the sequence to change the style of dimensions selectively.
43. Can dimvars be changed during a dimensioning command?
44. If dimensions are hidden by other objects in the drawing, what can you do to see them better?
45. How would you quickly bring all dimensions to the top of the display order?

# CHAPTER 13

## Applying Dimensional Constraints

AutoCAD's dimensions report the size of objects, as you saw in the previous chapters. A special type of dimension, called the dimensional constraint, controls the size of objects and the distance between them. In this chapter, you learn about dimensional constraints; in the following chapter, you learn how to control the geometry in and between objects using geometric constraints.

Typically, you apply geometric constraints to objects first, and then apply dimensional constraints. But I find it best to learn about dimensional constraints first.

In this chapter, you learn the following commands:

**DC...** commands apply dimensional constraints within and between objects.

**DELCONSTRAINT** erases dimensional constraints.

**PARAMETERS** edits links between dimensional constraints.

**CONSTRAINTSETTINGS** specifies default conditions for dimensional constraints.

---

**NEW TO AUTOCAD 2011** IN THIS CHAPTER
- Each dimensional constraint now has its own command.

## ABOUT DIMENSIONAL CONSTRAINTS

*Dimensional constraints* define the size of objects, such as the length of lines and radius of circles. For example, when you change the value of the linear dimensional constraint shown in the figure below, the line is forced to change its length. This is not something you can do with regular dimensions, because they only update their values when the associated geometry is changed.

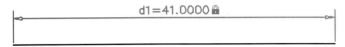

Next to the text is a 🔒 padlock icon. This indicates that the dimension is constraining. (If you find the icon cluttering, it can be turned off.) The padlock reminds you that you cannot edit the line's length directly: it can only be done by editing the value of the dimensional constraint (d1=41.0).

Dimensional constraints can be applied between objects. For example, the figure below indicates that the distance between the center of the ellipse and circle is dimensionally constrained. If you move the circle, the ellipse moves along, keeping the same distance between them.

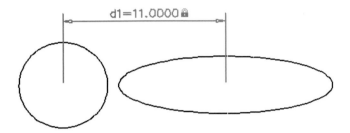

Dimensional constraints contain *values* (numbers) just like regular dimensions, such as d1=11.0. What's really powerful, however, is that they can contain *formulas* (parameters) and *references* to other dimensional constraints. (In the figures above, the 'd1=' is a simple formula. Note that AutoCAD copies the constraint naming convention used by most other parametric modelers.) You use the Parameters Manager palette to edit formulas and define relationships between parameters, as described near the end of this chapter.

Dimensional constraints are used together with *geometric constraints* to create *parametric drawings*. These drawings change their design and shape automatically, according to the values you enter. (Geometric constraints fix objects geometrically, such as horizontal or parallel to another object; you learn about them in the following chapter.) While AutoCAD is limited to 2D constraints, they can be used in 3D to a limited extent, such as in controlling the elevation and thickness of objects.

(*Historical note*: Parametric CAD was invented by two Russian computer programmers in the mid-1980s. In the USA, they formed a software company named Parametric Technology Corp to market and sell the software, called Pro/Engineer, to mechanical engineers. Some 15 years later, they left PTC to develop Revit, parametric software for architects. Revit is now owned by Autodesk. While Pro/E continues to use its own parametric system, AutoCAD and most other CAD packages license the technology from a single company, the D-Cubed division of Siemens.)

 **DC... COMMANDS**

The collection of commands that begin with "DC" applies dimensional constraints between objects, as well as to geometric features within objects.

These commands operate almost exactly like the dimensioning commands you learned in earlier chapters. The difference is that these commands apply dimensional constraints. For example, the DCLINEAR command applies linear dimensional constraints, just like the DIMLINEAR command.

The DC... commands support the following dimensional constraints. (These commands are new to AutoCAD 2011, having replaced the all-in-one DIMCONSTRAINT command of AutoCAD 2010.)

| DC... Command | Dimension Constraint | Equivalent Dimensioning Command |
| --- | --- | --- |
| DCLINEAR | Linear | DIMLINEAR |
| DCHORIZONTAL | Horizontal | DIMHORIZONTAL * |
| DCVERTICAL | Vertical | DIMVERTICAL * |
| DCALIGNED | Aligned | DIMALIGNED |
| DCANGULAR | Angular | DIMANGULAR |
| DCRADIUS | Radial | DIMRADIUS |
| DCDIAMETER | Diameter | DIMDIAMETER |

*) *Commands undocumented by Autodesk.*

Other commands convert regular dimensions into dimensional constraints, and toggle their display. A system variable changes the forms of dimensional constraints between dynamic and annotational. Dimensional constraints cannot be converted back to regular dimensions.

| DC... Command | Constraint Function |
| --- | --- |
| DCCONVERT | Converts associative dimensions to dimensional constraints. |
| DCDISPLAY | Toggles the display of dimensional constraints. |

## Types of Dimensional Constraints

AutoCAD has four types of dimensional constraints:

**Dynamic constraints** display in model tab only; they do not display in layouts and they do not plot. They are displayed in a unique and uneditable dimension style to ensure that they always appear the same size, no matter the zoom level. This is the default style of dimensional constraint, and the type you use the most in this chapter.

*Left: Dynamic constraint independent of zoom level.*
*Right: Annotational constraint taking on current dimensional style.*

**Annotational constraints** act just like regular dimensions. They are displayed in both model and layout tabs. They are plotted, are affected by dimension styles, and their size changes with zoom levels. To create annotational dimensions from scratch, set system variable CCONSTRAINTFORM=1. You can switch dimensional constraints between dynamic and annotational with the PROPERTIES palette's **Constraint Form** property.

**Reference constraints** refer to constraints that do not constrain geometry on their own. They are created automatically when a drawing is *over constrained* — when there are too many dimensional constraints applied to the same set of objects. Reference constraints are identified by the parentheses that surround the text, as illustrated by **d4** in the figure below.

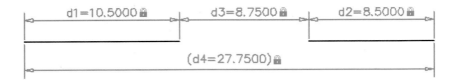

Reference constraints are useful for reporting the overall widths of models, the value being the sum of two or more dynamic constraints. As the values of dynamic constraints change, the values reported by references are updated; you cannot edit the value of reference constraints.

**Parametric constraints** are like dynamic constraints, except that they are created in the Block Editor with the BCPARAMETERS command. They sport user-definable grips for controlling the size of the related dynamic blocks.

**Note** If you are unsure which form to use, follow these guidelines:

- Generally, you should use dynamic constraints (the default format).
- If you intend to plot drawings with dimensional constraints, then you should use annotational ones.
- Reference constraints are a by-product of over-constrained drawings, and so are generally not used.
- Parametric constraints are only used with dynamic blocks.

Autodesk, unfortunately, made the nomenclature confusing by using the work "parameter" in dynamic blocks before they implemented the parametric functionality. Technically, all dimensional and geometric constraints are parametric; better yet, the "parametric constraints" of dynamic blocks can also be parametric.

### TUTORIAL: PLACING DIMENSIONAL CONSTRAINTS

The best way to understand dimensional constraints is to try them out. In this tutorial, you place a dimensional constraint on one circle, making its radius the same as that of another circle.

1. Start AutoCAD with a new drawing.
2. With the **CIRCLE** command, draw two circles similar to the ones illustrated below. Their sizes do not matter, except that one should be one larger than the other.

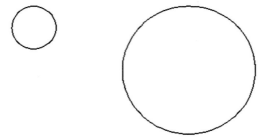

3. Dimension one of the circles with the **DIMRADIUS** command. (In the following step, you will convert this dimension into a dimensional constraint.)
   Command: **dimradius**
   Select arc or circle: *(Pick one arc, circle, or polyarc.)*

Specify dimension line location or [Mtext/Text/Angle]: *(Pick a point to position the dimension leader.)*

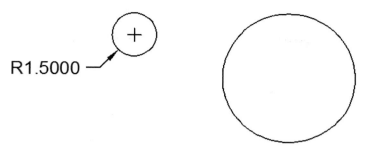

### Converting Associative Dimensions to Constraints

4. Convert the associative dimension to a dimensional constraint by starting the **DCCONVERT** command as follows:
   - From the ribbon's Parametrics tab, click the **Convert** button.
   - Enter the **dcconvert** command, as follows:

     Command: **dcconvert**

5. At the prompt, pick the radial dimension that you drew earlier:

   Select associative dimensions to convert: *(Select the radial dimension.)*
   
   Select associative dimensions to convert: *(Press* **ENTER** *to continue.)*
   
   1 associative dimensions converted

Notice that a padlock icon appears and that the dimension text changes to **rad1=1.5**. (The number you see may differ.)

The word "rad" is short for *radial*, and the "1" means that this is the *first* radial dimensional constraint in the drawing.

The other item to note is that the dimension style changes to one unique to dimensional constraints: its color is gray and the scale is annotative, which means the arrowhead and text display at the same size, no matter the zoom level of the drawing. You cannot edit this dimension style.

### Placing Dimensional Constraints

6. Place a constrained radial dimension on the second circle. Start the **DCRADIUS** command, as follows:
   - From the ribbon's Parametrics tab, choose **Radius**.
   - Enter the **DCRADIUS** command:

     Command: **dcradius**

7. Follow the prompt to place the dimensional constraint:

    Select arc or circle: *(Pick the second circle.)*

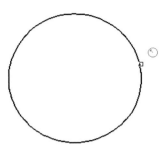

Notice the icon that appears near the cursor. It reminds you that you are placing a radial constraint.

   Specify dimension line location: *(Pick a point to position the dimension leader.)* *(Press* ENTER *to end the command.)*

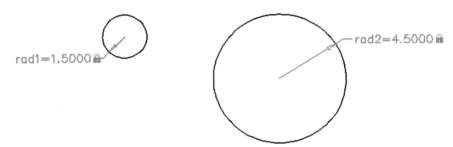

The dimension text reads "rad2," meaning this is the second radial constraint.

**Note** AutoCAD automatically assigns names according to the type of dimensional constraint, which you may edit with different names. The predefined names are as follows:

| Name | Meaning |
|---|---|
| d | Distance (length) |
| rad | Radius |
| dia | Diameter |
| ang | Angle |
| user | User-defined |

## Changing Object Size Through Constraints

8. Try changing the radius of one of the circles:

    a. Select the second circle, the one attached to **rad2**.

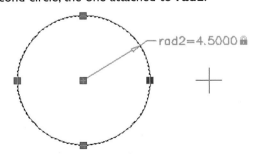

b. Select a grip, and then attempt to drag it to make the circle larger. Notice that the circle won't change. That's because the dimensional constraint locks the radius, preventing you from changing the circle's size — as indicated by the padlock icon.

**Note** To "unlock" the circle's radius temporarily, follow these steps:
a. Select the circle, and then click one of the quadrant grips; it turns red.
b. *Tap* the **Ctrl** key while the cursor hovers over the grip.
c. Move the cursor; do not click or drag the grip!
d. Pick the new location. Notice that the circle's movement is not ghosted, as is normally the case.

The circle changes size, and the dimensional constraint updates its value. Expressions are converted to numerical values.

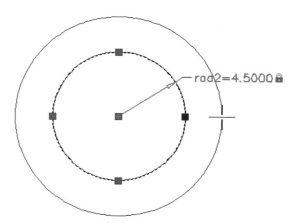

AutoCAD reacts differently when you *hold down* the **CTRL** key while *dragging* the grip: it makes a copy of the circle, complete with constraint(s). The constraint identifier is updated, such as **rad3** or **rad4**.

The **CTRL** key similarly affects all other types of dimensional constraints.

9. To change the circle's radius, you edit the value of the dimensional constraint:
   a. Double-click the text of the dimensional constraint. Notice that you can edit it.

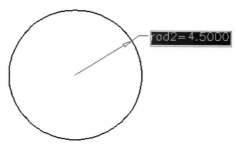

   b. Change the radius to a smaller radius, leaving alone the "rad2=" prefix.

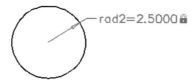

   c. Press **ESC** to exit editing mode. Notice that the circle changes its radius. (You can use the **TEXTEDIT** command to edit text of dimensional constraints.)

**Note** The text displayed by dimensional constraints comes in two parts: a *name* and an *expression*.
- *Name* identifies the constraint dimension (such as **rad2**).
- *Expression* is a value that reports the distance (such as **=2.5**).
- *Parameter* is the name (rad2) and expression (=2.5) together, like this: **rad2=2.5**. It can also be a formula that calculates the distance, such as **dia1 = 2 x dia2** — one circle's diameter is 2 times the diameter of another.

The Constraint Settings dialog box determines whether the drawing shows names, expressions, or both.

Leave the drawing open for the next tutorial.

As you can see from this tutorial, the word "constraint" and the padlock icon are somewhat misleading, for both imply that dimensional constraints are locked up tight. They are not, for they can be edited both directly and indirectly through the Parameters Manager palette.

### TUTORIAL: LINKING DIMENSIONAL CONSTRAINTS

In the previous tutorial, you saw how dimensional constraints control the size of objects; constraints can also link size *between* objects. (Geometric constraints link the *position* of objects.)

You continue the tutorial from above by linking the radius of one circle with the other: when you change one, the other will too.

1. Double-click the text of the **rad1** circle.

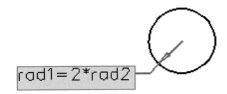

2. Edit the text to read:
   rad1=**2*rad2**

3. Press **ENTER**. Notice that one circle is half the size of the other: circle **rad1** is twice the radius of second circle.

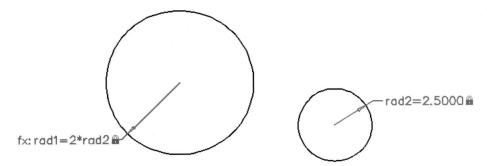

4. Edit the value of the second circle, such as doubling its value, and then press **ENTER**. Notice that the first circle again changes size to accommodate the second one.

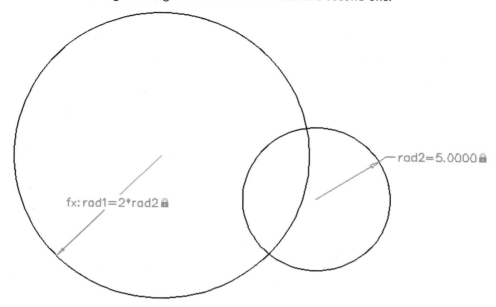

5. Move one of the circles through dragging its center grip. Notice that you can move it, and that the radius (size) remains linked.

### TUTORIAL: LINKING OBJECTS WITH CONSTRAINTS

In the previous tutorial, you saw how one circle can control the size of the other through the use of parametrics (**rad1=2\*rad2**). In this tutorial, you link the two circles with a linear dimension, so that they maintain a constant distance apart.

1. Place a constrained linear dimension between the two circles. Start the **DCLINEAR** command, as follows:
   - From the ribbon's Parametrics tab, choose ▫ **Linear**.
   - Enter the **DCLINEAR** command:

     Command: **dclinear**

2. Hover the cursor over one circle. Notice the two icons that appear in the drawing:

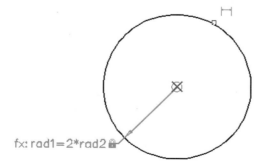

   - Near the cursor is a ⊢⊣ blue linear constraint icon. Its purpose is to remind you of the current dimensional constraint mode.
   - At the center of the circle a ⊠ red marker appears. It previews the linear constraint's starting point.

3. Click the circle:
   Specify first constraint point or [Object] <Object>: *(Choose a circle.)*
4. Repeat for the other circle:
   Specify second constraint point: *(Choose another circle.)*

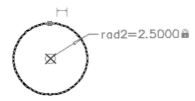

5. Pick a point to locate the dimension line, and then press **ENTER** to end the command:
   Specify dimension line location: *(Pick a point.)*
   Dimension text = 10.5000 *(Press* **ENTER.***)*

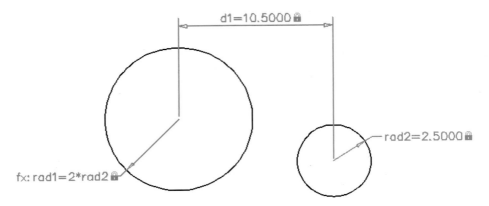

6. To see that the circles are now linked, move one of them. Notice that the other circle travels along.

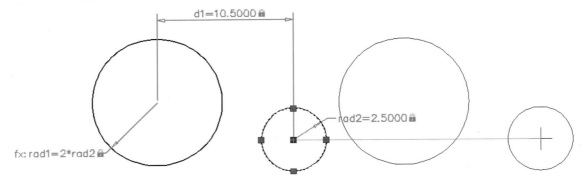

Linear dimensional constraints are identified by **d**, as in **d1**=10.5. The "d" is short for *distance*.

7. When you turn the linear constraint into a parameter, then you can make the distance a function of the size of the circles, like this:
   a. Double-click the text of the linear constraint.

b. Change the text to read as follows:

   d1=rad2/2

   This makes the **d1** distance between the circles equal to one-half the size of circle **rad2**.

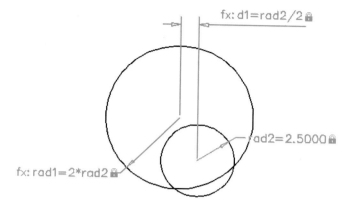

8. Change the value of **rad2** to see the other circle and the distance between them change.

## DELCONSTRAINT

The DELCONSTRAINT command erases dimensional constraints from drawings. (It also erases geometric constraints.)

Constraints can also be removed with the ERASE command, but cannot be exploded with the EXPLODE or XPLODE commands. If the constraints are referred to by others, AutoCAD removes the links.

### TUTORIAL: ERASING CONSTRAINTS

1. To remove dimensional constraints from the drawing, start the **DELCONSTRAINT** command, as follows:
   - From the ribbon's Parametric tab, choose **Delete Constraints**.
   - Enter the **delconstraint** command, as follows:

      Command: **delconstraint**
   - Alternatively, enter the **delcon** alias.

2. At the prompt, choose one of the radial dimensions from the previous tutorial:
   All constraints will be removed from selected objects...
   Select objects: *(Choose one of the radial dimensions.)*
   Select objects: *(Press ENTER to exit the command.)*
   1 constraint removed

You can enter **All** at the 'Select objects' prompt to remove all constraints from the drawing. If you make a mistake, reverse it with the U command.

## $fx$ PARAMETERS

The **PARAMETERS** command creates links between dimensional constraints through the Parameters Manager palette.

Parameters are interactive formulas that control the size and position of objects. When you change the name of a parameter, AutoCAD automatically changes all related occurrences.

The Parameters Manager palette defines and edits parameters associated with dimensional constraints. Through it, AutoCAD maintains the relationships between dimensional constraints correctly. (There is no relationship between dimensional and geometric constraints.)

The palette also allows you to edit parameters. You can change names, expressions, and values; types cannot, however, be edited.

### TUTORIAL: EDITING PARAMETER VALUES

In this tutorial, you apply dimensional constraints to a rectangle, and then edit them through the Parameters Manager palette.

1. Begin this tutorial by drawing a rectangle with the **RECTANGLE** command.
2. Apply dimensional constraints to two sides of the rectangle with the ribbon's Parametrics tab: **Horizontal** for the horizontal one, and **Vertical** for vertical.

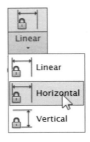

```
Command: _DcHorizontal
Specify first constraint point or [Object] <Object>: o
Select object: (Pick a horizontal on the rectangle.)
Specify dimension line location: (Pick a point to locate the dimension line.)
Dimension text = 13.0000 (Press ENTER.)
```

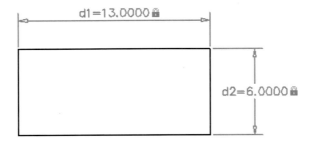

3. Open the Parameters Manager palette with the **PARAMETERS** command.
   - In the ribbon's Parametric tab, choose **Parameters Manager**.
   - Enter the **PARAMETERS** command:

      Command: **parameters**

Notice the palette, in which the dimensional constraints appear as **d1** and **d2**.

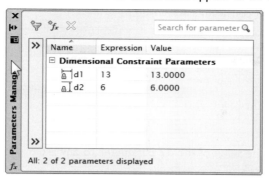

4. Change the expression of **d1** to this formula:

   **d2*2**

   This means that width **d1** is always twice the height **d2**.

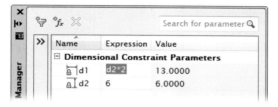

Notice that the parameter changes for **d1**:

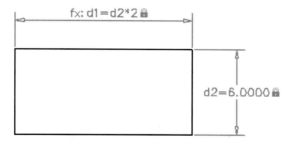

5. Select the rectangle, and then move a grip to change the shape of the rectangle.

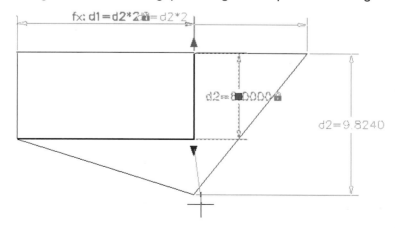

Notice that the parameters update in the palette. This shows that the parameters in the palette and dimensional constraints are linked.

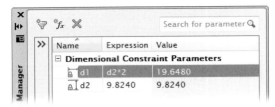

6. In the palette, change the value of the **d2** parameter to any arbitrary number. Notice that the rectangle changes its size automatically.

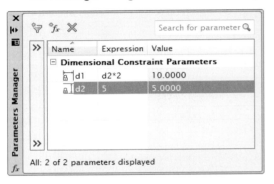
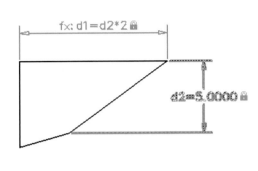

Also notice that AutoCAD highlights the dimensional constraint linked to **d2** in the Parameters Manager palette.

The -PARAMETERS command operates at the command line, creating new expressions, as well as editing, renaming, and deleting them.

Let's examine the palette in greater detail.

### Parameters Manager Palette

The Parameters Manager palette contains the following toolbar buttons, headers, and shortcut menu.

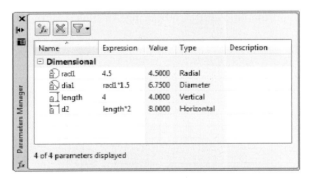

**Add** creates new user-defined expressions. The default names look like **user1** and can be renamed by you. (Renaming is almost mandatory in large drawings.) You can enter numeric values (such as 4.5), expressions (such as **rad1*1.5**), or formulas (such as **length*2**). Do not use the equals sign ( = ). See the boxed text "Expressions in Parameters" for a list of all expressions supported by AutoCAD.

**Remove** deletes the selected expression.

**Filter** filters the display of parameters; displays the Filters pane. Filters let you create groups of related parameters.

**Search** searches for parameter names; this is useful for complex drawings that contain many dimensional constraints.

## Headers

Click the name of a header to sort its content alphabetically; click again to sort in reverse order.

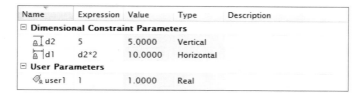

**Name** heads the list of variables names. Although AutoCAD assigns names automatically, you can give them names more meaningful than "d2" and "dia1" — such as "Length" and "Diameter."

**Expression** specifies the parametric expression, and consists of constants, names, and expressions (formulae). See boxed text.

**Value** reports the value of the expression.

**Type** reports the type of parameter, such as Vertical or Real (not user editable).

**Description** provides space to describe the parameter. Describing the purpose of the parameter to make it easier for others to follow your reasoning.

## Shortcut Menu

Double-click an expression, and then right-click to see the following shortcut menu:

**Cut** copies the expression to the Clipboard, and then deletes the expression.

**Copy** copies the expression to the Clipboard.

**Paste** pastes the contents from the Clipboard into the field.

**Delete** deletes the expression.

**Expression** provides a submenu of functions. In addition to those listed by the submenu, you can also use algebraic expressions, including +, -, *, and /. See boxed text.

## EXPRESSIONS IN PARAMETERS

The following expressions can be used in parameters. To access a list of the expressions, double-click an Expression text entry box, and then right-click.

| Expression | Meaning |
|---|---|
| + | Add |
| - | Subtract |
| / | Divide |
| * | Multiply |
| cos | Cosine |
| sin | Sine |
| tan | Tangent |
| acos | Arccosine |
| asin | Arcsine |
| atan | Arctangent |
| cosh | Hyperbolic cosine |
| sinh | Hyperbolic sine |
| tanh | Hyperbolic tangent |
| atanh | Hyperbolic arctangent |
| sqrt | Square root |
| sign | Changes positive to negative and vice versa |
| exp | Exponent |
| floor | Rounds down |
| ceil | Ceiling (rounds up) |
| round | Rounds to the nearest integer, upwards or down |
| abs | Absolute value (makes negative numbers positive) |
| max | Maximum of two expressions |
| min | Minimum of two expressions |
| ln | Natural logarithm |
| log | Logarithm |
| pow | Raises one expression to the power of the second expression |
| exp10 | Exponent to the power of 10 |
| r2d | Converts radians to degrees |
| d2r | Converts degrees to radians |
| trunc | Truncates (removes decimal portion of number) |
| PI | 3.141... |
| E | 2.17... |

```
cos(expr)
sin(expr)
tan(expr)
acos(expr)
asin(expr)
atan(expr)
cosh(expr)
sinh(expr)
tanh(expr)
acosh(expr)
asinh(expr)
atanh(expr)
sqrt(expr)
sign(expr)
exp(expr)
floor(expr)
ceil(expr)
round(expr)
abs(expr)
max(expr1;expr2)
min(expr1;expr2)
ln(expr)
log(expr)
pow(expr1;expr2)
exp10(expr)
r2d(expr)
d2r(expr)
trunc(expr)
PI
E
```

### Undocumented Expression

| random | Returns a random number between 0 and 1. |
|---|---|

## CONSTRAINTSETTINGS

The CONSTRAINTSETTINGS command specifies default conditions for dimensional and geometric constraints through a dialog box. In this chapter, we look at the dialog box's Dimensional tab of options; the next chapter covers the remaining two tabs.

1.  To set options for dimensional constraints, start the **CONSTRAINTSETTINGS** command, as follows:

    - From the 2D ribbon's Parametric tab, click the arrow.
    - Enter the **constraintsettings** command, as follows:

        Command: **constraintsettings**

    - Alternatively, enter the **csettings** alias.

2.  If you entered the command name, then you may need to choose the **Dimensional** tab.

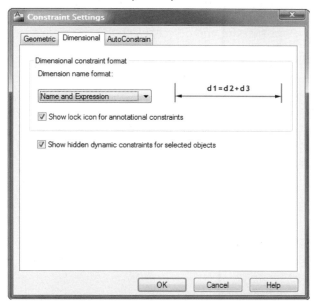

**Dimension Name Format** — droplist displays (1) name only, (2) value only, or (3) name and expression, as illustrated below (**CONSTRAINTNAMEFORMAT**).

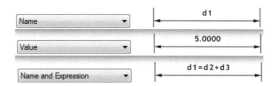

*From top to bottom:*
Name only, value only, name and expression.

**Show Lock Icon for Annotational Constraints** — toggles the display of the padlock icon (**DIMCONSTRAINTICON**).

**Show Hidden Dynamic Constraints for Selected Objects** — toggles the display of hidden constraints between displayed and not displayed when related objects are selected (**DYNCONSTRAINTMODE**).

3.  Change settings, and then click **OK**.

## CONSTRAINTS SETTINGS: ADDITIONAL METHODS

The Dimensional panel of the ribbon's Parametrics tab contain a few more buttons worth knowing about.

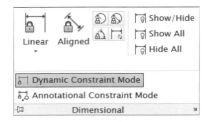

The Show and Hide buttons toggle the display of dimensional constraints:

**Show/Hide** — prompts you to select objects, and then prompts you to decide if they should be visible (show) or invisible (hide):

Select objects: *(Choose one or more objects or dimensional constraints.)*
Select objects: *(Press* ENTER *to end object selection.)*
Enter an option [Show/Hide]<Show>: *(Type* **S** *or* **H**.*)*

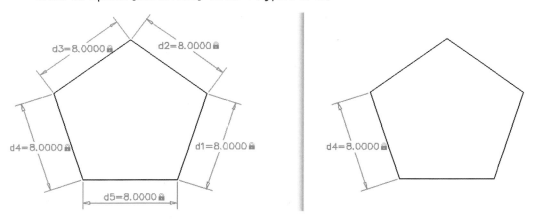

**Show All** — displays all dimensional constraints.

**Hide All** — hides all dimensional constraints.

The **Constraint Mode** buttons switch the form between dynamic and annotational:

**Dynamic** — constructs dynamic dimensional constraints.

**Annotational** — constructs annotational dimensional constraints.

## EXERCISES

1. Draw a line of any length and angle, and then apply a horizontal dimensional constraint to line **d1** — similar to the one illustrated below:

2. Continuing with the same drawing:
   a. Draw a circle.
   b. Use dimensional constraints to connect the size of objects. Apply a diameter constraint to circle (**dia1**).
   c. Edit the circle's text to reference the line:

   **dia1=d1/3**

   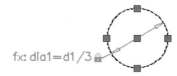

   d. Write out the meaning of this formula:

3. Continuing with the drawing from the previous exercise:
   a. Resize the line by selecting the dimensional constraint, and then dragging a triangular handle. Describe what happens:

   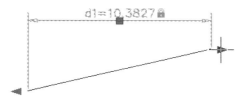

   b. Resize the circle, and then describe what happens.

4. Select the circle, and then drag one of its grips. Describe what happens:

## CHAPTER REVIEW

1. What is the purpose of dimensional constraints?
2. List three ways in which dynamic dimensional constraints differ from regular dimensions:
    a.
    b.
    c.
3. What does the padlock icon indicate?
    What do parentheses around dimension text indicate?
4. Describe the differences between the following forms of dimensional constraints:
    a. Dynamic
    b. Annotational
    c. Reference
    d. Parametric
5. Can dimensional constraints switch between forms?
    If so, how?
6. Can regular dimensions be converted to dimensional constraints?
    If so, how?
7. Define "overconstrained."
8. Describe what happens when you add a dimensional constraint to a fully constrained drawing.
9. Match the following abbreviations:
    a. ang           i. distance
    b. rad           ii. diameter
    c. dia           iii. user-defined
    d. d             iv. radius
    e. user          v. angle
10. When objects are constrained, are they permanently fixed in place?
11. How can you unlock a constraint temporarily?
12. What happens when you hold down the **CTRL** key while dragging a grip of an constrained object?
13. Write out the formulas for the following constraint conditions:
    a. Link the length of line 1 to the radius of arc 2.
    b. Make the diameter of circle 3 half the length of line 4.
    c. Make the length of line 5 four times the length of line 6.
    d. Link the radius of circle 7 to the diameter of circle 8.
14. When entering formulas into the Parameters Manager palette, when would you use the equals ( = ) sign?
15. Name the ribbon tab that contains commands for placing dimensional constraints?
16. Can constraints be erased?
    If so, how?

17. Match the text displayed by the dimensional constraint with the type of text:

a. 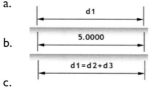 i. Value

b. ii. Name and expression

c. iii. Name

# CHAPTER 14

## Using Geometric Constraints

Dimensional constraints control the sizes of objects, as you saw in the previous chapter. Related to them are geometric constraints, which control the position of objects in drawings and the location of objects relative to each other.

In this chapter, you learn the following commands:

> **AUTOCONSTRAIN** applies geometric constraints automatically.
> **GC...** commands apply geometric constraints to selected objects (new to AutoCAD 2011).
> **CONSTRAINTSETTINGS** specifies default conditions for automatic and geometric constraints.
> **CONSTRAINTBAR** toggles the visibility of constraint bars.
> **CONSTRAINTINFER** toggles the inference of geometric constraints during drawing and editing (new to AutoCAD 2011).

---

**NEW TO AUTOCAD 2011** IN THIS CHAPTER
- Geometric constraints can now be inferred during some drawing and editing commands.
- Each constraint now has its own command.

## ABOUT GEOMETRIC CONSTRAINTS

*Geometric constraints* keep objects together or in place.

For example, when you apply a Fix constraint to an object, it is pinned to a fixed position in the drawing — like a pin in a piece of paper. Or when you apply a Concentric constraint to two circles, they remain concentric. When one circle is moved, and the other one moves with it, as illustrated below:

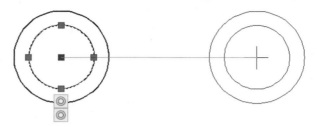

Geometric constraints are like "sticky" object snaps. For instance, the Tangent osnap exists only for the split-second interval in which AutoCAD finds and then applies the tangency location; in contrast, the Tangency constraint forces lines to remember that they are tangent to arcs — even when the two do not touch, because they are connected through the constraint.

There some geometric constraints that match object snap modes, such as Intersect and Perpendicular. Then there are a few constraints with no osnap equivalent, such as Equal and Colinear. In total, AutoCAD provides these geometric constraints:

| Geometric Constraint | Command | Equivalent Drafting Mode |
|---|---|---|
| Horizontal | GCHORIZONTAL | Ortho mode |
| Vertical | GCVERTICAL | Ortho mode |
| Perpendicular | GCPERPENDICULAR | PERpendicular osnap |
| Tangent | GCTANGENT | TANgent osnap |
| Coincident | GCCOINCIDENT | ENDpoint osnap |
| Concentric | GCCONCENTRIC | CENter osnap |
| Parallel | GCPARALLEL | PARallel osnap |
| Symmetric | GCSYMMETRIC | MIRROR command |
| Smooth | GCSMOOTH | ... |
| Colinear | GCCOLLINEAR | ... |
| Equal | GCEQUAL | ... |
| Fix | GCFIX | ... |

AutoCAD's constraints operate primarily in 2D; programs like Autodesk's Inventor also have 3D constraints. Spline frames cannot be constrained.

Objects do not need to touch for geometric constraints to work. For example, a line in a front view can have equal length to a line in the top view, so that changing one line changes the other; one end of the front-view line can always be vertical to the end top-view line. In this way, the views stay in step.

Geometric and dimensional constraints work together, and together are known as *parametrics*. (Recall from the previous chapter that dimensional constraints define the size of geometric features.) Sometimes, the two types of constraints can replace each other. For example, the Perpendicular geometric constraint is the same as an Angular dimensional constraint set to an angle of 90 degrees.

AutoCAD does not allow you to add *overconstrain*, or add too many or conflicting constraints. Normally, you want drawings to be *fully constrained*, which means there are no missing constraints.

While individual constraints can be applied with the GEOMCONSTRAINT command, it is easier and faster to let the AUTOCONSTRAIN command find all possible geometric constraints for you, after which you remove and add individual ones as necessary. Once the geometric constraints are in place, you typically add the dimensional constraints described in the previous chapter.

# AUTOCONSTRAIN

The **AUTOCONSTRAIN** command automatically constrains selected objects.

This command is applied after objects are drawn. It finds probable constraints between geometric objects that are within the tolerance distance specified by the Constraint Settings dialog box. (The dialog box also controls which constraints are applied, and in which order of priority.)

**TUTORIAL: CONSTRAINING GEOMETRY AUTOMATICALLY**

In this tutorial, you open an existing drawing, and then automatically constrain all geometry.

1. Open the *geom-constraint.dwg* file, a drawing of a part made of lines and circles.

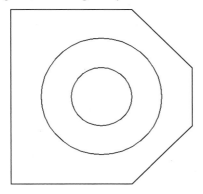

2. To apply geometric constraints to the entire drawing, start the **AUTOCONSTRAIN** command, as follows:
   * In the ribbon's Parametric tab, choose the **Auto Constrain** button.
   * Enter the **autoconstrain** command:

      Command: **autoconstrain**

3. At the prompt, enter 'All' to select all objects:
   Select objects or [Settings] <Settings>: *(Enter **all** to select all objects.)*
   Select objects or [Settings] <Settings>: *(Press ENTER.)*

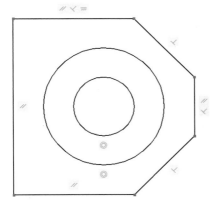

Notice the many toolbar-like "tags" that appear next to the lines and circles. These are called "constraint tags"; their icons report the constraint applied to adjacent geometry.

**Notes** You don't need to select everything in the drawing; you can select specific objects using any selection mode.

This command's **Settings** option displays the Constraint Settings dialog box's AutoConstrain tab, as described later.

Constraint bars can be turned off with the **CONSTRAINTBAR** command, or through buttons on the ribbon's Parametric tab.

4. There are many constraints in the drawing! (I count 11.) Most of them connect at least two objects, and so it becomes difficult to determine the connections, even in a drawing as simple as this one. For this reason, AutoCAD lets you view objects related to constraints in two ways: by constraint and by object.

   a. **By Constraint.** First, view constrained objects by constraint. Hover the cursor over one of the Concentric tags, as shown below.

   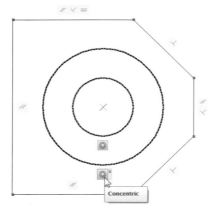

   Notice that AutoCAD displays the following information:
   - The two ⊚ concentric tags become opaque to help them stand out from other tags. The two relate to one another, because the inner circle is concentric with the outer one, and vice versa.
   - The two circles constrained by concentricity are highlighted, and the center point of concentricity is indicated by the X.

   b. **By Object.** Alternatively, you can view connected objects by object. Hover the cursor over the short vertical line, as shown below.

   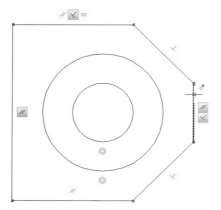

Notice that AutoCAD highlights the related constraints: the ⟨⟩ perpendicular and ∥ parallel constraints on this and other lines.

5. To prove that constraints hold the otherwise-individualistic circles together, try moving one of them by dragging it by its grips.

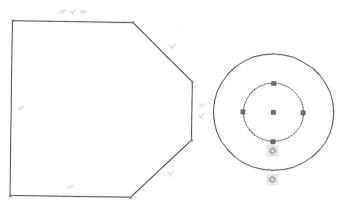

Notice that the two circles move as a whole — independently of the polygon. (The polygon and circles are not constrained to each other, nor are they fixed in place. See the exercises at the end of this chapter.)

6. When you attempt to move one of the lines making up the polygon, it stays connected to the others. The polygon, however, may become distorted, depending on how constraints were applied.

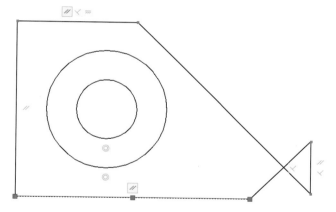

7. You can unlock constraints temporarily by following these steps:
   a. Select an object, such as one of the lines.
   b. Click a grip so that it turns red, and then hover the cursor over the grip.
   c. Tap the **CTRL** key; don't hold it down *.
   d. Move the cursor, and then click for the new location. Notice that the line separates from the rest of the polygon.

   *) AutoCAD reacts differently when you hold down the **CTRL** key while dragging the grip: a copy of the object is made, complete with constraint(s).

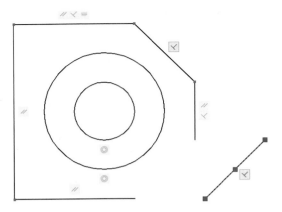

## TUTORIAL: MODIFYING GEOMETRIC CONSTRAINTS

After AutoCAD adds constraints to drawings automatically, you may wish to modify them. The types of modifications available to you include hiding and deleting constraints, or adding new ones. Unlike dimensional constraints, geometric constraints cannot be edited — because there is no need to.

1. To hide all constraint bars, click the **Hide All** button found in the ribbon's Parametric tab. (Alternatively, you can enter the **CONSTRAINTBAR** command, and then specify the **Hideall** option; the **Showall** option displays them all again.)
2. To hide an individual constraint bar, click the **x** in its upper right corner.

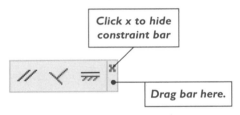

3. To move a constraint bar, drag it around by its tab. (See figure above.)
4. To delete an individual constraint from the bar, right-click it, and then choose **Delete** from the shortcut menu.

   (As an alternative, you can hover the cursor over it until it turns blue, and then press **DEL**.)
5. To delete all constraints associated with an object, use the **DELCONSTRAINT** command. See the previous chapter.
6. To add more constraints, see the **GEOMCONSTRAINT** command described below.

 **Notes** The **DELCONSTRAINT** command erases all geometric and dimensional constraints from a selection set of objects — both visible and hidden.

The Parameters Manager palette (**PARAMETERS** command) does not apply to geometric constraints.

# GEOMCONSTRAINT

The GEOMCONSTRAINT command applies geometric relationships between objects or to points on objects.

## TUTORIAL: APPLYING CONSTRAINTS MANUALLY

In this tutorial, you force the circle to remain centered in the keyhole shape with the GEOMCONSTRAINT command's **Concentric** option.

1. Begin AutoCAD with a new drawing.
2. Ensure that the (or INFER) **Inferred Constraints** button is turned off (looks gray). It's the first button on the status bar.
3. Draw a shape with the **PLINE** command similar to the figure below. With the **CIRCLE** command, add a circle; its location in the drawing is unimportant.

4. Start the **GEOMCONSTRAINT** command, as follows:
   - In the ribbon's Parametrics tab, choose the **Concentric** button.
   - Enter the **gcconcentric** command:

      Command: **gcconcentric**

5. AutoCAD prompts you to select an object. Choose a polyarc at one end of the polyline.
    Select first object: *(Choose a polyarc.)*

6. At the next prompt, choose the circle.
    Select second object: *(Choose the circle.)*

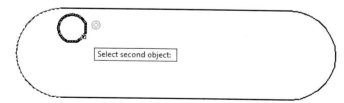

Notice that AutoCAD displays constraint bars to indicate the type and approximate location of the applied constraint.

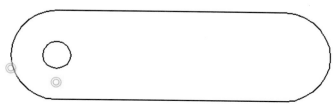

7. With the constraint in place, let's see what happens when the objects are moved. First, use the **ROTATE** command to rotate the keyhole about one end. (The base point is centered on the polyarc furthest from the circle). Notice that the circle moves with the keyhole, due to the concentric constraint.

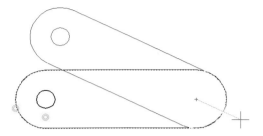

8. Now drag the circle. Notice that the keyhole stretches itself to keep the large polyarc centered around the circle.

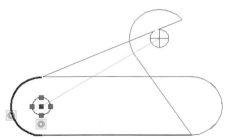

### GEOMETRIC CONSTRAINT CATALOG

AutoCAD provides many types of geometric constraints. They are applied through commands that begin with GC, or through buttons on the ribbon. Below, all of them are listed in alphabetical order.

Be aware that some constraints act differently, depending on the option you select. For example, the Fix constraint allows the selected object to be scaled and rotated about a lock point — when you use its 'Select point:' option. But when you select the same object with Fix's 'Object:' option, it is truly fixed in place: it cannot be scaled, rotated, or moved.

#### GcCoincident

The **GCCOINCIDENT** command places the **Coincident** constraint. It forces a point of the second object to lock to an end point, midpoint, or any other point you pick on the first object, in a manner similar to Endpoint, Midpoint, and Nearest object snaps. The objects can be lines, polylines, polyarcs, circles, arcs, ellipses, splines, and any two valid points on an object.

Select first point or [Object/Autoconstrain] <Object>: *(Pick a constraint point, or enter an option.)*

Select second point or [Object]<Object>: *(Pick another constraint point or type* **O**.*)*

**First point** — specifies the endpoint of the first object.

**Second point** — specifies the endpoint of the second object, which will be made coincident to the first one, as illustrated below.

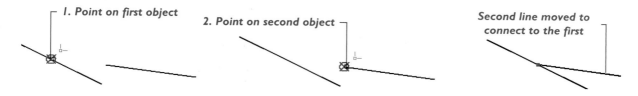

**Autoconstrain** — constrains multiple objects automatically like the **AUTOCONSTRAIN** command.

**Object** — selects an object instead of a constraint point; prompts you:
Select object: *(Choose an object.)*
Select point or [Multiple]: *(Pick a point on the object, or type* **M**.*)*

**Multiple** — selects more than one additional point to coincide with the first.

> **Note** The **Object** option can be used to slide a circle or polygon along a line. It constrains the center of the circle to the line linearly.

##  GcCollinear

The GCCOLLINEAR command places the **COLlinear** constraint. (Notice that it is spelled with two Ls.) It forces one object to lie in the same line as a second. It works with lines, polyline segments, ellipses, and multiline text, and displays the same prompts as GCEQUAL.

**Left:** Two lines selected by GCCOLLINEAR command.
**Right:** Second line made collinear to the first.

## ◎ GcConcentric

The GCCONCENTRIC command places the **CONcentric** constraint. It forces two curved objects to share the same center point, like Center osnap mode. It works with circles, arcs, polyarcs, and ellipses. The prompts are the same as those displayed by GCPERPENDICULAR:

Select first object: *(Choose the reference object.)*
Select second object: *(Choose the object to be constrained.)*

**Left:** Two circles selected by GCCONCENTRIC command.
**Right:** Second circle made concentric to the first.

The second object is made concentric to the first one.

## = GcEqual

The GCEQUAL command places the **Equal** constraint. It forces one line or polyline segment to be the same length as a second; also, it forces an arc, polyarc, or circle to have the same radius as a second one:

   Select first object or [Multiple]: *(Choose an object, or type **M**.)*
   Select second object: *(Choose a second object.)*

**First object** — chooses the reference object.

**Second object** — chooses the object to be constrained, and then scales it to be the same size as the first.

**Multiple** — selects additional objects to be constrained by repeating the 'Select object to make equal to the first' prompt until you press **ENTER**.

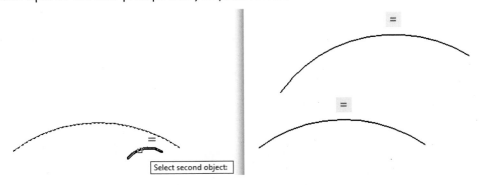

**Left:** *Two arcs selected by* GCEQUAL *command.*
**Right:** *Second arc made the same size as the first.*

The second object is resized to be the same size as the first one.

## GcFix

The GCFIX command places the **Fix** constraint. It locks objects in position. This constraint works with all objects.

   Select point or [Object] <Object>: *(Pick a point on an object, or type **O**.)*

**Select point** — allows the object to be scaled (resized) and rotated about the lock point.
**Object** — locks the object in place; it cannot be resized, rotated, or moved.

The object is fixed in place, depending on how it was selected, by point or by object.

## GcHorizontal

The GCHORIZONTAL command places the **Horizontal** constraint. It forces geometry to be horizontal, like ortho mode:

   Select an object or [2Points]: *(Choose an object, or type **2P**.)*

**Object** — forces the selected object to be horizontal. If the object is an arc, then its endpoints are extended to make the two pick points horizontal.

**2Points** — picks two points on an object; the orientation of the two pick points is made horizontal, as illustrated below. This is useful for orienting an arc or polyline. Works with lines, polyline segments, arcs and circles.

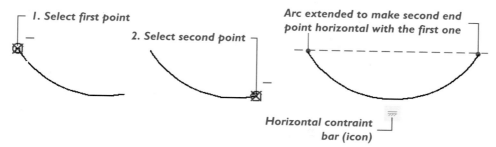

The two points need not be on the same object.

## GcParallel

The GCPARALLEL command places the **PArallel** constraint. It forces one object to be parallel to a second, like the Parallel osnap. It works with lines, polyline segments, ellipses, and multiline text.

Select first object: *(Choose the reference object.)*
Select second object: *(Choose the object to be constrained.)*

**First object** — picks the reference object.
**Second object** — chooses the objects to be made parallel.

**Left:** *Two objects selected by* GCPARALLEL *command.*
**Right:** *Mtext made the same angle as the line.*

The second object is rotated to be parallel to the first one.

## GcPerpendicular

The GCPERPENDICULAR command places the **Perpendicular** constraint. It forces the first object to be perpendicular to the second, like Perpendicular osnap. The second is rotated to be perpendicular to the first, as illustrated below. Works with lines, polyline segments, ellipses, and multiline text.

Select first object: *(Choose the reference object.)*
Select second object: *(Choose the object to be constrained.)*

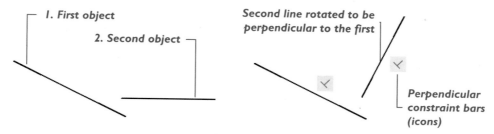

The second object is made perpendicular to the first one.

## GcSmooth

The **GCSMOOTH** command places the **SMooth** constraint. It forces the connection between a spline and another open curve to be smooth; endpoints become coincident. The first object must be a spline; the second can be another spline, a line, an arc, or a polyarc.

```
Select first spline curve: (Choose the reference spline.)
Select second curve: (Choose another object to be constrained to the first.)
```

**First curve** — must be a spline.

**Second curve** — can be another spline, a line, an arc, or a polyarc.

**Left:** *Spline and arc selected by* **GCSMOOTH** *command.*
**Right:** *Arc smoothly connected to the spline.*

The second object is constrained smoothly to the first.

**Note** AutoCAD applies a G2 curvature continuous condition with the SMooth constraint. "G2 curvature" specifies that the second derivative of the curves is equal at the point of contact. This ensures that the radius of curvature is continuous across the spline and arc or polyarc. The "G" is short for *geometry*; there are also G1 and G3 curves that apply the first (tangent) and third derivatives across curves and surfaces. G2 is the practical maximum in CAD.

## GcSymmetric

The **GCSYMMETRIC** command places the **Symmetric** constraint. It forces two objects to have mirror-like symmetry about a line of symmetry, like the **MIRROR** command. Depending on which of its options you select, the object selected second may be scaled to be the same size as the first. It works with lines, polylines and polyarcs, circles, arcs, and ellipses. Before starting this command, make sure you have a line of symmetry ready!

```
Select first object or [2Points] <2Points>: (Choose an object, or type 2P.)
Select second object: (Choose another object.)
Select symmetry line: (Pick an object.)
```

**First object** — chooses the object to be the reference.

**Second object** — chooses the object to be made symmetric, and then scales the first one to be a mirror image of the second.

**2Points** — chooses pick points on two objects; the objects maintain their size.

**Symmetry line** — chooses an object to use as the mirror (a.k.a. symmetry) line.

**Note** The symmetry constraint makes the angles of lines symmetrical; in arcs and circles, the center points and radii are made symmetrical. Polylines and other objects may become distorted to match the constraint conditions.

### GcTangent

The GCTANGENT command places the **Tangent** constraint. It forces one straight or curved object to be tangent to a curved object, like the Tangent osnap. Objects need not touch one another. This option works with line, polyline segments and polyarcs, circles, arcs, and ellipses. Displays the same prompts as GCPERPENDICULAR:

> Select first object: *(Choose the reference object.)*
> Select second object: *(Choose the object to be constrained.)*

The second object is made tangent to the first.

### GcVertical

The Vertical constraint forces geometry to be vertical, like ortho mode. Displays the same prompt as Horizontal:

> Select an object or [2Points]: *(Choose an object, or type **2P**.)*

The object is made vertical.

## CONSTRAINTINFER

The CONSTRAINTINFER system variable determines whether geometric constraints are applied automatically as you draw and edit. (The AUTOCONSTRAIN command applies constraints *after* you create a drawing; this system variable applies them *during*.)

Applying constraints as you draw is tricky work for AutoCAD, and so it follows a number of rules to determine when it is appropriate to add a geometric constraint to the object you just drew:

- AutoCAD applies constraints that match the object snap modes that are currently turned on. For instance, the Perpendicular constraint is applied to objects that PERpendicular osnap recognizes.
- The **LINE** and **PLINE** commands automatically apply the Coincident constraint to endpoints, and when the Close option is employed; osnap modes are ignored during these commands.
- The **RECTANG** command applies parallel and perpendicular constraints.
- The **FILLET** command applies tangent and coincident constraints between the arc and the lines.
- The **CHAMFER** applies coincident constraints between the lines.
- Some constraints are applied during editing operations. For example, a Coincident constraint is applied when you stretch a line so that it meets the endpoint of another. The Horizontal and Vertical constraints are applied during the **COPY**, **MOVE**, and **STRETCH** commands (or related grips editing) when object tracking is turned on.

The following constraints are not applied: Collinear, Equal, Fix, Smooth, and Symmetric.

The following object snaps are ignored by constraint inference: APParent intersection, EXTension, INTersection, and QUAdrant.

Autodesk notes that the following editing commands do not infer constraints: ARRAY, BREAK, EXTEND, MIRROR, OFFSET, TRIM, and SCALE.

## TUTORIAL: INFERRING CONSTRAINTS

In this tutorial, you practice drawing without and with inferred constraints.

1. Begin a new drawing.
2. Ensure that the status bar's (or INFER) **Inferred Constraints** button is off (looks gray).
3. Turn on ortho mode: click the (or ORTHO) button on the status bar (looks blue).

### Inferred Constraints on Lines

In this tutorial, you see the effect that ortho mode has on lines drawn without and with inferred constraints.

4. With the **LINE** command, draw a line of any size.
5. Grips-edit the line, as follows:
   a. Select the line; notice the blue grips.
   b. Select a grip at the end of the line; notice that it turns red.
   c. Drag the red grip; notice that the line changes its angle and length.

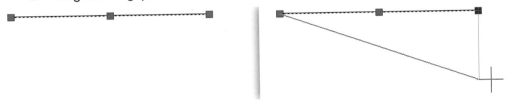

*Left:* Horizontal line drawn with ortho mode turned on.
*Right:* Line's angle and length changed by grips-editing.

6. Repeat the exercise with inferred constraints turned on:
   a. Click the (or INFER) **Inferred Constraints** button on the status bar (looks blue). Ensure ortho mode is still on.
   b. Draw a line. Notice the horizontal constraint.
   c. Use grips editing to stretch the line, as above. Notice that you can only move the line, and that it maintains its length and remains horizontal.

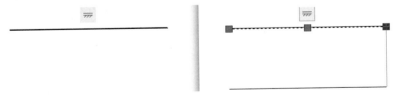

*Left:* Horizontal line drawn with inferred constraints and ortho mode turned on.
*Right:* Line's position changed by grips-editing.

When ortho mode is on, AutoCAD infers horizontal and vertical constraints on objects being drawn.

### Inferred Constraints on Rectangles

In this tutorial, you see the effect inferred constraints has on rectangles.

7. Turn off the status bar's ![icon] (or INFER) **Inferred Constraints** button (looks gray).
8. With the **RECTANG** command, draw a rectangle of any size.
9. Select the rectangle, and then use the grips to stretch it. Notice that the rectangle becomes distorted.

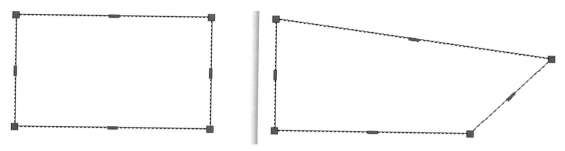

*Left: Rectangle drawn without constraints.*
*Right: Rectangle distorted by grips-editing.*

10. Click the ![icon] button to turn on inferred constraints.
11. Draw a new rectangle. Notice the ∥ parallel and ⊢ perpendicular constraint tags that appear.

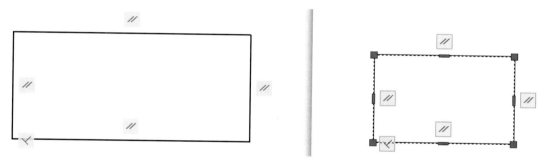

*Left: Rectangle drawn with constraints.*
*Right: Rectangle resized by grips-editing.*

12. Use grips to drag a corner of the rectangle. Notice that the grips force the rectangle to maintain its shape even as it changes its size.

When you draw with the **LINE**, **PLINE**, or **RECTANG** command, AutoCAD infers appropriate constraints. (Constraints are not applied to polylines drawn by the **POLYGON**, **DONUT**, or **SKETCH** commands.)

### Inferred Constraints on Fillets

In this tutorial, you see the effect that inferred constraints have on fillets.

13. Repeat the exercise with a fillet, as follows:
    a. Turn off inferred constraints.
    b. Draw two lines, and then apply a fillet with the **FILLET** command.
    c. Edit the fillet with grips; notice that the arc pulls away from the lines.

*Unconstrained fillet arc being moved.*

    d. Erase the fillet arc.
    e. Turn on inferred constraints.
    f. Reuse the **FILLET** command to apply another fillet to the two lines. Two Tangency constraint tags appear, as well as two small, blue, square dots, which indicate coincident constraints:

   ⌀ Tangency constraints keep the ends of the fillet arc tangent with the two lines.
   ■ Coincident constraints keep the ends of the arc attached to the lines' ends.

    g. Edit the fillet with grips; notice that this time the entire assembly moves — lines and arc together.

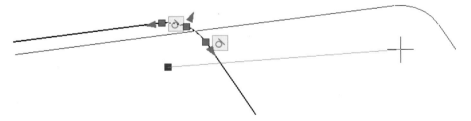

When you apply fillets and chamfers, AutoCAD infers appropriate constraints to keep the fillet arc and chamfer line in place.

# CONSTRAINTSETTINGS

The CONSTRAINTSETTINGS command specifies default conditions for geometric and dimensional constraints through a dialog box.

In this chapter, we look at the dialog box's Geometric and AutoConstrain tabs; the previous chapter covered the other tab for dimensional constraints.

1. To set options for dimensional constraints, start the **CONSTRAINTSETTINGS** command, as follows:

    - From the ribbon's Parametric tab, click the arrow. Clicking the arrow in the Geometric pane displays the dialog box with the Geometric tab.
    - Enter the **constraintsettings** command, as follows:

        Command: **constraintsettings**

    - Alternatively, enter the **csettings** alias.

2. Choose the **Geometric** or **AutoConstrain** tab.
3. Change settings, and then click **OK**.

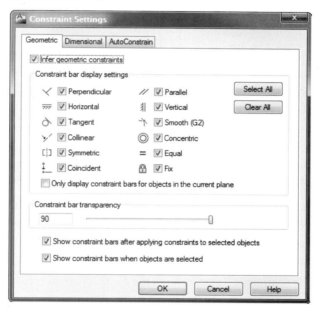

## Geometric Tab

The **Geometric** tab controls how geometric constraints are added to objects in drawings.

- **Infer Geometric Constraints** — toggles the use of inferred constraints. (This setting is stored in system variable **CONSTRAINTINFER**.)

    **Constraint Bar Settings** — toggles the display of individual geometric constraints (**CONSTRAINTBARMODE**). **Select All** turns on all geometric constraints; **Clear All** turns off all geometric constraints.

    **Only Display Constraint Bars for Object in the Current Plane** — toggles the display of constraints (1) always or (2) only when in the current UCS.

    **Constraint Bar Transparency** — adjusts the translucency of constraint bars from 10 (almost opaque) to 90, almost transparent (**CBARTRANSPARENCY**).

**Show Constraint Bars After Applying Constraints to Selected Objects** — toggles the default visibility of constraint bars between invisible and visible (**CONSTRAINTBARDISPLAY**).

**Show Constraint Bars When Objects are Selected** — toggles the display of constraint bars when you select objects. When you have constraint bars turned off, this option ensures you can see them when you select an object.

### AutoConstrain Tab

The **AutoConstrain** tab controls how geometric constraints are added automatically with the AUTOCONSTRAIN command. AutoCAD 2011 added Equal to the list of constraint types that can be automatically constrained.

**Priority** — specifies the order in which constraints are applied automatically with the AUTOCONSTRAIN command; use the **Move Up** and **Move Down** buttons to change the order. The check marks in the Apply column toggle individual geometric constraints.

**Select All** sets all geometric constraints as auto-constraints; **Clear All** turns off auto-constraining for all geometric constraints; **Reset** returns the AutoConstrain settings to default values.

**Tangent Objects Must Share an Intersection Point** — determines how objects must relate when applying the Tangent constraint:

    **On**: Tangent constraints are applied only when objects touch.

    **Off**: Tangent constraints are applied if objects would touch after being theoretically extended.

**Perpendicular Objects Must Share an Intersection Point** — determines how two objects relate for applying the Perpendicular constraint:

    **On**: Objects must touch.

    **Off**: Objects needn't touch.

**Tolerances Distance** — specifies the distance within which the constraints are applied automatically by the **AUTOCONSTRAIN** command.

**Tolerances Angle** — specifies the angle within which angle-related constraints are applied automatically.

## TUTORIAL: COMBINING DIMENSIONAL AND GEOMETRIC CONSTRAINTS

In this tutorial, you use dimensional and geometric constraints to create a piston simulator. One end of a shaft rotates about a crankshaft, the other end slides in and out of the cylinder. This mechanism is used by external combustion engines, such as steam locomotives, and internal ones, such as automobiles.

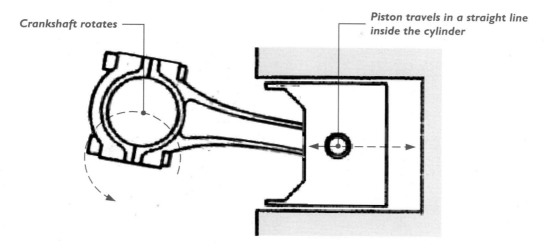

This mechanism translates linear motion (the piston) into rotational motion (the crankshaft).

This tutorial draws the three elements symbolically: the piston and cylinder are represented by lines; the crankshaft by a circle.

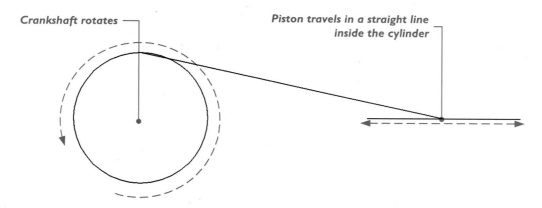

1. Begin a new drawing in AutoCAD.
2. With the **CIRCLE** and **LINE** commands, draw the three elements as shown by the figure. The size of the circle and lines does not matter; osnap is useful for attaching the diagonal line to the other two.

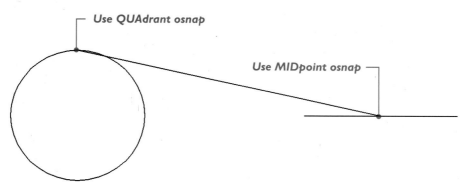

3. The right end of the diagonal line represents the piston inside the cylinder. It travels back and forth in a line. Use the Coincident constraint to fix the end of the diagonal line along the horizontal one, as follows:
   a. From the ribbon's Parametric tab, click the **Coincident** button.
   b. At the prompt, choose the horizontal line using the **Object** option. This is the reference object, along which the diagonal line will slide.
      Command: _GcCoincident
      Select first point or [Object/Autoconstrain] <Object>: o
      Select object: *(Pick horizontal line.)*
   c. At the second prompt, hover the cursor near the diagonal line's right end:
      Select point or [Multiple]: *(Pick right end of diagonal line.)*
   
   When you see the red marker appear, click!

   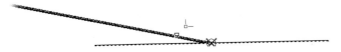

   Notice the blue square; it indicates the presence of the Coincidence constraint.

4. Fix the length of the diagonal line using Aligned dimensional constraint, as follows:
   a. From the Parametrics tab, choose **Aligned**.
   b. Using the **Object** option to select the diagonal line:
      Command: _DcAligned
      Specify first constraint point or [Object/Point & line/2Lines] <Object>: **o**
      Select object: *(Choose diagonal line.)*
      Specify dimension line location: *(Pick a point.)*
   c. Accept whatever length the line is:
      Dimension text = 20 3/8 *(Press* ENTER.*)*

   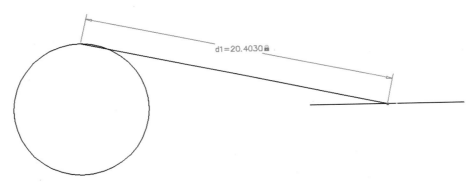

5. You can now try out the rudimentary beginnings of the piston action:
   a. Pick the diagonal line.
   b. Choose the blue grip at the line's left end.
   c. Drag the grip. As you do, notice that the right end travels along the horizontal line.
   d. Press ESC.

6. The next step is to fix the diagonal line's left end to the circle. This can be accomplished by again using the Aligned dimensional constraint. Fix the distance between the line and the circle's center point, as follows:

   a. Choose **Aligned** from the Parametrics tab.
   b. For the first constraint point, choose the diagonal line's left end.

   ```
   Command: _DcAligned
   Specify first constraint point or [Object/Point & line/2Lines] <Object>: (Pick the diagonal line's left end.)
   ```

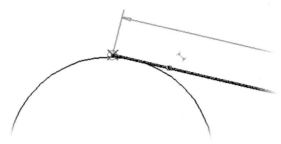

   c. For the second constraint point, choose the circle's center.

   ```
   Specify second constraint point: (Pick the circle's center.)
   ```

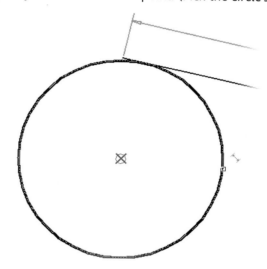

   d. And then place the dimension.

   ```
   Specify dimension line location: (Pick a point.)
   Dimension text = 4 1/2 (Press ENTER.)
   ```

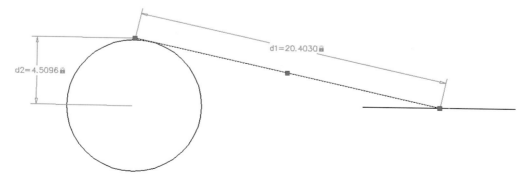

7. Select the diagonal line, and then move it by dragging the grip at its left end. Notice that the circle moves along. It needs to be fixed in place.
8. To fix objects in place, use the Fix constraint, as follows:
    a. From the Parametrics tab, choose 🔒 **Fix**.
    b. Use the **Point** option, not the Object option. (*Point* fixes the circle in place, but still allows it to rotate.) Hover the cursor over the circle; when the red marker appears at the circle's center point, click!

    Command: _GcFix

    Select point or [Object] <Object>: *(Pick the circle.)*

Notice the 🔒 Fix tag that appears near the circle.

9. The mechanism is now properly constrained. Test it by selecting the diagonal line, and then move it by dragging the grip at its left end. Notice that the left end travels around the circle, while the right end moves back and forth along the horizontal line.

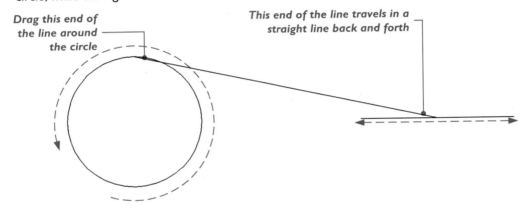

Drag this end of the line around the circle

This end of the line travels in a straight line back and forth

## EXERCISES

1. Open the *parametrics-2.dwg* file.

   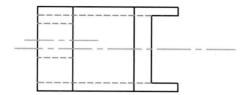

   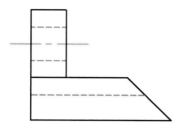 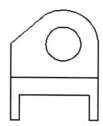

   a. Apply geometric constraints. Hints:
      - Apply geometric constraints first, and then apply dimensional ones.
      - Use AutoConstrain in small batches; otherwise you may have unintended results, such as diagonal lines forced perpendicular.
      - Initially accept all default values for dimensional constraints, and then edit them afterwards.
   b. Apply dimensional constraints.

2. In this exercise, you continue from the "Constraining Geometry Automatically" tutorial, found at the beginning of this chapter.
   a. Open the *geom-constraint.dwg* file.
   b. Apply the **AUTOCONSTRAIN** command.
   c. Recall that you were able to move the circles out of the polygon. What kind of a constraint would keep the circles positioned in the polygon? Apply the constraint to the drawing.
   d. What kind of constraint keeps the drawing from moving? Apply the constraint to the drawing.

## CHAPTER REVIEW

1. What is the purpose of geometric constraints?
2. How do geometric constraints differ from object snaps?
3. When are drawings full constrained?
4. Describe the primary difference between the **AUTOCONSTRAIN** and **GEOMCONSTRAINT** command.
5. Define *tolerance distance*:
6. What are *constraint tabs*?
7. List two ways that you can identify which constraints and objects are connected with each other:
    a.
    b.
8. Describe the function of the following constraints:
    a. CONcentric
    b. Equal
    c. Parallel
    d. Symmetric
9. Must objects touch to be constrained to each other?
10. Indicate the difference between using the Fix constraint's 'Select point' option and 'Object' option:
    a. Select point
    b. Object
11. What is the effect of the Horizontal constraint's '2Points' option on arcs?
12. Which object cannot be constrained?
13. How might you turn off the display of constraint bars?
14. Which type of constraint should be applied to drawings first — dimensional or geometric?

**Bonus Question**

15. The SMooth constraint applies G2 curvature between curves. Define the parts of the name "G2":
    a. G
    b. 2

# CHAPTER 15
## Reporting on Drawings

A benefit of drawing with CAD is that you can obtain information from the drawing. AutoCAD's inquiry commands provide the following kinds of information:

**ID** reports the x, y, z coordinates of single points.

**MEASUREGEOM** reports distances, angles, areas, and volumes.

**MASSPROP** reports the area, perimeter, centroid, and so on of regions and solids.

**LIST** and **PROPERTIES** report information about objects.

**TABLE** creates tables of data.

**TABLEDIT** edits data in tables.

**TABLESTYLE** specifies table styles.

**DWGPROPS** reports and stores information about the drawing.

**TIME** reports on time spent editing the drawing.

**SETVAR** reports the settings of system variables.

**Note** AutoCAD always works to 14 digits. All reporting commands display accuracy and units according to the current setting of the **UNITS COMMAND**; sometimes, the actual values are not returned. For instance, the value 3.128362563 is stored by AutoCAD, but could be displayed as 3.13, 3 1/8, or 3 — depending on the setting of **UNITS**.

---

**NEW TO AUTOCAD 2011** IN THIS CHAPTER

- Commands such as **LIST** and **PROPERTIES** now report object transparency.

# ID

The ID command reports the x, y, z coordinates of single points. The command can be used transparently during other commands.

**TUTORIAL: REPORTING POINT COORDINATES**

In this tutorial, you learn to find the x, y, z coordinates of points in drawings.

1. To find the coordinates of points, start AutoCAD with a new drawing, and then enter the **ID** command:
   - In the ribbon's Home tab, choose the **ID Point** button in the Utilities panel.
   - At the 'Command:' prompt, enter the **id** command.

     Command: **id** *(Press ENTER.)*

2. In all cases, AutoCAD prompts you to pick a point in the drawing.
   Specify point: *(Pick a point.)*

AutoCAD reports the x, y, z coordinates of the point in the command prompt bar:
X = 13.3006   Y = 4.4596   Z = 0.0000

AutoCAD does not actually report the z coordinate, because all drawing is done in the x,y-plane. Instead, this command reports the current value of the elevation, which is always 0.0000 — unless changed with the ELEV command.

To capture points of geometric features accurately, use an object snap mode, such as ENDpoint or CENter.

## LastPoint and @

AutoCAD stores the x, y, z coordinates recorded by ID and other commands in the LASTPOINT system variable. You can access the coordinates in the next command by entering **@** at the prompt. This shorthand tells AutoCAD to read the value stored by LASTPOINT.

Here is an example of using @ with the LINE command:

   Command: **line**
   Specify first point: **@**
   Specify next point or [Undo]: *(Pick a point.)*

And with the CIRCLE command:

   Command: **circle**
   Specify center point for circle or [3P/2P/Ttr (tan tan radius)]: **@**
   Specify radius of circle or [Diameter]: *(Pick a point.)*

LASTPOINT stores the coordinates only until the next 'Specify point' prompt. As soon as you pick another point, or enter coordinates at the keyboard, AutoCAD stores the new coordinates in the LASTPOINT system variable.

## MEASUREGEOM

The MEASUREGEOM command reports distances and angles between points, and radii, areas, and volumes of objects.

### TUTORIAL: REPORTING DISTANCE AND ANGLE MEASUREMENTS

In this tutorial, you draw several objects, and then measure them with the MEASUREGEOM command.

1. Start a new drawing with the *acad.dwt* template file, and then draw a line at an angle. It can be of any length.
2. To ensure that measurements are accurate, turn on ENDpoint object snap.
3. To find the length and angle, start the **MEASUREGEOM** command:
   - In the ribbon's Home tab, choose the **Measure** button in the Utilities panel.
   - At the 'Command:' prompt, enter the **measuregeom** command:

      Command: **measuregeom** *(Press ENTER.)*

   - Alternatively, enter the **mea** alias at the 'Command:' prompt.
4. In all cases, AutoCAD prompts you to choose the type of measurement. To find the length (i.e. distance) of the line, Enter **d**, as follows:

   Enter an option [Distance/Radius/Angle/ARea/Volume] <Distance>: **d**
5. Pick the two endpoints of the line:

   Specify first point: *(Pick point 1.)*

   Specify second point: *(Pick point 2.)*

Notice that AutoCAD reports more than just the length of the line. It also reports the horizontal and vertical distances along the x and y axes, and the angle of the line to the positive x axis.

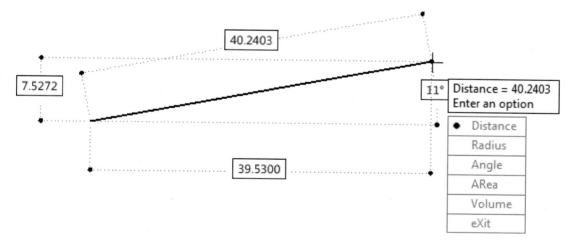

The measurements shown in the drawing area are temporary, however; as soon as you leave this command, the numbers vanish. But not to worry, for the information is repeated in the command bar, along with some additional information:

   Distance = 9.7188,  Angle in XY Plane = 213,  Angle from XY Plane = 0
   Delta X = -8.1564,  Delta Y = -5.2847,   Delta Z = 0.0000

To copy this data for use elsewhere, select the text, and then press CTRL+C.

The x- and y-related information has the following meaning:

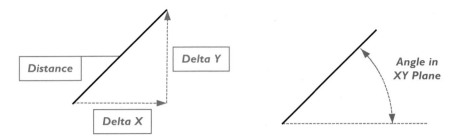

*Left:* The distance, delta X and delta Y reported by the MEASUREGEOM command.
*Right:* The angle in the XY plane.

AutoCAD stores the distance in the DISTANCE system variable, which you can reuse for the next command.

You can find the shortest distance between two objects, such as a line and a circle, by using the PERpendicular object snap:

Command: **measuregeom**
Enter an option [Distance/Radius/Angle/ARea/Volume] <Distance>: **d**
Specify first point: **per**
to *(Pick a point on the line.)*

 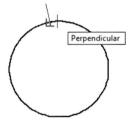

Specify second point: **per**
to *(Pick a point on the circle.)*

AutoCAD measures the shortest distance between the two objects, whether or not they would pass by each other. In the figure above, there is a gap between the line and the point of perpendicularity, because AutoCAD "extends" the line segment to find the point of perpendicularity.

When there are more than two objects in the drawing, AutoCAD cannot find the perpendicular until you pick the next point. An ellipsis ( ... ) appears next to the Perpendicular icon to indicate the deferment. You always need to two pick two points.

### Angle Measurement

The MEASUREGEOM command's **Angle** option measures angles in a manner different from what you might expect given the tutorial above.

Command: **measuregeom**
Enter an option [Distance/Radius/Angle/ARea/Volume] <Distance>: **a**
Select arc, circle, line, or <Specify vertex>: *(Enter an option.)*

Each of these options is handled differently:

**Arc** option measures the angle between the two endpoints of the arc.

**Circle** option measures the angle between two points that you pick on the circle.

**Line** option measures the angle between two lines.

**Specify Vertex** option measures the angle between any two points in the drawing, after you pick the vertex point.

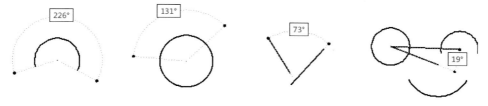

*Left to right:* Measuring the angles between...
Endpoints of an arc.   Two points on a circle.   Between two lines.   Of a vertex formed by center points on arcs and circle.

## TUTORIAL: DETERMINING AREAS AND PERIMETERS

The **Area** option measures the area and perimeter of closed objects. The list includes circles, ellipses, open and closed splines, open and closed polylines, polygons, regions, hatch patterns, and 3D solids — as well as arbitrary areas of picked points. You can add and subtract many areas, and have AutoCAD report the total.

1. For this tutorial, draw a rectangle that overlaps a circle.

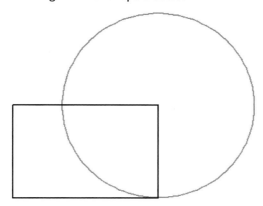

2. To find the area and perimeters of both objects, start the **MEASUREGEOM** command, and then enter **ar** for the **Area** option:

   Command: **measuregeom**

   Enter an option [Distance/Radius/Angle/ARea/Volume] <Distance>: **ar**

   (Remember to enter **ar**, and not just **a**, which is the abbreviation for the **Angle** option.)

3. For this tutorial, enter **o** for the **Objects** option. This allows you to choose any closed object, except self-intersecting ones.

   Specify first corner point or [Object/Add area/Subtract area/eXit] <Object>: **o**

4. AutoCAD prompts you to select objects. Choose the rectangle:

   Select objects: *(Select the rectangle.)*

   Notice that the interior of the rectangle turns green; this is AutoCAD's way of confirming that it measured the area.

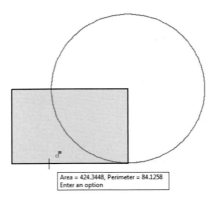

5. Press **ESC** to exit the command.

### Adding and Subtracting Areas

6. To add areas, repeat the command, and then use the **Area** option's **Add area** option. (You have to add the first selection to nothing before you can add more to it.)

    Command: **measuregeom**

    Enter an option [Distance/Radius/Angle/ARea/Volume] <Distance>: **ar**

    Specify first corner point or [Object/Add area/Subtract area/eXit] <Object>: **a**

7. Specify the **Object** option:

    Specify first corner point or [Object/Subtract area/eXit]: **o**

8. Select the rectangle, and then the circle. As you do, notice that AutoCAD reports the individual areas and perimeters and total area in the command bar.

    (ADD mode) Select objects: *(Select the rectangle.)*

    Area = 26.5451, Perimeter = 21.2725

    Total area = 26.5451

    (ADD mode) Select objects: *(Select the circle.)*

    Area = 50.2668, Circumference = 25.1331

    Total area = 76.8118

    Notice that the overlapping area is colored a darker green.

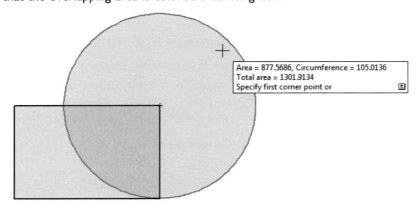

9. To subtract areas, follow these steps:

    a. Press **ENTER** to exit add mode:

(ADD mode) Select objects: *(Press ENTER.)*

b. Enter **s** to specify **Subtract** area mode:

Specify first corner point or [Object/Subtract area/eXit]: **s**

c. This time, you'll pick points instead of objects:

Specify first corner point or [Object/Add area/eXit]: *(Pick a point, such as at the intersection of the rectangle and circle.)*

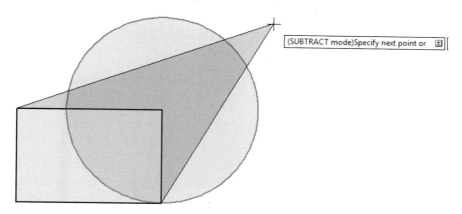

(SUBTRACT mode) Specify next point or [Arc/Length/Undo]: *(Pick another point.)*

(SUBTRACT mode) Specify next point or [Arc/Length/Undo]: *(Pick another point.)*

Notice that AutoCAD is coloring the subtraction area in pink.

10. Press **ESC** to end the command.

AutoCAD stores the area and perimeter measurements in the AREA and PERIMETER system variables. (AREA is one of a few system variables with the same name as a related command, and so you must use the SETVAR command to access the value stored in the system variable. Entering AREA at the 'Command:' prompt executes the command.)

Command: **setvar**

Enter variable name or [?]: **area**

AREA = 2.0316 (read only)

You can measure open objects like open polylines, polyarcs, and splines (but not arcs or connected lines). AutoCAD measures the area (through the Area option's Object option) by temporarily "closing" the open object, as illustrated below.

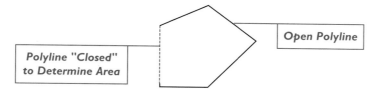

When it comes to perimeters, however, open objects have shorter perimeters than closed ones of the same area. That's because AutoCAD measures the length of the open objects without closing its perimeter.

The table below lists the data reported by the MEASUREGEOM's Area option:

| Objects | Area Measurement |
|---|---|
| Circles | Area, circumference. |
| Ellipses, polygons | Area, perimeter. |
| Regions | Area, perimeter. |
| Closed polylines and closed splines | Area, perimeter. |
| Open polylines and open splines | Area, length. |
| Wide polylines | Area and length/perimeter along center line. |
| 3D solids | Area of all faces. |
| Hatch patterns | Area, perimeter. |

**Notes** Here's a quick way to find the area of multiple irregular areas: hatch them with a single pattern, and then look up the area with the Properties palette.

Alternatively, use the **BOUNDARY** command and then find the area from the Properties palette.

### TUTORIAL: REPORTING VOLUMES

The **Volume** option measures the volume of 2D and 3D objects. In the case of 2D objects, you are prompted to specify the height. This make the command useful for measuring room volumes on 2D floor plans, for example: measure the area of the room, and then specify a height of 8' to arrive at the room's volume.

1. For this tutorial, draw a 2D rectangle and a 3D sphere with the **SPHERE** command.

2. The rectangle represents the floor of a room. To find its volume, start the **MEASUREGEOM** command, and then enter **v** for the Volume option:

    Command: **measuregeom**

    Enter an option [Distance/Radius/Angle/ARea/Volume] <Distance>: **v**

3. Use the **Object** option to select the rectangle, as follows:

    Specify first corner point or [Object/Add volume/Subtract volume/eXit] <Object>: **o**

    Select objects: *(Choose the rectangle.)*

4. Notice that AutoCAD prompts you for the height:

    Enter height: **8**

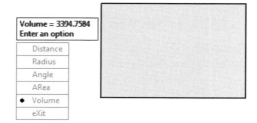

AutoCAD reports the volume in the tooltip and in the command bar.

Volume = 166.0767

5. Press **ESC** to end the command.

### Volumes of 3D Objects

6. To find the volume of 3D objects, repeat the command and its **Volume** option.
   Command: **measuregeom**
   Enter an option [Distance/Radius/Angle/ARea/Volume] <Distance>: **v**
   Specify first corner point or [Object/Add volume/Subtract volume/eXit] <Object>: **o**
7. Select the sphere.
   Select objects: *(Choose the sphere.)*
8. AutoCAD reports the volume — without asking for a height.

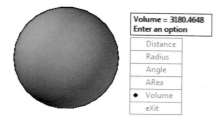

9. Press **ESC** to end the command.

## MASSPROP

The MASSPROP command reports the area, perimeter, centroid, and so on, of 2D regions and 3D solid models. It does not work with 3D mesh objects or other 2D or 3D objects not made with ShapeManager kernel.

The boundary must be made from a single region object. You can use the SUBTRACT command to remove islands from regions.

### TUTORIAL: REPORTING INFORMATION ABOUT REGIONS

1. For this tutorial, draw a closed polyline shaped like the one illustrated above.
2. Use the **REGION** command to turn the polyline into a region object.
3. To find out information about the region, start the **MASSPROP** command:
   - At the 'Command:' prompt, enter the **massprop** command.
     Command: **massprop** *(Press ENTER.)*
4. Notice that AutoCAD prompts you to select objects.
   Select objects: *(Select region objects.)*
   Select objects: *(Press ENTER to end object selection.)*

   AutoCAD switches to the Text window, to display a report similar to the following:
   ---------------- REGIONS ----------------

```
Area:              1.7211
Perimeter:         14.3193
Bounding box:   X: 1.2616  --  3.7327
                Y: 0.9440  --  3.0999
Centroid:       X: 2.6010
                Y: 2.2224
Moments of inertia:  X: 9.0893
                     Y: 12.3840
Product of inertia:  XY: 9.9040
Radii of gyration:   X: 2.2981
                     Y: 2.6825
Principal moments and X-Y directions about centroid:
                I: 0.5772 along [0.9653 -0.2612]
                J: 0.7529 along [0.2612 0.9653]
```

All values are relative to the origin in the current UCS.

5. AutoCAD asks if you wish to save the data to a file.

   Write analysis to a file? [Yes/No] <N>: **y**

6. If you answer **Y**, AutoCAD displays the Create Mass and Area Properties File dialog box. Enter a file name, and then click **Save**. AutoCAD creates the *.mpr* mass properties report file.

7. Press **F2** to return to the drawing window.

The same command works with 3D solid models, and provides a similar report.

### Finding the Center

The centroid value is useful for finding the geographic center of irregularly shaped objects, like maps, such as that of Greenland illustrated below. The bounding box is the smallest rectangle that completely surrounds the region. The area and perimeter values are the same as those reported by the AREA, LIST, and PROPERTIES commands.

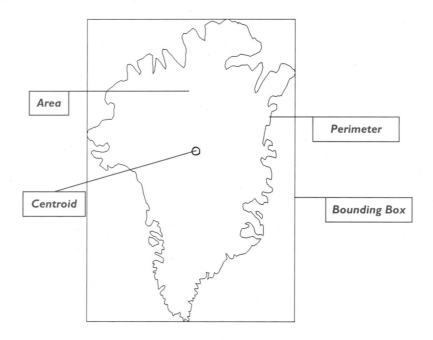

 **LIST**

The LIST command reports information about selected objects in the command bar.

## TUTORIAL: REPORTING INFORMATION ABOUT OBJECTS

1. To find out information about objects, start the LIST command:
   - In the ribbon's Home tab, choose the **List** button in the Properties panel.
   - At the 'Command:' prompt, enter the **list** command:
   
   `Command: **list** (Press ENTER.)
   
   - Alternatively, enter the **ls** alias at the 'Command:' prompt.

2. In all cases, AutoCAD prompts you to select objects. While you can select more than one object, it is better to pick just one, because of the amount of information generated.
   
   Select objects: *(Select one object.)*
   
   Select objects: *(Press ENTER to end object selection.)*
   
   AutoCAD reports on the object(s) in the Text window.
   
   ```
   CIRCLE    Layer: "0"
       Space: Model space
       Handle = 15f22
    center point, X=1231.9752  Y=-1589.5176  Z= 0.0000
       radius  593.9697
    circumference 3732.0214
           area 1108353.756431.9752  Y=-1589.5176  Z= 0.0000
   ```

3. You can select the text, and then use CTRL+C to copy it to the Clipboard for pasting into other documents.

4. Press F2 to return to the drawing window.

AutoCAD reports information that is specific to each object. For instance, circles return the following information:

| Line # | Example | Information Reported |
|---|---|---|
| 1 | CIRCLE | Type of object. |
|   | "0" | Layer on which the object resides. |
| 2 | Model space | Space in which the object resides: model or paper space. |
| 3 | 9D | Handle (unique hexadecimal identifier assigned by AutoCAD). |
| 4 | 0.5,7.5,0.0 | X, y, z coordinates of the object's location. |
| Other |  | Additional geometric information specific to the object: Area, perimeter, length. Color, linetype, and lineweight, if not set to BYLAYER. Thickness, if not zero. |

The geometric information varies among objects, as illustrated by the table below. For complex objects, the listing can get very long, because AutoCAD reports on every subpart.

| Object | Information Reported |
|---|---|
| Arc | Center point, Radius, Start angle, End angle, Length. |
| Box | Position, Length, Width, Height, Rotation. |

| | |
|---|---|
| Circle | Center point, Radius. |
| Ellipse | Center, Major Axis, Minor Axis, Radius Ratio. |
| Hatch | Pattern name, Scale, Angle, Associativity, Area, Origin. |
| Helix | Position, Length, Base radius, Top radius, Turns, Step Dist, and Axis. |
| Line | From point, To point, Length, Angle in XY Plane, Delta X, Delta Y, Delta Z. |
| Multileader | Leader Type, Content Type, Landing, Leader Number, Vertex, Annotative scale. |
| Plane Surface | U-isolines, V-isolines, Bounding Box. |
| Point | Point. |
| Polyline | Constant width (or Starting width and Ending width), Area, Perimeter. |
| Polyarc | Bulge, Center, Radius, Start angle, End angle. |
| Sphere | Position, Radius. |
| Spline | Area, Circumference, Order, Properties, Parametric Range, Number of control points, Control Points, Number of fit points, User Data, Start Tangent, End Tangent. |
| Text | Style, Annotative, Font file, Start point, Height, Text, Rotation angle, Width scale factor, Obliquing angle, Generation. |
| etc. | |

**Note** The **LIST** command is more useful than **AREA** for individual objects, because it provides more information. For example, **AREA** reports the area and circumference of circles, but **LIST** also reports their radius. It is also easier for finding the area and length of ellipses, and open and closed splines and polylines.

## PROPERTIES

The **PROPERTIES** command reports on objects in a format easier to digest and more complete than **LIST**. Compare the data for the circle shown at right with that listed on the previous page.

When you select all objects in the drawing, the Property window reports the total number.

A drawback is that the data listed by the Properties palette cannot be copied to the Clipboard.

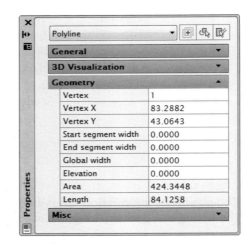

**TABLE**

The TABLE command creates the outline structure of tables, following which you fill cells with text, formulas, and drawings.

Tables consist of rows and columns. AutoCAD descriptively names three kinds of rows, names that are used when creating table styles: title, header, and data row. Columns don't get special names.

The topmost row is usually reserved for the title of the table, such as "Parts List" illustrated below. AutoCAD calls this the "title row."

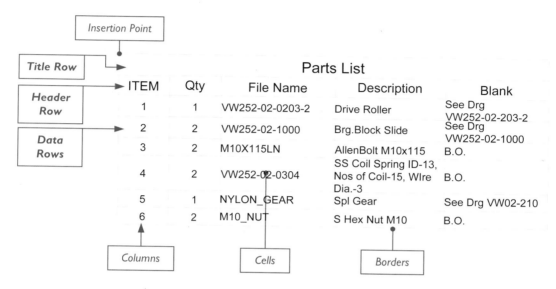

The second row typically labels the columns, such as "ITEM" and "Qty" (short for "quantity"). AutoCAD calls this the "header row."

All other rows in the table are called "data rows."

The rows and columns of tables define cells, just as in spreadsheets. Each cell contains one piece of information, either some text (or numbers or formulas) or an image of a block or drawing.

Tables are not static; you can change them at any time. You can add and remove rows and columns, merge cells to make them longer or taller, format cells to add color or more spacing, and export the table data for use by other software. Tables can receive data, from attributes in blocks or external files. The title, header, and data rows can be formatted differently from each other, as can individual cells. Tables are changed through grips editing, the Properties palette, and commands found on shortcut menus.

Tables have insertion points, just like text and blocks. The insertion point is the anchor from which the table grows and shrinks. The insertion point is typically located at the table's upper-left corner. When you add rows, the table grows downward from its insertion point.

In some drawings, it makes more sense to have the insertion point at the bottom corner, so that the table grows upwards. Below, the title and header rows are at the bottom of the table.

### BASIC TUTORIAL: PLACING TABLES

1. To place tables in drawings, start the **TABLE** command with one of these methods:
   - In the ribbon's Home tab, choose the **Table** button in the Annotation panel.
   - At the 'Command:' prompt, enter the **table** command.

     Command: **table** *(Press* ENTER.*)*
   - Alternatively, enter the **tb** alias at the 'Command:' prompt.

   In all cases, AutoCAD displays the Insert Table dialog box.

   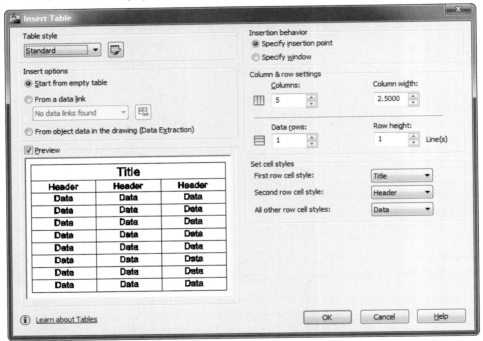

2. Ignore all the options, and click **OK**.
3. AutoCAD prompts you for the insertion point:

   Specify insertion point: *(Pick a point.)*

   Notice that AutoCAD displays either the Text Editor tab on the ribbon:

   The title row is highlighted. This means that when you start typing, text is entered in that cell.

   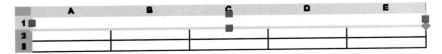

4. Enter a title for the table:
   **Table of Parts**
5. Press the **TAB** key. Notice that AutoCAD highlights the next cell.
6. Enter a header name for the column:
   **Item Number**

|   | A | B | C | D | E |
|---|---|---|---|---|---|
| 1 |   |   | Table of Parts |   |   |
| 2 | Item Number |   |   |   |   |
| 3 |   |   |   |   |   |

7. Press the **TAB** key, and the highlight jumps to the next cell. (Press **SHIFT+TAB** to move to the previous cell. You can also use the cursor keys to move to other cells.)
   Enter the next header name:
   **Quantity** (Press **TAB**.)
8. AutoCAD auto-fill cells when you drag the cursor across cells in a specific manner. Here's how:
   a. Enter "1" (one) in the cell under Item Number.
   b. Press **ENTER** to exit cell editing. Notice that the table is no longer selected.
   c. Single-click the cell containing the number "1." (Use a single click.) Notice the cell is highlighted in color and sports five grips:
      **Square** grips resize the cell.
      **Diamond** grip auto-fills adjacent cells.

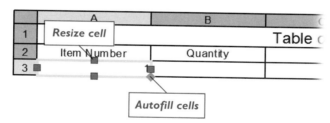

   d. Drag the diamond grip across blank cells. Notice that the tooltip displays the number that will be auto-filled.

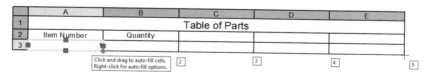

   Numbers are incremented; if you auto-fill with a cell containing a word, the same word is repeated in each cell.
   e. Release the mouse button. Notice that the numbers 2 - 5 fill the cells.

|   | A | B | C | D | E |
|---|---|---|---|---|---|
| 1 |   |   | Table of Parts |   |   |
| 2 | Item Number | Quantity |   |   |   |
| 3 |   | 1 | 2 | 3 | 4 | 5 |

9. When done filling the table's cells, click outside of the table to exit the command.

**Notes** When you press TAB in the last cell, AutoCAD automatically creates another row.

To move around the table, you can also press the four cursor keys on the keypad. Pressing **ENTER** moves the highlight down the table.

To create line breaks (where text is forced to start on the line below), press **ALT+ENTER**. To change the text style of a cell quickly, select a text style from the **Text Style** droplist on the Standard toolbar.

### ADVANCED TUTORIALS: INSERTING TABLES

The Insert Table dialog box contains several options for controlling the initial size of the table.

### Insertion Behavior

The Insertion Behavior section of the dialog box provides two basic methods of specifying the overall size of the table:

**Specify Insertion Point** — you specify where the table must start; AutoCAD determines the size of the table.

**Specify Window** — you specify the area into which the table must fit; AutoCAD determines how to size the cells to fit.

### Column & Row Settings

The numbers you specify for the table's rows and column are not critical, because you can always add and remove rows and columns later, as well as change the width and height of cells. Still, the Column & Row Settings section provides two ways of changing the size of the cells that make up the rows and columns:

**Columns** and **Data Rows** — you specify the number of columns and rows; AutoCAD determines the size based on column width and row height.

**Column Width** and **Row Height** — you specify the width of columns and height of rows; AutoCAD determines the number of rows and columns to create.

**Note** The **Row Height** is measured in "lines." A *line* is defined as the text height + top margin + bottom margin. The default value is 0.3 units, which comes from 0.18 (text height) + 0.06 (top margin) + 0.6 (bottom margin).

### Table and Cell Styles

When the drawing already has table styles defined, you can select one from the **Table Style Name** droplist. (You can use DesignCenter to borrow table styles from other drawings.)

If you want to create or modify a table style, click the button, which displays the Table Style dialog box.

### Set Cell Styles

You can specify style override with the Set Cell Styles options. Although called "cell styles," the styles apply to entire rows: the first row uses the "Title" style by default, the second row the "Header" style, and remaining rows the "Data" style.

These styles can be modified with the TABLESTYLE command, as detailed later in this chapter.

### Data Source

Tables can be created from data in the drawing or from external data files, such as spreadsheets. This is an advanced topic not covered by this book.

Tables can be inserted from spreadsheet programs: copy the data (using CTRL+C), and then paste them (CTRL+V) into the AutoCAD drawing. When spreadsheets are copied from Excel XP, they are pasted as table objects. For Excel, use the PASTESPEC command, and then select "AutoCAD Entities"; the spreadsheet data are pasted as a table. For other spreadsheets, the data are pasted as mtext.

### Preview

The preview image, unfortunately, does not reflect the number of rows, columns, and sizes specified; it reflects only the visual characteristics of the current table style, such as fonts and colors.

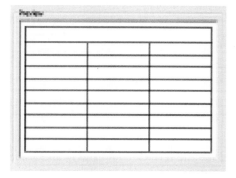

### EDITING TABLES

AutoCAD provides many options for editing tables. The methods of editing tables and cells include:

- **TABLEDIT** command edits the text of the selected cell.
- **TABLETOOLBAR** system variable toggles display of the table editing toolbar.
- **TINSERT** command inserts blocks and drawings in cells.
- **PROPERTIES** command changes many properties of tables.
- **TABLEEXPORT** command exports the table data as a comma-delimited text file.

I find it curious that a command named TABLEDIT edits cells (and then just the text in cells) and not the table itself. Another curiosity: the spelling of the TABLEDIT command has one "e," while TABLEEXPORT has two.

Many other table and cell editing and property commands are "hidden" in shortcut menus.

### TablEdit Command

The TABLEDIT command (one "e") edits the text of selected cells. It edits a cell at a time.

1. For this tutorial, open the *table.dwg* drawing.
2. To edit text in a table cell, start the **TABLEDIT** command with one of these methods:
   - At the 'Command:' prompt, enter the **tabledit** command. AutoCAD prompts you to select a cell:

     Command: **tabledit** (Press ENTER.)
     Pick a table cell: (Select a single cell.)

   - Double-click the cell you wish to edit, which means you need not enter the command.
   - Alternatively, select a cell, and then click the right mouse button. From the shortcut menu, select **Edit Text**.

   In all cases, AutoCAD highlights the selected cell in gray:

|   | A | B | C | D | E |
|---|---|---|---|---|---|
| 1 | | | | Parts List | |
| 2 | Item | Qty | Part Number | Description | Comments |
| 3 | 1 | 1 | VW02-210 | Spl. Gear | See Drg#VW202-210 |
| 4 | 3 | 1 | Helix_Helical gear | Helical Gear | See DrgHelix_Helical_Gear |
| 5 | 9 | 2 | Vw252-02-1101 | Upper Plate | See Drg#VW252-02-1100_U |
| 6 | 10 | 2 | VW252-02-1102 | Upper Bush | See Drg#VW252-02-1100_U |
| 7 | 12 | 1 | VW252-02-0201-N | M.S. Pipe | See Drg#VW252-02-0203-2 |
| 8 | 13 | 1 | VW252-02-0202 | Upper End SS Shaft | See Drg#VW252-02-0203-2 |
| 9 | 14 | 1 | VW252-02-0203N | Rubber Coating | See Drg#VW252-02-0203-2 |

3. Make your editing changes; see the **MTEXT** command in an earlier chapter. You can edit the text, or change properties, for example, make the text boldface, insert a symbol, or change its justification. (These are called "local overrides.")
4. Use the **TAB** key or the cursor keys to edit other cells.
5. When you are finished editing the text, press the **ESC** key, or click outside of the table.

### TInsert Command

The TINSERT command inserts blocks and drawings in cells (short for "table insert"). Cells can contain blocks or drawings together with text.

1. To insert blocks into cells, start the **TINSERT** command with one of these methods:
   - At the 'Command:' prompt, enter the **tinsert** command. AutoCAD prompts you to select a cell:

     Command: **tinsert** *(Press ENTER.)*

     Pick a table cell: *(Select a single cell.)*

   - Alternatively, select a cell, and then click the right mouse button. From the shortcut menu, select **Insert Block**.

   In both cases, AutoCAD displays the Insert a Block in a Table Cell dialog box.

   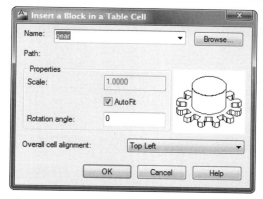

2. Change the dialog box settings, as follows:

   | | |
   |---|---|
   | Name | **gear** |
   | AutoFit | (turned on) |
   | Rotation Angle | **0** |
   | Overall Cell Alignment | **Middle Center** |

   "Gear" is the name of a block definition provided in the *table.dwg* file.

   **Browse** lets you insert an entire drawing, including *.dwg* and *.dxf* files.

   **Overall Cell Alignment** is like text justification; the **Middle Center** cell alignment option centers the block in the cell; you can choose any other alignment.

   **AutoFit** determines the size of the block:

   ☑ Block scaled automatically to fit the cell.

   ☐ Cell resized to fit the block; the **Scale Factor** option becomes available.

   **Rotation angle** rotates the block; 90 degrees, for example, turns the block on its left side.

3. Click **OK**. Notice that AutoCAD places the block next to the text in the cell.

| Parts List | | | | |
|---|---|---|---|---|
| Item | Qty | Part Number | Description | Comments |
| 1 | 1 | VW02-210 | Spl. Gear | See Drg#VW202-210 |
| 3 | 1 | Helix_Helical gear | Helical Gear | See DrgHelix_Helical_Gear |
| 9 | 2 | Vw252-02-1101 | Upper Plate | See Drg#VW252-02-1100_U |
| 10 | 2 | VW252-02-1102 | Upper Bush | See Drg#VW252-02-1100_U |
| 12 | 1 | VW252-02-0201-N | M.S. Pipe | See Drg#VW252-02-0203-2 |
| 13 | 1 | VW252-02-0202 | | See Drg#VW252-02-0203-2 |
| 14 | 1 | VW252-02-0203N | Rubber Coating | See Drg#VW252-02-0203-2 |

**Note** To make it appear as if a single cell bridges two cells (yet remaining two cells), change the border property so that the line between two cells is invisible.

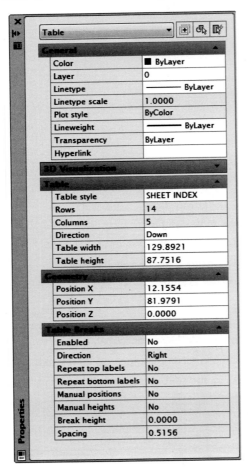

## Properties Command

The **PROPERTIES** command changes many (but not all) properties of tables and cells.

*Left: Properties for tables.*
*Right: Properties for cells.*

The Properties palette has different displays, depending on what part of the table you select.

### Grips Editing

Grips editing changes the size of the table, rows, and columns — depending on where you click, and where you grip. Once the cell(s) or table is selected, blue grips appear. You can use the grips to change the size of the table, the rows, and the columns — again, depending on which grips you drag.

### Managing Cell Content

In the past, AutoCAD restricted cells to one piece of content, either text or a block. As of AutoCAD 2008, cells can contain different types of content. There appears to be no command for accessing the dialog box: when a cell contains two or more objects, right-click, and then choose **Manage Content** from the shortcut menu. Notice the dialog box that controls the display order.

The Cell Content dialog box lists the contents of the selected cell. Use the **Move Up** and **Move Down** buttons to change the display order. Click **Delete** to erase the selected item.

> **Flow** — determines where to place the cell contents according to the width of the cell; some might be next to each other; others might be stacked on top of each other.
>
> **Stacked Horizontal** — lays out cell content horizontally.
>
> **Stacked Vertical** — stacks cell content vertically.
>
> **Content Spacing** — specifies the width between cell content items.

### TableExport Command

The TABLEEXPORT command (two "e"s) exports the table data as CSV-format files (short for "comma-separated values").

This means the table data are saved as an ASCII text file, with data separated by commas, one line of data per table row. Here is an example:

"VW252 - Washing Unit Drawing Sheets"
"Sheet Number","Sheet Title"
"00","Cover Sheet"
"01","VW252-Washing-Unit"
"03","Drive Roller Sub Assy"

The .csv files can be imported into just about any spreadsheet and database program, as well as some word processors and other software.

AutoCAD exports the table data literally. If the table's direction is up, you may want to use the PROPERTIES command to change the table's Direction property to **Down** before exporting the table — otherwise the title and header rows appear at the end of the .csv file for tables with an Up direction.

1. To export tables from drawings, start the **TABLEEXPORT** command with one of these methods:
   - At the 'Command:' prompt, enter the **TABLEEXPORT** command. AutoCAD prompts you to select a table:

     Command: **tableexport** *(Press ENTER.)*

     Select a table: *(Select a single table.)*

   - Alternatively, select a table, and then click the right mouse button. From the shortcut menu, select **Export**.

   In either case, AutoCAD displays the Export Data dialog box.

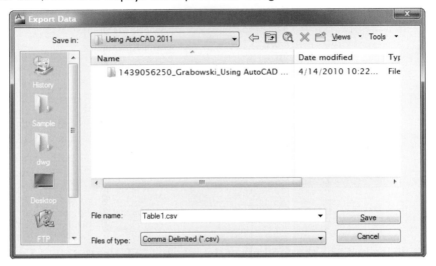

2. Specify the path, enter a file name, and then click **OK**.

   AutoCAD exports the data. You can open the file in a spreadsheet or database program.

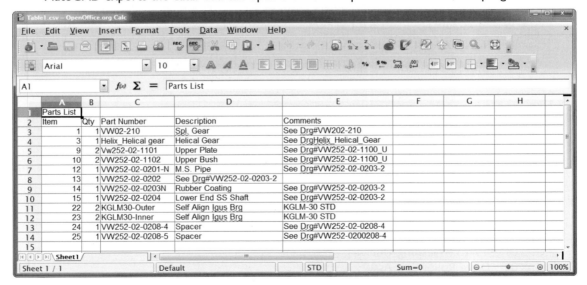

# TABLESTYLE

The **TABLESTYLE** command defines named table styles, allowing you to change the look of tables quickly, for example giving them thicker lines, color, and other fonts.

Tables have styles similar to text and dimensions. I find the default style included with AutoCAD, called "Standard," quite ugly; you'll want to change it right away with this command — or use DesignCenter to borrow table styles from other drawings.

As of AutoCAD 2008, table styles include cell styles, which define sub-styles for individual cells. AutoCAD includes predefined cell styles named Data, Header, and Title. You can edit these cell styles, as well as create new ones.

This command allows you to change the properties of cells, header rows, and title rows independently of each other. In addition, you can specify the color and thickness of lines, color of text and color behind cells, margin widths, justification within cells, and more.

### BASIC TUTORIAL: CREATING TABLE STYLES

1. Open the *tablestyle.dwg* file, a drawing of a table with the Standard style.

   | \multicolumn{5}{c}{Parts List} |
   
   | Item | Qty | Part Number | Description | Comments |
   |---|---|---|---|---|
   | 1 | 1 | VW02-210 | Spl. Gear | See Drg#VW202-210 |
   | 3 | 1 | Helix_Helical gear | Helical Gear | See DrgHelix_Helical_Gear |
   | 9 | 2 | VW252-02-1101 | Upper Plate | See Drg#VW252-02-1100_U |
   | 10 | 2 | VW252-02-1102 | Upper Bush | See Drg#VW252-02-1100_U |
   | 12 | 1 | VW252-02-0201-N | M.S. Pipe | See Drg#VW252-02-0203-2 |
   | 13 | 1 | VW252-02-0202 | Upper End SS Shaft | See Drg#VW252-02-0203-2 |
   | 14 | 1 | VW252-02-0203N | Rubber Coating | See Drg#VW252-02-0203-2 |
   | 15 | 1 | VW252-02-0204 | Lower End SS Shaft | See Drg#VW252-02-0203-2 |
   | 22 | 2 | KGLM30-Outer | Self Align Igus Brg | KGLM-30 STD |
   | 23 | 2 | KGLM30-Inner | Self Align Igus Brg | KGLM-30 STD |
   | 24 | 1 | VW252-02-0208-4 | Spacer | See Drg#VW252-02-0208-4 |
   | 25 | 1 | VW252-02-0208-5 | Spacer | See Drg#VW252-0200208-4 |

2. To modify the table style, start the **TABLESTYLE** command with one of these methods:
   - In the ribbon's Home tab, choose the **Table Style** button in the Annotation panel.
   - At the 'Command:' prompt, enter the **tablestyle** command.

     Command: **tablestyle** *(Press ENTER.)*

   - Alternatively, enter the **ts** alias at the 'Command:' prompt.

In all cases, AutoCAD displays the Table Style dialog box.

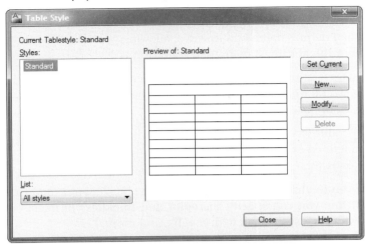

3. To improve the look of the Standard style, click **Modify**. AutoCAD displays the Modify Table Style dialog box.

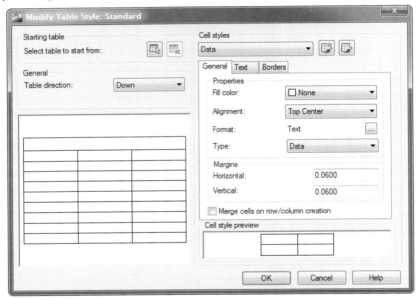

The most important item in this dialog box is the **Cell Styles** droplist, because it displays the names of cell styles — the styles that affect the look of cells.

**Data** — generic style for data cells, all cells that aren't headers or titles.

**Header** — cell style for header cells, usually in the table's second row.

**Title** — cell style for title cells, usually the table's first row.

**Create New Cell Style** — displays a dialog box for naming new cell styles. Note that new styles are based on existing ones. Clicking **Continue** returns you to the Table Style dialog box.

**Manage Cell Styles** — displays a dialog box for renaming and deleting style names.

4. Let's change the Standard style to make tables more attractive, starting with the **Data** cell style. Make the following changes:
   a. In the Modify Table Style dialog box, choose the **Text** tab.

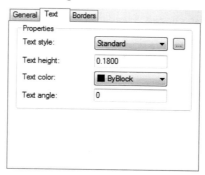

   b. Click the [...] button next to Text Style. Notice the Text Style dialog box.
   c. In the Text Style dialog box, select "Arial" from the **Font Name** list. (Note that the Upside Down and Backwards options do not work for text in tables.)

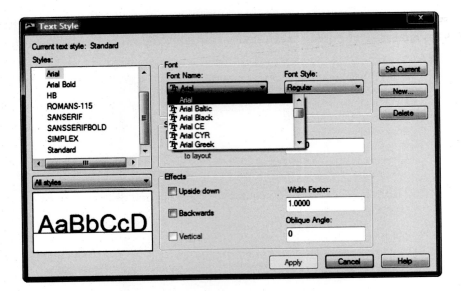

d. Click **Set Current**, and then **Close**. Notice that the fonts change in the Cell Style Preview window.

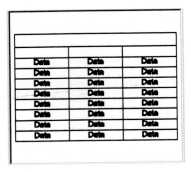

5. From the **Cell Styles** droplist, choose "Header."
    a. Choose the **General** tab.
    b. From the **Fill Color** droplist, select "Yellow."

6. From the **Cell Styles** droplist, choose "Title." Make these changes to the style:
    a. Choose the **Text** tab.

b. Change the text color to white: (1) from the **Text Color** droplist, select "Select Color," (2) in the Select Color dialog box, select the **True Color** tab, (3) change Luminance to "100."

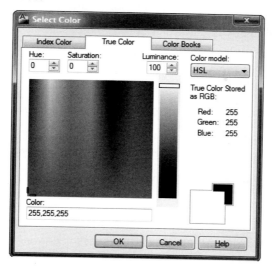

Click **OK**. (This trick works only when the AutoCAD screen color is white.)

c. Choose the **General** tab, and then change the **Fill Color** to "Black."

Notice the changes to the preview window.

7. Click **OK**, and then **Close** to exit the dialog boxes.

You have modified the Standard table style. To see what it looks like, use the TABLE command to create a table, and then fill its cells with text.

**Note** Changing a table style definition retroactively changes existing tables made with the same style.

# DWGPROPS

The DWGPROPS command reports and stores information about drawing files.

This command displays a dialog box of data related to the file, such as its size and attributes. As well, you can add information about the drawing, such as the name of the drafter and general comments. This data can be searched with AutoCAD's Design Center (ADCENTER) command and the Windows Explorer's Find (or Search) options.

## TUTORIAL: REPORTING INFORMATION ABOUT DRAWINGS

1. To find out information about drawings, start the **DWGPROPS** command:
   - In the application menu, choose **Drawing Utilities** and then **Drawing Properties**.
   - At the 'Command:' prompt, enter the **dwgprops** command.

      Command: **dwgprops** *(Press ENTER.)*

   Notice that AutoCAD displays the Properties dialog box — which is different from the Properties palette (illustrated earlier). This dialog box is for drawing properties, whereas the palette is for object properties.

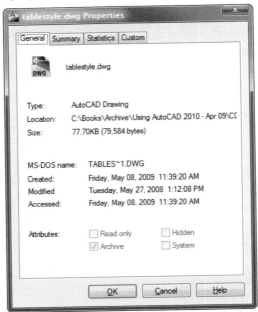

The General tab shows information about the drawing file's name, location, size, and dates. (If the drawing has not yet been saved, most fields are blank.) Attributes can be changed only by the Windows Explorer's Property option.

2. Click the **Summary** tab.

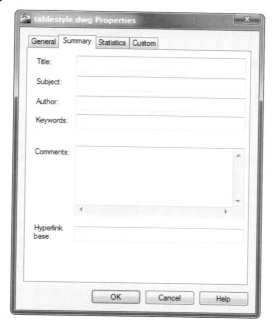

You can add data to the fields, such as your name, the subject of the drawing, and general comments.

**Title** — can differ from the drawing's file name, such as the project name.

**Keywords** — useful when drawings are searched by database programs.

**Hyperlink Base** — base URL address used by relative hyperlinks in the drawing.

3. Click the **Statistics** tab.

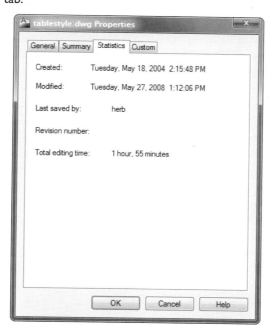

**Created** — date and time the drawing was first created with the **NEW** command; the date and time are retrieved from the **TDCREATE** system variable.

**Last Modified** — date and time the drawing was last edited or had objects added, retrieved from the **TDUPDATE** system variable.

**Last Saved By** — user's login name, retrieved from **LOGINNAME** system variable, which in turn is obtained from Windows.

**Revision Number** — AutoCAD software revision number.

**Total Editing Time** — time the drawing has been open. The name is misleading, because it includes time when the drawing is displayed by AutoCAD without being edited, including coffee breaks, tea time, and lunch. The value is retrieved from the **TDINDWG** system variable.

**Note** If the drawing was last saved by a non-Autodesk software product, an additional message appears in this tab:
This file may have been last saved by a program other than Autodesk software.

4. Click the **Custom** tab.

   You can enter text in up to ten fields, along with a value for each. The custom fields are searchable by the **Find** option of DesignCenter.

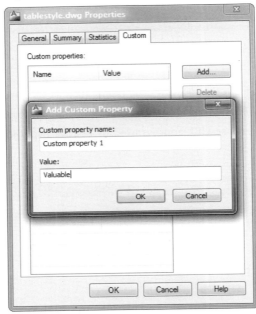

   To add a field, click **Add**; enter a name for the field and optionally a value. Click **OK**.
   To delete a field, select it, and then click the **Delete** button.

5. Click **OK** to close the dialog box.

## 'SETVAR

The **SETVAR** command (short for "set variables," or sysvars, for short) reports the settings of system variables. You have used this command many times in this and previous chapters, perhaps without realizing it.

System variables store the state of the drawing and control numerous commands. You can change the value of most system variables, thus changing the state of the drawing (snap, grid, and so on) and the actions of commands (mirrored text, current layer, and so on). Some system variables cannot be changed; they are called "read-only."

This command can be used transparently during other commands.

(*History*: In old versions of AutoCAD, system variables could only be changed with the **SETVAR** command. But then Autodesk made it possible to enter system variables directly at the 'Command:' prompt. The only time you *must* use this command is to examine the list of system variables names and values, or when the name of the system variable is the same as the command, such as **AREA**.)

All of AutoCAD's system variables are listed in Appendix C of this book.

### TUTORIAL: LISTING SYSTEM VARIABLES

1. To list the values of system variables, start the **SETVAR** command:
   - At the 'Command:' prompt, enter the **setvar** command:

     Command: **setvar** *(Press ENTER.)*

   - Alternatively, enter the **set** alias at the 'Command:' prompt.

2. In all cases, AutoCAD prompts you to enter the name of a system variable, or else enter **?** to list the names and current values of sysvars:

   Enter variable name or [?] <AREA>: **?**

3. AutoCAD asks which variables you wish to see. You can use the following wildcard characters to list related groups of sysvars, such as **grid*** (means all names starting with "grid").

   **\*** — means all variables.

   **?** — means a single character.

   Enter variable(s) to list <*>: *(Press ENTER to list all variables.)*

   Notice that AutoCAD flips to the Text window, and displays the value of many system variables.

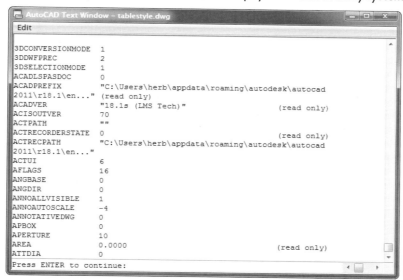

4. With nearly 800 system variables, the listing is very long. Press **ESC** to end it early.

## 'TIME

The TIME command reports the time spent editing drawings, and provides timer functions.

This command can be used transparently in the middle of other commands.

### TUTORIAL: REPORTING ON TIME

1. To find out time-related information about drawings, start the TIME command:
   - At the 'Command:' prompt, enter the **time** command.

        Command: **time** *(Press ENTER.)*

   AutoCAD switches to the text screen and reports time-related data similar to the following:

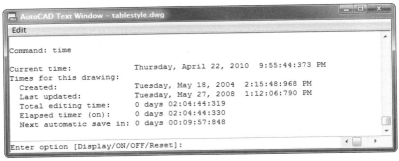

2. Press ENTER to exit the command:
        Enter option [Display/ON/OFF/Reset]: *(Press ENTER.)*

The accuracy of the time depends on your computer's clock. The meaning of the data reported by this command is as follows:

| Time | Meaning |
| --- | --- |
| Current Time | Current date and 24-hour time to the nearest millisecond. |
| Created | Date and time drawing was created. |
| Last Updated | Latest use of the SAVE and QSAVE commands. |
| Total Editing Time | Cumulative time drawing has been open in AutoCAD, excluding plotting time, and sessions when drawing was exited without saving. |
| Elapsed Timer | Stopwatch-like timer. |
| Next Automatic Save In | Minutes remaining until the next automatic save. |

### REPORTING ON TIME: ADDITIONAL METHODS

The TIME command includes options to redisplay its output and act as a stopwatch.

- **Display** redisplays the time report.
- **ON** starts the timer.
- **OFF** turns off the timer.
- **Reset** resets the timer.

Let's look at each.

### Display

The **Display** option redisplays the time report, with updated times.

### ON, OFF, and Reset

The **ON** option starts the timer. The **OFF** option stops the time, and the **Reset** option sets the timer back to 0 days and 0 time (0 hours, 0 minutes, and 0 seconds).

## EXERCISES

1. Start AutoCAD with a new drawing. Use this drawing for all the exercises in this chapter.
   a. Draw a line with endpoints at 2,2 and 6,6.
   b. With the **ID** command and ENDpoint object snap, select the first endpoint of the line.
   c. What coordinates does the command return?
   d. Repeat for the other end. Are the coordinates what you expect?

2. Draw a line, circle, and arc on the screen.
   a. Select the line with the **LIST** command.
   b. What is the line's angle in the XY Plane?
   c. Select the circle with **LIST** command.
   d. What is the circle's radius?
   e. Select the arc with **LIST** command.
   f. What is the arc's starting angle?

3. Continue with the drawing from exercise #2.
   a. Select the line with the **PROPERTIES** command.
   b. Deselect the line, and then select the arc.
   c. How do the data displayed by the Properties window change?
   d. Select all three objects. Does the Properties window display a smaller or larger amount of information?

4. Continue with the same drawing, and enter the **DBLIST** command.
   a. What do you notice about the information listed by AutoCAD?
   b. Use the **COPY** command's **Multiple** option to copy the objects three times.
   c. Use the **DBLIST** command again, and notice the length of the listing.

5. Continuing with the same drawing, use the **DIST** command with ENDpoint object snap.
   a. Select two endpoints of one of the lines on the screen.
   b. Is the length the same as reported by the **LIST** command?
   c. Repeat the **DIST** command, but use PERpendicular object snap to find the distance between a line and a circle.

6. Draw two rectangles of different sizes with the **RECTANG** command.
   a. Use the **AREA** command to determine the area of one rectangle.
   b. Which option of the **AREA** command did you use?
   c. Repeat the same procedure on the second rectangle. Add the two areas together, and obtain the total area.
   d. Use the **AREA** command, but this time use the **Add** option to find the area of both rectangles.
   e. Does AutoCAD arrive at the same answer as you?

7. Open the *area.dwg* file, a drawing of geometric shapes.
    a. Find the area of each object, and then write down the figures.
    b. Using the **Add** option, find the total area of all objects.

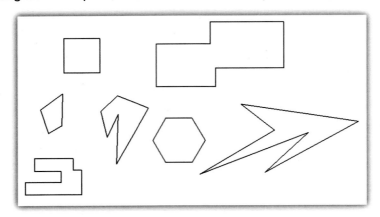

8. Use the **TIME** command to check the current time in the drawing.
    a. Is the time reported by AutoCAD the same as on your wristwatch?
    b. The same as shown on your computer?

9. Use the elapsed timer option of the **TIME** command to determine the time needed to complete the following exercise. Before starting, though, write down the time you estimate it will take you to complete exercise #10.

10. Draw the base plate shown below.
    a. Use the **AREA** command to find the area. *Hints:* Use the **RECTANGLE** command with its **Fillet** option. Remember to use the **Subtract** option to remove the holes from the total.

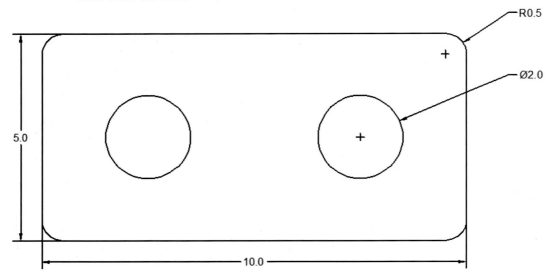

    b. Turn off the timer. How close was your estimate?

11. Open the *34486.dwg* file, a drawing of a 50' x 25' basement floor plan. Each unit in the drawing represents a foot.

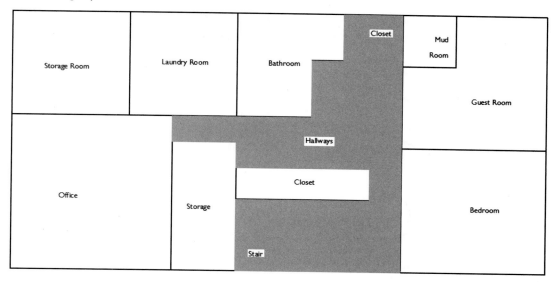

   a. What is the total square footage of the basement?
   b. Create a new layer called "Areas."
   c. Draw rectangles that match each room.
   d. Find the area of each room:
   
   Storage Room
   
   Laundry Room
   
   Bathroom
   
   Mud Room
   
   Guest Room
   
   Bedroom
   
   Closet
   
   Storage
   
   Office
   
   e. What is the area of the hallways, including stairs?

12. Start a new drawing with *acad.dwt*. Find the value of the following system variables:

    **OFFSETGAPTYPE**
    
    **LASTPOINT**
    
    **AREA**
    
    **MAXSORT**
    
   If you do not understand the meaning of a system variable, refer to Appendix C of this book.

13. Open the *15_44.dwg* file, a drawing of a gasket.
    a. Ensure the **DELOBJ** system variable is set to 1.
    b. Start the **BOUNDARY** command, and set object type to **Region**. Click **Pick Points**, and select a point inside the gasket. How many regions is the gasket composed of?

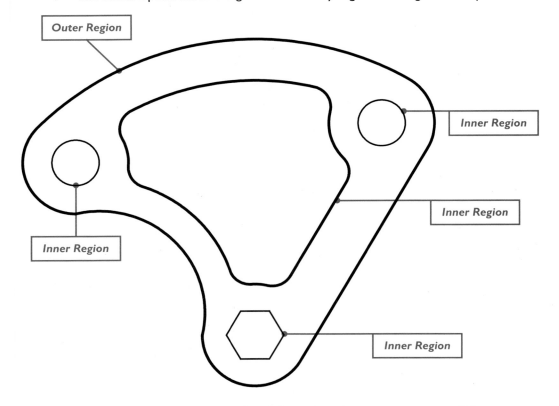

    c. Start the **SUBTRACT** command, and then select the outermost region. When prompted "Select solids and regions to subtract," select the four inner regions (also called "islands").
    d. Apply the **MASSPROP** command to the gasket, and then save the data to an *.mpr* file.
    e. Write down the values from the report:

Area: _____

Perimeter: _____

Bounding box:    X: _____    Y: _____

Centroid:    X: _____    Y: _____

Moments of inertia:   X: _____    Y: _____

Product of inertia:   XY: _____

Radii of gyration:   X: _____    Y: _____

Principal moments and X-Y directions about centroid:   I: _____

                                                                                  J: _____

    f. Use the **POINT** command to place a point at the centroid of the gasket.

14. Open the *15_43.dwg* file, and then find the centroid of the gasket.

15. Employ AutoCAD's **TABLE** command to create a table with seven rows and five columns.

16. Use the **TABLE** command to recreate the table of reference drawings illustrated below. (A copy of the table is available in the *refdwg.tif* file.)

| B-46858 | CONV. C-12, LAYOUT |
|---------|-----------------------|
| A-50918 | CHUTE AT DISCHARGE END, ARRGT & DET. |
|  |  |
|  |  |
|  |  |
|  |  |
| **REFERENCE DRAWINGS** ||

17. With the **TABLE** command, recreate the table of general notes illustrated below. (A copy of the table is available in the *generalnotes.tif* file.)

| 1. | ALL FIELD WELDED CONSTRUCTION (EXCEPT LADDER) |
|----|---|
| 2. | PAINT NOTE: 1 COAT OF RED PRIMER AND 1 COAT OF EQUIPMENT GREY (SHOP PAINT LADDER, FIELD PAINT PLATFORM AFTER INSTALLATION) |
| 3. | ALL WELDING MUST CONFORM TO THE REQUIREMENTS OF THE LATEST ISSUE OF C.S.A. STANDARD W.59 EXCEPT THE REFERENCE TO C.S.A. STANDARD W.47 DOES NOT APPLY. |
| **GENERAL NOTES** ||

18. Recreate the bill of materials table illustrated below. (A copy of the table is available in the *table-scan.tif* file.) Use auto-fill to fill the cells under "No. PER UNIT."

| No. | No. OF UNITS | No. PER UNIT | DESCRIPTION | MATERIAL | STOCK No. | REM |
|---|---|---|---|---|---|---|
| 3-1 | 2 | | WASTE OIL DISPOSAL UNIT | | | |
| -2 | 2 | 2 | 3/16" PLATE x 1'-1½" ⌀ | STEEL | 2502001 | |
| -3 | | 2 | 3/16" x 3" FLAT BAR x 3'-6½" | " | 2500205 | |
| -4 | | 2 | 3" x 3" BUTT HINGE | " | 1970323 | |
| -5 | | 2 | 3/8" x 1½" FLAT BAR x 3" | " | 2500503 | |
| -6 | | 2 | 12" STD. BLACK PIPE x 1'-6" | " | 3010003 | |

19. Open the *chapter16.dwg* file, a drawing of a parts list table.

    a. Edit the table to add the values in the Qty column (short for "quantity"). What approach did you take to enter the values?

| Parts List | | | |
|---|---|---|---|
| Item | Qty | Part Number | Description |
| 1 | | PC-15 | Paper clips |
| 2 | | P-24 | Pencils |
| 3 | | PP-33 | Paper Pads |
| 4 | | R-42 | Rulers |
| 5 | | CB-51 | Clipboard |

    b. Split the table into two sections.

20. Create the parts list table illustrated below. In the **Description** column, insert the appropriate blocks. Describe the approach you took to insert the blocks.

| Parts List | | | |
|---|---|---|---|
| Item | Qty | Part | Illustration |
| 1 | 5 | Air Injector (Dynamic Pump) | |
| 2 | 4 | Evaporator - Circular Fin | |
| 3 | 3 | Heater Feed with Air Outlet | |
| 4 | 2 | Pressure Gauge | |
| 5 | 1 | Thermometer - Type 2 | |

21. Open the *table.dwg* file.
    a. Use the **TABLEEXPORT** command to create a .csv file.
    b. If you have access to a spreadsheet program, open the .csv file and view its contents.
    c. Are the data in the spreadsheet the same as in the table in AutoCAD?

22. Open the *tablestyle.dwg* file. Use the **TABLESTYLE** command to make a new style based on the Standard style:

| \multicolumn{5}{c}{Parts List} | | | | |
|---|---|---|---|---|
| Item | Qty | Part Number | Description | Comments |
| 1 | 1 | VW02-210 | Spl. Gear | See Drg#VW202-210 |
| 3 | 1 | Helix_Helical gear | Helical Gear | See DrgHelix_Helical_Gear |
| 9 | 2 | Vw252-02-1101 | Upper Plate | See Drg#VW252-02-1100_U |
| 10 | 2 | VW252-02-1102 | Upper Bush | See Drg#VW252-02-1100_U |
| 12 | 1 | VW252-02-0201-N | M.S. Pipe | See Drg#VW252-02-0203-2 |
| 13 | 1 | VW252-02-0202 | Upper End SS Shaft | See Drg#VW252-02-0203-2 |
| 14 | 1 | VW252-02-0203N | Rubber Coating | See Drg#VW252-02-0203-2 |
| 15 | 1 | VW252-02-0204 | Lower End SS Shaft | See Drg#VW252-02-0203-2 |
| 22 | 2 | KGLM30-Outer | Self Align Igus Brg | KGLM-30 STD |
| 23 | 2 | KGLM30-Inner | Self Align Igus Brg | KGLM-30 STD |
| 24 | 1 | VW252-02-0208-4 | Spacer | See Drg#VW252-02-0208-4 |
| 25 | 1 | VW252-02-0208-5 | Spacer | See Drg#VW252-0200208-4 |

   a. Change font to Times New Roman.
   b. Change fill color of data rows to light gray.
   c. Change border line color to white.
   d. Name the style.

## CHAPTER REVIEW

1. List the two kinds of actions the **ID** command performs:
    a.
    b.
2. Where are the **ID** coordinates stored?
   How do you access them in other drawing and editing commands?
3. What is the difference between the **PROPERTIES** and **LIST** commands?
4. What are the six distances returned by the **MEASUREGEOM** command?
5. What option of the **MEASUREGEOM** command's **Area** option calculates the area of a circle?
6. After using the **MEASUREGEOM** and **LIST** commands on the circle, which command do you find more useful?
7. How do you determine the last time a drawing was updated?
8. Can the **MEASUREGEOM** command return the length of an open polyline?
9. Why would you want to subtract areas?
10. Describe a purpose of the **DWGPROPS** dialog box.
11. List two reasons why the Total Editing Time might be inaccurate for drawings:
    a.
    b.
12. Can you change the value of all system variables?
13. Examine the bill of material below, and then answer the following questions. (Also available as *bofm.tif*.)

| PART No. | No. OF UNITS | No. PER UNIT | DESCRIPTION | MATERIAL | REQ. No. | REMARKS |
|---|---|---|---|---|---|---|
| C-10102-1 | 1 | | LUB DRIP TRAY ASSY. | | | |
| -2 | | 1 | 16 GA. × 1'-8 1/2" × 7'-3" SH. MET. TRAY | M.S. | 2503101 | |
| -3 | | 1 | 2" Ø STD. PIPE NIPPLE, 4" LG | " | 3252405 | |
| -4 | | 1 | 1/4" N.C. HX. HD. BOLT, 1" LG | " | PE-CONTR | TACK-WELD TO TRAY |
| -5 | | 2 | 5/16" Ø FLAT WASHER | " | -"- | ONE AS SHOWN |
| -6 | | 1 | 1/4" N.C. WING NUT | " | 2320501 | |
| -8 | | 1 | 3/8" × 2" - 2" L, 2'-0" LG | M.S. | 2501300 | |
| -9 | | 1 | 3/8" × 1 1/2" × 1 1/2" L, 8'-4" LG | " | 2501002 | |
| -10 | | 1 | -"- , 10" LG | " | -"- | |

   a. How many columns does the table have?
   b. How many rows?
   c. What is the wording of the title row?
   d. What does AutoCAD call the row containing the phrase "PART No."?
   e. Where is the insertion point located?

14. Examine the bill of material below, and then answer the following questions. (Also available as *revisions.tif*.)

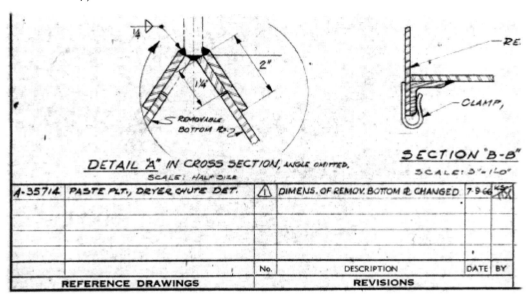

   a. How many columns does the table have?
   b. How many rows?
   c. Is the table's direction up or down?
   d. What might be the purpose of the Reference Drawings table?
   Of the Revisions table?

15. Explain the reason behind tables having title, header, and data rows.
16. Can individual cells be formatted differently from each other?
17. Where is the title row located when the table's direction is up?
18. After creating a table, can rows be added?
    Removed?

750

# V

## 3D Modeling

# CHAPTER 16

## Introduction to 3D Solid Modeling

In 2D drafting, you draw and edit with 2D objects — such as lines, arcs, and circles — to construct drawings. In 3D modeling, the process is similar: you draw and edit with basic boxes, spheres, and other 3D objects to produce complete drawings.

The difference between 2D and 3D lies in the nature of the basic objects, the kinds of editing tools available, and the added dimension (z, or depth). Editing consists of adding, subtracting, and manipulating solids. A hole, for example, is made by subtracting a cylinder from the model.

Solid modeling is used primarily in mechanical engineering, because models are precise and can be accurately analyzed. Solid modeling is also sometimes used by architects to represent complex intersections, such as the roofs on buildings, and to find interferences, such as between plumbing and the building's walls.

After constructing 3D models, designers often "convert" them to standard sets of 2D drawings for construction. The word "model" is often used to distinguish 3D from 2D drawings.

In this chapter, you learn the following commands for constructing solid models:

**BOX** creates basic 3D solid models.

**PRESSPULL** extrudes 2D objects into 3D models.

**REVOLVE** creates round solid bodies from 2D designs.

**UNION** joins two or more 3D solids into bodies.

**FILLET** rounds off edges of 3D solids.

**Subobject** selection selects vertices, edges, faces, and subobjects.

**ViewCube** rotates 3D viewpoints in real time.

**UCS** and **Dynamic UCS** change the orientation of the x, y workplane.

## HISTORY OF 3D IN AUTOCAD

AutoCAD provides several methods of creating 3D objects, each of which has its advantages and disadvantages. It is rumored that the very first release of AutoCAD contained 3D, but a dispute with the programmer led to the feature remaining unusable. Over the following decades, Autodesk added new 3D technologies frequently, such as polyfaces (useful for surveying and digital terrain modeling), solid modeling (useful for mechanical designs), surfaces (useful for conceptual modeling in architecture), and mesh modeling (useful for industrial design). The catch to this ad-hoc development is that older commands that seemed to create surfaces and meshes, do not — such as TABSURF and 3DMESH. Nor does SOLID command create 3D solids.

Historically, each method was added to AutoCAD as subsequent releases improved their ability to work in three dimensions:

**Thickness.** AutoCAD v2.1 gave certain 2D objects thickness and elevation using the ELEV command. This is sometimes called "2-1/2D," because the extrusions always result in flat tops. Adding thickness is easy to understand and edit, but is limited in capability, and sometimes renders incorrectly.

**Wireframes.** AutoCAD Release 9 could create true 3D *wireframes* with the 3DLINE command. In a later release, the 3DPOLY command was added, and then many editing commands were extended to allow 3D editing. Wireframe objects, however, are difficult to construct, are even more difficult to edit, and do not render at all. A better form of 3D was needed.

**Surface meshes.** As far back as 1989, Autodesk described Release 10 as "The 3D Release," because the new UCS (user-defined coordinate systems) allowed users to draw and edit on any plane in 3D space. It added 3D objects made from *polyface meshes* using commands like 3DMESH and RULESURF. Polyface meshes are triangular and rectangular planes joined to create 3D surfaces. The mesh renders well, but is difficult to edit, and AutoCAD provides little assistance in manipulating this type of surface.

**Solid modeling.** AutoCAD Release 11 added a solid modeling kernel for creating 3D solid objects with commands such as BOX and EXTRUDE. In the following releases, Autodesk switched the solid modeling engine from PADL to ACIS, and then to ShapeManager. Solids render well, can be edited in limited ways, and are popular for designing mechanical devices. In this chapter, you work with solid models.

**Surface modeling.** AutoCAD 2007 added commands to create 3D surfaces, such as LOFT and SWEEP. These are "true" surfaces, not simulated ones created by polyface meshes.

**Mesh modeling.** AutoCAD 2010 added 3D meshes, popular for styling automobiles and consumer products. To distinguish them from pre-2010 3D mesh objects that were made of polyfaces, Autodesk calls the new ones "full featured mesh objects" and "subdivision meshes"; in this book, they are named "3D mesh objects."

**NURBS surfacing.** AutoCAD 2011 adds NURBS splines and meshes that interoperate with each other, and produce far more complex looking surfaces than before. Interactive editing just about eliminates the need for commands.

**Custom 3D objects.** AutoCAD Release 13's ObjectARx programming interface made it possible for Autodesk and third-party developers to create custom 3D objects, such as the landscape objects created by the LSNEW command (since removed from AutoCAD).

**Photorealism.** AutoCAD Release 12 added the RENDER, LIGHT, RMAT, and related commands for creating realistic images of 3D models. This was advanced in AutoCAD 2007 with improved photorealism and material definitions. With AutoCAD 2011, Autodesk unifies the way materials are defined across all of its software products.

**Visual styles.** AutoCAD 2007 added visual styles that allowed users to edit 3D drawings in simulated rendering modes. Visual styles can be customized to output drawings suitable for technical documentation or for appearing hand-sketched. AutoCAD 2011 adds several more predefined styles, such as X-ray and grayscale looks.

---

In this chapter, you learn about 3D solid modeling, and in the following chapter, 3D mesh modeling. Solid modeling is better for precise prismatic shapes, such as machine parts, while 3D meshes are better for organic flowing shapes, such as hair dryers and cell phones.

## PARTS OF 3D SOLID MODELS

Three-dimensional models have an entire vocabulary to define *features* (a.k.a. subobjects), as illustrated below.

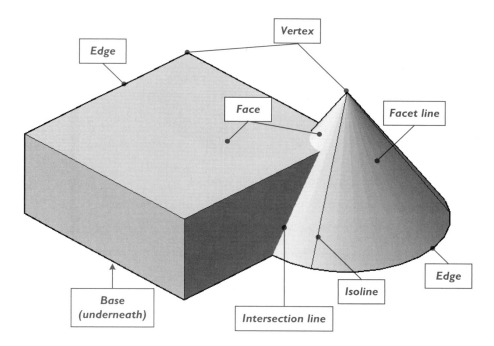

**Faces** define the surfaces of solid models. Faces can be flat (as on boxes) or curved, as on spheres and cones. When drawings are in 2D or 3D wireframe mode, faces are invisible.

**Edges** outline faces. On boxes, each face has four edges; a cone has just one edge, around the base; spheres have no edges. The look of the edge can be modified with the VISUALSTYLE command.

**Vertices** are at all intersections of edges and faces. On boxes, a vertex is located at each corner, eight in all. Cones have just one vertex, at the tip; spheres have no vertices.

**Bases** form the bottoms of most 3D models. When creating primitives like boxes and cylinders, you first define the size of the base, followed by the height. The bases of boxes are squares or rectangles; the bases of cone are circles or ellipses; spheres have no bases.

**Isolines** illustrate the surface of curved 3D models, such as spheres and cones. Since curved surfaces have few edges, isolines help you visualize the curves. The number of isolines is adjusted with the ISOLINES system variable; the default is 4 isolines, but a better value is 12.

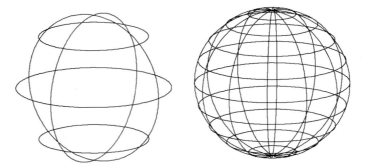

*Left to right:* Isolines = 0 (sphere is invisible), 4 (default value), and 12.

**Facet lines** appear when curved portions of 3D models are not smoothly rendered, as illustrated below. Unsmooth renderings are faster to display than smoothed renderings.

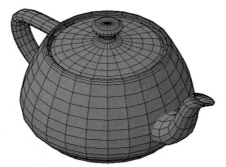

**Intersection lines** appear at the intersections of 3D models.

When creating 3D drawings, you usually construct them in one of two ways:

- By using basic 3D elements called "primitives" like boxes, cones, and spheres.
- By turning 2D elements like circles and polygons into 3D elements through actions like extruding and revolving.

# TUTORIAL: PREPARING TO MODEL IN THREE DIMENSIONS

In this tutorial, you prepare AutoCAD to model the 3D part illustrated below. It is a solid model made of a box, with a fillet (the rounded end) and a hole.

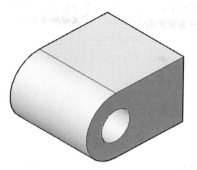

Start AutoCAD with a new drawing (*acad3d.dwt*), and then follow these steps. See the figure below for the locations of options and controls.

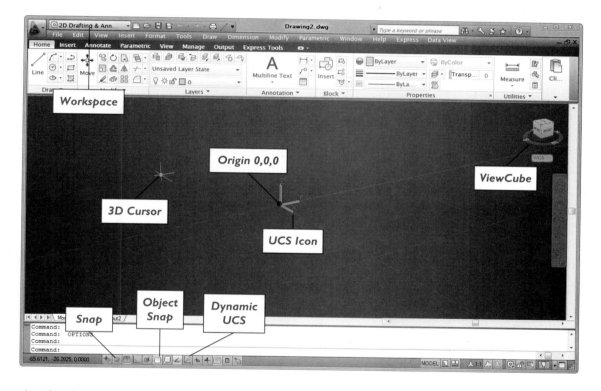

1. Set the workspace for 3D modeling, as follows:
    a. On the Quick Access toolbar or the status bar, click the **Workspace** droplist.
    b. Choose "3D Modeling."

Notice that the ribbon changes to emphasize 3D drawing and editing tasks. (If palettes appear, such as Tools or Materials, close them.)

2. Even though AutoCAD is a 3D CAD program, almost all "3D" drawing is actually done in 2D on the x, y-plane. The *dynamic UCS* (user-defined coordinate system) causes AutoCAD automatically to rotate the x, y-plane to match the selected face, as you will see later in this tutorial. (More on UCSs later in this chapter.)

   Ensure that the **Allow Dynamic Ucs** button is turned on. On the status bar, the button should look blue.

3. Ensure that **Object Snap** is turned on, and that **Center** is the only active snap mode.

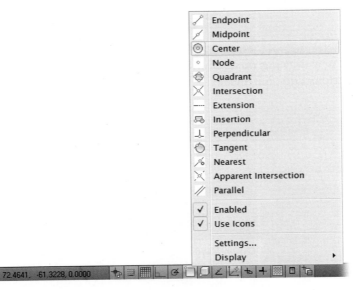

4. Turn on **Snap** mode, and then set the snap spacing to the smallest distance in this model, **0.25**, as follows:

   a. Right-click the **Snap** button, and then choose **Settings** from the shortcut menu.
   b. In the Drafting Settings dialog box, enter **0.25** for the X Spacing, and then press **Tab**.

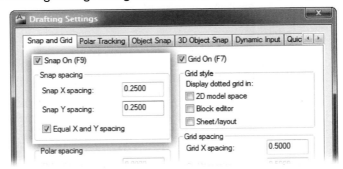

   c. Click **OK** to dismiss the dialog box.

5. Change the viewpoint from plan to isometric view. This viewpoint lets you see three sides of the model at the same time: front, side, and top.

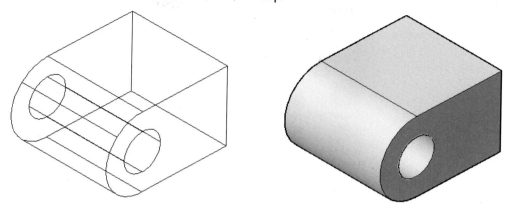

*Left:* A wireframe 3D model in plan view.
*Right:* Same model viewed from an isometric viewpoint in Conceptual rendering.

   a. Click the ribbon's **View** tab.
   b. In the Views panel, click the **Views** button, and then choose **SW Isometric**.

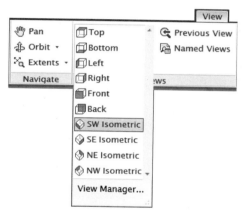

Notice that the cursor and UCS icon rotate to show the three axes (x, y, and z) in isometric mode.

*Left:* UCS icon.
*Right:* 3D cursor.

6. Change the visual style to **Conceptual**. This is a real-time rendering mode known as a "visual style" that makes it easier to view 3D models than would be possible in traditional CAD wireframe mode.

a. Click the ribbon's **Home** tab.
b. In the View panel, click the **2D Wireframe** droplist, and then choose "Conceptual."

You won't see a change until you begin drawing.

7. With this preliminary work accomplished, save the drawing, as follows:
   a. Click the red **A**.
   b. Choose **Save**.

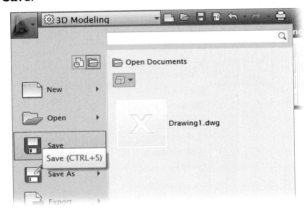

   c. In the Save Drawing As dialog box, name the drawing "First 3D Model."

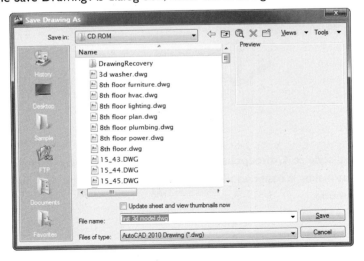

   d. Click **Save**.

## TUTORIAL: DRAWING SOLID PRIMITIVES

In this part of the tutorial, you draw a 3D box shape; later, you turn a circle into a cylinder.

1. In the Home tab's Modeling panel, choose **Box**.

   Notice that AutoCAD prompts you in the command bar. If dynamic input is turned on, you see the prompts in the drawing area.

2. Boxes are defined by the two corners of their rectangular base and their height. Start the box at the origin of the drawing, which is indicated by the UCS icon:
   Specify corner of box or [CEnter] <0,0,0>: *(Pick a point.)*

   Recall that the origin has x, y coordinates of 0,0. (With snap mode turned on, it is easy to pick the 0,0 point precisely.)

3. For the opposite corner of the box's rectangular base, drag the preview rectangle to 5,4:
   Specify corner or [Cube/Length]: *(Drag the base to 5,4.)*

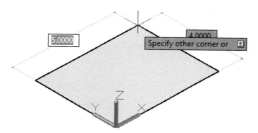

4. For the height, drag the rectangle to 3.
   Specify height: *(Drag the box to a height of 3.)*

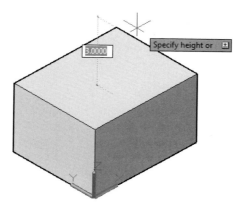

   Specifying the height is the same as specifying the z coordinate.

5. You have drawn a box as a 3D solid model. Press **CTRL+S** to save your work.

If you find it too finicky to *show* AutoCAD the size of the box, you can enter numbers for the x,y coordinates and the height like this:

> Specify corner of box or [CEnter] <0,0,0>: **0,0**
>
> Specify corner or [Cube/Length]: **5,4**
>
> Specify height: **3**

Other commands for creating 3D primitives include SPHERE, TORUS (donut-like shapes), WEDGE, CYLINDER, and CONE. You can practice them in the exercises at the end of this chapter.

 **Note** To zoom the 3D model, place the cursor over the 3D model, and then roll the mouse's roller wheel. In one direction, the model zooms larger; in the other, smaller. By having the cursor over the model, you keep it centered in the drawing area while zooming.

### TUTORIAL: MODIFYING 3D MODELS

Once you draw a basic 3D model, you can modify it with generic editing commands, like FILLET and CHAMFER, and with commands specific to 3D, such as SOLIDEDIT and PRESSPULL. You can also edit 3D models directly with grips.

In this tutorial, you round one end of the box with the FILLET command. The command operates in 3D somewhat differently from what you may be used to from 2D drafting.

1. In the Home tab's Modify panel, choose ⌐ **Fillet**.

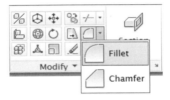

2. When AutoCAD prompts you to select an object, click the edge shown in the figure:
   Select first object or [Undo/Polyline/Radius/Trim/Multiple]: *(Pick the edge.)*

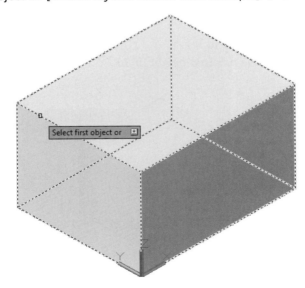

3. When prompted, specify a radius of 1.5 units (half of the box's height of 3 units):
   Enter fillet radius: **1.5**
4. Select the lower edge, as shown below:
   Select an edge or [Chain/Radius]: *(Pick lower edge.)*

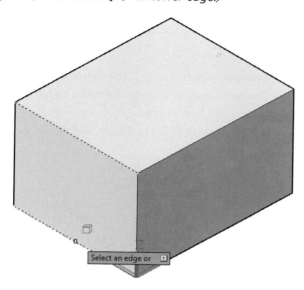

5. Press **ENTER** to end the command:
   Select an edge or [Chain/Radius]: *(Press ENTER.)*

   Notice that the one end is instantly redrawn with curvature.

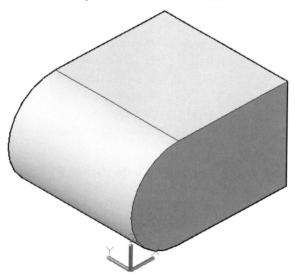

## TUTORIAL: CREATING HOLES

One of the most common actions in 3D modeling is making holes. This is done indirectly by removing a cylinder shape from the model in two steps: (1) draw a cylinder, and then (2) remove it from the model.

You could use the CYLINDER command, but I find it easier to place a circle, and then extrude it into a cylinder shape. This is an example of the second method of creating parts of 3D models that I mentioned earlier: extruding 2D objects.

1. You will be placing the circle on a side face of the now-rounded box. Selecting faces can be tricky, and so AutoCAD provides *subobject filtering*. It makes it easier to select a specific part of a 3D object, such as a single face or edge.
   In the Home tab's Subobject panel, click **No Filter**, and then choose **Face**.

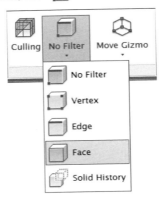

Notice the square face icon near the cursor. It indicates that subselection mode is set to Face. The complete list of subobject filters is shown in the following table:

| Subobject Filter | Icon | Meaning |
| --- | --- | --- |
| None | ... | Selects the 3D model. |
| Vertex |  | Selects a vertex. |
| Edge |  | Selects an edge. |
| Face |  | Selects a face. |
| Solid History |  | Selects a primitive used to make the 3D model. |

2. In the Home tab's Draw palette, choose **Circle** (Center, Radius).

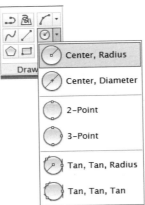

3. Move cursor near the semicircle of the fillet; notice that AutoCAD's object snap finds the center of the fillet.

    Specify center point for circle or [3P/2P/Ttr (tan tan radius)]: *(Pick the filleted curve.)*

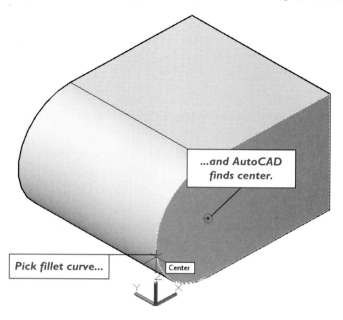

4. Enter the radius of 0.75 units:

    Specify radius of circle or [Diameter]: **.75**

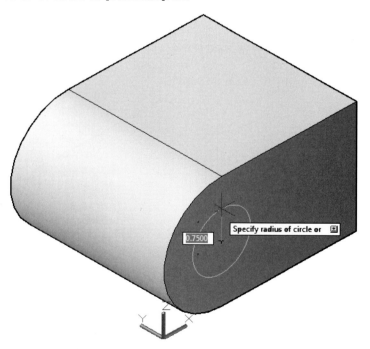

5. With the circle drawn, you extrude it into a cylinder shape. In the Home tab's Modeling panel, choose 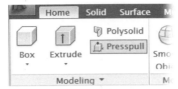 **Press/Pull**.

6. Turn the circle into a cylinder like this:
   a. Click inside the circle, and then drag it. Notice that the 2D circle elongates into a 3D cylinder.
   b. Drag the length of the cylinder until it clears the other end of the model; the precise length does not matter.
   c. Click to end the command.

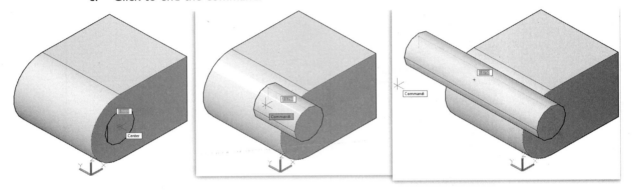

**Left to right:** (a) Select the circle, (b) drag it, and (c) click.

Notice that the model now has a hole through it.

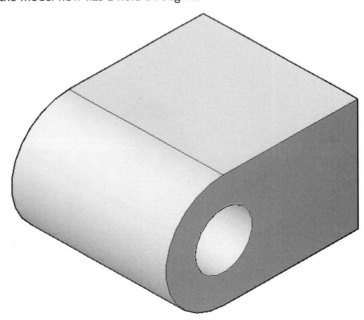

7. You can get a better view of the hole by rotating the model like this:
   a. Grab any part of the ViewCube.

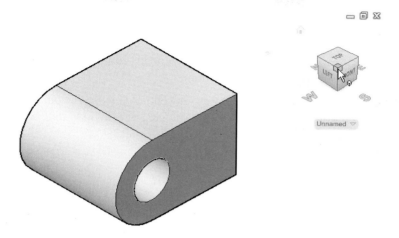

   b. Drag the cube. As you do, notice that the 3D model rotates.

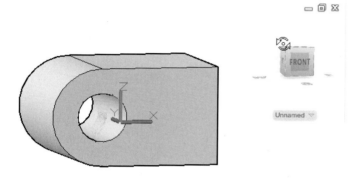

8. Press CTRL+S to save your work.

In this tutorial, you completed the most common solid modeling tasks:

> Created a 3D primitive (box).
> Modified a primitive (fillet).
> Extruded a 3D shape from a 2D object (circle to cylinder).
> Rotated the 3D viewpoint (view cube).

## REVOLVED MODELS

In the earlier tutorial, you created a 3D object (the cylinder) by extruding a circle. A second way to generate a 3D object is to revolve a 2D outline. The REVOLVE command creates symmetrically round objects, like wineglasses and chair legs.

To revolve (convert) 2D objects into 3D solids, AutoCAD needs to know two things: (1) the objects to revolve, and (2) the axis about which to revolve them.

Objects that can be revolved include circles, ellipses, closed polylines, polygons, donuts, closed splines, and regions. (AutoCAD cannot revolve objects within blocks, or polylines that cross over themselves.) These objects are known as the profile.

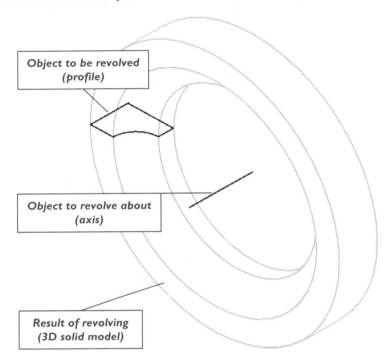

The axis can be the x or y axis, or an object, such as a line or polyline segment. The length of the axis is unimportant; AutoCAD uses only the axis' orientation in space to determine the location and angle of the revolution.

You can specify full 360-degree revolutions, or partial revolutions, with positive or negative angles.

Introduction to 3D Solid Modeling    769

## TUTORIAL: REVOLVING OBJECTS

In this tutorial, you create the small washer illustrated below. Since it is round, it makes sense to create it by revolving the profile of a cross-section.

1. Open the *3D washer.dwg* file, the drawing of a rubber washer's profile. This washer design has an inner fillet.

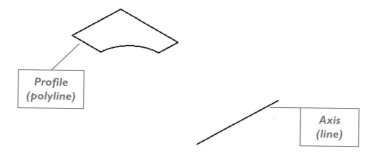

   Notice that there is a *profile* drawn from a polyline, and a single line, which is the *axis* about which you will revolve the profile.

2. In the Home tab's Modeling panel, click the ▼ button below Extrude, and then choose **Revolve** from the flyout.

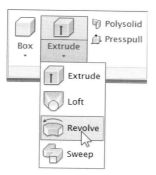

3. When prompted, select the profile (polyline outline), and then press ENTER:
   Select objects to revolve or [MOde]: *(Choose the profile.)*
   Select objects to revolve or [MOde]: *(Press ENTER.)*

4. Select the axis (line object):
   Specify axis start point or define axis by [Object/X/Y/Z] <Object>: **o**
   Select an object: *(Choose the line.)*

5. Finally, specify the angle of revolution. For a full washer, specify 360 degrees:
   Specify angle of revolution or [STart angle] <360>: *(Press ENTER.)*

Notice that AutoCAD instantly draws the washer in 3D as a solid model.

When you specify a smaller angle of revolution, you get a washer that's been cut (at left). You can use the PRESSPULL command to extend parts of the washer (at right) to create a horseshoe-like shape.

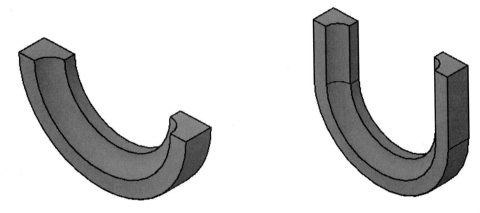

**Left:** *Profile revolved by 180 degrees.*
**Right:** *Endcaps of the 3D model extruded.*

You get different results, depending on the axis you choose. The figures below show the result of revolving the same profile about the x and y axes — instead of the line. The resulting shape differs, as does the size. (So that you can see the profile, I revolved them by 270 degrees and colored the area of the profile white.) This shows that the distance of the profile from the axis affects the shape and size, and so it is crucial to place the profile correctly relative to the axis.

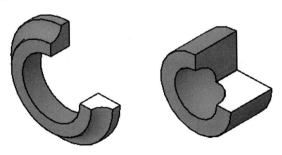

**Left:** *Profile revolved about x axis.*
**Right:** *Profile revolved about y axis; washer hole is closed.*

# UNDERSTANDING THE UCS

To create and edit 3D drawings, you must become comfortable with changing the point of view, and working at any angle in 3D space. AutoCAD provides a number of tools that create and save 3D viewpoints. One of the commands is UCS — short for "user-defined coordinate system".

## WORLD COORDINATE SYSTEM (WCS)

If there are coordinate systems that can be defined by users, then there must be an absolute system to which they refer. There is: the *world coordinate system* (WCS, for short).

When you start AutoCAD with a template file like *acad.dwt*, it begins in WCS plan mode. You look straight down onto the x, y-plane, with the z axis coming out of the monitor towards your face. You can change the viewpoint, and it is still the WCS.

### User-defined Coordinate Systems (UCS)

But when you change the coordinate system, you define a new one — a UCS. Why would you do this? To solve a problem that AutoCAD has; namely, it is 2D-oriented, and so it is difficult to draw and edit in 3D.

The workaround is to create UCSes that rotate the x, y-plane in 3D space to match the plane in which you want to drawn and edit. Think about drawing on the face of a sloped roof: the task is hard if you could not reorient the drawing plane to match the roof. A tutorial later in this section does exactly that.

### UCS Icons

The UCS icon helps orients you in 3D space. When you first start AutoCAD, it shows the direction of the x (red) and y (green) axes in the WCS:

The small yellow square at the intersection of the axes means that AutoCAD is in WCS (world coordinate system). You are looking straight down on the x,y-plane, so you do not see the z axis. When you invoke a user-defined coordinate system, the UCS icon loses the square.

When you change to a 3D viewpoint, however, the z (blue) axis appears. For example, enter VPOINT **1,2,3** to see the following:

The UCS icon is usually positioned at the origin (0,0); if the origin is outside the viewport, then the icon is located in the lower left corner of the viewport.

When the viewpoint goes underneath the x, y-plane, the z axis is negative. The z portion of the UCS icon becomes dotted. Enter VPOINT **1,2,-3** to see an example of this:

The icon can change its look. The original UCS icon showed only the direction of the x and y axes, and no z axis. It was introduced in 1990 when "full 3D" was added to AutoCAD Release 10. (See figure below.)

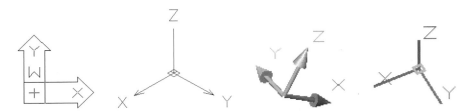

**Left to right:** Development of the UCS icon through the decades:
original 2D UCS icon; earliest 3D UCS icon; redesign of 3D UCS icon; current design.

When you switch to layouts, the icon is strictly 2D and strictly ornamental. (Its design represents the plastic 30-60-90 triangle used by hand drafters.) There is no 3D in layouts, because they are meant to represent flat sheets of paper.

You can adjust the look of the UCS icon with the UCSICON command's **Properties** option.

### TUTORIAL: CHANGING FROM WCS TO UCS

In this tutorial, you draw an object in the WCS, and then draw the same object in a UCS to see how the two coordinate systems affect 2D drafting and editing in 3D space.

1. Start AutoCAD with a new drawing using the *acad.dwt* template file. AutoCAD always starts in WCS mode, the world coordinate system.
2. With the **CIRCLE** command, draw a circle with a radius of 2. The circle is drawn in the x, y-plane of the WCS.

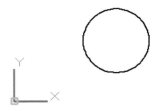

3. Rotate the 3D viewpoint: in the View tab's Views panel, choose  **SW Isometric**. Notice that the circle and UCS icon rotate to match the new viewpoint; the circle looks like an ellipse because the viewpoint is rotated.

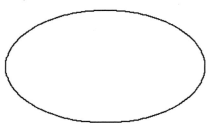

4. The drawing is still in WCS; only the viewpoint has changed (to -1,-1,1). To see this, draw another circle with any size of radius.

5. Turn on the grid, which always lies in the x, y-plane of the current UCS (actually, the WCS in this case).

   Notice that the second circle and the grid lie in the same plane as the first circle.

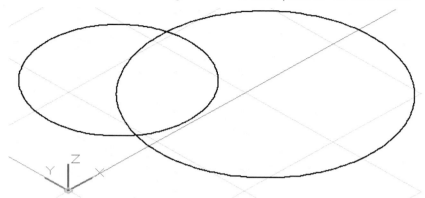

6. UCSs are created with the **ucs** command. In the View tab's Coordinates panel, choose **UCS**. (Alternatively, enter **ucs** at the 'Command:' prompt.)

7. Enter **n** to create a new UCS:
   Enter an option
   [New/Move/orthoGraphic/Prev/Restore/Save/Del/Apply/?/World] <World>: **n**

8. To align the coordinate system with the current viewpoint, enter **v**:
   Specify origin of new UCS or [ZAxis/3point/OBject/Face/View/X/Y/Z] <0,0,0>: **v**

   Notice that the UCS icon rotates to show just the x and y axes, just as if you were looking at the drawing in plan view. In addition, the grid "faces you."

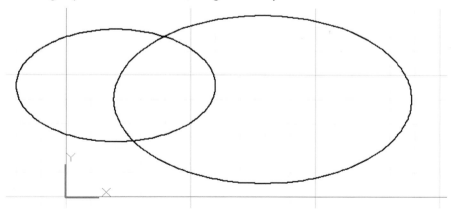

   This aligns the UCS to the current view, and is probably the easiest option to use — but also the least useful, because it does not necessarily align with features on 3D objects. (You'll see how to do this in the next tutorial.)

9. Draw a third circle. Note that you are drawing it in the new x,y-working plane.

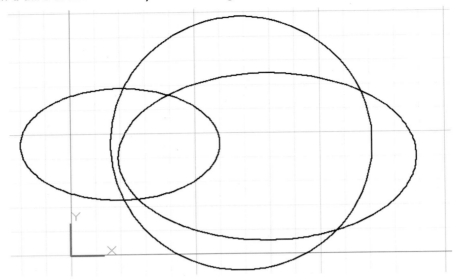

10. You can return to the WCS by clicking the ◎ **World** button in the View tab's Coordinates panel. Alternatively, you can enter the **UCS** command's **World** option:

    Command: **ucs**

    Enter an option [New/Move/orthoGraphic/Prev/Restore/Save/Del/Apply/?/World] <World>: *(Press* ENTER.*)*

11. Return to plan view with the **PLAN** command.

By changing the orientation of the coordinate system to match the viewpoint of 3D objects, you can draw just as in plan view.

### TUTORIAL: USING DYNAMIC UCS

The UCS command is good enough for setting up specific coordinate systems that you want to reuse frequently. In other cases, however, you want the coordinate system to match the current working plane on-the-fly.

This is where *dynamic UCS* becomes useful. When it is turned on, AutoCAD matches the UCS to the selected face, as described in this tutorial.

1. Open the *Dynamic UCS.dwg* file, a drawing of a wedge-shaped structure. You will place text on three sides using dynamic UCS.

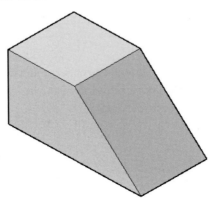

2. Ensure Dynamic UCS is turned on: its  button on the status bar should glow blue.
3. When working with UCSes, it makes sense to turn on cursor labels.

These display X, Y, and Z on the three axes of the cross hair cursor, just as with the UCS axes. To do so, follow these steps:
   a. Enter the **OPTIONS** command, and then choose the **3D Modeling** tab.
   b. In the 3D Crosshairs section, turn on the **Show Labels for Dynamic UCS**, and then click **OK**.

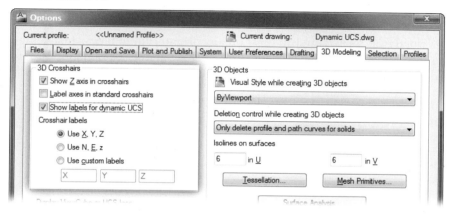

4. To see dynamic UCS in action, start the **TEXT** command.
5. At the 'Specify start point of text or [Justify/Style]:' prompt, move the cursor over the faces of the wedge. Notice that the cursor changes its orientation, depending on the angle of the face. In the figure below, I superimposed the cursor on all three faces:

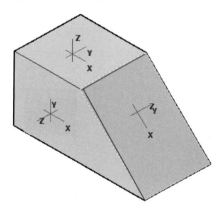

In each case, the x, y axes of the cursor orient themselves so that they lie *in the plane* of the face. The z axis changes its orientation so that it always points *away from the face*.

6. Place text on each of the three faces:
   Command: **text**
   Specify start point of text or [Justify/Style]: (Pick a point on the side face.)
   Specify height <0.2000>: *(Press ENTER.)*
   Specify rotation angle of text <0>: *(Press ENTER.)*
   **Side** *(Press ENTER.)*

7. Repeat the TEXT command for the top and sloping faces. The result will look similar to the figure below.

Dynamic UCS is wonderful for letting you draw on any orientation of faces. Just remember, however, that it kicks in only *during* commands; it does not operate at the 'Command:' prompt. As well, dynamic UCS does not operate at all on curved faces, like those on spheres and cones; flat faces only!

## JOINING SOLID MODELS

Sometimes you need to join two or more solid models. This is accomplished with the UNION command, and the result is known as a body.

This command joins 2D regions into a single region, and 3D solid objects into a single solid model; regions and 3D solid models cannot, however, be mixed. The objects do not need to be touching or intersecting to be unioned.

To execute this command, AutoCAD needs to know just one thing from you: the objects to be joined.

### TUTORIAL: UNIONING SOLIDS

In this tutorial, you join three solid models into a single body.

1. Open the *Union.dwg* file, a drawing of three solid models.
2. Move the cursor over the parts to confirm they are independent of each other; AutoCAD highlights just one part at a time.

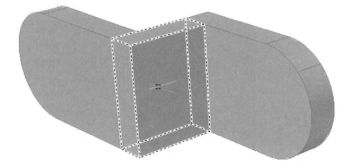

3. Join the three parts as follows:
    a. In the Home tab's Solid Editing panel, choose **Union**.
    b. In response to the prompt, enter **All** to select all objects in the drawing.

    Select objects: **all**

    Select objects: *(Press ENTER.)*

    c. Press ENTER to exit the command.

4. Move the cursor over the parts to confirm they are part of a single body; AutoCAD highlights all of them at once.

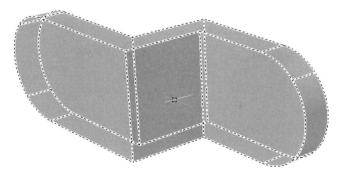

5. Even though the three are joined into one, you can still edit the originals with the SHOW-HIST system variable. Change the value to **2**, as follows:

   Command: **showhist**

   Enter new value for SHOWHIST <1>: **2**

   Notice that AutoCAD ghosts the outlines of the original solid models, including the boxes before their ends were filleted.

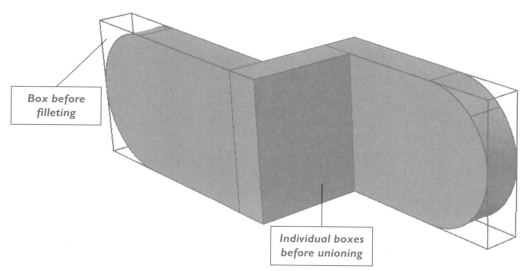

*Box before filleting*

*Individual boxes before unioning*

6. To edit the *subobjects* (individual parts of the solid body), follow these steps:
   a. In the Home tab's Subobject panel, select **Solid History**.

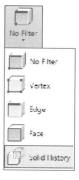

   b. Hold down the **CTRL** key, and then select one of the subobjects. The cursor shows the icon, indicating it is in subobject selection mode.

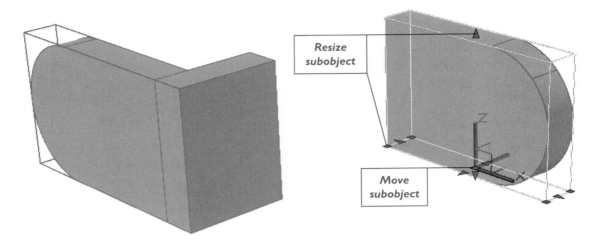

   c. Notice the grips; you can drag them to change the size of the subobject, even separate them by moving them away from each other.
   d. Press ESC to exit direct editing mode.

The subobjects still comprise a single solid body, even when you separate sub-objects. The solid is not contiguous, but still a single object.

## EXERCISES

### 3D SOLID MODEL PRIMITIVES

Create 3D primitives with these modeling commands:

1. **SPHERE** command
    Specify center point or [3P/2P/Ttr]: *(Pick a point.)*
    Specify radius or [Diameter] <1.0>: **2**

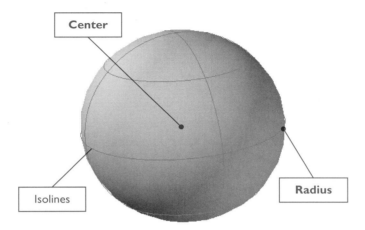

2. **BOX** command
    Specify first corner or [Center]: *(Pick a point.)*
    Specify other corner or [Cube/Length]: **1,4**
    Specify height or [2Point] <2.0>: **9**

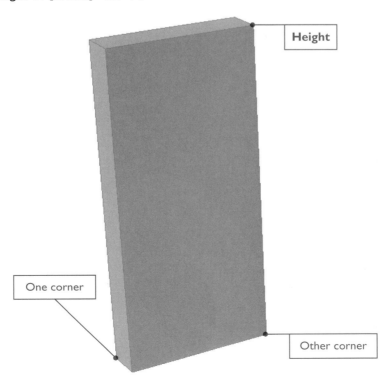

3. **CYLINDER** command

    Specify center point of base or [3P/2P/Ttr/Elliptical]: *(Pick a point.)*
    Specify base radius or [Diameter] <2.0000>: **2**
    Specify height or [2Point/Axis endpoint] <9.0>: **2**

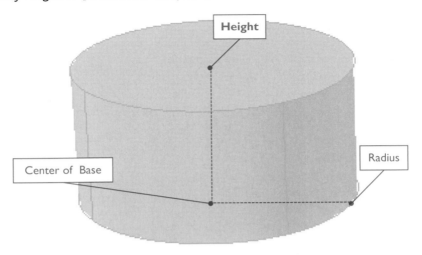

4. **WEDGE** command

    Specify first corner or [Center]: *(Pick a point.)*
    Specify other corner or [Cube/Length]: **@4,6**
    Specify height or [2Point] <2.0>: **2**

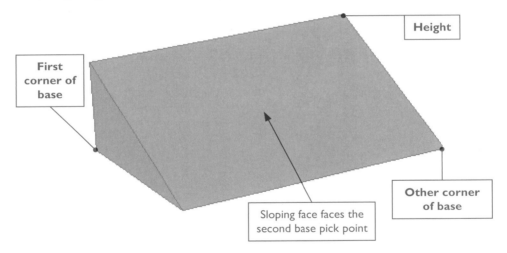

5. **CONE** command

    Specify center point of base or [3P/2P/Ttr/Elliptical]: *(Pick a point.)*
    Specify base radius or [Diameter] <2.0000>: **2**
    Specify height or [2Point/Axis endpoint/Top radius] <2.0000>: **3**

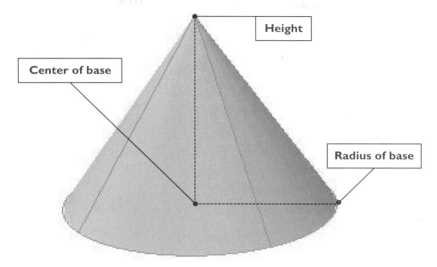

6. **TORUS** command

    Specify center point or [3P/2P/Ttr]: *(Pick a point.)*
    Specify radius or [Diameter] <2.0000>: **2**
    Specify tube radius or [2Point/Diameter]: **.5**

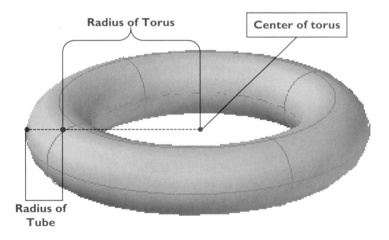

## OTHER SOLIDS OPERATIONS

Construct 3D models from 2D objects using these modeling commands:

7. **REVOLVE** command
   a. Open the *Revolve Exercise.dwg* file, a drawing of a spool profile.

   b. Use the **REVOLVE** command to create the spool.
   c. Change the visual style to **Conceptual**.
   d. Drag the ViewCube to rotate the view to see the spool from another angle.

8. Create a 3D model of the flow reducer using these dimensions:

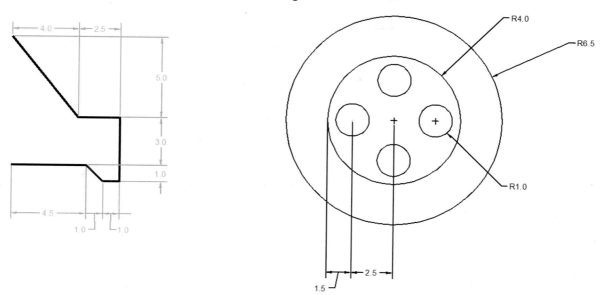

The result should look like the following views of the 3D model:

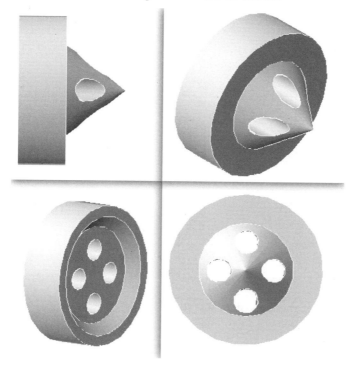

## CHAPTER REVIEW

1. Where is most "3D drafting" performed in AutoCAD?
2. Are faces only flat?
3. Label the parts of the solid models illustrated below:

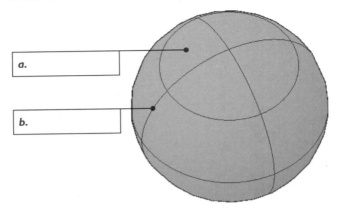

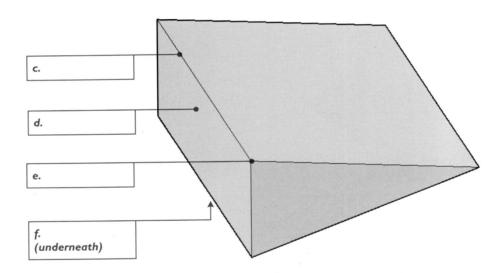

4. How do isolines affect the look of curved surfaces?
5. What are *subobjects*?
6. How do you select a subobject?
7. How does AutoCAD enable drafters to easily work on x, y-planes in 3D space?
8. Does dynamic UCS work on curved surfaces?
9. Describe the advantage of working in *isometric mode*.
10. What is the meaning of the following abbreviations?
    a. WCS
    b. UCS

11. Describe three functions performed by the UCS icon:
    a.
    b.
    c.
12. How is the ViewCube useful?
13. What is a *primitive*?
14. Describe two ways to begin creating 3D models:
    a.
    b.
15. How are holes made in 3D models?
16. Which command joins two or more solid models?
    Must the models be touching to be joined?
17. What is the name for a single solid model made up of two or more other solids?
18. Can bodies be edited?
    If so, how?_____
19. Identify the following primitives:

    a.

    b.

    c.

    d.

    e.

    f.

    g.

# CHAPTER 17

## Introduction to 3D Mesh Modeling

In the last chapter, you experienced the precision of 3D solid modeling; in this chapter, you learn about a second way of producing 3D models, through mesh modeling.

Three-dimensional mesh modeling is better than solid modeling for organic flowing shapes, such as hair dryers and cell phones. Mesh modeling is especially popular for styling automobiles and consumer products.

To distinguish them from pre-2010 3D mesh objects made of polyfaces, Autodesk calls the new ones "full featured mesh objects" and "subdivision" meshes; in this book, they are named "3D mesh objects."

In this chapter, you learn the following commands for constructing mesh models:

> **MESH** creates 3D mesh primitives.
>
> **MESHOPTIONS** specifies the number of tesselation lines for mesh primitives.
>
> **MESHSMOOTHMORE** and **MESHSMOOTHLESS** increase and decrease the smoothness levels of 3D mesh models.
>
> **MESHCREASE** and **MESHUNCREASE** add and remove creases from smoothed mesh models.
>
> **MESHCAP** caps open areas of meshes (new to AutoCAD 2011).

---

**NEW TO AUTOCAD 2011** IN THIS CHAPTER

- **MESHCAP** command closes open areas of meshes.

---

## PARTS OF 3D MESH MODELS

Mesh models differ from solid models in that they are hollow; you cannot analyze mesh models, for example to find their mass or centroids. Their advantage is that you can easily manipulate the surfaces of meshes to create unique shapes.

Visually, 3D mesh models are covered by tessellation lines, whereas solid models sport lines only along their edges and isolines. Tessellation lines delineate the meshes that cover surfaces; each face has one or more meshes. Three-dimensional mesh models use much of the same vocabulary as solid models, as the figure below illustrates.

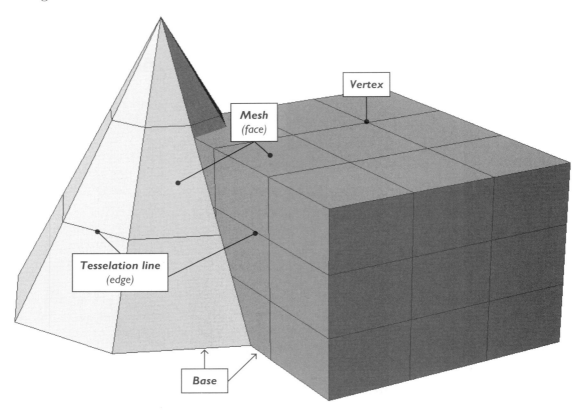

**Meshes** cover the surfaces of mesh models. Generally, each mesh has four edges; those on the bases of cylinders and cones are triangular. During subobject selection, AutoCAD considers each mesh to be a face.

**Tessellation lines** define the edges of flat and curved meshes. During subobject selection, AutoCAD considers each tessellation line to be an edge.

**Vertices** appear at the intersections of tessellation lines.

**Bases** form the bottoms of 3D models. When creating primitives like boxes and cylinders, you first define the size of the base and then the height. The bases of boxes are squares or rectangles; those of cones are circles or ellipses.

## MULTIPLYING MESHES

The number of meshes on mesh models is whatever number you require for modeling. When first drawing a mesh model primitive, such as box or sphere, you specify the number of tessellation lines (which define meshes) with the MESHPRIMITIVEOPTIONS command. The default is three meshes in each direction.

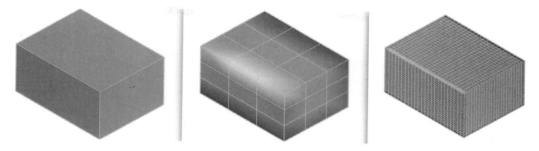

**Left to right:**
*Mesh density = 1 (minimum), 3 (default), and 24 tesselation divisions.*

A greater number of meshes allows you model a greater number of details at the cost of increased processing time and slower computer response. The solution is to selectively add more meshes in one area of the model through *subdivision* with the MESHREFINE command, as illustrated below.

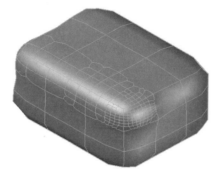

### Increasing Smoothness

Mesh models are big on *smoothness*, a term appearing frequently in this chapter. Indeed, the entire point to meshes is that they can be rounded off (smoothed) to create organic shapes. Once you smooth mesh models with the MESHSMOOTHMORE command, you can selectively sharpen edges by adding *creases* with the MESHCREASE command.

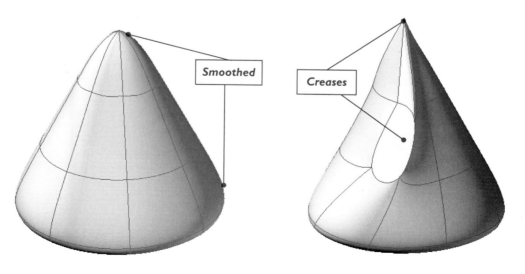

Smoothing can be applied more than once, up to a maximum level of 4. Each application progressively rounds off the corners to the point that a cube looks nearly like a sphere. (Actually, smoothing can go all the way to level 255, but your computer might not have the processing power required, subsequently slowing down operations, and so Autodesk limits it to 4.)

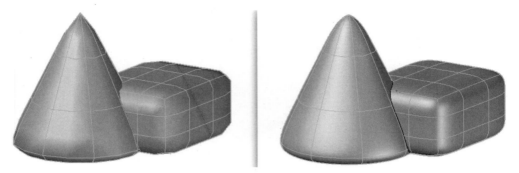

**Left:** *Pyramid and box with smoothness level = 1.*
**Right:** *Smoothness level = 4.*

You can "un-smooth" smoothed mesh models with the MESHSMOOTHLESS command all the way back to level 0 where the hard edges reappear.

Unfortunately, AutoCAD uses the term "smoothness" for two different operations. In one, the MESHSMOOTHMORE command smooths hard-edged mesh models; in another, the MESHSMOOTH command converts 3D solid models and surfaces into mesh models, even though they won't look smooth after conversion.

### Manipulating Mesh Models

Mesh models were meant to be manipulated directly. You just grab a face, vertex, or edge, and then drag — no command needed. (You use the same subobject selection method as with 3D solid models.)

As you drag, adjacent faces drag along, as illustrated below. To create the organic shape shown at right, I maximized the smoothness of the cube (to level 4), and then dragged two meshes — one on the left face and another on the top face. Total modeling time: under ten seconds.

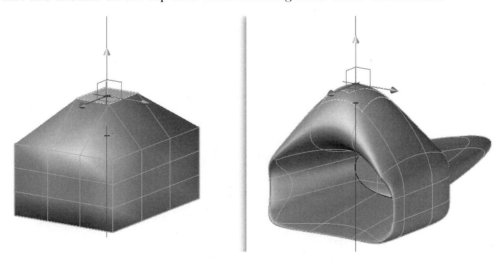

**Left:** *Dragging a face on an unsmoothed cube.*
**Right:** *Dragging faces on a smoothed cube.*

Use the 3D gizmo to move, rotate, and scale faces in the x, y, or z directions. In the figures above, I selected the z axis (blue rod) of the 3D Move gizmo to ensure my dragging action was precisely vertical.

## Creating and Editing 3D Mesh Models

3D models can be created in several ways: from primitives, as described next; from other objects, such as solid models and surfaces; and from mesh creation commands, like TABSURF and LOFT. Some of these are described in this chapter.

In addition to editing mesh models by smoothing, creasing, and dragging subobjects, you can edit them using Boolean operations, filleting and chamfering corners, and so on. The catch is that these operation cannot be applied directly to mesh models; they must first be converted to solid models.

### TUTORIAL: CREATING 3D MESH PRIMITIVES (MESH)

The MESH command creates 3D mesh primitives: square and rectangular boxes, pointy and flat top cones, round and elliptical cylinders, multisided pyramids, spheres, wedges, and tori.

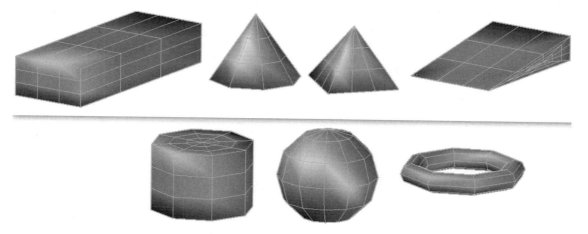

*Top row, left to right:* Mesh box, cone, pyramid, and wedge.
*Bottom row:* Mesh cylinder, sphere, and torus.

In this tutorial, you create a shape that looks like a key cap on a computer keyboard. Starting with a box, you smooth it, depress the top, and then crease the bottom edges. The finished 3D model will look something like the figure illustrated below.

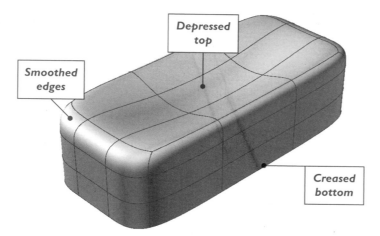

1. Start a new drawing with AutoCAD using the *acad3d.dwt* template drawing.
2. Turn off the grid by clicking the **Grid** button on the status bar.
   Ensure the workspace is set to "3D Modeling."

3. Drawing meshes is easier using the ribbon than entering commands at the keyboard. To do so, click the **Mesh Modeling** tab in the ribbon.

4. To draw the box, choose **Mesh Box** in the Primitives panel of the Mesh Modeling tab.

When you click the button, AutoCAD starts the MESH command and enters the **B** (box) option on your behalf:

Current smoothness level is set to : 0

Enter an option [Box/Cone/CYlinder/Pyramid/Sphere/Wedge/Torus/SEttings] <Box>: B

5. At the following prompt, pick a point anywhere in the drawing:

Specify first corner or [Center]: *(Pick a point to locate the first corner of the box.)*

(The current smoothness level is 0, which means the box will have sharp edges.)

6. Pick a second point to define the size of the base:

Specify other corner or [Cube/Length]: *(Pick another point to specify the area of the base.)*

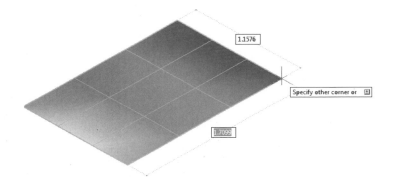

Notice that the base is made of a 3x3 grid of meshes.

7. Drag the cursor to indicate the height:
    Specify height or [2Point]: *(Pick a point to specify the height.)*

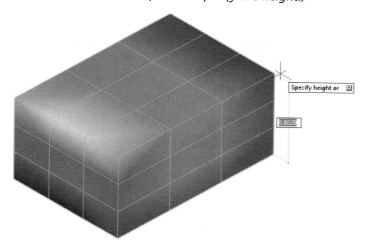

The box is complete. In the next tutorial, you apply smoothing to the box.

### Setting Tessellation Lines

The **MESH** command uses a default number of tessellation lines, which define the number of mesh "rectangles" on the surface of mesh models. AutoCAD provides no fewer than three ways to change default values.

Here we look at two methods; the third method uses system variables and is much too cumbersome.

### MeshOptions

The **MESHOPTIONS** command presets the number of tessellation lines for all primitives through a dialog box. To access the dialog box, click the ⬊ diagonal arrow on the Primitives panel, as illustrated below.

The Mesh Primitive Options dialog box previews each of the primitives as you change the number of tessellation lines.

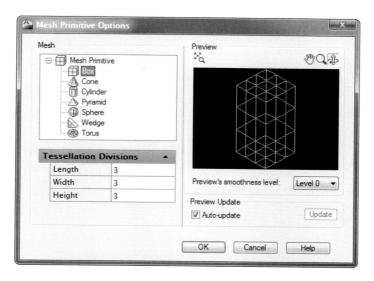

The minimum number is usually 1, the maximum 256:

**Length** — specifies the tessellation lines along the length of boxes (x axis); default = 3.

**Width** — specifies the tessellation lines along the width of boxes (y axis).

**Height** — specifies the tessellation lines along the height of boxes (z axis).

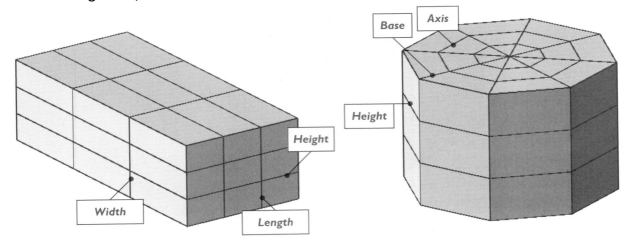

For cones, the divisions are specified by different names:

**Axis** — specifies the tessellation lines around the perimeter of the cylinder's base. The minimum value is 3, which draws a cylinder with a triangular base; the same restriction applies to the axes of cones and spheres, as well as to all parameters of tori.

**Base** — specifies the tessellation lines between the perimeter and center of the cylinder's base.

**Height** — specifies the tessellation lines between the base and top of the cylinder.

Remember to avoid entering large values for Tessellation Division, because they slow down AutoCAD.

### MESHSMOOTHMORE AND MESHSMOOTHLESS

The **MESHSMOOTHMORE** command rounds off the corners of 3D mesh models. Each time you apply the command, the model is smoothed some more — up to the maximum level set by **SMOOTHMESHMAXLEV**, which is set to 4 by default. When you go beyond the limit, AutoCAD warns you with this dialog box:

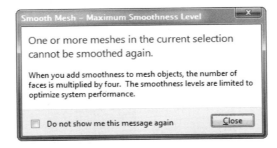

The complementary command is **MESHSMOOTHLESS**, which reduces the level of smoothness by one step until it reaches 0.

**Note** You can draw mesh primitives with any level of smoothness with the **MESH** command's **SEttings** option, as follows:

Command: **mesh**
Current smoothness level is set to : 0
Enter an option [Box/Cone/CYlinder/Pyramid/Sphere/Wedge/Torus/SEttings] <Box>: **se**
Specify level of smoothness or [Tessellation] <0>: *(Enter a value between 0 and 4.)*

The **Tessellation** option displays the Mesh Primitive Options dialog box described earlier.

## TUTORIAL: SMOOTHING MESH MODELS

In this tutorial, you apply smoothing to the box created in the earlier tutorial.

1. From the 3D Modeling ribbon's Mesh Modeling tab, choose the **Smooth More** button in the Mesh panel. (Don't click Smooth Object, which has a different function!)

2. Pick the box. Notice that its edges are somewhat rounded off (smoother).
   Select mesh objects to increase the smoothness level: *(Pick the box.)*
   Select mesh objects to increase the smoothness level: *(Press **Enter** to exit the command.)*

3. Press **SPACEBAR** to repeat the command, and then choose the box again.
4. Repeat until AutoCAD presents its warning dialog box.

**Note** The dialog box indicates the box has reached the maximum level of smoothness (4). If you have a really powerful computer, you can increase the limit to 255 with the **SMOOTHMESHMAXLEV** system variable — although Autodesk recommends using 5 as the maximum. The technical editor finds that any value above 5 looks no different; at 8, AutoCAD complained the computer lacked sufficient memory.

5. Click **OK** to exit the dialog box. At this point, the box should look something like the one illustrated below.

In the next tutorial, you depress the top of the box.

### MOVING, ROTATING, AND SCALING MESH FACES, EDGES, AND VERTICES

There is no command for manipulating the surface of 3D mesh objects. Rather, you use the cursor interactively to modify surfaces. You choose a face, edge, or vertex, and then drag. You can use 3D gizmos to move, rotate, and scale faces, as illustrated below.

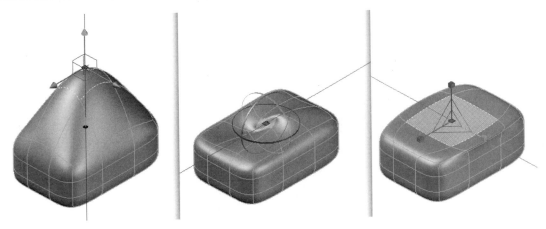

*Left: Moving a face.*
*Center: Rotating a face.*
*Right: Scaling a face larger (highlighted in light color).*

The dynamic UCS icon lets you choose an axis or plane along which to perform the transformation, such as along the z axis or in the x, y-plane.

*Introduction to 3D Mesh Modeling* **797**

**TUTORIAL: MOVING FACES**

In this tutorial, you move a face to create a depression in the top of the box.

1. Before manipulating a 3D mesh object, you need to decide on the subobject to be manipulated. For this tutorial, you will be moving a face, so you should set the subobject selection mode to Face.
2. In the Subobject panel (Mesh Modeling tab), choose **Face**.

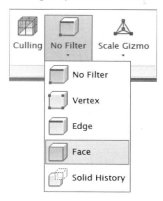

Notice the  Face icon next to the square pickbox cursor: AutoCAD confirms it is in face subobject selection mode.

3. Move the cursor over the box. Notice that AutoCAD highlights individual faces.

4. *Click* the center face on the top of the box. Notice the 3D Move gizmo (which looks like the UCS icon).

Somewhat harder to see is the ▬ red dot at the center of the face. This is AutoCAD's feedback indicating which face has been selected.

**Note** You can select more than one face for editing by clicking on each face. As you choose additional faces, AutoCAD places a small red dot in each one.

**SHIFT+click** removes subobjects from the selection set. (If you were choosing edges, AutoCAD marks them with short red lines; selected vertices are also marked with red dots. You can freely select faces, edges, and vertices by choosing the appropriate button in the Subobject panel.)

5. Move the cursor over the blue z axis of the 3D Move gizmo. Notice that it turns gold, and that a blue line shoots laser-like from each end of the axis. Click to choose the z axis.

**Note** If you want to rotate or scale the face, now is the time to switch modes. Right-click the z-axis, and then choose the mode from the shortcut menu.

6. With the z axis selected, move the cursor. As you do, notice that the top of the box bulges out or inward, depending on the cursor movement. For this tutorial, move the cursor down (negative z direction) to create a mild depression. This is the top of the key cap.

Notice also that adjacent faces are also moved, but to a lessor extent. This ensures that the 3D mesh model remains smooth.

7. Click to position the depression.

8. Press **ESC** to exit move-face mode.

In the next tutorial, you sharpen the base of the box.

### MESHCREASE AND MESHUNCREASE

You've seen how 3D mesh models are all about smoothness. But there are times you may want certain edges to be sharp — known as *creases* in this business. The **MESHCREASE** command applies creases to subobjects, not to entire mesh models.

There is a tricky aspect to creasing: crease values between 1 and 6 cause the crease to disappear when its value is less than the smoothness level of the mesh. Fortunately, there is a solution to the head scratcher: enter A for Always, and the crease won't disappear.

The **MESHUNCREASE** command removes creases from subobjects.

### TUTORIAL: ADDING CREASES

In this tutorial, you add creases to the base of the box.

1. Rotate the model so that you can see the bottom more clearly. Drag the corner ViewCube until you see the word Bottom.

**Left:** *Dragging the ViewCube by a corner.*
**Right:** *ViewCube rotated to show the bottom of the 3D model.*

2. You will be creasing the edges (tessellation lines) of the bottom of the box, so change subobject selection to **Edge**. Notice the 🔲 edge icon near the cursor.

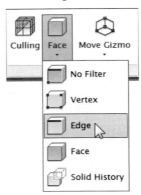

3. In the Mesh panel, choose **Add Crease**.

4. AutoCAD prompts you to select subobjects. Pick an edge making up the base of the box.
   Select mesh subobjects to crease: *(Choose an edge.)*

   Notice that AutoCAD highlights selected edges in white. (No red dots or lines appear this time.)

5. Continue, picking one by one the remaining edges making up the base of the box. In total, you pick 12 edges.

If you pick an incorrect edge, enter **u** to undo the last chosen point. (**Undo** is a hidden option in this command.)

6. When done, press **ENTER**.
   Select mesh subobjects to crease: *(Press* **Enter***.)*

7. AutoCAD asks for the crease value. While the value can range from 0 to 6, enter **A** for "Always" to keep the crease there, no matter how the smoothness level of the rest of the model may change. (Entering **A** is the same as entering 0.)

    Specify crease value [Always] <Always>: **a**

8. Use the ViewCube to rotate the view of the key cap.

    (As a shortcut, click the tiny **house** icon near the ViewCube: this rotates the model back to the "home view," which is the default SW Isometric viewpoint.)

The tutorial is complete. Save your work under the name of *Key cap.dwg*.

 **MESHCAP**

The **MESHCAP** command adds a mesh face that connects open edges.

Sometimes mesh objects are not *water tight*, which means that they have gaps, as illustrated by the figure below. Gaps can appear when mesh objects are improperly translated from other programs, or when the **ERASE** command removes faces from meshes. Leaky mesh objects cannot be converted to solids. Using this command repairs non-watertight meshes.

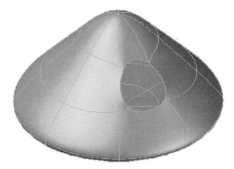

This command prompts you to select the edges to be capped, and then AutoCAD generates the cap out of a single multi-edged face.

**Note** To delete a face, select it using ▇ face subobject selection mode, and then press **DEL**.

### TUTORIAL: CAPPING LEAKING MESH OBJECTS

In this tutorial, you cap an opening in a mesh object.

1. Open the *meshcap.dwg* drawing file. Notice the large hole in the mesh object.

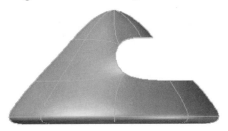

2. Use the ViewCube to rotate the viewpoint to an isometric view that lets you clearly see the edges of the hole.

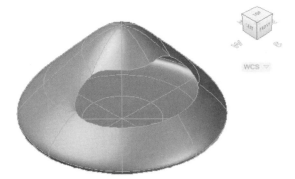

3. To close the opening, start the **MESHCAP** command using one of the following methods:
   - In the ribbon's Mesh tab, choose the **Close Hole** button in the Mesh Edit panel.
   - Or, at the 'Command:' prompt, enter the **MESHCAP** command.

       Command: **meshcap** *(Press ENTER.)*

4. Notice that AutoCAD switches subselection mode to ▯ Edge automatically. Use the pick cursor to select the edges around the hole:

   Select connecting mesh edges to create a new mesh face: *(Choose all edges surrounding the opening.)*

5. When done, press **ENTER**.

   Select connecting mesh edges to create a new mesh face: *(Press ENTER.)*

   Notice that AutoCAD fills in the opening with a single face.

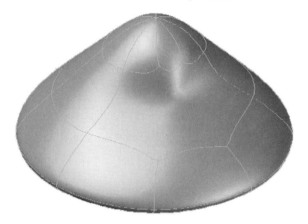

# EXERCISES

## 3D SURFACE MODEL PRIMITIVES

Create 3D primitives with these modeling commands:

1. **MESH** command's **Sphere** option:
   Current smoothness level is set to : 0
   Specify center point or [3P/2P/Ttr]: *(Pick a point anywhere in the drawing.)*
   Specify radius or [Diameter] <1.0>: **2**

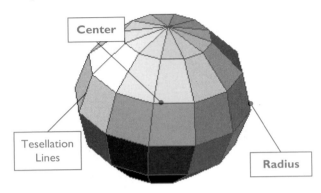

2. **MESH** command's **Box** option:
   Specify first corner or [Center]: *(Pick a point anywhere in the drawing.)*
   Specify other corner or [Cube/Length]: **3,4**
   Specify height or [2Point] <1.0>: **2**

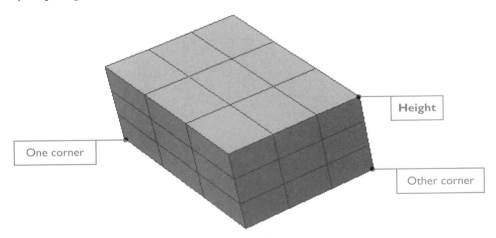

3. **MESH** command's **Cylinder** option:

   Specify center point of base or [3P/2P/Ttr/Elliptical]: *(Pick a point anywhere in the drawing.)*

   Specify base radius or [Diameter] <3>: **2**

   Specify height or [2Point/Axis endpoint] <1">: **2**

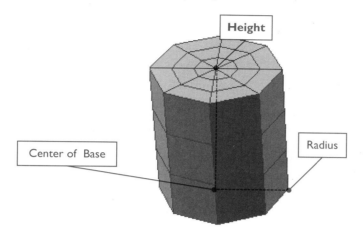

4. **MESH** command's **Wedge** option:

   Specify first corner or [Center]: *(Pick a point anywhere in the drawing.)*

   Specify other corner or [Cube/Length]: **@4,6**

   Specify height or [2Point]: **2**

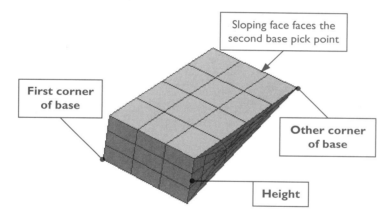

5. MESH command's **Cone** option:
   Specify center point of base or [Edge/Sides]: *(Pick a point anywhere in the drawing.)*
   Specify base radius or [Inscribed]: **3**
   Specify height or [2Point/Axis endpoint/Top radius]: **5**

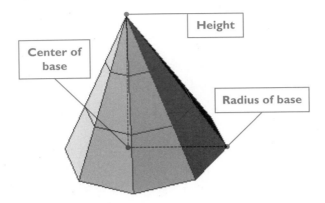

6. MESH command's **Torus** option:
   Specify center point or [3P/2P/Ttr]: *(Pick a point anywhere in the drawing.)*
   Specify radius or [Diameter] <1.0>: **4**
   Specify tube radius or [2Point/Diameter]: **1**

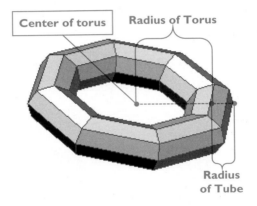

## OTHER MESH OPERATIONS

Modify 3D surface models.

7. In a new drawing, draw a cube.
   a. With which command did you draw the cube?

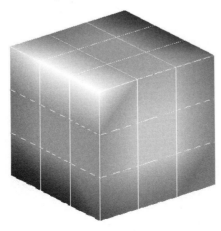

b. Smooth the box.

   Which command smooths 3D mesh models?

c. Repeat until the box is at its maximum level of smoothness.

d. Use the **MESHSMOOTHLESS** command to unsmooth the box.

8. Create a 3D mesh model of the deflated basketball using these commands:

   a. **MESH** command's **Sphere** option.
   b. **MESHSMOOTHMORE** command.
   c. Subobject selection set to **Face**.
   d. Rotate face.

9. Create a 3D mesh model of the topless ball cap using these commands:
   a. **MESH** command's **Torus** option.
   b. **MESHSMOOTHMORE** command.
   c. Subobject selection set to **Face**.
   d. Scale face.

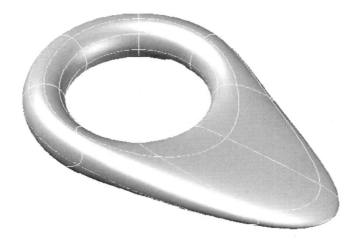

10. Recreate the apple as a 3D mesh model. It is approximately 2.5" (64mm) in diameter, and sits on a flat base.

11. Create the airplane body as a 3D mesh model.

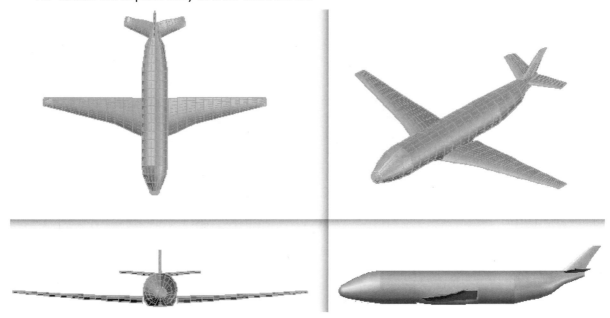

## CHAPTER REVIEW

1. List two ways in which 3D mesh models differ from 3D solid models:
   a.
   b.
2. Label the parts of the mesh models illustrated below:

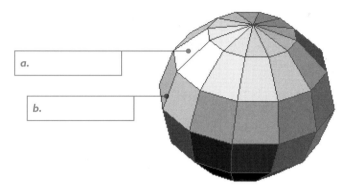

3. What is the purpose of *tessellation lines*?
4. Must meshes always have four straight sides?
5. Which command distorts meshes?
6. How do you select a subobject on a 3D mesh model?
7. What is a *crease*?
   How do creases affect meshes?
8. What happens when mesh smoothness exceeds 4?
   What happens when tessellation density has a large value?
9. Which command increases the smoothness of 3D mesh models?
10. What is a *primitive* in terms of 3D mesh models?
11. Can 3D mesh models be edited?
12. Identify the following primitives:

     a.

     b.

     c.

     d.

     e.

     f.

     g.

13. What happens when you click the Home icon in the ViewCube?
14. Can more than one face or edge be selected for editing? If so, how?
15. Why might you want to increase the number of meshes?

# CHAPTER 18

## Additional 3D Techniques

The previous two chapters introduced you to 3D modeling with solids and meshes. In this chapter, you learn about other types of 3D modeling, such as creating surfaces and editing 3D models.

**3DOSNAP** snaps to 3D features of solid and surface models (new to AutoCAD 2011).

**EDGESURF** fits meshes between four connected edges.

**RULESURF** stretches meshes between two disconnected edges.

**TABSURF** raises meshes from path curves and direction vectors.

**REVSURF** revolves meshes from path curves and axes of rotation.

**EXTRUDE** and **SWEEP** create 3D objects from 2D designs.

**UNION**, **SUBTRACT**, and **INTERSECT** create new 3D bodies through Boolean operations.

**INTERFERE** creates 3D bodies from intersecting solids.

**CHAMFER** and **CHAMFEREDGE** (new to AutoCAD 2011) apply cutoffs to the edges of 3D models.

**FILLET** and **FILLETEDGE** (new to AutoCAD 2011) round the edges of 3D models.

**MASSPROP** reports the properties of 3D solids.

---

**NEW TO AUTOCAD 2011** IN THIS CHAPTER

- **3DOSNAP** sets 3D object snap modes for 3D models.
- Fewer 3D objects are now selected through culling.
- AutoCAD now supports 3D mice.
- **CHAMFEREDGE** and **FILLETEDGE** provide interactive chamfering and filleting.

## ADDITIONAL 3D DESIGN AIDS

The primary drawback to 3D design is that you work in a 2D work environment. The screen of your computer's monitor is flat; you cannot reach into the screen to manipulate 3D objects. The traditional mouse moves about on a flat surface, pushing the cursor in only the x and y directions.

Perhaps one day in the future, tactile-feedback 3D screens will become commonplace; in the meantime, Autodesk provides software tools that help mitigate the problem. In earlier chapters, you learned about subobject selection, whereby you can select just a single face, edge, or vertex of a 3D object. You also learned about UCSs, which reorient the x, y working plane anywhere in 3D space, and the related dynamic UCS that temporarily locates the x, y plane on the selected face of a 3D object.

AutoCAD 2011 introduces three more aids for 3D design:

- 3D object snaps snap to geometric features unique to 3D objects.
- Culling makes AutoCAD ignore objects that lie underneath other objects.
- AutoCAD now supports 3D mice from Logitech's 3dconnexion division.

Lets look at these items in greater detail.

###  3D OBJECT SNAPS

While 2D object snaps work with 3D models, they do not support geometry specific to 3D. Autodesk provides a separate set of 3D object snaps that find the following geometric features:

- □ ZVERtex — snaps to vertices and control vertices*.
- △ ZMIDpoint — snaps to the midpoints of edges.
- ○ ZCENter — snaps to the centers of faces.

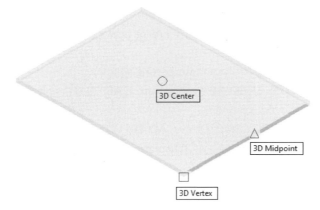

- ⊠ ZKNOt — snaps to knots of splines.
- ⊢ ZPERpendicular — snaps to the perpendiculars of planar faces.
- ⊠ ZNEAr — snaps to the object nearest a face.
- ZNONe — turns off 3D object snap modes.

*Control vertices and knots are found on splines.*

To turn on 3D object snapping, click the  (or 3DOSNAP) button on the status bar. Right-click the button to toggle individual modes. Choosing **Settings** displays the 3D Object Snap tab of the Drafting

Settings dialog box, which offers the same option as does the shortcut menu. (This dialog box can be accessed directly through the **3DOSNAP** command.)

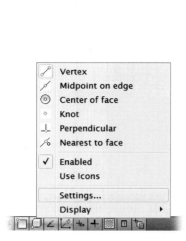
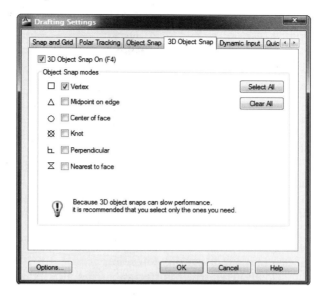

Autodesk warns that AutoCAD can slow down when too many 3D osnap modes are turned on at the same time. In particular, I recommend leaving ZNEAr turned off.

##   CULLED SELECTIONS

When you hover the cursor over overlapping 3D objects in a shaded visual style, all of them are highlighted through AutoCAD's preview feature. It can be difficult to select just the topmost object. Turning on *culling* forces AutoCAD to select just the topmost object.

(Culling has no effect in wireframe modes, since you can see through all objects to select the one of interest.)

### TUTORIAL: CULLING 3D OBJECT SELECTION

In this tutorial, you learn how culling affects the selection of 3D objects.

1. Start AutoCAD, and then open the *truck model.dwg* drawing file.
2. In the ribbon's Home tab, ensure that Culling is turned off (found in the Subobject panel).

3. Pass the cursor over the truck model. Notice that many objects are highlighted.

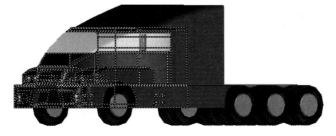

4. Click the Culling button (turns blue).
5. When you now pass the cursor over the truck model, just the topmost object is highlighted.

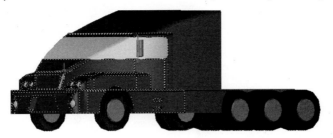

  3D MOUSE SUPPORT

AutoCAD supports the line of 3D mice from Logitech's 3dconnexion division. After you install the generic 3dconnexion driver on your computer, AutoCAD adds a 3D mouse icon to the navigation bar automatically. The driver is available from www.3dconnexion.com/service/drivers.html.

These 3D mice rely on a cylinder-shaped controller that navigates through the model with six degrees of freedom: zoom in/out, pan left/right, pan up/down, tilt, spin, and roll.

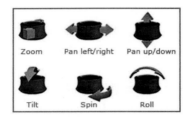

The Space Navigator is a compact model that provides just the basics: the 3D controller and two buttons; the Space Pilot Pro adds two dozen task-specific and customizable buttons, as well as a color display that reports the status of the 3D mouse and optional updates from your email account, social media sites, RSS feeds, and so on.

*Left: Bottom-of-the-line Space Navigator for $99.*
*Right: Top-of-the-line Space Pilot Pro for $399.*

Some users find these 3D mice remarkable, but others experience difficulty in keeping track of the 3D controller's six functions.

## 3D Mouse Navigation Bar

Once the driver is installed, the navigation bar adds the ![icon] icon for 3D mice:

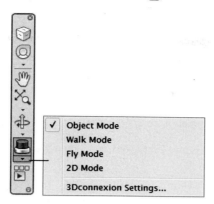

**Object Mode** — changes 3D viewpoint in same direction as controller movement.

**Walk Mode** — changes 3D viewpoint in opposite direction to controller movement, while maintaining orientation and height of the current view.

**Fly Mode** — changes 3D viewpoint in opposite direction to controller movement, while not maintaining orientation and current view height.

**2D Mode** — pans and zooms in same direction as controller movement like a 2D mouse.

**3Dconnexion Settings** — displays the 3Dconnexion Settings dialog box.

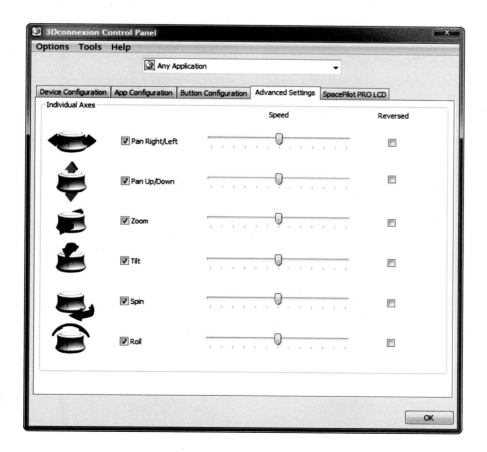

## 2D EDITING COMMANDS THAT WORK IN 3D

Some 2D editing commands you learned about earlier in this book can be used on 3D solid and mesh models. They work as you would expect. For instance, erasing and rotating 3D solid models are just like erasing and rotating 2D objects. These commands also work well with 3D models:

**ALIGN** moves, resizes (scales), and/or rotates solid models in 3D space.

**CHAMFER** cuts the edges of faces.

**ERASE** erases the selected solid models from the drawing.

**EXPLODE** converts 3D solid models into 2D regions, and 3D meshes into polyfaces.

**FILLET** rounds the edges of faces.

**MIRROR** and **MIRROR3D** make mirror copies of solid objects.

**MOVE** moves solid models in 3D space (by any distance in the x, y, and/or z directions).

**ROTATE** and **ROTATE3D** rotate solid objects in two and three dimensions.

**SCALE** changes the size of solid objects.

Some commands work partially with 3D objects. For instance, the **PROPERTIES** command changes only some properties of solid and mesh models. Some modes of the **OSNAP** command work well with 3D objects.

## 2D Commands That Don't Work

Not all 2D editing commands work with solid models. The **STRETCH** command only moves solid models, but does not stretch them. **EXTEND** and **LENGTHEN** apply only to open objects, which solid models are not; the workaround is to use the **SOLIDEDIT** command (not covered by this book.)

**HATCH**, **GRADIENT**, and **HATCHEDIT** don't work, because solid and mesh models cannot be hatched. The closest workaround is to change their color with the **COLOR** command; alternatively, use the **MATERIAL** command to apply a material to the 3D model, and then change the visual style to Realistic.

You cannot break solid models with the **BREAK** command. As a workaround, use the **SLICE** command to cut off a portion of the solid model.

**DIVIDE** and **MEASURE** do not apply to solid models. The **TRIM** command does not work with solid models; the workaround is to use the **INTERSECT** or **SLICE** command. 3D models cannot be offset with the **OFFSET** command; the workaround is to use the **COPY** or **SOLIDEDIT** command.

Most grips editing options work with 3D solids: Copy, Rotate, Move, and Scale. The Stretch option moves them, while the Mirror option rotates.

## AutoCAD LT

AutoCAD LT can view but not edit the shape of solid and mesh models; it can apply operations to them, such as copying, moving, rotating, and arraying.

 **EDGESURF**

The **EDGESURF** command (short for "edge surface") constructs a 3D mesh object between four edges.

The edges of the mesh are defined by open objects: lines, arcs, splines, and open 2D or 3D polylines. The figure below illustrates a mesh that was created between two lines and two arcs. (Technically, AutoCAD interpolates a bicubic mesh between four general space curves, creating a Coons patch.) The type of surface is determined by the **MESHTYPE** system variable, polyface meshes or 3D mesh objects.

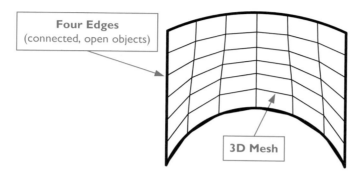

 **Notes** For the **EDGESURF** command to work correctly, there must be precisely four objects — no more, no less. Even more important, the four objects *must* connect at their endpoints, so use object snap or snap modes to ensure they do.

The "edge surf" name of this command implies that it creates surface objects. But here the word "surface" means that the object is not solidly filled.

(*History*: The Coons surface was developed by Steven Coons, codeveloper of the first CAD system. In 1963, he envisioned "the designer seated at the *console* [early term for keyboard], drawing a sketch of his proposed device on the screen of an *oscilloscope tube* [early form of monitor], with a light pen modifying his sketch into a perfect drawing." See design.osu.edu/carlson/history/lesson4.html. Until AutoCAD 2010, this command created Coons surfaces made of polyface meshes.)

### TUTORIAL: MESHING BETWEEN FOUR EDGES

In this tutorial, you create a surface between four connected edges.

1. Start AutoCAD with a new drawing (*acad3d.dwt*), and then ensure that ENDpoint object snap is turned on.
2. Using the **ARC** command, draw four shallow arcs, ensuring that their endpoints meet.

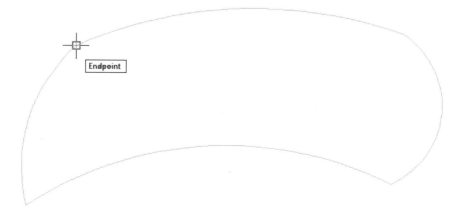

3. To construct a surface between the four arcs, start the **EDGESURF** command:
   - In the 3D ribbon's Mesh Modeling tab, choose **EdgeSurf** in the Primitives panel.
   - At the Command: prompt, enter the **edgesurf** command.

   Command: **edgesurf** *(Press* ENTER.*)*

   Notice that AutoCAD reports the current setting for the **SURFAB1** and **SURFAB2** system variables, which are discussed next.
   Current wire frame density: SURFTAB1=6  SURFTAB2=6

4. You are prompted to select the first object. Choose one of the arcs.
   Select object 1 for surface edge: *(Pick an object.)*

   The order in which you select the splines does not matter. Your selection must consist of a single pick, however; you cannot take selection shortcuts, such as windowing all four arcs or pressing **CTRL+A**.

5. And then choose the other three splines:
   Select object 2 for surface edge: *(Pick arc.)*
   Select object 3 for surface edge: *(Pick arc.)*
   Select object 4 for surface edge: *(Pick arc.)*

   AutoCAD instantly draws the mesh.

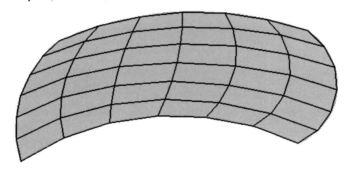

 **Note**  If one object is not connected to an adjacent one, AutoCAD complains "Edge x does not touch another edge." You can force objects to touch with the **FILLET** command, or the **TRIM** and **EXTEND** commands; then restart the **EDGESURF** command.

### DRAWING SURFACES: ADDITIONAL METHODS

The mesh resulting from the tutorial above is quite coarse, because it uses just 36 faces on the surface area — 6 x 6. You can improve the smoothness with these two system variables.

- **SURFTAB1** sets tabulations and mesh density in the m direction.
- **SURFTAB2** sets mesh density in the n direction.

The first object you selected defines the *m* direction; the object connected to the first boundary line defines the *n* direction. The letters "m" and "n" are used instead of x and y, because the direction of the meshes doesn't necessarily correspond with the x and y axes, and because the two directions are not necessarily at right angles.

These two system variables also affect the smoothness of the surface meshes created by the REVSURF, RULESURF, and TABSURF commands, as discussed later.

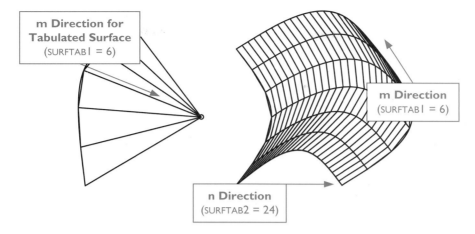

As you saw in the tutorial, the default value of 6 for SURFTAB1 and SURFTAB2 is far too low. In general, you should increase the mesh density for larger meshes. While the range is between 2 and 32766, I recommend you set a value of at least 16; the technical editor recommends using 32766 when you need a long lunch break.

    Command: **surftab1**

    Enter new value for SURFTAB1 <6>: *(Enter a higher value, such as **16**.)*

    Command: **surftab2**

    Enter new value for SURFTAB2 <6>: *(Enter a higher value, such as **16**.)*

Unfortunately, when you change the value of these system variables, it is not retroactive. You have to erase the mesh, and then reconstruct it. Shown below is the mesh using m and n values of 16.

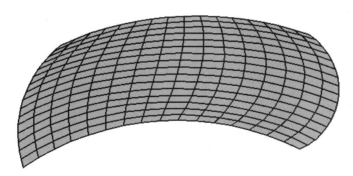

## RULESURF

The RULESURF command (short for "ruled surfaces") creates meshes that span two objects — lines, points, arcs, circles, open and closed 2D polylines, and 3D polylines. These objects are called "defining curves." (*Curve* is a general term in mathematics that includes straight lines and curves.)

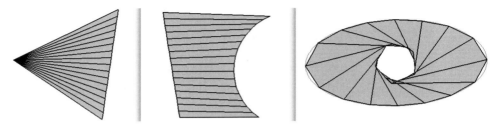

*Ruled surfaces:*
**Left**: *Between a point and line.* **Middle**: *Between a line and an arc.* **Right**: *Between a polygon and an ellipse.*

Using this command is just like using EDGESURF, except that you work with two curves, and they needn't touch.

    Command: **rulesurf** *(Press* ENTER.*)*
    Current wire frame density:  SURFTAB1=16
    Select first defining curve: *(Pick object.)*
    Select second defining curve: *(Pick object.)*

### Defining Starting Points

AutoCAD uses several rules to determine the starting points of ruled surfaces:

**For Closed Objects.** Circles begin at 0 degrees. This angle is affected by the direction of the x axis and the value of SNAPANG. Closed polylines begin at the last vertex, and then go backward to the first vertex. In effect, ruled surfaces are drawn on circles and closed polylines in opposite directions, creating the iris effect illustrated above. It's better to use round polylines than circles.

**For Open Objects.** AutoCAD draws the ruled surface from the endpoint nearest to the pick point on the object. Selecting opposite ends creates ruled surfaces that cross.

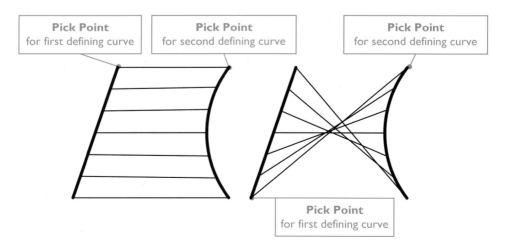

## TABSURF

The TABSURF command constructs meshes defined by paths and direction vectors. Constructing a tabulated surface takes these steps:

1. Draw a *path curve* from which the mesh is calculated. Path curves can be lines, arcs, circles, ellipses, splines, and 2D or 3D polylines — open or closed.
2. Draw a *direction vector* that defines the direction and length of the mesh. Direction vectors are lines and open 2D or 3D polylines. On polylines, AutoCAD uses only the first and last points to determine the direction vector; intermediate vertices, curves, and splines are ignored. (The direction vector need neither touch the path curve, nor be perpendicular to the plane of the path curve.)
3. Start the TABSURF command, and then pick the path curve and direction vector.

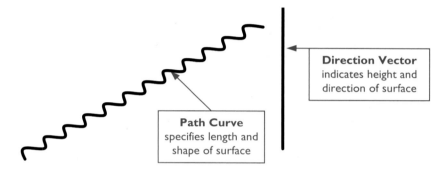

Command: **tabsurf** *(Press ENTER.)*
Current wire frame density: SURFTAB1=6
Select object for path curve: *(Pick object.)*
Select object for direction vector: *(Pick either end of object.)*

The point you pick on the direction vector determines the direction that the mesh is projected. Select a point closer to one end of the direction vector, and the tabulation is projected in that direction.

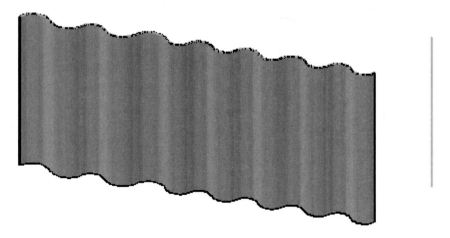

**Note** To draw vertical lines in plan view, use point filters, as follows:

Command: **line**
Specify first point: *(Pick a point.)*
Specify next point or [Undo]: **.xy**
of *(Pick another point, or even the same point.)*
(need Z): *(Enter a height, such as* **10**.*)*

The number of tabulations on the surface is controlled by the SURFTAB1 system variable, except when it comes to polyline segments. In this case, AutoCAD ignores this system variable, drawing a tabulation line at the end of each straight segments. Polyline arc segments are still divided by the number of tabulations specified by SURFTAB1.

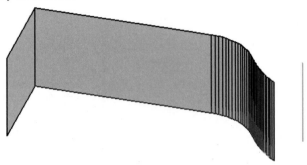

## REVSURF

The REVSURF command (short for "revolved surface") constructs 3D meshes by rotating objects about an axis. To create the revolved surfaces, you take these steps:

1. Draw a *path curve*, which is the object to be revolved. The path curves can be lines, circles, arcs, splines, and 2D or 3D polylines — they must be open. A sphere, for example, can be made from an arc rotated 360 degrees, while a torus (donut) is a circle that has been rotated through 360 degrees.
2. Draw an *axis of revolution*, about which the object revolves.
3. Use the **REVSURF** command to construct the meshed surface. The axes can be lines and open polylines (2D or 3D); the same considerations apply as with the **TABSURF** command.

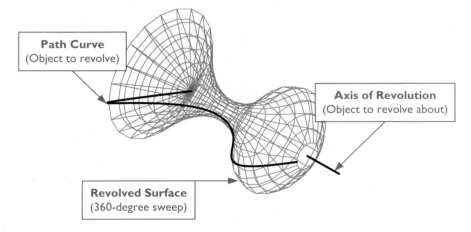

This command differs from the previous ones in that you also specify angles. The start and included angles allow the construction of a revolution of less than 360 degrees (full circle); this lets you see inside the 3D object

> Command: **revsurf** *(Press ENTER.)*
> Current wire frame density:  SURFTAB1=6  SURFTAB2=6
> Select object to revolve: *(Pick object.)*
> Select object that defines the axis of revolution: *(Pick object.)*
> Specify start angle <0>: *(Press ENTER, or specify an angle, such as **15**.)*
> Specify included angle (+=ccw, -=cw) <360>: *(Press ENTER, or specify an angle, such as **285**.)*

The start angle is defined by the location of the path curve, not by the x axis, as you would expect from other commands involving angles. The right-hand rule determines the direction: curl your fingers, and then point the thumb in the direction of the path curve. The direction of your fingers shows the direction the angle will take. (Specify a negative angle for the surface to revolve in the opposite direction.)

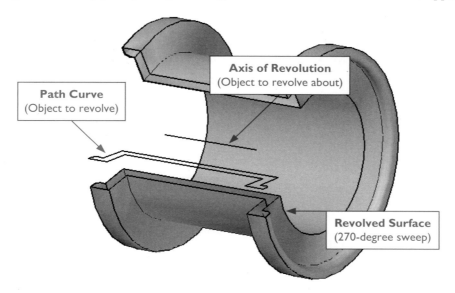

The related **REVOLVE** command creates solid and surface models, as discussed in Chapter 16, "Introduction to 3D Solid Modeling."

## SWEEP AND EXTRUDE

The **SWEEP** and **EXTRUDE** commands extrude 2D objects along paths to create 3D solid models or surfaces.

Both commands are more sophisticated than both the **PRESSPULL** command from Chapter 16 and the dragging of subobjects from Chapter 17, because you specify a path along which to direct the extrusion. This is handy for creating objects like handrails and snaking tubes.

The two commands operate similarly, though **EXTRUDE** has these differences from **SWEEP**:

The *extrusion* process takes place at right angles to the center line of the path: the profile must be perpendicular to the path; in contrast, the *sweep* process does not have this restriction. (The solution for extrusions is to draw the path in one UCS and the profile in a second UCS at an angle to the first. AutoCAD reorients the profile to be perpendicular to the path.)

EXTRUDE can specify an angle, so that the extrusion grows smaller or larger along the path; SWEEP cannot do this. (It would be nice if Autodesk would combine the best of both commands into one.) To create extruded or swept models, take these steps:

1. Draw the profile of objects to be extruded or swept. This includes just about any open or closed objects. The exceptions are objects within blocks, polylines that cross over themselves, and polylines with more than 500 vertices.

2. If you plan to extrude an object with holes in it, such as the profile of a gear with an axle hole, then you must convert them to a single region with the REGION command, and then use the SUBTRACT command to remove the hole areas.

3. Draw the path along which to extrude/sweep. The path is either an object — line, circle, arc, ellipse, elliptical arc, polyline, or spline — or a height that you specify.
   For extrusions, the path is optional. You can also drag the height with your cursor.

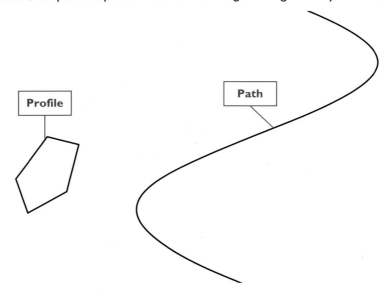

4. Enter the EXTRUDE or SWEEP command to create the extrusion.

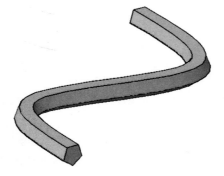

Note that extrude does not work when the profile and path lie in the same plane. sweep is exempt from this limitation, but it is not clear how it knows to align the profile with the path.

## TUTORIAL: SWEEPING OBJECTS INTO SOLIDS

In this tutorial, you sweep a circle along a spline to create a bent handrail shape.

1. Draw a circle with the **CIRCLE** command. This object will define the *profile* of the handrail.
2. Draw a polyline with the **PLINE** command. This object will define the length (*path*) of the handrail. The polyline and circle do not need to be near each other.

3. To sweep the 2D object into a 3D solid, start the **SWEEP** command:
   - In the 3D ribbon's Home tab, choose **Sweep** from the Modeling panel.
   - At the 'Command:' prompt, enter the **sweep** command.

      Command: **sweep** *(Press ENTER.)*

4. AutoCAD prompts you select the object(s) to sweep.

      Current wire frame density: ISOLINES=4, Closed profiles creation mode = Solid
      Select objects to sweep or [MOde]: *(Pick the circle.)*
      Select objects to sweep or [MOde]:*(Press ENTER to end object selection.)*

5. Indicate the height to which the object(s) should be extruded:
      Select sweep path or [Alignment/Base point/Scale/Twist]: *(Pick the polyline.)*

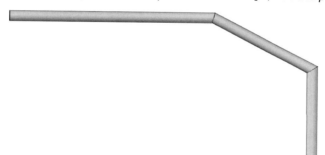

As an interesting aside, this command can twist the swept objects with the **Twist** option. AutoCAD asks for the twist angle and whether you want banking (in which the swept object is naturally banked along 3D paths). Below, I swept a rectangle along the polyline through 1000 degrees.

**Note** You can sweep subobjects, such as faces and edges of 3D solids and meshes: hold down the **CTRL** key, and then select the subobjects. This command also sweeps multiple profiles along a single path at once, useful for drawing a curving pair of railway tracks

The **EXTRUDE** command operates similarly:

Command: **extrude** *(Press ENTER.)*
Current wire frame density: ISOLINES=4, Closed profiles creation mode = Solid
Select objects to extrude or [MOde]: *(Select one or more objects.)*
Select objects to extrude or [MOde]: *(Press ENTER to end object selection.)*
Specify height of extrusion or [Direction/Path/Taper angle/Expression] <0.5>: *(Move the cursor, enter a value, or type an option.)*

**MOde** — determines if a 3D solid or surface is created by this command.

**Specify height of extrusion** — pick two points to indicate the height, or enter a distance, or drag the cursor.

**Direction** — pick two points to indicate the height and direction of extrusion.

**Path** — selects a line or other open object. Remember that the path object must be perpendicular to the object being extruded.

**Taper angle** — specifies positive and negative angles that must not allow the extruded 3D object to intersect itself. Although the allowable range is -90 to +90 degrees, shallow angles work best.

**Expression** — indicates the extrusion distance through a parametric formula.

**Note** This command creates either 3D surfaces or solids, depending on the setting of the MOde option. The **SURFACEMODELINGMODE** system variable determines if the surface is regular or NURBS.

## UNION

The **UNION** command merges 2D regions and 3D models into a single body, as you saw earlier in Chapter 16. It also works with 3D mesh objects, including those made with the **MESH** command and with other like **EDGESURF**.

With mesh objects, however, the process differs somewhat, because AutoCAD turns the meshes into 3D solids. After the union is complete, you turn the 3D solids back into meshes. You see how this works in the following tutorial.

**Notes** The **UNION**, **INTERSECT**, and **SUBTRACT** commands are collectively known as "Boolean operations," named after George Boole, a 19th-century mathematician who developed Boolean logic.

You may be familiar with using the terms **or** (union), **and** (intersect), and **not** (subtract) in narrowing down searches on the Internet. These are based on the work of Mr. Boole. You can learn more about this topic at www.kerryr.net/pioneers/boole.htm.

## TUTORIAL: JOINING 3D MESHES

In this tutorial, you draw a box and a sphere as 3D mesh objects, and then join them into a single object.

1. In a new drawing, draw a box and a sphere with the **MESH** command, as shown below. For the box, enter 0,0,0 for the **Center** option, and then specify a **Cube** with lengths of 1. For the sphere, use a radius of 0.75.

2. To join two or more 3D mesh objects, start the **UNION** command:
   - From the ribbon's Home tab, choose **Union** from the Solid Editing panel.
   - At the 'Command:' prompt, enter the **union** command.

     Command: **union** *(Press* ENTER.*)*

   - Alternatively, enter the alias **uni** at the keyboard.

3. In all cases, AutoCAD prompts you to select the objects to join. Select them all:
   Select objects: *(Press* CTRL+A.*)*

4. Press **ENTER** to end object selection:)
   Select objects: *(Press* ENTER *to end object selection.)*

5. Notice the dialog box:

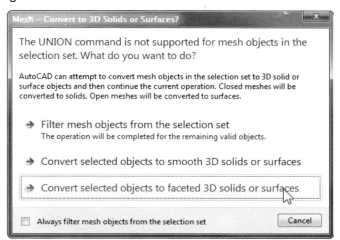

AutoCAD gives you three options, the first two of which you don't want:

   **Filter mesh objects** — does not convert mesh objects to solids, and so they cannot be unioned.

**Smooth 3D solids** — converts meshes to solids, but into a format that does not convert back well to meshes.

**Faceted 3D solids** — converts meshes into a format that converts back to meshes.

Choose the third option, **Convert Selected Objects to Faceted 3D Solids or Surfaces**. The selected objects are merged into a single object, but they do not look any different from before; this is normal.

6. Use the **MESHSMOOTH** command to convert the unioned 3D solid back to a 3D mesh.

   Command: **meshsmooth**

   Select objects to convert: *(Select the 3D solid.)*

   Select objects to convert: *(Press* ENTER *to end object selection.)*

7. After you press ENTER, AutoCAD displays this dialog box that gives a false warning message — well, false in our case. Recall that we asked AutoCAD to convert the 3D meshes into *faceted* solids; if we'd picked *smoothed* solids, the results would now look atrocious (as illustrated later).

   Choose **Create Mesh**.

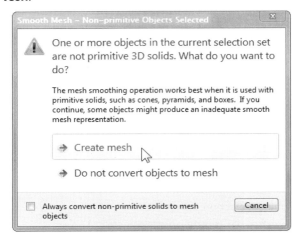

Again, the converted mesh model looks no different, but the **PROPERTIES** command will report now that it is a single mesh object.

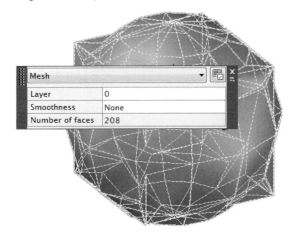

### Smoothed Conversion

So, how bad does it look when 3D meshes are converted to smooth solids? First, here is the result after conversion to smooth solids (at left):

*Left: 3D mesh converted to smooth 3D solid.*
*Right: 3D solid converted to 3D mesh.*

In some cases, this may be exactly what you want. After all, the U command lets us play around with what-ifs in AutoCAD. The problem gets worse after you use **MESHSMOOTH** to convert the 3D solid back to mesh form. Take a look (at right, above).

## SUBTRACT

The **SUBTRACT** command removes intersecting parts of 2D regions, 3D solid models, surfaces, and meshes. As with the **UNION** command, AutoCAD must first convert 3D meshes into solids.

Regions, 3D solids, and so on cannot be mixed: like subtracts from like. The objects must be intersecting to be subtracted. AutoCAD needs to know just two things from you: (1) the objects to be subtracted, and (2) the objects doing the subtraction.

### TUTORIAL: SUBTRACTING SOLIDS

1. Draw a box (with the **BOX** command) and four cylinders (**CYLINDER** command) similar to those illustrated below.

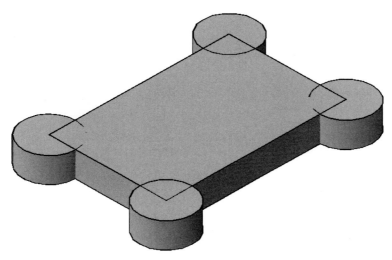

2. To remove 3D solids from each other, start the **SUBTRACT** command:
   - From the ribbon's Home tab, choose **Subtract** from the Solid Editing panel.
   - At the 'Command:' prompt, enter the **subtract** command.

     Command: **subtract** *(Press ENTER.)*
   - Alternatively, enter the alias **su** at the keyboard.

3. In all cases, AutoCAD prompts you to select the objects that will be subtracted.

   Select solids and regions to subtract from ...

   Select objects: *(Select one or more objects.)*

   Select objects: *(Press ENTER to end object selection.)*

4. Select the objects that will do the subtracting.

   Select solids and regions to subtract ...

   Select objects: *(Select one or more objects.)*

   Select objects: *(Press ENTER to end object selection.)*

The results may not be what you expect, because the order in which you select objects is important:

1. First, select the objects from which the subtracting takes place. These are the objects to be made smaller.
2. Second, select the objects that do the subtracting.

If the result looks wrong, enter the U command to undo, and then pick objects in the opposite order. In the figure below, I first selected the box ("subtract from"), and then the four cylinders ("to subtract"):

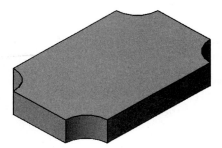

I repeated the command, and this time I selected the four cylinders ("subtract from"), and then the box ("to subtract"):

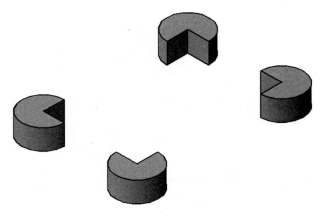

The "Pacman"-shaped cylinders form a single object, because AutoCAD first unions the objects before subtracting them.

## INTERSECT

The **INTERSECT** command removes all but intersecting parts of 2D regions, 3D solids, surfaces, and meshes.

This command determines the overlapping area of regions or the common volume of 3D objects. (Object types, such as solids and regions, cannot be mixed.)

### TUTORIAL: INTERSECTING SOLIDS

1. Draw the same box and four cylinders, as with the earlier tutorial.
2. To remove all but the intersections of the objects, start the **INTERSECT** command:
   - From the ribbon's Home tab, choose **Intersect** from the Solid Editing panel.
   - At the 'Command:' prompt, enter the **intersect** command.

        Command: **intersect** *(Press ENTER.)*
   - Alternatively, enter the alias **in** at the keyboard.
3. Notice that AutoCAD prompts you to select the objects from which to make the intersection. For this part of the tutorial, select all objects, as follows.

    Select objects: *(Press CTRL+A to select all objects.)*

    Select objects: *(Press ENTER to end object selection.)*

The result of this command is — nothing! The four cylinders intersect the box, but not each other. Because you asked this command to find the volume common to *all* objects, the result is nothing.

4. Repeat the command. This time select the box and only one of the cylinders. The result is a pie-shaped portion of cylinder, the part that was the intersection of the two objects. (The three cylinders remain in the drawing, because you did not make them part of the selection set.)

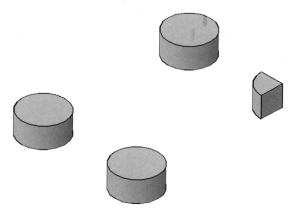

 **INTERFERE**

The **INTERFERE** command creates 3D bodies from intersecting solids.

This command is more general than the **INTERSECT** command, because it finds *all* intersections of *all* 3D solids. It deals with interferences in two ways: first, it highlights the solids that interfere (not the interference volumes themselves, unfortunately), and then it gives you the option to turn the interferences into independent solids. You can, if you like, move them away from the rest of the solid model, and then analyze them separately.

I think you can see that these two features are immediately useful. By highlighting interfering pairs of solids, AutoCAD shows you which parts of a model collide or interfere — for example, pipes that go through other pipes when they ought not to, or heating ducts that penetrate walls.

### TUTORIAL: CHECKING INTERFERENCE BETWEEN SOLIDS

For this tutorial, continue using the drawing of the box and four cylinders.

1. To determine if 3D solids interfere with each other, start the **INTERFERE** command:
   - From the ribbon's Home tab, choose **Intersect** from the Solid Editing panel.
   - At the 'Command:' prompt, enter the **interfere** command.
     
     Command: **interfere** *(Press ENTER.)*
   - Alternatively, enter the alias **inf** at the keyboard.

2. AutoCAD prompts you to select two sets of solids:
   
   Select first set of solids: *(Select one or more objects.)*
   
   Select objects: *(Press ENTER to end object selection.)*
   
   The prompts provide no clue that there is an alternative method to select objects. At the first prompt, enter **all** to select all objects in the drawing, and then press ENTER at the second prompt. AutoCAD filters out all non-solids, and then checks them against each other for interference.

3. Optionally, select a second set of objects.
   
   Select second set of solids: *(Press ENTER to skip, or select objects for the second selection set.)*

4. Notice that AutoCAD displays the intersecting solids in red, and the original objects in wireframe display mode.

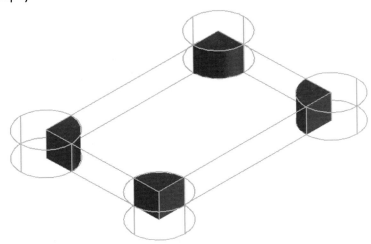

When you click the **Next** button, the Interference Checking dialog box guides you to each interference. This is particularly useful in complex drawings of many elements. Click the **Zoom** button when you need an even closer look.

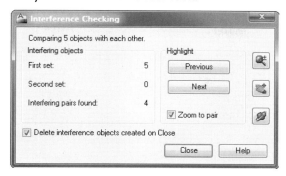

5. The **Delete Interference Objects Created on Close** option determines if the interference solids stay in the drawing after you exit this dialog box:

   **Yes** — interference objects are erased (default).

   **No** — interference objects remain.

6. Click **Close** to close the dialog box and exit the command.

The objects created by this command usually disappear from the drawing when you end this command. That's because they clutter the drawing, and so are not useful.

If you wish them to remain in the drawing, turn off the **Delete Interference Objects Created on Close** option. The objects are added to the selection set, and can be accessed with the **Previous** selection option.

## CHAMFER AND CHAMFEREDGE

The CHAMFER command chamfers the edges of 3D objects, in addition to the intersections of 2D objects. The CHAMFEREDGE command interactively bevels 3D objects.

When you select a 3D solid or mesh model, the commands automatically switch to 3D editing mode that presents prompts different from those you saw in 2D editing. They operate on a single 3D model at a time, but allow you to chamfer as many edges as you need during a single operation. (It does not work with surface models.)

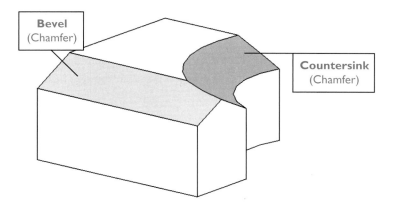

This command also works on curved portions of solid models. You can pick an isoline, the lines that define curved surfaces. When you select the isoline of a hole, for example, then AutoCAD creates a countersink.

## TUTORIAL: CHAMFERING OBJECTS

In this tutorial, you draw a rectangular box, and then chamfer its edges.

1. Draw a rectangular box with the **BOX** command.

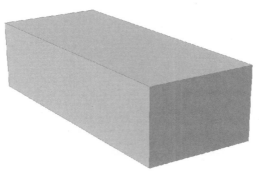

2. To chamfer the edge (intersection of two faces) of the solid model, start the **CHAMFER** command:
   - In the ribbon's Home tab, choose **Chamfer** from the Modify panel.
   - At the 'Command:' prompt, enter the **chamfer** command.

        Command: **chamfer** *(Press* ENTER.*)*

   - Alternatively, enter the **cha** alias at the 'Command:' prompt.

3. Notice that AutoCAD displays prompts familiar to you from 2D editing:
   (TRIM mode) Current chamfer Dist1 = 0.0000, Dist2 = 0.0000

   Select first line or [Polyline/Distance/Angle/Trim/Method/mUltiple]: *(Pick an edge of the solid model.)*

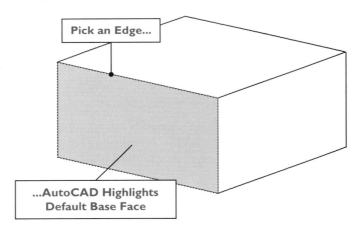

4. Notice that AutoCAD highlights one of the two surfaces adjoining the selected edge. The following prompt allows you to select a base face other than the one highlighted:
    Base surface selection...

    Enter surface selection option [Next/OK (current)] <OK>: *(Type* **N** *or* **OK**.*)*

   Your options are as follows:
   - Enter **N** (next) to select the *other* face as the base surface.
   - Enter **OK** when the two chamfer distances are the same, or when AutoCAD correctly guesses the base surface.

When you select an Edge, AutoCAD does two things: (1) highlights one face by outlining its edges with dashed lines, and (2) prompts you to select the *base surface* (a.k.a. the first face). The first chamfer distance applies to the base face, which AutoCAD calls the "base surface chamfer distance."

Because the two chamfer distances can be unequal, AutoCAD needs to know which distance applies to which face. (Recall that every edge has two adjoining faces.)

As an alternative, you can select an isoline. In this case, AutoCAD reverses the order: first it asks for the chamfer distances, and then for the other face.

5. Enter the chamfer distance(s). (You cannot specify an angle, as with 2D chamfering.) Notice the reference to the "base surface chamfer distance":

    Specify base surface chamfer distance or [Expression]: **0.5**

    Specify other surface chamfer distance or [Expression] <0.5>: *(Press* ENTER, *or enter a value.)*

If the chamfer distance is too large, AutoCAD complains, "Cannot blend edge with unselected adjacent tangent edge. Finding connected blend set failed. Failure while chamfering," and exits the command. Restart the command, and then enter a smaller value.

(The **Expression** option allows you to enter a parametric equation to specify the chamfer distances.)

6. AutoCAD can chamfer just the selected edge, or a *loop*. A "loop" consists of all the edges that touch the base surface. A rectangular surface, as illustrated below, has a loop with four edges.

    Select an edge or [Loop]: *(Pick an edge, or type* **L**.*)*

    Select an edge or [Loop]: *(Press* ESC.*)*

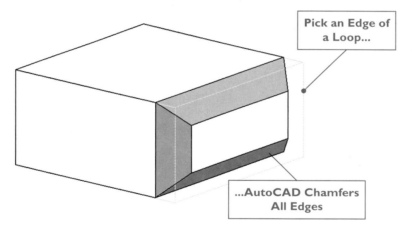

7. The command repeats until you press ESC.

## Chamfering Meshes

When you use the CHAMFER command on 3D mesh objects, AutoCAD will change them to 3D solids. Remember to follow the same process as described under the UNION command:

1. Upon AutoCAD's prompting, change the 3D meshes to faceted solids (no smoothed solids).
2. Use **MESHSMOOTH** to convert them back to 3D mesh objects.

**INTERACTIVE CHAMFERING**

The **CHAMFEREDGE** command chamfers the edges of 3D solids interactively. You can enter chamfer distances, or else drag the cursor to show the size of the chamfer.

- From the ribbon's Solid tab, choose **Chamfer Edge** from the Solid Editing panel.
- At the 'Command:' prompt, enter the **chamferedge** command.

Command: **chamferedge** *(Press ENTER.)*

The command turns on Edge subobject selection mode automatically, and jumps directly to the 'Select a loop' prompt, as follows:

Distance1 = 1.0000, Distance2 = 1.0000

Select an edge or [Loop/Distance]: *(Choose an edge.)*

Select an edge *[that]* belongs to the same face or [Loop/Distance]: *(Press ENTER to end object selection.)*

At this point, arrow grips appear on the face of the chamfer. You can drag them interactively to adjust the size of the chamfer, as illustrated by the figure:

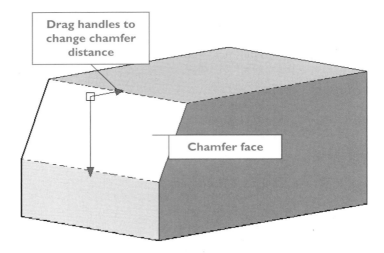

(The square grip does nothing.)

Press Enter to accept the chamfer or [Distance]: *(Press ENTER.)*

When done, press ENTER to exit the command.

## FILLET AND FILLETEDGE

Applying the FILLET and FILLETEDGE commands to 3D objects is similar to using the CHAMFER and CHAMFEREDGE commands, except that they create rounded corners. AutoCAD is limited to constant radius fillets; it cannot apply a variable radius to 3D fillets.

### TUTORIAL: CHAMFERING AND FILLETING 3D OBJECTS

1. In this tutorial, you draw a cylinder on top of a rectangular box, and then add fillets. Begin by drawing the box, as follows:

    Command: **box**

    Specify corner of box or [CEnter]: **0,0,0**

    Specify corner or [Cube/Length]: **l**

    Specify length: **9**

    Specify width: **4**

    Specify height: **1**

2. If necessary, use the **ZOOM Extents** command to see the box, and then use the ViewCube to set the 3D viewpoint.

3. Add the cylinder:

    Command: **cylinder**

    Specify center point of base or [3P/2P/Ttr/Elliptical]: **4.5,2,1**

    Specify radius for base of cylinder or [Diameter]: **1**

    Specify height or [2Point/Axis endpoint] <1.0>: **3**

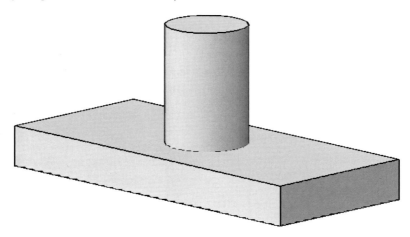

4. Start the **FILLET** command, and fillet the four edges of the end of the box:

    Command: **fillet**

    Current settings: Mode = TRIM, Radius = 0.0

    Select first object or [Undo/Polyline/Radius/Trim/Multiple]: *(Pick an edge of the box.)*

    Enter fillet radius <0.0>: **0.2**

    Select an edge or [Chain/Radius]: *(Pick another edge.)*

    Select an edge or [Chain/Radius]: *(Pick a third edge.)*

    Select an edge or [Chain/Radius]: *(Pick a fourth edge.)*

Select an edge or [Chain/Radius]: *(Press* ENTER.*)*
4 edge(s) selected for fillet.

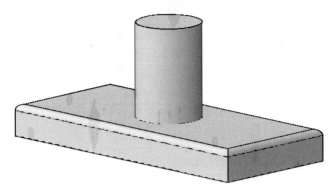

5. Another fillet is to be placed at the base of the cylinder. But the **FILLET** command cannot fillet between two solid objects. The proper procedure is to use the **UNION** command first to join the box and the cylinder into a single object.

    Command: **union**

    Select objects: *(Pick the box.)*

    Select objects: *(Pick the cylinder.)*

    Select objects: *(Press* ENTER.*)*

    The objects won't look any different.

6. Now you can apply the fillet between the box and the cylinder:

    Command: **fillet**

    Current settings: Mode = TRIM, Radius = 0.2

    Select first object or [Undo/Polyline/Radius/Trim/Multiple]: *(Pick the bottom edge of the cylinder.)*

    Enter fillet radius <0.2>: *(Press* ENTER *to reuse the radius.)*

    Select an edge or [Chain/Radius]: *(Press* ENTER *to exit the command.)*

    1 edge(s) selected for fillet.

    Notice the curve at the base of the cylinder.

7. Apply one last fillet, on top of the cylinder.

    Command: **fillet**

    Current settings: Mode = TRIM, Radius = 0.2

    Select first object or [Undo/Polyline/Radius/Trim/Multiple]: *(Pick the top edge of the cylinder.)*

    Enter fillet radius <0.2>: *(Press* ENTER *to reuse the radius.)*

    Select an edge or [Chain/Radius]: *(Press* ENTER.*)*

1 edge(s) selected for fillet.

Notice that the top of the cylinder is gracefully curved.

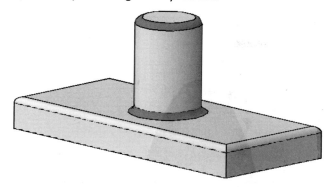

  **INTERACTIVE FILLETING**

The **FILLETEDGE** command fillets the edges of 3D solids interactively. You can enter a fillet radius as before, or else drag the cursor to show the radius of the fillet.

- From the ribbon's Solid tab, choose **Fillet Edge** from the Solid Editing panel.
- At the 'Command:' prompt, enter the **filletedge** command.

Command: **filletedge** *(Press ENTER.)*

The command turns on Edge subobject selection mode automatically, and jumps directly to the 'Select a loop' prompt, as follows:

Radius = 1.0000

Select an edge or [Chain/Radius]: *(Choose an edge.)*

Press Enter to accept the fillet or [Radius]: *(Press ENTER.)*

An arrow grip appears, which you can drag interactively to adjust the radius of the filter, as illustrated below.

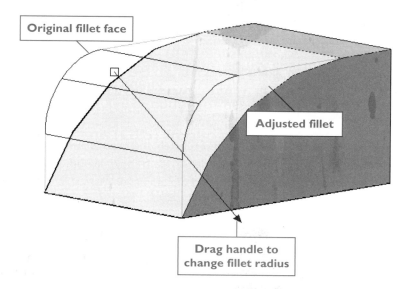

(The square grip does nothing.) When done, press ENTER to exit the command.

# MASSPROP

The MASSPROP command reports information about regions and solid models, such as their area, perimeter, centroid, and so on (short for "mass properties"). This command does not work with 3D meshes.

## TUTORIAL: REPORTING PROPERTIES

For this tutorial, continue with the solid model you filleted in the previous tutorial.

1. To find information about solids, start the **MASSPROP** command:
   - At the 'Command:' prompt, enter the **massprop** command.

   Command: **massprop** *(Press ENTER.)*

2. AutoCAD prompts you to select objects. While you can select more than one object, it is better to pick just one, because of the amount of information generated.

   Select objects: *(Select solid object.)*

   Select objects: *(Press ENTER to end object selection.)*

   AutoCAD switches to the Text window, and displays the report:

   ```
   ---------------- SOLIDS ----------------
   Mass:              45.2095
   Volume:            45.2095
   Bounding box:      X: 0.0000 -- 9.0000
                      Y: 0.0000 -- 4.0000
                      Z: 0.0000 -- 4.0000
   Centroid:          X: 4.5000
                      Y: 2.0000
                      Z: 0.9115
   Moments of inertia:   X: 307.5583
                         Y: 1235.5372
                         Z: 1389.0395
   Products of inertia: XY: 406.8852
                        YZ: 82.4135
                        ZX: 185.4304
   Radii of gyration:    X: 2.6082
                         Y: 5.2277
                         Z: 5.5430
   Principal moments and X-Y-Z directions about centroid:
       I: 89.1620 along [1.0000 0.0000 0.0000]
       J: 282.4871 along [0.0000 1.0000 0.0000]
       K: 292.7100 along [0.0000 0.0000 1.0000]
   ```

   The data are relative to the current UCS. Unfortunately, the mass is always equal to the volume because AutoCAD does not allow the density to be anything other than 1.0 — contrary to what we would expect from this kind of command.

3. AutoCAD asks if you wish to save the data to a file.

   Write analysis to a file? [Yes/No] <N>: **y**

If you answer **Y**, AutoCAD displays the Create Mass and Area Properties File dialog box. Enter a file name, and then click **Save**. AutoCAD creates the *.mpr* mass properties report file, which can be opened in Notepad or other text editor.

4. Press **F2** to return to the drawing window.

### MASSPROP DEFINITIONS

**Bounding Box** is the x, y, z-coordinate triple of the smallest 3D box enclosing 3D solids or bodies.

**Centroid** is the center of mass of 3D solids and bodies.

**Mass** equals the volume, because density = 1.

**Moment of Inertia** is Mass * $Radius^2$ for 3D bodies.

**Product of Inertia** is Mass * Distance (centroid to y, z axis) * Distance (centroid to x, z axis) for 3D bodies.

**Radius of Gyration** is $(MomentOfInertia / Mass)^{1/2}$ for 3D bodies.

**Volume** is the 3D space occupied by 3D bodies.

## EXERCISES

1. Draw four connected splines using the **SPLINE** command.
    a. Apply a Coons patch between them.
    b. Which command did you use?

2. Use the **REVSURF** command to draw a sphere with a diameter of 2.5 units. What object do you use as the defining curve?

3. Use the **TABSURF** command to draw a mesh with a hole in it, as illustrated below. *Hint:* The defining curves are both polylines. How must you draw the polylines to keep the tabulations from twisting?

4. Draw the chamfered 12mm-radius washer using the profile illustrated below. Use these dimensions:
    Outer height = 4mm
    Inner height = 2mm
    Width at bottom = 3mm
    Width at top = 1mm

   With which command did you draw it?

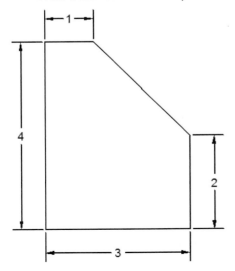
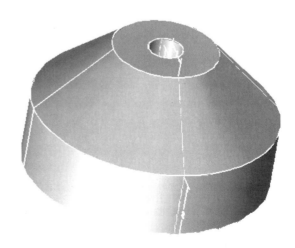

5. Open the *Ch10RevSurf.dwg file*, a partial drawing of a plastic wheel. Use the **REVSURF** command to generate a 360-degree 3D surface model.

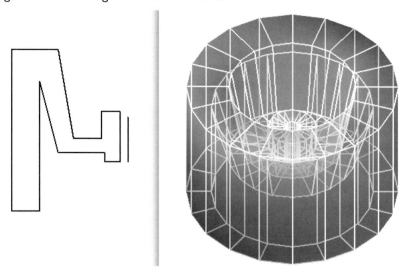

6. Open the *Ch11Glass.dwg* drawing file, a profile of a goblet. Use the **REVOLVE** command to create the goblet.

7. Open the *Ch11Extrude.dwg* drawing file, a profile of a gear.
   a. Use the **EXTRUDE** command to make the gear 0.5 units thick.
   b. Repeat, but with a taper angle of 15 degrees.

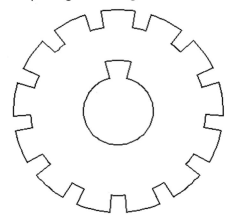

8. Open the *Ch11Path.dwg* drawing file, a pentagon profile and a path. Use the **SWEEP** command to sweep the pentagon along the spline path.

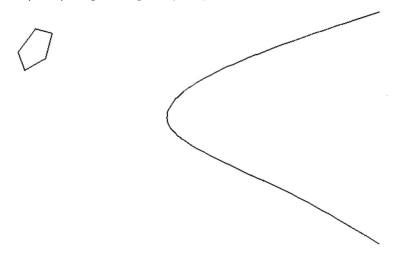

9. Create a profile of your own, and then apply the **SWEEP** command.

10. Open the *Ch11Pump.dwg* file, a drawing of a pump. Use the **INTERFERE** and **MOVE** commands to isolate the interference solids.

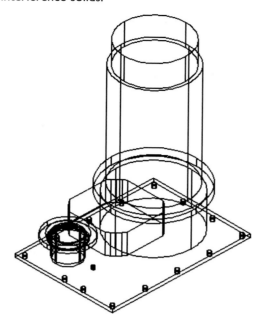

11. Open the *Ch11Sandpile.dwg* file, a drawing of a pile of sand, from which a bulldozer removed a chunk.

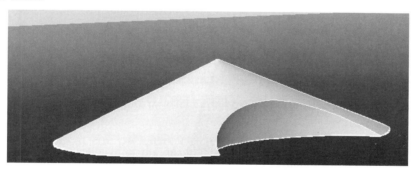

    a. Use the **MASSPROP** command to find the volume of sand (in cubic feet) in the pile.

    b. Assuming sand weights 90 pounds per cubic foot, how much does the pile weigh?

## ADDING BEVELS TO SOLIDS' EDGES

Before applying the CHAMFER command, first create a simple solid model of a hole inside a cube.

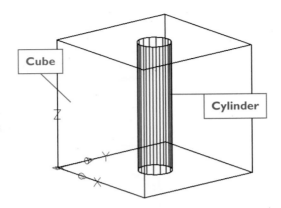

1. Start AutoCAD with a new drawing.
2. Draw a cube, as follows:

    Command: **box**

    Specify corner of box or [CEnter] <0,0,0>: *(Press ENTER.)*

    Specify corner or [Cube/Length]: **c**

    Specify length: **1**

3. Add the cylinder, as follows:

    Command: **cylinder**

    Current wire frame density: ISOLINES=16

    Specify center point for base of cylinder or [Elliptical] <0,0,0>: **0.5,0.5,0**

    Specify radius for base of cylinder or [Diameter]: **0.125**

    Specify height of cylinder or [Center of other end]: **1**

4. To view the cube better, rotate the viewpoint, as follows:

    Command: **vpoint**

    Current view direction: VIEWDIR= 0,0,0

    Specify a view point or [Rotate] <display compass and tripod>: **2.5,-2,1**

    Regenerating model.

    (To further improve the image, click the **LWT** button to thicken the lines, set ISOLINES to **16**, zoom to the extents, and then ZOOM **0.5x**.)

5. Subtract the cylinder from the box to create the hole:

    Command: **subtract**

    Select solids and regions to subtract from ..

    Select objects: *(Pick the cube.)*

    1 found Select objects: *(Press ENTER.)*

    Select solids and regions to subtract ...

    Select objects: *(Pick the cylinder.)*

    1 found Select objects: *(Press ENTER.)*

    (Following the subtraction operation, the model will look exactly the same.)

6. Start the **CHAMFER** command, and then select an edge of the cube.

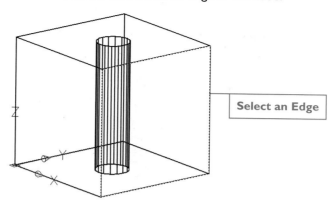

Command: **chamfer**

(TRIM mode) Current chamfer Dist1 = 0.5000, Dist2 = 0.5000

Select first line or [Polyline/Distance/Angle/Trim/Method]: *(Pick any edge of the cube.)*

Base surface selection...

(To highlight a different face, enter **N** — short for "next" — at the prompt; otherwise press **ENTER**.)

Enter surface selection option [Next/OK (current)] <OK>: *(Press* **ENTER**.*)*

7. Specify the chamfer distances here. (The **Distance** option earlier in this command is only for 2D entities.)

Specify base surface chamfer distance <0.5000>: **0.2**

Specify other surface chamfer distance <0.5000>: **0.2**

8. Select the edges you want chamfered: pick any two edges, and then press **ENTER** to exit the command.

Select an edge or [Loop]: *(Pick edge 1.)*

Select an edge or [Loop]: *(Pick edge 2.)*

Select an edge or [Loop]: *(Press* **ENTER**.*)*

Notice that AutoCAD performs a "double" chamfer: the intersection between the two chamfered edges is another 45-degree angle.

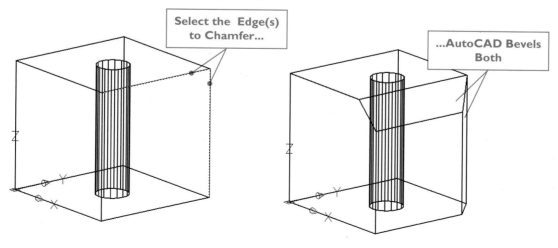

## ADDING COUNTERSINKS

In this tutorial, you continue from the previous one to create a second type of chamfer, one that countersinks the hole.

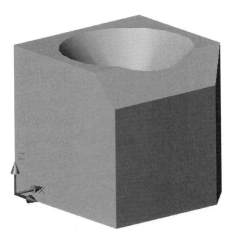

Continue with the drawing from the previous tutorial.

1. Start the **CHAMFER** command again, and then select a top edge on the cube (adjacent to the hole).

    Command: **chamfer**

    (TRIM mode) Current chamfer Dist1 = 0.2000, Dist2 = 0.2000

    Select first line or [Polyline/Distance/Angle/Trim/Method]: *(Pick an edge on the top face of the cube, but not one of the chamfer edges.)*

    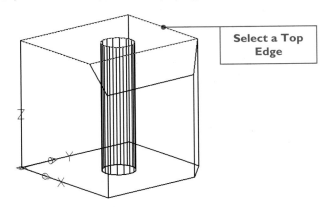

    If the "wrong" face is highlighted by AutoCAD, use the **Next** option to pick the top face:

    Enter surface selection option [Next/OK (current)] <OK>: **n**

    Enter surface selection option [Next/OK (current)] <OK>: *(Press ENTER.)*

2. Specify a chamfer distance of 0.3 units:

    Specify base surface chamfer distance <0.2000>: **0.3**

    Specify other surface chamfer distance <0.2000>: **0.3**

3. Select the top of the cylindrical hole:
   Select an edge or [Loop]: *(Select the top of the cylindrical hole.)*
   Select an edge or [Loop]: *(Press* ENTER.*)*

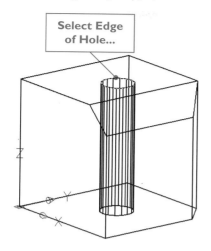
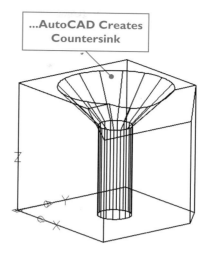

Select Edge of Hole... ...AutoCAD Creates Countersink

## CHAPTER REVIEW

1. What is an *extrusion*?
2. Describe two ways that extrusions differ from sweeps:
    a.
    b.
3. Describe a use for the following commands in constructing surfaces:

    **RULESURF**

    **EDGESURF**

    **TABSURF**

    **REVSURF**
4. The density of two-direction 3D surfaces is described by the values of *m* and *n*. Which system variables control the values of each?
5. Which command creates "Coons patches"?
6. How do **TABSURF** and **REVSURF** differ?

    How are they the same?
7. Can **CHAMFER** operate on 3D mesh models?

    If so, describe the process:
8. Does the **INTERSECT** command remove everything in common from two solid objects?
9. Does the **UNION** command combine two solid objects to create a region?
10. Why are the **UNION, INTERSECT**, and **SUBTRACT** commands called "Booleans"?
11. Describe how to create bolt holes in solid objects.
13. Does the **INTERFERE** command create solid objects from the intersection of two or more solid objects?

# VII

## Plotting Drawings

# CHAPTER 19
## Working with Layouts

In this chapter, you learn how to view drawings in more than one mode. Up until now, you have been drawing and editing drawings in model space. Models are drawn in model space; drawings are laid out for plotting in layouts. *Layouts* determine the arrangement and views of 2D and 3D models on the plot sheet.

In this chapter, you find the following commands and options:

**TILEMODE** switches drawings between model and layout tabs.
**ZOOM Xp** scales models relative to layouts.
**LAYER Current VP Freeze** freezes layers in selected viewports.
**Viewport overrides** override layer properties on a per-viewport basis.
**VPMAX** and **VPMIN** maximize and minimize the active viewport.
**QVLAYOUT** displays thumbnail views of layouts in filmroll fashion.
**SPACETRANS** scales model-space distances relative to layouts.
**PSLTSCALE** scales model-space linetypes relative to layouts.
**LAYOUTWIZARD** steps through the layout creation process.
**LAYOUT** creates and modifies layouts.
**VIEWPORTS** creates rectangular viewports.
**-VPORTS** and **VPCLIP** create and change polygonal viewports.
**EXPORTLAYOUT** exports layouts to model space.

## ABOUT LAYOUTS

Two of AutoCAD's drafting environments are known as *model space* and *layouts*.

Model space is where you draw the models full scale, 1:1. Recall from earlier chapters that unless you used annotative scaling, you had to calculate manually the correct scale factor for text, linetypes, and hatch patterns. Whether you employed annotative scaling or worked out the scale factors on your own, all this was to ensure text and other elements did not appear too small when printed.

Layouts solve the scaling problem in another way. In this environment, you arrange models as if they were being placed on sheets of paper. In layouts, the paper is full size (1:1), and the model is scaled to fit the paper. This solves the problem of working out scale factors for text, linetypes, and hatch patterns — they are all drawn full size in layouts — although this system introduces a different problem, in that dimensions cannot be seen in model tab.

The figure below illustrates the elements of a layout. It features two viewports, each showing the model at different scales.

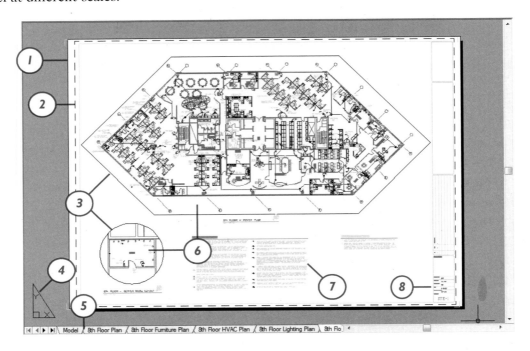

1: Edge of "paper" (solid line).
2: Printer margins (dashed line).
3: Polygonal and round viewports.
4: Paper space UCS icon.
5: Model and layout tabs.
6: The model (scaled).
7: Text placed in layout.
8: Title block placed in layout.

Showing a model in a layout takes two steps: (1) creating a viewport in a layout, which acts like a window into model space, and then (2) setting the viewport scale factor to size the model correctly.

You access layouts through tabs or buttons — depending on how you have AutoCAD configured. Sometimes, layout tabs are located just under the drawing, as illustrated above. Other times, layout viewing controls are stored in buttons on the status bar, as illustrated below.

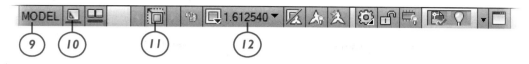

**9:** Toggle between model and paper space.  **11:** Maximize the viewport.
**10:** Quick views of layouts.  **12:** Viewport scale factor.

Every drawing starts with one model space (named "Model") and two or more layouts named "Layout1" and "Layout2." Each layout represents a drawing sheet to be plotted/printed. Each can have the same or different paper sizes and orientation; each can be assigned the same or a different printer.

You can think of layouts as interactive plot previews. You can create, move, and delete viewports, the entire model, and other elements at your whim. Layouts show the overall plan as well as zoomed-in details — all at the same time.

Each layout needs at least one viewport through which to show some or all of the model. But layouts often have two or more viewports, each showing details or overall views. Each viewport can show the model at a different scale factor, and, in 3D drawings, show different views, such as the top, side, and front. (Layouts themselves are strictly 2D.)

Layouts typically have drawing borders and title blocks. (The viewports fit inside the drawing border.) You can add dimensions, text, and hatch patterns to layouts. A command is available that automatically translates linetype scales between the model scale factor and the layout.

New layouts are created from scratch or imported from other drawings, and can be moved, resized, and deleted. The order in which layouts appear can be changed. As well, each viewport can have a different set of layers visible, and have distinctive plot styles and visual styles. While layouts are always rectangular, viewports can be any shape.

### Exploring Layouts

To experience layouts, open the *8th floor.dwg* file, a floor plan drawing. If necessary, select the **Model** tab. You see the entire model, but without notes, drawing borders, or title blocks.

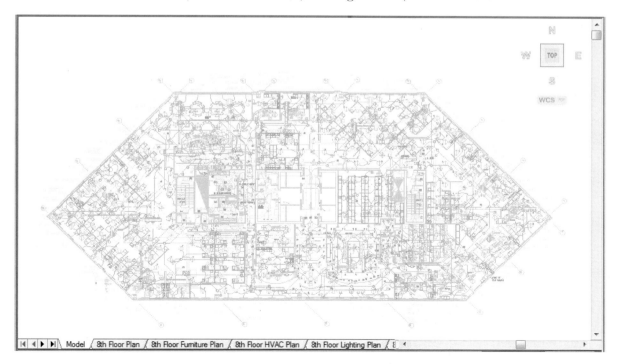

From among the layout tabs, select the **8th Floor Plan** tab by clicking on its name. The view changes dramatically. You see only the floor plan of the model, surrounded by the title block and drawing border. This illustrates how layers can be frozen in layouts to show details selectively.

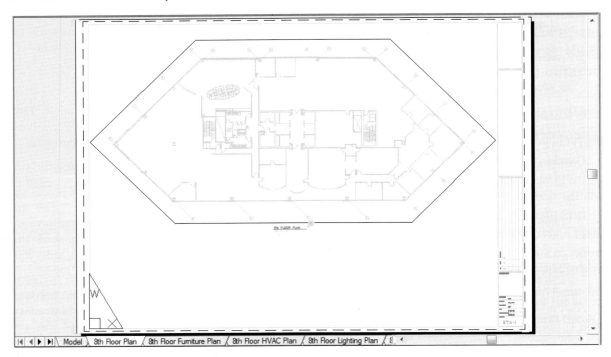

The drawing is shown on a white rectangle, which represents the paper. In this case, the selected paper is "Architectural E1" size, which works out to 42" wide by 30" high. The dashed rectangle represents the printer's margin, the unprintable area along the four edges of the paper.

Select the **8th Furniture Plan** tab. The view changes again. In this layout, layers are turned on to show the floor plan with the furniture — desks, chairs, and so on.

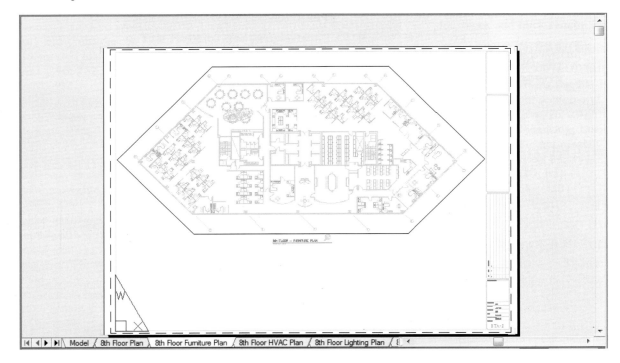

Select the other layout tabs, working your way through the HVAC plan (heating, ventilation, air conditioning), lighting plan, power plan, and plumbing plan. As an alternative to clicking tabs with the cursor, you can press CTRL+PGUP and CTRL+PGDN to switch between layouts.

When drawings have many layouts, an additional set of controls becomes useful. These "VCR" buttons take you to the next and previous layouts, as well as to the first and last layouts.

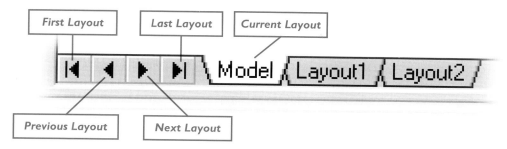

### "Hiding" Layout Tabs

Layout tabs can be removed, but I don't see much point to this option. The tabs share space with the horizontal scroll bar, which I find useful for panning, and so the tabs don't take up any extra room in AutoCAD's user interface.

Still, here's how to do it: right-click any tab. From the shortcut menu, choose **Hide Layout and Model Tabs**. AutoCAD doesn't hide the tabs, but relocates them as buttons and shortcut menus on the status bar, replacing the MODEL/PAPER button. To move the tabs back to the scroll bar, right-click the model or layout button, and then choose **Display Layout and Model Tabs**.

displays model space.

displays the last-accessed layout.

displays a list of layout names and preview thumbnail images. Select a thumbnail to go to that layout.

> **Note** To switch between model and paper mode, double-click the layout, as follows:
> Double-click *inside* the viewport to enter model mode; or click the **VsMax** button.
> Double-click *outside* the viewport to return to paper mode.

---

#### HISTORY: **TILEMODE**

Using the TILEMODE system variable was the original method of switching between model and paper space, as layouts were originally called. With the introduction of layout tabs in AutoCAD 2000, the need for TILEMODE as a "command" disappeared.

I mention the system variable to point out that when you select model or layout tabs, AutoCAD still executes TILEMODE **0** or **1** silently in the background.

### TUTORIAL: WORKING WITH LAYOUTS

This tutorial takes you through some aspects of working with layouts.

1. Open the *layout.dwg* file, the drawing of a streetscape.
2. Select the **Layout1** tab by either method:
   - If layout tabs are present, click the one labeled "Layout1."
   - If layout tabs are not present, click the layout button.

   Notice that the cityscape drawing appears on a sheet of paper with two rectangles. The inner rectangle is the viewport, while the dashed rectangle is the printer margin.

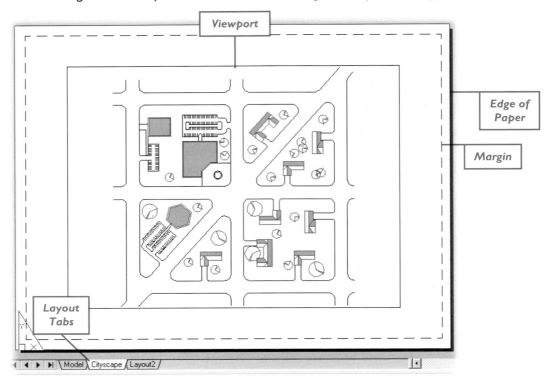

In layout mode, you work only with the "paper," not the model. "Working with the paper" includes drawing on the paper or page, manipulating viewports and their content, and plotting. You'll find that you cannot touch the model at all.

1. Try selecting the model. Notice that you can't, because the layout is in "paper" mode. (On the status bar, the indicator reads PAPER.)
2. Click the viewport border. Notice that you can select it, unlike the model.
3. Grab a grip, and then make the viewport smaller by dragging. Notice what happens to the model: it has become "cropped" — part of the model is cut off visually. This illustrates how you can selectively show details of the model in layouts. Layouts need not show the entire model.

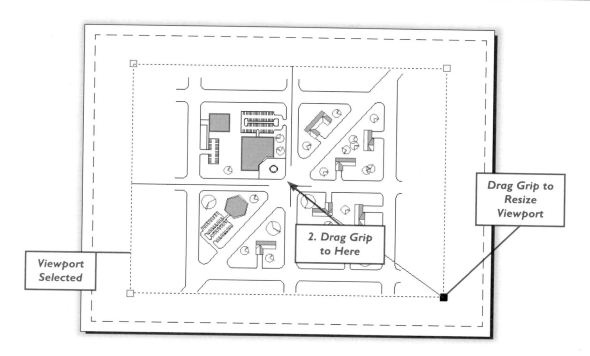

4. You can move viewports: grab an edge (not a handle), and then drag the viewport to a new position. Notice that the cropped model view moves with the viewport.

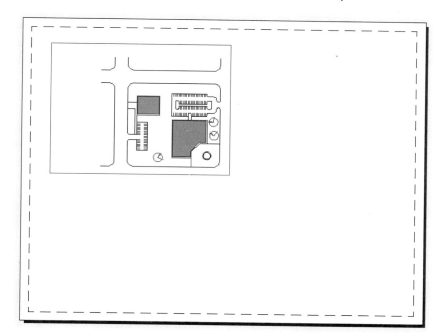

You can apply all of the usual grip editing commands to viewports: stretch, move, copy, scale (resize), rotate, and mirror.

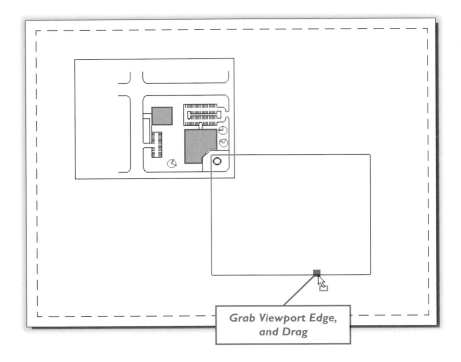

5. Use the **COPY** command, for example, to copy viewports. Notice that the copy of the viewport contains exactly the same model view as the original. And, as the figure below illustrates, viewports can overlap.

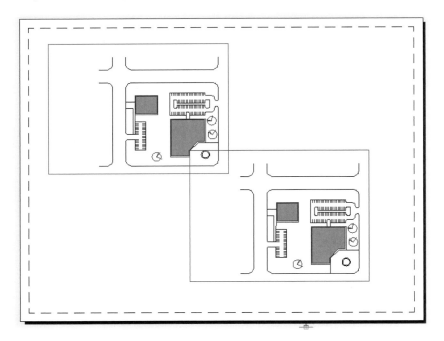

You can change the color of viewport borders, or make them invisible by placing them on a layer of their own, and then change the color or freeze the layer. The rest of the model is unaffected. (You cannot apply linetypes or lineweights to viewport borders.) To see viewport borders in the drawing, but not plot them, change the layer to be non-plotting.

6. The **ERASE** command deletes viewports. Other editing commands, however, do not work with viewports, including **OFFSET**, **TRIM**, **EXPLODE**, and **FILLET**.

## Hiding Viewport Borders

If you don't want the viewport borders to appear in drawings, use this trick to hide them: assign viewports their own layer, and then freeze the layer. This makes the viewport borders invisible, but the content remains visible.

1. With the **LAYER** command, create a new layer named "VPorts."
2. Select the viewport borders, and from the Layers toolbar or ribbon panel, click the **Layer Control** droplist to change the layer to "VPorts."

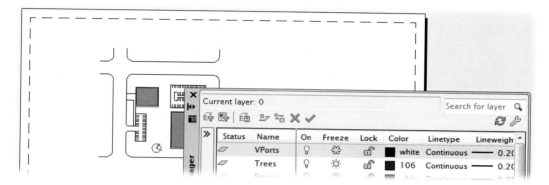

3. Freeze layer "VPorts." Notice that the viewport border disappears, but that the model view remains.
4. Thaw layer "VPorts" to bring back the viewport borders.

## Switching from Paper to Model Mode

I lied. There are three ways to view drawings: (1) in the Model tab, (2) in a layout tab, and (3) in model mode of a layout tab. You can edit the model inside the viewport.

Until now, you have been manipulating viewports. Let's work with the model inside the viewports.

1. On the status bar, click **PAPER**.

Notice that the button changes its name to MODEL. This means that you are now working with the model instead of the layout (paper) — yet the drawing is still in layout mode.

(If you cannot see words PAPER or MODEL on the status bar, then double-click inside the viewport to enter model space.)

2. Either way, notice that one viewport has a heavy border. This is the *active* viewport, the one in which you can manipulate the model.

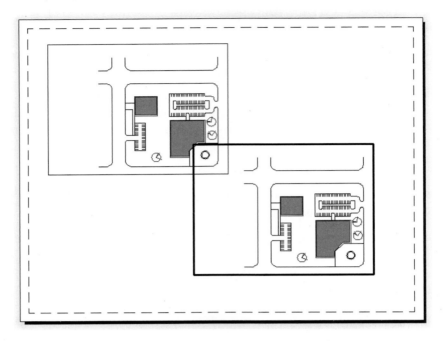

To make another viewport active, simply select it. Alternatively, you can press **CTRL+R** to cycle through viewports.

3. Select an object in the model. Notice that the grips and highlighting appear in both viewports. This confirms that you are seeing two images of the same model.

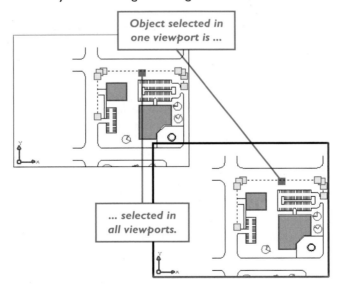

*Object selected in one viewport is ...*

*... selected in all viewports.*

### SCALING MODELS IN VIEWPORTS

You can show models at a different scale in every viewport. As well, you can hide and show layers, and override layer properties, such as color and linetype, and toggle the grid — all on a per-viewport basis. In this section, we look at the important topic of scaling models inside viewports.

#### VP Scale

The easiest way to scale models in viewports is to use the VP Scale droplist found on the status bar. (See illustration at right.)

1. Select a viewport.
2. Click the **VP Scale** list, and then choose a scale factor.

Notice that the model changes its size inside the viewport. You can choose a different scale factor for each viewport — or use the same one for all. (See figure on the following page.)

Not sure which scale factor to choose? Read ahead to the section on the ZOOM XP command.

(This droplist does not work if the viewport is locked, either through the Properties palette's **Display Locked** or the MIVEW and -VPORTS commands' **Lock** options.)

The scale factors listed by the droplist can be edited with the SCALELISTEDIT command. This command lets you add and remove scale factors; you can access it through the **Custom** item on the VP Scale droplist. In addition, the HIDEXREFSCALE system variable hides scale factors imported from externally-referenced drawings. See Chapter 20, "Plotting Drawings," for more on both of these.

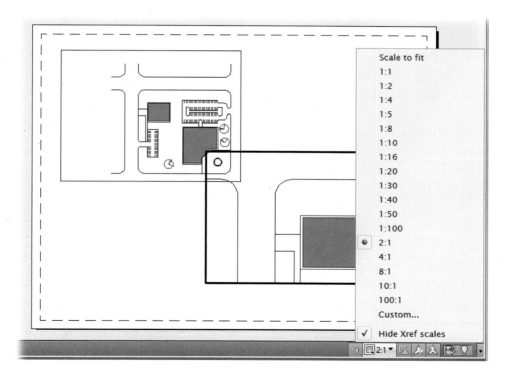

#### Zoom XP

AutoCAD's two primary visibility commands are ZOOM and LAYER. In model mode, AutoCAD commands apply to the active viewport only; all other viewports are ignored. The ZOOM command changes the size of the model in each viewport.

1. Enter the **ZOOM** command, and then use the **Extents** option. You should see the entire model in the active viewport. (Note that this changes the viewport scale.)

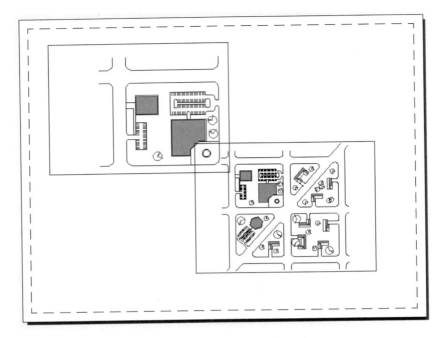

2. The **ZOOM** command's **XP** option is meant for use in layout viewports (short for "multiple in paper"). This option scales the model view relative to the layout. This is important: *You use ZOOM XP to scale models in layout viewports.*

3. Apply the scale factor of 1:2 by using it with the XP suffix, as follows:
   Command: **zoom**
   Specify corner of window, enter a scale factor (nX or nXP), or
   [All/Center/Dynamic/Extents/Previous/Scale/Window/Object] <real time>: **1/2xp**

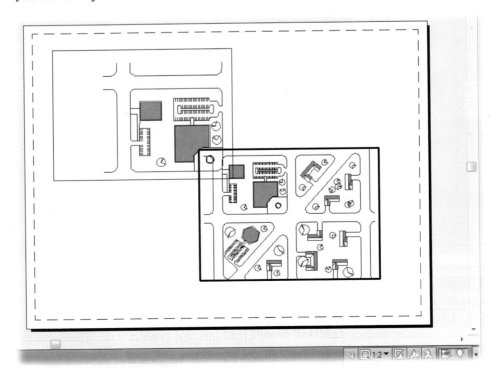

Notice that the viewport scale factor is updated on the status bar.

**Notes** Use the **PAN** command and the scroll bars to position models inside viewports.

You can lock viewports to prevent anyone from changing the view of the model, such as through zooming or panning: (1) In paper mode, select the viewport border, or (2) In the Properties palette, change the **Display Locked** property to **Yes**.

To enter model space without messing up the viewport scale, click the **Maximize Viewport** button.

The viewport scale factor can also be set with the Properties palette. It has the advantage of providing access to the QuickCalc calculator, which is useful when you need to work out the scale factor.

### SELECTIVELY DISPLAYING DETAILS IN VIEWPORTS

Earlier in this chapter, you saw how it was possible to make the viewport border invisible by freezing its layer. The LAYER command also allows you to change what is displayed inside the viewports on a per-viewport basis:

**Freezing layers** hides parts of the drawing; thaw the layers to make the parts visible again.

**Overriding layer properties** displays parts of drawings with different colors, linetypes, and so on.

The "ByLayer" properties that can be overridden in each viewport are color, linetype, lineweight, transparency, and plot style. Whether freezing or overriding, the process is the same:

1. Select a viewport.
2. Use the **LAYER** palette to freeze layers or override properties.
3. Repeat for other viewports.

Let's look at this in some detail.

### Freezing Layers

The LAYER command freezes layers in the current viewport through the **Current VP Freeze** column.

1. Ensure the *layout.dwg* drawing is open. Select one of the viewports.
2. Open the Layer Properties Manager palette with the **LAYER** command.

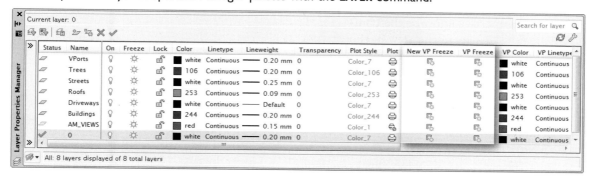

Towards the center are two columns you have not yet used:

**VP Freeze** column freezes the selected layers in the current viewport (a.k.a. "VP") — as soon as you make the change in the palette (or click **Apply** in the dialog box).

**New VP Freeze** column freezes the selected layers — whenever a new viewport is created.

These two options have no effect on Model tab and in unselected layouts and viewports.

3. Hold down the **CTRL** key, and then select the names of the **Buildings**, **Roofs**, and **Trees** layers, as illustrated below.

| Status | Name | sparency | Plot Style | Plot | New VP Freeze | VP Freeze | VP Color | VP Linetype | VP Lineweight | VP Transparency | VP Plot Style |
|---|---|---|---|---|---|---|---|---|---|---|---|
| | VPorts | | Color_7 | 🖨 | 🗔 | 🗔 | ■ white | Continuous | — 0.20 mm | 0 | Color_7 |
| | Trees | | Color_106 | 🖨 | 🗔 | 🗔 | ■ 106 | Continuous | — 0.20 mm | 0 | Color_106 |
| | Streets | | Color_7 | 🖨 | 🗔 | 🗔 | ■ white | Continuous | — 0.25 mm | 0 | Color_7 |
| | Roofs | | Color_253 | 🖨 | 🗔 | 🗔 | □ 253 | Continuous | — 0.09 mm | 0 | Color_253 |
| | Driveways | | Color_7 | 🖨 | 🗔 | 🗔 | ■ white | Continuous | — Default | 0 | Color_7 |
| | Buildings | | Color_244 | 🖨 | 🗔 | 🗔 | | | — 0.20 mm | 0 | Color_244 |
| | AM_VIEWS | | Color_1 | 🖨 | 🗔 | 🗔 | | | — 0.15 mm | 0 | Color_1 |
| ✓ | 0 | | Color_7 | 🖨 | 🗔 | 🗔 | | | — 0.20 mm | 0 | Color_7 |

*Click icon in **VP Layer** column*

4. Let go of the **CTRL** key, and then click a highlighted icon in the **Current VP Freeze** column. Notice that all selected suns turn to snowflakes, indicating the layers will be frozen (hidden from view).

### EDITING VIEWPORTS

You can make many kinds of changes to viewports through the Properties palette. Some do not, however, work, such as linetypes, lineweights, and transparency.

#### Viewport Border Properties

The following properties can be applied to viewport borders:

**Color** changes the color of the viewport border.
**Layer** specifies the border's layer name.
**Plot Style** specifies the plot style, if enabled.
**Hyperlink** activates a link when the border is clicked.
**Geometry** specifies the center coordinates and size.

#### Viewport Content Properties

The following properties apply to the content of viewports:
**On** toggles the display of the view in the viewport.
**Display Locked** prevents view changes, such as zooms.
**Annotation Scale** specifies the annotative scale factor.
**Standard Scale** sets views using factors from SCALELISTEDIT.
**Custom Scale** specifies other scale factors.
**Shade Plot** specifies shading mode during plotting.

#### Other Viewport Commands

These properties are controlled by other commands:
**Clipped** accessed through the MVIEW and -VPORTS commands.
**UCS Per Viewport** accessed through the UCS command.
**Visual Style** accessed through the VSCURRENT system variable; cannot be set in paper space.
**Linked to Sheet View** accessed through the SHEETSET command.

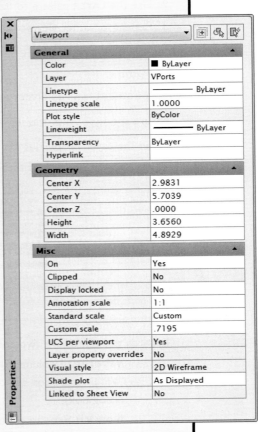

5. Choose the **OK** button. Notice that the buildings, roofs, and trees disappear from the viewport. The other viewport, however, is unaffected. This shows that viewports can have their layers frozen independently of each other.

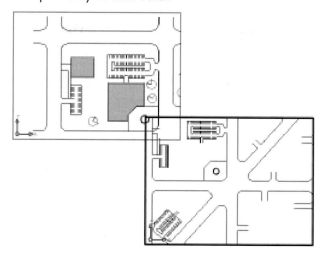

### Overriding Layer Properties

You can override layer properties with commands like COLOR and LINETYPE, but these overrides affect all viewports equally. The LAYER command can override properties in each viewport.

For example, you might have some layers colored light gray in one viewport but black in others. Some viewports might show layers with the Hidden linetype, while others show the Continuous linetype, as illustrated below.

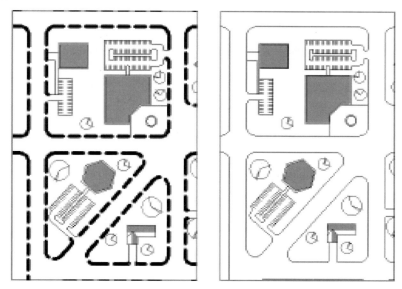

*Left: Streets in hidden linetype and heavy lineweight overrides.*
*Right: Streets displayed normally (ByLayer).*

The LAYER command overrides properties with the VP Color, VP Linetype, VP Lineweight, and VP Plot Style columns.

1. Continuing with the *layout.dwg* drawing, double-click inside one of the two viewports to enter model space.

2. Enter the **U** command to undo the layer changes from the previous tutorial.

3. (If the layers palette is closed, open it again with the **LAYER** command.) Towards the right of

the palette are the columns that we are interested in:

**VP Color** column overrides the ByLayer color in the current viewport.

**VP Linetype** column overrides the ByLayer linetype.

**VP Lineweight** column overrides the ByLayer lineweight.

**VP Transparency** column overrides the ByLayer transparency (new to AutoCAD 2011).

**VP Plot Style** column overrides the ByLayer plot style, if plot styles are enabled.

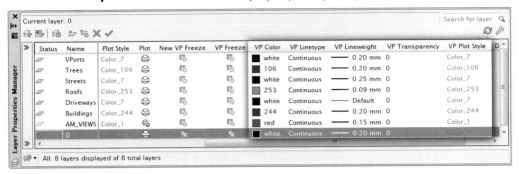

4. Select the name of the **Streets** layer, as illustrated below.
5. In the VP Lineweight column, select **2.11mm**.

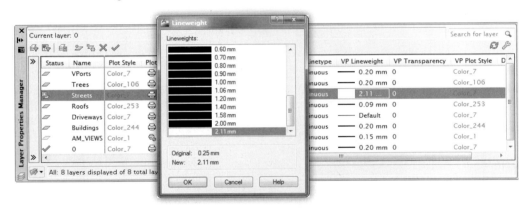

6. Notice that the streets are displayed with heavy lines in the current viewport; the other viewport is unaffected by the change.

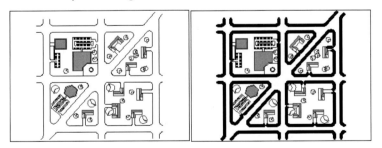

**Left:** *Streets displayed normally (ByLayer).*
**Right:** *Streets in heavy lineweight override.*

**Note** The **VPLAYEROVERRIDES** system variable reports whether *any* viewport contains layer property overrides. When "1," at least one does; when "0," none does. (This sysvar is read-only).

The information is also provided by the Properties palette. Select a viewport border, and then check whether the value of **Layer Property Overrides** is Yes or No. Like the sysvar, this setting is read-only.

### VpOverrideMode — Toggling Properties Overrides

You can toggle property overrides with the VPOVERRIDEMODE system variable. It has two settings:

- **1** — overrides are displayed and plotted.
- **0** — overrides are neither displayed nor plotted.

This sysvar is handy for temporarily turning off overrides. To turn overrides off "permanently," see the next section.

### Reverting Properties to ByLayer

After you've overridden the layer properties of several viewports, you may want to change them back to "normal" (ByLayer). Autodesk provides three (!) ways to do this: through the LAYER, -VPORTS, and MVIEW commands.

The LAYER command's palette indicates overridden properties by highlighting them in cyan (light blue). In addition, a different icon appears in the Status column; pausing the cursor over the icon displays a yellow tooltip that reports the override value(s).

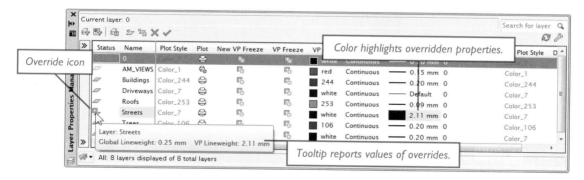

To remove overrides, follow these steps:

1. In the Layer dialog box, right-click any layer name.
2. From the shortcut menu, select **Remove Viewport Overrides For**. Notice the options:

   **Selected Layers** — removes overrides from selected layers only.

   **All Layers** — removes overlays from all layers.

   For both options, overrides can be removed from the current viewport or all viewports.

3. Select an option. Notice that the overrides and cyan highlights disappear.

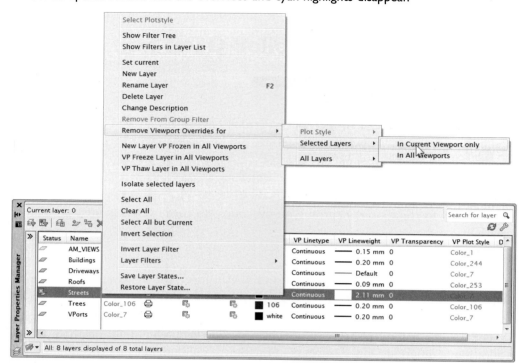

The MVIEW and -VPORTS commands operate identically, curiously enough. Both have a LAyer option that removes the overrides. Here are the prompts for MVIEW:

   Command: **mview**
   Specify corner of viewport or [ON/OFF/Fit/.../LAyer/2/3/4] <Fit>: **la**
   Reset viewport layer property overrides back to global properties? [Yes/No]: **y**
   Select objects: *(Select one or more viewports.)*
   Select objects: *(Press* ENTER *to exit object selection.)*

Notice that the overrides are removed.

### INSERTING TITLE BLOCKS

Layout mode is where you place scale-dependent objects, such as drawing notes and title blocks. In this tutorial, you add a title block and border to a drawing. Continue with the *Layout.dwg* file:

1. Ensure that Layout1 is in PAPER mode (no viewport with a heavy outline).
   If necessary, use the ZOOM All command to see the entire layout.
2. Use the layer palette to create a new layer called "Titleblock," and then make it current.
3. Save your work with the QSAVE command.
4. To access title blocks, open DesignCenter by pressing CTRL+2.
5. In the DesignCenter's Folder List, go to the folder holding the template drawings (not in the \template folder, which is empty):
   C:\Users\<*login name*>\AppData\Local\Autodesk\AutoCAD 2011\R18.1\enu\Template
   Replace <*login name*> with your Windows login name, such as "administrator."

*Working with Layouts* **871**

**Note** The folder is inconveniently hidden by Windows, and so it might not show up in DesignCenter. Here is one way to access it: (1) enter the **ADCNAVIGATE** command, and (2) provide the folder path listed above. DesignCenter immediately displays the contents of the *template* folder.

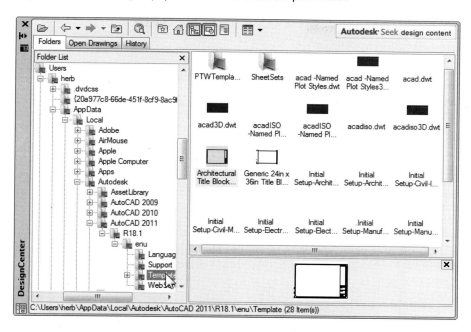

6. Drag the *Architectural Title Block.dwg* file from DesignCenter into the drawing.
7. When AutoCAD prompts you for the insertion point, pick the lower-left corner of the margin.

    Command: _-INSERT Enter block name or [?] <A$C27FF03C9>: "Architectural Title Block.dwg"

    Specify insertion point or [Scale/X/Y/Z/Rotate/PScale/PX/PY/PZ/PRotate]: *(Pick the lower left corner of the margin.)*

8. Press **ENTER** at each of the remaining prompts.

    Enter X scale factor, specify opposite corner, or [Corner/XYZ] <1>: *(Move the cursor to set the scale factor interactively.)*

    Enter Y scale factor <use X scale factor>: *(Press **ENTER** to keep scale factor at 1.0.)*

    Specify rotation angle <0>: *(Press **ENTER** to keep angle at 1.0.)*

9. You may need to move the viewports so that they do not interfere with the title block or border.

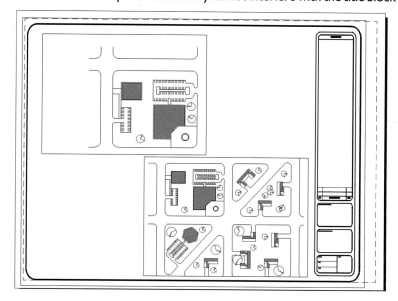

### USING TILEMODE

At the start of this chapter, I mentioned the TILEMODE system variable, which has been replaced by the layout tabs. For completeness, here is how to use it:

    Command: **tilemode**
    Enter new value for TILEMODE <0>: **1**

Entering **1** returns to the model tab, while entering **0** returns to the last active layout tab.

### VPMAX AND VPMIN

The VPMAX command maximizes the size of the current viewport (short for "viewport maximize"). This command is useful for increasing the size of small viewports. Conversely, the VPMIN command returns the viewport to its normal size. (As an alternative, you can double-click viewport borders to maximize and minimize them.)

Using this pair of commands is faster than switching back and forth between paper and model space, because AutoCAD automatically switches to model space at the same viewing scale as the layout. Another advantage is that these two commands do not affect model/paper space zoom scales.

The VMPAX and VPMIN commands work in layouts only, and not in Model tab. Drawings cannot be plotted or published while the viewport is maximized.

*Left: Maximize Viewport button.*
*Right: Minimize Viewport button surrounded by the Previous and Next buttons.*

When the layout has more than one viewport, two arrow buttons appear, as illustrated above at right. These buttons allow you to move to the next and previous viewport, in the order in which they were created. To return the viewport to its normal size, click the **Minimize Viewport** button.

### TUTORIAL: MAXIMIZING AND MINIMIZING VIEWPORTS

1. For this tutorial, open the *vpmax.dwg* file.
   Click the **Layout** tab to ensure the drawing displays the viewports in layout mode.

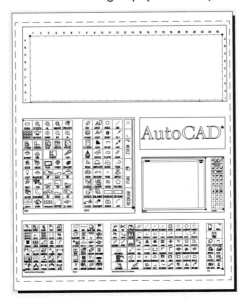

2. To maximize viewports, start the **VPMAX** command:
   - In paper space, double-click a viewport border.
   - In paper space, select a viewport. Right-click, and then select **Maximize Viewport** from the shortcut menu.
   - On the status bar, click the **Maximize Viewport** button.
   - At the 'Command:' prompt, enter the **vpmax** command.

        Command: **vpmax** *(Press ENTER.)*

3. If the layout contains more than one viewport, AutoCAD asks you to select one of them:
        Select a viewport to maximize: *(Select a viewport.)*

   Notice that AutoCAD does the things illustrated in the figure below:

   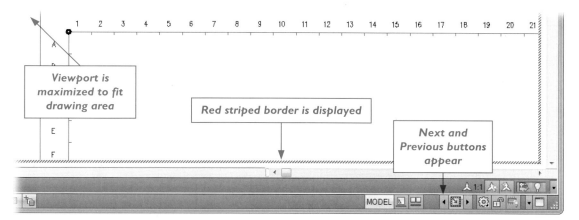

   - Viewport is made as large as the drawing area.
   - Viewport has a red striped border.
   - **Previous** and **Next** buttons appear on the status bar, if the layout contains more than one viewport.
   - The Maximize Viewport button turns into the Minimize Viewport button.
   - Space changes from paper to model.

4. To see other viewports in their maximized state, click the **Maximize Previous Viewport** or **Maximize Next Viewport** button. Notice that AutoCAD displays the next viewport every time you click the button.

5. When done editing, minimize the viewport using one of these methods:
   - Double-click the red striped border.
   - Right-click anywhere inside the viewport, and then select the **Minimize Viewport** option from the shortcut menu.
   - On the status bar, click the **Minimize Viewport** button.
   - At the 'Command:' prompt, enter the **vpmin** command.

        `Command: **vpmin** *(Press ENTER.)*

   AutoCAD returns the viewport to its natural size.

**Note** Here's how to maximize a viewport but remain in paper space: select a viewport, and then use the **ZOOM** command with its **Object** option. AutoCAD maximizes the viewport.

 **QVLAYOUT**

The **QVLAYOUT** command displays a filmstrip-like interface for switching layouts (short for "quick view layouts").

Autodesk added this interface to AutoCAD 2009 as a "quicker" and more visual way of accessing layouts. Frankly, I think that layout tabs are better, but the decision is up to you. The best way to understand quick views is to experience them.

### TUTORIAL: SWITCHING LAYOUTS THROUGH QUICK VIEWS

1. For this tutorial, open a drawing that has many layouts: *8th floor.dwg*.
2. To switch to another layout, start quick views through one of the following methods:
   - On the status bar, click the [icon] Quick View Layouts button.
   - At the 'Command:' prompt, enter the **qvlayout** command.

      Command: **qvlayout**

   Notice the filmstrip of thumbnails.

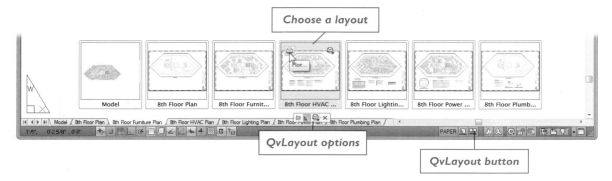

3. To switch to a layout, click on its thumbnail image.

This user interface has a number of features. The small toolbar-like rectangle contains these options:

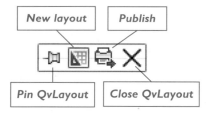

**Pin** — "pins" the quick view layout, keeping it open until you click the **x** button; when un-pinned, the quick view closes as soon as you pick any point outside of it.

**New** — creates new layouts; see the **LAYOUT** command later in this chapter.

**Publish** — publishes the layout; see the **PUBLISH** command in the next chapter.

**Close** — closes the quick view interface; alternatively, use the **QVLAYOUTCLOSE** command.

 **Notes** If there are quick views beyond the width of your monitor, move the mouse past the edge of the image strip to see additional thumbnails.

To change the size of the thumbnails, hold down the **CTRL** key while rotating the mouse's roller button back and forth. This makes the images larger and smaller.

If no images appear in the thumbnails, enter the **UPDATETHUMBSNOW** command to force a redraw.

When the cursor passes over a thumbnail image, the following controls become available:

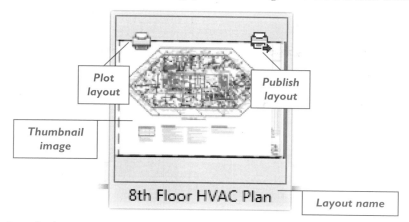

**Plot** — plots the layout; see **PLOT** command in the next chapter.
**Publish** — publishes the layout.

 **'SPACETRANS**

The SPACETRANS command converts distances between model space units and paper space units (short for "space translation").

The command automatically determines the height of text placed in model mode of layouts, scaling it appropriately for paper mode. The command is meant to be used *transparently* during other commands when AutoCAD asks for heights, lengths, and distances. However, a bug in SPACETRANS allows it to operate transparently only after it has been run normally (not during another command) at least once.

Instead of using this command, you may find it easier to use annotative scaling to scale text, because the annotative feature does the scaling for you, automatically; SPACETRANS requires you to enter the command each and every time you want to place scaled text. SPACETRANS may only be used in layouts — either in paper or model mode; attempting to use it in model space results in this complaint:

```
** Command not allowed in Model Tab **
```

### TUTORIAL: DETERMINING TEXT HEIGHT IN LAYOUTS

In this tutorial, you use the SPACETRANS command during the TEXT command to have AutoCAD scale the text height relative to the viewport scale. Continue with the *layout.dwg* drawing.

1. In a layout tab, double click inside a viewport. (This ensures AutoCAD is displaying model space through the viewport in the layout.)
2. Start the **TEXT** command, and then enter the **SPACETRANS** command transparently, as follows:
   Command: **text**
   Specify start point of text or [Justify/Style]: *(Pick a point.)*
   Specify height <0.2>: **'spacetrans**
3. You can enter fractions or decimals. For example, you may enter 1/8" or 0.125" (a typical height for text in drawings).
   >>Specify paper space distance <1.000>: **1/8**

4. Press ENTER. Notice that AutoCAD resumes the TEXT command, and that it displays the calculated height: 0.125 x viewport scale factor:

    Resuming TEXT command.

    Specify height <0.2000>: 0.075783521943615

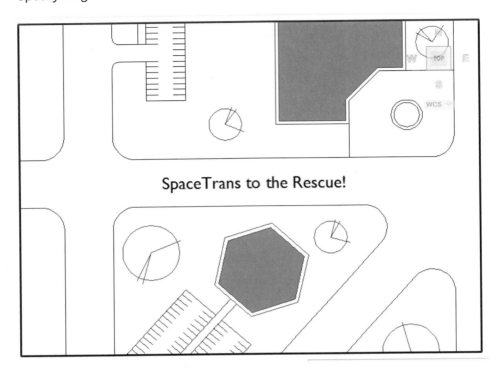

5. Continue with the TEXT command's other prompts:

    Specify rotation angle of text <0>: *(Press ENTER.)*

    Enter text: **SpaceTrans to the Rescue!** *(Press ENTER.)*

    Enter text: *(Press ENTER.)*

    AutoCAD places the text in the model with a height scaled to 0.075 units — instead of 0.125 units.

**Notes** Use SPACETRANS together with the SCALETEXT command to change existing text. Start the SCALETEXT command, and then enter 'SPACETRANS to specify the new height. AutoCAD scales the text correctly.

Because SPACETRANS cannot be used with the MTEXT command, you can instead, use SPACETRANS as a calculator at the 'Command:' prompt to find the scaled text height; then enter the height manually in the mtext editor, as follows:

   Command: **spacetrans**

   Specify model space distance <0.1250>: **1.0**

   0.663080708963838

When the viewport scale changes, the text also changes size, depending on the mode in which it was placed: When text is placed in PAPER mode:
- It is not visible in model space.
- It does not change size when viewport scale changes.
- It issues the following prompts, noting the word "model":

    Select a viewport: *(Select a viewport border.)*

    Specify model space distance <1.0>: *(Enter a value.)*

When the text is placed in MODEL mode:
- It is visible in model space.
- Its apparent size changes as seen from the layout.
- It issues the following single prompt, noting the word "paper":
  Specify paper space distance <1.000>: *(Enter a distance.)*

## PSLTSCALE

The **PSLTSCALE** system variable matches linetype scaling to viewport scaling, so that linetypes drawn in the model appear the correct size (short for "paper space line type scale").

Command: **psltscale**

Enter new value for PSLTSCALE <1>: **0**

When set to **1** (the default), the viewport scale factor (set by ZOOM **Xp** or the viewport properties) controls linetype scaling. The lengths of dashes and gaps are based on paper space units — for objects drawn in model space and in layouts. The advantage is that viewports can have different zoom levels (scale factors), yet the linetypes look the same.

When set to **0**, the linetype scale in model space is independent of layout scale factors.

 **Note** After changing the value of **PSLTSCALE**, you need to use the **REGENALL** command to update linetype scaling in all viewports.

### TUTORIAL: MAKING LINETYPE SCALES UNIFORM

1. Open the *psltscale.dwg* file, a drawing of objects with several linetypes in which **PSLTSCALE** is turned off (set to **0**). Notice that the linetypes in the two viewports have different scales, because the viewports have different scales.

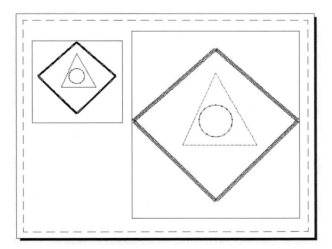

2. To make linetype scales the same in all layout viewports, start the **PSLTSCALE** system variable:
   Command: **psltscale** *(Press ENTER.)*

3. AutoCAD prompts you to change the value:
   Enter new value for PSLTSCALE <0>: **1**

4. The change to **PSLTSCALE** has no effect until you regenerate all the viewports:
   Command: **regenall** (Press ENTER.)

   Notice that the linetypes have the same scale factor in both viewports.

   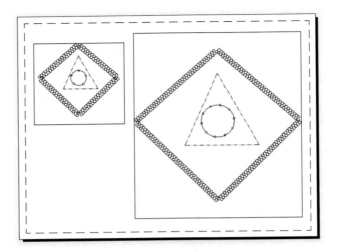

5. Change the linetype scale:
   Command: **ltscale**
   Enter new linetype scale factor <0.2500>: **.75**
   Regenerating layout. Regenerating model.

   The linetype scale changes in both viewports after AutoCAD automatically regenerates them. **VPMAX** shows the correct scale, whereas the Model tab does not.

## LAYOUTWIZARD

The **LAYOUTWIZARD** command takes you through the steps of creating new layouts.

### TUTORIAL: MANAGING LAYOUTS

1. To manage layouts, start the **LAYOUTWIZARD** command:
   • At the 'Command:' prompt, enter the **layoutwizard** command.

      Command: **layoutwizard** (Press ENTER.)

   Notice that AutoCAD displays this first dialog box:

   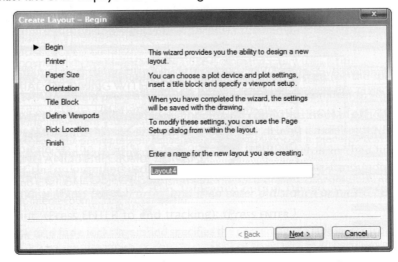

2. Enter a name for the layout, which will appear on the tab.
   Click **Next**.

3. Select a printer from the list of Windows system printers and AutoCAD HDI printer drivers (shown as *.pc3* files). You can select a different printer later, if need be.

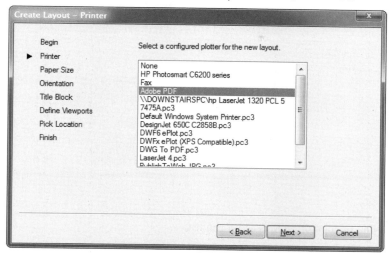

   Click **Next**.

4. Select a paper size from the droplist. Only those sizes supported by the printer are listed.

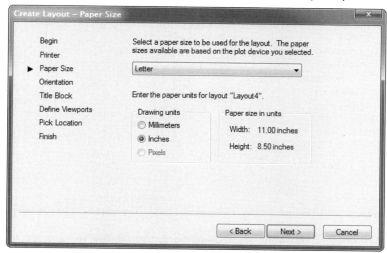

   **Drawing Units:** Select the drawing units — inches or mm.
   Click **Next**.

5. Select an orientation for the paper — landscape or portrait.

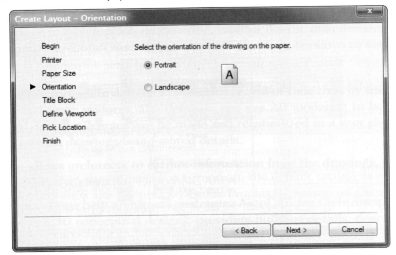

Click **Next**.

6. Select a drawing border/title block from the list.

(For A-size drawings, make sure the orientation of the border matches the orientation of the page you selected in the previous setup.)

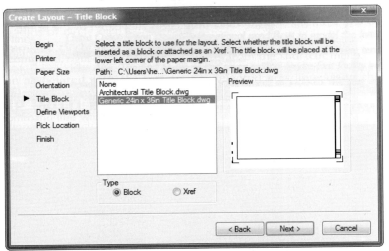

**Type:** Specifies whether you want the drawing border to be inserted as a block, or attached as an externally-referenced drawing (xref). If you are not sure, select **Block**.

The following table provides you with the pros and cons of each:

| Type | Pros | Cons |
| --- | --- | --- |
| Block | Border and title block are part of the drawing, making it complete and easier to transport, such as by email. | Drawing size is larger.<br><br>Blocks cannot be as readily updated as xrefs. |
| Xref | Drawing size is smaller.<br><br>All xref'ed title blocks can be easily changed. | When sending the drawing to another office or client, you must ensure the xref is packaged with the drawing file. |

7. **Viewport Setup:** Select the number and style of viewports. If you are not sure, select **Single**.

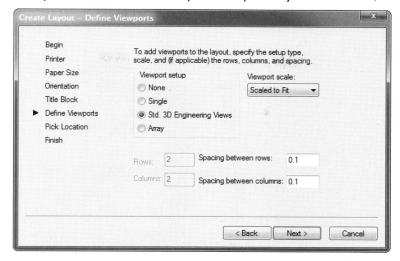

| Viewport Setup | Comments |
| --- | --- |
| None | No viewport is created. |
| Single | One viewport is created. |
| Std. 3D Engineering Views | Four viewports are created, and then adjusted to show the standard engineering viewpoints of 3D drawings: front, side, top, and isometric. |
| Array | Rows x columns array of viewports is created. |

For engineering drawings, you would choose Std. 3D Engineering Views, which sets up the viewports and viewpoints illustrated below (at left). The standard engineering viewpoints are as follows:

    Top view        Isometric view
    Front view     Side view.

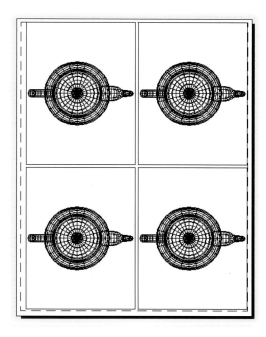

***Left:*** *Four standard engineering views.*
***Right:*** *Four views in an array.*

The **Array** option sets up the viewports as a rectangular array. You can specify the number of rows and columns, such as the 2x2 illustrated above, at right.

**Viewport Scale** selects the viewport scale, which ranges from 100:1 to 1:100, and from 1/128"=1' to 1'=1'.

If you are not sure of the scale, select the **Scaled to Fit** option. (You can change it later by right-clicking the viewport border, selecting **Properties**, and then changing the value of **Custom Scale**.)

Optionally, you can also specify the gap between viewports; 0.1 units is the default.

8. Click the **Select Location** button to position the viewport(s).

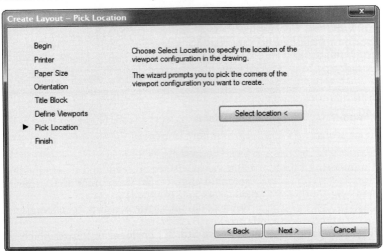

AutoCAD prompts you at the command line:
   Specify first corner: (Pick point 1.)
   Specify opposite corner: (Pick point 2.)

Pick two points to form a rectangle. AutoCAD later fits the viewport(s) to the rectangle.

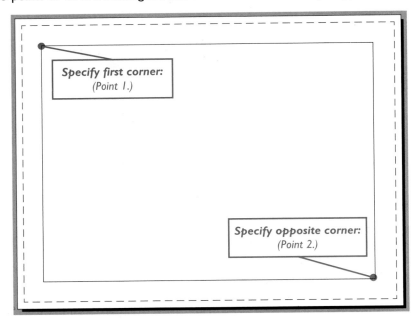

Click **Next**.

9. Click **Finish**.

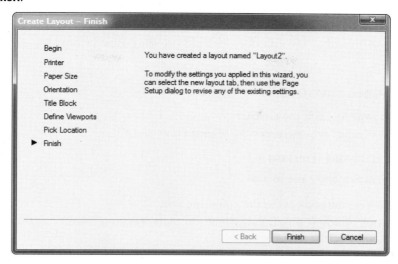

AutoCAD creates the viewport, which allows the model to show through into the layout, the elements of which are illustrated below. (The title block and number of viewports vary, depending on the options you selected during the wizard.)

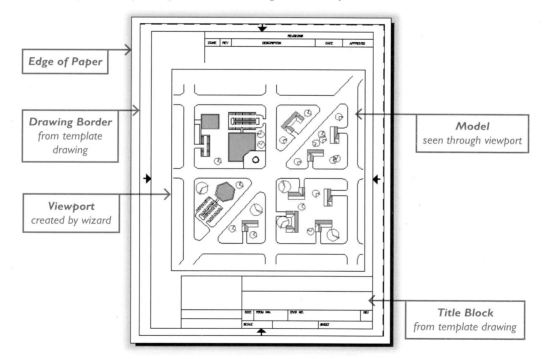

 **LAYOUT**

The LAYOUT command manages layouts, for example creating new layouts based on ones stored in other drawings.

### TUTORIAL: MANAGING LAYOUTS

1. To manage viewports, start the **LAYOUT** command:
   - At the 'Command:' prompt, enter the **layout** command.

     Command: **layout** *(Press ENTER.)*

   - As an alternative, enter the **lo** alias.

2. AutoCAD displays the prompts at the command line:
   Enter layout option [Copy/Delete/New/Template/Rename/SAveas/Set/?] <set>: *(Enter an option.)*

   **Copy** — copies layouts; prompts for the name of a layout to copy:
   Enter name of layout to copy: *(Enter a name, such as* layout1.*)*

   **Delete** — removes layouts from the drawings. The Model tab cannot be deleted. Prompts for the name of the layout to erase:
   Enter name of layout to delete: *(Enter a name, such as* layout1.*)*

   **New** — creates new layouts. You can enter a names of up to 255 characters long, but only the first 31 are shown on the tab, fewer if there is less room. Prompts for a name:
   Enter name of new layout: *(Enter a name, such as* layout2.*)*

   **Template** — creates new layouts based on .*dwt* template, .*dwg* drawing, or .*dxf* interchange files. Displays the Insert Layout(s) dialog box (illustrated below), and then inserts the selected layouts *and all its objects* into the drawing.

   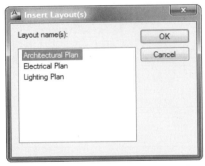

   **Rename** — changes the name of layouts; prompts for the old and new names
   Enter name of layout to rename: *(Enter a name, such as* layout1.*)*
   Enter new layout name: *(Enter a name, such as* layout3.*)*

   **Saveas** — saves layouts as .*dwt* template files; prompts for the name of the layout to save:
   Enter layout to save to template: *(Enter a name, such as* layout3.*)*

   **Set** — makes a layout current; prompts for the name of the layout to bring to the foreground:
   Enter layout to make current: *(Enter a name, such as* layout3.*)*

   **?** — lists the names of layouts.

**Note** A shortcut menu provides several more options not found in the **LAYOUT** command. Right-click any layout tab, if one is visible, to see the following shortcut menu:

```
New layout
From template...
Delete
Rename
Move or Copy...
Select All Layouts

Activate Previous Layout
Activate Model Tab

Page Setup Manager...
Plot...

Import Layout as Sheet...
Export Layout to Model...

Hide Layout and Model tabs
```

Layout1 / Layout2 / Layout3 \ Architectural Plan /

The added options are as follows:

**Select All Layouts** — selects all layouts; next command affects all of them.
**Activate Previous Layout** — returns to the previously-accessed layout, which is useful when a drawing has many layouts.
**Activate Model Tab** — returns to the Model tab.
**Page Setup Manager** — executes the **PAGESETUP** command.
**Plot** — executes the **PLOT** command.
**Import Layout as Sheet** — adds the current layout as a sheet; available only when the Sheet Set Manager palette is open.
**Export Layout to Model** exports the contents of the layout to model space, flattening 3D objects in the process (**EXPORTLAYOUT** command).
**Hide Layout and Model Tabs** — "hides" the tabs by moving them to the status bar.

 **VIEWPORTS**

The **VIEWPORTS** command creates and merges (removes) rectangular viewports.

When you switch to a layout for the first time, AutoCAD creates a single viewport automatically, which you can copy, move, and resize. To create additional viewports, use the **VIEWPORTS** command.

You can create non-rectangular viewports and even convert objects into viewports, but to do so you need the **-VPORTS** command instead, which I describe later. Both commands work in model space and in layouts, but non-rectangular viewports are available only in layouts.

The **VIEWPORTS** and **-VPORTS** commands need two pieces of information: (1) the number of viewports, and (2) their locations.

### TUTORIAL: CREATING RECTANGULAR VIEWPORTS

In this first tutorial, you standard rectangular viewports.

1. To create one or more rectangular viewports, start the **VIEWPORTS** command:
   - At the 'Command:' prompt, enter the **viewports** command.

     Command: **viewports** *(Press ENTER.)*

   - Alternatively, enter the **vports** alias at the 'Command:' prompt.

   In all cases, AutoCAD displays the Viewports dialog box. (If necessary, click the **New Viewports** tab.)

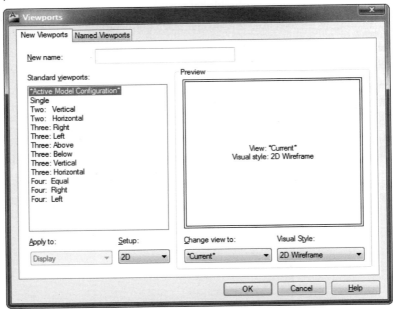

2. From the list of names under Standard Viewports, select a viewport style, as illustrated below.

| | |
|---|---|
| Single | [Single viewport] |
| Two: Vertical | [Two vertical viewports] |
| Two: Horizontal | [Two horizontal viewports] |
| Three: Right | [Three viewports, large on right] |
| Three: Left | [Three viewports, large on left] |
| Three: Above | [Three viewports, large on bottom] |
| Three: Below | [Three viewports, large on top] |
| Three: Vertical | [Three vertical viewports] |
| Three: Horizontal | [Three horizontal viewports] |
| Four: Equal | [Four equal viewports] |
| Four: Right | [Four viewports, large on left] |
| Four: Left | [Four viewports, large on right] |

3. Click **OK**. AutoCAD prompts you for the location of the viewports:
   Specify first corner or [Fit] <Fit>: *(Press* ENTER *to fit the viewports within the display area.)*
4. Press **ENTER**, and AutoCAD fits the viewports within the display area.
   Or, you can pick two points (forming a rectangle), and AutoCAD fits the viewports between them.

**Notes** Viewports created in layouts are called "floating," because they can be moved and edited in other ways. Each viewport is independent of the others. Polygonal (non-rectangular) viewports and viewports converted from objects can only be created in layouts.

Viewports created in model space are called "tiled," because they stick together as tightly as tiles. They cannot be moved or copied. You can only merge and erase them with the **-VPORTS** command.

 **-VPORTS** AND **VPCLIP**

The **-VPORTS** command creates non-rectangular viewports, while **VPCLIP** *clips* viewports.

Layouts are not limited to viewports that are rectangular or square. You can create polygonal (non-rectangular) and circular viewports. These are sometimes called "clipped viewports," because they usually hide portions of models.

You can create clipped viewports with the -vports command, or else convert closed objects into viewports, including closed polylines, circles, ellipses, splines, and regions. (Closed polylines can be made of line and arc segments, and can even intersect themselves.) There is one drawback: Viewports created as polygons and from objects cannot have their scales locked.

### TUTORIAL: CREATING POLYGONAL VIEWPORTS

In this tutorial, you create a non-rectangular viewport with the *layout.dwg* drawing.

1. Before creating polygonal viewports, make a new layout, for the following actions cannot be carried out in model space:
   a. Right-click any layout tab.
   b. Choose **New layout** from the shortcut menu.
   c. Click on the new layout tab to make it current.
   d. If a viewport appears automatically, remove it by selecting it and then pressing **Del**.

2. Start the **-VPORTS** command:
   - At the 'Command:' prompt, enter the **-vports** command.

      Command: **-vports** *(Press* ENTER.*)*

   Notice that this command displays its prompts at the command line.

3. To draw the outline of a polygonal viewport, enter **P**:
   Specify start point or
   [ON/OFF/Fit/Shadeplot/Lock/Object/Polygonal/Restore/LAyer/2/3/4] <Fit>: **p**

4. Pick a point in the layout to start the polygonal viewport.
   Specify start point: *(Pick a point.)*

5. The prompts that follow are exactly like those of the **PLINE** command. You can draw line and arc segments. You must pick a minimum of three points, after which you can close the polyline.

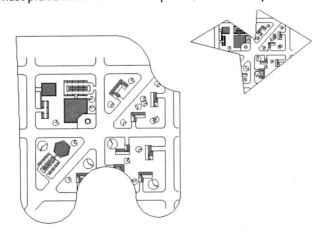

An example of what is possible with polygonal viewport construction is illustrated above. There are two viewports: the larger one is made with line and arc segments; the smaller one in the upper right is made with the minimum of three line segments.

 **Note** In addition to clipping viewports, AutoCAD can also clip externally-referenced drawings and other attached objects with the **CLIP** command. If you prefer, you can use the following commands for specific file types:

> Xrefs — **XREFCLIP**
> Image files — **IMAGECLIP**
> DWF files — **DWFCLIP**
> DGN files — **DGNCLIP**
> PDF files — **PDFCLIP**

### TUTORIAL: CREATING VIEWPORTS FROM OBJECTS

Unusually-shaped viewports can be made from objects. As in the previous tutorial, create a new layout.

1. Draw a polyline, circle, ellipse, spline, or region. If drawing a polyline or spline, use the **Close** option to ensure the object is closed. The figure below illustrates a self-intersecting polyline drawn with the **PLINE** command, and then smoothed with the **PEDIT** command.

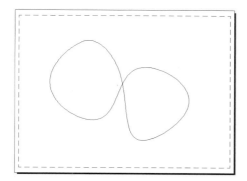 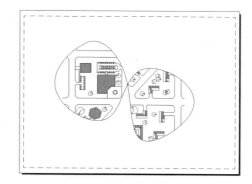

*Left:* Closed spline drawn in paper space.
*Right:* Object converted to viewport boundary.

2. Start the **-VPORTS** command:
   - At the 'Command:' prompt, enter the **-vports** command.

      Command: **-vports** (*Press* ENTER.)

3. Enter **O** to convert a closed object into a viewport:
   Specify corner of viewport or
   [ON/OFF/Fit/Shadeplot/Lock/Object/Polygonal/Restore/2/3/4] <Fit>: **o**

4. Pick the object to convert.
   Select object to clip viewport: *(Select the closed object.)*

Notice that AutoCAD converts the object into a viewport, and the model shows through, as illustrated above.

Once the viewport is in place, you can use grips editing to change the border, as illustrated by the figure below.

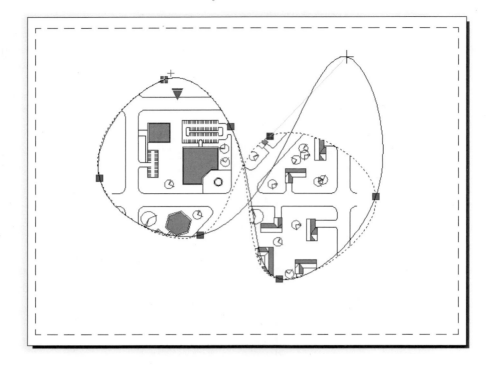

### VPCLIP

The **VPCLIP** command operates in a manner similar to **-VPORTS**: it clips rectangular viewports with polylines or objects. In addition, it can further clip a clipped viewport. Better than **-VPORTS**, it converts clipped viewports back into rectangular ones via the **Delete** option.

> Command: **vpclip** *(Press* ENTER.*)*
>
> Select viewport to clip: *(Pick a viewport.)*
>
> Select clipping object or [Polygonal] <Polygonal>: *(Enter an option; see table below.)*

Only when you select a viewport previously clipped by this command does the **Delete** option appear in the prompt line. The options of this command are as follows:

| VpClip Option | Comment |
|---|---|
| Clipping object | Selects a closed object to be converted into a viewport boundary. |
| Polygonal | Picks points to designate line and arc segments for the viewport boundary. |
| Delete | Deletes the clipped viewport, and restores the rectangular viewport. |

## EXPORTLAYOUT

The **EXPORTLAYOUT** command copies the content of layouts to model space.

The command works by saving the layout as a *.dwg* file on your computer's hard drive, and then opens it as a new drawing in Model tab. This works only in layout mode, and not when a viewport is maximized or in Model tab.

This command is useful for freezing layouts, since they are saved to separate drawings. Another case is when another CAD packages cannot read AutoCAD's layouts; use this command to export layouts to model space, which then makes them available to these packages.

1. Open the *union.dwg* drawing file.

   If necessary, switch to the **Layout1** tab.

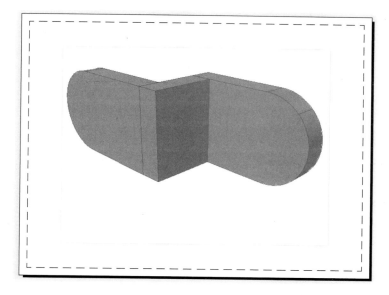

2. Start the **EXPORTLAYOUT** command:
   * At the 'Command:' prompt, enter the **exportlayout** command.

     Command: **exportlayout** *(Press* ENTER.*)*

   Notice the Export Layout to Model Space Drawing dialog box. The drawing has "_layout1" appended to its file name.

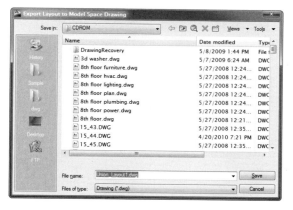

3. Click **Save**, and then wait while layout is converted to model space and saved on disk.

4. Click **Open**. AutoCAD opens the drawing in its own window.

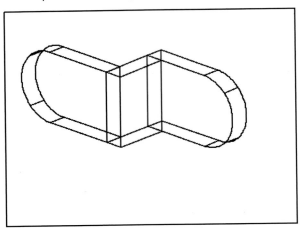

The 3D objects are still 3D, but the viewport is mimicked by a rectangle made of a polyline. Portions of the drawing that extend beyond the viewport are not clipped. I have found that extremely complex layouts sometimes export incorrectly.

What has changed is that the drawing is scaled to be the same size as in layout mode, such as 10" wide. This could create a problem, warns the technical editor, when a circle of diameter 100' shown in a 1:2 viewport ends up having a diameter of 50' after being exporting to model space. This could cause problems when the exported *.dwg* file is sent to other CAD packages.

The reason? He thinks Autodesk is allowing for viewports of different scales. The solution is to ensure the viewport is at 1:1 scale before running the **EXPORTLAYOUT** command.

## EXERCISES

1. Open the *grader.dwg* file, the 2D drawing of a grader.
    a. Click the **Layout1** tab, and create two viewports on an A0- or E-size sheet of paper:
    b. The first viewport showing the entire grader, without the drawing border.
    c. The second viewport showing a detail of the air intake.
    d. The result should look similar to the figure below. (All figures in these exercises are courtesy of Autodesk, Inc.)
    e. Save the drawing as *layout1.dwg*.

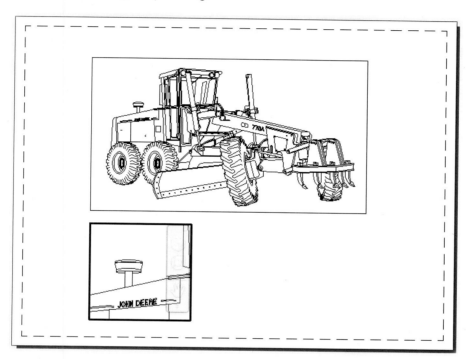

2. Start a new drawing, and then use the Layout Wizard to create a viewport with the following options:

    | | |
    |---|---|
    | Name | **Practice Layout** |
    | Plotter | **DWF 6** |
    | Paper Size | **ANSI B** |
    | Drawing Units | **Inches** |
    | Orientation | **Landscape** |
    | Title Block | **ANSI B** |
    | Viewport Setup | **Single** |
    | Viewport Scale | **Scaled to Fit** |

    Save the result as *layout2.dwg*.

3. Open the *langer.dwg* file, a drawing detailing sump pumps.
    a. Create four layout tabs of the following names, which match the names of the details.
       Pump Connection Detail
       Branch Pipe Support Detail
       Duplex Ejector Pumps
       Air Handler Unit Hanging Detail
    b. In each layout, place the namesake detail.
    c. What is the viewport scale factor?
    d. Save the drawing as *layout3.dwg*. The result should look similar to the figure below.

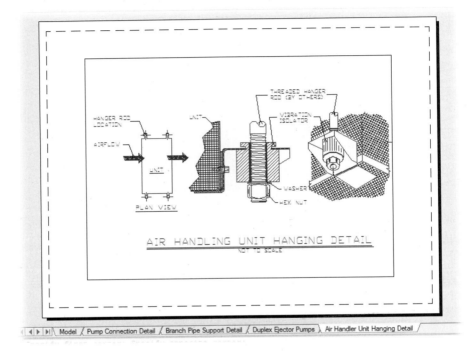

4. Use the **VIEWPORTS** command to create viewports with the following arrangements:
    a.

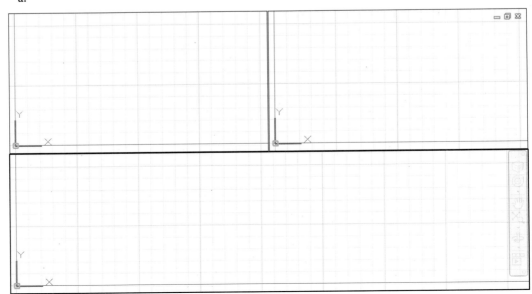

b.

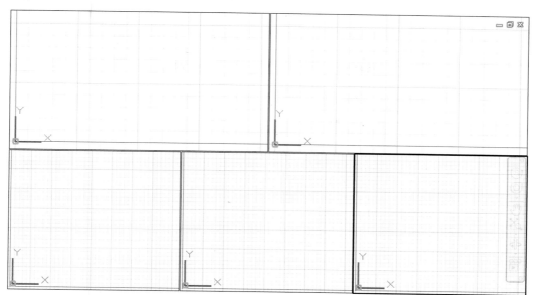

c.

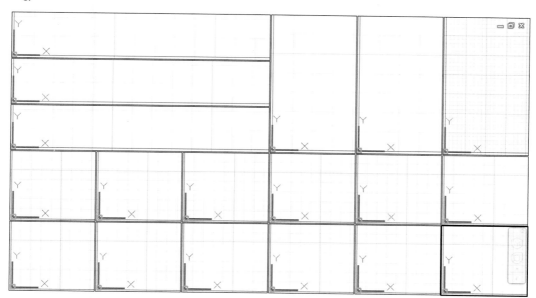

5. Open the *wright.dwg* file, a floor plan and elevation of a house designed by architect Frank Lloyd Wright.

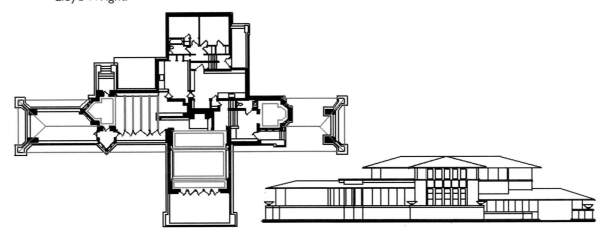

Create a pair of clipped viewports that show the two views independently, as illustrated below. Save the drawing as *layouts5.dwg*.

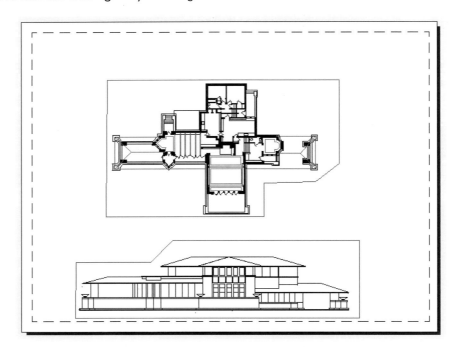

## CHAPTER REVIEW

1. Explain the purpose of layouts.
2. Describe two ways to switch between model and layout mode:
   a.
   b.
3. In which tab are the following elements drawn full size?
   The model
   Title block
4. What does the layout represent?
5. Can a drawing have more than one layout?
6. How do you see the model in layouts?
7. Can viewports be modified?
   Can viewports be copied?
8. Name the parts of the layout:
   a.
   b.
   c.
   d.

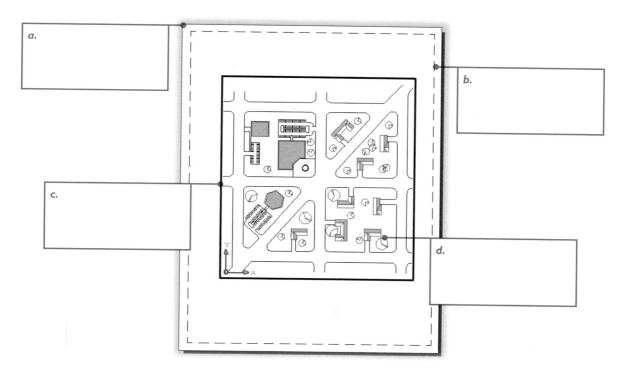

9. A layout has two viewports, one of which has a heavier border.
   What is the significance of the heavy border?
   What mode is AutoCAD in?
10. Explain the function of the following keystrokes when used with layout tabs:
    **CTRL+R**
    **CTRL+PGDN**

11. Describe the kind of drafting you can do when the layout is in the following mode:
    PAPER
    MODEL
12. Describe what happens when you resize a viewport in a layout?
    When you copy a viewport.
    When you select an object in one viewport.
13. Describe how to hide the viewport border.
14. What are the differences between the two viewports shown below? Explain a possible reason for the differences.

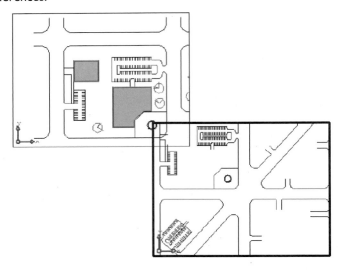

15. What are two ways to scale models in layouts?
    a.
    b.
16. What happens when you double-click a viewport?
17. Calculate the scale factor for the following viewport:
    Drawing of automobile = 12', bumper to bumper length
    Width of viewport = 10"
18. How would you move the model in the viewport?
19. Explain the difference between the **LAYER** command's **Current VP Freeze** and **New VP Freeze** options.
20. What does the snowflake mean in the Layer Properties Manager dialog box?
21. In which mode would you insert a title block?
    PAPER mode of layout tabs.
    MODEL mode of layout tabs.
    Model tab.
22. What advantage does the **VIEWPORTS** command have over the **-VPORTS** command?
    **-VPORTS** command over **VIEWPORTS** ?
23. Must viewports be rectangular?
    If no, in what way?
    If yes, why?

24. Describe the difference between the following:
    Tiled viewports
    Floating viewports
25. Briefly explain the function of the **PSLTSCALE** command.
26. When does **PSLTSCALE** take effect?
27. When is the **SPACETRANS** command commonly used?
28. What scale factor does **SPACETRANS** calculate?
29. Which command maximizes a viewport and also
    Switches to model space?
    Stays in paper space?
30. Describe one or more benefits to using the **VPMAX** command.
31. How can you add and remove scale factors from the VP Scale list?
32. When might you want to override layer properties in viewports?
33. List at least two ways of turning off layer property overrides.
34. What is the purpose of the **QVLAYOUT** command?
35. How might the **EXPORTLAYOUT** command be useful?

# CHAPTER 20
## Plotting Drawings

The end product of CAD drafting is typically the drawing plotted on paper. In this chapter, you learn to plot drawings with these commands:

**PREVIEW** previews drawings before plotting.
**PLOT** plots drawings.
**OPTIONS** specifies default settings for plotting.
**AUTOPUBLISH** saves drawings in DWG and DWF or PDF simultaneously.
**SCALELISTEDIT** customizes lists of plot scales.
**HIDEXREFSCALES** hides plot scales imported from xrefs.
**VIEWPLOTDETAILS** reports on successful and failed plots.
**PLOTSTAMP** stamps plots with information about drawings and plotting.
**PAGESETUP** assigns plotters to layouts.
**PUBLISH** creates and plots drawing sets.
**PLOTTERMANAGER** creates and edits plotter configurations.
**STYLESMANAGER** creates and edits plot style tables.

---

**NEW TO AUTOCAD 2011** IN THIS CHAPTER
- **PLOT** and **PAGESETUP** now support transparency.
- New **PLOTTRANSPARENCYOVERRIDE** system variable determines how transparent objects are plotted.

## ALL ABOUT PLOTTERS AND PRINTERS

AutoCAD works with any printer connected to your computer and your network; it checks automatically for all system printers registered with Windows.

### SYSTEM PRINTERS

A system printer is any local and network printer recognized by Windows. To see the list of system printers for your computer: (1) choose the **Start** button on the Windows Taskbar, and then (2) select **Printers** (or Devices and Printers). Windows opens the Printers window, which lists the names of your computer's system printers.

The figure illustrates the Printers window for the author's computer. From left to right, the icons indicate:

**Adobe PDF** — the unadorned printer icon indicates a local printer. In this specific case, Adobe PDF creates PDF files from any document by "printing" it to a file.

**HP LaserJet...** — the check mark on this icon indicates the default printer. AutoCAD and other programs use this printer automatically, unless you specify a different one. To select a different printer as the default, right-click one of the other printer icons, and then select **Default Printer** from the shortcut menu.

Windows Vista and earlier also display a tube badge that represents "sharing," meaning that the printer can be accessed by other computers on the network. (Windows 7 does not show which printers are shared.) To allow network access to one of your printers: (1) right-click its icon, and then select **Sharing**; (2) in the Properties | Sharing dialog box, select the **Shared As** radio button, and then (3) choose **OK**.

To change the properties of a system printer, right-click its icon, and then select **Properties**. The content of the Properties dialog box varies, depending on the printer's capabilities. Commonly, though, you can set the default resolution, paper size and source, color management, and so on.

### Nontraditional Printers

AutoCAD also works with nontraditional printers, the most common today being Adobe Acrobat PDF (short for "portable document format"). AutoCAD creates PDF files of drawings with the EXPORTPDF command.

Autodesk promotes the DWF format for sharing drawings over the Internet. The 3DDWF, PUBLISH, and EXPORTDWF commands create *.dwf* files from drawings. In addition, Autodesk also provides the free DWF Writer software that creates *.dwf* files from any application; you can download it from www.autodesk.com/dwfwriter.

Less popular today are faxes. If computers have modems, then fax capability is included. Windows lets you fax from any software program, including AutoCAD, provided the computer is connected to the telephone system. (Not all versions of Windows include the fax software, so you may have to purchase it separately.) The process is as simple as selecting the fax as the printer. The drawback to faxing is that AutoCAD typically splits large drawings onto multiple A- or A4-size sheets of paper (roughly 8" x 11" each).

## LOCAL AND NETWORK PRINTER CONNECTIONS

Printers usually connect to computers through USB (universal serial bus) connections. Computers can theoretically have up to 128 USB ports, but two to eight are typical. To support older printers, some computers still have parallel ports; the very oldest printers and pen plotters use serial ports, which operate slowly and are difficult to configure.

Network printers are connected *directly* or *indirectly* to the network. If connected directly, the printer contains its own network card; if indirectly, the printer is connected to a computer, and then through the computer's network card to the network.

Computers connected to networks can print to any network printer — provided the computers have been given permission to access the network printers. This works in reverse, too: you can restrict other networked computers from using your computer's printer.

### Differences Between Local and Network Printers

The primary difference between *local* and *network* printers is how Windows sees them. During printer setup, you tell Windows whether the printer is Local or Network, so that it knows whether to search for it on your computer's local ports, or along the network.

Another difference is that local printers are typically more available. A network printer might be inaccessible because the network is down, or because too many other computers are sending files.

The advantage to networking printers is that everyone in the office can share all printers. If the printer attached to your computer breaks down (through mechanical failure, lack of paper or ink, and so on), you can easily access another one.

## ABOUT DEVICE DRIVERS

A term you will occasionally read about is "device driver." Without device drivers, software programs cannot communicate with printers, graphics boards, and other computer hardware — collectively known as "peripherals."

To understand the role of device drivers, it helps to know their history. In the early days of personal computers, there were very few choices for external peripherals, and the peripherals had very few abilities. For instance, early graphics boards and printers only output monochrome text — no color, no graphics!

As personal computers became popular in the late 1980s, more and more companies began inventing add-ons for computers. Graphics boards and printers gained the ability to output graphics, and then added color. Varieties of input devices flourished, such as digitizing tablets and 100-button keypads.

Peripherals vendors typically took one of two approaches to make their products work with operating systems. One approach was to mimic a competitor's protocol, such as the popular HPGL (Hewlett-Packard Graphics Language) for plotters and TIGA (Texas Instruments Graphics Architecture) for graphics boards. If the standard lacked the desired capabilities, then the second approach was to create a new standard.

Even with standards in place, peripherals still needed a way to communicate with the operating system. Initially, each peripheral vendor wrote its own software, called a "device driver," software that drives the devices — or, "drivers" for short.

When Microsoft began marketing Windows, they decided to include as many drivers as possible to reduce the resistance to switching from DOS to Windows. Microsoft undertook the thankless job of writing thousands of drivers for nearly every peripheral on the market — albeit in many cases, one device driver served several related hardware products.

Initially, Autodesk undertook the same approach. The company wrote AutoCAD drivers for all peripherals, and even boasted about the long list. By the mid-1980s, however, the job became overwhelming, and so Autodesk began to require that peripheral vendors write the drivers themselves using the ADI kit supplied by Autodesk — the AutoCAD Device Interface. In the 1990s, the capabilities were expanded by the enhanced HDI kit — Heidi Device Interface.

(HEIDI is short for "HOOPS Extended Immediate-mode Drawing Interface." HOOPS, in turn, is short for "Hierarchical Object-Oriented Picture System." So really, HDI is short for Hierarchical Object-Oriented Picture System Extended Immediate-mode Drawing Interface Device Interface — whew!)

**Drivers Today**

Today, Autodesk provides drivers for a number of plotters and two graphics board standards, OpenGL and DirectX. It relies on Microsoft to provide drivers for common peripherals, such as mice, and on hardware vendors for specialized peripherals.

From time to time, hardware companies update device drivers to fix bugs and enhance capabilities. They typically provide free downloads of the latest drivers. If plots don't work correctly, the first thing to suspect is a device driver that needs updating. On the other hand, if the driver is working well, don't upgrade, because sometimes upgrades can cause new problems.

To see the list of drivers installed on your computer: (1) right-click the Computer icon on the desktop, (2) choose the **Properties** item on the shortcut menu, and then (3) click **Device Manager** to see a list similar to that illustrated at right.

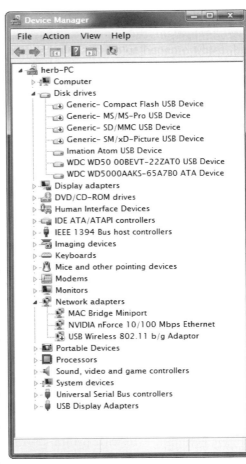

With this background information in place, let's see how AutoCAD handles the plotting of drawings. In general, the process is as follows:

1. Start the **PLOT** command and set the plotting options.
2. Use the **PREVIEW** command to check that the plot will turn out correctly.
3. Send the drawing to the printer or, if necessary, go back to the Plot dialog box to correct settings.

The rest of this chapter concerns itself with setting up plotters by a variety of methods.

## PREVIEW

The PREVIEW command shows how drawings will be plotted.

Previewing is important, because it ensures that the drawings will be plotted as you expect. The preview shows whether all drawing elements appear on the paper correctly. It provides a visual check of settings, such as whether the drawing is centered — or not, as illustrated below.

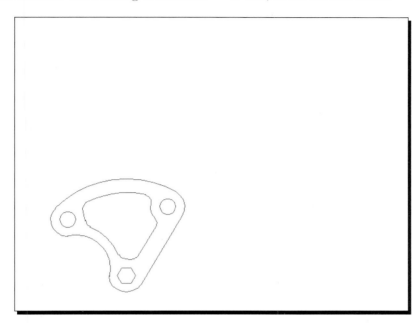

This saves you time and money, because previewing is much, much faster than plotting and costs nothing, except for a few extra seconds of your time — and the electricity to run things.

### TUTORIAL: PREVIEWING PLOTS

1. To preview drawings before plotting, start the **PREVIEW** command:
   - In the ribbon's Output tab, click the **Preview** button in the Plot panel.
   - Or, at the 'Command:' prompt, enter the **preview** command:

     Command: **preview** *(Press* ENTER.*)*
   - Alternatively, enter the **pre** alias at the 'Command:' prompt.

    **Note** Sometimes the preview process does not go smoothly because of "missing" plotters.
   When a plotter is not assigned to the drawing automatically, AutoCAD complains:
      No plotter is assigned. Use Page Setup to assign a plotter to the current Layout.
   When a plotter is assigned, but is not available from your computer, AutoCAD complains:
      The selected layout has an invalid hardcopy configuration.

   The first solution is to use the **PAGESETUP** command to assign a printer or plotter to the layout. Another is to modify *.dwt* template files by assigning the system printer to Model tab and all layout tabs. It would be much easier if Autodesk again did this for us, as it did in earlier releases of AutoCAD.

2. Once the preview is generated, AutoCAD displays it in the preview window. (If the drawing is complex, a dialog box may appear showing AutoCAD's progress in generating the preview image.)

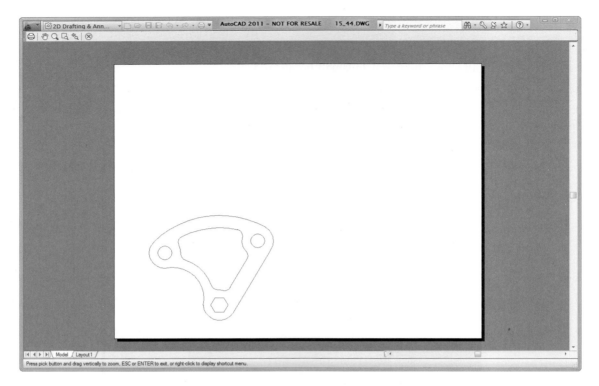

The preview shows the drawing on a white rectangle that represents the paper. The color of the area that represents the sheet of paper can be changed: use the **OPTIONS** command's **Display | Colors** button, and then select **Plot Preview** from the Context list.

3. Press **ESC** to exit preview mode and return to the drawing.

## PREVIEWING PLOTS: ADDITIONAL METHODS

Once in the preview window, you can zoom and pan the image.

### Controlling the Preview Image

The cursor initially looks like a magnifying glass. As you drag the cursor up and down (by holding down the left mouse button), the preview zooms in and out.

For other viewing options, you can use the toolbar, or else right-click the drawing for a shortcut menu of viewing choices:

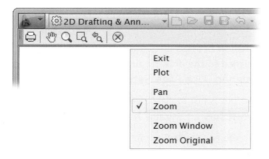

When you select **Pan** from the toolbar or shortcut menu, the cursor changes to a hand. Hold down the left mouse button and the view moves (pans) as you move the mouse around.

The **Zoom** button returns you to the real-time zoom mode.

The **Zoom Window** option operates identically to the regular ZOOM command's **Window** option. The cursor changes to ⌖; pick two points to identify the area to enlarge.

The **Zoom Original** option returns the original view, much like the ZOOM **Extents** command.

To plot, click the **Plot** button.

Otherwise, click ⊗ **Cancel** to return to AutoCAD. Alternatively, press ENTER, the spacebar, or ESC to exit the preview window.

(Although the preview screen has Model and Layout tabs, they do not operate.)

### PLOT

The PLOT command displays the Plot dialog box, and then plots the drawing.

The dialog box lists all of the options AutoCAD needs resolved before outputting drawings on paper or to files. Drawings can be plotted to printers, plotters, or to files.

### TUTORIAL: PLOTTING DRAWINGS

1. For this tutorial, open the *15_44.dwg* file.
2. To plot the drawing, start the **PLOT** command using one of the following methods:
   - On the keyboard, press the **CTRL+P** shortcut.
   - Right-click a layout tab, and select **Plot** from the shortcut menu.
   - In the ribbon's Output tab, click the **Plot** button in the Output panel.
   - On the Quick Access toolbar, click the **Plot** button.
   - From the A button's application menu, choose **Print**, and then **Plot.**
   - At the 'Command:' prompt, enter the **plot** command.

        Command: **plot** *(Press ENTER.)*
   - Alternatively, enter the aliases **print** or **dwfout** (an older command that was integrated into plotting) at the 'Command:' prompt.

   Notice that AutoCAD displays the Plot dialog box.

   The Plot dialog box has many options, and it can be confusing to navigate. For your first plots, I recommend you specify the following information as a minimum:

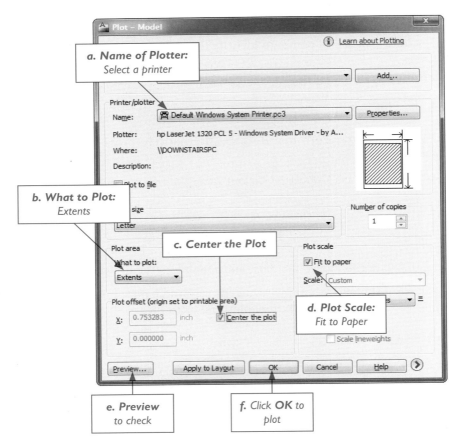

a. **Printer/plotter** – select the name of a printer or plotter. In many cases, this is whatever printer is hooked up to your computer (the "System" or "Default" printer).

b. **What to Plot** — select **Extents** to ensure that every object is included in the plot.

c. **Plot Offset** — select **Center the Plot**. (If you find that the plot is not centered, it may be because What to Plot is set to Display.)

d. **Plot Scale** — select **Fit to Paper** so that nothing is printed off the edge of the page.

e. Click the **Preview** button to check that the plot will turn out.

f. Click the **Plot** button to start the plot. (In the Plot dialog box, click the **OK** button to start the plot.) Notice that AutoCAD plots the drawing.

3. If you wish only to set up the plotting parameters, but not actually plot, then click **Apply to Layout** to save the changes, and then click **Cancel** to return to the drawing editor.

**Notes** AutoCAD does not plot layers that are frozen or have the "No Plot" property. As well, AutoCAD does not plot the Defpoints layer. If part of your drawing does not plot, it could be because you drew on the Defpoints layer.

To create 2D DWF files of drawings, use the **EXPORTDWF** command; to create multi-sheet 2D DWF files, use **PUBLISH** and its **DWF File** option; to create 3D versions of DWF files, use the **3DDWF** command.

### Select a Plotter/Printer

Your computer may have two or more printers available for plotting. The Printer/Plotter droplist contains the list of printers available to you — local and networked, as well as special drivers for plotting to files, such as PDF and DWF.

From the **Name** droplist, select a printer name.

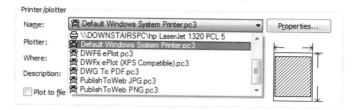

If the printer you want is not listed here, read how to add it with the PLOTTERMANAGER command, described later.

To change properties specific to the printer, such as its resolution or color management, choose the **Properties** button.

### Plot to File

The **Plot to File** option saves drawings to files. Plot files can be imported into other software. For example, many brands of plotters and graphics programs can import files created by HPGL plotter drivers, usually meant for Hewlett-Packard brand plotters.

Some "plotters" listed by the Plot dialog box are designed only to plot to files. These include "DWF6," "DWG to PDF," and "PublishtoWeb." For them, the **Plot to File** option is turned on automatically and grayed-out, because it cannot be turned off.

To make PDF files of drawings, choose the "DWG To PDF" plotter. These can be viewed by the free Acrobat Reader software from Adobe. (Alternatively, use the EXPORTPDF command.)

(The technical editor provides this tip: Plot-to-file is handy when you are on the road and need access to someone else's printer. Connecting to foreign printers directly can be a hassle, because it involves cables and installing printer drivers, which might not be available for your computer and its operating system. Instead, plot to a USB stick on your computer, and then on the other computer, copy the file to the printer.)

Here's how to create plot files:

1. Choose a format, such as PDF or DWF. Notice that the **Plot to file** option is turned on automatically.

   (If plotting to a printer file, such as HP LaserJet, then click **Plot to File** so that a check mark appears.)
2. Click **Plot**.
3. Instead of plotting the drawing, AutoCAD displays the Browse for Plot File dialog box. Select a folder, and if necessary change the file name.
4. Click **Save**.

AutoCAD generates the plot file, saving it under the name of the drawing and using the appropriate file extension, such as *.pdf* or *.dwf*. If plotting with a printer driver, then the file extension is *.plt*.

### Select the Paper Size

**Paper Size** specifies the dimensions of the paper (media) used by printers. AutoCAD reads the appropriate media sizes from the selected printer. Thus the list of sizes you see here varies according to the printer. Make your selection from the droplist, and ensure the same size of paper is loaded into the printer.

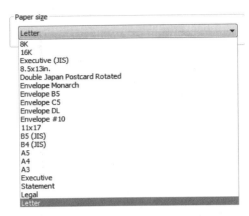

Most desktop printers use Letter paper in North America; drafters call it "A size." It measures 8.5x11". In Europe and other parts of the world that use metric sizes, A4 paper is the equivalent; it measures 8.26x11.7".

If the paper size you need is not listed, you can add it with the **PLOTTERMANAGER**'s **Custom Paper Sizes** option.

Sometimes printers will not print drawings. This may occur when drawings are set up with a size of paper that is unavailable.

### Select the Plot Area

Most times you want to plot the entire drawing — known as the "extents." In some cases, however, you may want to plot smaller areas of drawings.

AutoCAD offers the Plot Area droplist with several options for determining how much of the drawing to plot:

| Plot Area | Comments |
|---|---|
| Display | Plots the current view of drawings — the "what you see is what you get" plot. |
| Extents | Plots the extents, a rectangle that encompasses every part of the drawing containing objects; like performing a ZOOM **Extents**, and then plotting with the **Display** option. |
| Limits | Plots the limits of model space, as set by the LIMITS command. |
| Window | Plots a rectangular area identified by two windowed picks. After clicking **Window**, you are prompted to pick the two corners of a rectangle: |
| | Specify first corner: *(Pick a point or type X,Y coordinates.)* |
| | Specify opposite corner: *(Pick another point or type X,Y coordinates.)* |
| Layout | Plots the extents of the page in layouts (available only when plotting layouts). |
| View | Plots named views; available only when the drawing contains named views (created with the VIEW command). Select **View**, and then select a view name. |

## Plot Offset

The plot origin is at the lower left corner of the media for most plotters, just as it is for AutoCAD drawings. Some plotters, however, locate their origin at the center of the media; AutoCAD normally adjusts for them.

Sometimes you need to shift the plot on the media, for example to avoid a preprinted title block. Use the **Plot Offset** section of the Plot dialog box:

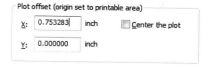

- Positive values shift the plot to the right and up.
- Negative values shift the plot to the left and down.

In most cases, however, you will probably choose the **Center the plot** option to center the plot on the media.

 **Note** Changing the plot's offset may result in a *clipped* plot, where part of the drawing is not plotted, because it extends beyond the edge of the media's margin.

## Number of Copies

Most times, you want just one copy of the plot. If you need more, however, increase the **Number of Copies** counter.

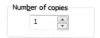

The largest number you can dial in is 99. When plotting to file, number of copies is limited to one.

## Plot Scale

To plot the drawing at a specific scale, select one of the predefined scale factors provided in the Plot Scale list box. These range from 1:1 to 1:100 and 1/128"=1'.

For draft plots, select **Fit to Paper**. Together with a plot area of **Extents**, this ensures your entire drawing fits whatever size of media you select.

Another choice is **Custom**, where you specify the scale factor. Specify the number of inches (or millimeters) on the paper to match the number of drawing units to be plotted. For example, a scale of 1" = 8'-0" means that 1" of paper contains 8'-0" (or 96") of drawing.

To set the custom scale correctly, enter the following:

- Inches      **1"**
- Units       **96"**

This is the same as a scale of 1/8" = 1'-0". The example assumes that the drawing units are set to architectural units. You can customize the list of plot scales with the SCALELISTEDIT command. (This command also affects annotative scale factors and the viewport scale list in the Properties palette.)

### Preview the Plot

You should preview the plot to see the area of the paper on which the plot appears. AutoCAD's plot preview allows two types of preview: partial and full.

#### Partial Preview

The partial preview does not show the drawing elements but rather the position of the plot on the paper. It's called "partial," because it shows only the basics. (*History*: this option was more desirable in decades past, because slow computers took a long time to generate full previews.)

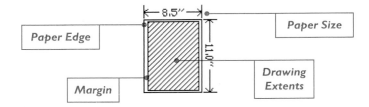

The area covered by the drawing is shown by the diagonal hatching. The paper's edge is shown by the outer rectangle. The *margin* is the non-printing area outside of the drawing; its size varies according to the capabilities of the printer and the settings in the Plot dialog box: Plot area, Plot offset, and Plot scale.

#### Full Preview

Click the **Preview** button to see the full preview, which shows the drawing as it would appear as a final plot on paper. See the PREVIEW command earlier in this chapter.

### Save the Settings

If you plan to use the same plot parameters again, save them. In the Page Setup Name section (near the top of the Plot dialog box), choose the **Add** button.

AutoCAD displays the User Defined Page Setups dialog box:

Enter a descriptive name, and then choose **OK**. The next time you plot this drawing, you can select the page setup name from the **Name** droplist, and all of the dialog box's parameters change to those of the saved page setup.

(As an alternative, you can select **Import** from the droplist. AutoCAD displays the Select Page Setup From File dialog box; choose a *.dwg* drawing file that you know contains page setups.)

## Plot the Drawing

When you have completed these steps, make sure the plotter is ready, and then choose the **OK** button at the bottom of the Plot dialog box.

(To save the settings but not plot the drawing, click **Apply to Layout**, and then **Cancel**.)

AutoCAD displays a dialog box indicating its progress plotting each layout.

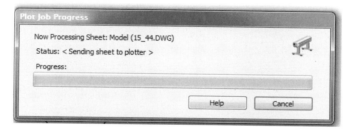

In the tray at the right end of the status bar, look for the tiny pen plotter icon moving a sheet of paper back and forth. (The technical editor notes that pen plotters hardly exist any more — kind of like the rarely-used diskettes that inspired the icon for the Save and SaveAs buttons.)

Right-click the plotter icon to see additional options:

**Cancel Sheet** — cancels the current sheet being plotted.

**Cancel Entire Job** — cancels multi-sheet plot jobs.

**View Plot and Publish Details** — displays the dialog box discussed in the next section.

**View DWF File** — opens the *.dwf* file in the DWF Viewer, if drawings are published in DWF format.

**Enable Balloon Notification** — toggles the display of the balloon described below.

# VIEWPLOTDETAILS

The VIEWPLOTDETAILS command reports on successful and failed plots.

Once the plot is complete, AutoCAD displays a yellow balloon at the right end of the status bar. It reports whether (or not) the plot was successful.

Should the plotter be next to your desk, you don't need yellow balloons telling you that the drawing has been plotted. But when the plotter is located elsewhere, it's good to know the status without having to consume precious calories by getting up from your desk and walking over to it. "This is so tempting," muses the copy editor. "Although," adds the technical editor, "it's the only exercise some people get." (To dismiss the balloon, click the **x** in the upper right corner.)

Click the blue underlined text to view a full report on the plot. AutoCAD displays the Plot and Publish Details dialog box.

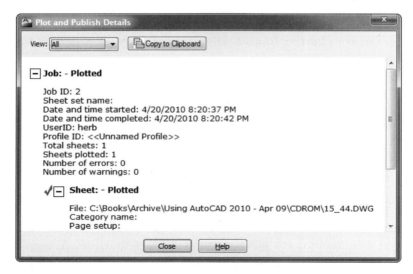

> **View** droplist — displays all reports or just those containing errors.
>
> **Copy to Clipboard** button — copies the reports to the Clipboard; use the **Edit | Paste** command in a word processor to capture the report in a document.

Click **Close** to close the dialog box.

(Alternative methods of accessing this dialog box include entering the VIEWPLOTDETAILS command, or right-clicking the plotter icon on the status bar, and then selecting **View Plot and Publish Details**.)

**PLOTTING DRAWINGS: ADDITIONAL METHODS**

The PLOT command's dialog box sports a **More Options** button that expands to show more options. Click the ⊙ button, or press ALT+>.

Let's look at each, except for **Plot Stamp On**, which is discussed later in this chapter.

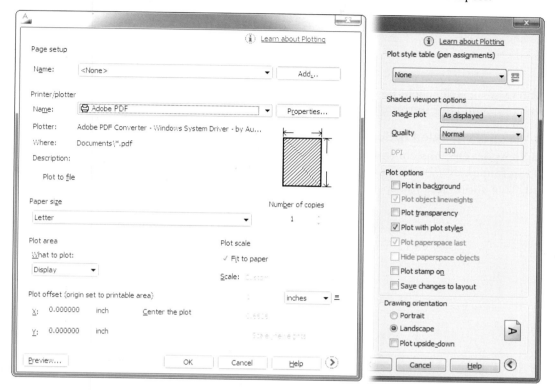

### About Plot Style Tables

The word "pen" in Pen Assignments derives from an earlier age when plotters plotted with actual pens — commonly ink pens, but also ballpoint pens, felt pens, and even pencils and knives (for cutting signs). The term has carried over to today's non-pen plotters that use laser and inkjet technology; now, it refers to varying widths, colors, and shades of gray.

Early releases of AutoCAD assigned colors of objects to pens. For example, all objects colored red (either by layer or by object) were plotted by a specific pen; those colored blue plotted with another pen, and so on. The pens didn't need to contain red or blue ink; typically they contained black ink and had tips of varying widths, such as 0.1" or 0.05".

Plot styles today assign plotter-specific properties to layers and objects, such as widths, colors, line-end capping, and patterns.

### Selecting a Plot Style

Select one of the preassigned plot style tables from the list. To create a new style, select **New** from the droplist.

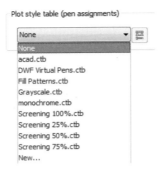

To edit an existing style, click the **Edit** button. You learn more about plot styles under the STYLES-MANAGER command later in this chapter.

### Select Viewport Shade Options

**Shaded Viewport Options** control plotting of viewports. Viewports typically display drawings with 2D and 3D wireframe views, but sometimes you want 3D models plotted as renderings or with hidden-lines removed. This option lets you specify the type and quality of shading.

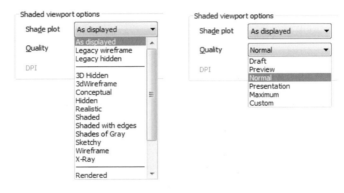

*Left: Shade Plot options.*
*Right: Quality options.*

You can check the effect of changing **Shade plot** in the preview window; **Quality** and **DPI** don't show up until you plot on paper. Although this option has the word "viewport" in its name, it applies equally to drawings plotted from the Model tab.

Under some conditions, the options are unavailable:

**Shade plot** is unavailable when at least one viewport in layout mode has a Shade Plot other than "As Displayed" assigned to it by the Properties palette. This option is always available when plotting from the Model tab.

**DPI** (dots per inch) is available only when **Quality** is set to "Custom."

### Shade Plot Options

The **Shade Plot** droplist applies visual and rendering styles, such as those applied to the samples illustrated below. Even if your graphics board cannot display all visual styles effects, such as shadows and materials, AutoCAD can make the plotter simulate the effects in the plotted output.

**Note** If a viewport does not plot, it may be that its display has been turned off. You can turn it back on with the **PROPERTIES** command: in the Misc section, change the **On** option to "Yes."

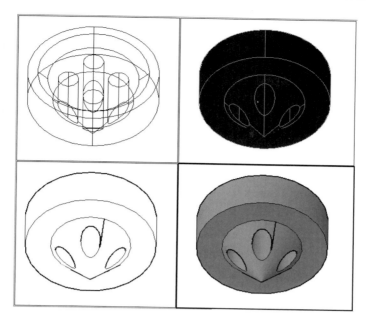

*Top left:* 3D drawing displayed in wireframe. *Top right:* Gooch shaded (conceptual visual style).
*Bottom left:* 3D drawing with hidden lines removed. *Bottom right:* 3D drawing rendered.

| Shade Plot | Comment |
|---|---|
| **As Displayed** | Plots objects as displayed in model or layout tabs. If shaded with the VISUALSTYLES command or rendered with the RENDER command, then the drawing is plotted that way. |
| **Legacy Wireframe** | Plots objects in wireframe, regardless of viewport display mode. |
| **Legacy Hidden** | Plots objects with hidden lines removed, regardless of display mode. |
| *Visual Styles\** | |
| **3D Hidden** | Plots 3D objects with hidden lines removed. |
| **3dWireframe** | Plots 3D objects in wireframe. |
| **Conceptual** | Plots 3D objects using Gooch rendering. |
| **Hidden** | Plots objects with hidden lines removed. |
| **Realistic** | Plots 3D objects in a realistic mode. |
| **Shaded** | Plots 3D objects with shading. |
| **Shaded with edges** | Plots 3D objects shaded with edges. |
| **Shades of Gray** | Plots 3D objects with shades of gray, no matter the original colors. |
| **Sketchy** | Plots objects with shaky lines, simulating a hand-drawn effect. |
| **Wireframe** | Plots objects in wireframe (edges only). |
| **X-Ray** | Plots 3D objects with see-through shades and edges. |
| *Rendered\** | |
| **Rendered** | Plots 3D objects rendered, regardless of display mode. |
| **Draft** | Plots 3D objects at the lowest quality. |
| **Low** | Plots 3D objects... |
| **Medium** | ....at varying levels... |
| **High** | ...of quality. |
| **Presentation** | Plots objects at the highest quality. |

*\*If the drawing has other modes defined, their names are included in this list.*

### Quality and DPI Options

The **Quality** droplist specifies the resolution for shaded and rendered plots, because they are raster images. (Wireframe and hidden-line-removed are plotted as vectors.)

| Quality | Meaning |
| --- | --- |
| Draft | Plotted at the output device's lowest resolution. |
| Preview | Plotted at a maximum resolution of 150 dpi. |
| Normal | Plotted at a maximum of resolution 300 dpi. |
| Presentation | Plotted at a maximum resolution of 600 dpi. |
| Maximum | Plotted at the output device's highest resolution. |
| Custom | Plotted at the resolution specified in the **DPI** text box |

DPI is short for "dots per inch."

 **Note** Two-dimensional drawings cannot be rendered, or have hidden lines removed or visual styles applied.

You can see the effect of shading and hidden-line removal on 3D drawings in the plot preview window. Note that applying visual styles and renderings to drawings takes longer, and can slow the preview display and increase the plotting time.

### Plot Options

The **Plot Options** area lists miscellaneous options.

#### Plot in Background

When the **Plot in Background** option is turned on, AutoCAD plots the drawing "in the background," allowing you to return to editing the drawing more quickly. (This is a replacement for autospooling, discussed later in this chapter.)

Always turn this option on, unless AutoCAD slows down unacceptably.

#### Plot Object Lineweights and Plot with Plot Styles

The **Plot with Lineweights** and **Plot with Plot Styles** options toggle each other; you can have one or the other, neither, but not both:

**Plot with Lineweights** plots drawings with lineweights, *if* lineweights are turned on, and *if* lineweights have been assigned to layers and objects.

**Plot with Plot Styles** plots drawings with plot styles, again only if defined and turned on.

Lineweights are described in Chapter 7, "Properties and Layers," and plot styles are discussed later in this chapter.

### Plot Transparency

The **Plot Transparency** option plots objects with transparency. When this option is turned on, AutoCAD 2011 rasterizes the entire drawing to create the transparency effect; the rasterization process can increase the plotting time, resulting in slower plot production.

The related PLOTTRANSPARENCYOVERRIDE system variable determines when translucent objects are plotted:

0 – Does not plot translucency, regardless of object or layer transparency settings.
1 – Uses the settings of the PageSetup and Plot commands (default).
2 – Plots objects with transparency.

### Plot Paperspace Last

AutoCAD normally plots paper space objects before model space objects. Model space objects include viewport borders and title blocks. The **Plot Paperspace Last** option reverses the order.

Why does this matter? Perhaps it has to do with how old color plotters would lay down the ink. This option is not available when plotting from the model tab, because it has no paperspace elements.

The technical editor notes that other options also affect the plotted output: there are differences between how Hidden modes hide circles with thickness (some looking like a can, others like a pipe), and how values of the DISPSILH system variable create various looks.

### Hide Paperspace Objects

The **Hide Paperspace Objects** option determines if hidden-line removal applies to 3D objects in paper space. This allows you to create drawings without hidden-line removal, yet have hidden lines removed for the plot. The technical editor is unsure why anyone would want 3D objects in paper space, but here's how to do it:

1. In model space, create a 3D solid model with internal details, such as slots and holes.
2. From the **Edit** menu, select **Copy**, and then copy the entire model.
3. Switch to a layout tab, and then paste the model (**Edit | Paste**).
4. Use the PREVIEW command to notice the differences:

    **Hide Paperspace Objects = off** — solids in paper space appear as wireframes, showing internal details.

    **Hide Paperspace Objects = on** — solids in paper space have hidden lines removed.

This option is overridden by the **Shade plot** option of viewports, and is not available when plotting from the model tab.

### Plot Stamp On

The **Plot Stamp On** option toggles plot stamping on plotted drawings. See the PLOTSTAMP command later in this chapter.

### Save Changes to Layout

When turned on, the **Save Changes to Layout** option stores the changes made in this dialog box with the current layout.

The next time you plot from this layout tab, the same settings appear in the Plot dialog box. Select a different layout or drawing, and the settings are different.

## Drawing Orientation

Larger CAD drawings are usually plotted in landscape mode, with the long edge of the page oriented horizontally.

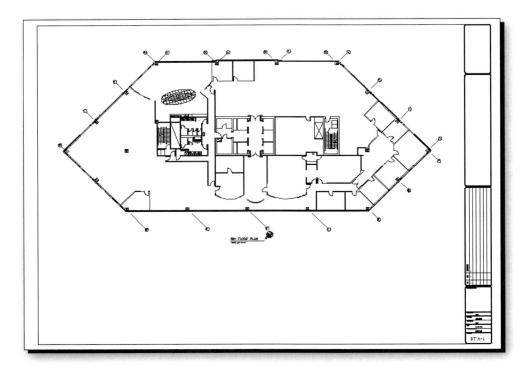

*Above*: D-size drawing plotted in landscape mode.
*Below*: A-size drawing plotted in portrait mode.

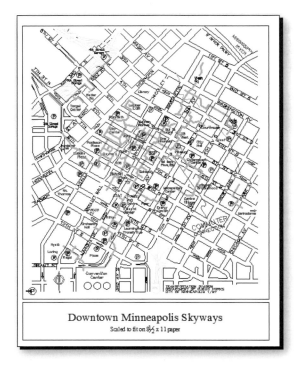

Smaller drawings are often printed in portrait mode, where the long edge is upright.

To switch between the two orientations, choose the **Landscape or Portrait** buttons in the Drawing Orientation area.

The **Plot upside-down** option is handy if you stuck the paper with the title block the wrong way around in the printer.

### AUTOSPOOL

AutoCAD has an "autospool" option that lets it spool plots as an alternative to background plots. Spooling is set up through the Add a Plotter wizard's Port page.

*Spooling* is a technique that speeds up printing jobs. When printing a document or plotting a drawing, the print data is sent to a file on disk; shortly thereafter, the spooling software starts up automatically, and sends the print data from the file to the printer. It is faster to redirect the plot job to a file on disk than to send it directly to the printer; this trick allows applications to finish prints job faster.

(Spooling is short for "simultaneous peripheral operations online," and is more commonly known as *buffering*. In Windows, spooling is handled by the Print Manager. When a printer won't print, the problem often lies with a bug in the Printer Manager and its *.spl* spooling files.)

AutoCAD allows you to use independent software for plot spooling. Before using the software, however, you must set up AutoCAD:

1. Install and configure the spooler software according to the vendor's instructions.
2. AutoCAD needs to know the name of the spooler, as well as the folder in which to place the spool files.

   From the Tools menu, select **Options**, and then the **Files** tab:
   a. Open **Print File, Spooler, and Prolog Section Names**, and in **Print Spool Executable**, specify the spooler program name.
   b. Open **Printer Support File Path**, and in **Print Spool File Location**, specify the folder name.

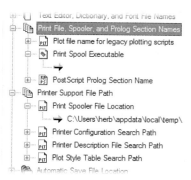

   c. Choose **OK**.
3. Set up the plotter by selecting **Print | Plotter Manager** from the application menu, and then working through the wizard.
4. Start the **PLOT** command, and then select the plotter from the Plotter Configuration droplist.

 **OPTIONS**

One of the many tasks of the OPTIONS command is to set defaults for plotting. These are available in two of the dialog box's tabs: the Plot and Publish tab, and the Files tab.

### PLOT AND PUBLISH TAB

The Plot and Publish tab contains defaults specific to plotting and publishing drawings. "Plotting" means printing one drawing at a time, while "publishing" means printing sets of drawings (two or more).

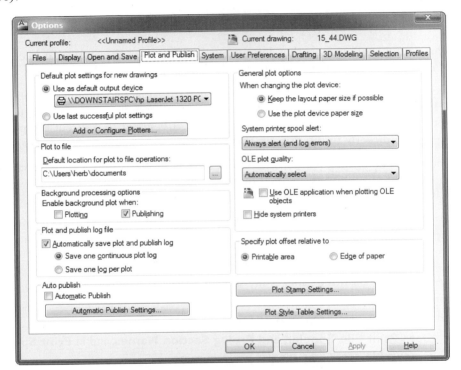

### Default Plot Settings for New Drawings

Here you specify the default plotter:

- ⊙ **Use As Default Output Device** selects the default printer from a list of system printers and devices defined by AutoCAD's *.pc3* (printer configuration) files. It will be used for new drawings, as well as for those saved in versions prior to AutoCAD 2000.
- ○ **Use Last Successful Plot Setting**s selects the last-used plotter settings, instead of the default printer.

Click the **Add or Configure Plotters** button to display the Plotter Manager window; see the PLOTTER-MANAGER command later in this chapter.

### Plot To File

**Default Location for Plot to File Operations** specifies the drive and folder for storing drawings plotted to files.

Type the path, or click the **...** button to display the Select Default Location for All Plot-to-File Operations dialog box.

## Background Processing Options

Background plotting lets you work on drawings while AutoCAD generates the plot. When PLOT, -PLOT, PUBLISH, and -PUBLISH are used in scripts, this setting is ignored and drawings are plotted and published in the foreground.

- ☐ **Plotting** performs background plot jobs; see the **PLOT** command in this chapter.
- ☑ **Publishing** performs background publish jobs; see the **PUBLISH** command.

## Plot and Publish Log File

The log file saves plotting and publishing results in a CSV text file (comma-separated value) that can be imported into spreadsheet and database programs. Data include Job ID, Job name, Sheet set name, Category name, Date and time started and completed, Sheet name, Full file path, Selected layout name, Page setup name, Named page setup path, Device name, Paper size name, and Final status.

- ☑ **Automatically Save Plot and Publish Log** creates log files:
    - ⦿ **Save One Continuous Plot Log** stores all plot data in a single log file.
    - ○ **Save One Log File Per Plot** stores data about each plot in a separate log file.

## AutoPublish

AutoCAD can save the drawing in DWF or PDF formats at the same time as it is saved in DWG format. Turn on this option when you work with these electronic plot files, and need them to be up to date. Otherwise, leave this option turned off, because it prolongs the save-file process.

- ☐ **Automatic Publish** saves drawings in DWF or PDF format during the **SAVE** and **CLOSE** commands. (You can also toggle auto-DWF/PDF publishing with the **AUTOPUBLISH** command.)

Click the **Automatic Publish Settings** button to specify whether drawings are saved in DWF, DWFx, or PDF format.

## General Plot Options

When changing the plot device:

- ⦿ **Keep the Layout Paper Size If Possible** uses the paper size specified by the plot setup. When the printer cannot handle the specified paper size, AutoCAD displays warning message, and then uses the size of paper specified by the PC3 plotter configuration file, or by the system printer's settings.
- ○ **Use the Plot Device Paper Size** ignores the size specified by the layout, and uses the paper size specified by the plotter's PC3 configuration file or the default system settings.

**System Printer Spool Alert** determines the type of alert displayed by port conflicts during plot spooling:

- **Always Alert (And Log Errors)** displays an alert and records the error to the log file.
- **Alert First Time Only (And Log Errors)** displays an alert the first time only, and then records the error to the log file.
- **Never Alert (And Log First Error)** never displays alerts, but records error to log files.
- **Never Alert (Do Not Log Errors)** never displays alerts and does not record to log files.

**OLE Plot Quality** specifies the quality of plotted OLE objects:

- **Monochrome** is suitable for black text.
- **Low Graphics** is suitable for colored test and simple graphics.
- **High Graphics** is meant for photographs.
- **Automatically Select** allows AutoCAD to choose the setting.
- ☐ **Use OLE Application When Plotting OLE Objects** launches the application that created the OLE object in order to improve the plot quality.
- ☐ **Hide System Printers** toggles the display of Windows system printers. System printers configured by the Add-a-Plotter wizard are always displayed in the Plot dialog box.

## Specify Plot Offset Relative To

The **Plot Offset** option in the Plot dialog box is controlled by the following setting:

- ⊙ **Printable Area** offsets the plot relative to the printable area.
- ○ **Edge of Paper** offsets the plot relative to the edge of the paper.

## Other Options

**Plot Stamp Settings** button opens the Plot Stamp dialog box; see the PLOTSTAMP command.

**Plot Style Table Settings** button opens the Plot Style Table Settings dialog box; see the STYLESMANAGER command.

## FILE TAB

The Options dialog box's File tab specifies the default location for plot-related files, as illustrated below:

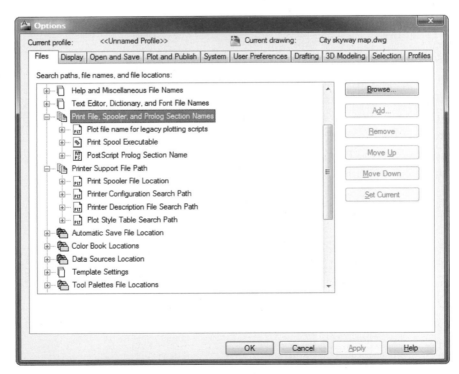

## SCALELISTEDIT

AutoCAD comes with a long list of standard scale factors, and so you use the SCALELISTEDIT command to remove unused ones and to add your own custom factors. The command customizes the list of scale factors used by plots, viewports, page layouts, and annotative scales, as illustrated below.

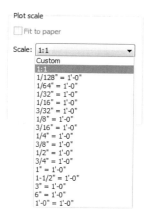
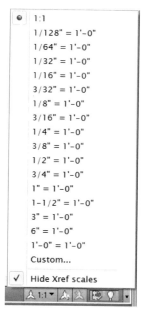

**Left**: Scale factors listed in Plot dialog box.
**Right**: The same scale factors available in the drawing bar.

1. To add and remove scale factors from AutoCAD, start the **SCALELISTEDIT** command:
   - From the Format menu, choose **Scale List**.
   - In the ribbon's Annotate tab, choose **Scale List** from the Annotation Scaling panel.
   - At the 'Command:' prompt, enter the **scalelistedit** command.

   Command: **scalelistedit** *(Press* ENTER.*)*

   AutoCAD displays the Edit Scale List dialog box, which lists the scale factors currently defined.

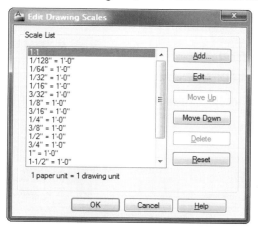

2. To add a new scale factor, click **Add**.

   To edit the name of an existing scale factor, select it, and then click **Edit**.

**Name Appearing in Scale List** — provides descriptive names that appear in place of scale factors; alternatively, enter an actual scale factor, such as 1:10.

**Paper Units** — specifies the size of objects on paper. This value is usually 1 (one) for drawings larger than the paper.

**Drawing Units** — specifies the size of objects in the drawing. This value is usually 1 for drawings smaller than the paper.

3. Change the order in which scale factors appear with the **Move Up** and **Move Down** buttons. This lets you place frequently-used scales at the top of the list.

4. You can remove unused factors with the **Delete** button. If you make a big mistake, just click the **Reset** button to restore the default list; but be careful, because this action also deletes all custom scales you added.

5. Click **OK** to exit the dialog box.

The -SCALELISTEDIT command performs the same functions at the command line:

   Command: **-scalelistedit**

   Enter option [?/Add/Delete/Reset/Exit] <Add>: *(Enter an option.)*


### HIDEXREFSCALES

The HIDEXREFSCALES system variable usefully hides all scale factors imported from xref'ed drawings when set to 1:

   Command: **hidexrefscales**

   Enter new value for HIDEXREFSCALES <0>: **1**

To view imported scale factors, change this system variable to 0. Alternatively, uncheck the option in the drawing bar:

The MEASUREINIT system variable specifies Imperial or metric scale factors automatically when new drawings are created without a template.

# PLOTSTAMP

The **PLOTSTAMP** command stamps plots with information about the drawing and plot.

When you plot sets of drawings, it sometimes becomes difficult to determine which set was plotted most recently, or which drawing files produced the plots. To identify the plots more readily, apply plot stamps.

A plot stamp consists of one or two lines of text that list useful information, such as the drawing file name, the date and time the drawing was plotted, the plotter used to make the plot, the plot scale, the name of the computer that generated the plot, and so on. In addition, AutoCAD lets you specify two custom pieces of data.

Plot stamps are often placed along the edge of the drawing, as illustrated above. You can change the location, as well as the font and size of text. After you have set up a plot stamp, its parameters can be saved to a *.pss* plot stamp settings file for use with other AutoCAD systems.

## TUTORIAL: STAMPING PLOTS

1. To apply a stamp to the plotted drawing, start the **PLOTSTAMP** command:
   - In the Plot or Publish dialog boxes, choose the **Plot Stamp Settings** button.

   - At the 'Command:' prompt, enter the **plotstamp** command.

     Command: **plotstamp** *(Press ENTER.)*

   - Alternatively, enter the **ddplotstamp** (the command's old name) alias at the 'Command:' prompt.

   In all cases, AutoCAD displays the Plot Stamp dialog box.

   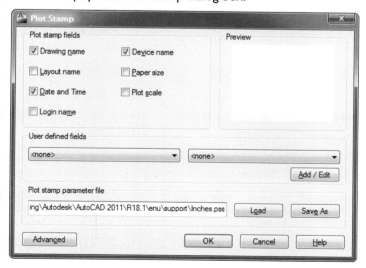

2. In the Plot Stamp Fields area, select the text you wish stamped on the plot:

| Plot Stamp Fields | Examples | Comments |
|---|---|---|
| Drawing Name | C:\17_22.DWG | Full path and file name of the drawing. |
| Layout Name | Model | Layout name; "Model" if plotted from model space. |
| Date and Time | 6/26/2003 10:24:42 AM | Date and time of the plot; format is determined by the Regional Settings dialog box of the Control Panel. |
| Login Name | Administrator | Windows login name, as stored in the LOGINNAME sysvar. |
| Device Name | Lexmark Optra R+ | Name of the plotting device. |
| Paper Size | Letter 8 ½ x 11 in | Size of the paper. |
| Plot Scale | 1:0.8543125 | Plot scale factor; when you see an unusual scale factor, such as that shown in the example, this means the drawing was probably scaled to fit the margins. |

3. To define your own fields, click the **Add/Edit** button. AutoCAD displays the User Defined Fields dialog box.

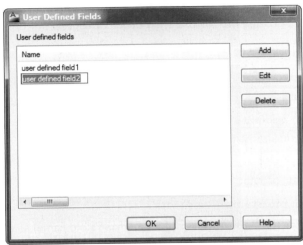

Click **Add**, and then enter text. You can add as many user-defined fields as you wish, but AutoCAD allows you to use only two per plot stamp. When done, click **OK**. (To include a user-defined field with the plot stamp, select it from the droplist under User Defined Fields.)

4. To reuse the settings with other drawings, or have fun swapping them with friends, click **Save As** to save the settings to the *.pss* file. AutoCAD displays the Plotstamp Parameter File Name dialog box:
    a. Enter a file name.
    b. Click **Save**. AutoCAD saves the data in a *.pss* (plot stamp parameter) file.

5. Click **OK** to exit the dialog box.

**Notes** When options of the Plot Stamp dialog box are grayed out, this means that the *inches.pss* or *mm.pss* file in AutoCAD's \*support* folder is set to read-only. To change the setting with Windows Explorer: (1) right-click the *.pss* file, (2) select **Properties**, (3) uncheck **Read-only**, and (4) click **OK**.

The plot stamp data are not saved with the drawing, but are generated anew with each plot. Plot stamps are plotted with color 7 (on raster plotters) or pen 7 (on pen plotters).

## STAMPING PLOTS: ADVANCED OPTIONS

The Advanced Options dialog box provides additional options for locating the stamp on the plot. In the Plot Stamp dialog box, click **Advanced**.

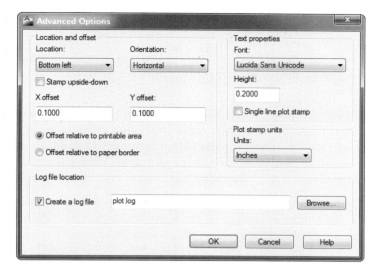

### Location and Offset

The options in the Location and Offset area positions the stamp on the plot. The preview, unfortunately, is not located in this dialog box, so you need to go through a cycle: change settings, click **OK** to get out of this dialog box, check the plot preview in the first dialog box, and then return to this one.

The **Stamp Upside-down** option is useful, because otherwise the stamp can be confused for other text in the drawing. Also, some filing systems work better with upside-down plot stamps.

 **Note** A large offset value could position the plot stamp text beyond the plotter's printable area, which may cause the text to be cut off. To prevent this, use the **Offset relative to printable area** option.

### Text Properties

The **Text Properties** section selects the font and size of text. I recommend a narrow font, such as Arial Narrow or Future Condensed, to ensure that all the plot stamp text fits the page. The default size, 0.2", tends to be too large; I recommend 0.1" or smaller.

### Plot Stamp Units

The **Plot Stamp Units** section defines the units for the offset and height numbers used by this dialog box. Select from inches, millimeters, or pixels.

### Log File Location

The **Log File Location** section lets you specify the name and folder for the plots' log files. The *.log* file is in ASCII format, and can be useful for billing clients and checking who is hogging the plotters. The file contains exactly the same information as is stamped on the plot, such as:

    C:\17_22.DWG,Model,6/26/2003 12:34:48 PM,Administrator,Lexmark Optra R plus.

Click **OK** to exit the dialog boxes.

# PAGESETUP

The **PAGESETUP** command assigns plotters to drawings.

AutoCAD doesn't assign any plotters or printers to new drawings, unfortunately, even though older releases assigned the default Windows system printer. Right before plotting the drawing, you have to select a printer in the **PLOT** command's dialog box. Or, you can assign a default printer, such as your office's networked large-format plotter, as part of the initial setup for template drawings. This is done with the **PAGESETUP** command.

*Page setups* are lists of instructions that tell AutoCAD how to plot model space, layouts, and sheets. Autodesk calls these "pages," because AutoCAD prints them onto pages of paper. The flexibility of page setups means that you can have a different page setup for model tab and for every layout and sheet in every drawing: each could be plotted by a different plotter on different size paper, and at different scales and orientations — in color or in monochrome.

The Page Setup Manager lists the layouts to which the page setup will be applied. This list can consist of a single layout name, or of all the layouts in the current drawing. (When opened from the Sheet Set Manager, it displays the name of the current sheet set.) Page setup data are stored in the drawing's *.dwg* file.

### TUTORIAL: PREPARING LAYOUTS FOR PLOTTING

1. To prepare drawings for plotting, start the **PAGESETUP** command:
   - From the ribbon's Output tab, choose **Page Setup** from the Plot panel.
   - In the application menu, choose **Print** and then **Page Setup**.
   - Right-click a layout tab, and then select **Page Setup Manager** from the shortcut menu.
   - In the Sheet Set Manager, click the **Publish** button, and then select **Manage Page Setups** from the menu.
   - At the 'Command:' prompt, enter the **pagesetup** command.

   Command: **pagesetup** *(Press ENTER.)*

   AutoCAD displays the Page Setup Manager listing page setups, if any are already defined.

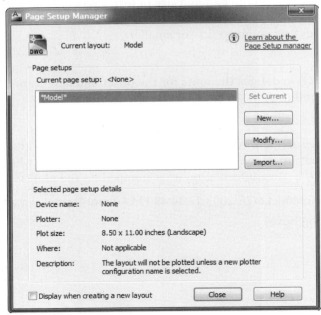

(The **Display When Creating a New Layout** option displays this dialog box each time you create a new layout tab.)

2. Look in the lower half of the dialog box. Notice that no plotter has been selected for this layout. To specify a plotter, click **Modify**.
3. Notice the Page Setup dialog box, which looks almost identical to the Plot dialog box.

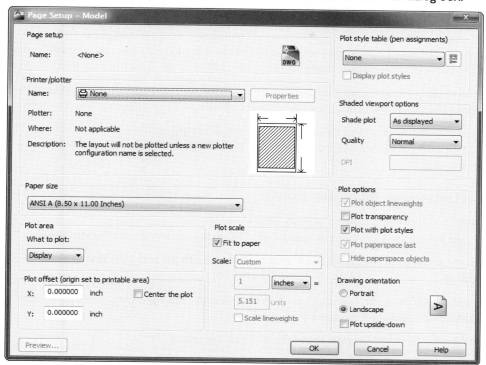

In this dialog box, you can preset many options for future plotting. The most important, however, is specifying the printer or plotter to use.

4. In the Printer/Plotter area, select a printer from the Name droplist. The names of printers in the droplist vary with every computer, and include Windows printer drivers, AutoCAD plotter and export drivers, and attached network printer drivers.

5. Set other parameters, if you wish. If you are unsure which settings to use, I recommend these:

    | | |
    |---|---|
    | Plot Area | **Extents** |
    | Plot Offset | **Center the Plot** |
    | Plot Scale | **Fit to Paper** |

6. Click **OK**. Your drawing is now ready to plot.

## PUBLISH

The PUBLISH command plots *sets* of drawing: one or more drawings and/or combinations of model and layout tabs. (In contrast, the PLOT command plots just one drawing at a time.) A drawing set consists of one or more drawings and layouts that are plotted, in order, at one time — these are called "sheets" by Autodesk.

Engineering and architectural offices often use drawing sets to combine all drawings belonging to a particular project. This makes it easier to plot many drawings at once, such as at specific stages of large projects. Alternatively, drawing sets can be saved as *.dwf* or *.pdf* files for archiving or plotting later.

This command selects the drawing files and layouts needed for sets, and then lets you reorder, rename, and copy them. The command then plots or exports drawings — in effect, it is a SuperPlot command.

Lists of drawing sets can be saved as *.dsd* drawing set description files for later reuse. (The BATCHPLT utility found in earlier releases was removed from AutoCAD; its functions are now performed by the PUBLISH command.)

### TUTORIAL: PUBLISHING DRAWING SETS

1. To create a drawing set, start the PUBLISH command:
   - From the ribbon's Output tab, choose the **Publish** button from the Publish panel.
   - In the application menu, choose **Publish**. (Choosing **Print** and then **Batch Plot** does the same thing.)
   - Hold down the CTRL key, select two or more layout tabs, right-click, and then choose **Publish Selected Layouts** from the shortcut menu.
   - At the 'Command:' prompt, enter the **publish** command.

        Command: **publish** *(Press ENTER.)*

   In all cases, AutoCAD displays the Publish Drawing Sheets dialog box. Initially, the dialog box lists the names of the drawing(s) and layout(s) currently open in AutoCAD.

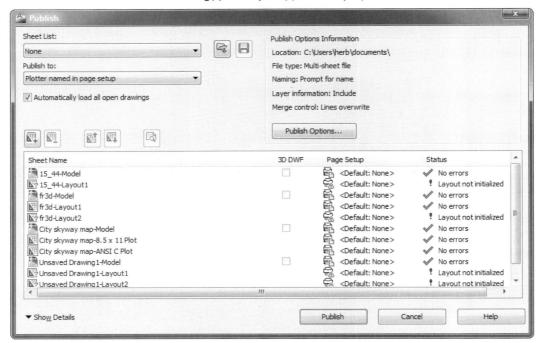

2. AutoCAD provides a number of ways to create lists of drawings (sheets) to print (plot):

   a. **Automatically Load All Open Drawings** option adds all open drawings and layouts to the list of sheet names. If you don't want AutoCAD to do this for you, turn off the option.

   b. Click the curiously-named **Add Sheets** button to add drawing files from your computer's hard drive or network locations. You can choose among DWG drawing, DWS drawing standards, DWT drawing template, and DXF drawing exchange format. You can include model or layouts or both.

   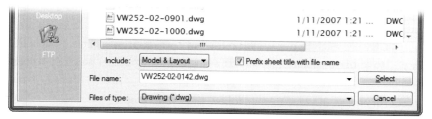

   As AutoCAD adds drawings to the list, it also extracts the layouts. Each is listed separately by file name, along with one of the following suffixes:

   - **-Model** — indicates the model tab (model space view) of the drawing.
   - **-Layout***n* or *LayoutName* — indicates the layout tab number or name. If sheets of the same name already exist in the list, AutoCAD will prompt you to change their names.

   c. Alternatively, click the **Load Sheet List** button to bulk-load drawings through .dsd or .bp3 files.

**Note** DSD (drawing set descriptions) files are created by the **PUBLISH** command's button and simply contain the lists of drawings that will be published. Typically, this is a list of project drawings that you may need to print on a regular basis.

BP3 (batch plot v3) files are predecessors to DSD files, and are no longer used. They were created by the Batch Plot utility program provided with AutoCAD 2000 through 2004.

   d. If there are some sheets you want to remove from the list, select them, and then click the **Remove** button.
      - Hold down the SHIFT key to select a range of layout names.
      - Hold down the CTRL key to select nonconsecutive layout names.

      AutoCAD removes the selected layouts without asking if you're sure — even if you're not.

3. Drawing sets are usually plotted in a specific order. Use the **Move Up** and **Move Down** buttons to change the order of the layouts.

   Some plotters stack output pages in reverse order, and thus you can reverse the plot order so that the last layout is plotted first. To do so, click **Show Details**, and then click the **Reverse Plot Order** button.

4. The drawing sets can be "published" to printers, or saved as electronic plots for transmittal by email or other electronic forms. Make your selection from the Publish To droplist among DWF, DWFx, PDF, or another format set up with the **PAGESETUP** command:

The **Plotter Named in Page Setups** option sends all of the layouts to a plotter — but the layouts might not end up being plotted by the printer you expect. Prior to pressing the Publish button, ensure each layout is set up for the correct printer. This complexity is the necessary result of the flexibility that allows a single drawing set to be plotted to a variety of printers.

To save model sheets in 3D *.dwf* format, follow these two steps:

   a. Choose **DWF** from the Publish To droplist.
   b. Click the check boxes in the column under **3D DWF**.

Layouts and sheets cannot be plotted as 3D DWFs, since they are supposed to represent the 2D world.

5. Click **Save List** to save the list of layouts for later reuse.

6. Click **Publish**. AutoCAD loads each drawing, and then generates the plot or the *.dwg* file. A dialog box notes its progress, as plotting many sheets takes a long time.

When AutoCAD cannot find a drawing or layout, it skips it and carries on with the next one. If errors are found, AutoCAD generates a report and saves it as a *.csv* comma-separated value log file, which you can read after importing it into a spreadsheet program, such as OpenOffice Calc.

When done, AutoCAD asks if you wish to view the output in an external viewer, if an appropriate one is installed on your computer. DWF and DWFx output is shown in Design Review (provided with AutoCAD), and PDF in Acrobat Reader (free from www.adobe.com/acrobat).

### SHOW DETAILS

The Publish dialog box places some commands in the Details section; you see this section after clicking the **Show Details** button.

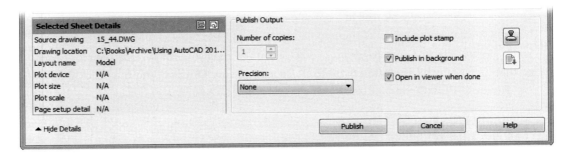

## PRECISION PRESETS MANAGER DIALOG BOX

The Precision Presets Manager dialog box is accessed through the PUBLISH command's **Precision** option, or through the EXPORTSETTINGS command's Export to DWF/PDF Options dialog box's **Override Precision** option.

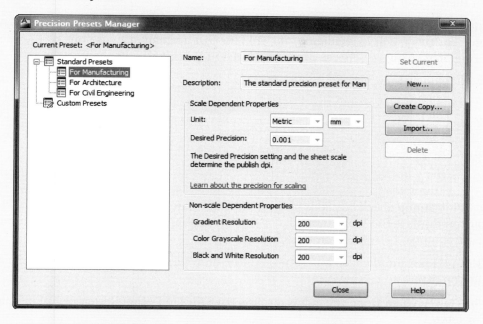

**Current Preset** specifies the name of the current precision preset.
**Name** names the preset.
**Description** describes the purpose of the preset.

### Scale Dependent Properties options

**Unit** chooses Metric or Imperial units, and subunits (mm, m, feet, inches, etc).
**Desired Precision** specifies the precision at which the drawing will be exported.

### Non-scale Dependent Properties

A lower resolution setting reduces file size but make gradients look postertized and images look grainier.

**Gradient Resolution** specifies the resolution of gradient hatch areas.
**Color Grayscale Resolution** specifies the resolution of color and grayscale images.
**Black and White Resolution** specifies the resolution of monochrome images.

### Buttons

**Set Current** sets the selected preset as the working preset.
**New** creates new presets; displays the New Precision Preset dialog box which prompts for a name and description.
**Create Copy** copies the selected preset; displays the Copy Precision Preset dialog box, which prompts you for a name and description.
**Import** imports .*dsd* Drawing Set Description files created by the PUBLISH command; displays the Select List of Sheets dialog box.
**Delete** erases the preset.

### Publish Output

The **Publish Output** section contains additional options for generating sheets.

You can have the plotter output more than one copy of the sheets, as determined by the **Number of copies** option. It is unavailable when outputting to PDF and DWF files, because they can print multiple copies from their viewer programs.

The **Precision** droplist selects a predefined level of precision, the units and subunits, and resolution of non-vector entities (such as gradients) for DWF or PDF formats. Choose one of the presets (For Manufacturing, and so on), or choose **Manage Precision Presets** to edit them.

The **Include Plot Stamp** check box adds plot stamp data to the edge of each plot. The **Plot Stamp Settings** button displays the Plot Stamp dialog box for customizing the plot stamp information; see the PLOTSTAMP command elsewhere in this chapter.

The **Publish in Background** check box returns to the drawing editor more quickly by generating the sheets while you return to editing.

The **Open in Viewer When Done** check box displays DWF files in Design Review or PDF files in Acrobat Viewer v7 or later.

The **Currently publishing in default order** button reverses the order in which layouts are plotted. AutoCAD normally publishes the drawing set in the order listed by the dialog box. Some plotters output pages face-up, which means the set ends up in reverse order; this button fixes the problem.

### PUBLISH OPTIONS

Before publishing the drawing list in DWF or PDF format, click the **Publish Options** button. It displays the Publish Options dialog box, which controls output for DWF and PDF files.

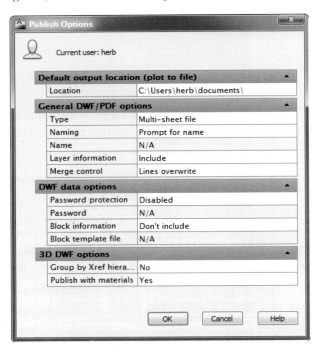

### Current User

The name that appears next to the icon in the upper-left corner reports the Windows login name of the current user, such as "Administrator."

### Default Output Location

The **Default Output Location** (formerly named Directory) section determines where plot files are saved when you publish drawing sheets. Enter a different drive and folder name, or click the ⋯ button to display the Select Folder for Generated Files dialog box.

### General DWF/PDF Options

The **General DWF/PDF Options** section specifies how drawings are published as PDF and DWF files.

#### Type

AutoCAD can publish all sheets in a single file, or each sheet in its own file. Click the droplist and choose one of the following:

- **Single-Sheet File** outputs each sheet to its own file. Use this option for compatibility with older viewing software that does not accept electronic files in multisheet format.
- **Multi-Sheet File** outputs all sheets to a single file. Use this option for the convenience of publishing to a single file.

#### Naming

When AutoCAD creates a single-sheet DWF, the *.dwf* file takes on the drawing's (sheet's) name. But what name should AutoCAD use when publishing more than one drawing? Here you specify how the file is named:

- **Specify Name** specifies the path and file name here in the dialog box.
- **Prompt for Name** delays specifying the path and file name until later, after you click the Publish button.

#### Name

The **Name** option specifies the name of the file (valid with the Specify Name option only).

#### Layer Information

This option determines whether layers are included in the file. Choose one of the following:

- **Include** includes layer names and properties, and leaves objects on layers.
- **Don't Include** does not include layer names, so all objects are layer-less. This can be useful for limiting the amount of information provided to clients.

#### Merge Control

When lines and other objects cross each other in the drawing, AutoCAD can output the intersections in one of two ways:

- **Lines merge** blends the colors of overlapping lines.
- **Lines overwrite** draws the last line overtop of other lines.

### DWF Data Options

The following options apply to DWF files only, even though they are also displayed when you choose PDF as the output format.

#### Password Protection

DWF files are often distributed by email or on CD; adding a password prevents unauthorized people from viewing the files. The recipients must have the password to open the file; how they receive the password is up to you. If you forget the password, the DWF file cannot be recovered, although you can generate another DWF file from the original drawings.

- **Disabled** — password protection is turned off.
- **Specify Password** — allows you to specify the password here.
- **Prompt for Password** — AutoCAD prompts you for the password later when you click the **Publish** button.

### Password

The **Password** option specifies the password (when Specify Password option is enabled). You can use letters, numbers, punctuation, and non-ASCII characters for the password. The password is case-sensitive, which means that "autocad" is not the same password as "AutoCAD."

### Block Information

This option determines how blocks are exported:

**Include** — includes blocks and attribute data in the file.

**Don't Include** — explodes blocks to remove block and attribute information.

### Block Template File

You can use a DXE or BLK block template file to specify which blocks and attributes should be included in the file.

- **Create** creates a new *.blk* block template file; displays the Publish Block Template dialog box.
- **Edit** edits existing *.blk* files; displays the Publish Block Template dialog box.

The ⋯ button opens the Select Block Template file, from which you can choose a *.blk* file.

### 3D DWF Options

Usually, the PUBLISH command generates 2D DWF files. It can output 3D ones when the check box is selected from the Publish dialog box's **3D DWF** column (available only from the Model "sheet" of drawings, and when DWF is the Publish To format.) AutoCAD does not generate 3D DWFs from layouts, because they are 2D only.

Once 3D DWF is selected for a Model sheet, the Publish Options dialog box makes the 3D DWF options available:

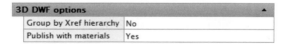

**Group by Xref Hierarchy** lists objects under each xref name, if externally-referenced drawings are linked to the sheet.

**Publish with Materials** includes material with objects; materials are displayed by Design Review.

 **Note** The technical editor suggests setting up the DWF plotter with a very large paper size. He finds that this dramatically increases the resolution at the cost of a slight increase in file size.

# PLOTTERMANAGER

The **PLOTTERMANAGER** command creates and edits plotter configurations.

AutoCAD plots to any printer connected to your computer. This includes printers found in your office, large-format plotters used in engineering and architectural offices, and files on disk. Large-format plotters create D- and E-size plots that are roughly three to four feet, or one meter, across.

Most software relies on Windows to provide device drivers. The problem is that the device drivers provided by Microsoft are not accurate or flexible enough for AutoCAD's high-precision plotting. For this reason, Autodesk includes its own set of drivers, known as HDI.

To plot drawings with HDI drivers, first run the Plotter Manager. This configures in great detail how the plotters should print the drawings.

## TUTORIAL: CREATING PLOTTER CONFIGURATIONS

1. To create configurations for plotters, start the **PLOTTERMANAGER** command:
    - From the 2D ribbon's Output tab, select **Plotter Manager** from the Plot tab.
    - In the application menu, choose **Print** and then **Manage Plotters**.
    - At the 'Command:' prompt, enter the **plottermanager** command.

        Command: **plottermanager** *(Press* ENTER.*)*

    AutoCAD displays the contents of the *C:\Users\username\AppData\Roaming\Autodesk \Auto-CAD 2011\R18.1\enu\Plotters folder.*

2. Select a PC3 file to edit (consulting the "Tutorial: Editing Plotter Configurations" that follows), or double-click **Add-a-Plotter Wizard** to create a new plotter definition.

3. To create a new PC3 file, choose **Add-a-Plotter Wizard**. Notice that AutoCAD starts the *addplwiz.exe* program, and then displays the Introduction Page.

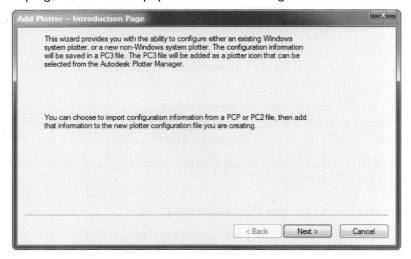

4. Choose **Next**.
5. In the Begin page, set up AutoCAD with one of these styles of printer:

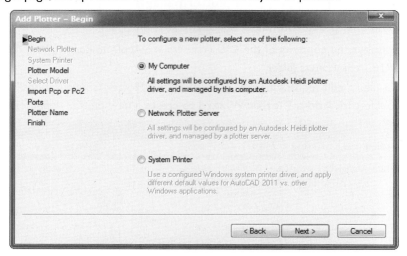

- **My Computer** consists of local printers and plotters controlled by AutoCAD's plotter drivers.
- **Network Plotter Server** consists of printers located on the network, also controlled by AutoCAD's plotter drivers.
- **System Printer** consists of local and network printers controlled by Windows printer drivers.

Unless you have a reason to do otherwise, choose **My Computer,** and then click **Next**.

6. AutoCAD displays a list of printer and plotter drivers provided by Autodesk.

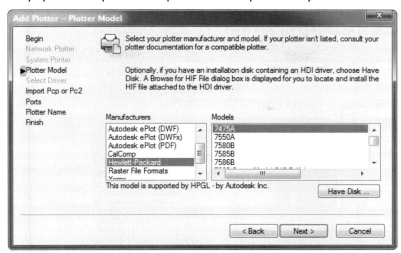

   a. From the Manufacturers list, select the brand name of the plotter.
   b. From the Models list, select the specific model number. (If your plotter comes with AutoCAD-specific drivers on a CD, choose the **Have Disk** button.)

      If you had selected Network Plotter Server, AutoCAD would have prompted you for the network location of the server.

      If you had selected System Printer, AutoCAD would have prompted you to select one.
   c. Choose **Next**.

**Notes** If you do not see the name of your plotter's manufacturer, check the documentation. Often it lists brand names and model numbers of compatible plotters. If you cannot find this information, try selecting Adobe for PostScript printers and Hewlett-Packard for large-format inkjet plotters.

To plot the drawing in a raster format, select **Raster File Formats**, and then a specific format:

| Format | Color Depths | File Extension |
|---|---|---|
| Independent JPEG Group JFIF | Gray, RGB | .jpg |
| Portable Network Graphics PNG | Bitonal, gray, indexed, RGB, RGBA | .png |
| TIFF Uncompressed | Bitonal, indexed, gray, RGB, RGBA | .tif |
| TIFF Compressed | Bitonal, indexed, gray, RGB, RGBA | .tif |
| CALS MIL-R-28002A Type 1 | Bitonal | .cal |
| Dimensional CALS Type 1 | Bitonal | .cal |
| MS-Windows BMP Uncompressed | Bitonal, gray, indexed, RGB | .bmp |
| TrueVision TGA 2.0 Uncompressed | Indexed, gray, RGB, RGBA | .tga |
| Z-Soft PC Paintbrush PCX | Indexed, RGB | .pcx |

7. The **Import PCP or PC2** page is important only if you created plotter configuration files with AutoCAD Release 13 (*.pcp*) and 14 (*.pc2*). Here you import these files for use with newer releases of AutoCAD (as *.pc3* files).

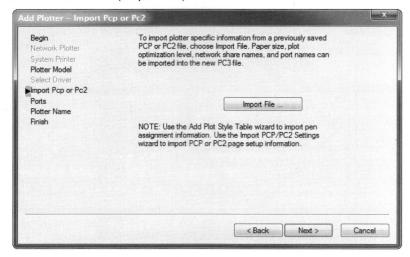

Choose **Next**.

8. The Ports page can be a difficult step for some users. Here you select the *port* to which the plotter is connected — or no port at all.

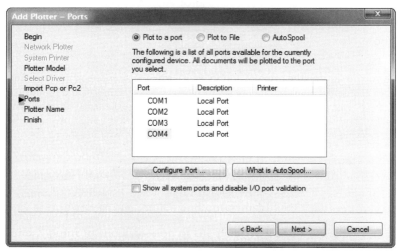

Your choices are:

**Plot to a Port** — AutoCAD sends the drawing to the plotter through a local port or network port, such as a parallel port (designated LPT), serial port (designated COM), or USB port. Select this option for drawings plotted by printers and plotters.

**Plot to File** — AutoCAD sends the drawing to a file on disk. Select this option to save the drawing to disk as a file in both plotter and raster formats.

**AutoSpool** — AutoCAD sends the drawing to a file in a specified folder (defined by AutoCAD's Options dialog box), where another program processes the file. Select this option only if you know what you are doing. (If you need to ask, then you don't know.)

9. Some ports have further options. If necessary, choose the **Configure Port** button:

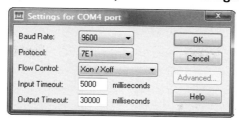

- **Serial** ports specify communications settings.
- **Parallel** and **USB** ports specify the transmission retry time.
- **Network** ports specify nothing, curiously enough.

Change settings, and then click **OK**.
Choose **Next**.

10. In the Plotter Name page, give the plotter configuration a name.

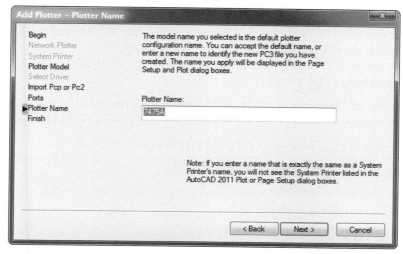

Enter a descriptive name, and then choose **Next**.

11. The Finish page has three buttons:

> **Edit Plotter Configuration** — Displays AutoCAD's Plotter Configuration Editor dialog box, which allows you to specify options, such as media source, type of paper, type of graphics, and initialization strings.
>
> **Calibrate Plotter** — calibrates the plotter. This allows you to confirm, for example, that a ten-inch line is indeed plotted 10.0000000 inches long.
>
> **Finish** — completes the plotter configuration.

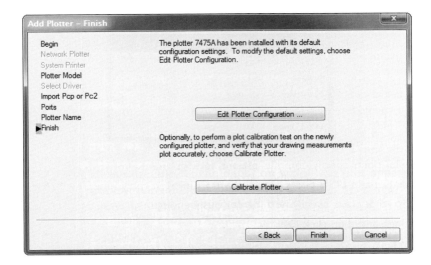

12. Choose **Finish**.

This configures the new plotter for AutoCAD. When you next use the PLOT command, this configuration appears in the list of available plotters.

### TUTORIAL: EDITING PLOTTER CONFIGURATIONS

1. To edit a plotter configuration, start the **PLOTTERMANAGER** command:
   - From the 2D ribbon's Output tab, select **Plotter Manager** from the Plot tab.
   - At the 'Command:' prompt, enter the **plottermanager** command.

      Command: **plottermanager** *(Press* ENTER.*)*

   Again, AutoCAD requests that Windows display the Plotters window, which lists all HDI plotter configurations as *.pc3* files.

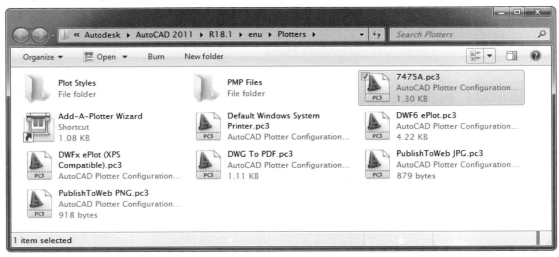

2. Double-click any *.pc3* icon, except *Default Windows System Printer.pc3*, because it must be edited through Windows.

Windows displays the Plotter Configuration Editor dialog box.

3. Click the **Device and Settings** tab.

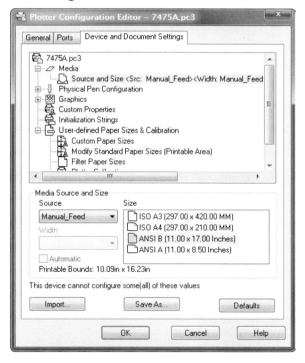

The available settings vary, depending on the capabilities of the plotter. Some settings are unavailable because the specific plotter model does not support them. Other settings are handled through the **Custom Properties** button.

Selected options are noted in angle brackets, <like this>. When a change is made to a setting, AutoCAD places a red checkmark in front of it.

4. Change settings, and then press **OK**.

## THE PLOTTER CONFIGURATION EDITOR OPTIONS

The Plotter Configuration Editor lists features and capabilities that AutoCAD takes advantage of. Not all settings listed below are available for all printers and plotters. Some features must be set through Windows: select the **Custom Properties** node, and then click the **Custom Properties** button.

### Media Source and Size

The **Media Source and Size** section specifies the paper source and size and type of paper.

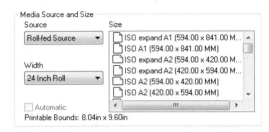

**Source** specifies the location of the paper, such as trays, sheet feeds, and roll feeds.

**Width** specifies the width of the paper (applies only to roll-fed sources of paper.)

☐ **Automatic** means that the printer determines the paper source.

**Size** lists the standard and custom paper sizes the printer is capable of handling.

### Media Type

The **Media Type** option determines the type of the paper, since this can affect how the plotter lays down its ink — not all printers support this.

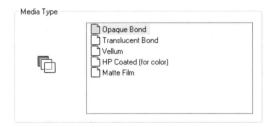

### Duplex Printing

The **Duplex Printing** option determines whether the paper is printed on both sides — if the printer is capable of it.

- ⊙ **None** turns off double-sided printing.
- O **Short Side** sets the binding margin on the short edge.
- O **Long Side** sets the binding margin on the long edge.

### Physical Pen Configuration

The **Physical Pen Configuration** option is for pen plotters only; it is largely obsolete, or nearly so. It controls the actions of each pen.

- ☐ **Prompt for Pen Swapping** forces AutoCAD to pause the plot so that you can change pens; available for single-pen plotters only.
- ☐ **Area Fill Correction** forces AutoCAD to plot filled areas by half-a-pen width narrower, so as not to draw filled areas too wide.

**Pen Optimization Level** optimizes pen motion to reduce total plotting time. This setting is not effective with very slow computers and slow plotters.

### Physical Pen Characteristics

The **Physical Pen Characteristics** option allows you to specify the color, width, and manufacturer-recommended top speed of each pen (in inches or millimeters per second).

## Graphics

The **Graphics** options specify options for vector and raster graphics and for TrueType fonts. The settings here vary depending on the capabilities of each printer and plotter.

### Resolution and Color Depths / Vector Graphics

The **Resolution and Color Depths** section (a.k.a. Vector Graphics) specifies options for color depth (number of colors), resolution (dpi), and dithering (simulation of a color by mixing two colors). Some printers trade off fewer colors for higher resolution.

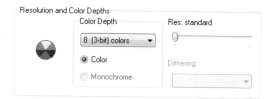

The **Raster Images** option trades off plotting speed for output quality (for plotters without pens).

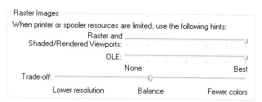

The **Merge Control** option specifies the look of crossing lines. This option is not valid when printing all colors as black, when using PostScript printers, and when the printer does not support merge control.

- **Lines Overwrite** — only topmost lines are visible at intersections.
- **Lines Merge** — colors of crossing lines are merged.

## Custom Properties

The **Custom Properties** option displays the Custom Properties button.

Click the button to view the dialog boxes that control the printer. The dialog box varies wildly, depending on the printer model.

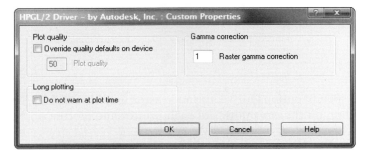

## Initialization Strings

The **Initialization Strings** option sets control codes for non-system printers, including those for pre-initialization, post-initialization, and termination. These codes allow you to use plotters and printers not supported entirely by AutoCAD.

 **Note** Use the backslash ( \ ) to emulate escape strings. For example, **\27** is sent as the escape character, and **\10** as the line-feed character.

## User-Defined Paper Sizes and Calibration

The **User-Defined Paper Sizes and Calibration** option allows you to specify custom paper sizes (those not recognized by AutoCAD) and to calibrate the printer.

To help you define custom paper sizes, AutoCAD runs the Custom Paper Size wizard. Click **Add** to start it.

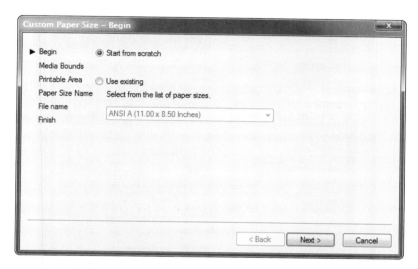

 **Note** Calibration ensures drawings are plotted accurately. Autodesk recommends, "If your plotter provides a calibration utility, it is recommended that you use it instead of the AutoCAD utility."

Following printer calibration, AutoCAD stores the calibration data in a *.pmp* (plotter model parameter) file. The *.pmp* file must be attached to the printer's *.pc3* file, unless it was created during the Calibration stage of the Add-a-Plotter wizard.

## Save As

To save the settings, click the **Save As** button. You can share the *.pc3* file with other AutoCAD users through the **Import** button.

## STYLESMANAGER

The **STYLESMANAGER** command creates and edits plot style tables.

Plot styles control the printing properties of objects. Depending on the capabilities of the printer, plot styles control color or grayscale, dithering and screening, pen number and virtual pens, linetype and lineweight, line end and join styles, and fill style. Plot style tables are collections of plot styles assigned to layouts and the model tab.

AutoCAD works with two types of plot style table: *color-dependent* and *named*.

### COLOR-DEPENDENT PLOT STYLE TABLES (.*CTB* FILES)

Color-dependent plot style tables (.*ctb* files) assign plot styles by the color of the layer or the objects, if ByLayer color has been overridden.

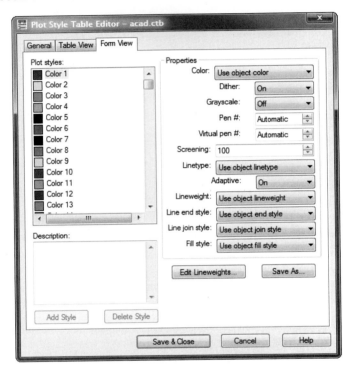

(*History:* Prior to AutoCAD 2000, this was the only way to control printing. During the old **PLOT** command, the CAD operator would assign AutoCAD color numbers to pen numbers. For example, blue might be assigned to pen #3. During plotting, AutoCAD instructs the plotter to plot all blue-colored objects with whatever pen was in pen holder #3. Pens could be assigned linetypes, colors, and widths, depending on the capabilities of the plotter.)

Color-dependent plot style tables have exactly 256 plot styles, one for each ACI color (short for "AutoCAD Index Color"). AutoCAD cannot accurately plot true colors (16.7 million) or color books (DIC, Pantone, and RAL) with these style tables; if you must have true colors, use the **Use Entity Color** option to plot with the ACI color nearest to the true color.

 **Note** If you are not sure whether a drawing uses color-dependent or named plot styles, look at the Plot Style list in the layers palette. If its style names are grayed out, color-dependent styles are in effect; if not, named plot styles are in effect.

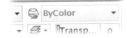

## NAMED PLOT STYLE TABLES (.STB FILES)

Named plot style tables (*.stb* files) contain named plot styles assigned to layers and objects. This is much more flexible than color-dependent styles, because it is not limited to controlling objects by their color. Instead, every layer and object can have its own plot style.

You can assign a different named STB to each layout in the drawing.

## CONVERTCTB AND CONVERTPSTYLES

You can switch drawings between color-dependent and named plot style tables with the CONVERTPSTYLES command (short for "convert plot styles").

### From Color-dependent to Named

When a drawing changes from color-dependent to named plot styles, AutoCAD removes CTB color-dependent plot style tables from layouts, and replaces them with named ones. Color-dependent plot style tables should first be converted using these steps:

> **Step 1.** First, use the **CONVERTCTB** command to convert color-dependent plot style tables to named tables. Select a *.ctb* file from the dialog box, and then specify the name for the *.stb* file. The 256 converted plot styles are given generic names, such as Style1 and Style2.
>
> **Step 2.** Next, use the **STYLESMANAGER** command to rename the generic plot styles. Styles should be renamed *before* being attached to layouts.
>
> **Step 3.** Finally, use the **CONVERTPSTYLES** command to switch the drawing to named plot styles. The command has no options, other than displaying a warning:

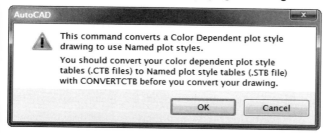

Click **OK**, and a moment later AutoCAD confirms the conversion:

Drawing converted from Named plot style mode to Color Dependent mode.

### From Named to Color-dependent

When a drawing changes from named plot styles to color-dependent, AutoCAD erases all STB plot style names assigned to layers and objects in the drawing. Using the CONVERTPSTYLES command results in a message similar to that shown above, along with the confirmation message:

Drawing converted from Color Dependent plot style mode to Named mode.

Although it appears that the UNDO command reverses the conversion, named plot styles are labeled "missing" and are **not** returned.

### Setting Default Plot Styles

In almost all cases, your drawings should use named plot styles, because they have more benefits than color-based ones. This comparison lists the benefits:

| Color-based Tables | Named Tables |
| --- | --- |
| 256 plot styles | Only as many plot styles as your drawings need. *Benefit:* Fewer plot styles to deal with. |
| Generic style names, such as "Color1" and "Color2" | Descriptive style names, such as "50% Screening." *Benefit:* Easier to understand style names. |
| Object colors control for plots | Object color independent of plotting. *Benefit:* More versatile use of colors in drawings. |
| Single plot style for all layouts | Every layout having its own plot style. *Benefit:* Use of different layouts for different plotters. |

Despite the drawbacks to color-based plot styles, they remain the default setting for new drawings. To change the default plot style table for new drawings through a template file, follow these steps:

1. Enter the **OPTIONS** command.
2. Choose the **Plot and Publish** tab.
3. Click **Plot Style Table Settings**. Notice the Plot Style Table Settings dialog box.

4. Select **Use Named Plot Styles**, and then choose a default plot style table (*.stb* file).
5. Click **OK** twice to exit the dialog boxes, and then save the drawing as a template file with the **SAVEAS** command.

## Assigning Plot Styles

Named plot styles can be assigned to entire layouts, layers, and individual objects.

Like colors and linetypes, plot styles can be ByLayer and ByBlock. *ByLayer* means objects take on the plot style assigned to the layer; change the plot style, and all objects on the layer change. *ByBlock* means objects in a block take on the plot style assigned to the block. In addition, a plot style can be *Default*, which means objects take on the plot style assigned to the layout.

### Assigned Plot Styles to Layouts

Assigning a plot style to all layouts takes these steps:

1. Ensure the drawing is set to use named plot styles. Enter the **New** command, and then choose the *acad -Named Plot Styles.dwt* template file.
2. Start the **PLOT** command, and then click the **More** button to expand the dialog box.
3. In the Plot Style Table droplist, choose an *.stb* file (plot style name).

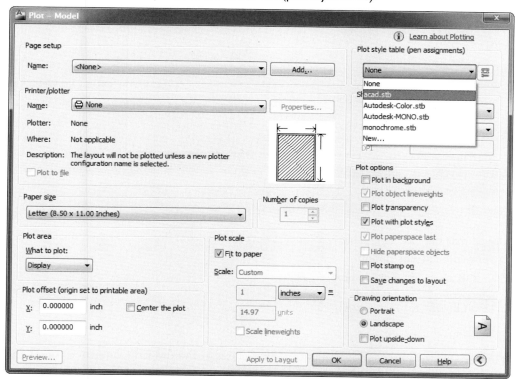

4. Notice that AutoCAD asks whether to apply the table to all layouts:

You can answer yes or no:

> **Yes:** the same plot style table is assigned to the model tab and all layouts in the drawing.
>
> **No:** this plot style table is assigned only to the model tab.

5. Click **OK** to exit the dialog box.

If you want different a plot style table for each layout, then you have to take some more steps:

6. Select a layout tab. If AutoCAD does not automatically display the Page Setup Manager again, right-click the layout tab, and select **Page Setup Manager**.

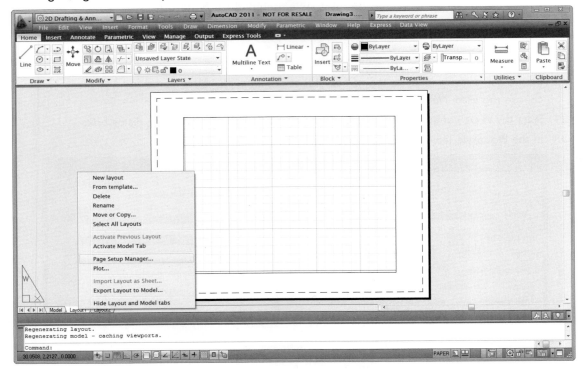

7. Notice that the Plot style table is "None." Select a table from the list, and then turn on the **Display plot styles** option.
8. Click **OK** to exit the dialog box.

Repeat for each layout tab.

### Assigned Plot Styles to Objects

To assign a plot style to one or more objects, select them in the drawing, and then select the named plot style from the Plot Style Control droplist on the Properties palette.

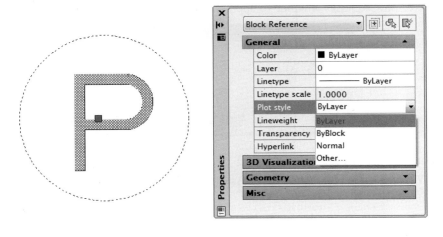

Choose **Other** to see a list of all plot styles in the table.

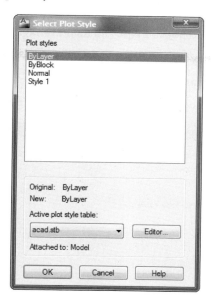

## Assigned Plot Styles to Layers

To assign a plot style to layers, start the LAYER command.

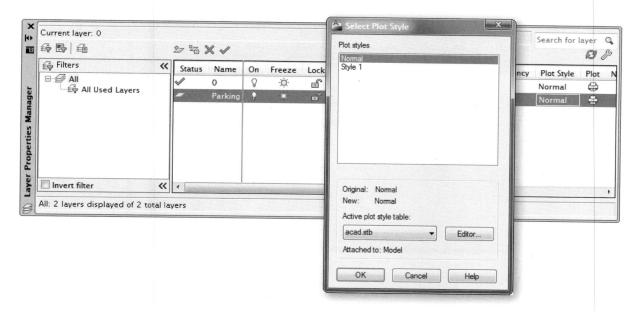

In the palette, select one or more layers, and then select a named plot style from the **Plot Style** column. The Select Plot Style dialog box appears; choose a named plot style, and then click **OK** to exit the dialog box. All objects assigned to that layer now take on the same plot style.

## TUTORIAL: CREATING NEW PLOT STYLES

1. To create new plot styles, start the **STYLESMANAGER** command:
   - From the application menu, choose **Print** and then **Manage Plot Styles**.
   - At the 'Command:' prompt, enter the **stylesmanager** command:

     Command: **stylesmanager** *(Press* ENTER.*)*

2. In the window that appears, double-click **Add-A-Plot Style Table Wizard**.

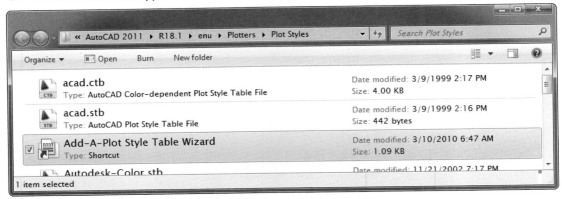

In all cases, AutoCAD runs the Add Plot Style Table "wizard," which leads you through the steps for creating a new plot style table.

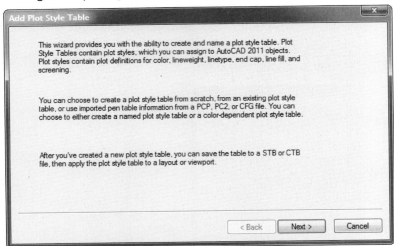

Click **Next**.

3. In the Begin dialog box, AutoCAD asks for the kind of plot style table to create. Select **Start from Scratch**, and then click **Next**.

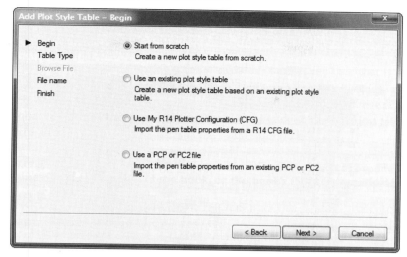

4. In the Table Type dialog box, select **Named Plot Style Table**.

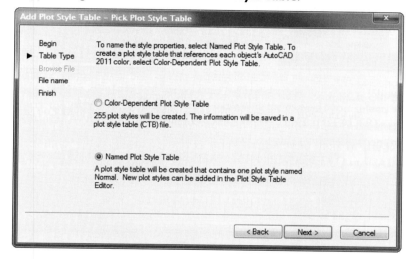

Click **Next**.

5. In the File Name dialog box, enter a descriptive name for the style table.

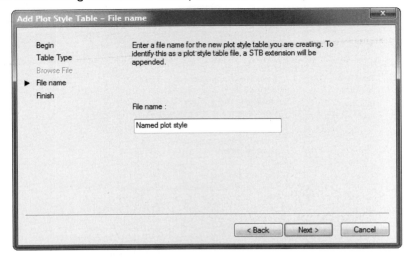

Click **Next**.

6. In the Finish dialog box, click **Finish**.

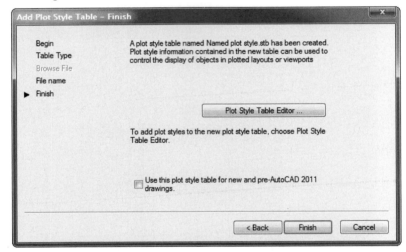

AutoCAD adds the *.stb* style table file to its collection stored in the \*plot styles* folder.

## TUTORIAL: EDITING PLOT STYLES

1. To edit plot styles, start the **stylesmanager** command:
   - From the application menu, choose **Print** and then **Manage Plot Styles**.
   - At the 'Command:' prompt, enter the **stylesmanager** command.

   Command: **stylesmanager** *(Press* ENTER.*)*

2. In both cases, AutoCAD requests that Windows display the Plot Styles window — the same one you saw before. In this case, however, you double-click a *.ctb* or *.stb* file name:

   **CTB**: color-based plot style table files.

   **STB**: named plot style table files.

   Notice that AutoCAD opens the Plot Style Table Editor.

3. Click the **Form View** tab, which lists plot style names on the left and the associated plot style properties on the right.

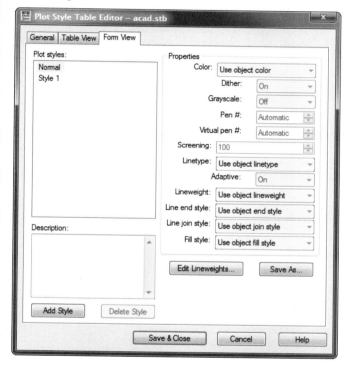

4. Under Plot styles, select a plot style name — other than Normal. Notice the plot style properties at the right.

   Edit the properties, as required.

    **Note** The first plot style in every named plot style table is called "NORMAL." It lists the object's default properties, so that you can see its properties with no plot style applied. The NORMAL style cannot be edited or removed.

5. To add a plot style, click **Add Style**.

   Name the style, and then edit its properties. (You cannot add plot styles to, or delete them from, color-based tables.)

6. To delete a style, select it, and then click **Delete Style**.

7. To save the plot style table, click **Save As**, and then name the table file.
8. When done, click **Save and Close**.

| *Plot Style Property* | *Comment* |
|---|---|
| Color | Specifies plotted color of objects; plot color overrides object color. Default: **Use Object Color**. |
| Dither | Toggles dithering to approximate colors with dot patterns. Default: **Off** for .stb files; **On** for .ctb files. |
| Grayscale | Converts object colors to grayscale during printing. Default: **Off** |
| Pen # | Specifies pen #; ranges from 0 to 32 (available for pen plotters only). Default: **Automatic** (pen #0). |
| Virtual Pen # | Simulates pen plotters for non-pen plotters; ranges from 0 to 255. Default: **Automatic** (pen #0). |
| Screening | Specifies color intensity; ranges from 0 (white) to 100. Selecting a value other than 100 turns on Dithering. Default: **100**. |
| Linetype | Overrides object linetype with selected linetype. Default: **Use Object Linetype**. |
| Adaptive | Adjusts scale of the linetype to complete the pattern. Turn off if linetype scale is crucial. Default: **On**. |
| Lineweight | Overrides object lineweight with selected lineweight. Default: **Use Object Lineweight**. |
| Line End Style | Assigns a style to the end of object lines: Butt, Square, Round, and Diamond. Default: **Use Object End Style**. |
| Line Join Style | Assigns a style to the intersection of object lines: Miter, Bevel, Round, and Diamond. Default: **Use Object Join Style**. |
| Fill Style | Overrides object fill style with the following: Solid, Crosshatch, Diamonds, Horizontal Bars, Slant Left, Slant Right, Square Dots, Checkerboard, and Vertical Bar. Default: **Use Object Fill Style**. |

## EXERCISES

1. Open the *edit4.dwg* file, a landscape drawing.

   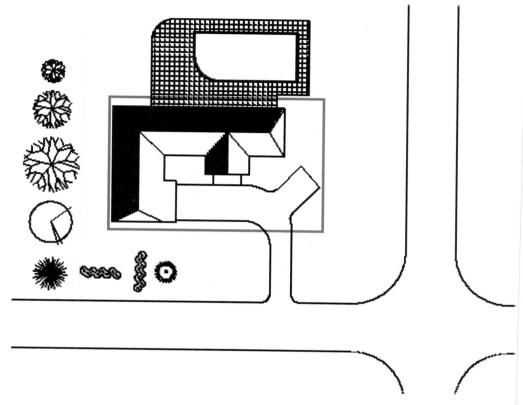

   a. Start the **PLOT** command, and assign a printer, if necessary.
   b. Plot the drawing at a scale factor of "Scaled to Fit."
   c. Does the drawing fit the page?

2. Use a ruler to measure the house's longest side. Assume the length represents 50'. At what scale was the house plotted?

3. Plot the drawing again, using both landscape or portrait orientation.

4. Using the same drawing, set the plot scale factor to 1:1.
   Preview the drawing.
   Is the drawing larger or smaller?

5. Plot the drawing again, but this time use the **Window** option, and window the area shown above by the blue rectangle.

6. Use the **PLOTSTAMP** command to attach a stamp to the plot.

7. Use the **PUBLISH** command to create a drawing set of the three drawings listed below:
   15_43.dwg
   15_44.dwg
   15_45.dwg

8. Create a plotter configuration for the printer attached to your computer.
   Name the configuration "UsingAutoCAD."

9. Create a plotter configuration for outputting compressed TIFF files with the **PLOT** command.
   Test the configuration by plotting a drawing, and then viewing the *.tif* file with AutoCAD's **REPLAY** command.

10. Calibrate the printer attached to your computer.
    How inaccurate was it before calibration?

11. Create a named plot style table with the following plot styles:
    50% screening.
    Checkerboard fill.
    Diamond join style.
    0.02 lineweight.

    Attach the plot style table to a drawing, and then plot it.

# CHAPTER REVIEW

1. What is the purpose of plotting?
2. What part of the drawing is plotted when plotted to the extents of the drawing?
3. Describe how to plot only a portion of drawings.
4. How do you rotate the plot on the paper?
5. Why would you plot drawings to file?
6. Can you plot a drawing without first assigning a plotter or printer?
7. Explain the benefit to using the **PREVIEW** command.
8. Briefly describe the purpose of *device drivers*.
9. Does AutoCAD plot objects on frozen layers?
10. What kinds of plotters and printers does AutoCAD work with?
11. How large is a drawing plotted when the scale factor is
    a. Scaled to Fit?
    b. 1:1?
    c. 1/128" = 1?
12. Name an advantage to using full preview.
    To using partial preview.
13. What is the benefit of background plotting?
14. When might you want to offset the drawing to the paper?
    What is the drawback of offsetting?
15. Can AutoCAD create rendered plots?
    Can 2D drawings be rendered?
16. How do you instruct AutoCAD to plot with hidden lines removed in model space?
    In a layout?
17. Describe how plot stamps are useful.
18. When a plot stamp indicates the scale factor is 1:0.97531864, what might this indicate?
19. What are *drawing sets*?
    When might you want to use them?
20. Must all layouts in a drawing set come from the same drawing?
21. Explain the purpose of the **PUBLISH** command.
    Of the **PLOTTERMANAGER** command.
22. Can AutoCAD plot to printers located on networks?
    Connected via USB ports?
23. Explain the meaning of the check mark on the printer icon:

24. What is the purpose of the *.pc3* file?
    The *.stb* file?
    The *.ctb* file?

25. Can AutoCAD print on both side of the paper?
26. Explain the purpose of the **STYLESMANAGER** command.
27. Of the two lists of plot style names shown below, which are color-based names and which are named styles?

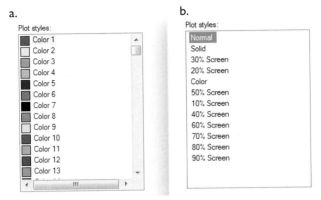

    a.

    b.

28. Describe two advantages of named plot styles over color-based ones:

    a.

    b.

29. Can named plot styles be applied to layers?

    Layouts?

    Views?

30. Explain the purpose of the **SCALELISTEDIT** command.
31. Describe how to create a PDF file of a drawing.
32. Why can 3D DWF files be created only from model tab?
33. If a plotted drawing does not fit the paper, what can you do to correct the situation?

    a.

    b.

    c.

34. What might be some reasons for plots failing to show lineweights?

    a.

    b.

    c.

# VII

## Appendices

# APPENDIX A

# AutoCAD, Computers, and Windows

The Windows operating system is the foundation on which AutoCAD and other programs run. Windows loads automatically when you start the computer. (You might notice operating system messages and version numbers when your computer starts.)

AutoCAD 2011 operates on these dialects of Windows:

- XP with service pack 2
- Vista with SP1
- 7

You install a separate version of AutoCAD for 32- or 64-bit operating systems. The 64-bit version handles larger drawing files in memory.

## HARDWARE OF AN AUTOCAD SYSTEM

CAD systems run on computers; peripherals connect to computers, as discussed later in this chapter. From smallest to largest, the five primary categories of computer are as follows:

**Handheld** computers are the smallest of computers, fitting in your hand. Typically, data are input by writing with a stylus or your finger on the computer's screen; handwriting recognition software translates writing to text and graphics. Handheld computers can view and edit CAD drawings, although AutoCAD itself is not available.

**Netbook** computers can run AutoCAD, even though they are not supported by Autodesk. Because of their slower speed and lower graphics resolution, netbooks are suitable for 2D drawings only; 3D graphics overwhelm this class of portable computer.

**Slate** computers are like netbooks, but without the keyboard. Like handhelds, input is via stylus or finger. Like netbooks, their slow processors and limited resolution make them suitable only for viewing and simple editing of CAD drawings. At time of writing, it was not known if Autodesk will support this class of computer, such as Apple's iPad.

**Notebook** computers are the smallest practical computers for running AutoCAD. They are the size of a notebook (hence the name), making them transportable. Notebook computers usually consist of a case that contains everything, except the printer. Almost all current notebook computers are powerful enough to run AutoCAD.

**Desktop** computers are most commonly seen on (or under) desks. These computers usually consist of a case that contains slots for additional adapters, such as for high-end video. The monitor, keyboard, mouse, and output devices (such as printers) are all external to the case.

**Workstations** are larger, faster, and more expensive than desktop computers. They typically contain two or more fast processors, much more memory, and much larger storage space. This class of computer is meant for running larger and more sophisticated programs than personal computers typically handle. For example, the special effects of Hollywood movies are created on workstations; some specific brands of CAD software run best on workstations.

**Mainframe** computers are the largest and most powerful of computers. They are meant for processing huge amounts of data, and are used by governments and companies to process data, such as in making weather forecasts and processing credit card transactions. These computers are not used for CAD.

## COMPONENTS OF COMPUTERS

Computers are made up of several essential parts. Let's look at some of them.

### System Board

The system board (sometimes call the "motherboard") is a fiberglass board that holds most of the computer's chips and expansion slots. The CPU (central processing unit), RAM memory chips, ROM (read-only memory), drive controllers, and other parts are mounted on this board.

In addition, the board contains slots into which expansion boards can be mounted, such as video capture cards and drive adapters. These are the types of expansion slots:

- **ISA** (industry standard architecture) is the oldest type of expansion slot. The newest computers no longer have these slots.
- **PCI** (peripheral connect interface) is the most common slot in today's computers, found in both PC and Macintosh computers.
- **AGP** (advanced graphics port) is designed for high-speed data transfer between the CPU and graphics boards.
- **Dedicated Memory** slots are designed for adding more memory to desktop and notebook computers. They access memory extremely quickly.
- **Memory Slots** are external to computers and read memory cards used by digital cameras and other portable devices.

Internal expansion slots are becoming less important as computer designers integrate most functions into the motherboard, including sound, networking, graphics, and drive interfaces. At the same time, high-speed external interfaces allow you to attach external peripherals to computers, such as hard drives, scanners, and additional monitors.

USB, FireWire, and external memory slots have replaced internal expansion slots, making it possible to add peripherals without ever opening the computer case. USB (universal serial bus) has the advantage of being ubiquitous; it can even recharge portable electronic devices. FireWire has the advantage of providing faster data throughput than USB because it imposes no load on the computer's CPU. Memory cards placed in external memory slots can provide ReadyBoost memory to Windows, which helps speed up operations by reducing hard disk access.

Other external slots include the following:

- **PC Card** or **PCMCIA** (personal computer memory card interface adapter; also known in jest as "people can't memorize confusing industry acronyms") is an older interface designed for notebook computers. The adapter card looks like a thick credit card, and might contain a FireWire interface, wireless network interface, or even a tiny disk drive.

- **ExpressCard** is a smaller and faster replacement for the PC Card interface, and is provided in many notebook computers today. The slot comes in two widths, the rectangular ExpressCard/34 (34 mm wide) and the L-shaped ExpressCard/54 (54 mm wide).
- **SD** and **miniSD** (secure digital) are designed for handheld computers and notebook computers. These cards most often contain memory, but also can provide wireless networking, GPS (global positioning system), and Bluetooth communications capabilities.

### Central Processing Unit

The central processing unit (or CPU) is the "brain" of the computer. This is where the software is processed; and then the CPU sends instructions to the graphics board, printer, plotter, and other peripherals. With a few exceptions, all information passes through the central processing unit. To speed up calculations and access to memory, CPUs are usually packaged with cache memory, a very fast form of memory.

Physically, the central processor is a large computer chip mounted on the motherboard. Because of the heat generated by CPUs, it is common to have heat sinks (which look like a series of metal fins) and fans mounted on the chips. AMD and Intel manufacture the CPUs commonly used by AutoCAD.

CPUs are also located in other parts of the computer. The computer's graphics board has its own dedicated CPU for processing graphical images, called the "GPU" (graphics processing unit). Laser printers and networked hard drives incorporate CPUs to process data.

AutoCAD can make use of the graphics board's GPU for processing visual styles and materials; this is known as "hardware acceleration," because GPUs are typically faster than CPUs at these tasks.

### Memory

Computer memory can be divided into two categories: ROM (read-only memory) and RAM (random access memory). ROM memory is contained in preprogrammed chips mounted on the motherboard. ROM stores basic sets of commands for the computer, such as the instructions for starting the computer.

Random access memory (RAM) is what most people mean when they refer to "computer memory." It stores nearly all the information used by the computer while running; it is temporary, because all data in RAM are lost when the computer is turned off or when the operating system freezes up.

Software programs (such as AutoCAD) require a certain amount of RAM to run the program. This amount is given in gigabytes. AutoCAD 2011 needs at least 2GB, meaning two gigabytes of RAM. A gigabyte is 1,024MB; a megabyte is 1,024KB (kilobytes); a kilobyte is 1,024 bytes.

Hard drive manufacturers often round down these values to make their products appear to have larger capacities, by counting only 1 million bytes in a megabyte — a loss of 2.4%.

### Disk Drives

Disk drives store large amounts of data, such as application programs, the Windows operating system, drawings, and other documents. Disk drives are either *non-removable* (fixed) or *removable*.

In almost all cases, fixed drives are hard drives. Today's desktop hard drives typically hold 500GB to 2TB (terrabyte = 1,024GB) of data.

Removable drives include diskettes, USB data keys, CDs and DVDs, and even tape drives. The now-rarely-used 3-1/2" diskette holds 1.44MB, a relatively small amount of data; some computers support 2.88MB of data on these diskettes.

USB data keys hold 32MB to 32GB of flash memory, and connect to the USB ports of computers. They are small and expensive relative to other forms of data storage, but have the advantage of being portable and a form of memory that doesn't lose its data when not attached to a power source.

CD (short for "compact disc, read-only memory") and DVD ("digital versatile disc") discs are common for distributing software. They are the same discs that hold music and movies, but use a different format for coding data. These discs can be read, but not written to. This makes the data secure from accidental erasure and attack from computer viruses. Your copy of AutoCAD may have arrived on a DVD.

All CDs allow you to store up to 700MB of data. DVDs hold 4.7GB of data; dual-layer DVDs allow twice as much data to be recorded, while BluRay discs have a capacity of 27GB - 54GB. Two types of CDs and DVDs can be written to (called "burning discs"):

- **CD-R and DVD-R** discs can be written to, but not erased.
- **CD-RW and DVD-RW** discs can be written to and erased.

DVD drives can read CDs, but CD drives cannot read DVDs. Similarly, BluRay drives can read DVDs and CDs, but DVD and CD drives cannot read BluRay discs.

Tape drives were the original backup medium, because they were relatively cheap. They suffer, however, from slow access speed. Today's tape drives can store 40GB or more data on tapes similar to those used by 8mm digital video cameras.

### DISK CARE

All of your work is recorded to disk, so caring for disks is very important. Disks sometimes becomes damaged, through which you lose your work! Frequent backing-up (copying files to a second disk drive, or recordable DVD-R) minimizes the possibility of file loss. AutoCAD includes the ability to back up drawings automatically every few minutes to your computer's hard disk, or safer yet, to the drive on another computer.

Today's hard drives have self-parking heads that move to an area that does not have stored data; this helps protect disks from damage when moved. Some have sensors that automatically protect the head when a strong gravitational force is felt, such as that of a falling notebook computer.

When the hard drive is on, do not move or tilt the computer or the external drive. This is particularly important with notebook computers. A hard drive can be damaged by shock. If you move the computer or the drive, then move it gently. Research by Google has shown that heating and cooling causes hard drives to break down more quickly, so it is best to leave your computer's hard drives powered up all the time.

SSD drives are solid-state drives made of the same memory as used for USB data keys. These drives read data very quickly, and do not suffer damage, as do hard disks. There is some question, however, of how long the memory cells last; when one cell breaks down, the entire SSD drive no longer works.

CD-ROMs and DVDs are relatively sturdy forms of data storage, but should also be handled with care. Avoid touching either surface of the discs; handle them by their centers and edges. When not in use, keep the discs in cases or protective sleeves; these protect the surfaces from which the laser reads data. Write on the label side with soft-tip pens only; damaging the label with a ballpoint pen damages the data on the disc. Keep discs out of the sun and out of high humidity areas.

There is some controversy about the lifetime of recorded CDs and DVDs. Some claim they last just ten years before the aluminum inside the disc begins to fail; others claim discs can last 100 years or more. In any case, it makes sense to copy important data to new storage formats as they become common.

### Write-Protecting Data

Normally, disks can be both read and written to (new data saved on it). But you can write-protect some disks to preserve what is already stored on them. This allows the disk to be read by the computer, but not written to. Windows allows you to set read-only status for files, folders, and entire drives.

CD-ROMs and DVDs cannot be overwritten, so the data are always safe — except when damaged physically. Some hard drives include utility software that electronically locks the drive, preventing misuse.

Some drives can erase data from CD-R and DVD-R discs, while all DVD-RW and CD-RW discs can have data erased and added. SD memory cards have a slider switch to prevent data from being overridden.

## PERIPHERAL HARDWARE

Peripheral hardware consists of add-on devices that perform specific functions. A properly-equipped CAD station consists of the following peripheral devices.

### PLOTTERS

Plotters produce "hard" copies of your drawings on paper, vinyl sticker material, cloth, and other media. The two primary types of plotter today are inkjet plotters and laser printers. Some offices still have older pen plotters.

---

**AUTOCAD 2011**

**HARDWARE & OS REQUIREMENTS**

Windows XP SP2:

    1.6GHz CPU with SSE2*;
    2.6GHz dual core for 3D modeling

    2GB RAM or more

    1.8GB free disk space for installation;
    (DVD-ROM drive for installation)

    1024x768 16-bit graphics board

    Internet Explorer 7.0 or higher

Windows 7 and Vista SP1:

    3.0GHz CPU with SSE2*

    2GB RAM or more

    1.8GB free disk space for installation;
    (DVD-ROM drive for installation)

    1024x768 16-bit graphics board
    1280x1024 32-bit for 3D modeling

    Internet Explorer 7.0 or higher

(64-bit AutoCAD cannot be installed on 32-bit Windows operating systems.)

*) **SSE2** allows AutoCAD to use the CPU's hardware for certain math functions and vector transformations (short for "streaming single-instruction multiple-data extension 2").

---

### Inkjet Plotters

The most common plotter today is the inkjet plotter, which prints by sending an electrical current to the print head. Inside the print head, electricity heats up the ink, causing it to squirt out the end of dozens of nozzles, hitting the paper and soaking in. Typical print heads have 48 nozzles lined up vertically. Print heads in large-format inkjet plotters can have as many as 512 nozzles.

For color plots, there are at least four print heads: one each for cyan (light blue), magenta (pink), yellow, and black. By combining these four colors using a process called dithering, inkjet plotters seem to produce 256 or more colors. For higher quality output, specialty inkjet plotters use seven or more colors of ink.

Desktop inkjet printers produce plots of A (8.5" x 11") or B-size (11" x 17"), while floor models plot up to E-size (36" x 48"). Inkjet printers have a typical resolution of 360 dots per inch (dpi), but some are capable of 1440 dpi and even 2880 dpi. Higher resolutions are generated by squirting smaller droplets of ink, and placing the droplets closer together.

Inkjet plotters can use normal paper, but specially-coated paper results in a cleaner looking print with longer lasting and brighter colors.

When you need paper plots to last many years, make sure your inkjet printer uses pigment inks and archival-quality paper.

### Laser Printers

Laser printers create plots with a process similar to photocopying: using a laser beam, the printer "paints" the image of text and drawings on its internal drum. This creates an electrostatic charge on the drum. The drum rotates, picking up black toner particles that stick to the electrically-charged portions of the drum. The toner particles are transferred to the paper, which passes over a hot fuser to melt the toner onto the paper.

The laser-copy process produces sharp, clear prints. Although some laser printers have excellent graphics capabilities, they are usually restricted to a maximum of B-size. Resolution is typically 600 dpi, but some models boast 1200 dpi or higher.

## GRAPHICS BOARDS

Graphics boards convert the images generated by AutoCAD into what you see on the monitor. Just about any graphics board made in the last few years works with AutoCAD, but may not support all of its 3D capabilities.

Some of the realistic renderings and visual styles in AutoCAD require advanced capabilities in graphics boards — such as shadow casting and high quality transparency. Autodesk maintains a Web site that lists graphics boards it has tested: www.autodesk.com/autocad-graphicscard. Autodesk ranks each on three levels: Not Supported; Supported but not Recommended; and Supported and Recommended.

At the time of writing this book, only certain graphics boards from AMD ATI and nVidia are fully compatible with the graphic demands of AutoCAD's imaging. Boards from other vendors might not be supported.

You can test the graphics board in your computer by running the 3DCONFIG command, which determines its capabilities for the following:

**Smooth Display** — removes jagged effect from diagonal lines and curved edges through anti-aliasing.

**Gooch Hardware Shader** — softens the contrast between light and dark areas by substituting warm and cool colors.

**Per-pixel Lighting** — computes colors of individual pixels for smoother 3D objects and lighting.

**Full Shadow Display** — generates accurate shadows.

**Advanced Material Effects** — processes materials effects on the graphics board.

**Texture Compression** — compresses texture and image files to reduce memory usage on the graphics board.

**Enhanced 3D Performance** — uses the graphic board for displaying 3D objects.

**Enhanced 2D Precision** — removes artifacts that are sometimes displayed at extreme zoom levels.

The technical editor observes that AutoCAD tends to be very conservative in evaluating the graphics boards plugged into your computer. He has overridden the default values without problems. For truly problematic cases, AutoCAD has the **/nohardware** startup switch that overrides hardware acceleration.

### Monitors

Monitors display the work in progress. A monitor is sometimes referred to as a "display device," "LCD" (liquid crystal display), or "CRT" (cathode ray tube). CAD operators prefer monitors that measure 19 inches (diagonally) or larger, because they can see more of the drawing at a time.

The quality of the image on the monitor is determined by its resolution, which depends on the number of dots (pixels) displayed on the screen. The pixels make up the image. A typical 1600 x 900-resolution display contains 1,600 pixels horizontally and 924 pixels vertically.

Many professional CAD users prefer higher resolution displays, such as 2048x1152 or higher. The device driver, the operating system, the graphics board, and the monitor work together to support the high resolution.

A few graphics boards support two or even four monitors. The advantage to multi-monitor setups is that one monitor can display the drawing, while the second can contain AutoCAD's toolbars, palettes, and other user interface elements. The third and fourth monitors could be used to display other software, or simply spread out the AutoCAD window.

If your computer's graphics board supports only two monitors, you can add more using DisplayLink, an external graphics adapter that attaches to USB ports. The drawback to multiple monitors is the added cost and loss of desk space.

## INPUT DEVICES

Just as word processing programs require keyboards to input letters, numbers, and symbols, CAD programs require input devices to make and manipulate drawing elements. In addition to the keyboard, the most common input devices for CAD drawing are the mouse, digitizer, and scanner.

### Mouse

The *mouse* is an input device used for pointing and positioning the cursor. (The name "mouse" comes from its mouse-like appearance; cordless mice dispense with the tail, relying on a small transmitter to connect with the computer.) The mouse is used for command input and screen interaction; it cannot digitize drawings.

Mice usually sport three or more buttons, with one button disguised as a roller wheel. More buttons are better, because they can be used for additional functions; AutoCAD supports up to 15 buttons, each assigned to a different command. For example, the middle button can be made to issue a double-click, reducing by half the number of clicks. The roller wheel allows real-time zooms and pans in AutoCAD.

*Optical mice* use a laser to track movement over most surfaces, except glass table tops and other shiny surfaces. *Mechanical mice* have a ball under their housing. As the ball rolls, the mouse measures its motion, and then transmits the relative movement to the computer.

###  3D Mouse

AutoCAD 2011 supports the 3D mice from Logitech 3dconnexion. Once the driver is installed in Windows and the 3D mouse is attached to the computer, AutoCAD adds a 3D mouse icon to the navigation bar, from which you can determine how the 3D mouse interacts with the drawing.

The action of a 3D mouse can be tricky to learn, because the controller knob works six ways: pan left/right, pan up/down, zoom, tilt, spin, and roll. Once they become familiar with it, however, users can be much more productive than with just a 2D mouse.

### Digitizer

The digitizer is an electronic input device that transmits the absolute x and y location of the puck resting on its sensitized pad. Digitizers have a fine grid of wires sandwiched between glass layers. When the puck is moved across the pad, the absolute location is read and transmitted to the computer.

Digitizers can be used in two ways: as pointing devices to move the cursor around the screen (like a mouse), or as tracing devices to copy drawings into the computer at scale and in proper proportion. In "tablet" mode, the digitizing pad is calibrated to the actual absolute coordinates of the drawing. When used as a pointing device, the tablet is not calibrated.

Digitizers can be used with a stylus (similar in appearance to a pencil) or with a multi-button puck. Digitizers come in several sizes, ranging from a few square inches, to very large. Small pads can be used to digitize large drawings: moving the drawing on the pad and then recalibrating. This becomes annoying when you frequently work with large-scale drawings. In that case, the expense of a large-format digitizer (36" x 24" or larger) may be justified. Digitizers are now very rare, and used primarily for tracing existing maps.

### Scanners

Scanners are used to "read" paper drawings and maps into computers. The scanner works by shining a bright light at the paper, while a head moves across the paper, measuring the amount of reflected light. (The head is made of many CCDs, the same charge couple devices used by digital cameras). Bright areas are the paper, while dark areas are lines.

The scanner sends data to the computer in raster format, where they can be imported by AutoCAD with the IMAGEATTACH command, or else converted to vector files by specialized software. Scanners are available in a variety of sizes, ranging from very small units that read business cards to E-size scanners that read large engineering drawings.

## DRIVES AND FILES

Windows names the computer's disk drives with letters of the alphabet, followed by the colon symbol ( : ), such as A: and C:.

> **A:** and **B:** are reserved for two floppy disk drives, even when your computer has no floppy drives.
> **C:** almost always designates the first hard drive.
> **D:** and subsequent letters are for all other drives, including additional hard drives, network drives, CD and DVD drives, and USB, and other removable drives.

Computers can access *network drives* over networks. Computers must be connected with a network cable or a wireless network connection to share drawings and other files. The names of network drives are usually prefixed by a double backslash and their computer's name, such as \\**Kat**\**C**. The networked version of AutoCAD allows a single copy to be used by two or more operators, depending on the terms of the network license.

*Removable drives* are any disk drives from which the medium can be removed. When reading or writing using removable disks, you must be careful that the correct disk is inserted. (Removable drives are sometimes called "backup drives.")

You must also be careful when removing the drives, since Windows might not be finished writing data to it. Use the **Safely Remove Hardware** command to ensure the drive is ready to be removed.

### FILE NAMES, EXTENSIONS, AND PATHS

Files can have names of up to 255 letters and characters. At the end of the file name you often find a dot ( . ) and a three-letter code, called the *file extension*. This denotes the type of file. For example, AutoCAD drawing files have the *.dwg* extension, while its template files use the *.dwt* extension.

Many programs, including AutoCAD, add file extensions to file names automatically, while others require you to specify the extension. The display of file extensions is normally turned off by Microsoft. Because it is useful to identify files by their extension, I recommend that you turn on the display, as follows:

1. In Windows Explorer, select **View | Folder Options**.
2. In the dialog box, choose the **View** tab.
3. Uncheck the **Hide extensions for known file types** option.

At the same time, I recommend that you turn on **Hidden Files** so that you can see AutoCAD support files, such as templates, which are stored in hidden folders.

The full file name includes the *path*, which specifies the names of the drive and folders in which the file is stored. For example, an AutoCAD drawing named *widget* stored on the C: drive in the \autocad\drawings folder is fully named by its path as:

   c:\autocad\drawings\widget.dwg

(Sometimes you might see file names that are just eight characters long and perhaps end in ~1. Older operating systems, such as Windows 3.1 and DOS, limited file names to eight characters. When a file name of nine or more characters is copied to a computer running one of these older operating systems, the file name is truncated (chopped off) to six characters, and ~1 is added. For example, *houseplan.dwg* becomes *housep~1.dwg*. Similar names are truncated to ~2, and so on.)

### Wild-Card Characters

In all file dialog boxes, as well as in some Windows functions and certain AutoCAD commands, you can use *wildcard characters* to specify groups of files. The two most common wildcard characters are the question mark (?) and the asterisk (*).

The **question mark** ( ? ) fills in for any *single* character, even if the character is nothing. By using one or more question marks, you set an upper limit to the number of characters. When you enter:

   car??

you refer to variations on "car...", such as:

   car     cars     carts     card2

But not "cardindex," because it has more than five characters.

The **asterisk** ( * ) represents *any* number of characters. For example, when you enter:

   *.dwg

you are referring to all files with the "....dwg" file extension, such as

   floorplan.dwg     my drawing 1.dwg     8th floor.dwg

Alternately, when you type

   floorplan.*

you refer to all files named "Floorplan..." regardless of their file extension, as well as to those with no extension:

   floorplan.dwg     floorplan.bak     floorplan.dwf     floorplan.

### DISPLAYING FILES

Windows provides several ways to display and manipulate files on your computer.

### Windows Explorer

Windows Explorer is the "official" method to view files in Windows. Not to be confused with Internet Explorer, Windows Explorer lets you view files in all folders on all drives, including on computers networked to yours.

### File-Related Dialog Boxes

Almost all of AutoCAD's file-related dialog boxes — such as those opened by the OPEN, SAVEAS, and IMPORT commands — let you view and manipulate files. This is a handy shortcut that obviates the need to switch to Explorer.

This makes a lot of sense, because you're often saving and opening file from inside AutoCAD. File-related dialog boxes are displayed when you use the OPEN, SAVEAS, EXPORT, and other commands that involve files.

AutoCAD's File dialog box provides the following functions:

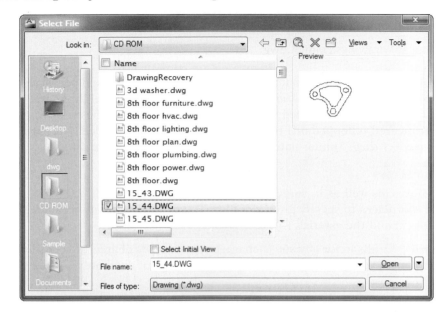

**Look In** — selects a different folder by several methods: (1) the **Look in** list box lets you move directly to another folder, (2) single-click the **Up One Level** icon to move up the folder tree one folder at a time; double-click the folder to move down into it, or (3) click the **Back** button to return to the previous folder.

**Create a New Folder** — creates a new folder in the current folder. As an alternative, press ALT+5.

**Delete** — erases the selected file(s); you can also press the DEL key. If you delete a file by accident, you can go to the Recycle Bin and retrieve the file.

**Views** — lists files in different views, such as icons or with details (size, type, and modified date-time). Click the column headers to sort in alphabetical order or by size, date, and so on. Click a second time to sort in reverse order.

**Open** — select a file name, and then click **Open.** Alternatively, you can double-click the file name to load it without choosing Open.

**Places List** — provides convenient access to frequently used folders. Drag a folder from the File list to the Places list to add more locations.

Right-click a file or folder to bring up a context-sensitive menu that performs additional functions, such as these:

**Rename** — renames files. You can also click twice (slowly) on the file name, and then type a new name.

**Cut** *or* **Copy** — moves or copies files. You can also drag the file to another folder. Hold down the CTRL key to copy the file; hold down the SHIFT key to move a file.

## DOS Session

The oldest method is to start a DOS session, and then type in commands — such as the DIR, TREE, COPY, MOVE, and DELETE commands — together with parameters, such as /s, *.*, and a:.

This approach appeals to power users, because Windows Explorer cannot perform some file-related tasks, such as renaming all files in a folder (**ren** *filename1.ext filename2.ext*), or copying files to devices (**copy** *filename.ext* **lpt:**).

## AUTOCAD FILE EXTENSIONS

AutoCAD uses many types of files, the contents of which are described by the file's extension. For example, linetypes are stored in files that end with .lin. Extensions used by AutoCAD are listed here.

### Drawing Files

| Extension | Description |
|---|---|
| .bak | Backup drawing files. |
| .dwf | Drawing Web format files. |
| .dwg | AutoCAD drawing files. |
| .dws | CAD standards files. |
| .dwt | Drawing template files. |
| .dxb | AutoCAD binary drawing interchange files. |
| .dxf | AutoCAD drawing interchange files. |
| .dwfx | XPS compatible version of DWF. |

### Support Files

| Extension | Description |
|---|---|
| .$ac | Temporary files created by AutoCAD. |
| .aws | AutoCAD workspace files. |
| .blk | Block template files. |
| .cfg | Configuration files. |
| .chm | HTML format help files. |
| .cui | Customize User Interface files. |
| .cus | Custom dictionary files. |
| .dct | Dictionary files. |
| .dst | Sheet set data files. |
| .dsd | Drawing set description files. |
| .err | Error log files. |
| .fdc | Field catalog files. |
| .fmp | Font mapping files. |
| .hlp | Windows-format help files. |
| .ies | Illumination distribution data files. |
| .lin | Linetype definition files. |
| .lli | Landscape libraries. |
| .log | Log files created by the LOGFILEON command. |
| .mli | Rendering material library files. |
| .mln | Multiline library files. |
| .mnc | Compiled menu files (obsolete as of AutoCAD 2006). |
| .mnd | Uncompiled menu files containing macros (obsolete). |
| .mnl | AutoLISP routines used by AutoCAD menus (obsolete). |
| .mns | AutoCAD-generated menu source files (obsolete). |
| .mnu | Menu source files (obsolete as of AutoCAD 2006). |
| .msg | Message files. |
| .pat | Hatch pattern definition files. |
| .pgp | Program parameters files (external commands and aliases). |
| .scr | Script files. |
| .shp | Shape and font definition files. |
| .shx | Compiled shape and Autodesk font files. |
| .ttf | Microsoft font files. |
| .xpg | Xml-format tool palette group files. |

## AutoCAD Program Files

| Extension | Description |
|---|---|
| .arx | ObjectARx (AutoCAD Runtime eXtension) program files. |
| .dll | Dynamic link library files. |
| .exe | Executable files, such as AutoCAD itself. |

## Plotting Support Files

| Extension | Description |
|---|---|
| .ctb | Color-table based plot parameter files. |
| .pcp | Plot configuration parameters files for AutoCAD Release 14. |
| .pc2 | Plot configuration parameters files for AutoCAD 2000. |
| .pc3 | Plot configuration parameters files since AutoCAD 2000i. |
| .plt | Plot files. |
| .stb | Style-table based plot parameter files. |

## Import-Export Files

| Extension | Description |
|---|---|
| .3ds | 3D Studio files. |
| .bmp | Windows raster files (device-independent bitmap). |
| .cdf | Comma delimited files (created by EATTEXT). |
| .dgn | MicroStation V8 and V7 design files (created by DGNEXPORT). |
| .dxe | Data extraction files (created by DATAEXTRACTION). |
| .dxx | DXF files (created by ATTEXT). |
| .kml | Google Earth files (keyhole markup language). |
| .kmx | Compressed DML files (read by GEOGRAPHICLOCATION). |
| .pcx | Raster format file. |
| .png | Portable Network Graphics raster files. |
| .sat | ACIS solid object files (short for "Save As Text"). |
| .slb | Slide library files. |
| .sld | Slide files. |
| .stl | Solid object stereo-lithography files (solids modeling). |
| .tif | Raster format files (Tagged image file format). |
| .tga | Raster format files (Targa). |
| .txt | Space delimited files (created by EATTEXT). |
| .wmf | Windows metafile files. |
| .xls | Excel spreadsheet files (created by EATTEXT). |
| .xml | DesignXML format files. |

## Miscellaneous Files

| Extension | Description |
|---|---|
| .css | Cascading style sheet files. |
| .html | HTML files. |
| .htt | HTML template files. |
| .ini | Initialization files. |
| .js | JavaScript files. |
| .map | Map files displayed by GEOGRAPHICLOCATION. |
| .xls | XML style sheet. |
| .xml | Extended markup language files. |
| .xmx | External messages files. |

## LISP and ObjectARX Programming Files

| Extension | Description |
|---|---|
| .actm | Active macro source code files. |
| .cpp | ObjectARX source code files. |
| .dcl | Dialog control language descriptions of dialog box files. |
| .dvb | Visual Basic for Applications program files. |
| .def | ObjectARX definition files. |
| .fas | AutoLISP fast load programs. |
| .frm | VBA form definition files. |
| .h | ADS and ObjectARX function definitions files. |
| .lib | ObjectARX function library files. |
| .lsp | AutoLISP program files. |
| .mak | ObjectARX make files. |
| .rx | Lists of ObjectARX applications that load automatically. |
| .tlb | ActiveX Automation type library files. |
| .unt | Unit definition files. |
| .vlx | Compiled Visual LISP files. |

## REMOVING TEMPORARY FILES

While it is working, AutoCAD creates "temporary" files. When you exit AutoCAD, these temporary files are erased. If, however, your computer crashes, these files might be left on the hard disk. You can erase temporary files, but only when AutoCAD is not running. Never erase files with these extensions while AutoCAD is still running:

. $$$  .$a  .ac$  .dwk  .dwl  .ef$  .sv$  .swr

These files can be erased to free up disk space when AutoCAD is not running. Alternatively, the DRAW-INGRECOVERY and RECOVERALL commands help you select which of these files can be reused, following a crash.

A-980

# APPENDIX B

# AutoCAD Commands, Aliases, and Keyboard Shortcuts

## AUTOCAD COMMANDS

The following commands are available in AutoCAD. Commands prefixed with - (hyphen) display their prompts at the command line only and not in a dialog box. Shortcut keystrokes are indicated as CTRL+F4.

This icon indicates the 86 commands that are new in AutoCAD 2011.

*The Illustrated AutoCAD 2011 Quick Reference* by author Ralph Grabowski provides a complete reference guide to all of AutoCAD's commands, and is available through Autodesk Press.

| Command | Description |
|---|---|
| **A** | |
| About | Displays an AutoCAD information dialog box that includes version and serial numbers. |
| AcisIn | Imports ASCII-format ACIS files into the drawing, and then creates 3D solids, 2D regions, or body objects. |
| AcisOut | Exports AutoCAD 3D solids, 2D regions, or bodies as .sat ASCII-format ACIS files. |
| ActBasePoint | Inserts base points in action macros. |
| ActManager | Manages action macro files. |
| ActRecord | Starts recording user actions as a macro. |
| ActStop | Stops recording the macro, and then displays the Action Macro dialog box, where you fill in the macro name and other details. |
| -ActStop | Stops recording the macro, and then prompts you at the command line for the macro's name and other details. |
| ActUserInput | Adds user input to existing macros. You use this command while recording a macro. |
| ActUserMessage | Adds dialog-box messages to macros being recorded; displays the Insert User Message dialog box. You use this command while recording a macro. |
| -ActUserMessage | Adds dialog-box messages to macros being recorded; prompts you at the command line. You use this command while recording a macro. |
| AdcClose | CTRL+2: Closes the DesignCenter palette. |
| AdCenter | CTRL+2: Opens DesignCenter; manages AutoCAD content. |
| AdcNavigate | Directs DesignCenter to the file name, folder, or network path you specify. |

| Command | Description |
|---|---|
| AddSelected | Creates new objects based on the object type and properties of selected objects. |
| Adjust | Changes the fade, contrast, and other parameters of images and underlays; parameters vary according to the object type (a super command). |
| AecToAcad | Explodes AEC objects and adds them to a new drawing file. |
| Ai_EditCustFile | Edits custom files (undocumented). |
| Ai_Publish | Prompts you for an email and then asks to quit (undocumented). |
| Ai_Send_Feedback | Opens the default Web browser at Autodesk's Contact Us page (undocumented). |
| Align | Uses three pairs of 3D points to move and rotate (align) 3D objects. |
| AllPlay | Plays all show motion views in sequence, without prompting. Useful for running animations in front of an audience (undocumented). |
| AmeConvert | Converts drawings made with AME v2.0 and v2.1 into ACIS solid models. |
| AnalysisCurvature | Displays color curvature gradients on 3D models. |
| AnalysisDraft | Displays color spacing gradients between parts and molds. |
| AnalysisOptions | Changes display options for curvature, draft, and zebra analysis; displays a dialog box. |
| AnalysisZebra | Displays black and white surface continuity stripes on 3D models. |
| AniPath | Creates animations of drawings along paths. |
| AnnoReset | Resets the location of all scale representations of an annotative object to that of the current scale representation. |
| AnnoUpdate | Updates annotative objects when styles have changed in the drawing. |
| Antz | Controls the ViewCube orientation from the command line; displays all SteeringWheels at once (undocumented). |
| Aperture | Adjusts the size of the target box used with object snap. |
| AppLoad | Displays a dialog box that lets you list AutoLISP, Visual Basic, and ObjectARX program names for easy loading into AutoCAD. |
| Arc | Draws arcs by a variety of methods. |
| Archive | Archives sheet sets in DWG or DWF format. |
| Area | Computes the area and perimeter of polygonal shapes. |
| Array | Makes multiple copies of objects. |
| Arx | Loads and unloads ObjectARX programs. Also displays the names of ObjectARX program command names. |
| Attach | Inserts external references, images, and DWF, DWFx, DGN, and PDF underlays in the current drawing. Xrefs can use geographic data if both the drawing and xref specify longitude/latitude (a super command). |
| -Attach | Inserts external files through the command line (undocumented). |
| AttachURL | Attaches hyperlinks to objects and areas. |
| AttDef | Creates attribute definitions. |
| AttDisp | Controls whether attributes are displayed. |
| AttEdit | Edits attributes. |
| AttExt | Extracts attribute data from drawings, and writes them to files for use with other programs. |
| AttIpEdit | Launches the in-place attribute editor. |
| AttReDef | Assigns existing attributes to new blocks, and new attributes to existing blocks. |
| AttSync | Synchronizes changed attributes with all blocks. |
| Audit | Diagnoses and corrects errors in drawing files. |
| AutoConstrain | Constrains objects within a preset distance of each other automatically. |
| AutoPublish | Toggles the simultaneous saving of drawings in DWG and DWF formats. |

| Command | Description |
| --- | --- |

**B**

| | |
| --- | --- |
| **Background** | Accesses the Background tab of the View dialog box directly (undocumented). |
| **Base** | Specifies the origin for inserting one drawing into another. |
| **BAttMan** | Edits all aspects of attributes in a block; short for Block Attribute Manager. |
| **BEdit** | Switches to the block editor; makes the following commands available: |

  **BAction** — adds actions to parameters in dynamic block definitions.

  **BActionBar** — toggles display of action bars.

  **BActionSet** — associates actions with selected objects within dynamic blocks.

  **BActionTool** — adds actions to dynamic block definitions.

  **BAssociate** — associates actions with parameters in dynamic blocks.

  **BAttOrder** — controls the order in which attributes are displayed in dynamic blocks.

  **BAuthorPalette** — opens the Block Authoring palette in the Block Editor.

  **BAuthorPaletteClose** — closes the Block Authoring palette.

  **BClose** — closes the Block Editor.

  **BConstruction** — converts objects between geometry and construction geometry.

  **BCParameter** — applies constraint parameters to selected objects, or converts dimensional constraints to parameter constraints; BC is short for Block Constraint.

  **BCycleOrder** — changes the cycling order of grips within dynamic block references.

  **BGripSet** — creates, repositions, resets, and deletes grips associated with parameters.

  **BLookupTable** — creates and edits lookup tables associated with dynamic blocks.

  **BParameter** — adds parameters and grips to dynamic blocks.

  **BSave** — saves changes to dynamic blocks.

  **BSaveAs** — saves dynamic blocks by another name.

  **BTable** — defines variations of dynamic blocks; you can copy'n paste an Excel spreadsheet into the table.

  **BTestBlock** — displays a window within the Block Editor to test the dynamic block being edited.

  **BvHide** — hides objects that have been assigned the state of invisibility.

  **BvShow** — shows objects assigned invisibility.

  **BvState** — assigns visibility states to objects in dynamic blocks.
  **BwBlockAs** — saves the block as a *.dwg* file on disc (undocumented).

| | |
| --- | --- |
| **BESettings** | Displays the Block Editor Settings dialog box. |
| **BHatch** | *See Hatch command.* |
| **Blipmode** | Toggles display of marker blips. |
| **Block** | Creates symbols from groups of objects. |
| **BlockIcon** | Generates preview images for blocks created with AutoCAD Release 14 and earlier. |
| **BmpOut** | Exports selected objects from the current viewport to raster *.bmp* files. |
| **Boundary** | Draws closed boundary polylines. |
| **Box** | Creates 3D solid boxes and cubes. |
| **Break** | Erases parts of objects; breaks objects in two. |
| **BRep** | Removes construction history from 3D solid models. |
| **Browser** | Launches your computer's default Web browser with the URL you specify. |

| Command | Description |
|---|---|
| **C** | |
| Cal | Runs a geometry calculator that evaluates integer, real, and vector expressions. |
| Camera | Sets the camera and target locations. |
| Chamfer | Trims intersecting lines, connecting them with a chamfer. |
| ChamferEdge | Bevels the edges of 3D solids and surfaces interactively. |
| Change | Permits modification of an object's characteristics. |
| CheckStandards | Compares the settings of layers, linetypes, text styles, and dimension styles with those of another drawing. |
| ChProp | Changes properties (linetype, color, and so on) of objects. |
| ChSpace | Moves and scales objects from model to paper space. |
| Circle | Draws circles by a variety of methods. |
| ClassicImage | Replaces the IMAGE command dialog box. |
| ClassicLayer | Replaces the LAYER command for the modal layer dialog box, the one where you need to click OK before continuing to edit. |
| ClassicXref | Replaces the XREF command dialog box (undocumented). |
| CleanScreenOff | CTRL+0: Turns on toolbars, title bar, and window borders. |
| CleanScreenOn | CTRL+0: Maximizes drawing area by turning off toolbars, title bar, and other elements. |
| Clip | Crops external references, images, viewports, and DWF, DWFx, DGN, and PDF underlays to a specified boundary; can invert clips for all reference files (a super command). |
| Close | CTRL+F4: Closes the current drawing. |
| CloseAll | Closes all open drawings; keeps AutoCAD open. |
| Color | Sets new colors for subsequently-drawn objects. |
| CommandLine | CTRL+9: Displays the command line palette. |
| CommandLineHide | CTRL+9: Hides the command line palette. |
| Compile | Compiles shapes and *.shp* and *.pfb* font files. |
| Cone | Creates 3D solid cones. |
| ConstraintBar | Displays the available geometric constraints on objects; a toolbar-like element. |
| ConstraintSettings | Controls the display of geometric constraints on constraint bars through a dialog box. |
| +ConstraintSettings | Specifies which tab of the Constraint Settings dialog box to display (undocumented). |
| Convert | Converts 2D polylines and associative hatches in pre-AutoCAD Release 14 drawings to the "lightweight" format to save on memory and disk space. |
| ConvertCTB | Converts drawings from plot styles to color-based tables. |
| ConvertOldLights | Converts light blocks in drawings created by AutoCAD 2006 and older. |
| ConvertOldMaterials | Converts material definitions in drawings created by AutoCAD 2006 and older. |
| ConvertPStyles | Converts drawings from color-based tables to plot styles. |
| ConvToMesh | Converts 3D solids and surfaces into mesh objects. |
| ConvToNurbs | Converts 3D solids and surfaces into NURBS surfaces. |
| ConvToSolid | Converts closed polylines and circles to extruded 3D solids. |
| ConvToSurface | Converts 2D solids, regions, lines, polylines, arcs, and flat 3D faces to surfaces. |
| Copy | Makes one or more copies of objects. |
| CopyBase | CTRL+SHIFT+C: Copies objects with a specified base point. |
| CopyClip | CTRL+C: Copies selected objects to the Clipboard in several formats. |
| CopyHist | Copies Text window text to the Clipboard. |
| CopyLink | Copies all objects in the current viewport to the Clipboard in several formats. |
| CopyToLayer | Copies selected objects to another layer through a dialog box. |
| Cui | Opens the Customize User Interface dialog box. |

| Command | Description |
| --- | --- |
| CuiExport | Exports customization settings to *.cui* files on disk. |
| CuiImport | Imports customization settings from *.cui* files. |
| CuiLoad | Loads *.cui* files. |
| CuiUnload | Unloads *.cui* files. |
| CuiXAppend | Appends *.bmp*, *.dib*, and *.rle* image files to *.cuix* files (undocumented). |
| CuiXCreate | Creates new *.cuix* files, and then prompts you to add *.cui* and *.bmp* files (undocumented). |
| CuiXList | Lists root parts and images stored in *.cuix* files (undocumented). |
| Customize | Customizes tool palette groups. |
| CutClip | CTRL+X: Cuts selected objects from the drawing to the Clipboard in several formats. |
| CvAdd | Adds control vertices to NURBS surfaces and splines. |
| CvHide | Hides control vertices for all NURBS surfaces and curves. |
| CvRebuild | Rebuilds the shapes of NURBS surfaces and curves; displays a dialog box. |
| -CvRebuild | Rebuilds the shapes of NURBS surfaces and curves at the command line. |
| CvRemove | Removes control vertices from NURBS surfaces and curves. |
| CvShow | Shows control vertices of selected NURBS surfaces and curves. |
| Cylinder | Creates 3D cylinders. |

## D

| Command | Description |
| --- | --- |
| DataExtraction | Extracts attribute data and drawing information to spreadsheets or tables. |
| DataLink | Links tables with external spreadsheet files. |
| DataLinkUpdate | Updates linked tables. |
| DbcClose | CTRL+6: Closes the dbConnect Manager (undocumented). |
| DbConnect | CTRL+6: Connects objects in drawings to tables in external database files. |
| DbList | Provides information about all objects in drawings. |
| DcAligned | Constrains the distance between two points on objects. |
| DcAngular | Constrains the angle between objects. |
| DcConvert | Converts associative dimensions to dimensional constraints. |
| DcDiameter | Constrains the diameters of circular objects. |
| DcDisplay | Toggles the display of dynamic constraints. |
| DcForm | Toggles the creation of dimensional constraints between dynamic and annotational. |
| DcHorizontal | Constrains the horizontal distance between objects. |
| DcLinear | Constrains objects horizontally, vertically, or rotationally. |
| DcRadius | Constrains the radii of curved objects. |
| DcVertical | Constrains the vertical distance between two objects or two points on an object. |
| DdEdit | *See the TextEdit command.* |
| DdPType | Specifies the style and size of points. |
| DdVPoint | Sets 3D viewpoints. |
| Delay | Creates a delay between operations in a script file. |
| DelConstraint | Removes all geometric and dimensional constraints from a selection set of objects. |
| DetachURL | Removes hyperlinks from objects. |
| DgnAdjust | Adjusts the contrast of attached DGN files. |
| DgnAttach | Attaches MicroStation V7 and V8 2D DGN files as underlays. |
| DgnClip | Clips attached DGN files. |
| DgnExport | Exports the current drawing as a DGN file. |

| Command | Description |
|---|---|
| -DgnExport | Command line version of the equivalent DGNEXPORT command (undocumented). |
| DgnImport | Imports MicroStation V7 and V8 2D DGN files as AutoCAD entities. |
| -DgnImport | Provides a command line version of the equivalent DGNIMPORT command (undocumented). |
| DgnLayers | Toggles the display of layers (levels) in DGN files placed as underlays in AutoCAD. |
| DgnMapping | Displays the DGN Mapping Setups dialog box for matching layers, linetypes, lineweights, and colors between MicroStation and AutoCAD. |
| Dist | Computes the distance between two points. |
| DistantLight | Places distant lights in drawings. |
| Divide | Divides objects into an equal number of parts, and places specified blocks or point objects at the division points. |
| Donut | Constructs solid filled circles and doughnuts. |
| Dragmode | Toggles display of dragged objects. |
| DrawingRecovery | Opens the Drawing Recovery palette. |
| DrawingRecoveryHide | Closes the Drawing Recovery palette. |
| DrawOrder | Changes the order in which objects are displayed: selected objects and images are placed above or below other objects. |
| DSettings | Specifies drawing settings for snap, grid, polar, and object snap tracking. |
| DsViewer | Opens the Aerial View palette. |
| DView | Displays 3D views dynamically. |
| DwfAdjust | Changes the contrast, fade, and color of *.dwf* files. |
| DwfAttach | Attaches *.dwf* files as uneditable underlays with a dialog box. |
| DwfClip | Clips DWF underlays. |
| DwfFormat | Toggles the default format — DWF or DWFx. |
| DwfLayers | Toggles the display of layers of DWF underlays. |
| DwgProps | Sets and displays the properties of the current drawing. |
| DxbIn | Imports binary drawing interchange files. |

## DIMENSIONS

| Command | Description |
|---|---|
| Dim | Specifies semi-automatic dimensioning capabilities. |
| Dim1 | Executes a single AutoCAD Release 12-style dimension command. |
| DimAligned | Draws linear dimensions aligned to objects. |
| DimAngular | Draws angular dimensions. |
| DimArc | Draws arc dimensions along the circumference of arcs. |
| DimBaseline | Draws linear, angular, and ordinate dimensions that continue from baselines. |
| DimBreak | Breaks extension and dimension lines where they cross other objects. |
| DimCenter | Draws center marks on circles and arcs. |
| DimConstraint | Applies dimensional constraints to selected objects or points on objects. |
| DimContinue | Draws linear, angular, and ordinate dimensions that continue from the last dimension. |
| DimDiameter | Draws diameter dimensions on circles and arcs. |
| DimDisassociate | Removes associativity from dimensions. |
| DimEdit | Edits the text and extension lines of associative dimensions. |
| DimInspect | Adds inspection text to dimensions. |
| DimJogged | Draws radial dimensions with jogged leaders. |
| DimJogLine | Jogs linear dimension lines. |
| DimLinear | Draws linear dimensions. |
| DimOrdinate | Draws ordinate dimensions in the x and y directions. |

| Command | Description |
|---|---|
| DimOverride | Overrides current dimension variables to change the look of selected dimensions. |
| DimRadius | Draws radial dimensions for circles and arcs. |
| DimReassociate | Associates dimensions with objects. |
| DimRegen | Updates associative dimensions. |
| DimSpace | Spaces two or more linear dimensions evenly. |
| DimStyle | Creates, names, modifies, and applies named dimension styles. |
| DimTEdit | Moves and rotates text in dimensions. |

### E

| Command | Description |
|---|---|
| EAttEdit | Starts the enhanced attribute editor. |
| Edge | Changes the visibility of 3D face edges. |
| EdgeSurf | Draws edge-defined 3D meshes. |
| EditShot | Prompts for a view name at the command line, and then displays the Shot Properties tab of the New View / Shot Properties dialog box (undocumented). |
| EditTableCell | Prompts you to select a table, a cell range, and an operation (undocumented). |
| Elev | Sets current elevation and thickness. |
| Ellipse | Constructs ellipses and elliptical arcs. |
| Erase | Removes objects from drawings. |
| eTransmit | Packages the drawing and related files for transmission by email or courier. |
| ExAcReload | Reloads xrefs, and reports whether they have changed (undocumented). |
| Explode | Breaks down blocks into individual objects; reduces polylines to lines and arcs. |
| Export | Exports drawings in several file formats. |
| -Export | Exports drawings in DWF, DWFx, or PDF format from the command line (undocumented). |
| ExportDwf | Exports drawings as DWF files, and allows you to override individual page setups on a sheet-by-sheet basis through a dialog box. |
| ExportDwfx | Exports drawings as DWFx files, and allows you to override individual page setups on a sheet-by-sheet basis through a dialog box. |
| ExportLayout | Copies layouts to model space. |
| ExportPdf | Exports drawings as PDF files, and allows you to override individual page setups on a sheet-by-sheet basis through a dialog box. |
| ExportSettings | Exports drawings as DWF, DWFx, or PDF files, and allows you to set export options through an integrated dialog box. |
| Extend | Extends objects to meet boundary objects. |
| ExternalReferences | Displays the External References palette for attaching drawings, images, and .dwf files. |
| ExternalReferencesClose | Closes the External References palette; the longest command name in AutoCAD. |
| Extrude | Extrudes 2D closed objects into 3D solid objects. |

### F

| Command | Description |
|---|---|
| FbxExport | Creates .fbx files (Motionbuilder FilmBoX format) of selected objects through dialog box. |
| -FbxExport | Creates .fbx files of selected objects in the current drawing at the command line. |
| FbxImport | Imports .fbx files containing objects, lights, camera and materials through a dialog box. |
| -FbxImport | Imports .fbx files at the command line. |
| Field | Inserts updatable field text in drawings. |
| Fill | Toggles the display of solid fills. |
| Fillet | Connects two lines with an arc. |
| FilletEdge | Rounds and fillets the edges of solid objects interactively. |

| Command | Description |
| --- | --- |
| Filter | Creates selection sets of objects based on their properties. |
| Find | Finds and replaces text. |
| FlatShot | Creates 2D views of 3D models. |
| Fog | Invokes the RENDERENVIRONMENT command. |
| FreePoint | Places target-less point lights in drawings (undocumented). |
| FreeSpot | Places target-less spot lights in drawings. |
| FreeWeb | Places target-less web lights in drawings. |

### G

| Command | Description |
| --- | --- |
| GcCoincident | Constrains selected objects to points. |
| GcCollinear | Constrains selected objects to lie along the same line. |
| GcConcentric | Constrains two arcs, circles, or ellipses to the same center point. |
| GcEqual | Constrains selected objects to the same size or radius. |
| GcFix | Locks selected objects in one position. |
| GcHorizontal | Constrains selected objects to be horizontal. |
| GcParallel | Constrains selected objects to be parallel to each other. |
| GcPerpedicular | Constrains selected objects to be perpendicular to one another. |
| GcSmooth | Constrains splines to be contiguous (and maintain G2 continuity) with other splines, lines, arcs, and polylines. |
| GcSymmetric | Constrains selected objects to be symmetrically constrained about a line of symmetry. |
| GcTangent | Constrains two curves to maintain tangency to each other, or their extensions. |
| GcVertical | Constrains selected objects to be vertical. |
| GeographicLocation | Specifies the sun position through date, time, city, longitude, and latitude. |
| GeomConstraint | Applies geometric relationships between objects or points on objects. |
| GoToURL | Links to hyperlinks |
| Gradient | Opens the Hatch and Gradient dialog box at the Gradient tab. |
| GraphicsConfig | Sets options for 3D display performance through a dialog box; shortcut for the 3DCONFIG command's Manual Tune button. |
| -GraphicsConfig | Sets options for 3D display performance through the command line. |
| GraphScr | F2: Switches to the drawing window from the Text window. |
| Grid | F7 and CTRL+G: Displays the grid at a specified spacing. |
| Group | Creates named selection sets of objects. (CTRL+SHIFT+A toggles group selection style.) |

### H

| Command | Description |
| --- | --- |
| Hatch | Opens the Hatch and Gradient dialog box (formerly BHATCH.) |
| HatchEdit | Edits associative hatch patterns. |
| HatchGenerateBoundary | Creates objects from the boundaries of hatch patterns. |
| HatchSetBoundary | Creates new hatch boundaries from existing objects, and then fills it with a pattern selected from another object. |
| HatchSetOrigin | Relocates the origin of selected hatch patterns. |
| HatchToBack | Sets the draw order for all hatches in the drawing behind all other objects. |
| Helix | Draws 3D helix objects (spirals). |
| Help | ? and F1: Displays a list of AutoCAD commands with detailed information. |

| Command | Description |
| --- | --- |
| Hide | Removes hidden lines from the currently-displayed view. |
| HidePalettes | CTRL+SHIFT+H: Hides all open palettes. |
| HlSettings | Specifies properties for hidden line removal. |
| HyperLink | CTRL+K: Attaches hyperlinks to objects, or modifies existing hyperlinks. |
| HyperLinkOptions | Controls the visibility of the hyperlink cursor and the display of hyperlink tooltips. |

## I

| Command | Description |
| --- | --- |
| Id | Describes the position of a point in x, y, z coordinates. |
| Image | Activates the EXTERNALREFERENCES command. |
| -Image | Controls the insertion of raster images through the command line. |
| ImageAdjust | Controls the brightness, contrast, and fading of raster images. |
| ImageAttach | Attaches raster images to the current drawing. |
| ImageClip | Places rectangular or irregular clipping boundaries around images. |
| ImageQuality | Controls the display quality of images. |
| ImpliedFaceX | Extracts loops from 3D solids to create regions (undocumented). |
| Import | Imports a variety of file formats into drawings. |
| Imprint | Imprints 2D edges onto the faces of 3D models. |
| InitialSetup | Runs the Initial Setup wizard (undocumented). |
| Insert | Inserts blocks and other drawings into the current drawing. |
| InsertObj | Inserts objects generated by another Windows application. |
| Interfere | Determines the interference of two or more 3D solids; displays dialog box with options. |
| Intersect | Creates a 3D solid or 2D region from the intersection of two or more 3D solids or 2D regions. |
| IsolateObjects | Displays selected objects across layers; unselected objects are hidden. |
| Isoplane | F5 and CTRL+E: Switches to the next isoplane. |

## J

| Command | Description |
| --- | --- |
| Join | Unifies individual lines, arcs, splines, and elliptical arcs. |
| JpgOut | Exports views in JPEG format. |
| JustifyText | Changes the justification of text. |

## L

| Command | Description |
| --- | --- |
| Layer | Displays the Layer palette; creates and changes layers; toggles the state of layers; assigns linetypes, lineweights, plot styles, colors, and other properties to layers. |
| LayerClose | Closes the Layer palette. |
| LayerPalette | Displays the Layer palette. |

## LAYER

| Command | Description |
| --- | --- |
| LayCur | Changes the layer of selected objects to the current layer. |
| LayDel | Erases all objects from the specified layer; purges layer from drawing. |
| LayerP | Displays the previous layer state. |
| LayerPMode | Toggles the availability of the LAYERP command. |
| LayerState | Displays the Layer State dialog box for creating and editing layer states. |
| LayFrz | Freezes the layers of the selected objects. |
| LayIso | Turns off all layers except ones holding selected objects. |
| LayLck | Locks the layer of the selected object. |
| LayMch | Changes the layers of selected objects to that of a selected object. |

| Command | Description |
|---|---|
| LayMCur | Makes the selected object's layer current, like the AI_MOLC command. |
| LayMrg | Moves objects to another layer; purges the original layer from the drawing. |
| LayOff | Turns off the layer of the selected object. |
| LayOn | Turns on all layers. |
| LayThw | Thaws all layers. |
| LayTrans | Translates layer names from one space to another. |
| LayULk | Unlocks the layer of a selected object. |
| LayUnIso | Turns on layers that were turned off with the last LAYISO command. |
| LayVpi | Isolates the selected object's layer in the current viewport. |
| LayWalk | Displays objects on selected layers. |
| Layout | Creates a new layout and renames, copies, saves, or deletes existing layouts. |
| LayoutWizard | Designates page and plot settings for new layouts. |
| Leader | Draws leader dimensions. |
| Lengthen | Lengthens or shortens open objects. |
| Light | Creates, names, places, and deletes "lights" used by the RENDER command. |
| LightList | Edits light parameters in the Lights In Model palette. |
| LightListClose | Closes the Lights In Model palette. |
| Limits | Sets drawing boundaries. |
| Line | Draws straight line segments. |
| Linetype | Lists, creates, and modifies linetype definitions; loads them for use in drawings. |
| List | Displays database information for selected objects. |
| LiveSection | Activates section planes. |
| Load | Loads shape files into drawings. |
| Loft | Creates 3D surfaces and solids by lofting two or more curves. |
| LogFileOff | Closes the .log keyboard logging file. |
| LogFileOn | Writes the text of the 'Command:' prompt area to .log log files. |
| LtScale | Specifies the scale factor for linetypes. |
| LWeight | Sets the current lineweight, lineweight display options, and lineweight units. |

## M

| Command | Description |
|---|---|
| Markup | CTRL+7: Reviews and changes the status of marked-up .dwf files. |
| MarkupClose | CTRL+7: Closes the Markup Set Manager palette. |
| MassProp | Calculates and displays the mass properties of 3D solids and 2D regions. |
| MatBrowserClose | Closes the new Materials Browser palette. |
| MatBrowserOpen | Opens the new Materials Browser palette. |
| MatchCell | Copies the properties from one table cell to other cells. |
| MatchProp | Copies properties from one object to other objects. |
| MatEditorClose | Closes the new Materials Editor palette. |
| MatEditorOpen | Opens the new Materials Editor palette. |
| MaterialAssign | Attaches materials to objects. |
| MaterialAttach | Attaches materials to objects by layer name. |
| MaterialMap | Maps materials to selected objects. |
| Measure | Places points or blocks at specified distances along objects. |

| Command | Description |
| --- | --- |
| MeasureGeom | Measures the distance, radius, angle, area, and volume of selected objects or sequence of points. |
| Menu | Loads *.cui* and *.mnu* menu files of AutoCAD commands into the menu area (obsolete). |
| MenuLoad | Loads *.cui* and *.mnu* menu files of AutoCAD commands into the menu area (obsolete). |
| MenuUnload | Loads *.cui* and *.mnu* menu files of AutoCAD commands into the menu area (obsolete). |
| Mesh | Creates 3D mesh primitive objects: boxes, cones, cylinders, pyramids, spheres, wedges, and tori. |
| MeshCap | Creates mesh faces that connect open edges. |
| MeshCollapse | Merges vertices of mesh faces and edges. |
| MeshCrease | Sharpens the edges of selected mesh subobjects. |
| MeshExtrude | Extends mesh faces into 3D space. |
| MeshMerge | Merges adjacent mesh faces into a single face. |
| MeshOptions | Displays the Mesh Tessellation Options dialog box, which controls default settings for converting existing objects to mesh objects; previews converted objects. |
| MeshPrimitiveOptions | Displays the Mesh Primitive Options dialog box, which sets the tessellation defaults for primitive mesh objects created with the MESH command. |
| MeshRefine | Multiplies the number of faces in selected mesh objects or faces. |
| MeshSmooth | Converts objects into 3D mesh objects: 3D faces, legacy polygonal meshes, polyface meshes, regions, closed polylines, 3D solids, and 3D surfaces. |
| MeshSmoothLess | Decreases the level of smoothness for mesh objects by one level; minimum level is 0. |
| MeshSmoothMore | Increases the level of smoothness for mesh objects by one level. |
| MeshSpin | Rotates adjoining edges of triangular mesh faces. |
| MeshSplit | Splits a mesh face into two faces. |
| MeshUncrease | Removes creases from mesh objects. |
| MigrateMaterials | Converts legacy materials in Tool Palettes to new style of materials, and then adds these to the Materials Browser. |
| MInsert | Inserts arrays of blocks. |
| Mirror | Creates mirror images of objects. |
| Mirror3D | Creates mirror images of objects rotated about a plane. |
| MLeader | Creates multiline leaders. |
| MLeaderAlign | Aligns two or more mleaders. |
| MLeaderCollect | Collects two or more mleaders with block annotations into a single mleader with multiple annotations. |
| MLeaderEdit | Adds and removes leaders from mleaders. |
| MLeaderStyle | Specifies mleader styles. |
| MlEdit | Edits multilines. |
| MLine | Draws multiple parallel lines (up to 16). |
| MlStyle | Defines named mline styles, including color, linetype, and endcapping. |
| Model | Switches from layout tabs to Model tab. |
| Move | Moves objects. |
| MRedo | Redoes more than one undo operation. |
| MSlide | Creates *.sld* slide files of the current display. |
| MSpace | Switches to model space. |
| MtEdit | Edits mtext. |
| MText | Places formatted paragraph text inside a rectangular boundary. |
| Multiple | Repeats commands. |

| Command | Description |
|---|---|
| MView | Creates and manipulates viewports in paper space. |
| MvSetup | Sets up new drawings. |

### N

| Command | Description |
|---|---|
| NavBar | Provides access to navigation and orientation tools from a single interface. |
| NavSMotion | Displays the show motion interface. |
| NavSMotionClose | Closes the show motion interface. |
| NavSWheel | SHIFT+W: Displays the navigation steering wheel. |
| NavVCube | Toggles the display of the ViewCube. |
| NetLoad | Loads .Net files. |
| New | CTRL+N: Creates new drawings. |
| NewSheetSet | Creates a new sheet set. |
| NewShot | Creates named views with show motion options; displays the Shot Properties tab of the New View / Shot Properties dialog box. |
| NewView | Shortcut to the VIEW command's New View dialog box. |

### O

| Command | Description |
|---|---|
| ObjectScale | Adds and removes scale factors from annotative objects. |
| Offset | Constructs parallel copies of objects. |
| OleConvert | Converts OLE (object linking and embedding) objects to other formats (undocumented). |
| OleLinks | Controls objects linked to drawings. |
| OleOpen | Opens the source OLE file and application (undocumented). |
| OleReset | Resets the OLE object to its original form (undocumented). |
| OleScale | Displays the OLE Properties dialog box. |
| Oops | Restores objects erased by the previous command. |
| Open | CTRL+O: Opens existing drawings. |
| OpenDwfMarkup | Opens marked-up .dwf files. |
| OpenSheetset | Opens .dst sheet set files. |
| OptChProp | Changes the styles of objects at the command line: text, dimension, table, or multiline style (undocumented). |
| Options | Customizes AutoCAD's settings. |
| Ortho | F8 and CTRL+L: Forces lines to be drawn orthogonally or at an angle. |
| OSnap | F3 and CTRL+F: Locates geometric points of objects. |

### P

| Command | Description |
|---|---|
| PageSetup | Specifies the layout page, plotting device, paper size, and settings for new layouts. |
| Pan | Moves the view within the current viewport. |
| Parameters | Defines and edits parameters associated with dimensional constraints through a palette. |
| -Parameters | Controls parameters from the command line. |
| ParametersClose | Closes the Parameters Manager palette. |
| PartiaLoad | Loads additional geometry into partially-opened drawings. |
| -PartialOpen | Loads geometry from a selected view or layer into drawings. |
| PasteAsHyperlink | Pastes objects as hyperlinks. |
| PasteBlock | CTRL+SHIFT+V: Pastes copied block into drawings. |
| PasteClip | CTRL+V: Pastes objects from the Clipboard into the upper left corner of drawings. |
| PasteOrig | Pastes copied objects from the Clipboard into new drawings using the coordinates from the original. |

| Command | Description |
| --- | --- |
| PasteSpec | Controls the format of objects pasted from the Clipboard. |
| PcInWizard | Imports .pcp and .pc2 configuration files of plot settings. |
| PdfAdjust | Adjusts the fade, contrast, and monochrome settings of PDF underlays; can be replaced by the ADJUST super command. |
| PdfAttach | Inserts PDF files as an underlay into the current drawing; can be replaced by the ATTACH super command. |
| PdfClip | Crops the display of selected PDF underlays to a user-specified boundary; inverts clipping boundaries; can be replaced by the CLIP super command; |
| PdfLayers | Toggles the display of layers in PDF underlays; can be replaced by the ULAYERS super command. |
| PEdit | Edits polylines and polyface objects. |
| PFace | Constructs polygon meshes defined by the location of vertices. |
| Plan | Returns to the plan view of the current UCS. |
| PlaneSurf | Creates flat surface areas. |
| PLine | Creates connected lines, arcs, and splines of specified width. |
| Plot | CTRL+P: Plots drawing to printers and plotters. |
| PlotStamp | Adds information about the drawing to the edge of the plot. |
| PlotStyle | Sets plot styles for new objects, or assigns plot styles to selected objects. |
| PlotterManager | Launches the Add-a-Plotter wizard and the Plotter Configuration Editor. |
| PngOut | Exports views as PNG files (portable network graphics format). |
| Point | Draws points. |
| Pointcloud | Displays options for attaching indexed point cloud files through the command line. |
| PointcloudAttach | Inserts indexed point cloud files through a dialog box. |
| -PointcloudAttach | Inserts indexed point cloud files into the current drawing through the command line. |
| PointcloudIndex | Creates indexed point clouds files from PCG or ISD scan files. |
| PointLight | Places point lights in drawings. |
| Polygon | Draws regular polygons with a specified number of sides. |
| PolySolid | Converts lines, arcs, circles, and polylines into 3D solids with thickness. |
| PressPull | CTRL+SHIFT+E: Presses and pulls bounded areas. |
| Preview | Provides a Windows-like plot preview. |
| ProjectGeometry | Projects points, lines, or curves onto 3D solids and surfaces. |
| Properties | CTRL+1: Displays and changes the properties of existing objects. |
| PropertiesClose | CTRL+1: Closes the Properties palette. |
| PSetUpIn | Imports user-defined page setups into new drawing layouts. |
| PSpace | Switches to paper space (layout mode). |
| Publish | Plots one or more drawings as a drawing set, or exports them in DWF format. |
| PublishToWeb | Creates a Web page from one or more drawings in DWF, JPEG, or PNG formats. |
| Purge | Deletes unused blocks, layers, linetypes, and so on. |
| Pyramid | Creates 3D solid pyramids and cones. |

## Q

| Command | Description |
| --- | --- |
| QcClose | CTRL+8: Closes the Quick Calc palette. |
| QDim | Creates continuous dimensions quickly. |
| QLeader | Creates leaders and leader annotation quickly. |
| QNew | Starts new drawings based on template files. |

| Command | Description |
|---|---|
| QSave | CTRL+S: Saves drawings without requesting a file name. |
| QSelect | Creates selection sets based on filtering criteria. |
| QText | Redraws text as rectangles with the next regeneration. |
| QuickCalc | CTRL+8: Opens the Quick Calc palette. |
| QuickCui | Opens the Customize User Interface dialog box partially collapsed. |
| QuickProperties | CTRL+SHIFT+P: Toggles the display of the Quick Properties palette (undocumented). |
| Quit | ALT+F4 and CTRL+Q: Exits AutoCAD. |
| QvDrawing | Turns on the thumbnail strip of drawing and layout views. |
| QvDrawingClose | Closes the thumbnail strip. |
| QvLayout | Turns on the thumbnail strip of layout views. |
| QvLayoutClose | Closes the thumbnail strip. |

**R**

| Command | Description |
|---|---|
| Ray | Draws semi-infinite construction lines. |
| Recover | Attempts to recover corrupted or damaged files. |
| RecoverAll | Recovers and updates drawings and all attached reference files. |
| Rectang | Draws rectangles. |
| Redefine | Restores AutoCAD's definition of undefined commands. |
| Redo | CTRL+Y: Restores the operations changed by the previous UNDO command. |
| Redraw | Cleans up the display of the current viewport. |
| RedrawAll | Redraws all viewports. |
| RefClose | Saves or discards changes made during in-place editing of xrefs and blocks. |
| RefEdit | Selects references for editing. |
| RefSet | Adds and removes objects from a working set during in-place editing of references. |
| Regen | Regenerates the drawing in the current viewport. |
| RegenAll | Regenerates all viewports. |
| RegenAuto | Controls whether drawings are regenerated automatically. |
| Region | Creates 2D region objects from existing closed objects. |
| Reinit | Reinitializes the I/O ports, digitizer, display, plotter, and the *acad.pgp* file. |
| Rename | Renames blocks, linetypes, layers, text styles, views, and so on. |
| Render | Renders 3D objects. |
| RenderCrop | Renders a rectangular area of the drawing. |
| RenderEnvironment | Sets the fog options; displays the Render Environment dialog box. |
| RenderExposure | Changes the brightness and contrast of renderings. |
| -RenderOutputSize | Presets the size of the next rendering in pixels (undocumented). |
| RenderPresets | Specifies the parameters for renderings; displays the Render Presets Manager dialog box. |
| -RenderPresets | Prompts you for rendering settings at the command prompt. |
| RenderWin | Displays the Render window, which lists and displays the current and previous renderings and parameters. |
| ResetBlock | Resets dynamic blocks to their original definitions. |
| Resume | Continues playing a script file that had been interrupted by the ESC key. |
| RevCloud | Draws revision clouds. |
| Reverse | Reverses the order of vertices in lines, polylines, splines, and helices; the Reverse option is also added to the PEDIT command. |
| Revolve | Creates 3D solids by revolving 2D closed objects around an axis. |
| RevSurf | Draws revolved 3D meshes. |

| Command | Description |
|---|---|
| Ribbon | Displays the ribbon portion of the user interface. |
| +Ribbon | Opens tabs by menu group name and element id (undocumented). |
| RibbonClose | Closes the ribbon user interface |
| Rotate | Rotates objects about specified center points. |
| Rotate3D | Rotates objects about a 3D axis. |
| RPref | Sets preferences for renderings; displays the Render Settings palette. |
| RPrefClose | Closes the Render Settings palette. |
| RScript | Restarts scripts. |
| RuleSurf | Draws ruled 3D meshes. |

## S

| Command | Description |
|---|---|
| Save / SaveAs | CTRL+SHIFT+S: Saves the current drawing by a specified name. |
| +SaveAs | Specifies the format in which to save drawings through the command line: in the current version of DWG, DWT, DWS, or DXF, or in an earlier version as specified by a number (undocumented). |
| SaveImg | Saves the current rendering in bitmap, Targa, or TIFF formats. |
| Scale | Changes the size of objects equally in the x, y, z directions. |
| ScaleListEdit | Adds and removes scale factors used by PLOT and other commands. |
| ScaleText | Resizes text. |
| Script | Runs script files in AutoCAD. |
| Section | Creates 2D regions from 3D solids by intersecting the solid with a plane. |
| SectionPlane | Creates section objects independently of the 3D model. |
| SectionPlaneJog | Adds a jogged segment to a section object (replaces the JOGSECTION command). |
| SectionPlaneSettings | Sets display options for the selected section plane. |
| SectionPlaneToBlock | Saves selected section planes as 2D or 3D blocks (replaces the GENERATESECTION command). |
| SecurityOptions | Sets up passwords and digital signatures for drawings. |
| Seek | Opens a Web browser and displays the Autodesk Seek home page for finding blocks and drawings; new name for Content Search and AdCenter's DC Online. |
| Select | Preselects objects to be edited; CTRL+A selects all objects. |
| SelectSimilar | Adds similar objects to the selection set based on their properties. |
| SequencePlay | Plays all show motion views in a view category (undocumented). |
| SetByLayer | Resets properties to those specified by layers. |
| SetiDropHandler | Specifies how to treat i-drop objects when dragged into drawings from Web sites. |
| SetVar | Views and changes AutoCAD's system variables. |
| -ShadeMode | Shades 3D objects in the current viewport. |
| Shape | Places shapes from shape files into drawings. |
| ShareWithSeek | Uploads blocks and drawings to Autodesk's Seek Web site. |
| Sheetset | CTRL+4: Opens the Sheet Set Manager palette. |
| SheetsetHide | CTRL+4: Closes the Sheet Set Manager palette. |
| Shell | Runs other programs outside of AutoCAD. |
| ShowMat | Reports the material definition assigned to selected objects. |
| ShowPalettes | CTRL+SHIFT+H: Shows all palettes that were previously closed with the HIDEPALETTES command. |
| SigValidate | Displays digital signature information in drawings. |
| Sketch | Allows freehand sketching. |
| Slice | Slices 3D solids with a plane. |

| Command | Description |
|---|---|
| Snap | F9 and CTRL+B: Toggles snap mode on or off, changes the snap resolution, sets spacing for the X- and Y-axis, rotates the grid, and sets isometric mode. |
| SolDraw | Creates 2D profiles and sections of 3D solid models in viewports created with the SOLVIEW command. |
| Solid | Draws filled triangles and rectangles. |
| SolidEdit | Edits faces and edges of 3D solid objects. |
| SolProf | Creates profile images of 3D solid models. |
| SolView | Creates viewports in paper space of orthogonal multi- and sectional view drawings of 3D solid model. |
| SpaceTrans | Converts length values between model and paper space. |
| Spell | Checks the spelling of text in the drawing. |
| Sphere | Draws 3D spheres. |
| Spline | Draws NURBS (spline) curves. |
| SplinEdit | Edits splines. |
| Spotlight | Places spotlights in drawings. |
| Standards | Compares CAD standards between two drawings. |
| Status | Displays information about the current drawing. |
| StlOut | Exports 3D solids to *.stl* files, in ASCII or binary format, for use with stereolithography. |
| Stretch | Moves selected objects while keeping connections to other objects unchanged. |
| Style | Creates and modifies text styles. |
| StylesManager | Displays the Plot Style Manager. |
| Subtract | Creates new 3D solids and 2D regions by subtracting one set of objects from a second. |
| SunProperties | Sets the properties of the Sun light; opens the Sun Properties palette. |
| SunPropertiesClose | Closes the Sun Properties palette. |
| SurfBlend | Adds continuous blend surfaces between two existing surfaces. |
| SurfExtend | Lengthens surfaces by specified distances. |
| SurfFillet | Adds surface fillets between pairs of surfaces. |
| SurfNetwork | Creates surfaces between several curves defined by surfaces and edges of solids (in the U and V directions). |
| SurfOffset | Creates parallel surfaces at a specified distance from the original. |
| SurfPatch | Creates new surfaces by fitting caps over surface edges that form closed loops. |
| SurfSculpt | Creates 3D solids by trimming and combining surfaces that bound watertight areas. |
| SurfTrim | Trims at the intersections of surfaces and other geometry. |
| SurfUntrim | Replaces surface areas removed by the SURFTRIM command. |
| Sweep | Creates 3D solids and surfaces by sweeping 2D curves along paths. |
| SysWindows | CTRL+TAB: Controls the size and position of windows. |

## T

| Command | Description |
|---|---|
| Table | Inserts tables in drawings. |
| TablEdit | Edits the content of table cells. |
| TableExport | Exports tables as comma-separated text files. |
| TableStyle | Specifies table styles. |
| Tablet | Aligns digitizers with existing drawing coordinates; operates only when a digitizer is connected. |
| TabSurf | Draws tabulated 3D meshes. |
| TargetPoint | Places target point lights in drawings. |

| Command | Description |
|---|---|
| Text | Places text in drawings. |
| TextEdit | Edits dimensional constraints, dimensions, and text objects (formerly DDEDIT). |
| TextScr | F2: Displays the Text window. |
| TextToFront | Displays text and/or dimensions in front of all other objects. |
| Thicken | Creates 3D solids by thickening surfaces. |
| TifOut | Exports views as TIFF files (tagged image file format). |
| Time | Tracks time. |
| TInsert | Inserts blocks and drawings in table cells. |
| Tolerance | Selects tolerance symbols. |
| Toolbar | Opens the Customize User Interface dialog box (obsolete). |
| -Toolbar | Controls the display of toolboxes. |
| ToolPalettes | CTRL+3: Opens the Tool palette. |
| ToolPalettesClose | CTRL+3: Closes the Tool palette. |
| Torus | Draws doughnut-shaped 3D solids. |
| TpNavigate | Sets the default tab for the Tools palette. |
| Trace | Draws lines with width. |
| Transparency | Toggles the background of bi-level images between transparent and opaque. |
| TraySettings | Specifies options for commands operating from the tray (right end of status bar). |
| TreeStat | Displays information on the spatial index. |
| Trim | Trims objects by defining other objects as cutting edges. |

## U

| Command | Description |
|---|---|
| U | CTRL+Z: Undoes the effect of commands. |
| UCS | Creates and manipulates user-defined coordinate systems. |
| UCSicon | Controls the display of the UCS icon. |
| UcsMan | Manages user-defined coordinate systems. |
| ULayers | Controls the display of layers in DWF, DWFx, DGN, and PDF underlays; a super command. |
| Undefine | Disables commands. (Don't undefine the REDEFINE command!) |
| Undo | Undoes several commands in a single operation. |
| Union | Creates new 3D solids and 2D regions from two solids or regions. |
| UnisolateObjects | Displays objects hidden with the ISOLATEOBJECTS command. |
| Units | Selects the display format and precision of units and angles. |
| UpdateField | Updates the contents of field text. |
| UpdateThumbsNow | Updates the preview images in sheet sets. |

## V

*All VBA-related commands will be removed from AutoCAD 2012.*

| Command | Description |
|---|---|
| VbaIDE | ALT+F8: Launches the Visual Basic Editor. |
| VbaLoad | Loads VBA projects into AutoCAD. |
| VbaMan | Loads, unloads, saves, creates, embeds, and extracts VBA projects. |
| VbaRun | ALT+F11: Runs VBA macros. |
| VbaStmt | Executes VBA statements at the command prompt. |
| VbaUnload | Unloads global VBA projects. |
| View | Saves the display as a view; displays named views. |
| ViewGo | Displays the named view, along with the associated background setting, layer state, live section, UCS, and visual style. This is a shortcut to the -VIEW command's Restore option. |

| Command | Description |
| --- | --- |
| ViewPlay | Plays the animation associated with a named view. |
| ViewPlotDetails | Reports on successful and unsuccessful plots. |
| ViewRes | Controls the fast zoom mode and resolution for circle and arc regenerations. |
| VisualStyles | Defines and edits visual styles used for renderings; displays the Visual Styles Manager palette. |
| VisualStylesClose | Closes the Visual Styles Manager palette. |
| VLisp | Launches the Visual LISP interactive development environment. |
| VpClip | Clips viewport objects. |
| VpLayer | Controls the independent visibility of layers in viewports. |
| VpMax | Maximizes the current viewport (in layout mode). |
| VpMin | Restores the maximized viewport. |
| VPoint | Sets the viewpoint from which to view 3D drawings. |
| VPorts | CTRL+R: Sets the number and configuration of viewports. |
| VsCurrent | Applies a named visual style to the current viewport. |
| VSlide | Displays .sld slide files created with the MSLIDE command. |
| VsSave | Saves a visual style by name in model space only. |
| VtOptions | Specifies options for visual transitions during the ZOOM command. |

## W

| Command | Description |
| --- | --- |
| WalkFlySettings | Specifies the settings for walk and fly animations; displays the Walk and Fly Settings dialog box. |
| WBlock | Writes objects to drawing files. |
| WebLight | Places web lights in drawings; these use .ies files to define light distribution. |
| Wedge | Creates 3D solid wedges. |
| WelcomeScreen | Displays the Welcome Screen dialog box; requires WelcomeScreen to be installed (undocumented). |
| WhoHas | Displays ownership information for opened drawing files. |
| WipeOut | Creates blank areas in drawings. |
| WmfIn | Imports .wmf files into drawings as blocks. |
| WmfOpts | Controls how .wmf files are imported. |
| WmfOut | Exports drawings as .wmf files. |
| WorkSpace | Creates, modifies, and saves workspaces; sets the current workspace. |
| WsSave | Saves the current user interface as a workspace. |
| WsSettings | Specifies options for workspaces through a dialog box. |

## X

| Command | Description |
| --- | --- |
| XAttach | Attaches externally-referenced drawing files to the drawing. |
| XBind | Binds externally-referenced drawings; converts them to blocks. |
| XClip | Defines clipping boundaries; sets the front and back clipping planes. |
| XEdges | Creates wireframe geometry from the edges of 3D solids and surfaces, as well as 2D regions. |
| XLine | Draws infinite construction lines. |
| XOpen | Opens externally-referenced drawings in independent windows. |
| Xplode | Breaks compound objects into component objects, with user control. |
| -Xref | Places externally-referenced drawings into drawings. |

## Z

| Command | Description |
| --- | --- |
| Zoom | Increases and decreases the viewing size of drawings. |

| Command | Description |
| --- | --- |
| **3** | |
| 3D | Draws 3D objects of polygon meshes (boxes, cones, dishes, domes, meshes, pyramids, spheres, tori, and wedges). |
| 3dAlign | Functions similarly to the ALIGN command, but with different prompts. |
| 3dArray | Creates 3D arrays. |
| 3dClip | Switches to interactive 3D view, and opens the Adjust Clipping Planes palette. |
| 3dConfig | Configures the 3D graphics system in a dialog box. |
| 3dCOrbit | Switches to interactive 3D view, and sets objects into continuous motion. |
| 3dDistance | Switches to interactive 3D view, and makes objects appear closer or farther away. |
| 3dDWF | Exports drawings as 3D DWF files; displays the Export 3D DWF dialog box. |
| 3dDwfPublish | Exports drawings as 3D *.dwf* files (undocumented). |
| 3dEditBar | Reshapes, scales, and edits the tangency of NURBS surfaces. |
| 3dFace | Creates 3D faces. |
| 3dFly | Navigates 3D drawings in fly-through mode; displays the Position Locator palette. |
| 3dFOrbit | Displays an arc ball for real-time view changes in 3D. |
| 3dMesh | Draws 3D meshes. |
| 3dMove | Displays the move handle on selected objects, so that they can be moved in 3D space. |
| 3dOrbit | Controls the interactive 3D viewing. |
| 3dOrbitCtr | Sets the center of rotations for the 3DORBIT command. |
| 3dOsnap | F4: Sets 3D object snap modes in the Drafting Settings dialog box. |
| -3dOsnap | Sets 3D object snap modes at the command line. |
| 3dOrbitTransparent | Changes the 3D view in real time (undocumented). |
| 3dPan | Invokes interactive 3D view to drag the view horizontally and vertically. |
| 3dPanTransparent | Pans 3D views in real time (undocumented). |
| 3dPoly | Draws 3D polylines. |
| 3dPrint | Sends 3D models to 3D printing services in STL format. |
| 3dRotate | Displays the rotate handle on selected objects so that they can be rotated in 3D space. |
| 3dScale | Scales in 3D, including 3D mesh faces; has a shortcut menu; works with point filters. |
| 3dsIn | Imports 3D Studio geometry and rendering data. |
| 3dSwivel | Switches to interactive 3D view, simulating the effect of turning the camera. |
| 3dSwivelTransparent | Swivels the 3D viewpoint in real time (undocumented). |
| 3dWalk | Enters Walk mode; accesses the Walk Mode right-click menu. |
| 3dZoom | Switches to interactive 3D view to zoom in and out. |
| 3dZoomTransparent | Zooms the 3D view in real time; an undocumented transparent command. |

## COMMAND ALIASES

Aliases are shortened versions of command names. They are defined in the *acad.ppg* file or "hardwired" by AutoCAD. Those listed below in italics can be used only in the Block Editor. The icon indicates the aliases that are new to AutoCAD 2011.

| Command | Aliases |
|---|---|
| **A** | |
| ActRecord | arr |
| ActStop | ars |
| -ActStop | -ars |
| ActUserInput | aru |
| ActUserMessage | arm |
| -ActUserMessage | -arm |
| AdCenter | adc, dcenter, dc, content |
| Align | al |
| AllPlay | aplay |
| *AnalysisCurvature* | curvatureanalysis |
| *AnalysisDraftangle* | draftangleanalysis |
| *AnalysisZebra* | zebra |
| AppLoad | ap |
| Arc | a |
| Area | aa |
| Array | ar |
| -Array | -ar |
| AttDef | att, ddattdef |
| -AttDef | -att |
| AttEdit | ate, ddatte |
| -AttEdit | -ate, atte |
| AttExt | ddattext |
| AttIpEdit | ati |
| **B** | |
| BAction | ac |
| BClose | cc |
| BCParameter | cparam |
| BEdit | be |
| Block | b, bmake, bmod, acadblockdialog |
| -Block | -b |
| Boundary | bo, bpoly |
| -Boundary | -bo |
| BParameter | param |
| Break | br |
| BSave | bs |
| BvState | bvs |
| **C** | |
| Camera | cam |
| Chamfer | cha |
| Change | -ch |
| CheckStandards | chk |
| Circle | c |
| Color | col, colour, ddcolor |
| CommandLine | cli |
| ConstraintBar | cbar |
| ConstraintSettings | csettings |
| Copy | cp, co |
| CTableStyle | ct |
| Cui | toolbar, to, tbconfig |
| CuiLoad | menuload |
| CuiUnload | menuunload |
| *CvAdd* | insertcontrolpoint |
| *CvHide* | pointoff |
| *CvrRbuild* | rebuild |
| *CvRemove* | removecontrolpoint |
| *CvShow* | pointon |
| Cylinder | cyl |
| **D** | |
| DataExtraction | dx, eattext |
| DataLink | dl |
| DataLinkUpdate | dlu |
| DbConnect | dbc, ase, aad, aex, asq, ali, aro |
| DdGrips | gr |
| DdVpoint | vp |
| DelConstraint | dcon, delcon |
| Dist | di |
| Divide | div |
| Donut | doughnut, do |
| DrawingRecovery | drm |
| Draworder | dr |
| DSettings | ds, se, ddrmodes, osnap, os, ddosnap |

| Command | Aliases |
|---|---|
| DsViewer | av |
| DView | dv |

**DIMENSIONS**

| Command | Aliases |
|---|---|
| DimAligned | dal, dimali |
| DimAngular | dan, dimang |
| DimArc | dar |
| DimBaseline | dba, dimbase |
| DimCenter | dce |
| DimContinue | dco, dimcont |
| DimDiameter | ddi, dimdia |
| DimDisassociate | dda |
| DimEdit | dimed, ded |
| DimJogged | Jog |
| DimJogLine | djl |
| DimLinear | dli, dimlin, dimhorizontal, dimvertical, dimrotated |
| DimOrdinate | dor, dimord |
| DimOverride | dov, dimover |
| DimRadius | dra, dimrad |
| DimReassociate | dre |
| DimStyle | d, dimsty, dst, ddim |
| DimTEdit | dimted |

**E**

| Command | Aliases |
|---|---|
| EditShot | eshot |
| EditText | ed, ddedit |
| Ellipse | el |
| Erase | e |
| Explode | x |
| Export | exp |
| -Export | -qpub |
| ExportDwf | qdwf |
| ExportDwfx | qdwfx |
| ExportPdf | qpdf |
| -ExportToAutoCAD | aectoacad |
| Extend | ex |
| ExternalReferences | er, image, im, xref, xr |
| Extrude | ext |

**F**

| Command | Aliases |
|---|---|
| Fillet | f |
| Filter | fi |
| FlatShot | fshot |

**G**

| Command | Aliases |
|---|---|
| GeographicLocation | geo, north, northdir |
| GeomConstraint | gcon |
| Gradient | gd |
| Group | g |
| -Group | -g |

**H**

| Command | Aliases |
|---|---|
| Hatch | bhatch, h, bh |
| -Hatch | -h |
| HatchEdit | he |
| HatchToBack | hb |
| Hide | hi |
| HidePalettes | poff |

**I**

| Command | Aliases |
|---|---|
| -Image | -im |
| ImageAdjust | iad |
| ImageAttach | iat |
| ImageClip | icl |
| Import | imp |
| Insert | i, inserturl, ddinsert |
| -Insert | -i |
| Insertobj | io |
| Interfere | inf |
| Intersect | in |
| IsolateObjects | isolate |

**J**

| Command | Aliases |
|---|---|
| Join | j |

**L**

| Command | Aliases |
|---|---|
| Layer | la, ddlmodes |
| -Layer | -la |
| LayerState | las, lman |
| -Layout | lo |
| Leader | lead |
| Lengthen | len |
| Line | l |
| Linetype | ltype, lt, ddltype |
| -Linetype | -ltype, -lt |
| List | ls, li |
| LtScale | lts |
| Lweight | lineweight, lw |

| Command | Aliases | | Command | Aliases |
|---|---|---|---|---|
| **M** | | | **P** | |
| Markup | msm | | Pan | p |
| MatBrowserOpen | mat | | -Pan | -p |
| MatchProp | ma, painter | | Parameters | par |
| MaterialMap | setuv | | -Parameters | -par |
| Materials | mat, finish, rmat | | -PartialOpen | partialopen |
| Measure | me | | PasteSpec | pa |
| MeasureGeom | mea | | PEdit | pe |
| MeshCrease | crease | | PLine | pl |
| MeshRefine | refine | | Plot | print, dwfout |
| MeshSmooth | smooth, convtomesh | | PlotStamp | ddplotstamp |
| MeshSmoothLess | less | | Point | po |
| MeshSmoothMore | more | | Pointcloud | pc |
| MeshSplit | split | | Pointcloudattach | pcattach |
| MeshUncrease | uncrease | | Pointcloudindex | pcindex |
| Mirror | mi | | PointLight | freepoint |
| Mirror3D | 3dmirror | | Polygon | pol |
| MLeader | mld | | PolySolid | psolid |
| MLeaderAlign | mla | | Preview | pre |
| MLeaderCollect | mle | | Properties | props, pr, mo, ch, ddchprop, ddmodify |
| MLeaderEdit | mle | | PropertiesClose | prclose |
| MLeaderStyle | mls | | PSpace | ps |
| MLine | ml | | PublishToWeb | ptw |
| Move | m | | Purge | pu |
| MSpace | ms | | -Purge | -pu |
| MText | t, mt | | Pyramid | pyr |
| -MText | -t | | **Q** | |
| MView | mv | | QLeader | le |
| **N** | | | QuickCalc | qc |
| NavSMotion | motion | | QuickCui | qcui |
| NavSMotionClose | motioncls | | Quit | exit |
| NavSWheel | wheel | | QvDrawing | qvd |
| NavVCube | cube | | QvDrawingClose | qvdc |
| NewShot | nshot | | QvLayout | qvl |
| NewView | nview | | QvLayoutClose | qvlc |
| **O** | | | **R** | |
| Offset | o | | Rectang | rec, rectangle |
| Open | openurl, dxfin | | Redraw | r |
| Options | op, preferences, ddselect, ddgrips, gr | | RedrawAll | ra |
| -OSnap | -os, ddosnap | | Regen | re |
| | | | RegenAll | rea |
| | | | Region | reg |

| Command | Aliases |
|---|---|
| Rename | ren |
| -Rename | -ren |
| Render | rr |
| RenderCrop | rc |
| RenderEnvironment | fog |
| RenderPresets | rp, rfileopt |
| RenderWin | rw, rendscr |
| Revolve | rev |
| Ribbon | dashboard |
| RibbonClose | dashboardclose |
| Rotate | ro |
| RPref | rpr |

**S**

| Command | Aliases |
|---|---|
| Save | saveurl |
| SaveAs | dxfout |
| Scale | sc |
| Script | scr |
| Section | sec |
| SectionPlane | splane |
| SectionPlaneJog | jogsection |
| SectionPlaneToBlock | generatesection |
| SequencePlay | splay |
| SetVar | set |
| ShadeMode | sha |
| Sheetset | ssm |
| Shell | sh |
| ShowPalettes | pon |
| Slice | sl |
| Snap | sn |
| Solid | so |
| Spell | sp |
| Spline | spl |
| SplinEdit | spe |
| Standards | sta |
| Stretch | s |
| Style | st, ddstyle |
| Subtract | su |
| SurfBlend | blendsrf |
| SurfExtend | extendsrf |
| SurfFillet | filletsrf |
| SurfNetwork | networksrf |
| SurfOffset | offsetsrf |
| SurfPatch | patch |
| SurfSculpt | createsolid |

**T**

| Command | Aliases |
|---|---|
| Table | tb |
| TableStyle | ts |
| Tablet | ta |
| Text | dt, dtext |
| TextEdit | tedit |
| Thickness | th |
| Tilemode | ti, tm |
| Tolerance | tol |
| Toolbar | to |
| ToolPalettes | tp |
| Torus | tor |
| Trim | tr |

**U**

| Command | Aliases |
|---|---|
| Ucs | dducs |
| UcsMan | uc, dducs, dducsp |
| Union | uni |
| UnisolateObjects | unhide, unisolate |
| Units | un, ddunits |
| -Units | -un |

**V**

| Command | Aliases |
|---|---|
| View | v, ddview |
| -View | -v |
| ViewGo | vgo |
| ViewPlay | vplay |
| VisualStyles | vsm |
| -VisualStyles | -vsm |
| VLisp | vlide |
| VPoint | -vp |
| VPorts | viewports |
| VsCurrent | vs, shademode, sha |

**W**

| Command | Aliases |
|---|---|
| Wblock | w, acadwblockdialog |
| -Wblock | -w |
| Wedge | we |

**X**

| Command | Aliases |
|---|---|
| XAttach | xa |
| XBind | xb |
| -XBind | -xb |
| XClip | xc |
| XLine | xl |
| Xplode | xp |
| -XRef | -xr |

| Command | Aliases | Command | Aliases |
|---|---|---|---|
| **Z** | | 3dMove | 3m |
| Zoom | z | 3dOrbit | 3do, orbit |
| | | 3dPoly | 3p |
| **3** | | 3dPrint | 3dp, 3dplot, rapidprototype |
| 3dAlign | 3al | 3dRotate | 3r |
| 3dArray | 3a | 3dScale | 3s |
| 3dDwf | 3ddwfpublish | 3dWalk | 3dw, 3dnavigate |
| 3dFace | 3f | | |

## KEYBOARD SHORTCUTS

Msot keyboard shortcuts are customized through the CUI command; some are hardcoded into Auto-CAD. The ⊕ icon indicates the shortcuts that are revised in AutoCAD 2011.

### FUNCTION KEYS

| Function Key | Meaning |
|---|---|
| F1 | Calls up the help window. |
| ⊕ SHIFT+F1 | Selects no subobjects. |
| F2 | Toggles between the graphics and text windows. |
| ⊕ SHIFT+F2 | Selects vertex subobjects. |
| F3 | Toggles object snap on and off. |
| ⊕ SHIFT+F3 | Selects edge subobjects. |
| ⊕ F4 | Toggles 3D object snap modes on and off (toggles Tablet mode before AutoCAD 2011). |
| ⊕ SHIFT+F4 | Selects face subobjects. |
| CTRL+F4 | Closes the current drawing. |
| ALT+F4 | Exits AutoCAD. |
| F5 | Switches to the next isometric plane when in iso mode: left, top, and right. |
| ⊕ SHIFT+F5 | Toggles solid history. |
| F6 | Toggles dynamic UCS on and off. |
| F7 | Toggles grid display on and off. |
| F8 | Toggles ortho mode on and off. |
| ALT+F8 | Displays the Macros dialog box. |
| F9 | Toggles snap mode on and off; to override temporarily, hold down F9 while selecting objects. |
| F10 | Toggles polar tracking on and off. |
| F11 | Toggles object snap tracking on and off. |
| ALT+F11 | Starts Visual Basic for Applications editor. |
| F12 | Toggles dynamic input. |

### ALTERNATE KEYS

| Alt-key | Meaning |
|---|---|
| ALT | Accesses keystroke shortcuts on menus. |
| ALT+TAB | Switches to the next application. |

## CONTROL KEYS

| Ctrl-key | Meaning |
|---|---|
| CTRL | Unlocks locked toolbars and windows temporarily; selects a face when picking 3D solid models; (NEW 2011) cycles through editing modes while a polyline grip is hot (red). (NEW 2011) selects polyline segments. |
| CTRL+0 | Toggles clean screen. |
| CTRL+1 | Toggles the Properties palette. |
| CTRL+SHIFT+1 | Toggles inference of constraints |
| CTRL+2 | Toggles the AutoCAD DesignCenter palette. |
| CTRL+3 | Toggles the Tool palettes. |
| CTRL+4 | Toggles Sheetset Manager palette. |
| CTRL+6 | Launches dbConnect. |
| CTRL+7 | Toggles Markup Set Manager palette. |
| CTRL+8 | Toggles QuickCalc palette. |
| CTRL+9 | Toggles Command Line palette. |
| CTRL+A | Selects all objects in the current model or layout space. |
| CTRL+SHIFT+A | Toggles group selection mode. |
| CTRL+B | Turns snap mode on or off. |
| CTRL+C | Copies selected objects to the Clipboard. |
| CTRL+SHIFT+C | Copies selected objects with a base point to the Clipboard. |
| CTRL+D | Toggles dynamic UCS (formerly toggled coordinate mode on the status bar). |
| CTRL+E | Switches to the next isoplane. |
| CTRL+SHIFT+E | Starts the PRESSPULL command (formerly CTRL+ALT). |
| CTRL+F | Toggles object snap on and off. |
| CTRL+G | Turns the grid on and off. |
| CTRL+H | Toggles pickstyle mode. |
| CTRL+SHIFT+H | Toggles the visibility of all open palettes (HIDEPALETTES and SHOWPALETTES commands). |
| CTRL+I | Cycles through coordinate displays (CTRL+D in AutoCAD 2008 and earlier). |
| CTRL+K | Creates hyperlinks. |
| CTRL+L | Turns ortho mode on and off. |
| CTRL+N | Starts new drawings. |
| CTRL+O | Opens drawings. |
| CTRL+P | Prints the drawing. |
| CTRL+SHIFT+P | Toggles Quick Properties palette. |
| CTRL+Q | Quits AutoCAD. |
| CTRL+R | Switches to the next viewport. |
| CTRL+S | Saves the current drawing. |
| CTRL+SHIFT+S | Displays the Save Drawing As dialog box. |
| CTRL+T | Toggles tablet mode. |
| CTRL+V | Pastes from the Clipboard into the drawing or to the command prompt area. |
| CTRL+SHIFT+V | Pastes with an insertion point. |
| CTRL+X | Cuts selected objects to the Clipboard. |
| CTRL+Y | Performs the REDO command. |
| CTRL+Z | Performs the U command. |
| CTRL+TAB | Switches to the next drawing. |

## TEMPORARY OVERRIDE (SHIFT) KEYS

| Shift-key | Meaning |
|---|---|
| SHIFT | Toggles orthogonal mode; selects 3D faces, edges, and vertices; activates 3DORBIT command; switches between trim and extend modes during the EXTEND and TRIM commands. |
| SHIFT+A | Overrides object snap (osnap) mode. |
| SHIFT+C | Overrides CENter osnap. |
| SHIFT+D | Disables all snap modes and tracking. |
| SHIFT+E | Overrides ENDpoint osnap. |
| SHIFT+J | Overrides osnap tracking mode. |
| SHIFT+L | Disables all snap modes and tracking. |
| SHIFT+M | Overrides MIDpoint osnap. |
| SHIFT+P | Overrides ENDpoint osnap. |
| SHIFT+Q | Overrides osnap tracking mode. |
| SHIFT+S | Enables osnap enforcement. |
| SHIFT+V | Overrides MIDpoint osnap. |
| SHIFT+W | Toggles the navigation wheel. |
| SHIFT+X | Overrides polar mode. |
| SHIFT+' | Overrides osnap mode. |
| SHIFT+, | Overrides CENter osnap. |
| SHIFT+. | Overrides polar mode. |
| SHIFT+; | Enables osnap enforcement. |

## COMMAND LINE KEYSTROKES

These keystrokes are used in the command-prompt area and the Text window.

| Keystroke | Meaning |
|---|---|
| left arrow | Moves the cursor one character to the left. |
| right arrow | Moves the cursor one character to the right. |
| HOME | Moves the cursor to the beginning of the line of command text. |
| END | Moves the cursor to the end of the line. |
| DEL | Deletes the character to the right of the cursor. |
| BACKSPACE | Deletes the character to the left of the cursor. |
| INS | Switches between insert and typeover modes. |
| up arrow | Displays the previous line in the command history. |
| down arrow | Displays the next line in the command history. |
| PGUP | Displays the previous screen of command text. |
| PGDN | Displays the next screen of command text. |
| CTRL+V | Pastes text from the Clipboard into the command line. |
| ESC | Cancels the current command. |
| TAB | Cycles through command names. |

## MOUSE AND DIGITIZER BUTTONS

| Button # | Mouse Button | Meaning |
|---|---|---|
| ... | Wheel | Zooms or pans. |
| 1 | Left | Selects objects. |
| 3 | Center, wheel | Displays object snap menu (when MBUTTONPAN = 1); hold down wheel button to pan. |
| 2 | Right | Displays shortcut menus. |
| 4 | | Cancels command. |
| 5 | | Toggles snap mode. |
| 6 | | Toggles orthographic mode. |

| Button # | Mouse Button | Meaning |
|---|---|---|
| 7 | | Toggles grid display |
| 8 | | Toggles coordinate display. |
| 9 | | Switches to isometric plane. |
| 10 | | Toggles tablet mode. |
| SHIFT+1 | SHIFT+Left | Toggles cycle mode. |
| SHIFT+2 | SHIFT+Left | Displays object snap shortcut menu. |
| CTRL+2 | CTRL+Right | Displays object snap shortcut men |

## MTEXT EDITOR SHORTCUT KEYS

| Shortcut | Meaning |
|---|---|
| TAB | Moves cursor to the next tab stop, or creates the next indent level. |
| SHIFT+TAB | Outdents. |
| CTRL+A | Selects all text. |
| CTRL+B | **Boldfaces**. |
| CTRL+I | *Italicizes.* |
| CTRL+U | Underlines. |
| CTRL+C | Copies selected text to the Clipboard. |
| CTRL+X | Cuts text from the editor and sends it to the Clipboard. |
| CTRL+V | Pastes text from the Clipboard. |
| CTRL+Y | Redoes last undo. |
| CTRL+Z | Undoes last action. |
| CTRL+F | Inserts field text; displays Field dialog box. |
| CTRL+R | Finds and replaces text. |
| CTRL+SPACE | Removes formatting from selected text. |
| CTRL+SHIFT+SPACE | Inserts non-breaking space. |
| CTRL+SHIFT+U | Converts selected text to all UPPERCASE. |
| CTRL+SHIFT+L | Converts selected text to all lowercase. |

# APPENDIX C

# AutoCAD System Variables

AutoCAD stores information about its current state, the drawing and the operating system in over 700 system variables. Those variables allow you to change the state of AutoCAD, and help programmers — who often work with menu macros and AutoLISP — to determine the state of the AutoCAD system.

## CONVENTIONS

The following pages list all documented system variables, plus several more not documented by Autodesk. The listing uses the following conventions:

| | |
|---|---|
| **Bold** | System variable is documented by Autodesk in AutoCAD 2011. |
| ***Bold Italic*** | System variable is listed neither by the **SETVAR** command nor by Autodesk's documentation. |
| 🖼 | System variable must be accessed through the **SETVAR** command; otherwise, the equivalent command is executed. |
| **Default** | System variable has this default value, as set in the *acad.dwg* prototype drawing; other template files may have different default values. |
| (R/O) | System variable is read-only (cannot be changed by the user). |
| **Toggle** | System variable has one of two values: 0 (off, closed, disabled) or 1 (on, open, enabled). |
| **ACI color** | System variable takes on one of AutoCAD's 255 colors; ACI is short for "AutoCAD color index." |
| 🌀 | System variable is new to AutoCAD 2011. |

| Variable Name | Default Value | Meaning |
|---|---|---|
| _AuthorPalettePath | varies | Path to the customized Tool palette folder. |
| _PkSer (r/o) | varies | Software serial number, such as "117-69999999". |
| _Server (r/o) | 0 | Network authorization code. |
| _ToolPalettePath | varies | Path to the Tool palette folder. |
| _VerNum (r/o) | varies | Internal program build number, such as "Z.54.01". |

## A

| Variable Name | Default Value | Meaning |
|---|---|---|
| AcadLspAsDoc | 0 | Controls whether *acad.lsp* is loaded into:<br>  **0** The first drawing only.<br>  **1** Every drawing. |
| AcadPrefix (r/o) | varies | Specifies paths used by AutoCAD search in Options \| Files dialog box. |
| AcadVer (r/o) | "18.0" | Specifies the AutoCAD version number. |
| AcisOutVer | 70 | Controls the ACIS version number; values are 15, 16, 17, 18, 20, 21, 30, 40, or 70. |
| ActPath | "" | Specifies paths to action macro files. |
| ActRecorderState | 0 | Reports the status of the Action Recorder:<br>  **0** Inactive.<br>  **1** Actively recording a macro.<br>  **2** Actively playing back a macro. |
| ActRecPath | "C:\users\..." | Specifies current path to recorded action macros. |
| ActUi | 6 | Controls state of Active Macro panel on the ribbon:<br>  **0** No change.<br>  **1** Expands during playback.<br>  **2** Expands during recording.<br>  **4** Prompts for name and description when recording is complete. |
| AdcState (r/o) | 0 | Toggle: reports if DesignCenter is active. |
| AFlags | 0 | Controls the default attribute display mode:<br>  **0** No mode specified.<br>  **1** Invisible.<br>  **2** Constant.<br>  **4** Verify.<br>  **8** Preset.<br>  **16** Lock position in block.<br>  **32** Multiple-line attributes. |
| AngBase | 0 | Controls the direction of zero degrees relative to the UCS. |
| AngDir | 0 | Controls the rotation of positive angles:<br>  **0** Clockwise.<br>  **1** Counterclockwise. |
| AnnoAllVisible | 1 | Toggles display of annotative objects at the current scale:<br>  **0** Displays only objects matching VP scale.<br>  **1** Displays all annotative objects (default). |

| Variable Name | Default Value | Meaning |
|---|---|---|
| **AnnoAutoScale** | 4 | Updates annotative objects when the annotation scale is changed:<br>  **1** Adds the new annotation scale to annotative objects (except on layers that are off, frozen, locked, or have viewports set to freeze).<br>  **2** As above, but includes objects on locked layers (excludes objects on layers that are off, frozen, or have viewports set to freeze).<br>  **3** The opposite of 2: applies to annotative objects on all layers, except objects on locked layers.<br>  **4** Applies to all objects regardless of layer status.<br>  **-1** Same as 1, but turned off.<br>  **-2** Same as 2, but turned off.<br>  **-3** Same as 3, but turned off.<br>  **-4** Same as 4, but turned off. |
| **AnnotativeDwg** | 0 | Toggles whether the drawing acts like an annotative block when inserted into other drawings:<br>  **0** Non-annotative block.<br>  **1** Annotative block.<br>(This sysvar is read-only when drawing contains annotative text.) |
| **ApBox** | 0 | Toggle: displays the AutoSnap aperture box cursor. |
| **Aperture** | 10 | Controls the object snap aperture in pixels:<br>  **1** Minimum size.<br>  **50** Maximum size. |
| *AppFrameResources* (r/o) | *"pack://application:,,,/AcWindows; etc..."* | *Specifies resource files used by applications.* |
| **ApplyGlobalOpacities** | | Toggles all palettes between (0) opaque and (1) translucent. |
| **ApState** | 0 | Toggle: reports the state of the block Authoring Palette in the Block Editor. |
| **Area** (r/o) | 0.0 | Reports the area measured by the last Area command. |
| **AttDia** | 0 | Controls the user interface for entering attributes:<br>  **0** Command-line prompts.<br>  **1** Dialog box. |
| **AttIpe** | 0 | Toggles display of in-place attribute text editor's toolbar:<br>  **0** Reduced-function toolbar.<br>  **1** Full-function toolbar. |
| **AttMode** | 1 | Controls the display of attributes:<br>  **0** Off.<br>  **1** Normal.<br>  **2** On; displays invisible attributes. |
| **AttMulti** | 1 | Toggles creation of multi-line attributes:<br>  **0** Attributes can be single-line only.<br>  **1** Attributes can be multi-line. |
| **AttReq** | 1 | Toggles attribute values during insertion:<br>  **0** Uses default values.<br>  **1** Prompts user for values. |

| Variable Name | Default Value | Meaning |
|---|---|---|
| **AuditCtl** | 0 | Toggles creation of *.adt* audit log files:<br>  **0** File not created.<br>  **1** File created. |
| **AUnits** | 0 | Controls the type of angular units:<br>  **0** Decimal degrees.<br>  **1** Degrees-minutes-seconds.<br>  **2** Grads.<br>  **3** Radians.<br>  **4** Surveyor's units. |
| **AUPrec** | 0 | Specifies number of decimal places displayed by angles; range is 0 - 8. |
| **AutoDwfPublish** | 0 | Toggles automatic publishing of DWF Files when drawings are saved or closed (to be removed in a future release):<br>  **0** DWF are not published.<br>  **1** DWF are published. |
| **AutomaticPub** | 0 | Toggles automatic exporting of drawing as DWF and PDF files when drawings are saved or closed:<br>  **0** DWF and PDF files are not published.<br>  **1** DWF and PDF are published. |
| **AutoSnap** | 63 | Controls the AutoSnap display (sum):<br>  **0** Turns off all AutoSnap features.<br>  **1** Turns on marker.<br>  **2** Turns on SnapTip.<br>  **4** Turns on magnetic cursor.<br>  **8** Turns on polar tracking.<br>  **16** Turns on object snap tracking.<br>  **32** Turns on tooltips for polar tracking and object snap tracking. |
| **B** | | |
| **BackgroundPlot** | 2 | Controls background plotting and publishing (ignored during scripts):<br>  **0** Plot foreground; publish foreground.<br>  **1** Plot background; publish foreground.<br>  **2** Plot foreground; publish background.<br>  **3** Plot background; publish background. |
| **BackZ** | 0.0 | Controls the location of the back clipping plane offset from the target plane. |
| **BActionBarMode** | 1 | Toggles between action objects and action bars in the Block Editor:<br>  **0** Displays legacy action objects (compatible with AutoCAD 2009 and earlier).<br>  **1** Displays action bars. |
| **BActionColor** | "7" | Specifies ACI text color for actions in Block Editor. |
| **BConStatusMode** | 0 | Toggles the display of the constraint status. |

| Variable Name | Default Value | Meaning |
|---|---|---|
| **BDependencyHighlight** | 1 | Toggles highlighting of dependent objects when parameters, actions, or grips selected in Block Editor:<br>**0** Not highlighted.<br>**1** Highlighted. |
| **BGripObjColor** | "141" | Specifies ACI color of grips in Block Editor. |
| **BGripObjSize** | 8 | Controls the size of grips in Block Editor; range is 1 to 255. |
| *BgrdPlotTimeout* | *20* | *Controls the time out for failed background plots; ranges from 0 to 300 seconds.* |
| **BindType** | 0 | Controls how xref names are converted when being bound or edited:<br>**0** From *xref\|name* to *xref$0$name*.<br>**1** From *xref\|name* to *name*. |
| **BlipMode** | 0 | Toggles the display of blip marks. |
| **BlockEditLock** | 0 | Toggles the locking of dynamic blocks being edited:<br>**0** Unlocked.<br>**1** Locked. |
| **BlockEditor** (r/o) | 0 | Reports whether Block Editor is open. |
| **BlockTestWindow** (r/o) | 0 | Reports whether the Block Test window is open. |
| **BParameterColor** | "7" | Specifies ACI color of parameters in the Block Editor. |
| **BParameterFont** | "Simplex.shx" | Controls the font used for parameter and action text in the Block Editor. |
| **BParameterSize** | 12 | Controls the size of parameter text and features in Block Editor:<br>**1** Minimum.<br>**255** Maximum. |
| **BPTextHorizontal** | 1 | Toggles the alignment of text with dimension lines in Block Editor:<br>**0** Text is aligned with dimension lines.<br>**1** Text is horizontal. |
| **BTMarkDisplay** | 1 | Controls the display of value set markers:<br>**0** Unlocked.<br>**1** Locked. |
| **BVMode** | 0 | Controls the display of invisible objects in the Block Editor:<br>**0** Invisible.<br>**1** Visible and dimmed. |

## C

| Variable Name | Default Value | Meaning |
|---|---|---|
| **CalcInput** | 1 | Controls how formulas and global constants are evaluated in dialog boxes:<br>**0** Not evaluated.<br>**1** Evaluated after pressing Alt+Enter. |
| **CameraDisplay** | 0 | Toggles the display of camera glyphs. |
| **CameraHeight** | 0 | Specifies the default height of cameras. |
| **CAnnoScale** | "1:1" | Names the current annotative scale for the current viewport. |
| **CAnnoScaleValue**(r/o) | 1.0 | Reports the current annotative scale. |

| Variable Name | Default Value | Meaning |
|---|---|---|
| **CaptureThumbnails** | 1 | Determines how thumbnails are generated by NavSWheel's Restore option: |
| | |   **0**  No preview thumbnails generated. |
| | |   **1**  Generated when bracket is positioned over empty frames. |
| | |   **2**  Generated automatically after every view change. |
| **CBarTransparency** | 50 | Controls the transparency of constraint bars; range is 10 to 90. |
| **CConstraintForm** | 0 | Determines the type of new constraint applied to objects by the DimConstraint command: |
| | |   **0**  Applies dynamic constraints. |
| | |   **1**  Applies annotative constraints. |
| **CDate** (R/O) | varies | Specifies the current date and time in the format YyyyMmDd.HhMmSsDd, such as 20080503.18082328 |
| *CDynDisplayMode* | *0* | *Toggles display of dynamic user interface.* |
| **CeColor** | "BYLAYER" | Controls the current color. |
| **CeLtScale** | 1.0 | Controls the current linetype scaling factor. |
| **CeLType** | "BYLAYER" | Controls the current linetype. |
| **CeLWeight** | -1 | Controls the current lineweight in millimeters; valid values are 0, 5, 9, 13, 15, 18, 20, 25, 30, 35, 40, 50, 53, 60, 70, 80, 90, 100, 106, 120, 140, 158, 200, and 211, plus the following: |
| | |   **-1**  BYLAYER. |
| | |   **-2**  BYBLOCK. |
| | |   **-3**  Default, as defined by LwDefault. |
| **CenterMT** | 0 | Controls how corner grips stretch uncentered multiline text: |
| | |   **0**  Center grip moves in same direction; opposite grip stays in place. |
| | |   **1**  Center grip stays in place; side grips move in direction of stretch. |
| **CeTransparency** | ByLayer | Sets the level of translucency for new objects: |
| | | **ByLayer** Determined by the object's layer (default). |
| | | **ByBlock** Determined by the block in which the object resides. |
| | |   **0**  Fully opaque (not translucent). |
| | |   **1-90**  Range of translucency, as a percentage. |
| **ChamferA** | 0.0 | Specifies the current value of the first chamfer distance. |
| **ChamferB** | 0.0 | Specifies the current value of the second chamfer distance. |
| **ChamferC** | 0.0 | Specifies the current value of the chamfer length. |
| **ChamferD** | 0 | Specifies the current value of the chamfer angle. |
| **ChamMode** | 0 | Toggles the chamfer input mode: |
| | |   **0**  Chamfer by two lengths. |
| | |   **1**  Chamfer by length and angle. |
| *CipMode* | *0* | *Toggles Customer Involvement Program.* |
| **CircleRad** | 0.0 | Specifies the most-recent circle radius. |
| **ClassicKeys** | 0 | Toggles meaning of Ctrl+C: (0) copies objects, or (1) cancels commands. |
| **CLayer** | "0" | Specifies name of current layer. |
| **CleanScreenState** (R/O) | 0 | Toggle: reports whether clean screen mode is active. |
| **CliState** (R/O) | 1 | Reports the state of the command line palette. |

| Variable Name | Default Value | Meaning |
|---|---|---|
| **CMaterial** | "ByLayer" | Sets the name of the current material. |
| **CmdActive** (r/o) | 1 | Reports the type of command currently active (used by programs):<br>  **1** Regular command.<br>  **2** Transparent command.<br>  **4** Script file.<br>  **8** Dialog box.<br>  **16** Dynamic data exchange.<br>  **32** AutoLISP command.<br>  **64** ARX command. |
| **CmdDia** | 1 | Toggles QLeader's in-place text editor. |
| **CmdEcho** | 1 | Toggles AutoLISP command display:<br>  **0** No command echoing.<br>  **1** Command echoing. |
| **CmdInputHistoryMax** | 20 | Controls the maximum command input items stored; works with InputHistoryMode. |
| **CmdNames** (r/o) | *varies* | Reports name of the command currently active, such as "SETVAR". |
| **CMleaderStyle** | "Standard" | Reports name of current mleader style. |
| **CMLJust** | 0 | Controls the multiline justification mode:<br>  **0** Top.<br>  **1** Middle.<br>  **2** Bottom. |
| **CMLScale** | 1.0 | Controls the scale of overall multiline width:<br>  *-n* Flips offsets of multiline.<br>  **0** Collapses to single line.<br>  *n* Scales by a factor of *n*. |
| **CMLStyle** | "STANDARD" | Specifies the current multiline style name. |
| **Compass** | 0 | Toggles the display of the 3D compass. |
| **ConstraintBarDisplay** | 1 | Toggles the display of constraint bars. |
| **ConstraintBarMode** | 4031 | Determines the geometric constraints displayed by constraint bars:<br>  **1** Horizontal<br>  **2** Vertical<br>  **4** Perpendicular<br>  **8** Parallel<br>  **16** Tangent<br>  **32** Smooth<br>  **64** Coincident<br>  **128** Concentric<br>  **256** Colinear<br>  **512** Symmetric<br>  **1024** Equal<br>  **2048** Fix<br>  **4031** All displayed |

| Variable Name | Default Value | Meaning |
|---|---|---|
| ⊙ *ConstraintCursorDisplay* | 1 | Toggles the display of the constraint icons that appear next to the cursor: (0) off or (1) on. |
| ⊙ **ConstraintInfer** | 0 | Toggles the inference of geometric constraints as objects are drawn and edited: (0) not inferred or (1) inferred. |
| **ConstraintNameFormat** | 2 | Determines how constraint text is displayed:<br>  **0** By variable names.<br>  **1** By values.<br>  **2** By expressions (formulas). |
| **ConstraintRelax** | 0 | Toggles constraints between enforced and relaxed during editing. |
| **ConstraintSolveMode** | 1 | Determines how constraints behave when applied or edited:<br>  **0** ⊙ Geometry size not retained when constraints are applied or modified.<br>  **1** ⊙ Geometry size is retained, even when constraints are applied or modified. |
| **Coords** | 1 | Controls the coordinate display style:<br>  **0** Updated by screen picks.<br>  **1** Continuous display.<br>  **2** Polar display upon request. |
| **CopyMode** | 0 | Toggles whether the Copy command repeats itself:<br>  **0** Command repeats copying.<br>  **1** Command makes one copy, then exits. |
| **CPlotStyle** | "ByColor" | Specifies the current plot style; options for named plot styles are: ByLayer, ByBlock, Normal, and User Defined. |
| **CProfile** (r/o) | "<<Unnamed Profile>>" | Specifies the name of the current profile. |
| **CrossingAreaColor** | 100 | Specifies the ACI color of crossing rectangle. |
| **CShadow** | 0 | Specifies how shadows are cast by 3D objects (if graphics board is capable; see 3dConfig command):<br>  **0** Casts and receives shadows.<br>  **1** Casts shadows.<br>  **2** Receives shadows.<br>  **3** Ignores shadows. |
| **CTab** | "Model" | Specifies the name of the current tab. |
| **CTableStyle** | "Standard" | Specifies the name of the current table style name. |
| ⊙ **CullingObj** | 1 | Toggle: determines which 3D objects are highlighted during selection: (0) all or (1) topmost object only. |
| ⊙ **CullingObjSelection** | 0 | Toggle: hidden objects are selected when a selection window is dragged: (0) not culled so hidden objects are selected, or (1) culled so hidden objects are not selected. |
| **CursorSize** | 5 | Controls the cursor size as a percent of the viewport size.<br>  **1** Minimum size.<br>  **100** Full viewport. |
| **CVPort** | 2 | Specifies the current viewport number. |

| Variable Name | Default Value | Meaning |
|---|---|---|

## D

| Variable Name | Default Value | Meaning |
|---|---|---|
| **DataLinkNotify** | 2 | Controls reporting on data links:<br>  0  All notifications disabled.<br>  1  Displays data link icon in tray.<br>  2  Displays the icon and a warning balloon in the tray. |
| **Date** (r/o) | varies | Reports the current date in Julian format, such as 2448860.54043252. |
| **DbcState** (r/o) | 0 | Toggle: specifies whether dbConnect Manager is active. |
| **DblClkEdit** | 1 | Toggles editing by double-clicking objects. |
| **DBMod** (r/o) | 0 | Reports how the drawing has been modified:<br>  0  No modification since last save.<br>  1  Object database modified.<br>  2  Symbol table modified.<br>  4  Database variable modified.<br>  8  Window modified.<br>  16  View modified.<br>  32  Field modified. |
| **DctCust** | "sample.cus" | Specifies the name of custom spelling dictionary. |
| **DctMain** | "enu" | Controls the code for spelling dictionary:<br>  **enu**  American English<br>  **eng**  British English (ise)<br>  **enc**  Canadian English<br>  **cat**  Catalan<br>  **csy**  Czech<br>  **dan**  Danish<br>  **nld**  Dutch (primary)<br>  **fin**  Finnish<br>  **fra**  French (accented capitals)<br>  **frc**  French (unaccented capitals)<br>  **deu**  German (post-reform)<br>  **deo**  German (pre-reform)<br>  **ita**  Italian<br>  **nor**  Norwegian (Bokmal)<br>  **ptb**  Portuguese (Brazilian)<br>  **ptg**  Portuguese (Iberian)<br>  **rus**  Russian<br>  **esp**  Spanish<br>  **sve**  Swedish |
| **DefaultGizmo** | 0 | Determines which gizmo is the default when sub-objects are selected: 3D Move, 3D Rotate, 3D Scale, or none. |
| **DefaultLighting** | 1 | Toggles distant lighting. |
| **DefaultLightingType** | 0 | Toggles between new (1) and old (0) type of lights. |
| *DefaultViewCategory* | "" | *Specifies the default name for View Category in the View command's New View dialog box.* |
| **DefLPlStyle** | "ByColor" | Reports the default plot style for layer 0. |

| Variable Name | Default Value | Meaning |
|---|---|---|
| **DefPlStyle** | "ByColor" | Reports the default plot style for new objects. |
| **DelObj** | 1 | Toggles the deletion of source objects:<br>  **-2** Users are prompted whether to erase all defining objects.<br>  **-1** Users are prompted whether to erase profiles and cross sections.<br>  **0** Objects retained.<br>  **1** Profiles and cross sections erased.<br>  **2** All defining objects erased. |
| **DemandLoad** | 3 | Controls application loading when drawing contains proxy objects:<br>  **0** Applications not demand-loaded.<br>  **1** Apps loaded when drawing opened.<br>  **2** Apps loaded at first command.<br>  **3** Apps loaded when drawing is opened or at first command. |
| **DgnFrame** | 0 | Controls display of DGN underlay frame:<br>  **0** Does not display frame.<br>  **1** Displays and plots frame.<br>  **2** Displays but does not plot frame. |
| **DgnImportMax** | 1000000 | Limits maximum MicroStation elements to import into AutoCAD; 0 = no limit. |
| **DgnMappingPath** (r/o) | "C:\users\..." | Specifies path to folder holding the *dgnsetups.ini* file. |
| **DgnOsnap** | 1 | Toggles object snapping to DGN elements:<br>  **0** Does not osnap.<br>  **1** Osnaps. |
| **DiaStat** (r/o) | 0 | Reports whether user exited dialog box by clicking:<br>  **0** Cancel button.<br>  **1** OK button. |
| **Digitizer** (r/o) | 0 | Reports the type of digitizer attached to AutoCAD (read-only):<br>  **0** None.<br>  **1** Integrated touch.<br>  **2** External touch.<br>  **4** Integrated pen.<br>  **8** External pen.<br>  **16** Multiple input.<br>  **128** Input devices are ready. |
| **Dimension Variables** | | |
| **DimADec** | 0 | Controls angular dimension precision:<br>  **-1** Use DimDec setting (default).<br>  **0** Zero decimal places (minimum).<br>  **8** Eight decimal places (maximum). |
| **DimAlt** | Off | Toggles alternate units:<br>  **On** Enabled.<br>  **Off** Disabled. |
| **DimAltD** | 2 | Controls alternate unit decimal places. |
| **DimAltF** | 25.4 | Controls alternate unit scale factor. |

| Variable Name | Default Value | Meaning |
| --- | --- | --- |
| **DimAltRnd** | 0.0 | Controls rounding factor of alternate units. |
| **DimAltTD** | 2 | Controls decimal places of tolerance alternate units; range is 0 to 8. |
| **DimAltTZ** | 0 | Controls display of zeros in alternate tolerance units:<br>  **0**  Zeros not suppressed.<br>  **1**  All zeros suppressed.<br>  **2**  Includes 0 feet, but suppresses 0 inches.<br>  **3**  Includes 0 inches, but suppresses 0 feet.<br>  **4**  Suppresses leading zeros.<br>  **8**  Suppresses trailing zeros. |
| **DimAltU** | 2 | Controls display of alternate units:<br>  **1**  Scientific.<br>  **2**  Decimal.<br>  **3**  Engineering.<br>  **4**  Architectural; stacked.<br>  **5**  Fractional; stacked.<br>  **6**  Architectural.<br>  **7**  Fractional.<br>  **8**  Windows desktop units setting. |
| **DimAltZ** | 0 | Controls the display of zeros in alternate units:<br>  **0**  Suppresses 0 ft and 0 in.<br>  **1**  Includes 0 ft and 0 in.<br>  **2**  Includes 0 ft; suppresses 0 in.<br>  **3**  Suppresses 0 ft; includes 0 in.<br>  **4**  Suppresses leading 0 in decimal dimensions.<br>  **8**  Suppresses trailing 0 in decimal dimensions.<br>  **12**  Suppresses leading and trailing zeroes. |
| **DimAnno** (R/O) | 0 | Reports whether current dimstyle is:<br>  **0**  Not annotative.<br>  **1**  Annotative. |
| **DimAPost** | "" | Specifies the prefix and suffix for alternate text. |
| **DimArcSym** | 0 | Specifies the location of the arc symbol:<br>  **0**  Before dimension text.<br>  **1**  Above the dimension text.<br>  **2**  Not displayed. |
| *DimAso* | *On* | *Toggles associative dimensions:*<br>  ***On***  *Dimensions are created associative.*<br>  ***Off***  *Dimensions are not associative.* |
| **DimAssoc** | 2 | Controls how dimensions are created:<br>  **0**  Dimension elements exploded.<br>  **1**  Single dimension object, attached to defpoints.<br>  **2**  Single dimension object, attached to geometric objects. |
| **DimASz** | 0.18 | Controls the default arrowhead length. |

| Variable Name | Default Value | Meaning |
|---|---|---|
| **DimAtFit** | 3 | Controls how text and arrows are fitted when there is insufficient space between extension lines (leader is added when DimTMove = 1):<br>**0** Text and arrows outside extension lines.<br>**1** Arrows first outside, then text.<br>**2** Text first outside, then arrows.<br>**3** Either text or arrows, whichever fits better. |
| **DimAUnit** | 0 | Controls the format of angular dimensions:<br>**0** Decimal degrees.<br>**1** Degrees.Minutes.Seconds.<br>**2** Grads.<br>**3** Radians.<br>**4** Surveyor units. |
| **DimAZin** | 0 | Controls the display of zeros in angular dimensions:<br>**0** Displays all leading and trailing zeros.<br>**1** Suppresses 0 in front of decimal.<br>**2** Suppresses trailing zeros behind decimal.<br>**3** Suppresses zeros in front and behind the decimal. |
| **DimBlk**<br>  Closed filled<br>  Closed blank<br>  Closed<br>  Dot<br>  Architectural tick<br>  Oblique<br>  Open<br>  Origin indicator<br>  Origin indicator 2<br>  Right angle<br>  Open 30<br>  Dot small<br>  Dot blank<br>  Dot small blank<br>  Box<br>  Box filled<br>  Datum triangle<br>  Datum triangle filled<br>  Integral<br>  None | "" | Specifies the name of the arrowhead block:<br>Architectural tick: "Archtick"<br>Box filled: "Boxfilled"<br>Box: "Boxblank"<br>Closed blank: "Closedblank"<br>Closed filled: "" (default)<br>Closed: "Closed"<br>Datum triangle filled: "Datumfilled"<br>Datum triangle: "Datumblank"<br>Dot blanked: "Dotblank"<br>Dot small: "Dotsmall"<br>Dot: "Dot"<br>Integral: "Integral"<br>None: "None"<br>Oblique: "Oblique"<br>Open 30: "Open30"<br>Open: "Open"<br>Origin indication: "Origin"<br>Right-angle: "Open90" |
| **DimBlk1** | "" | Specifies the name of first arrowhead's block; uses same list of names as under DimBlk.<br>**.** No arrowhead. |
| **DimBlk2** | "" | Specifies name of second arrowhead block. |
| **DimCen** | 0.09 | Controls how center marks are drawn:<br>**-n** Draws center lines.<br>**0** No center mark or lines drawn.<br>**+n** Draws center marks of length *n*. |

| Variable Name | Default Value | Meaning |
| --- | --- | --- |
| **DimClrD** | 0 | ACI color of dimension lines:<br>   **0** BYBLOCK (default).<br>  **256** BYLAYER. |
| **DimClrE** | 0 | Specifies ACI color of extension lines and leaders. |
| **DimClrT** | 0 | Specifies ACI color of dimension text. |
| **DimConstraintIcon** | 2 | Toggles the display of the lock icon next to dimensional constraints:<br>  **0** Does not display the lock icon.<br>  **1** Displays icon next to dynamic constraints.<br>  **2** Displays the icon next to annotational constraints.<br>  **3** Displays the icon next to dynamic and annotational constraints. |
| **DimDec** | 4 | Controls the number of decimal places displayed; range is 0 to 8. |
| **DimDLE** | 0.0 | Controls the length of the dimension line extension. |
| **DimDLI** | 0.38 | Controls the increment of the continued dimension lines. |
| **DimDSep** | "." | Specifies the decimal separator (must be a single character). |
| **DimExe** | 0.18 | Controls the extension above the dimension line. |
| **DimExO** | 0.0625 | Specifies the extension line origin offset. |
| **DimFrac** | 0 | Controls the fraction format when DimLUnit set to 4 or 5:<br>  **0** Horizontal.<br>  **1** Diagonal.<br>  **2** Not stacked. |
| **DimFXL** | 1 | Specifies default length of fixed extension lines. |
| **DimFxlOn** | 0 | Toggles fixed extension lines. |
| **DimGap** | 0.09 | Controls the gap between text and the dimension line. |
| **DimJogAngle** | 45 | Specifies default angle for jogged dimension lines. |
| **DimJust** | 0 | Controls the positioning of horizontal text:<br>  **0** Center justify.<br>  **1** Next to first extension line.<br>  **2** Next to second extension line.<br>  **3** Above first extension line.<br>  **4** Above second extension line. |
| **DimLdrBlk** | "" | Specifies the name of the block used for leader arrowheads; same as DimBlk.<br>  **.** Suppresses display of arrowhead. |
| **DimLFac** | 1.0 | Controls the linear unit scale factor. |
| **DimLim** | Off | Toggles the display of dimension limits. |
| **DimLtEx1** | "" | Specifies linetype for the first extension line. |
| **DimLtEx2** | "" | Specifies linetype for second extension line. |
| **DimLtype** | "" | Specifies linetype name for dimension line. |

| Variable Name | Default Value | Meaning |
|---|---|---|
| **DimLUnit** | 2 | Controls dimension units (except angular); replaces DimUnit:<br>  **1** Scientific.<br>  **2** Decimal.<br>  **3** Engineering.<br>  **4** Architectural.<br>  **5** Fractional.<br>  **6** Windows desktop. |
| **DimLwD** | -2 | Controls the dimension line lineweight; valid values are BYLAYER, BYBLOCK, or integer multiples of 0.01mm. |
| **DimLwE** | -2 | Controls the extension lineweight. |
| **DimPost** | "" | Specifies the default prefix or suffix for dimension text (maximum 13 characters):<br>  "" No suffix. |
| **DimRnd** | 0.0 | Controls the rounding value for dimension distances. |
| **DimSAh** | Off | Toggles separate arrowhead blocks:<br>  **Off** Use arrowheads defined by DimBlk.<br>  **On** Use arrowheads defined by DimBlk1 and DimBlk2. |
| **DimScale** | 1.0 | Controls the overall dimension scale factor:<br>  **0** Value is computed from the scale between current model space viewport and paper space.<br>  **>0** Scales text and arrowheads. |
| **DimSD1** | Off | Toggles display of the first dimension line:<br>  **On** First dimension line is suppressed.<br>  **Off** Not suppressed. |
| **DimSD2** | Off | Toggles display of the second dimension line:<br>  **On** Second dimension line is suppressed.<br>  **Off** Not suppressed. |
| **DimSE1** | Off | Toggles display of the first extension line:<br>  **On** First extension line is suppressed.<br>  **Off** Not suppressed. |
| **DimSE2** | Off | Toggles display of the second extension line:<br>  **On** Second extension line is suppressed.<br>  **Off** Not suppressed. |
| *DimSho* | *On* | *Toggles dimension updates while dragging:*<br>  **On** *Dimensions are updated during drag.*<br>  **Off** *Dimensions are updated after drag.* |
| **DimSOXD** | Off | Toggles display of dimension lines outside of extension lines:<br>  **On** Dimension lines not drawn outside extension lines.<br>  **Off** Are drawn outside extension lines. |
| **DimStyle** (R/O) | "STANDARD" | Reports the current dimension style. |

| Variable Name | Default Value | Meaning |
| --- | --- | --- |
| **DimTAD** | 0 | Controls the vertical position of text:<br>  **0** Centers text between extension lines.<br>  **1** Places text above dimension line, except when dimension line not horizontal and DimTIH = 1.<br>  **2** Places text on side of dimension line farthest from the defining points.<br>  **3** Conforms to JIS (Japanese Industry Standard). |
| **DimTDec** | 4 | Specifies number of decimal places for primary tolerances; range 0 - 8. |
| **DimTFac** | 1.0 | Controls the scale factor for tolerance text height. |
| **DimTFill** | 0 | Toggles background fill color for dimension text. |
| **DimTFillClr** | 0 | Specifies background color for dimension text. |
| **DimTIH** | On | Toggles alignment of text placed inside extension lines:<br>  **Off** Text aligned with dimension line.<br>  **On** Text is horizontal. |
| **DimTIX** | Off | Toggles placement of text inside extension lines:<br>  **Off** Text placed inside extension lines, if room.<br>  **On** Text forced between the extension lines. |
| **DimTM** | 0.0 | Controls the value of the minus tolerance. |
| **DimTMove** | 0 | Controls how dimension text is moved:<br>  **0** Moves dimension line with text.<br>  **1** Moves text, and adds a leader to the text.<br>  **2** Move text anywhere without leader. |
| **DimTOFL** | Off | Toggles placement of dimension lines:<br>  **Off** Dimension lines not drawn when arrowheads are outside.<br>  **On** Dimension lines drawn, even when arrowheads are outside. |
| **DimTOH** | On | Toggles text alignment when outside of extension lines:<br>  **Off** Text aligned with dimension line.<br>  **On** Text is horizontal. |
| **DimTol** | Off | Toggles generation of dimension tolerances:<br>  **Off** Tolerances not drawn.<br>  **On** Tolerances are drawn. |
| **DimTolJ** | 1 | Controls vertical justification of tolerances:<br>  **0** Bottom.<br>  **1** Middle.<br>  **2** Top. |
| **DimTP** | 0.0 | Specifies the value of the plus tolerance. |
| **DimTSz** | 0.0 | Controls the size of oblique tick strokes:<br>  **0** Arrowheads.<br>  **>0** Oblique strokes. |
| **DimTVP** | 0.0 | Controls the vertical position of text when DimTAD = 0:<br>  **1** Turns on DimTAD (=1).<br>  **>-0.7** *or* **<0.7** Splits dimension line for text. |
| **DimTxSty** | "STANDARD" | Specifies the dimension text style. |

| Variable Name | Default Value | Meaning |
| --- | --- | --- |
| **DimTxt** | 0.18 | Controls the text height. |
| **DimTxtDirection** | Off | Toggles the reading direction of dimension text:<br>　0　Left to right.<br>　1　Right to left. |
| **DimTZin** | 0 | Controls the display of zeros in tolerances:<br>　**0**　Suppresses 0 ft and 0 in.<br>　**1**　Includes 0 ft and 0 in.<br>　**2**　Includes 0 ft; suppresses 0 in.<br>　**3**　Suppresses 0 ft; includes 0 in.<br>　**4**　Suppresses leading 0 in decimal dimensions.<br>　**8**　Suppresses trailing 0 in decimal dimensions.<br>　**12**　Suppresses leading and trailing zeroes. |
| **DimUPT** | Off | Controls user-positioned text:<br>　**Off**　Cursor positions dimension line.<br>　**On**　Cursor also positions text. |
| **DimZIN** | 0 | Controls the display of zero in feet-inches units:<br>　**0**　Suppresses 0 ft and 0 in.<br>　**1**　Includes 0 ft and 0 in.<br>　**2**　Includes 0 ft; suppress 0 in.<br>　**3**　Suppresses 0 ft; include 0 in.<br>　**4**　Suppresses leading 0 in decimal dim.<br>　**8**　Suppresses trailing 0 in decimal dim.<br>　**12**　Suppresses leading and trailing zeroes. |
| **DispSilh** | 0 | Toggles the silhouette display of 3D solids. |
| **Distance** (r/o) | 0.0 | Reports the distance last measured by the Dist command. |
| **DivMeshBoxHeight** | 3 | Specifies the default number of subdivisions along the height of boxes; range is 1 to 256. |
| **DivMeshBoxLength** | 3 | Specifies the default number of subdivisions along the length of boxes; range is 1 to 256. |
| **DivMeshBoxWidth** | 3 | Specifies the default number of subdivisions along the width of boxes; range is 1 to 256. |
| **DivMeshConeAxis** | 8 | Specifies the default number of subdivisions around the base of cones; range is 1 to 256. |
| **DivMeshConeBase** | 3 | Specifies the default number of subdivisions on the base of cones; range is 1 to 256. |
| **DivMeshConeHeight** | 3 | Specifies the default number of subdivisions between the base and the tip of cones; range is 1 to 256. |
| **DivMeshCylAxis** | 8 | Specifies the default number of subdivisions around the trunk of cylinders; range is 1 to 256. |
| **DivMeshCylBase** | 3 | Specifies the default number of subdivisions on the end caps of cylinders; range is 1 to 256. |
| **DivMeshCylHeight** | 3 | Specifies the default number of subdivisions along the trunk of cylinders. |
| **DivMeshPyrBase** | 3 | Specifies the default number of subdivisions on the base of pyramids; range is 1 to 256. |

| Variable Name | Default Value | Meaning |
|---|---|---|
| **DivMeshPyrHeight** | 3 | Specifies the default number of subdivisions between the base and the tip of pyramids; range is 1 to 256. |
| **DivMeshPyrLength** | 3 | Specifies the default number of subdivisions along the base of pyramids; range is 1 to 256. |
| **DivMeshSphereAxis** | 12 | Specifies the default number of subdivisions between the two polar ends of spheres; range is 1 to 256. |
| **DivMeshSphereHeight** | 6 | Specifies the default number of subdivisions around spheres; range is 1 to 256. |
| **DivMeshTorusPath** | 8 | Specifies the default number of subdivisions along paths swept by tori profiles; range is 1 to 256. |
| **DivMeshTorusSection** | 8 | Specifies the default number of subdivisions around paths swept by tori profiles; range is 1 to 256. |
| **DivMeshWedgeBase** | 3 | Specifies the default number of subdivisions on the base of wedges; range is 1 to 256. |
| **DivMeshWedgeHeight** | 3 | Specifies the default number of subdivisions along the height of wedges; range is 1 to 256. |
| **DivMeshWedgeLength** | 4 | Specifies the default number of subdivisions along the length of wedges; range is 1 to 256. |
| **DivMeshWedgeSlope** | 3 | Specifies the default number of subdivisions along the slope of wedges; range is 1 to 256. |
| **DivMeshWedgeWidth** | 3 | Specifies the default number of subdivisions along the width of wedges; range is 1 to 256. |
| **DonutId** | 0.5 | Controls the inside diameter of donuts. |
| **DonutOd** | 1.0 | Controls the outside diameter of donuts. |
| **DragMode** | 2 | Controls the drag mode:<br>  0 No drag.<br>  1 On if requested.<br>  2 Automatic. |
| **DragP1** | 10 | Controls the regen drag display; range is 0 to 32767. |
| **DragP2** | 25 | Controls the fast drag display; range is 0 to 32767. |
| **DragVs** | "" | Specifies the default visual style when 3D solids are created by dragging the cursor; disabled when visual style is 2D wireframe. |
| **DrawOrderCtl** | 3 | Controls the behavior of draw order:<br>  0 Draw order not restored until next regen or drawing reopened.<br>  1 Normal draw order behavior.<br>  2 Draw order inheritance.<br>  3 Combines options 1 and 2. |
| **DrState** | 0 | Toggles the Drawing Recovery palette. |
| **DTextEd** | 2 | Controls the user interface of Text command:<br>  0 Uses in-place text editor.<br>  1 Displays Edit Text dialog box.<br>  2 Click elsewhere in drawing to start new text string. |

| Variable Name | Default Value | Meaning |
|---|---|---|
| **DwfFrame** | 2 | Display of the frame around DWF overlays: |
| | |   **0** Frame is turned off. |
| | |   **1** Frame is displayed and plotted. |
| | |   **2** Frame is displayed but not plotted. |
| **DwfOsnap** | 1 | Toggles o'snapping of the DWF frame. |
| **DwgCheck** | 0 | Toggles checking of whether drawing was edited by software other than AutoCAD: |
| | |   **0** Suppresses dialog box. |
| | |   **1** Displays warning dialog box. |
| | |   **2** Prints warning in command bar. |
| **DwgCodePage** (r/o) | *varies* | Same values as SysCodePage. |
| **DwgName** (r/o) | *varies* | Reports the current drawing file name, such as "drawing1.dwg". |
| **DwgPrefix** (r/o) | *varies* | Reports the drawing's drive and folder, such as "d:\acad 2011\". |
| **DwgTitled** (r/o) | 0 | Reports whether the drawing file name is: |
| | |   **0** "drawing1.dwg". |
| | |   **1** User-assigned name. |
| **DxEval** | 12 | Controls which commands cause AutoCAD to check for changed external data files (sum): |
| | |   **0** None. |
| | |   **1** Open. |
| | |   **2** Save. |
| | |   **4** Plot. |
| | |   **8** Publish. |
| | |   **12** Plot and Publish. |
| | |   **16** eTransmit and Archive. |
| | |   **32** Save with automatic update. |
| | |   **64** Plot with automatic update. |
| | |   **128** Publish with automatic update. |
| | |   **256** eTransmit/Archive with automatic update. |
| **DynConstraintDisplay** | 1 | Toggles the display of dynamic constraints. |
| **DynConstraintMode** | 1 | Toggles the display of hidden constraints: |
| | |   **0** Constraints remain hidden when objects are selected. |
| | |   **1** Constraints are displayed. |
| **DynDiGrip** | 31 | Controls the dynamic dimensions displayed during grip stretch editing (DynDiVis =2): |
| | |   **0** Displays none. |
| | |   **1** Resulting dimension. |
| | |   **2** Length change dimension. |
| | |   **4** Absolute angle dimension. |
| | |   **8** Angle change dimension. |
| | |   **16** Arc radius dimension. |
| | |   **31** Displays all. |

| Variable Name | Default Value | Meaning |
|---|---|---|
| **DynDiVis** | 1 | Controls dynamic dimensions displayed during grip stretch editing:<br>  **0** First (in the cycle order).<br>  **1** First two (in the cycle order).<br>  **2** All (as specified by DynDiGrip). |
| **DynMode** | 3 | Controls dynamic input features. (Click DYN on status bar to turn on hidden modes.)<br>  **-3** Both on hidden.<br>  **-2** Dimensional input on hidden.<br>  **-1** Pointer input on hidden.<br>  **0** Off.<br>  **1** Pointer input on.<br>  **2** Dimensional input on.<br>  **3** Both on. |
| **DynPiCoords** | 0 | Toggles pointer input coordinates:<br>  **0** Relative.<br>  **1** Absolute. |
| **DynPiFormat** | 0 | Toggles pointer input coordinates:<br>  **0** Polar.<br>  **1** Cartesian. |
| **DynPiVis** | 1 | Controls when pointer input is displayed:<br>  **0** When user types at prompts for points.<br>  **1** When prompted for points.<br>  **2** Always. |
| **DynPrompt** | 1 | Toggles display of prompts in Dynamic Input tooltips. |
| **DynToolTips** | 1 | Toggles which tooltips are affected by tooltip appearance settings:<br>  **0** Only Dynamic Input value fields.<br>  **1** All drafting tooltips. |

## E

| Variable Name | Default Value | Meaning |
|---|---|---|
| **EdgeMode** | 0 | Toggles edge mode for the Trim and Extend commands:<br>  **0** Does not extend.<br>  **1** Extends cutting edge. |
| **Elevation** | 0.0 | Specifies the current elevation, relative to current UCS. |
| **EnterpriseMenu** (r/o) | "." | Reports the path and *.cui* file name. |
| **ErHighlight** | 1 | Toggle: external references are (0) not highlighted or (1) highlighted when selected in the External References dialog box – and vice versa. |
| *EntExts* | *1* | *Controls how drawing extents are calculated:*<br>  *0 Extents calculated every time; slows down AutoCAD, but uses less memory.*<br>  *1 Extents of every object are cached as a two-byte value (default).*<br>  *2 Extents of every object are cached as a four-byte value (fastest but uses more memory).* |
| **ErrNo** (r/o) | 0 | Reports error numbers from AutoLISP, ADS, and Arx. |
| **ErState** (r/o) | 0 | Toggle: reports display of the External References palette. |

| Variable Name | Default Value | Meaning |
|---|---|---|
| **Expert** | 0 | Controls the display of prompts:<br>  **0** Normal prompts.<br>  **1** 'About to regen, proceed?' and 'Really want to turn the current layer off?'<br>  **2** 'Block already defined. Redefine it?' and 'A drawing with this name already exists. Overwrite it?'<br>  **3** Linetype command messages.<br>  **4** UCS Save and VPorts Save.<br>  **5** DimStyle Save and DimOverride. |
| **ExplMode** | 1 | Toggles whether the Explode and Xplode commands explode non-uniformly scaled blocks:<br>  **0** Does not explode.<br>  **1** Explodes. |
| **ExportEPlotFormat** | 2 | Specifies the format in which to export drawings:<br>  **0** PDF.<br>  **1** DWF.<br>  **2** DWFx. |
| **ExportModelSpace** | 0 | Specifies which part of model space to export in DWF/x or PDF format:<br>  **0** Drawing as currently displayed.<br>  **1** Extents of the drawing.<br>  **2** Windowed area. |
| **ExportPageSetup** | 0 | Specifies which page setup to use in exporting drawings in DWF/x or PDF format:<br>  **0** Default setup.<br>  **1** Overridden setup. |
| **ExportPaperSpace** | 0 | Specifies which part of paper space export in DWF/x or PDF format:<br>  **0** Current layout only.<br>  **1** All layouts. |
| **ExtMax** (r/o) | $1.0E+20, 1.0E+20, 1.0E+20$ | Upper-right coordinate of drawing extents. |
| **ExtMin** (r/o) | $-1.0E+20, -1.0E+20, -1.0E+20$ | Lower-left coordinate of drawing extents. |
| **ExtNames** | 1 | Controls the format of named objects:<br>  **0** Names are limited to 31 characters, and can include A - Z, 0 - 9, dollar ($), underscore (_), and hyphen (-).<br>  **1** Names are limited to 255 characters, and can include A - Z, 0 - 9, spaces, and any characters not used by Windows or AutoCAD for special purposes. |

**F**

| Variable Name | Default Value | Meaning |
|---|---|---|
| **FaceterDevNormal** | 40 | Specifies the maximum angle between adjacent mesh surfaces and surface normals; range is any positive angle. |
| **FaceterDevSurface** | 0.0010 | Specifies the accuracy of meshes converted from surfaces and solids; range is any positive number:<br>  **0** Turns off this option. |

| Variable Name | Default Value | Meaning |
|---|---|---|
| **FaceterGridRatio** | 0.0000 | Specifies the maximum aspect ratio for mesh subdivisions; range is 0 to 100. Values <1 are inverted: 0.5 = 2. |
| **FaceterMaxEdgeLength** | 0.0000 | Specifies the maximum edge length for meshes converted from surfaces and solids; range is any positive number. |
| **FaceterMaxGrid** | 4096 | Specifies the maximum grid lines in the U and V directions for meshes converted from surfaces and solids; range is 0 to 4096. |
| **FaceTerMeshType** | 0 | Specifies the type of mesh created from solids and surfaces:<br>  **0** Optimized mesh type.<br>  **1** Quadrilateral type.<br>  **2** Triangular type. |
| **FaceterMinUGrid** | 0 | Specifies the minimum number of grid lines in the U direction for meshes converted from solids and surfaces; range is 0 to 1023:<br>  **0** Turns off this option. |
| **FaceTerMinVGrid** | 0 | Specifies the minimum number of grid lines in the V direction for meshes converted from solids and surfaces; range is 0 to 1023:<br>  **0** Turns off this option. |
| **FaceterPrimitiveMode** | 1 | Toggles smoothness settings for meshes converted from solids and surfaces:<br>  **0** Applies settings from the Mesh Tesselation Options dialog box.<br>  **1** Applies settings from the Mesh Primitive Options dialog box. |
| **FaceterSmoothLev** | 1 | Specifies the level of smoothness for meshes converted from solids and surfaces:<br>  **<0** Does not smooth objects.<br>  **0** Does not smooth objects.<br>  **1** Applies smoothness level 1.<br>  **2** Applies smoothness level 2.<br>  **3** Applies smoothness level 3. |
| **FacetRatio** | 0 | Controls the aspect ratio of facets on rounded 3D bodies:<br>  **0** Creates an *n* by 1 mesh.<br>  **1** Creates an *n* by *m* mesh. |
| **FaceTRres** | 0.5000 | Controls the smoothness of shaded and hidden-line objects; range is 0.01 to 10. |
| *FbxImportLog* | *1* | *Toggles creation of log files that report the status of imported .fbx files: (0) off, or (1) on.* |
| **FieldDisplay** | 1 | Toggles background to field text:<br>  **0** No background.<br>  **1** Gray background. |
| **FieldEval** | 31 | Controls how fields are updated:<br>  **0** Not updated.<br>  **1** Updated with Open.<br>  **2** Updated with Save.<br>  **4** Updated with Plot.<br>  **8** Updated with eTransmit.<br>  **16** Updated with regeneration. |

| Variable Name | Default Value | Meaning |
|---|---|---|
| **FileDia** | 1 | Toggles user interface for file-access commands, like Open and Save:<br>**0** Displays command-line prompts.<br>**1** Displays file dialog boxes. |
| **FilletRad** | 0.0 | Specifies the current fillet radius. |
| **FilletRad3d** | 1.0 | Specifies current fillet radius for the FilletEdge and SurfFillet commands: |
| **FillMode** | 1 | Toggles the fill of solid objects, wide polylines, fills, and hatches. |
| **FontAlt** | "simplex.shx" | Specifies the font used for missing fonts. |
| **FontMap** | "acad.fmp" | Specifies the name of the font mapping file. |
| **Frame** | 3 | Toggles the visibility of frames on all xrefs, images, and DWF, DWFx, PDF, and DGN underlays; overrides the XClipFrame, ImageFrame, DwfFrame, PdfFrame, DgnFrame commands:<br>**0** Not visible; not plotted<br>**1** Visible; plotted.<br>**2** Visible; not plotted.<br>**3** Uses the settings of the XClipFrame, ImageFrame, DwfFrame, PdfFrame, DgnFrame commands. |
| **FrontZ** (r/o) | 0.0 | Reports the front clipping plane offset. |
| **FullOpen** (r/o) | 1 | Reports whether the drawing is:<br>**0** Partially loaded.<br>**1** Fully open. |
| **FullPlotPath** | 1 | Specifies the format of file name sent to plot spooler:<br>**0** Drawing file name only.<br>**1** Full path and drawing file. |

## G

| Variable Name | Default Value | Meaning |
|---|---|---|
| **GeoLatLongFormat** | 0 | Specifies format of latitude and longitude:<br>**0** Decimal degrees.<br>**1** Degrees, minutes, and seconds. |
| **GeoMarkerVisibility** | 1 | Toggles visibility of geographic markers. |
| **GfAng** | 0 | Controls the angle of gradient fill; 0 to 360 degrees. |
| **GfClr1** | "RGB 000,000,255" | Specifies the first gradient color in RGB format. |
| **GfClr2** | "RGB 255,255,153" | Specifies the second gradient color in RGB format. |
| **GfClrLum** | 1.0 | Controls the level of gray in one-color gradients:<br>**0** Black.<br>**1** White. |
| **GfClrState** | 1 | Specifies the type of gradient fill:<br>**0** Two-color.<br>**1** One-color. |

| Variable Name | Default Value | Meaning |
|---|---|---|
| **GfName** | 1 | Specifies the style of gradient fill:<br>  1 Linear.<br>  2 Cylindrical.<br>  3 Inverted cylindrical.<br>  4 Spherical.<br>  5 Inverted spherical.<br>  6 Hemispherical.<br>  7 Inverted hemispherical.<br>  8 Curved.<br>  9 Inverted curved. |
| **GfAShift** | 0 | Controls the origin of the gradient fill:<br>  0 Centered.<br>  1 Shifted up and left. |
| **GlobalOpacity** | 0 | Default level of translucency for all palettes when transparency is turned on:<br>  **0** Fully transparent.<br>  **100** Fully opaque. |
| *GlobCheck* | 0 | *Controls reporting on dialog boxes:*<br>  *-1 Turns off local language.*<br>  *0 Turns off.*<br>  *1 Warns if larger than 640x400.*<br>  *2 Reports size in pixels.*<br>  *3 Displays additional information.* |
| **GridDisplay** | 2 | Determines grid display (sum of bit codes):<br>  0 Grid restricted to area specified by the Limits command.<br>  1 Grid is infinite.<br>  2 Adaptive grid, with fewer grid lines when zoomed out.<br>  4 Displays more grid lines when zoomed in.<br>  8 Grid follows the xy plane of the dynamic UCS. |
| **GridMajor** | 5 | Specifies number of minor grid lines per major line. |
| **GridMode** | 1 | Toggles the display of the grid. |
| **GridStyle** | 0 | Switches grid between dots and lines (bit mode):<br>  0 Lined grid is displayed.<br>  1 Dotted grid displayed in 2D model space.<br>  2 Dotted grid displayed in block editor.<br>  4 Dotted grid displayed in sheets and layout tabs. |
| **GridUnit** | 0.5,0.5 | Controls the x, y spacing of the grid. |
| **GripBlock** | 0 | Toggles the display of grips in blocks:<br>  0 At block insertion point.<br>  1 Of all objects within block. |
| **GripColor** | 150 | Specifies ACI color of unselected grips. |
| **GripContour** | 251 | Specifies ACI color of the grip outline. |
| **GripDynColor** | 140 | Specifies ACI color of grips in dynamic blocks. |

| Variable Name | Default Value | Meaning |
|---|---|---|
| **GripHot** | 12 | Specifies ACI color of selected grips. |
| **GripHover** | 11 | Specifies ACI grip color when cursor hovers. |
| **GripMultifunctional** | 3 | Controls access to multifunctional grips:<br>  0  Disabled.<br>  1  Uses Ctrl cycling and the hot grip shortcut menu.<br>  2  Uses the dynamic menu and the hot grip shortcut menu.<br>  3  Uses Ctrl cycling, the dynamic menu, and the hot grip shortcut menu. |
| **GripObjLimit** | 100 | Controls the maximum number of grips displayed; range is 1 to 32767; 0 = grips never suppressed. |
| **Grips** | 1 | Toggles the display of grips:<br>  0  Grips are not displayed.<br>  1  Grips are displayed, except polyline midpoint segment grips.<br>  2  Midpoint grips are displayed on polyline segments. |
| **GripSize** | 5 | Specifies the size of grip; range 1 - 255 pixels. |
| **GripSubObjMode** | 1 | Controls the use of Shift to select subobjects:<br>  0  Grips not displayed as hot when subobjects selected.<br>  1  Face, edge, or vertex grips displayed as hot when subobjects of solids, surfaces, and meshes selected (default).<br>  2  Line and arc segment grips display hot on 2D polylines.<br>  3  Combines actions of 1 and 2. |
| **GripTips** | 1 | Toggles display of grip tips when cursor hovers over custom objects:<br>  0  Grip tips and Ctrl-cycling tooltips not displayed.<br>  1  Grip tips and Ctrl-cycling tooltips displayed (default). |
| **GtAuto** | 1 | Toggles display of grip tools. |
| **GtDefault** | 0 | Toggles which commands are the default commands in 3D views:<br>  0  Scale, Move, and Rotate.<br>  1  3dScale, 3dMove, and 3dRotate. |
| **GtLocation** | 0 | Controls location of grip tools:<br>  0  Aligns grip tool with UCS icon.<br>  1  Aligns with the last selected object. |

**H**

| Variable Name | Default Value | Meaning |
|---|---|---|
| **HaloGap** | 0 | Controls the distance by which haloed lines are shortened in 2D wireframe visual style; a percentage of 1 unit. |
| **Handles** (R/O) | On | Reports whether object handles (not grips) can be accessed by applications; obsolete. |
| **HatchBoundSet** (R/O) | 0 | Reports the contents of the hatch boundary set:<br>  0  Current viewport (default).<br>  1  Existing set. |
| **HatchType** (R/O) | 0 | Specifies the type of hatch pattern:<br>  0  Predefined (default).<br>  1  User defined.<br>  2  Custom. |

Appendix C: AutoCAD System Variables    C-1033

| Variable Name | Default Value | Meaning |
|---|---|---|
| HelpPrefix | "C:\Program Files\Autodesk\Autocad 2011 \Help\Index.Html" | Specifies path to HTML help file. |
| HidePrecision | 0 | Controls the precision of hide calculations in 2D wireframe visual style:<br>**0** Single precision, less accurate, faster.<br>**1** Double precision, more accurate, but slower (recommended). |
| HideText | On | Controls the display of text during Hide:<br>**On** Text is hidden but does not hide other objects, unless text object has thickness.<br>**Off** Text is not hidden and does not hide other objects. |
| *HideXrefScales* | *1* | *Toggles inclusion of xrefs in the list of scale factors.* |
| Highlight | 1 | Toggles object selection highlighting. |
| HPAng | 0.0 | Specifies current hatch pattern angle. |
| HpAnnotative | 0 | Toggles whether new hatch patterns take on annotative scale factors:<br>**0** Non-annotative.<br>**1** Annotative. |
| HpAssoc | 1 | Toggles associativity of hatches:<br>**0** Not associative.<br>**1** Associative. |
| HpBackgroundColor | "." | Specifies background fill color for hatch patterns:<br>**"None"** *or* **"."** No background color (default).<br>**1** *through* **255** AutoCAD Color Index<br>*Name* Name of the first seven colors, such as "Red."<br>**RGB:** *or* **HSL:** Red-green-blue or hue-saturation-luminosity.<br>*Colorbook* Color book name and number. |
| HpBound | 1 | Controls the object created by the Hatch and Boundary commands:<br>**0** Region.<br>**1** Polyline. |
| HpBoundRetain | 0 | Toggles whether boundary objects are (0) not created or (1) created for new hatches; the HpBound system variable determines whether the boundaries are region or polylines: |
| HpColor | "." | Specifies default color for new hatch patterns:<br>**"None"** *or* **"."** Uses current background color (default).<br>**1** *through* **255** AutoCAD Color Index<br>*Name* Name of the first seven colors, such as "Red."<br>**RGB:** *or* **HSL:** Red-green-blue or hue-saturation-luminosity.<br>*Colorbook* Color book name and number. |
| HpDlgMode | 2 | Determines when Hatch and Gradient and Hatch Edit dialog boxes are displayed by the Hatch, Gradient, and HatchEdit commands:<br>**0** Dialog boxes are never displayed; enter the seTtings option to display them.<br>**1** Hatch and Gradient dialog boxes are displayed; Hatch Edit is not.<br>**2** Dialog boxes are displayed when the ribbon is not active. |
| HpDouble | 0 | Toggles double hatching. |

| Variable Name | Default Value | Meaning |
|---|---|---|
| **HpDrawOrder** | 3 | Controls draw order of hatches and fills:<br>  **0** None.<br>  **1** Behind all other objects.<br>  **2** In front of all other objects.<br>  **3** Behind the hatch boundary.<br>  **4** In front of the hatch boundary. |
| **HpGapTol** | 0 | Controls largest gap allowed in hatch boundaries; ranges from 0 to 5000 units. |
| **HpInherit** | 0 | Toggles how MatchProp copies the hatch origin from source object to destination objects:<br>  **0** As specified by HpOrigin.<br>  **1** As specified by the source hatch object. |
| **HpIslandDetection** | 1 | Determines how islands are hatched:<br>  **0** Alternating islands are hatched (Normal mode).<br>  **1** Outermost island only is hatched (Outer mode; default).<br>  **2** Everything is hatched (Ignore mode). |
| **HpIslandDetectionMode** | 1 | Toggle: type of island detection (controls HpIslandDetection):<br>  0 Legacy island detection.<br>  1 Shiny new island detection method. |
| *HpLastPattern (R/O)* | "Ansi31" | *Reports the name of the last hatch pattern used.* |
| *HpLayer* | "Use Current" | *Specifies the name of the layer to use for new hatches and fills.* |
| **HpMaxlines** | 1000000 | Controls the maximum number of hatch lines; range is 100 - 10,000,000. |
| **HpName** | "ANSI31" | Specifies default hatch name. |
| **HpObjWarning** | 10000 | Specifies the maximum number of hatch boundaries that can be selected before AutoCAD flashes warning message; range is 1 to 1073741823. |
| **HpOrigin** | 0,0 | Specifies the default origin for hatch objects. |
| **HpOriginMode** | 0 | Controls the default hatch origin point:<br>  **0** Specified by HpOrigin.<br>  **1** Bottom-left corner of hatch's rectangular extents.<br>  **2** Bottom-right corner of rectangular extents.<br>  **3** Top-right corner of rectangular extents.<br>  **4** Top-left corner of rectangular extents.<br>  **5** Center of hatch's rectangular extents. |
| *HpOriginStoreAsDefault* | 0 | *Toggles whether user-defined hatch origin is stored as default value:*<br>  **0** *Does not store.*<br>  **1** *Stores specified origin as default.* |
| **HpQuickPreview** | On | Toggles whether preview hatches when the cursor is inside boundaries:<br>  **Off** Does not preview.<br>  **On** Preview. |
| *HpRelativePs (R/O)* | Off | *Toggles whether pattern scaling is relative to the paper space scale factor:*<br>  **Off** *Does not scale hatch patterns.*<br>  **On** *Hatch patterns are scaled* |

| Variable Name | Default Value | Meaning |
|---|---|---|
| **HpScale** | 1.0 | Specifies the current hatch scale factor; cannot be zero. |
| **HpSeparate** | 0 | Controls number of hatch objects made from multiple boundaries:<br>**0** Single hatch object created.<br>**1** Separate hatch object created. |
| **HpSpace** | 1.0 | Controls the default spacing of user-defined hatches; cannot be zero. |
| **HpTransparency** | "." | Sets default translucency percentage for new hatches; has no effect on existing patterns:<br>**"Use current"** *or* **"."** — Translucency specified by CeTransparency.<br>**"ByLayer"** *or* **"ByBlock"** — Translucency specified by the associated layer or block.<br>**0** *to* **90** — Range of translucency from 0 (none) to 90 (mostly transparent). |
| **HyperlinkBase** | "" | Specifies the path for relative hyperlinks. |

## I

| Variable Name | Default Value | Meaning |
|---|---|---|
| **ImageFrame** | 1 | Toggles the visibility of frames on images:<br>**0** Not visible; not plotted<br>**1** Visible; plotted.<br>**2** Visible; not plotted. |
| **ImageHlt** | 0 | Toggles image frame highlighting when raster images are selected. |
| **ImpliedFace** | 1 | Toggles detection of implied faces. |
| **IndexCtl** | 0 | Controls creation of layer and spatial indices:<br>**0** No indices created.<br>**1** Layer index created.<br>**2** Spatial index created.<br>**3** Both indices created. |
| **InetLocation** | "http://www.autodesk.com" | Specifies the default URL for Browser. |
| **InputHistoryMode** | 15 | Controls the content and location of user input history (bitcode sum of):<br>**0** No history displayed.<br>**1** Displayed at the command line, and in dynamic prompt tooltips accessed with Up and Down arrow keys.<br>**2** Current command displayed in the shortcut menu.<br>**4** All commands displayed in the shortcut menu.<br>**8** Blip mark for recent input displayed in the drawing.<br>**15** All on. |
| **InsBase** | 0.0,0.0,0.0 | Controls the default insertion base point relative to the current UCS for Insert and XRef commands. |
| **InsName** | "" | Specifies the default block name:<br>**.** No default. |

| Variable Name | Default Value | Meaning |
|---|---|---|
| **InsUnits** | 1 | Specifies drawing units of blocks dragged into drawings: |
| | |   **0**  Unitless. |
| | |   **1**  Inches. |
| | |   **2**  Feet. |
| | |   **3**  Miles. |
| | |   **4**  Millimeters. |
| | |   **5**  Centimeters. |
| | |   **6**  Meters. |
| | |   **7**  Kilometers. |
| | |   **8**  Microinches. |
| | |   **9**  Mils. |
| | |   **10**  Yards. |
| | |   **11**  Angstroms. |
| | |   **12**  Nanometers. |
| | |   **13**  Microns. |
| | |   **14**  Decimeters. |
| | |   **15**  Decameters. |
| | |   **16**  Hectometers. |
| | |   **17**  Gigameters. |
| | |   **18**  Astronomical Units. |
| | |   **19**  Light Years. |
| | |   **20**  Parsecs. |
| **InsUnitsDefSource** | 1 | Controls source drawing units value; ranges from 0 to 20; see above. |
| **InsUnitsDefTarget** | 1 | Controls target drawing units; see list above. |
| **IntelligentUpdate** | 20 | Controls graphics refresh rate in frames per second; range is 0 (off) to 100 fps. |
| **InterfereColor** | "1" | Specifies color of interference objects. |
| **InterfereObjVs** | "Realistic" | Specifies visual style of interference objects. |
| **InterfereVpVs** | "3d wireframe" | Specifies visual style during interference checking. |
| **IntersectionColor** | 257 | Specifies ACI color of intersection polylines in 2D wireframe visual style (257=ByEntity). |
| **IntersectionDisplay** | Off | Toggles display of 3D surface intersections during Hide command in 2D wireframe: |
| | |   **Off**  Does not draw intersections. |
| | |   **On**  Draws polylines at intersections. |
| **ISaveBak** | 1 | Toggles creation of *.bak* backup files. |
| **ISavePercent** | 50 | Controls the percentage of waste in saved *.dwg* file before cleanup occurs: |
| | |   **0**  Slower full saves. |
| | |   **>0**  Faster partial saves. |
| | |   **100**  Maximum. |
| **IsoLines** | 4 | Controls the number of contour lines on 3D solids; range is 0 - 2047. |

| Variable Name | Default Value | Meaning |
|---|---|---|
| **L** | | |
| **LargeObjectSupport** | 0 | Toggles support for large-object support in DWG files:<br>  **0** Creates objects compatible with AutoCAD 2009 and earlier.<br>  **1** Uses large objects in drawings. |
| **LastAngle** (r/o) | 0 | Reports the end angle of last-drawn arc. |
| **LastPoint** | 0,0,0 | Reports the x,y,z coordinates of the last-entered point. |
| **LastPrompt** (r/o) | "*varies*" | Reports the last string on the command line. |
| **Latitude** | "37.7950" | Specifies last-used angle of latitude; default = San Francisco. |
| **LayerDlgMode** | 1 | Toggles the command mapped to the Layer command:<br>  **0** ClassicLayer command.<br>  **1** LayerPalette command. |
| **LayerEval** | 1 | Controls when newly-added (unreconciled) layers are detected:<br>  **0** Never.<br>  **1** When xref layers are added to the drawing<br>  **2** When new layers are added to drawings and xrefs. |
| **LayerEvalCtl** | 1 | Controls unreconciled layer filters:<br>  **0** Disables notification of new layers.<br>  **1** Enables notification, as controlled by the LayerEval sysvar. |
| **LayerFilterAlert** | 2 | Controls the deletion of layer filters in excess of 99 filters *and* the number of layers:<br>  **0** Off.<br>  **1** Deletes all filters without warning, when layer dialog box opened.<br>  **2** Recommends deleting all filters when layer dialog box opened.<br>  **3** Displays dialog box for selecting filters to erase, upon opening the drawing. |
| **LayerManagerState** (r/o) | 0 | Reports the status of the Layer Properties Manager palette. |
| **LayerNotify** | 15 | Controls when AutoCAD displays alerts about unreconciled layers (bit code):<br>  **0** Off.<br>  **1** Plotting.<br>  **2** Opening drawings.<br>  **4** Loading, reloading, and attaching xrefs.<br>  **8** Restoring layer states.<br>  **16** Saving drawings.<br>  **32** Inserting blocks, etc. |
| **LayLockFadeCtl** | 90 | Controls the fading of locked layers during the LayIso command:<br>  **0** Not faded.<br>  **>0** Faded up to 90 percent.<br>  **<0** Not faded, but the value is saved. |
| **LayoutRegenCtl** | 2 | Controls display list for layouts:<br>  **0** Display-list regen'ed with each tab change.<br>  **1** Display-list is saved for model tab and last layout tab.<br>  **2** Display list is saved for all tabs. |

| Variable Name | Default Value | Meaning |
| --- | --- | --- |
| **LegacyCtrlPick** | 0 | Toggles function of Ctrl+pick:<br>  **0** Selects faces, edges, and vertices of 3D solids.<br>  **1** Cycles through overlapping objects.<br>  **2** Ctrl+click selects sub-objects when SubObjSelectionMode = 0; otherwise, selects sub-objects when Ctrl is not held down. |
| **LensLength** (r/o) | 50.0 | Reports perspective view lens length, in mm. |
| **LightGlyphDisplay** | 1 | Toggles display of light glyph. |
| **LightingUnits** | 2 | Controls the type of lighting used:<br>  **0** Generic lighting.<br>  **1** International units of photometric lighting.<br>  **2** American units of photometric lighting. |
| **LightListState** | 0 | Toggles display of Light List palette. |
| **LightsInBlocks** | 1 | Toggles use of lights in blocks:<br>  **0** Off.<br>  **1** On. |
| **LimCheck** | 0 | Toggles drawing limits checking. |
| **LimMax** | 12.0,9.0 | Controls the upper right drawing limits. |
| **LimMin** | 0.0,0.0 | Controls the lower left drawing limits. |
| **LinearBrightness** | 0 | Controls the overall brightness of generic lighting in renderings; range is -10 to 10. |
| **LinearContrast** | 0 | Controls the overall contrast of generic lighting in renderings; range is -10 to 10. |
| *LispInit* | *1* | *Toggles AutoLISP functions and variables:*<br>  **0** *Preserved from drawing to drawing.*<br>  **1** *Valid in current drawing only.* |
| **Locale** (r/o) | "enc" | Reports ISO language code; see DctMain. |
| **LocalRootPrefix** (r/o) | "d:\docume..." | Reports the path to folder holding local customizable files. |
| **LockUi** | 0 | Controls the position and size of toolbars and palettes; hold down Ctrl key to unlock temporarily (bit code sum):<br>  **0** Toolbars and palettes unlocked.<br>  **1** Docked toolbars locked.<br>  **2** Docked palettes locked.<br>  **4** Floating toolbars locked.<br>  **8** Floating palettes locked. |
| **LoftAng1** | 90 | Specifies angle of loft to first cross section; range is 0 to 359.9 degrees. |
| **LoftAng2** | 90 | Specifies angle of loft to second cross section. |
| **LoftMag1** | 0.0 | Specifies magnitude of loft at first cross section; range is 1 to 10. |
| **LoftMag2** | 0.0 | Specifies magnitude of loft at last cross section. |

| Variable Name | Default Value | Meaning |
|---|---|---|
| **LoftNormals** | 1 | Specifies location of loft normals:<br>  **0** Ruled.<br>  **1** Smooth.<br>  **2** First normal.<br>  **3** Last normal.<br>  **4** Ends normal.<br>  **5** All normal.<br>  **6** Use draft angle and magnitude. |
| **LoftParam** | 7 | Specifies the loft shape:<br>  **1** Minimizes twists between cross sections.<br>  **2** Aligns start-to-end direction of each crosssection.<br>  **4** Generates simple solids and surfaces, instead of spline solids and surfaces.<br>  **8** Closes the surface or solid between the first and last cross-sections. |
| **LogExpBrightness** | 65.0 | Controls the overall brightness of photometric lighting in renderings; range is 0 to 200. |
| **LogExpContrast** | 50.0 | Controls the overall contrast of photometric lighting in renderings; range is 0 to 100. |
| **LogExpDaylight** | 2 | Controls how daylight is displayed photometric renderings:<br>  **0** Off.<br>  **1** On.<br>  **2** Same as sun status. |
| **LogExpMidtones** | 1.00 | Controls the overall midtones of photometric lighting in renderings; range is 0.01 to 20. |
| **LogExpPhysicalScale** | 1500.000 | Scales photometric lights physically. |
| **LogFileMode** | 0 | Toggles writing command prompts to *.log* file. |
| **LogFileName** (R/O) | "...\Drawing1.log" | Reports file name and path for *.log* file. |
| **LogFilePath** | "d:\acad 2011\" | Specifies path to the *.log* file. |
| **LogInName** (R/O) | "*username*" | Reports user's login name; truncated after 30 characters. |
| **Longitude** | -122.3940 | Specifies current angle of longitude (default = San Francisco). |
| **LTScale** | 1.0 | Controls linetype scale factor; cannot be 0. |
| **LUnits** | 2 | Controls linear units display:<br>  **1** Scientific.<br>  **2** Decimal.<br>  **3** Engineering.<br>  **4** Architectural.<br>  **5** Fractional. |
| **LUPrec** | 4 | Controls decimal places (or inverse of smallest fraction) of linear units; range is 0 to 8. |
| **LwDefault** | 25 | Controls the default lineweight, in millimeters; must be one of the following values: 0, 5, 9, 13, 15, 18, 20, 25, 30, 35, 40, 50, 53, 60, 70, 80, 90, 100, 106, 120, 140, 158, 200, or 211. |
| **LwDisplay** | 0 | Toggles whether lineweights are displayed; setting saved separately for Model space and each layout tab. |

| Variable Name | Default Value | Meaning |
| --- | --- | --- |
| **LwUnits** | 1 | Toggles units used for lineweights:<br>  **0** Inches.<br>  **1** Millimeters. |
| **M** | | |
| *MacroTrace* | *0* | *Toggles diesel debug mode.* |
| **MatBrowserState** (R/O) | 0 | Reports state of Materials Browser palette: (0) closed or (1) open. |
| **MatEditorState** (R/O) | 0 | Reports state of Materials Editor palette): (0) closed or (1) open. |
| **MaterialsPath** (R/O) | "" | Reports the path to the material library folder. |
| **MaxActVP** | 64 | Controls the maximum number of viewports to display; range is 2 to 64. |
| *MaxObjMem* | *0* | *Controls the maximum number of objects in memory; object pager is turned off when value = 0, <0, or 2,147,483,647.* |
| **MaxSort** | 1000 | Controls the maximum names sorted alphabetically; range 0 - 32767. |
| **MaxTouches** (R/O) | 0 | Reports number of touch points supported by multi-touch digitizers. |
| **MButtonPan** | 1 | Toggles the behavior of the wheel mouse:<br>  **0** Behaves as defined by AutoCAD's *.cui* file.<br>  **1** Pans when dragging with wheel. |
| **MeasureInit** | 0 | Toggles drawing units for default drawings:<br>  **0** English.<br>  **1** Metric. |
| **Measurement** | 0 | Toggles current drawing units:<br>  **0** English.<br>  **1** Metric. |
| **MenuBar** | 0 | Toggles the display of the menu bar. |
| **MenuCtl** | 1 | Toggles the display of submenus in side menu:<br>  **0** Only with menu picks.<br>  **1** Also with keyboard entry. |
| **MenuEcho** | 0 | Controls menu and prompt echoing (sum):<br>  **0** Displays all prompts.<br>  **1** Suppresses menu echoing.<br>  **2** Suppresses system prompts.<br>  **4** Disables ^P toggle.<br>  **8** Displays all input-output strings. |
| **MenuName** (R/O) | "acad" | Reports path and file name of *.cui* file. |
| **MeshType** | 1 | Toggles the type of mesh used by the RevSurf, TabSurf, RuleSurf, EdgeSurf, and other commands:<br>  **0** Polyfaces (compatible with AutoCAD 2009 and earlier).<br>  **1** 3D mesh objects (AutoCAD 2010 and later). |
| **MirrHatch** | 0 | Toggles mirroring of hatches:<br>  **0** Retains hatch angle.<br>  **1** Mirrors hatch angle. |

| Variable Name | Default Value | Meaning |
|---|---|---|
| **MirrText** | 0 | Toggles text handling by Mirror command:<br>  **0** Retains text orientation.<br>  **1** Mirrors text. |
| **MLeaderScale** | 1 | Scales mleaders based on:<br>  **0.0** Makes scale the ratio of model space viewport to paper space.<br>  **>0** Specifies scale factor. |
| **ModeMacro** | "" | Invokes Diesel macros. |
| **MsLtScale** | 1 | Controls how linetypes are scaled:<br>  **0** Not scaled by the annotation scale.<br>  **1** Scaled by the annotation scale. |
| **MsmState** (R/O) | 0 | Specifies if Markup Set Manager is active:<br>  **0** No.<br>  **1** Yes. |
| **MsOleScale** | 1.0 | Controls the size of text-containing OLE objects when pasted in model space:<br>  **-1** Scales by value of PlotScale.<br>  **0** Scales by value of DimScale.<br>  **>0** Uses scale factor. |
| **MTextEd** | "Internal" | Controls the name of the MText editor:<br>  **.** Uses the default editor.<br>  **0** Cancels the editing operation.<br>  **-1** Uses the secondary editor.<br>  **"blank"** Uses MText internal editor.<br>  **"Internal"** Uses MText internal editor.<br>  **"oldeditor"** Uses the previous internal editor.<br>  **"Notepad"** Uses Windows Notepad editor.<br>  **":lisped"** Uses Built-in AutoLISP function.<br>  *string* Uses another editor with a name fewer than 256 characters long in the form of this syntax:<br>  *:AutoLISPtextEditorFunction#TextEditor.* |
| **MTextFixed** | 2 | Controls the mtext editor appearance:<br>  **0** Mtext editor is used.<br>  **1** Mtext editor remembers its location.<br>  **2** Difficult-to-read text is displayed horizontally at a larger size. |
| **MTJigString** | "abc" | Specifies the sample text displayed by mtext editor; maximum 10 letters; enter . for no text. |
| **MTextToolbar** | 1 | Toggles toolbar display during MText:<br>  **0** Does not display toolbar.<br>  **1** Displays Text Formatting toolbar from AutoCAD 2008 and earlier.<br>  **2** Displays ribbon's Text Formatting panel. |
| **MyDocumentsPrefix** (R/O) | "C:\Documents and Settings\*username*\My Documents" | Reports path to \*my documents* folder of the logged-in user. |

| Variable Name | Default Value | Meaning |
|---|---|---|
| **N** | | |
| **NavBarDisplay** | 1 | Toggles display of navigation bar; applies to all viewports and layouts (NavBar command applies to the current viewport or space):<br>  **0** Off.<br>  **1** On. |
| **NavSWheelMode** | 2 | Specifies style of steering wheel:<br>  **0** Big wheel for viewing objects.<br>  **1** Big wheel for touring buildings.<br>  **2** Big wheel for full navigation.<br>  **3** Big wheel for 2D navigation.<br>  **4** Mini wheel for viewing objects.<br>  **5** Mini wheel for touring buildings.<br>  **6** Mini wheel for full navigation. |
| **NavSWheelOpacityBig** | 50 | Changes translucency of big steering wheel. Range is 25% to 90% (almost opaque). |
| **NavSWheelOpacityMini** | 150 | Changes translucency of mini steering wheel. Range is 25% to 90% (almost opaque). |
| **NavSWheelSizeBig** | 1 | Specifies size of the big steering wheel:<br>  **0** Small size.<br>  **1** Normal size.<br>  **2** Large size. |
| **NavSWheelSizeMini** | 1 | Specifies size of the mini steering wheel:<br>  **0** Small size.<br>  **1** Normal size.<br>  **2** Large size.<br>  **3** Extra large. |
| *NavSWheelWalkSpeed* | *1.0* | *Sets speed of Walk mode; range is 0.1 to 10.* |
| **NavVCubeDisplay** | 3 | Toggles display of the viewcube:<br>  **0** Not displayed.<br>  **1** Displayed in 3D visual styles, but not 2D<br>  **2** Displayed in 2D visual styles, but not 3D.<br>  **3** Displayed in 2D and 3D visual styles. |
| **NavVCubeLocation** | 0 | Locates viewcube in current viewport:<br>  **0** Upper-right corner.<br>  **1** Upper-left corner.<br>  **2** Lower-left corner.<br>  **3** Lower-right corner. |
| **NavVCubeOpacity** | 50 | Sets translucency of viewcube; 0 = displayed only when cursor passes over it. |
| **NavVCubeOrient** | 1 | Specifies viewcube's orientation:<br>  **0** Relative to WCS.<br>  **1** Relative to current UCS. |

| Variable Name | Default Value | Meaning |
|---|---|---|
| **NavVCubeSize** | 1 | Sizes the viewcube: |
| | |   **0**  Small size. |
| | |   **1**  Normal size. |
| | |   **2**  Large size. |
| *NodeName* (R/O) | *"AC$"* | *Reports the name of the network node; range is one to three characters.* |
| **NoMutt** | 0 | Toggles display of messages (a.k.a. muttering) during scripts, LISP, macros: |
| | |   **0**  Displays prompt, as normal. |
| | |   **1**  Suppresses muttering. |
| **NorthDirection** | 0 | Specifies angle of sun relative to positive y axis, and N direction of viewcube. |
| *NwfState* | *1* | *Displays New Features Workshop at AutoCAD start.* |

## O

| Variable Name | Default Value | Meaning |
|---|---|---|
| **ObjectIsolationMode** | 0 | Toggles display of hidden and isolated objects between sessions: |
| | |   **0**  Hidden and isolated objects displayed for the current drawing session only. |
| | |   **1**  Hidden and isolated settings saved for the next drawing session. |
| **ObscuredColor** | 257 | Specifies ACI color of objects obscured by Hide in 2D wireframe visual style. |
| **ObscuredLtype** | 0 | Specifies linetype of objects obscured by Hide in 2D wireframe visual mode: |
| | |   **0**  Invisible. |
| | |   **1**  Solid. |
| | |   **2**  Dashed. |
| | |   **3**  Dotted. |
| | |   **4**  Short dash. |
| | |   **5**  Medium dash. |
| | |   **6**  Long dash. |
| | |   **7**  Double short dash. |
| | |   **8**  Double medium dash. |
| | |   **9**  Double long dash. |
| | |   **10**  Medium long dash. |
| | |   **11**  Sparse dot. |
| **OffsetDist** | -1.0 | Controls current offset distance: |
| | |   **<0**  Offsets through a specified point. |
| | |   **≥0**  Uses default offset distance. |
| **OffsetGapType** | 0 | Controls how polylines reconnect when segments are offset: |
| | |   **0**  Extends segments to fill gap. |
| | |   **1**  Fills gap with fillet (arc segment). |
| | |   **2**  Fills gap with chamfer (line segment). |
| **OleFrame** | 2 | Controls visibility of OLE frames: |
| | |   **0**  Frame is not displayed and not plotted. |
| | |   **1**  Frame is displayed and is plotted. |
| | |   **2**  Frame is displayed but is not plotted. |

| Variable Name | Default Value | Meaning |
|---|---|---|
| **OleHide** | 0 | Controls display and plotting of OLE objects:<br>  **0** All OLE objects visible.<br>  **1** Visible in paper space only.<br>  **2** Visible in model space only.<br>  **3** Not visible. |
| **OleQuality** | 3 | Specifies quality of displaying and plotting of embedded OLE objects:<br>  **0** Monochrome.<br>  **1** Low quality graphics.<br>  **2** High quality graphics.<br>  **3** Automatically selected mode. |
| **OleStartup** | 0 | Toggles loading of OLE source applications to improve plot quality:<br>  **0** Does not load OLE source application.<br>  **1** Loads OLE source app when plotting. |
| **OpenPartial** | 1 | Determines whether drawings can be edited when not fully loaded:<br>  **0** Drawings must be fully open.<br>  **1** Visible portions of drawings can be edited before file is fully open. |
| **OpmState** | 0 | Toggles whether Properties palette is active. |
| **OrthoMode** | 0 | Toggles orthographic mode. |
| **OsMode** | 4133 | Controls current object snap mode (sum):<br>  **0** NONe.<br>  **1** ENDpoint.<br>  **2** MIDpoint.<br>  **4** CENter.<br>  **8** NODe.<br>  **16** QUAdrant.<br>  **32** INTersection.<br>  **64** INSertion.<br>  **128** PERpendicular.<br>  **256** TANgent.<br>  **512** NEARest.<br>  **1024** QUIck.<br>  **2048** APPint.<br>  **4096** EXTension.<br>  **8192** PARallel.<br>  **16383** All modes on.<br>  **16384** Object snap turned off via OSNAP on the status bar. |
| **OSnapCoord** | 2 | Controls whether keyboard overrides object snap:<br>  **0** Object snap overrides keyboard.<br>  **1** Keyboard overrides object snap.<br>  **2** Keyboard overrides object snap, except in scripts. |
| *OSnapHatch* | *0* | *Toggles whether hatches are snapped:*<br>  ***0*** *Osnaps ignore hatches.*<br>  ***1*** *Hatches are snapped.* |
| **OSnapNodeLegacy** | 1 | Toggles whether osnap snaps to mtext insertion points. |

| Variable Name | Default Value | Meaning |
|---|---|---|
| **OsnapZ** | 0 | Toggles osnap behavior in z direction: |
| | |   **0** Uses the z-coordinate. |
| | |   **1** Uses the current elevation setting. |
| **OsOptions** | 3 | Determines when objects with negative z values are o'snapped: |
| | |   **0** Uses the actual z coordinate. |
| | |   **1** Substitutes z coordinate with the elevation of the current UCS. |

**P**

| Variable Name | Default Value | Meaning |
|---|---|---|
| **PaletteOpaque** | 0 | Controls transparency of palettes: |
| | |   **0** Turned off by user. |
| | |   **1** Turned on by user. |
| | |   **2** Unavailable, but turned on by user. |
| | |   **3** Unavailable, and turned off by user. |
| **PaperUpdate** | 0 | Toggles how AutoCAD plots layouts with paper size different from plotter's default: |
| | |   **0** Displays a warning dialog box. |
| | |   **1** Changes paper size to that of the plotter configuration file. |
| **ParameterCopyMode** | 1 | Determines how constraints and referenced variables are copied: |
| | |   **0** Does not copy constraints. |
| | |   **1** Copies constraints and variables; expressions are constants. |
| | |   **2** Copies constraints, variables, and expressions; references variables; replaces data with constants. |
| | |   **3** Copies constraints, variables, and expressions; references variables; adds variables, if necessary. |
| | |   **4** Copies all; adds variables, if necessary. |
| **ParametersStatus** (R/O) | 0 | Reports whether the Parameters Manager palette is open: |
| | |   **0** Closed. |
| | |   **1** Open. |
| **PdfFrame** | 1 | Toggles the display of the PDF frame for editing and plotting: |
| | |   **0** Not displayed; not plotted. |
| | |   **1** Displayed; plotted. |
| | |   **2** Displayed; not plotted. |
| **PdfOsnap** | 1 | Toggles o'snapping between frame and objects in attached PDF files: |
| | |   **0** Geometry not o'snapped. |
| | |   **1** Geometry o'snapped. |
| **PDMode** | 0 | Controls point display style (sum): |
| | |   **0** Dot. |
| | |   **1** No display. |
| | |   **2** +-symbol. |
| | |   **3** x-symbol. |
| | |   **4** Short line. |
| | |   **32** Circle. |
| | |   **64** Square. |

| Variable Name | Default Value | Meaning |
|---|---|---|
| **PDSize** | 0.0 | Controls point display size:<br>  **>0** Absolute size, in pixels.<br>  **0** 5% of drawing area height.<br>  **<0** Percentage of viewport size. |
| **PEditAccept** | 0 | Toggles display of the PEdit command's 'Object selected is not a polyline. Do you want to turn it into one? <Y>:' prompt. |
| **PEllipse** | 0 | Toggles object used to create ellipses:<br>  **0** True ellipses.<br>  **1** Polyline arcs. |
| **Perimeter** (r/o) | 0.0 | Reports perimeter calculated by Area, DbList, and List commands. |
| **Perspective** | 0 | Toggles perspective mode; not available in 2D wireframe visual mode. |
| **PerpectiveClip** | 5.0 | Controls position of eye point clipping, as a percentage; range is 0.01% - 10.0%. |
| **PFaceVMax** (r/o) | 4 | Reports the maximum vertices per 3D face. |
| **PickAdd** | 2 | Toggles meaning of Shift key on selection sets:<br>  **0** Adds to selection set.<br>  **1** Removes from selection set.<br>  **2** Persistent; keeps objects selected after the Select command ends (hold down Shift to remove objects from the selection set. |
| **PickAuto** | 1 | Toggles selection set mode:<br>  **0** Single pick mode.<br>  **1** Automatic windowing and crossing. |
| **PickBox** | 3 | Controls selection pickbox size; range is 0 to 50 pixels. |
| **PickDrag** | 0 | Toggles selection window mode:<br>  **0** Pick two corners.<br>  **1** Pick a corner; drag to second corner. |
| **PickFirst** | 1 | Toggles command-selection mode:<br>  **0** Enter command first.<br>  **1** Select objects first. |
| **PickStyle** | 1 | Controls how groups and associative hatches are selected:<br>  **0** Includes neither.<br>  **1** Includes groups.<br>  **2** Includes associative hatches.<br>  **3** Includes both. |
| **Platform** (r/o) | "varies" | Reports the name of the operating system. |
| **PLineConvertMode** | 0 | Determines how polylines are converted to splines:<br>  **0** Linear segments.<br>  **1** Arc segments. |
| **PLineGen** | 0 | Toggles polyline linetype generation:<br>  **0** From vertex to vertex.<br>  **1** From end to end. |

| Variable Name | Default Value | Meaning |
|---|---|---|
| **PLineType** | 2 | Controls automatic conversion and creation of 2D polylines by PLine:<br>**0** Does not convert; creates old-format polylines.<br>**1** Does not convert; creates optimized lwpolylines.<br>**2** Converts polylines in older drawings on open; PLine creates optimized lwpolyline objects. |
| **PLineWid** | 0.0 | Controls current polyline width. |
| **PlotOffset** | 0 | Toggles the plot offset measurement:<br>**0** Relative to edge of margins.<br>**1** Relative to edge of paper. |
| **PlotRotMode** | 2 | Controls the orientation of plots:<br>**0** Lower left = 0,0.<br>**1** Lower left plotter area = lower left of media.<br>**2** X, y-origin offsets calculated relative to the rotated origin position. |
| **PlotTransparencyOverride** | 1 | Controls when translucent objects are plotted:<br>**0** Does not plot translucency.<br>**1** Uses the settings of the PageSetup and Plot commands.<br>**2** Plots object transparency always; converts drawings to raster, which can take longer to complete. |
| **PlQuiet** | 0 | Toggles display during batch plotting and scripts (replaces CmdDia):<br>**0** Displays plot dialog boxes and nonfatal errors.<br>**1** Logs nonfatal errors; does not display plot dialog boxes. |
| **PointCloudAutoUpdate** | 1 | Toggle: automatic regeneration of point clouds during editing and real time panning, zooming, and orbiting:<br>0 Manual regeneration required; has no effect on PointCloudRtDensity.<br>1 Automatically regenerated. |
| **PointCloudDensity** | 15 | Specifies percentage of points to display simultaneously as a percentage of 1,500,000 (the maximum number of points per drawing). |
| **PointCloudLock** | 0 | Toggles locking of point clouds:<br>**0** Not locked.<br>**1** Locked; point clouds cannot be edited, moved, or rotated. |
| **PointCloudRtDensity** | 5 | Specifies percentage of points to display during real time zooming, panning, or orbiting. . |
| **PolarAddAng** | "" | Holds a list of up to 10 user-defined polar angles; each angle can be up to 25 characters long, each separated with a semicolon (;). For example: 0;15;22.5;45. |
| **PolarAng** | 90 | Controls the increment of polar angle; contrary to Autodesk documentation, you can specify any angle. |
| **PolarDist** | 0.0 | Controls the polar snap increment when SnapStyl is set to 1 (isometric). |

| Variable Name | Default Value | Meaning |
|---|---|---|
| **PolarMode** | 0 | Controls polar and object snap tracking: |
| | |   **0**  Measures polar angles based on current UCS (absolute), track orthogonally; does not use additional polar tracking angles; acquires object tracking points automatically. |
| | |   **1**  Measures polar angles from selected objects (relative). |
| | |   **2**  Uses polar tracking settings in object snap tracking. |
| | |   **4**  Uses additional polar tracking angles (via PolarAng). |
| | |   **8**  Acquires object snap tracking points by pressing Shift. |
| **PolySides** | 4 | Controls the default number of polygon sides; range is 3 to 1024. |
| **Popups** (R/O) | 1 | Reports display driver support of AUI (advanced user interface): |
| | |   **0**  Not available. |
| | |   **1**  Available. |
| **PreviewEffect** | 2 | Controls the visual effect for previewing selected objects: |
| | |   **0**  Dashed lines. |
| | |   **1**  Thick lines. |
| | |   **2**  Thick dashed lines. |
| **PreviewFaceEffect** | 1 | Toggles preview selection highlighting of face sub-objects: |
| | |   **0**  Faces not highlighted. |
| | |   **1**  Faces highlighted with texture fill. |
| **PreviewFilter** | 7 | Controls the exclusion of objects from selection previewing (bit code sum): |
| | |   **0**  No objects excluded. |
| | |   **1**  Objects on locked layers. |
| | |   **2**  Objects in xrefs. |
| | |   **4**  Tables. |
| | |   **7**  Objects on locked layers, in xrefs, and tables. |
| | |   **8**  Multiline text. |
| | |   **16**  Hatch patterns. |
| | |   **32**  Groups. |
| **PreviewType** | 0 | Specifies the view for drawing thumbnails: |
| | |   **0**  Use last saved view. |
| | |   **1**  Use Home view. |
| **Product** (R/O) | "AutoCAD" | Reports the name of the software. |
| **Program** (R/O) | "acad" | Reports the name of the software's executable file. |
| **ProjectName** | "" | Controls the project name of the current drawing; searches for xref and image files. |
| **ProjMode** | 1 | Controls the projection mode for Trim and Extend commands: |
| | |   **0**  Does not project. |
| | |   **1**  Projects to xy plane of current UCS. |
| | |   **2**  Projects to view plane. |
| **ProxyGraphics** | 1 | Toggles saving of proxy images in drawings: |
| | |   **0**  Not saved; displays bounding box. |
| | |   **1**  Image saved with drawing. |
| **ProxyNotice** | 1 | Toggles warning message displayed when drawing contains proxy objects. |

| Variable Name | Default Value | Meaning |
|---|---|---|
| **ProxyShow** | 1 | Controls the display of proxy objects:<br>  **0** Not displayed.<br>  **1** All displayed.<br>  **2** Bounding box displayed. |
| **ProxyWebSearch** | 0 | Toggles checking for object enablers:<br>  **0** Does not check for object enablers.<br>  **1** Checks for object enablers when an Internet connection is present. |
| **PsLtScale** | 1 | Toggles paper space linetype scaling:<br>  **0** Uses model space scale factor.<br>  **1** Uses viewport scale factor. |
| **PSolHeight** | 4.0 | Specifies default height of polysolid objects. |
| **PSolWidth** | 0.25 | Specifies default width of polysolid objects. |
| *PsProlog* | "" | *Specifies the PostScript prologue file name.* |
| *PsQuality* | 75 | *Controls resolution of PostScript display, in pixels:*<br>  *<0 Display as outlines; no fill.*<br>  *0 Displays no fills.*<br>  *>0 Displays filled.* |
| **PStyleMode** | 1 | Toggles the plot color matching mode of the drawing:<br>  **0** Uses named plot style tables.<br>  **1** Uses color-dependent plot style tables. |
| **PStylePolicy** (r/o) | 1 | Reports whether the object color is associated with its plot style:<br>  **0** Not associated.<br>  **1** Associated. |
| **PsVpScale** | 0.0 | Controls the view scale factor (ratio of units in paper space to units in newly-created model space viewports) 0 = scaled to fit. |
| **PublishAllSheets** | 1 | Determines which sheets (model space and layouts) are loaded automatically into the Publish command's list:<br>  **0** Current drawing only.<br>  **1** All open drawings. |
| **PublishCollate** | 1 | Controls how sheets are published:<br>  **0** Sheet sets are processed one sheet at a time; a separate plot is created for each sheet.<br>  **1** Sheets sets are processed as a single job and single plot file. |
| **PublishHatch** | 1 | Determines how hatch patterns are exported in DWF format to Impression:<br>  **0** Treated as separate objects.<br>  **1** Treated as a single object. |
| **PUcsBase** (r/o) | "" | Reports name of UCS defining the origin and orientation of orthographic UCS settings; in paper space only. |

| Variable Name | Default Value | Meaning |
|---|---|---|
| **Q** | | |
| QAFlags | 0 | Controls quality assurance flags: |
| | |   **0** Turned off. |
| | |   **1** ^C metacharacter cancels grips, just as if user pressed Esc. |
| | |   **2** Long text screen listings do not pause. |
| | |   **4** Error and warning messages displayed in command line, instead of in dialog boxes. |
| | |   **128** Screen picks accepted via the AutoLISP (command) function. |
| QcState | 0 | Toggles whether QuickCalc palette is open. |
| QpLocation | 0 | Locates the Quick Properties palette: |
| | |   **0** Near cursor as defined by DSettings command's Quick Properties tab. |
| | |   **1** Floating, in a constant location. |
| QpMode | -1 | Controls Quick Properties palette display: |
| | |   **0** (or negative numbers) Off. |
| | |   **1** On for all objects |
| | |   **2** On only for objects defined in CUI. |
| QTextMode | 0 | Toggles quick text mode. |
| QvDrawingPin | 0 | Toggles pinning of drawing quick views. |
| QvLayoutPin | 0 | Toggles pinning of layout quick views. |
| **R** | | |
| RasterDpi | 300 | Controls the conversion of millimeters or inches to pixels, and vice versa; range is 100 to 32767. |
| RasterPercent | 20 | Percentage of system memory to allocate to plotting raster images; range is 0 - 100%. |
| RasterPreview (R/O) | 1 | Toggles creation of BMP preview image. |
| RasterThreshold | 20 | Specifies amount of RAM to allocate to plotting raster images; range is 0 - 2000MB. |
| Rebuild2dCv | 6 | Specifies number of control vertices per spline; range is 2 to 32767. |
| Rebuild2dDegree | 3 | Specifies degree for splines; range is 1 to 11: |
| | |   **1** Splines are like straight lines (no bends). |
| | |   **2** Splines are parabolic (one bend). |
| | |   **3** Splines are cubic Beziers (two bends). |
| Rebuild2dOption | 1 | Toggle: rebuilding of splines: |
| | |   **0** Original curves not deleted. |
| | |   **1** Original curves are deleted. |
| RebuildDegreeU | 3 | Specifies U-direction degree for NURBS surfaces; range is 2 to 11. |
| RebuildDegreeV | 3 | Specifies V-direction degree for NURBS surfaces; range is 2 to 11. |
| RebuildOptions | 1 | Determines fate of original surfaces and trimmed areas on rebuilt surfaces: |
| | |   **0** Keeps original surfaces; trimmed areas are not applied. |
| | |   **1** Deletes original surfaces; trimmed areas are not applied. |
| | |   **2** Keeps original surfaces; trimmed areas applied to rebuilt objects. |
| | |   **3** Deletes original surfaces; trimmed areas are applied. |

| Variable Name | Default Value | Meaning |
| --- | --- | --- |
| RebuildU | 6 | Specifies default number of lines in the U direction for NURBS surfaces; range is 2 to 32767. |
| RebuildV | 6 | Specifies the default number of lines in the V direction for NURBS surfaces; range is 2 to 32767. |
| RecoverAuto | 0 | Displays recovery notifications of damaged drawing files:<br>**0** Displays a dialog box when opening damaged files; interrupts running scripts.<br>**1** Recovers damaged files automatically; displays dialog box report on recovered file; does not interrupt scripts.<br>**2** Recovers damaged files automatically; displays report on the recovered file at the command prompt. |
| RecoveryMode | 2 | Controls recording of drawing recovery information after software failure:<br>**0** Note recorded.<br>**1** Recorded; Drawing Recovery palette does not display automatically.<br>**2** Recorded, and Drawing Recovery palette displays automatically. |
| RefEditName | "" | Specifies the reference file name when in reference-editing mode. |
| RegenMode | 1 | Toggles regeneration mode:<br>**0** Regens with each view change.<br>**1** Regens only when required. |
| Re-Init | 0 | Controls the reinitialization of I/O (input/output) devices:<br>**1** Digitizer port.<br>*2 Plotter port; obsolete.*<br>**4** Digitizer.<br>*8 Plotter; obsolete.*<br>**16** Reloads PGP (program parameters) file. |
| RememberFolders | 1 | Toggles the path search method:<br>**0** Path specified in AutoCAD desktop properties is default for file dialog boxes.<br>**1** Last path specified by each file dialog box is remembered. |
| RenderPrefsState | 1 | Toggles display of the Render Preferences palette. |
| RenderUserLights | 1 | Determines which lights are rendered:<br>**0** Default *or* user-defined lights are rendered.<br>**1** Default *and* user-defined lights are rendered. |
| ReportError | 1 | Determines if AutoCAD sends an error report to Autodesk:<br>**0** Does not create error reports.<br>**1** Error reports are generated and sent to Autodesk. |
| RibbonContextSelect | 1 | Determines when context-sensitive ribbon tabs are displayed:<br>0 Never.<br>1 Single-clicking objects or selection sets..<br>2 Double-clicking objects or selection sets. |
| RibbonContextSelLim | 2500 | Determines the maximum number of objects that activate context-sensitive ribbon tabs; range is 0 to 32767. |

| Variable Name | Default Value | Meaning |
|---|---|---|
| **RibbonDockedHeight** | 0 | Determines the height of the ribbon when docked:<br>  **0** Ribbon sizes itself to height of the selected tab.<br>  **1** Minimum height in pixels.<br>  **500** Maximum height in pixels. |
| **RibbonSelectMode** | 1 | Determines fate of selections after command is selected from ribbon:<br>  **0** Pickfirst selection set is unselected.<br>  **1** Pickfirst selection set remains selected. |
| **RibbonState** (r/o) | 1 | Reports whether the ribbon is open. |
| **RoamableRootPrefix** (r/o) | "d:\documents and settings\*username*\application data\aut..." | Reports the path to the root folder where roamable customized files are located. |
| **RolloverOpacity** | 0 | Specifies the translucency of palettes when cursor moves over them; range is 0 to 100. |
| **RolloverTips** | 1 | Toggles display of rollover tips. |
| **RTDisplay** | 1 | Toggles raster display during real-time zoom and pan:<br>  **0** Displays the entire raster image.<br>  **1** Displays raster outline only. |

## S

| Variable Name | Default Value | Meaning |
|---|---|---|
| **SaveFidelity** | 1 | Toggles how annotative objects are translated to earlier releases:<br>  **0** Makes no changes.<br>  **1** Saves each scale representation to a separate layer. |
| **SaveFile** (r/o) | "" | Reports the automatic save file name. |
| **SaveFilePath** | "...\temp\" | Specifies the path for automatic save files. |
| **SaveName** (r/o) | "" | Reports the drawing's save-as file name. |
| **SaveTime** | 10 | Controls the automatic save interval, in minutes; 0 = disable auto save. |
| **ScreenBoxes** (r/o) | 0 | Reports the maximum number of menu items supported by display; 0 = screen menu turned off. |
| **ScreenMode** (r/o) | 3 | Reports the state of AutoCAD display:<br>  **0** Text screen.<br>  **1** Graphics screen.<br>  **2** Dual-screen display. |
| **ScreenSize** (r/o) | *varies* | Reports the current viewport size, in pixels, such as 719.0,381.0. |
| **SDI** | 0 | *Controls the multiple-document interface (SDI is "single document interface"):*<br>  *0 Turns on MDI.*<br>  *1 Turns off MDI. (Only one drawing may be loaded into AutoCAD.)*<br>  *2 Disables MDI for apps that cannot support MDI; read-only.*<br>  *3 (r/o) Disables MDI for apps that cannot support MDI.* |
| **SelectionAnnoDisplay** | 1 | Toggles how alternate scale representations are displayed when annotative objects selected:<br>  **0** Does not display them.<br>  **1** Displays them according to the value of XFadeCtl. |
| **SelectionArea** | 1 | Toggles use of colored selection areas. |

| Variable Name | Default Value | Meaning |
|---|---|---|
| **SelectionAreaOpacity** | 25 | Controls the opacity of color selection areas; range is 0 (transparent) to 100 (opaque). |
| **SelectionCycling** | 0 | Controls selection cycling:<br>**0** Off.<br>**1** On but does not display the list dialog box.<br>**2** On and displays the list dialog box of objects that you be selected. |
| **SelectionPreview** | 3 | Controls selection preview:<br>**0** Off.<br>**1** On when commands are inactive.<br>**2** On when commands prompt for object selection. |
| **SelectSimilarMode** | 130 | Controls matchable properties for SelectSimilar command (bit code):<br>**0** Object type.<br>**1** Color.<br>**2** Layer.<br>**4** Linetype.<br>**8** Linetype scale.<br>**16** Lineweight.<br>**32** Plot style.<br>**64** Text styles, dimension styles, table styles, and so on.<br>**128** Names of referenced objects, such as blocks, xrefs, and images.<br>**130** 2+128 = layer and name of referenced objects. |
| **SetBylayerMode** | 255 | Controls which properties are affected by the SetByLayer command:<br>**0** None.<br>**1** Colors.<br>**2** Linetypes.<br>**4** Lineweights.<br>**8** Materials.<br>**16** Plot styles.<br>**32** ByBlock is changed to Bylayer.<br>**64** Blocks are changed from ByBlock to ByLayer.<br>**128** Includes transparency property.<br>**255** All. |
| **ShadEdge** | 3 | Controls shading by Shade command:<br>**0** Shades only faces.<br>**1** Shades faces, and shows edges in background color.<br>**2** Show edges only in object color.<br>**3** Shows faces in object color, edges in background color. |
| **ShadeDif** | 70 | Controls percentage of diffuse to ambient light; range is 0 to 100%. |
| **ShadowPlaneLocation** | 0.0 | Specifies the default height of the shadow plane. |

| Variable Name | Default Value | Meaning |
|---|---|---|
| **ShortcutMenu** | 11 | Controls display of shortcut menus (sum):<br>  **0** Does not display shortcut menus.<br>  **1** Displays default shortcut menus.<br>  **2** Displays edit shortcut menus.<br>  **4** Displays command shortcut menus when commands are active.<br>  **8** Displays command shortcut menus only when options available at the command line.<br>  **16** Displays shortcut menus when the right button held down longer. |
| **ShowHist** | 1 | Toggles display of history in solids:<br>  **0** Does not display original solids.<br>  **1** Displays original solids depending on Show History property settings.<br>  **2** Displays all original solids. |
| **ShowLayerUsage** | 1 | Toggles layer-usage icons in Layers dialog box. |
| **ShowMotionPin** | 1 | Specifies pinning of Show Motion images. |
| **ShpName** | "" | Specifies the default shape name:<br>  **.** Set to no default. |
| **SigWarn** | 1 | Toggles display of dialog box when drawings with digital signatures are opened:<br>  **0** Displays only when signature is invalid.<br>  **1** Displays always. |
| **SketchInc** | 0.1 | Controls the Sketch command's recording increment. |
| **SkPoly** | 0 | Controls sketch line mode:<br>  **0** Records as lines.<br>  **1** Records as a polyline.<br>  **2** Records as a spline. |
| **SkTolerance** | 0.5 | Specifies how closely spline fits to freehand sketches; range is 0 to 1. |
| **SkyStatus** | 0 | Reports status of sun and sky background:<br>  **0** Sky is off.<br>  **1** Sky is on.<br>  **2** Sky and illumination are on. |
| **SmoothMeshConvert** | 0 | Controls how solids and surfaces are converted to meshes:<br>  **0** Converts to smooth models; optimizes or merges coplanar faces.<br>  **1** Converts to smooth models; retains original mesh faces.<br>  **2** Converts to flattened faces; optimizes or merges coplanar faces.<br>  **3** Converts to flattened faces; retains original mesh faces. |
| **SmoothMeshGrid** | 3 | Determines how the facet grid is displayed:<br>  **0** Does not display.<br>  **1** Displays for smoothing levels 0 and 1.<br>  **2** Displays for smoothing levels 0, 1, and 2.<br>  **3** Displays levels 0, 1, 2, and 3. |
| **SmoothMeshMaxFace** | 10000000 | Specifies maximum faces on meshes; range is 1 to 16000000. |
| **SmoothMeshMaxLev** | 4 | Specifies maximum smoothness level for meshes; range is 1 - 255. |

| Variable Name | Default Value | Meaning |
|---|---|---|
| **SmState** (R/O) | 0 | Reports state of Show Motion interface: (0) closed or (1) open. |
| **SnapAng** | 0 | Controls rotation angle for snap and grid; when not 0, grid lines are not displayed. |
| **SnapBase** | 0.0,0.0 | Controls current origin for snap and grid. |
| **SnapIsoPair** | 0 | Controls current isometric drawing plane:<br>  **0** Left isoplane.<br>  **1** Top isoplane.<br>  **2** Right isoplane. |
| **SnapMode** | 0 | Toggles snap mode. |
| **SnapStyl** | 0 | Toggles snap style:<br>  **0** Normal.<br>  **1** Isometric. |
| **SnapType** | 0 | Toggles snap for the current viewport:<br>  **0** Standard snap.<br>  **1** Polar snap. |
| **SnapUnit** | 0.5,0.5 | Controls x, y spacing for snap distances. |
| **SolidCheck** | 1 | Toggles solid validation. |
| **SolidHist** | 1 | Toggles retention of history in solids. |
| **SortEnts** | 127 | Controls object display sort order:<br>  **0** Off.<br>  **1** Object selection.<br>  **2** Object snap.<br>  **4** Redraw.<br>  **8** Slide generation.<br>  **16** Regeneration.<br>  **32** Plot.<br>  **64** PostScript output. |
| **SplDegree** | 3 | Specifies default degree of new splines created with control vertices; range is 1 to 5. |
| **SplFrame** | 0 | Toggles displays of helix and 3D mesh object controls. (Use CvShow and CvHide for polylines and polyface meshes.)<br>  **0** Does not display control polygon of helices; displays smoothed mesh objects, and does not display invisible edges of 3D faces or polyface meshes.<br>  **1** Displays control polygons of helices; displays unsmoothed mesh objects that are smoothed; displays edges of 3D faces and polyface meshes. |
| **SplineSegs** | 8 | Controls number of line segments that define splined polylines; range is -32768 to 32767.<br>  **<0** Draws with fit-curve arcs.<br>  **>0** Draws with line segments. |
| **SplineType** | 6 | Controls type of spline curve:<br>  **5** Quadratic Bezier spline.<br>  **6** Cubic Bezier spline. |

| Variable Name | Default Value | Meaning |
|---|---|---|
| ⊛ **SplKnots** | 0 | Specifies default knot setting when using fit points for new splines:<br>**0** Chords.<br>**1** Square root chords.<br>**2** Uniform. |
| ⊛ **SplMethod** | 0 | Toggles default type of new splines: (0) fit or (1) control vertices. |
| **SsFound** | "" | Specifies path and file name of sheet sets. |
| **SsLocate** | 1 | Toggles whether sheet set files are opened with drawing:<br>**0** Not opened.<br>**1** Opened automatically. |
| **SsmAutoOpen** | 1 | Toggles whether the Sheet Set Manager is opened with drawing (SsLocate must be 1):<br>**0** Not opened.<br>**1** Opened automatically. |
| **SsmPollTime** | 60 | Controls time interval between automatic refreshes of status data in sheet sets; range is 20 to 600 seconds (SsmSheetStatus = 2). |
| **SsmSheetStatus** | 2 | Controls refresh of status data in sheet sets:<br>**0** Does not refresh automatically.<br>**1** Refreshes when sheet set is loaded or updated.<br>**2** Also refreshes as specified by SsmPollTime. |
| **SsmState** (R/O) | 0 | Toggles whether Sheet Set Manager is open. |
| **StandardsViolation** | 2 | Controls whether alerts are displayed when CAD standards are violated:<br>**0** Displays no alerts.<br>**1** Displays alert when CAD standard violated.<br>**2** Displays icon on status bar when file is opened with CAD standards, and when non-standard objects are created. |
| **Startup** | 0 | Controls which dialog box is displayed by the New and QNew commands:<br>**0** Displays Select Template dialog box.<br>**1** Displays Startup and Create New Drawing dialog box. |
| **StatusBar** | 1 | Controls display of application and drawing status bars:<br>**0** Hides both status bars.<br>**1** Displays the application status bar.<br>**2** Displays both status bars.<br>**3** Displays the drawing status bar. |
| **StepSize** | 6.0 | Specifies length of steps in walk mode; range is 1E-6 to 1E+6. |
| **StepsPerSec** | 2 | Specifies speed of steps in walk mode; range is 1-30. |
| **SubObjSelectionMode** | 0 | Specifies which subobjects are selected by Ctrl+click:<br>**0** All.<br>**1** Only vertices.<br>**2** Only edges.<br>**3** Only faces.<br>**4** Only history subobjects. |
| **SunPropertiesState** | 0 | Toggles display of Sun Properties palette. |
| **SunStatus** | 0 | Toggles display of light by the sun. |

Appendix C: AutoCAD System Variables

| Variable Name | Default Value | Meaning |
|---|---|---|
| SurfaceAssociativity | 1 | Toggles surface associativity:<br>**0** Surfaces have no associativity to other surfaces.<br>**1** Surfaces adjust automatically to modifications made to related surfaces; DelObj is ignored. |
| SurfaceAssociativityDrag | 1 | Specifies how associative surfaces react during dragging (move) operations:<br>**0** No previews; display updated following the drag operations.<br>**1** Reviews first surface only; other surfaces updated following drag operation.<br>**2** Previews all surfaces. |
| SurfaceAutoTrim | 0 | Toggles automatic trimming of surfaces by projected geometry:<br>**0** Does not trim.<br>**1** Trims. |
| SurfaceModelingMode | 0 | Determines default 3D surface type:<br>**0** Creates procedural surfaces.<br>**1** Creates NURBS surfaces. |
| *SurfOffsetConnect* | *0* | *Toggles connection between offset surfaces (0) off or (1) on:* |
| *SurfTrimAutoExtend* | *1* | *Toggles automatic extension of trim geometry (0) off or (1) on:* |
| *SurfTrimProjection* | *0* | *Toggles projection of trim entities onto surfaces (0) off or (1) on:* |
| SurfTab1 | 6 | Controls density of m-direction surfaces and meshes; range 5 - 32766. |
| SurfTab2 | 6 | Specifies density of n-direction surfaces and meshes; range 2 - 32766. |
| SurfType | 6 | Controls smoothing of surface by PEdit:<br>**5** Quadratic Bezier spline.<br>**6** Cubic Bezier spline.<br>**8** Bezier surface. |
| SurfU | 6 | Controls surface density in m-direction; range is 2 to 200. |
| SurfV | 6 | Specifies surface density in n-direction; range is 2 to 200. |
| SysCodePage (R/O) | "ANSI_1252" | Reports the system code page; set by operating system. |

**T**

| Variable Name | Default Value | Meaning |
|---|---|---|
| TableIndicator | 1 | Toggles display of column letters and row numbers during table editing. |
| TableToolbar | 1 | Toggles display of cell editing toolbar:<br>**0** Does not display.<br>**1** Displays.<br>**2** Displays commands on ribbon. |
| TabMode | 0 | Toggles tablet mode. |
| Target (R/O) | 0.0,0.0,0.0 | Reports target coordinates in the current viewport. |
| *Taskbar* | *1* | *Toggles whether each drawing appears as a button on the Windows taskbar.* |
| TbCustomize | 1 | Toggles whether toolbars can be customized. |
| TDCreate (R/O) | *varies* | Reports the local date and time that the drawing was created, such as 2448860.54014699. |
| TDInDwg (R/O) | *varies* | Reports the duration since the drawing was loaded, such as 0.00040625. |

| Variable Name | Default Value | Meaning |
|---|---|---|
| **TDuCreate** (r/o) | *varies* | Reports the universal date and time when the drawing was created, such as 2451318. 67772165. |
| **TDUpdate** (r/o) | *varies* | Reports the date and time of last update, such as 2448860.54014699. |
| **TDUsrTimer** (r/o) | *varies* | Reports the decimal time elapsed by user-timer, such as 0.00040694. |
| **TDuUpdate** (r/o) | *varies* | Reports the universal date and time of the last save, such as 2451318.67772165. |
| **TempOverrides** | 1 | Toggles temporary overrides. |
| **TempPrefix** (r/o) | "d:\temp" | Reports the path for temporary files set by Temp variable. |
| **TextEd** (r/o) | 0 | Reports the status of the text editor:<br>　**0** Closed.<br>　**1** Open. |
| **TextEval** | 0 | Toggles the interpretation of text input during the -Text command:<br>　**0** Literal text.<br>　**1** Read ( and ! as AutoLISP code. |
| **TextFill** | 1 | Toggles the fill of TrueType fonts when plotted:<br>　**0** Outline text.<br>　**1** Filled text. |
| **TextOutputFileFormat** | 0 | Controls Unicodes for plot and text window log files:<br>　**0** ANSI format.<br>　**1** UTF-8 (Unicode).<br>　**2** UTF-16LE (Unicode).<br>　**3** UTF-16BE (Unicode). |
| **TextQlty** | 50 | Controls the resolution of TrueType fonts when plotted; range is 0 to 100. |
| **TextSize** | 0.2 | Controls the default height of text (2.5 in metric units). |
| **TextStyle** | "Standard" | Specifies the default name of text style. |
| **Thickness** | 0.0 | Controls the default object thickness. |
| **ThumbSize** | 1 | Specifies the size of thumbnail images:<br>　**0** 64x64 pixels.<br>　**1** 128x128 pixels.<br>　**2** 256x256 pixels. |
| **TileMode** | 1 | Toggles the view mode:<br>　**0** Displays layout tab.<br>　**1** Displays model tab. |
| **TimeZone** | -80000 | Specifies current time zone. |
| **ToolTipMerge** | 0 | Toggles the merging of tooltips during dynamic display. |
| **ToolTips** | 1 | Toggles the display of tooltips. |
| **TpState** (r/o) | 0 | Reports if Tool Palettes palette is open. |
| **TraceWid** | 0.0500 | Specifies current width of traces. |

| Variable Name | Default Value | Meaning |
|---|---|---|
| **TrackPath** | 0 | Controls display of polar and object snap tracking alignment paths:<br>**0** Displays object snap tracking path across the entire viewport.<br>**1** Displays object snap tracking path between the alignment point and "From point" to cursor location.<br>**2** Turns off polar tracking path.<br>**3** Turns off polar and object snap tracking paths. |
| **TransparencyDisplay** | 1 | Toggles display of translucency in objects: (0) off or (1) on. |
| **TrayIcons** | 1 | Toggles the display of the tray on status bar. |
| **TrayNotify** | 1 | Toggles service notifications displayed by the tray. |
| **TrayTimeout** | 0 | Controls length of time that tray notifications are displayed; range is 0 to 10 seconds. |
| **TreeDepth** | 3020 | Controls the maximum branch depth (in *xxyy* format):<br>*xx* Model-space nodes.<br>*yy* Paper-space nodes.<br>*>0* 3D drawing.<br>*<0* 2D drawing. |
| **TreeMax** | 10000000 | Controls the memory consumption during drawing regeneration. |
| **TrimMode** | 1 | Toggles trims during Chamfer and Fillet:<br>**0** Leaves selected edges in place.<br>**1** Trims selected edges. |
| **TSpaceFac** | 1.0 | Controls the mtext line spacing distance measured as a factor of "normal" text spacing; ranges from 0.25 to 4.0. |
| **TSpaceType** | 1 | Controls the type of mtext line spacing:<br>**1** At Least adjusts line spacing based on the height of the tallest character in a line of mtext.<br>**2** Exactly uses the specified line spacing; ignores character height. |
| **TStackAlign** | 1 | Controls vertical alignment of stacked text:<br>**0** Bottom aligned.<br>**1** Center aligned.<br>**2** Top aligned. |
| **TStackSize** | 70 | Controls size of stacked text as a percentage of the current text height; range is 25 to 125%. |

**U**

| Variable Name | Default Value | Meaning |
|---|---|---|
| **UcsAxisAng** | 90 | Controls the default angle for rotating the UCS around an axis (via the UCS command using the X, Y, or Z options); valid values limited to: 5, 10, 15, 18, 22.5, 30, 45, 90, or 180. |
| **UcsBase** | "" | Specifies name of UCS that defines the origin and orientation of orthographic UCS settings. |
| **UcsDetect** | 1 | Toggles dynamic UCS mode. |
| **UcsFollow** | 0 | Toggles view displayed with new UCSs:<br>**0** Does not change.<br>**1** Aligns UCS with new view automatically. |

| Variable Name | Default Value | Meaning |
|---|---|---|
| ⌨ **UcsIcon** | 3 | Controls display of the UCS icon:<br>  **0** Off.<br>  **1** On at lower-left corner.<br>  **2** On at UCS origin, if possible. |
| **UcsName** (r/o) | "" | Reports the name of current UCS view:<br>  "" Current UCS is unnamed. |
| **UcsOrg** (r/o) | 0.0,0.0,0.0 | Reports origin of current UCS relative to WCS. |
| **UcsOrtho** | 1 | Controls whether the related orthographic UCS settings are restored automatically:<br>  **0** Does not change UCS setting when orthographic view is restored.<br>  **1** Restores related ortho UCS automatically when an ortho view is restored. |
| **UcsView** | 1 | Toggles whether the current UCS is saved with a named view. |
| **UcsVp** | 1 | Toggles whether the UCS in active viewports remains fixed (locked) or changes (unlocked) to match the UCS of the current viewport. |
| **UcsXDir** (r/o) | 1.0,0.0,0.0 | Reports the x-direction of current UCS relative to WCS. |
| **UcsYDir** (r/o) | 0.0,1.0,0.0 | Reports the y-direction of current UCS relative to WCS. |
| **UndoCtl** (r/o) | 21 | Reports the status of undo (bit sum):<br>  **0** Disabled.<br>  **1** Enabled.<br>  **2** Limited to one command.<br>  **4** Set to auto-group mode.<br>  **8** Set to active group mode.<br>  **16** Set to combined zooms and pans. |
| **UndoMarks** (r/o) | 0 | Reports the number of undo marks. |
| **UnitMode** | 0 | Toggles the type of units display:<br>  **0** As set by Units command.<br>  **1** As entered by user. |
| **UOSnap** | 1 | Toggles object snapping for geometry in DWF, DWFx, PDF, and DGN underlays; overrides the DwfOsnap, PdfOsnap, and DgnOsnap system variables:<br>  **0** Disabled.<br>  **1** Enabled.<br>  **2** Setting dependent on DwfOsnap, PdfOsnap, and DgnOsnap. |
| **UpdateThumbnail** | 15 | Controls how thumbnails are updated (sum):<br>  **0** Thumbnail previews not updated.<br>  **1** Sheet views updated.<br>  **2** Model views updated.<br>  **4** Sheets updated.<br>  **8** Updated when sheets or views are created, modified, or restored.<br>  **16** Updated when the drawing is saved. |
| **UserI1** *thru* **UserI5** | 0 | Stores user-definable integer variables. |
| **UserR1** *thru* **UserR5** | 0.0 | Stores user-definable real variables. |

| Variable Name | Default Value | Meaning |
|---|---|---|
| **UserS1** *thru* **UserS5** | "" | Stores user-definable string variables; values are not saved. |

## V

| Variable Name | Default Value | Meaning |
|---|---|---|
| **ViewCtr** (R/O) | varies | Reports x, y, z coordinates of center of current view, such as 15,9,56. |
| **ViewDir** (R/O) | varies | Reports current view direction relative to UCS: 0,0,1=plan view. |
| **ViewMode** (R/O) | 0 | Reports the current view mode:<br>**0** Normal view.<br>**1** Perspective mode on.<br>**2** Front clipping on.<br>**4** Back clipping on.<br>**8** UCS-follow on.<br>**16** Front clip not at eye. |
| **ViewSize** (R/O) | varies | Reports the height of current view in drawing units. |
| **ViewTwist** (R/O) | 0 | Reports the twist angle of current view. |
| **VisRetain** | 1 | Controls xref drawing's layer settings:<br>**0** Does not save xref-dependent layer settings in the current drawing.<br>**1** Saves xref-dependent layer settings in the current drawing, and take precedence over settings in the xref'ed drawing the next time the current drawing is loaded. |
| **VpLayerOverrides** (R/O) | 0 | Reports whether VP (viewport) layer properties are overridden in the current viewport:<br>**0** None.<br>**1** At least one layer. |
| **VpLayerOverridesMode** | 1 | Controls display and plot of VP layer property overrides:<br>**0** Not displayed or plotted.<br>**1** Displayed and plotted. |
| **VpMaximizedState** (R/O) | 0 | Reports whether viewport is maximized by VpMax command. |
| **VpRotateAssoc** | 1 | Toggles whether the contents are rotated with viewports:<br>**0** Views are not rotated by viewports.<br>**1** Views are rotated by viewports. |
| **VsACurvatureHigh** | 1.0 | Specifies value above which surface curvature is displayed green; range is any real number. |
| **VsACurvatureLow** | -1.0 | Specifies value below which surface curvature is displayed blue; range is any real number. |
| **VsACurvatureType** | 0 | Controls type of curvature analysis used by AnalysisCurvature command:<br>**0** Gaussian evaluates areas of high and low curvature.<br>**1** Mean evaluates mean curvature of u and v surface values.<br>**2** Maximum evaluates maximum curvature of u and v surface values.<br>**3** Minimum evaluates minimum curvature of u and v surface values. |
| **VsADraftangleHigh** | 3 | Specifies draft angle above which portions of model are displayed green during draft analysis:<br>**-90** Surface is parallel to UCS with surface normal facing opposite direction from construction plane.<br>**0** Surface is perpendicular to the construction plane.<br>**90** Surface is parallel to construction plane with surface normal facing same direction as UCS. |

| Variable Name | Default Value | Meaning |
|---|---|---|
| ◉ **VsADraftangleLow** | -3 | Specifies draft angle below which portions of model are displayed with blue during draft analysis:<br>  **-90** Surface is parallel to UCS with surface normal facing opposite direction from construction plane.<br>    **0** Surface is perpendicular to construction plane.<br>  **90** Surface is parallel to construction plane with surface normal facing same direction as UCS. |
| ◉ **VsAZebraColor1** | "Rgb:255,255,255" | Specifies first color of zebra analysis stripes; uses ACI, RGB, HSL, or ColorBook values. |
| ◉ **VsAZebraColor2** | "Rgb:0,0,0" | Specifies second color of zebra analysis stripes. |
| ◉ **VsAZebraDirection** | 90 | Specifies angle of zebra stripes; range is 0 to 90 degrees. |
| ◉ **VsAZebraSize** | 45 | Specifies width of stripes in zebra analysis displays; range is 1 to 100 pixels. |
| ◉ **VsAZebraType** | 1 | Controls type of zebra analysis display:<br>  **0** Chrome ball effect.<br>  **1** Cylinder effect. |
| **VsBackgrounds** | 1 | Toggles display of backgrounds in visual styles. |
| **VsEdgeColor** | "ByEntity" | Specifies the edge color. |
| **VsEdgeJitter** | -2 | Specifies the level of pencil effect:<br>  **0** *or* **-*n*** None.<br>  **1** Low.<br>  **2** Medium.<br>  **3** High. |
| **VsEdgeOverhang** | -6 | Specifies extension of pencil lines beyond edges; range is 1 to 100 pixels; *-n* = none. |
| **VsEdges** | 1 | Specifies types of edges to display:<br>  **0** Displays no edges.<br>  **1** Display isolines.<br>  **2** Displays facets and edges. |
| **VsEdgeSmooth** | 1 | Specifies crease angle; range is 0 to 180 degrees. |
| ◉ **VsEdgeLEx** | -6 | Specifies length of line extensions of edges in visual styles. Negative number turns off extensions; range is 0 to 100 pixels. |
| **VsFaceColorMode** | 0 | Determines color of faces. |
| **VsFaceHighlight** | -30 | Specifies the size of highlights; range is -100 to 100. Ignored when VsMaterialMode is 1 or 2 and objects have materials attached. |
| **VsFaceOpacity** | -60 | Controls the transparency/opacity of faces; range is -100 to 100 (fully opaque). |
| **VsFaceStyle** | 0 | Determines how faces are displayed:<br>  **0** No rendering.<br>  **1** Real.<br>  **2** Gooch. |
| **VsHaloGap** | 0 | Specifies halo gap; range is 0 to 100 pixels. |
| **VsHidePrecision** | 0 | Toggles accuracy of hides and shades. |
| **VsIntersectionColor** | "7 (white)" | Specifies the color of intersecting polylines. |

| Variable Name | Default Value | Meaning |
|---|---|---|
| **VsIntersectionEdges** | 0 | Toggles the display of intersecting edges. |
| **VsIntersectionLtype** | 1 | Specifies the linetype for intersecting lines:<br>  **0** Off.<br>  **1** Solid.<br>  **2** Dashed.<br>  **3** Dotted.<br>  **4** Short dash.<br>  **5** Medium dash.<br>  **6** Long dash.<br>  **7** Double-short dash.<br>  **8** Double-medium dash.<br>  **9** Double-long dash.<br>  **10** Medium-long dash.<br>  **11** Sparse dot. |
| **VsIsoOntop** | 0 | Toggles whether isolines are displayed. |
| **VsLightingQuality** | 1 | Toggles the quality of lighting:<br>  **0** Displays facets.<br>  **1** Smooths facets.<br>  **2** Turns on per-pixel lighting. |
| **VsMaterialMode** | 0 | Controls the display of material finishes:<br>  **0** Does not display materials or textures.<br>  **1** Displays materials only.<br>  **2** Displays materials and textures. |
| **VSMax** (R/O) | *varies* | Reports the upper-right corner of virtual screen, such as 37.46,27.00,0.00. |
| **VSMin** (R/O) | *varies* | Reports the lower-left corner of virtual screen, such as -24.97, -18.00,0.0. |
| **VsMonoColor** | "RGB:255,255,255" | Specifies the monochrome tint. |
| **VsObscuredColor** | "ByEntity" | Specifies color of obscured lines. |
| **VsObscuredEdges** | 1 | Toggles display of obscured edges. |
| **VsObscuredLtype** | 1 | Specifies linetype of obscured lines; see VsIntersectionLtype. |
| **VsOccludedColor** | "Byentity" | Specifies color of hidden (occluded) lines in visual styles; uses ACI, RGB, HSL, or ColorBook values. |
| **VsOccludedEdges** | 1 | Toggles display of hidden (occluded) edges in visual styles: (0) off or (1) on. |
| **VsOccludedLtype** | 1 | Controls linetype of hidden (occluded) lines in visual styles. Changing this system variable creates a new unsaved visual style. Range is 1 to 11:<br>  **1** Solid lines (default for most visual styles).<br>  **2** Dashed lines (default for hidden and shaded with edges).<br>  **3** Dotted lines.<br>  **4** Short dashes.<br>  **5** Medium dashes.<br>  **6** Long dashes.<br>  **7** Double short dashes. |

| Variable Name | Default Value | Meaning |
|---|---|---|
| | | 8 Double medium dashes. |
| | | 9 Double long dashes. |
| | | 10 Medium long dashes. |
| | | 11 Sparse dots. |
| **VsShadows** | 0 | Determines the quality of shadows: |
| | | 0 Does not display shadows. |
| | | 1 Displays ground shadows. |
| | | 2 Displays full shadows. |
| **VsSilhEdges** | 0 | Toggles the display of silhouette edges. |
| **VsSilhWidth** | 5 | Specifies the width of silhouette edge lines; range is 1 to 25 pixels. |
| **VsState** | 0 | Toggles the Visual Styles palette. |
| **VtDuration** | 750 | Controls the duration of smooth view transition; range is 0 to 5000 seconds. |
| **VtEnable** | 3 | Controls smooth view transitions for pans, zooms, view rotations, and scripts: |
| | | 0 Turned off. |
| | | 1 Pan and zoom. |
| | | 2 View rotation. |
| | | 3 Pan, zoom, and view rotation. |
| | | 4 During scripts only. |
| | | 5 Pan and zoom during scripts. |
| | | 6 View rotation during scripts. |
| | | 7 Pan, zoom, and view rotation during scripts. |
| **VtFps** | 7 | Controls minimum speed for smooth view transitions; range is 1 to 30 frames per second. |
| **W** | | |
| *WbDefaultBrowser (R/O)* | 0 | *Specifies which Web browser is used for the new Web-based help system:* |
| | | *0 Default system browser.* |
| | | *1 Internet Explorer.* |
| *WbHelpOnline (R/O)* | 1 | *Toggles Web-based help is accessed from Autodesk when online: (0) local access only or (1) local and Autodesk online access.* |
| *WbHelpType (R/O)* | 1 | *Toggles the source of help files:* |
| | | *0 CHM (traditional compiled help files).* |
| | | *1 HTML (hypertext markup language).* |
| **WhipArc** | 0 | Toggles display of circular objects: |
| | | 0 Displays as connected vectors. |
| | | 1 Displays as true circles and arcs. |
| **WhipThread** | 1 | Controls multithreaded processing on two CPUs (if present) during redraws and regens: |
| | | 0 Performs single-threaded calculations. |
| | | 1 Makes regenerations multi-threaded. |
| | | 2 Makes redraws multi-threaded. |
| | | 3 Makes regens and redraws multi-threaded. |
| **WindowAreaColor** | 150 | Specifies ACI color of windowed selection area. |

| Variable Name | Default Value | Meaning |
|---|---|---|
| **WmfBkgnd** | Off | Toggles background of *.wmf* files:<br>**Off** Background is transparent.<br>**On** Background is same as AutoCAD's background color. |
| **WmfForegnd** | Off | Toggles foreground colors of exported WMF images:<br>**Off** Foreground is darker than background.<br>**On** Foreground is lighter than background. |
| **WorldUcs** (r/o) | 1 | Toggles matching of WCS with UCS:<br>**0** Current UCS does not match WCS.<br>**1** UCS matches WCS. |
| **WorldView** | 1 | Toggles view during 3dOrbit, DView, and VPoint commands:<br>**0** Current UCS.<br>**1** WCS. |
| **WriteStat** (r/o) | 1 | Toggle: reports whether *.dwg* file is read-only:<br>**0** Drawing file cannot be written to.<br>**1** Drawing file can be written to. |
| **WsAutosave** | 0 | Toggles automatic saving of changes to workspaces: (0) off or (1) on. |
| **WsCurrent** | "*varies*" | Controls name of current workspace. |

## X

| Variable Name | Default Value | Meaning |
|---|---|---|
| **XClipFrame** | 2 | Toggles visibility of xref clipping boundary:<br>**0** Hides and does not plot the xclip frame.<br>**1** Displays and plots xclip frames.<br>**2** Displays but does not plot xclip frames. |
| **XDwgFadeCtl** | 70 | Specifies the amount attached xrefs are shown faded:<br>**0** Not faded.<br>**>0** Increasingly faded, to a maximum of 90%.<br>**<0** Not faded, but fade value is retained. |
| **XEdit** | 1 | Toggles editing of xrefs:<br>**0** Cannot in-place refedit.<br>**1** Can in-place refedit. |
| **XFadeCtl** | 50 | Controls faded display of objects not being edited in-place:<br>**0** No fading; minimum value.<br>**90** 90% fading; maximum value. |
| **XLoadCtl** | 2 | Controls demand loading:<br>**0** Demand loading turned off; entire drawing is loaded.<br>**1** Demand loading turned on; xref file opened.<br>**2** Demand loading turned on; a *copy* of the xref file is opened. |
| **XLoadPath** | "...\temp" | Specifies path for storing temporary copies of demand-loaded xref files. |
| **XRefCtl** | 0 | Toggles creation of *.xlg* xref log files. |
| **XrefNotify** | 2 | Controls notification of updated and missing xrefs:<br>**0** Displays no alert.<br>**1** Displays icon indicating xrefs are attached; yellow alert indicates missing xrefs.<br>**2** Also displays balloon messages when an xref is modified. |

| Variable Name | Default Value | Meaning |
|---|---|---|
| **XrefType** | 0 | Toggles xrefs:<br>**0** Attached.<br>**1** Overlaid. |

**Z**

| Variable Name | Default Value | Meaning |
|---|---|---|
| **ZoomFactor** | 60 | Controls the zoom level via mouse wheel; range from 3 to 100. |
| **ZoomWheel** | 0 | Switches the zoom direction when mouse wheel is rotated forward:<br>**0** Zooms in (opposite of most other software, including Inventor).<br>**1** Zooms out. |

**3**

| Variable Name | Default Value | Meaning |
|---|---|---|
| **3dConversionMode** | 1 | Controls how material and light definitions are converted when pre-AutoCAD 2008 drawings are opened:<br>**0** None converted.<br>**1** Converted automatically.<br>**2** Converted after prompted. |
| **3dDwfPrc** | 2 | Level of precision in drawings exported as *.dwf* files:<br>**1** 1<br>**2** 0.5<br>**3** 0.2<br>**4** 0.1<br>**5** 0.01<br>**6** 0.001 |
| **3dOsMode** | 11 | Specifies the current 3D object snaps modes (bit code):<br>**0** Disables all 3D osnap modes.<br>**1** Temporarily disables all 3D osnap modes (ZNON).<br>**2** Snaps to vertices (ZVER).<br>**4** Snaps to midpoints of edges (ZMID).<br>**8** Snaps to centers of faces (ZCEN).<br>**16** Snaps to spline knots (ZKNO).<br>**32** Snaps to the perpendiculars of faces (ZPER).<br>**64** Snaps to objects nearest to faces (ZNEA).<br>**126** Turns on all 3D osnap modes.<br>**11** 1 + 2 + 8 = center of a face and vertex. |
| **3dSelectionMode** | 1 | Toggles how visually overlapping objects are selected (other than in 2D and 3D wireframe visual modes; this system variable is slated for removal in a future release of AutoCAD):<br>**0** Uses traditional 3D selection.<br>**1** Uses line-of-sight selection. |

# APPENDIX D

# Summary of Dimension Variables

The following sections summarize the meaning of dimension variables, grouped by purpose. (For more on dimension variables, see chapter 12 on editing dimensions.)

## General

**DimAnno** (<u>anno</u>tative) reports whether the current dimension style uses annotative scaling; read-only. When on (set to 1), DIMSCALE is set to 0.

**DimAso** (<u>asso</u>ciative) toggles dimensions between associative and non-associative.

**DimAssoc** (<u>assoc</u>iative) determines how dimensions are created: exploded, attached to defpoints, or attached to objects.

**DimScale** specifies the overall scale factor for dimensions. When set to zero, AutoCAD calculates the scale factor for dimensions by two methods:

- When DIMANNO > 0, AutoCAD divides the scale factor of the current model space viewport by that of paper space.
- When DIMANNO = 0, AutoCAD uses the current viewport scale factor.

**DimSho** (<u>sho</u>w) updates dimensions while dragging.

**DimStyle** specifies the current dimension style as set by the DIMSTYLE command. (Note that the system variable and the command share the same name, so use the SETVAR command to access the dimvar.)

## Dimension Lines

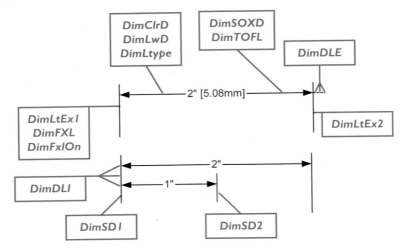

### Extension and Offset

**DimDLE** (dimension line extension) specifies how far the dimension line extends past the extension lines.

**DimLwD** (lineweight dimension) specifies the lineweight of dimension lines.

**DimDLI** (dimension line increment) specifies the offset for baseline dimensions.

### Suppression and Position of Dimension Lines

**DimSD1** (suppress dimension 1) suppresses the first dimension line.

**DimSD2** (suppress dimension 2) suppresses the second dimension line.

**DimSOXD** (suppress outside extension dimension) suppresses dimension lines outside extension lines.

**DimTOFL** (text outside forced line) forces dimension lines inside extension lines when text is outside the extension lines.

### Color and Lineweight

**DimClrD** (color dimension) specifies the color of dimension lines.

**DimLwD** (lineweight dimension) specifies the lineweight for dimension lines.

### Linetype

**DimLtype** (linetype) specifies the linetype for dimension lines.

## Extension Lines

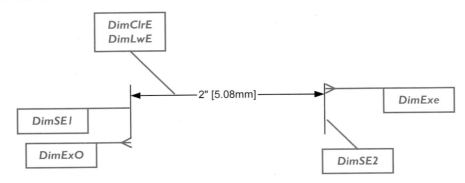

## Suppression and Linetype

**DimFXL** (fixed-length extension lines) specifies the default length of fixed-length extension lines.

**DimFxlOn** (fixed-length extension lines on) toggles the use of fixed-length extension lines.

**DimLtEx1** (linetype for extension line 1) specifies the linetype for the first extension line.

**DimLtEx2** (linetype for extension line 2) specifies the linetype for the second extension line.

## Color and Lineweight

**DimClrE** (color extension) specifies the color of extension lines.

**DimLwE** (lineweight extension) specifies the lineweight of extension lines.

## Extension and Offset

**DimExE** (extension extend) extends the extension lines above the dimension line.

**DimExO** (extension offset) offsets the extension line from the origin.

## Suppression and Position

**DimSE1** (suppress extension 1) suppresses the first extension line.

**DimSE2** (suppress extension 2) suppresses the second extension line.

### Arrowheads

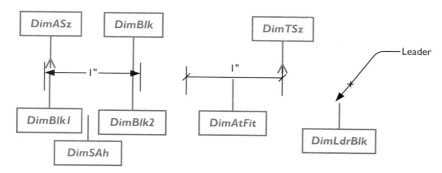

### Size and Fit

**DimASz** (arrow size) determines the length of arrowhead blocks.

**DimTSz** (tick size) specifies the size of tick strokes.

**DimATFit** (arrow text fit) determines under which conditions arrowheads and text are fitted between extension lines; see Alphabetical Listing of Dimvars later in this chapter.

### Names of Blocks

**DimBlk** (block) names the arrowhead block.

**DimBlk1** (block 1) names the first arrowhead block.

**DimBlk2** (block 2) names the second arrowhead block.

**DimLdrBlk** (leader block) names the leader arrowhead.

**DimSAh** (separate arrowheads) determines whether separate arrowhead blocks are used.

## Center Marks

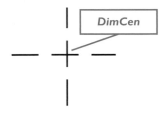

**DimCen** (<u>cen</u>ter) specifies the mark size and line.

## Arc Length Dimensions

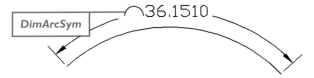

**DimArcSym** (<u>arc</u> <u>sym</u>bol) specifies the location of the arc length symbol.

## Jogged Dimensions

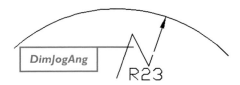

**DimJogAng** (<u>jog</u> <u>ang</u>le) specifies the default angle of jogged radial dimension lines.

## Text

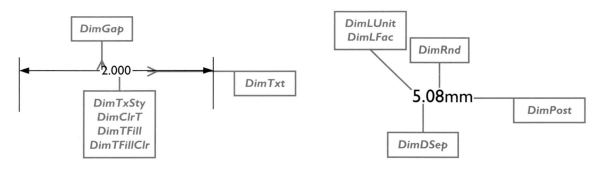

### Color and Format of Text

**DimClrT** (<u>col</u>or <u>t</u>ext) specifies the color of the text.

**DimGap** determines the gap between the dimension line and text.

**DimTxSty** (<u>t</u>ext <u>sty</u>le) specifies the text style.

**DimTxt** (<u>t</u>e<u>xt</u>) stores the text height.

**DimTxtDirection** specifies the reading direction of vertical and horizontal text.

### Background Fill

**DimTFill** (text fill) switches the background fill of dimension text among these: none, same as drawing background color, or as specified by DIMTFILLCLR.

**DimTFillClr** (text fill color) specifies the color of the background fill.

### Units, Scale, and Precision

**DimLUnit** (linear unit) specifies the format of linear units.

**DimPost** (postfix) specifies prefixes and suffixes of dimension text.

**DimLFac** (linear factor) specifies the linear unit scale factor.

**DimDSep** (decimal separator) specifies the decimal separator.

**DimRnd** (round) rounds distances.

**DimFrac** (fractions) determines how fractions are stacked: horizontal, diagonal, or not stacked.

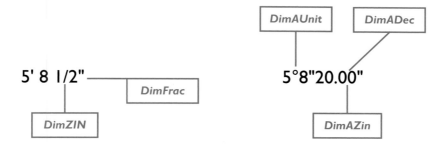

**DimZIn** (zero inches) suppresses zeroes in feet-inches units.

**DimADec** (angular decimal) specifies the precision of angular dimensions.

**DimAUnit** (angular unit) specifies the format for angular dimensions.

**DimAZIn** (angular zero inches) controls how zeros are suppressed in angular dimensions.

### Justification

**DimJust** (justify) specifies alignment (justification) of horizontal text: left, center, right.

**DimTAD** (text adjust dimension) positions dimension text vertically: top, center, outside, JIS.

**DimTVP** (text vertical position) specifies the vertical text position when DIMTAD is off.

**DimTIH** (text inside horizontal) determines whether text inside extensions is horizontal.

**DimTIX** (text in extension) forces text between extension lines.

**DimTMove** (text move) determines how dimension text is relocated: with dimension line, an added leader, or without a leader.

**DimTOH** (text outside horizontal) determines whether text outside extension lines is horizontal.

**DimUPT** (user position text) toggles whether user positions dimension line and/or text.

### Alternate Text

**DimAlt** (alternate) toggles alternate units.

**DimAltU** (alternate units) determines the format of alternate units.

**DimAltD** (alternate decimals) specifies decimal places of alternate units.

**DimAltF** (alternate factor) specifies the scale factor of alternate units.

**DimAltRnd** (alternate rounding) rounds off alternate units.

**DimAltZ** (alternate zero) suppresses zeros in alternate units.

**DimAPost** (alternate postfix) determines the prefixes and suffixes for alternate text.

Limits

**DimLim** (limits) toggles dimension limits.

Tolerance Text

**DimTol** (tolerance) toggles whether tolerances are drawn.

**DimTZIn** (tolerance zero inches) suppresses zeros in tolerances.

**DimTFac** (tolerance factor) scales the tolerance text height.

**DimTolJ** (tolerance justify) justifies tolerance text vertically.

**DimTP** (tolerance plus) specifies the plus tolerance value.

**DimTM** (tolerance minus) specifies the minus tolerance value.

Primary Tolerance

**DimDec** (decimals) specifies the decimal places for the primary tolerance.

**DimTDec** (tolerance decimals) specifies the decimal places for primary tolerance units.

Alternate Tolerance

**DimAltTD** (alternate tolerance decimals) specifies decimal places for tolerance alternate units.

**DimAltTZ** (alternate tolerance zero) suppresses zeros in tolerance alternate unit.

## Dimensional Constraints

**CConstraintForm** (current constraint form) controls whether annotational or dynamic constraints are applied to objects.

**ConstraintNameFormat** specifies the type of text displayed by dimensional constraints: variable names, values, or variable names and expressios.

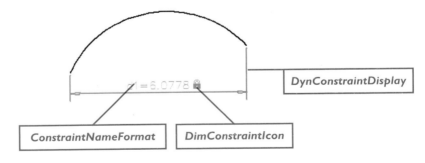

**DimConstraintIcon** toggles the lock icon next to dimensional constraint text.

**DynConstraintDisplay** displays and hides dynamic constraints.

## ALPHABETICAL LISTING OF DIMVARS

The table summarizes dimension variables, default values in the *acad.dwt* template, and their optional values.

| DimVar | Default | Settings and Options |
|---|---|---|
| **A** | | |
| DimADec | 0 | Specifies angular dimension precision: |
| | | -1 Uses **DimDec** setting (default). |
| | | 0 Uses zero decimal places (minimum). |
| | | 8 Uses eight decimal places (maximum). |
| DimAlt | Off | Specifies alternate units: |
| | | On Enables. |
| | | Off Disables. |
| DimAltD | 2 | Specifies alternate unit decimal places. |
| DimAltF | 25.4000 | Specifies alternate unit scale factor. |
| DimAltRnd | 0.0000 | Specifies rounding factor of alternate units. |
| DimAltTD | 2 | Specifies tolerance alternate unit decimal places. |
| DimAltTZ | 0 | Specifies how alternate tolerance units zeros are handled: |
| | | 0 Does not suppress zeros. |
| | | 1 Suppresses all zeros. |
| | | 2 Includes 0 feet, but suppresses 0 inches. |
| | | 3 Includes 0 inches, but suppresses 0 feet. |
| | | 4 Suppresses leading zeros. |
| | | 8 Suppresses trailing zeros. |
| DimAltU | 2 | Specifies format of alternate units: |
| | | 1 Scientific. |
| | | 2 Decimal. |
| | | 3 Engineering. |
| | | 4 Architectural; stacked. |
| | | 5 Fractional; stacked. |
| | | 6 Architectural. |
| | | 7 Fractional. |
| | | 8 Windows desktop units setting. |
| DimAltZ | 0 | Specifies zero suppression for alternate units: |
| | | 0 Suppresses 0 ft and 0 in. |
| | | 1 Includes 0 ft and 0 in. |
| | | 2 Includes 0 ft; suppress 0 in. |
| | | 3 Suppresses 0 ft; include 0 in. |
| | | 4 Suppresses leading 0 in decimal dims. |
| | | 8 Suppresses trailing 0 in decimal dims. |
| | | 12 Suppresses leading and trailing zeroes. |
| DimAnno | 0 | Reports whether the current dimension style is annotative. |
| DimAPost | "" | Specifies prefix and suffix for alternate text. |
| DimArcSym | "" | Specifies position of the arc length symbol: |
| | | 0 Places before the dimension text. |
| | | 1 Places above dimension text. |
| | | 2 Not displayed. |
| DimAso | On | Toggles associative dimensions: |
| | | On Specifies dimensions are associative. |
| | | Off Specifies dimensions are not associative. |
| DimAssoc | 2 | Controls creation of dimensions: |
| | | 0 Explodes dimension elements. |
| | | 1 Attaches dimension blocks to defpoints. |
| | | 2 Attaches dimension objects to geometry of objects. |
| DimASz | 0.1800 | Specifies arrowhead length. |

| DimVar | Default | Settings and Options |
|---|---|---|
| DimAtFit | 3 | Specifies how to handle text and arrows when space is insufficient between extension lines:<br>0  Places text and arrows outside extension lines.<br>1  Places arrows first outside, then text.<br>2  Places text first outside, then arrows.<br>3  Places either text or arrows, whichever fits better. |
| DimAUnit | 0 | Specifies the angular dimension format:<br>0  Sets decimal degrees.<br>1  Sets Degrees.Minutes.Seconds.<br>2  Sets grad.<br>3  Sets radian.<br>4  Sets surveyor units. |
| DimAZin | 0 | Suppresses zeros in angular dimensions:<br>0  Displays all leading and trailing zeros.<br>1  Suppresses 0 in front of decimal.<br>2  Suppresses trailing zeros behind decimal.<br>3  Suppresses zeros in front and behind the decimal. |

## B

| DimVar | Default | Settings and Options |
|---|---|---|
| DimBlk | "" | Specifies arrowhead block name:<br>Architectural tick:  "Archtick"<br>Box filled:  "Boxfilled"<br>Box:  "Boxblank"<br>Closed blank:  "Closedblank"<br>Closed filled:  ""  (default)<br>Closed:  "Closed"<br>Datum triangle filled:  "Datumfilled"<br>Datum triangle:  "Datumblank"<br>Dot blanked:  "Dotblank"<br>Dot small:  "Dotsmall"<br>Dot:  "Dot"<br>Integral:  "Integral"<br>None:  "None"<br>Oblique:  "Oblique"<br>Open 30:  "Open30"<br>Open:  "Open"<br>Origin indication:  "Origin"<br>Rightangle:  "Open90" |
| DimBlk1 | "" | Names the first arrowhead's block; uses same names as DimBlk.<br>.  Returns to default arrowhead. |
| DimBlk2 | "" | Names the second arrowhead's block.<br>.  Returns to default arrowhead. |

## C

| DimVar | Default | Settings and Options |
|---|---|---|
| DimCen | 0.0900 | Specifies the center mark size:<br>-n  Draws center lines.<br>0  Does not draw center mark or lines.<br>+n  Draws center marks of length n. |

## D

| DimVar | Default | Settings and Options |
|---|---|---|
| DimClrD | 0 | Specifies the dimension line color:<br>0  ByBlock (default).<br>1  Red<br>...<br>255  Dark gray<br>256  ByLayer |
| DimClrE | 0 | Specifies the extension line and leader color. |

| DimVar | Default | Settings and Options |
|---|---|---|
| DimClrT | 0 | Specifies the dimension text color. |
| DimDec | 4 | Specifies the number of primary tolerance decimal places. |
| DimDLE | 0.0000 | Specifies the length of the dimension line extension. |
| DimDLI | 0.3800 | Specifies the length of the dimension line continuation increment. |
| DimDSep | "." | Specifies the decimal separator (must be a single character). |

## E

| DimVar | Default | Settings and Options |
|---|---|---|
| DimExe | 0.1800 | Specifies the length of extension above dimension line. |
| DimExO | 0.0625 | Specifies the extension line origin offset. |

## F

| DimVar | Default | Settings and Options |
|---|---|---|
| DimFrac | 0 | Specifies fraction format when **DimLUnit** is set to 4 or 5:<br>0  Horizontal.<br>1  Diagonal.<br>2  Not stacked. |
| DimFXL | 1.0000 | Specifies the default length of fixed-length extension lines. |
| DimFxlOn | Off | Toggles fixed-length extension lines:<br>**On**  Displays extension lines at a fixed length.<br>**Off**  Displays extension lines at the user-drawn length |

## G

| DimVar | Default | Settings and Options |
|---|---|---|
| DimGap | 0.0900 | Specifies the gap between the dimension line and the text. |

## J

| DimVar | Default | Settings and Options |
|---|---|---|
| DimJogAng | 45 | Specifies the default angle for jogged radial dimension lines. |
| DimJust | 0 | Specifies the horizontal text positioning:<br>0  Sets center justification.<br>1  Places next to first extension line.<br>2  Places next to second extension line.<br>3  Places above first extension line.<br>4  Places above second extension line. |

## L

| DimVar | Default | Settings and Options |
|---|---|---|
| DimLdrBlk | "" | Specifies block name for leader arrowhead; same names as **DimBlock**.<br>.  Return to default. |
| DimLFac | 1.0000 | Specifies the linear unit scale factor. |
| DimLim | Off | Toggles generation of dimension limits. |
| DimLtEx1 | "" | Specifies the linetype used for the first extension line. |
| DimLtEx2 | "" | Specifies the linetype used for the first extension line. |
| DimLtype | "" | Specifies the linetype used for dimension lines. |
| DimLUnit | 2 | Specifies the linear dimension units (except angular):<br>1  Scientific.<br>2  Decimal.<br>3  Engineering.<br>4  Architectural.<br>5  Fractional.<br>6  Windows desktop. |
| DimLwD | -2 | Specifies the dimension line lineweight; valid values are BYLAYER, BYBLOCK, or an integer multiple of 0.01 mm. |
| DimLwE | -2 | Specifies the extension lineweight; valid values are BYLAYER, BYBLOCK, or an integer multiple of 0.01 mm. |

| DimVar | Default | Settings and Options |
|---|---|---|

**P**

| DimPost | "" | Specifies the default prefix or suffix for dimension text (maximum 13 characters): |

    ""      No suffix.
    <>mm  Millimeter suffix.
    <>Å   Angstrom suffix.

**R**

| DimRnd | 0.0000 | Specifies the rounding value for dimension distances. |

**S**

| DimSAh | Off | Specifies whether separate arrowhead blocks are used: |

    **Off**  Uses arrowhead defined by DimBlk.
    **On**  Uses arrowheads defined by DimBlk1 and DimBlk2.

| DimScale | 1.0000 | Specifies the overall scale factor for dimensions: |

    **0**  Value is computed from the scale between current model space viewport and paper space.
    **>0**  Scales text and arrowheads.

| DimSD1 | Off | Toggles suppression of the first dimension line: |

    **On**  First dimension line is suppressed.
    **Off**  Not suppressed.

| DimSD2 | Off | Toggles suppression of the second dimension line: |

    **On**  Second dimension line is suppressed.
    **Off**  Not suppressed.

| DimSE1 | Off | Toggles suppression of the first extension line: |

    **On**  First extension line is suppressed.
    **Off**  Not suppressed.

| DimSE2 | Off | Toggles suppression of the second extension line: |

    **On**  Second extension line is suppressed.
    **Off**  Not suppressed.

| DimSho | On | Toggles the updating of dimensions while dragging: |

    **On**  Dimensions are updated during drag.
    **Off**  Dimensions are updated after drag.

| DimSOXD | Off | Toggles suppression of dimension lines outside extension lines: |

    **On**  Dimension lines not drawn outside extension lines.
    **Off**  Are drawn outside extension lines.

| DimStyle | "STANDARD" | Names the current dimension style. |

**T**

| DimTAD | 0 | Vertical position of dimension text: |

    **0**  Centered between extension lines.
    **1**  Above dimension line, except when dimension line not horizontal and **DimTIH** = 1.
    **2**  On side of dimension line farthest from the defining points.
    **3**  Conforms to JIS.
    **4**  Places text below dimension line.

| DimTDec | 4 | Specifies the number of primary tolerance decimal places. |
| DimTFac | 1.0000 | Specifies the tolerance text height scaling factor. |
| DimTFill | 0 | Toggles the display of background fill: |

    **0**  Uses no fill.
    **1**  Uses background color of drawing.
    **2**  Uses color specified by DimTFillClr.

| DimTFillClr | 0 | Specifies the color used for the background of dimension text. |

| DimVar | Default | Settings and Options |
|---|---|---|
| DimTIH | On | Toggles the alignment of text inside extensions:<br>**Off** Text aligned with dimension line.<br>**On** Text is horizontal. |
| DimTIX | Off | Toggles the placement of text inside extensions:<br>**Off** Places text inside extension lines, if room.<br>**On** Forces text between the extension lines. |
| DimTM | 0.0000 | Specifies the minus tolerance. |
| DimTMove | 0 | Determines how dimension text is moved:<br>**0** Moves dimension line with text.<br>**1** Adds a leader when text is moved.<br>**2** Moves text anywhere; no leader. |
| DimTOFL | Off | Forces line inside extension lines:<br>**Off** Dimension lines not drawn when arrowheads are outside.<br>**On** Dimension lines drawn, even when arrowheads are outside. |
| DimTOH | On | Toggles the alignment of text outside extension lines:<br>**Off** Text aligned with dimension line.<br>**On** Text is horizontal. |
| DimTol | Off | Generate dimension tolerances:<br>**Off** Tolerances not drawn.<br>**On** Tolerances are drawn. |
| DimTolJ | 1 | Specifies the vertical justification of tolerance text:<br>**0** Bottom.<br>**1** Middle.<br>**2** Top. |
| DimTP | 0.0000 | Specifies the plus tolerance value. |
| DimTSz | 0.0000 | Specifies the size of oblique tick strokes:<br>**0** Arrowheads.<br>**>0** Oblique strokes. |
| DimTVP | 0.0000 | Specifies the text vertical position when **DimTAD**=0:<br>**1** Turns on **DimTAD**.<br>**>-0.7 or <0.7** Dimension line is split for text. |
| DimTxSty | "STANDARD" | Specifies the dimension text style. |
| DimTxt | 0.1800 | Specifies the text height. |
| DimTxtDirection | 0 | Specifies the reading direction of text:<br>**0** Horizontal text reads left to right;<br>Vertical text reads bottom to top.<br>**1** Horizontal text read right to left;<br>Vertical text reads top to bottom. |
| DimTZin | 0 | Specifies tolerance zero suppression:<br>**0** Suppresses 0 ft and 0 in.<br>**1** Includes 0 ft and 0 in.<br>**2** Includes 0 ft; suppress 0 in.<br>**3** Suppresses 0 ft; include 0 in.<br>**4** Suppresses leading 0 in decimal dim.<br>**8** Suppresses trailing 0 in decimal dim.<br>**12** Suppresses leading and trailing zeroes. |

## U

| DimVar | Default | Settings and Options |
|---|---|---|
| DimUPT | Off | User-positioned text:<br>**Off** Cursor positions dimension line.<br>**On** Cursor also positions text. |

| DimVar | Default | Settings and Options |
|---|---|---|

## Z

| DimZIN | 0 | Specifies the suppression of 0 in feet-inches units: |
|---|---|---|

    0   Suppresses 0 ft and 0 in.
    1   Includes 0 ft and 0 in.
    2   Includes 0 ft; suppress 0 in.
    3   Suppresses 0 ft; include 0 in.
    4   Suppresses leading 0 in decimal dim.
    8   Suppresses trailing 0 in decimal dim.
   12  Suppresses leading and trailing zeroes.

Two dimvars are considered obsolete by Autodesk, but remain in AutoCAD for compatibility reasons:

| Obsolete DimVar | Default | Comments |
|---|---|---|
| DimFit | 3 | Uses DimATfit and DimTMove instead. |
| DimUnit | 2 | Replaces DimLUnit and DimFrac. |

The following system variables are used by dimensional constraints:

| Sysvar | Default | Comments |
|---|---|---|
| CConstraintForm | 0 | Controls whether annotational or dynamic constraints are applied: |

    0   Applies dynamic constraints.
    1   Applies annotative constraints.

| ConstraintNameFormat | 3 | Specifies the type of text displayed by dimensional constraints: |
|---|---|---|

    0   Variable names.
    1   Values only.
    2   Variable names and expressions.

| DimConstraintIcon | 3 | Toggles the lock icon next to dimensional constraint text: |
|---|---|---|

    0   Does not display the lock icon.
    1   Displays icon next to dynamic constraints only.
    2   Displays the icon next to annotational constraints only.
    3   Displays the icon next to dynamic and annotational constraints.

| DynConstraintDisplay | 1 | Displays and hides dynamic constraints: |
|---|---|---|

    0   Off.
    1   On.

# IX

# Index

# INDEX

## A

a (add objects to the selection set)   293
accessing backup files   119
ACI (AutoCAD color index)   325
add to selection   298
adding
   areas   714
   creases   799
   dimensions from a baseline   583
   inspection text to dimensions   626
   leaders to mleaders (mleaderedit)   528
**ADDSELECTED**   302
advanced wizard   92
**AI_DIM**   627
**AI_DIM_TEXTABOVE**   630
**AI_DIM_TEXTCENTER**   630
**AI_DIM_TEXTHOME**   630
**AIDIM**   627
**AIDIMFLIPARROW**   618
**AIDIMPREC**   628
**AIDIMTEXTMOVE**   629
aliases   61, 1000
**ALIGN**   488
aligned
   dimensions   582
   mleaders   534
alignment point acquisition   215
all   441
   select all objects   288
   zoom   223
alphabetical listing of dimension variables   1073
alternate
   commands   65
   units   568, 649
always display text as wysiwyg   493

angle
   chamfering   394
   dimensions   580
   measurement   712
angular dimensions   594, 648
**ANNORESET**   550
annotation scaling   52, 109, 542
   controlling visibility   556
   dimensioning   599
   hatch   279
   linetype scales   338
   property   498
   text   490
   text style   556
   text, resetting and updating   556
**ANNOTATIVEDWG**   550
**ANNOUPDATE**   550
**APERTURE**   200
apparent intersection, osnap   202
application menu   17
applying constraints
   dimensional   663
   manually   691
applying layer states   361
**ARC**   147
arcs
   dimensions   569
   elliptical   158
   fillet   391
   grips   455
   length dimensions   592
   length symbol   641
   pline   154
   revcloud   399

Command and system variable names are shown in **BOLDFACE**.

architecture  lii
areas
    adding and subtracting  714
    perimeters  713
    plotting  910
    rectang  136
**ARRAY**  384
arrowheads  566, 638
    flipping  618
    size  638
ASCII characters  492
assigning plot styles  953
    to layers  955
    to layouts  953
    to objects  954
associative dimensions converted to dimensional
    constraints  667
associative hatch patterns  279
**ATTDIA**  65
attributes
    definitions, change  438
    text editor  537
au (select with the automatic option)  291
auto, undo  163
AutoCAD 2011  lxi
    classic workspace  21
    color system  325
    commands  981
    coordinates  70
    file extensions  977
    history of 3D  754
    LT, 3D  816
    overview  lx
    program files  978
    user interface  42
autocaps  508
autocomplete  60
**AUTOCONSTRAIN**  687
automatically
    backup  117
    constrain geometry  687
    scale text  543
autopublish  923
autospool  921
autotrack settings  215

## B

backgrounds
    color 0  326
    mask  499
    plot  918
    processing options  923
backup copies  117
**BASE**  262

base points
    blocks  257
    grips editing  454
baseline dimensions  583
    spacing  637
    quick dimensions  587
**BEDIT**  264
begin/end, undo  163
benefits of computer-aided drafting  li
bevel to solids' edges  846
binding external references  633
**BLOCK**  248
    change  438
    editor  264
    measure and divide  383
    template file  938
    templates  67
bold, mtext  498
borders
    hiding viewport  861
    viewport properties  866
**BOUNDARY**  285
boundaries
    defining  272
    recreating  273
    removing  273
    retention  274
box (select with a box)  290
**BREAK**  419
breaks
    dimensions  641
    dimension and extension lines  620
    first  420
    objects  419
bubbles, drawing mleaders with  529
bullets  503
buttons  62
    mouse and digitizer  1008
byblock color #0  326
byentity color #257  327
bylayer
    color #256  326
    reverting properties to  869

## C

c (select with a crossing window)  289
CAD applications  lii
canceling selection process  293
capping leaking mesh objects  802
cartesian coordinates  71
cells
    content  729
    styles  725
center marks  595
    for circles  640

centers
- ellipse 158
- finding the 718
- osnap 202

central processing unit 969

**CHAMFER** 393, 833

**CHAMFEREDGE** 833

chamfering
- 3D objects 837
- dimensions 573
- interactive 836
- meshes 835
- objects 393, 834
- rectangles 140

**CHANGE** 437

changing
- annotation scale 601
- colors 104, 324
- current layer 364
- dimstyle 653
- display precision 628
- draw order 306
- from wcs to ucs 772
- layer status 364
- linetype scaling 601
- linetypes 335
- lineweights 333
- object size through constraints 668
- objects 437
- properties 340
- screen colors 103
- selected items 292
- size 434
- text case 495
- windowed views 236
- workspaces 20

character set 509

characters, wild-card 975

check spelling 494, 539

checking interference between solids 832

**CIRCLE** 144

circles
- change 438
- dimensions 569
- fillet 391

circumscribed about circle 143

**CLAYER** 365

clean screen 53

**CLEANSCREENON** 231

clearing local overrides 656

closing
- line 133
- pline 154
- polyline editing 443

collecting bubbles into one mleader 532

**COLOR** 324
- 0 (background) 326
- 7 (white) 326
- books 329
- bylayer, byblock, and byentity 326
- dimstyles 636
- grips 450
- hatch and background color 276
- layer 349
- mtext 499
- names 326
- numbers 326
- xplode 441

color-dependent
- change to named 951
- plot style tables (*ctb* files) 950

column and row settings, tables 724

columns, mtext 505, 514

combining
- dimensional and geometric constraints 703
- paragraphs 504
- zooms and pans 164

commands 981
- aliases 1000
- bar 8
- in AutoCAD 8
- input considerations 55
- nomenclature lxiv
- undoing 162

companion files lxv

components of computers 968

cones, dimensions 571

configurations
- physical pen 947
- plotter 939

constraining geometry automatically 687

**CONSTRAINTINFER** 697

constraints
- dimensional 664
- geometric 685
- inferring 698

**CONSTRAINTSETTINGS** 679

constructing objects 66, 166, 373

construction lines 216

continuing
- arc 151
- dimensions 584
- line 134
- quick dimensions 586

control and alternate keys lxiv

controlling
- annotation visibility 556
- codes 491
- dimvars 631
- explosions 441

preview image 906
undo 163
conventions lxiv
converting
    field to text 495
    associative dimensions to constraints 667
**CONVERTCTB** 951
**CONVERTPSTYLES** 951
coordinate systems lx, 69, 77, 771
**COPY** 374
    grips editing 454
    plotting 911
    right-click 457
    rotate 434
    scale 435
countersinks 848
cp (select with a crossing polygon) 291
CPU 969
creases, adding 799
creating
    3D mesh models 791
    annotative text style 556
    blocks 249
    boundaries 285
    holes 764
    layer states 361
    layers 347
    new dimstyles 635
    new plot styles 956
    plotter configurations 939
    polygonal viewports 888
    rectangular viewports 886
    viewports from objects 889
crossing, trimming 423
*ctb* files (color-dependent plot style tables) 950
culled selections 813
culling 3D object selection 813
cursors 44
    icons 45
custom arrowheads 639
cylindrical coordinates 73

# D

data sources 725
data, write-protecting 970
datum point, quick dimensions 588
DC.. commands 665
**DDEDIT** 622
**DDPTYPE** 161
**DDPTYPE** 384
**DDSELECT** 296
decurve, polyline editing 446
defaults
    background colors 103
    lineweights 333
    plot settings for new drawings 922

defining
    boundaries by picking areas 272
    boundaries by selecting objects 273
    starting points 820
    text styles 515
**DELCONSTRAINT** 673
deleting
    dimstyles 652
    layers 361
    named views 234
delta, lengthen 428
Design Review 400
**DESIGNCENTER** 260, 632
designing blocks 252
destination, wblock 262
determining
    height of text 489
    text height in layouts 875
dialog boxes 63
    file-related 975
diameters
    circle 145
    quick dimensions 588
differences between local and network printers 903
different fillet results for different objects 390
digitizers 973
    buttons 1008
**DIMALIGNED** 581
**DIMANGULAR** 593
**DIMANNO** 550
**DIMARC** 592
**DIMBASELINE** 583
**DIMBREAK** 620
**DIMCEN** 596
**DIMCENTER** 595
**DIMCONTINUE** 583
**DIMDIAMETER** 589
**DIMEDIT** 622
dimensions
    aligned 582
    angular 648
    annotative 599
    arc length 592
    break 641
    breaking dimension and extension lines 620
    converting to dimensional constraints 667
    editing 617
    editing text 624
    editing through objects 657
    horizontal 575
    inspection text 626
    jogging 619
    lines 565
    modes 573
    moving text 629
    objects 568

obliquing extension lines   623
parts of   564
placing jogged radial   591
quick   585
radial   590
radius jog   641
rectang   135
repositioning text   623
rotated   577
spacing evenly   621
standards   574
text   566
variables and styles   574
variables, alphabetical listing   1073
variables, summary   1067
vertically   577
dimensional constraints   664
**DIMINSPECT**   625
**DIMJOGGED**   591
**DIMJOGLINE**   619
**DIMLINEAR**   575
**DIMRADIUS**   589
**DIMSPACE**   621
**DIMSTYLE**   633
dimension styles
   fit options   644
   global temporary changes   653
   modifying   635
   new   635
   overriding   652
   renaming and deleting   652
   setting the current   652
   sources   631
**DIMTEDIT**   622
direct distance entry   78, 458
direct editing with grips   449
disks
   care   970
   drives   969
displacement, copy   375
displaying
   details in viewports   865
   files   975
   menu bar   19
distance measurements   711
**DIVIDE**   381
dividing objects   382
**DONUT**   152
DOS (disk operating system) session   976
double hatching   277
dpi (dots per inch) plot options   918
draft-edit-plot cycle   66
drafting   66
   essentials   127
dragging, rotate   434
draw order, hatch   281

drawing
   arcs   148
   area   44
   basic objects   129
   circles   144
   construction lines   216
   donuts   152
   ellipses   157
   line segments   131
   mleaders with bubbles   529
   multiline leaders with text   526
   ortho mode   194
   points   160
   polygons   141
   polylines   153
   rectangles   134
   similar objects   302
   solid primitives   761
   surfaces   818
drawing
   files   lx, 977
   orientation   920
   plotting   907
   reporting information about   736
   with efficiency   247
   with polar mode   198
   with precision   185
   with snap mode   192
**DRAWORDER**   306
draworder for dimensions   656
drivers   904
drives   974
droplists   16
**DSETTINGS**   187
   osnap   208
**DTEXTED**   492
duplex printing   946
dwf (design Web format)
   3D options   938
   data options   937
   publish options   937
**DWGPROPS**   736
dynamic
   angles   58
   dimensions   58
   input   48
   input   9, 56
   lengthen   429
   ucs   774

# E

edge
   meshing between four   817
   moving, rotating, and scaling meshes   796
   polygon   142
   trimming   424

**EDGESURF** 817
edit dictionaries 507
editing 67
    blocks with the block editor 264
    content of blocks 249
    dimension text 624
    dimensions 617
    dimensions through objects 657
    drawings 319
    grips 450, 656
    hatch boundaries 284
    parameter values 674
    plot styles 959
    plotter configurations 944
    polyline 448
    tables 726
    tables with grips 729
    text 536
editor settings 509
    plotter configuration options 946
elevation, rectang 138
**ELLIPSE** 156
elliptical arc 158
endpoint, osnap 203
engineering liv
enhanced attribute editor 538
entering
    commands 54
    coordinates 57
entertainment lvii
eplots 68
equilateral triangles 166
**ERASE** 418
erasing
    constraints 673
    objects 418
    offset 380
    trimming 424
error messages and warnings 58
evenly spacing dimensions 621
executing commands with the ribbon 11
**EXIT** 22
exiting grips editing 455
**EXPERT** 65
**EXPLODE** 439
exploding
    blocks 259
    compound objects 439
**EXPORTLAYOUT** 891
express tools, dimensions 632
**EXTEND** 425
extending
    objects 425
    options 426
    polylines 426
    trimming 424

extension lines 565
    dimstyles 637
    obliquing 623
extension, osnap 203
extents, zoom 224
**EXTRUDE** 823

## F

f (select with a fence) 290
faces, moving, rotating, and scaling meshes 796
facilities management lvi
fence, trimming 423
fields, mtext 506
**FILEDIA** 64
file-related dialog boxes 975
files 974
    type 99
    AutoCAD program 978
    drawing 977
    extensions 977
    import-export 978
    LISP and ARX programming 979
    list 98
    miscellaneous 978
    names 974
    plot to 909
    plotting support 978
    removing temporary 979
    support 977
fill color, dimstyles 642
**FILLET** 389, 837
filleting
    3D objects 837
    inferred constraints on 700
    interactive 839
    objects 389
    rectang 139
    results for different objects 390
**FILLETEDGE** 837
**FILTER** 299
    layers 356
**FIND** 540
finding
    and replacing 508
    center 718
    finding files 102, 540
    replace, text 495
first line indent, mtext 511
first, break 420
fitting 488
    polyline editing 445
flip screen lxiv
flipping arrowheads 618
flyouts 15
**FONTALT** 520

fonts 483
    mtext 499
format, tolerances 650
fraction heights
    scale 643
    stack 500
frame around text 643
freezing
    layout modes 352
    layers 348, 865
from 206
ftp 112
full preview 912
full section hatches 266
function keys 61

## G

g (select the group) 292
gap tolerance 280
**GCCOINCIDENT** 692
**GCCOLLINEAR** 693
**GCCONCENTRIC** 693
**GCEQUAL** 694
**GCFIX** 694
**GCHORIZONTAL** 694
**GCPARALLEL** 695
**GCPERPENDICULAR** 695
**GCSMOOTH** 696
**GCSYMMETRIC** 696
**GCTANGENT** 697
**GCVERTICAL** 697
**GEOMCONSTRAINT** 691
geometric constraints 685
    modifying 690
global linetype scales 338
**GRADIENT** 282
graphics boards 972
**GRID** 49, 186
grid spacing 189
grips 449
    arcs 455
    color and size 450
    editing 656
    editing modes 451
    editing options 452, 657
    editing tables 729
    hatches 456
    polyline 456
group filters, layers 357
grouping, object 298

## H

halfwidth, pline 155
hardware of an AutoCAD system 967
**HATCH** 266

hatching
    angle 276
    grips 456
    hatching 277
    properties 275
    scale 276
    transparency 276
**HATCHEDIT** 284
**HATCHTOBACK** 307
headers 677
height 489
    mtext 513
**HIDEOBJECTS** 303
**HIDEXREFSCALES** 926
hiding
    layout tabs 857
    paperspace objects 919
    viewport borders 861
highlights and grips 47
history of 3D in AutoCAD 754
holes 764
    dimensions 572
horizontal, dimensions 580
how hatching works 267
**HPDRAWORDER** 306
hsl (hue, saturation, luminosity) 328
hyphen prefix 64

## I

icons, ucs 771
**ID** 710
implied windowing 298
importing
    text 508
    files 978
inferring constraints 698
info center 43
information about drawings 736
inherit from parent block 441
initial view 101
initialization strings, plotting 948
inkjet plotters 971
input devices 973
inscribed in circle 143
**INSERT** 255
inserting
    blocks with designcenter 260
    blocks with tool palettes 261
    field text 494
    right-click 457
    symbol 505
    tables 724
    title blocks 870
insertion, osnap 203
inspection text to dimensions 626

interactive
    chamfering   836
    filleting   839
**INTERFERE**   832
interference between solids   832
interior design   lv
**INTERSECT**   831
intersecting
    lines, fillet   390
    solids   831
intersection, osnap   204
introduction
    to 3D mesh modeling   787
    to dimensions   564
island detection   280
ISO (international organization of standards)
    patterns   275
    pen width   277
**ISOCIRCLE**   159
**ISOLATEOBJECTS**   303
isosceles triangles   167
italic, mtext   498

## J

jogged
    dimension lines   619
    linear dimension   641
    radius dimension   591, 641
**JOIN**   395
joining
    3D meshes   827
    objects   396
    polyline editing   443
    solid models   776
jointype, polyline editing   447
justification   502
justifying   486
    mtext   513
**JUSTIFYTEXT**   541

## K

keyboards
    coordinates   77
    layers   348
    shortcuts   60, 1005
keys   lxiv

## L

l (select the last object)   292
laser printers   972
**LASTPOINT**   710
**LAYER**   321, 346

layers
    assigning plot styles   955
    changing current   364
    changing status   364
    colors   349
    controls in layouts   352
    freezing   865
    freezing and thawing   348
    linetypes   349
    lineweights   349
    locking and unlocking   348
    names searching   355
    offset   380
    overriding properties   867
    plot style   350
    print toggles   350
    ribbon   363
    states manager   361
    tools   365
    transparency   350
    turning on and off   348
    xplode   441
**LAYERP**   365
**LAYERPMODE**   365
**LAYISO**   366
**LAYMCUR**   365
**LAYOUT**   884
layouts   51, 855
    assigning plot styles to   953
    freeze modes   352
    hiding tabs   857
    layer controls   352
    managing   884
    property modes   352
    save changes   919
    tabs and scroll bars   49
    text height   875
    working with   858
**LAYOUTWIZARD**   878
**LEADER**   521
leaders
    text editor   537
    to mleaders   528
leading and trailing   648
**LENGTHEN**   426
lengthening
    delta   428
    dynamic   429
    objects   426
    percent   428
    pline   156
    total   428
**LIMITS**   189
    text   568
**LINE**   130

lines
- construction 216
- change 437
- dimensions 568
- fillet intersecting 390
- fillet parallel 390
- inferred constraints on 698
- tessellation 793

line spacing 504
- mtext 513

linear dimensions 579, 646
- jog dimension 641

linear arrays 385
**LINETYPE** 335
linetypes
- dimstyles 636
- layer 349
- scales, uniform 877
- scaling 339
- scaling, dimensions 601

lineweights (see **LWEIGHT** command) 333
- dimstyles 636
- layer 349
- plot 918

linking
- dimensional constraints 670
- objects with constraints 671

LISP and ARX programming files 979
**LIST** 718
listing system variables 739
local
- dimstyle changes 655
- dimstyle overrides 655
- linetype scales 338
- printer connections 903

locking layers 348
log file location 929
**LTSCALE** 339
ltype gen, polyline editing 446
- xplode 441

**LWEIGHT** 332

## M

m (select through the multiple option) 292
**M2P** 462
managing
- cell content 729
- layouts 884

manipulating mesh models 790
manual, applying constraints 691
manufacturing lv
mapping lvi
mark/back, undo 164
**MARKUP** 400
- options 403

**MASSPROP** 717, 840
matching properties
- hatch 279
- properties 322

**MATCHPROP** 322
maximizing and minimizing viewports 872
**MEASURE** 382
"measuring" objects 382
**MEASUREGEOM** 711
measurement scale 647
measurements, distances and angles 711
media
- source and size 946
- type 946

memory 969
menu bars 44
- application menu 17

merge control 937
**MESH** 791
meshing
- between four edges 817
- chamfering meshes 835
- smoothing 795

**MESHCAP** 802
**MESHCREASE** 799
**MESHOPTIONS** 793
**MESHSMOOTHLESS** 794
**MESHSMOOTHMORE** 794
**MESHUNCREASE** 799
midpoint, osnap 204
**MIRROR** 376
mirrored grips editing 453
miscellaneous files 978
**MLEADER** 526
multileaders
- adding leaders to 528
- aligning 534
- collecting bubbles into one 532
- drawing with bubbles 529
- options 527

**MLEADERALIGN** 534
**MLEADERCOLLECT** 532
**MLEADEREDIT** 528
**MLEADERSTYLE** 529
mode indicators 50
models
- 3D 751
- 3D surface primitives 804
- switching from paper mode 861
- viewports scaling 863

modes, dimension 573
modifying
- 3D models 762
- dimstyles 635
- dynamic input 59
- geometric constraints 690

monitors 972
mouse 9, 973
    3D 973
    buttons 62, 1008
**MOVE** 431
moving
    command-less 431
    dimension text 629
    faces 797
    grips editing 452
    mesh faces, edges, and vertices 796
    right-click 457
    stretch 431
**MREDO** 165
**MSLTSCALE** 550
**MTEXT** 496
multiline text
    dimensions 579
    editor shortcut keys 1008
    shortcut menu 510
    text editor 537
    text entry area 511
multiline leaders 526
multiple 160
    chamfering 395
    fillet 392
    polyline editing 447
multiplying meshes 789
my network places 111

# N

named plot style tables (stb files) 951
named to color-dependent 952
**NAVBAR** 229
**NAVBARDISPLAY** 229
navigation bar 45, 229
    3D mouse 815
nearest, osnap 204
network 111
    printer connections 903
new
    dimstyles 635
    drawings 88
    in AutoCAD 2011 lxi
    plot styles 956
node, osnap 205
non-associative hatch patterns 279
none 207
nontraditional printers 902
noun/verb selection 297

# O

object
    grouping 298
    snaps 79

snaps, 3D 812
snaps, dimensions 574
tracking 460
objects
    assigning plot styles 954
    breaking 419
    chamfering 393, 834
    changing 437
    changing properties 340
    changing size through constraints 668
    colors 324
    constructing 373
    copies 374
    creating viewports from 889
    dimensioning 578
    dividing 382
    drawing similar 302
    editing dimensions through 657
    erasing 418
    exploding compound 439
    extending 425
    filleting 389
    hiding 304
    isolating 305
    joining 396
    lengthening 426
    linetype scaling 338
    lineweights 333
    linking with constraints 671
    "measuring" 382
    mirrored copies 377
    moving by stretching 431
    offsetting 379
    redlining 398
    revcloud 399
    reversing 397
    rotating 432
    selecting by location 286
    selecting by properties 294
    selecting similar 301
    shortening 422
    stretching 430
    sweeping into solids 825
    trimming 422
    unisolating 305
    zoom 224
**OBJECTSCALE** 548
oblique angle, tracking, and width factor 500
obliquing extension lines 623
off 207
**OFFSET** 378
offsettting
    hatches 266
    plot 911
**OFFSETGAPTYPE** 381
on and off, layers 348

online
    applications    lviii
    companion    lxvi
**OOPS**    418
opaque background    493
**OPEN**    96
opening
    documents    19
    drawings    96
    drawings with named views    235
    options, plot    918
        plot viewport shade    916
        shade plot    916
ordinate, quick dimensions    587
**ORTHO**    193
**OSNAP**    200
**-OSNAP**    211
osnapping
    with object tracking    460
other keys    62
**OTRACK**    214
    options    214
overline, mtext    498
overrides
    dimstyles    652
    layer properties    867
    layout properties    353
    properties    869

# P

p (select the previous selection set)    292
**PAGESETUP**    930
**PAN**    227
**-PAN**    228
panning the drawing    227
paper mode, switching to model mode    861
paper size    910
paperspace
    hide objects    919
    plot last    919
paragraph
    indent    512
    options    504
    panel    502
    text    496
parallel lines    390
    osnap    205
**PARAMETERS**    674
parameters manager palette    676
parametrics    lxii
partial open    100
partial preview    912
password protection    937
paste special    510
paths    974
pattern    275

PDF publish options    937
**PEDIT**    442
**PEDITACCEPT**    448
percent, lengthen    428
perimeters    713
peripheral hardware    971
perpendicular, osnap    205
physical pen characteristics    947
pick (select single object)    288
**PICKBOX**    295
pickbox size    296
placing
    blocks    255
    center marks    595
    dimensional constraints    666
    hatch patterns    268
    leaders    521
    text    481
    text all over drawings    485
**PLINE**    153
**PLINECONVERTMODE**    448
**PLOT**    907
plotting    68
    area    910
    background    918
    background processing options    923
    default settings for new drawings    922
    drawing    913
    file    909
    log file    923
    number of copies    911
    object lineweights    918
    offset    911
    offset relative to    924
    options    918
    paperspace last    919
    plot stamp units    929
    plot styles    918
    scale    911
    shade options    916
    stamp    927
    transparency    919
    with plot styles    918
plot style
    assigning    953
    assigning to layers    955
    assigning to layouts    953
    assigning to objects    954
    creating new    956
    editing    959
    layer    350
    selecting    915
    setting default    952
    switching    951
    tables    915
    tables, color-dependent    950
    tables, named    951

PLOTSTAMP   927
plotter configurations   939
PLOTTERMANAGER   939
plotters and printers   902
POINT   160
    dimensions   572
point filters   78
polar   194
    coordinates   72
    semicircular arrays   387
POLARANG   197
POLARDIST   197
POLARMODE   197
polyarcs, dimensions   569
POLYGON   140
polygonal
    creating viewports   888
polyline editing   448
    chamfering   394
    dimensioning segments   568
    extending   426
    fillet   390
    fillet   392
    grips   456
    trimming   423
popular justifications   502
precision, dimstyles   646
predefined views   232
prefix and suffix   647
preparing
    layouts for plotting   930
    models in three dimensions   757
press and drag   298
PREVIEW   905
previewing   99
    blocks   260
    partial plot   912
    plots   905
    selection   296
previous, zoom   224
primary units   646
primitives
    3D mesh primitives   791
    3D solid model   779
    3D surface models   804
print, layer toggles   350
printers   902
    local and network connections   903
    nontraditional   902
    system   902
project, trimming   424
PROPERTIES   321, 340, 656, 720
properties
    filter layers   359
    matching   322
    overridden layout   353

overrides   869
overriding layer properties   867
quick   343
selecting objects by   294
solid models   840
viewport border and content   866
protecting drawings   68
PSLTSCALE   877
PUBLISH   932
publishing
    drawing sets   932
    log file   923
    options   936
    output   936

## Q

QDIM   585
QLEADER   521
QNEW   90
QSAVE   117
QSELECT   294
quadrant osnap   205
quick   207
    continuous dimensions   585
    dimension text moves   630
    tour of AutoCAD 2011   3
quick access toolbar   17
QUICKPROPERTIES   343
QUIT   120
quitting AutoCAD   120
QVLAYOUT   874

## R

r (remove objects from the selection set)   293
radial dimensions   590
    placing jogged   591
    jog   641
radius
    fillet   392
    quick dimensions   587
rarely used zoom modes   226
RAY   216
read-only   100
real time zoom   226
recalling named views   232
recent
    drawings   18
    input   55
recovering erased objects   418
recreating boundaries   273
RECTANG   134
rectangles
    inferred constraints on   699
    viewports   886
    arrays   385

su and o (select subobjects and objects) 292
SUBTRACT 829
subtracting solids 829
sub-units 648
summary
    of all object snap modes 202
    of all dimension variables 1067
support files 977
surface model primitives 804
SWEEP 823
sweeping objects into solids 825
switching
    from paper to model mode 861
    layouts through quick views 874
symbols
    arc length 641
    dimensions 580
    dimstyles 638
    mtext 505
system
    board 968
    printers 902
system variables 64, 739, 1009

# T

TABLE 721
TABLEDIT 726
TABLEEXPORT 729
tables
    editing 726
    styles 724
TABLESTYLE 731
tabs
    hiding layout 857
    mtext 512
TABSURF 821
tan, tan, tan, circle 146
tangent, osnap 206
templates 89
    starting new drawings with 90
temporary files, removing 979
temporary osnaps 201
terminology lx
tessellation lines 793
TEXT 483
text
    adding inspection text to dimensions 626
    annotative style 556
    change 438
    dimensions 580
    dimstyles 642
    drawings 482
    editing 536
    editing dimension text 624
    editor 493
    finding 540
    height in layouts 875
    height in dimstyles 643
    height in mtext 498
    highlight color 494
    moving dimension text 629
    rejustifying 542
    repositioning dimension 623
    resizing 541
    single-line editor 537
    styles 515
text window 9
TEXTEDIT 536
TEXTFILL 520
TEXTQLTY 520
TEXTTOFRONT 307
TEXTTOFRONT 514
thawing layers 348
thickness, rectang 138
through offset 380
TILEMODE 872
TIME 740
TINSERT 726
title bar 43
title blocks, inserting 870
toggling
    drawing area 231
    grid 186
    ortho mode 193
    osnap mode 207
    polar mode 194
    snap mode 189
tolerance text 568
tolerances 650
TOOLBARS 21, 44
tool palettes 260
tools, layer 365
total, lengthen 428
touring the user interface 43
TRACKING 459
tracking 78
    lines 49
transparency 311
    hatch 276
    layer 350
    plotting 919
TRANSPARENCYDISPLAY 331
transparent
    panning 227
    redraws 222
    zoom 226
tray 52
tray options 53
triangles 166
TRIM 421

trimming
    chamfering   394
    closed objects   422
    fillet   392
    polylines   423
true color   327
tt (tracking)   207
ttr (tan tan radius), circle   146
turning layers on and off   348
types of dimensional constraints   665

# U

U   162, 419
u (undo the selected option)   293
UCS   771
user-defined coordinate systems   771
    changing from wcs   772
    dynamic   774
    icon   46, 771
underline, mtext   498
understanding CAD concepts   41
UNDO   163
undoing commands   162
    chamfering   394
    fillet   392
    grips editing   454
    line   133
    mtext   509
    pline   154
    polyline editing   446
    trimming   424
UNION   776, 826
UNISOLATEOBJECTS   304
UNITS   105
units
    dimstyles   646
    alternate   649
unlocking layers   348
updating
    annotative text   556
    field   495
upper and lower case   498
upper value and lower value   651
use shift to add to selection   298
user interface   lxi
    changes in AutoCAD 2011   42
user-defined
    coordinate systems   76, 771
    paper sizes and calibration   949
    patterns   275

# V

variables
    dimension   631, 1067
    system   1009
vertex, polyline editing   445

vertical, dimensions   580
vertices
    dimensions   572
    moving, rotating, and scaling meshes   796
VIEW   232
view cube   45
VIEWPLOTDETAILS   914
viewport borders
    content properties   866
    properties   866
    shade options, plotting   916
VIEWPORTS   886
viewports
    creating from objects   889
    creating polygonal   888
    creating rectangular   886
    hiding borders   861
    maximizing and minimizing   872
    scaling models   863
    selectively displaying details in   865
visibility, controlling annotation   556
visual selection cycling   288
volumes, measuring   716
VPSCALE   863
VPCLIP   888
VPMAX   872
VPMIN   872
-VPORTS   888
VPOVERRIDEMODE

# W

w (select within a window)   288
WBLOCK   262
wcs   771
wedges, dimensions   570
what's new in AutoCAD 2011   lxi
width
    mtext   514
    pline   154
    polyline editing   444
    rectang   137
wild-card characters   975
window
    controls   43
    implied   298
    resize   53
    zoom   225
Windows   967
    Explorer   975
wizards   92
working with
    dimension styles   633
    layers   348
    layouts   853
    layouts   858

workspaces 20, 52
world coordinate system (wcs) 771
wp (select with a windowed polygon) 291
write-protecting data 970

# X

x, y coordinates and elevation 50
XFADECTL 548
XLINE 216
XPLODE 441

# Z

zero suppression 647
ZOOM 222
ZOOM xp 863

# #

@ 710
0 feet and 0 inches 648
2D
    commands that don't work in 3D editing 816
    drafting lxii
    tutorial 23
2p (2 point) circle 145
3D
    chamfering and filleting 837
    culling object selection 813
    design aids 812
    dwf options 938
    modeling lxiii, 751
3D mesh models
    creating and editing 791
    joining 827
    parts 788
3D mouse 973
    navigation bar 815
    support 814
3D object snaps 812
3D solid modeling 753
    primitives 779
3D surface model primitives 804
3p (3 point) circle 145

# Index of Tutorials

## A
accessing droplists 16
accessing flyouts 15
adding creases 799
adding inspection text to dimensions 626
adding leaders to mleaders (mleaderedit) 528
aligning mleaders (mleaderalign) 534
annotative dimensioning 599
applying constraints manually 691
automatically scaling text 543

## B
breaking dimension and extension lines 620
breaking objects 419

## C
capping leaking mesh objects 802
chamfering and filleting 3D objects 837
chamfering objects 393
chamfering objects 834
changing all properties 340
changing colors 104
changing colors 324
changing display precision 628
changing from wcs to ucs 772
changing linetypes 335
changing lineweights 333
changing objects 437
changing size 434
changing the grid 192
changing windowed views 236
checking interference between solids 832
checking spelling 539
collecting bubbles into one mleader 532
combining dimensional and geometric constraints 703
constraining geometry automatically 687
continuing dimensions 584
controlled explosions 441
creating 3D mesh primitives (mesh) 791
creating and applying layer states 361
creating blocks 249
creating boundaries 285
creating custom arrowheads 639
creating group filters 357
creating holes 764
creating layers 347
creating new plot styles 956
creating plotter configurations 939
creating polygonal viewports 888
creating property filters 359
creating rectangular viewports 886
creating viewports from objects 889
culling 3D object selection 813

## D
defining text styles 515
determining areas and perimeters 713
determining text height in layouts 875
dimensioning 597
dimensioning objects 578
direct distance entry 458
dividing objects 382
drawing arcs 148
drawing circles 144
drawing construction lines 216
drawing donuts 152
drawing ellipses 157
drawing in ortho mode 194
drawing line segments 131
drawing mleaders with bubbles (mleaderstyle) 529
drawing multiline leaders with text (mleader) 526
drawing polygons 141
drawing polylines 153
drawing rectangles 134
drawing similar objects 302
drawing solid primitives 761
drawing with construction lines 218
drawing with polar mode 198
drawing with snap mode 192

## E
editing blocks with the block editor 264
editing dimension text 624
editing dimensions through objects 657
editing parameter values 674

# Index of Tutorials

editing plot styles 959
editing plotter configurations 944
editing text 536
editing with grips 450
erasing constraints 673
erasing objects 418
evenly spacing dimensions 621
executing commands with the ribbon 11
exploding compound objects 439
extending objects 425

## F
filleting objects 389
finding text 540
flipping arrowheads 618

## G

global retroactive dimstyle changes   653
global temporary dimstyle changes   653

## H

hiding and isolating objects   304

## I

inferring constraints   698
intersecting solids   831
inserting tables   724

## J

jogging dimension lines   619
joining 3D meshes   827
joining objects   396

## L

lengthening objects   426
linear and rectangular arrays   385
linking dimensional constraints   670
linking objects with constraints   671
listing system variables   739
local dimstyle changes   655
local dimvar overrides   655

## M

M2P   462
making backup copies   117
making copies   374
making linetype scales uniform   877
making mirrored copies   377
making offset copies   379
managing layouts   878
managing layouts   884
matching properties   322
maximizing and minimizing viewports   872
"measuring" objects   382
meshing between four edges   817
modifying 3D models   762
modifying geometric constraints   690
moving dimension text   629
moving faces   797
moving objects   431

## N

new — creating new dimstyles   635

## O

obliquing extension lines   623
opening drawings   96
opening drawings with named views   235

## P

panning the drawing   227
placing additional dimensions from a baseline   583
placing aligned dimensions   582
placing angular dimensions   594
placing arc length dimensions   592
placing blocks   255
placing center marks   595
placing dimensional constraints   666
placing dimensions horizontally   575
placing dimensions vertically and rotated   577
placing jogged radial dimensions   591
placing leaders   521
placing lines of text   483

### Index of Tutorials

placing paragraph text   496
placing radial dimensions   590
plotting drawings   907
polar and semicircular arrays   387
preparing layouts for plotting   930
preparing to model in three dimensions   757
previewing plots   905
publishing drawing sets   932

## Q

quick continuous dimensions   585
quitting AutoCAD   120

## R

redlining objects   398
redoing undos   165
rejustifying text   542
renaming and deleting named views   234
reporting distance and angle measurements   711
reporting information about drawings   736
reporting information about objects   719
reporting information about regions   717
reporting on time   740
reporting point coordinates   710
reporting properties   840
reporting volumes   716
repositioning dimension text   623
resizing text   541
reversing objects   397

revolving objects 769
rotating objects 432

## S

saving drawings 117
saving drawings by other names 110
saving drawings in other formats 114
saving drawings on other computers 111
selecting objects by location 286
selecting objects by properties 294
selecting similar objects 301
sending drawings by ftp 113
setting and changing colors 330
setting units 105
shortening objects 422
smoothing mesh models 795
specifying draworder 306
stamping plots 927
starting AutoCAD 4
starting AutoCAD with named views 235
starting new drawings 88
starting new drawings with templates 90
starting the pedit command 442
storing and recalling named views 232
stretching objects 430
subtracting solids 829
sweeping objects into solids 825
switching layouts through quick views 874

## T

through 380
toggling and selecting osnap mode 207
toggling ortho mode 193
toggling otrack mode 215
toggling polar mode 194
toggling snap mode 189
toggling the drawing area 231
toggling the grid 186
tracking 459

## U

undoing commands 162
unioning solids 776
using dynamic ucs 774
using object snap modes 211
using quick properties 343
using the advanced wizard 92
using the redraw command 221
using the zoom command 222

## W

working with layouts 858